PHYSICAL CHEMISTRY

Thermodynamics,
Statistical Mechanics, & Kinetics

PHYSICAL CHEMISTRY

Thermodynamics,

Statistical Mechanics, & Kinetics

ANDREW COOKSY

PEARSON

Boston   Columbus   Indianapolis   New York   San Francisco   Upper Saddle River
Amsterdam   Cape Town   Dubai   London   Madrid   Milan   Munich   Paris   Montréal   Toronto
Delhi   Mexico City   São Paulo   Sydney   Hong Kong   Seoul   Singapore   Taipei   Tokyo

Editor in Chief: *Adam Jaworski*
Executive Editor: *Jeanne Zalesky*
Senior Marketing Manager: *Jonathan Cottrell*
Project Editor: *Jessica Moro*
Editorial Assistant: *Lisa Tarabokjia*
Marketing Assistant: *Nicola Houston*
Director of Development: *Jennifer Hart*
Development Editor: *Daniel Schiller*
Media Producer: *Erin Fleming*
Managing Editor, Chemistry and Geosciences: *Gina M. Cheselka*
Full-Service Project Management/Composition: *GEX Publishing Services*
Illustrations: *Precision Graphics*
Image Lead: *Maya Melenchuk*
Photo Researcher: *Stephanie Ramsay*
Text Permissions Manager: *Joseph Croscup*
Text Permissions Research: *GEX Publishing Services*
Design Manager: *Mark Ong*
Interior Design: *Jerilyn Bockorick, Nesbitt Graphics*
Cover Design: *Richard Leeds, BigWig Design*
Operations Specialist: *Jeffrey Sargent*
Cover Image Credit: *Richard Megna/Fundamental Photographs*

Credits and acknowledgments borrowed from other sources and reproduced, with permission, in this textbook appear on the appropriate page within the text or in the back matter.

Many of the designations used by manufacturers and sellers to distinguish their products are claimed as trademarks. Where those designations appear in this book, and the publisher was aware of a trademark claim, the designations have been printed in initial caps or all caps.

Library of Congress Cataloging-in-Publication Data
Cooksy, Andrew.
  Physical chemistry : thermodynamics, statistical mechanics, & kinetics / Andrew Cooksy.
    pages cm
  Includes index.
  ISBN-13: 978-0-321-81415-9
  ISBN-10: 0-321-81415-0
  1. Chemistry, Physical and theoretical--Textbooks. 2. Thermodynamics--Textbooks. 3. Statistical mechanics--Textbooks. 4. Chemical kinetics--Textbooks. I. Title.
  QD453.3.C66 2013
  541--dc23
                          2012037313

9

www.pearsonhighered.com

ISBN-10: 0-321-81415-0
ISBN-13: 978-0-321-81415-9

Cragged, and steep, Truth stands, and he that will
Reach her, about must, and about must go,
And what the hill's suddenness resists, winne so.
—*John Donne (1572-1631), Satire III*

Thermodynamics, Statistical Mechanics, & Kinetics

Quantum Chemistry and Molecular Interactions

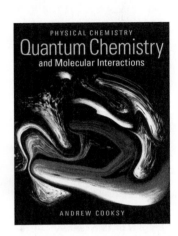

PHYSICAL CHEMISTRY
Thermodynamics, Statistical Mechanics, & Kinetics

I ask my students to review Ch. A at the beginning of the term to try and kick-start their rusty trigonometry, mechanics, calculus, and other basics. It's also a convenient summary of prereqs for students who aren't sure if they're ready for the course yet. This chapter also appears in the Quantum volume.

PART I
EXTRAPOLATING FROM MOLECULAR TO MACROSCOPIC SYSTEMS 37

Ch. 1 explains how the insights of 19th century scientists set the stage for the modern model of chemistry that will follow. A first pass at deriving the ideal gas law using the kinetic theory shows how macroscopic measurements were used to infer microscopic properties.

We present the main principles of stat mech, including the ergodic hypothesis and the nature of the ensemble. The entropy S is introduced right away as a measure of ensemble size, in hopes that this will help demystify S. We show that the Boltzmann and Gibbs definitions of entropy agree in the simplest ensembles. We conclude by deriving the ideal gas law again, but with fewer assumptions than we used in Ch. 1.

If we allow S to be defined by Boltzmann, then we can use S to define the temperature. This avoids the phenomenological definitions of either T or S required by a purely classical approach.

In Ch. 3 we begin to apply stat mech, using exp(−E/$k_B T$) to predict quantum state populations and using equipartition to predict the distribution of energy.

Ch. 4 covers selected, more advanced applications of stat mech, including a detailed derivation of the van der Waals eqn of gases, an approximate form of the pair correlation function (useful for many later expressions), and nuclear spin statistics.

This section includes a brief description of the theory underlying Bose-Einstein condensates.

Here we cover mean free path, collision frequency, Fick's laws, and other concepts we will need to describe molecular motion and chemical kinetics. A detailed derivation of the Gaussian distribution for diffusion is featured.

Using stat mech to obtain expressions for quantum state populations, we explain transition intensities, the principle of lasers, and other aspects of energy transfer.

PART II
NON-REACTIVE MACROSCOPIC SYSTEMS *231*

In Ch. 7, I introduce *G*, *H*, and Helmholtz free energy all at once, because the formal definitions clarify the relations between the parameters. The section on heat capacities then illustrates why, for example, enthalpy may be more useful than energy to describe a system at constant pressure.

Section 7.4 covers heat capacities of gases (with comparison to equipartition results), the Einstein and Debye heat capacities of crystals, and also the derivation of an approximate expression for the heat capacity of a liquid, showing why C (liquid) is greater than for the corresponding solid or gas.

Having introduced fundamentals of thermodynamics, this chapter focuses on energy, enthalpy, and mechanical work, clearing the way for the more *chemical* applications of thermodynamics that follow.

A novel derivation of the Sackur-Tetrode equation is given here, showing how the single-particle mass affects the number of ensemble states, and so affects the entropy of the many-particle gas.

Stat mech helps to establish a meaning of *S* separate from its relationship to other thermodynamic parameters. I believe this strengthens a student's understanding of the central role that *S* does play in thermodynamics. In particular, the 2nd law (which involves only *S*) will guide our discussion of the time-evolution of systems in the upcoming chapters.

"Tools" sections on bomb calorimetry, reaction calorimetry, and DSC point up the continuing importance of calorimetry in the practical analysis of chemical systems.

Phase transitions are marvelously counter-intuitive processes, where we need to carefully apply our understanding of microscopic molecular properties to bulk behavior. This chapter opens with a qualitative description of simple phase transitions in terms of square-well and other model potential energy functions, before moving into quantitative problems involving phase transitions, using the latent enthalpy, Clausius-Clapeyron eqn, and so on. The chapter concludes with a section on single-component phase diagrams.

11 Solutions 373

After presenting the Raoult's law and Henry's law solutions, we show in Sec. 11.2 that stat mech can help us obtain a quantitative understanding of trends in the Henry's law coefficients. Sec. 11.3 includes a discussion of multi-component phase diagrams; Sec. 11.4 covers basic elements of electrochem such as Nernst's law; 11.5 covers selected colligative properties; and 11.6 shows how we can predict properties of non-ideal solutions.

PART III
REACTIVE SYSTEMS 415

Chapters 10 and 11 describe processes (phase transitions, solvation) that inch us towards real chemical reactions. With Ch. 12 we arrive. Here the focus is on chemical thermodynamics, preceded by a microscopic perspective using reaction potential energy surfaces.

12 The Thermodynamics of Chemical Reactions 416

13 Chemical Kinetics: Elementary Reactions 449

We use simple collision theory and the Eyring equation to describe single-step chemical reactions. Particular attention is given to how well we can use these models to estimate the magnitude of the pre-exponential factor in the Arrhenius rate constant.

Section 14.1 demonstrates how, for relatively simple reaction mechanisms, we can integrate rate equations to obtain chemical concentrations as functions of time, but that as the complexity of the mechanism grows the analytical solution may become less valuable. We may use steady state, fast equilibrium, and other approximations when possible to regain some of the advantages of an analytical solution. The capstone idea of the text is that, by bringing together seemingly diverse principles from the microscopic and macroscopic realms, we can at last describe in detail chemical transformations in any system.

14 Chemical Kinetics: Multi-Step Reactions 477

This final section of the text surveys a number of applications to complex systems such as atmospheric and interstellar chemistry, combustion, and enzyme kinetics.

In addition to this answer list, detailed solutions for the Objectives Review are given in the solutions manual.

This book is intended to provide students with a detailed guide to the reasoning that forms the basis for physical chemistry—the framework that unites all chemistry. The study of physical chemistry gives us the opportunity to look at our science as an integrated whole, with each concept connected to the next. My goal has been to trace those connections, step-by-step whenever possible, to show how each new concept makes sense given its place in the framework.

Because its ideas build upon each other in this way, physical chemistry can serve as the foundation for an intuitive understanding of chemistry in all its forms, whether synthesizing new compounds, analyzing samples in a forensic laboratory, or studying the properties of novel materials. To that end, this book emphasizes the shared, fundamental principles of chemistry, showing how we can justify the form and behavior of complex chemical systems by applying the laws of mathematics and physics to the structures of individual particles and then extrapolating to larger systems. We learn physical chemistry so that we can recognize these fundamental principles when we run into them in our other courses and in our careers. The relevance of this discipline extends beyond chemistry to engineering, physics, biology, and medicine: any field in which the molecular structure of matter is important.

A key step toward cultivating an intuition about chemistry is a thorough and convincing presentation of these fundamentals. When we see not only what the ideas are, but also how they link together, those ideas become more discernible when we examine a new chemical system or process. The following features of this text seek to achieve that objective.

- My aim is to provide a rigorous treatment of the subject in a relaxed style. A combination of qualitative summaries and annotated, step-by-step derivations illuminates the logic connecting the theory to the parameters that we can measure by experiment. Although we use a lot of math to justify the theory we are developing, the math will always make sense if we look at it carefully. We take advantage of this to strengthen our confidence in the results and our understanding of how the math relates to the physics. Nothing is more empowering in physical chemistry than finding that you can successfully predict a phenomenon using both mathematics *and* a qualitative physical argument. The manifestation of atomic and molecular structure in bulk properties of materials is a theme that informs the unhurried narrative throughout the text.
- To illustrate how our understanding in this field continues to advance, we take the time to examine several tools commonly used in the laboratory, ("Tools of the Trade") while profiles of contemporary scientists

("Biosketches") showcase the ever-expanding frontiers of physical chemistry. Our intuition about chemistry operates at a deep level, held together by the theoretical framework, but these examples show how others are applying their understanding to solve real problems in the laboratory and beyond. They inspire us to think creatively about how the most fundamental chemical laws can answer our own questions about molecular structure and behavior.

- Our increasing appreciation and exploration of the interface between the molecular and the bulk scales has inspired a forward-looking coverage of topics that includes special attention to statistical mechanics throughout the volume.

Acknowledgments

I thank Kwang-Sik Yun and Andrew P. Stefani for providing the original inspiration and encouragement to carry out this project. My love of this field owes much to my mentors—William Klemperer, Richard J. Saykally, and Patrick Thaddeus—and to the many students and colleagues who have patiently discussed chemistry with me. I am particularly grateful to my fellow physical chemistry faculty—Steve Davis and Kwang-Sik Yun at Ole Miss, David Pullman and Karen Peterson at SDSU—for their many insights and limitless forbearance, and to William H. Green and the late John M. Brown for kindly hosting my sabbatical work in their research groups. An early prospectus for this book formed part of the proposal for an NSF CAREER grant, and I thank the agency for that support.

For helping me see this through, I thank my friends and mentors at Pearson, especially Nicole Folchetti, Adam Jaworski, Dan Kaveney, Jennifer Hart, Jessica Moro, and above all Jeanne Zalesky. A great debt is also owed to Dan Schiller for his patient and extensive work editing the manuscript. Thanks to Mary Myers for much work on the original manuscript, and to many in the open-source community for the tools used to assemble it. Finally, I thank all of our faculty and student reviewers for their careful reading and thoughtful criticisms. A textbook author could strive for no higher goal than to do justice to the fascination that we share for this subject.

Reviewers

Ludwik Adamowicz
University of Arizona

Larry Anderson
University of Colorado

Alexander Angerhofer
University of Florida

Matthew Asplund
Brigham Young University

Tom Baer
University of North Carolina

Russ Baughman
Truman State University

Nikos Bentenitis
Southwestern University

John Bevan
Texas A&M University

Charles Brooks
University of Michigan

Mark Bussell
Western Washington University

Beatriz Cardelino
Spelman College

Donna Chen
University of South Carolina

Samuel Colgate
University of Florida

Stephen Cooke
*Purchase College, the State University
of New York*

Paul Cooper
George Mason University

Phillip Coppens
*University at Buffalo, the State
University of New York*

Biamxiao Cui
Stanford University

Alfred D'Agostino
University of Notre Dame

Paul Davidovits
Boston College

Borguet Eric
Temple University

Michelle Foster
University of Massachusetts

Sophya Garashchuk
University of South Carolina

Franz Geiger
Northeastern University

Kathleen Gilbert
New Jersey Institute of Technology

Derek Gragson
California Polytechnic State University

Hua Guo
University New Mexico

John Hagen
California Polytechnic State University

Cynthia Hartzell
Northern Arizona University

Bill Hase
Texas Tech University

Clemens Heske
University of Nevada, Las Vegas

Lisa Hibbard
Spelman College

Brian Hoffman
Northeastern University

Xiche Hu
University of Toledo

Bruce Hudson
Syracuse University

David Jenson
Georgia Institute of Technology

Benjamin Killian
University of Florida

Judy Kim
University of California, San Diego

Krzysztof Kuczera
University of Kansas

Joseph Kushick
University of Massachusetts, Amherst

Marcus Lay
University of Georgia, Athens

Lisa Lever
University of South Carolina

Louis Madsen
*Virginia Polytechnic Institute and
State University*

Elache Mahdavian
Louisiana State University

Herve Marand
Virginia Polytechnic Institute and State University

Ruhullah Massoudi
South Carolina University

Gary Meints
Missouri State University

Ricardo Metz
University of Massachusetts

Kurt Mikkelson
University of Copenhagen

Phambu Nsoki
Tennessee State University

Jamiu Odutola
Alabama A&M University

Jason Pagano
Saginaw Valley State University

James Patterson
Brigham Young University

James Phillips
University of Wisconsin

Simon Phillpot
University of Florida

Rajeev Prabhakar
Miami University

Robert Quandt
Illinois State University

Ranko Richert
Arizona State University

Tim Royappa
University of West Florida

Stephen Sauer
University of Copenhagen

G. Alan Schick
Missouri State University

Charles Schmuttenmaer
Yale University

Rod Schoonover
California Polytechnic State University

Alexander Smirnov
North Carolina State University

J. Anthony Smith
Walla Walla University

Karl Sohlberg
Drexel University

David Styers-Barnett
Indiana University

James Terner
Virginia Commonwealth University

Greg Van Patten
Ohio University

John Vohs
University of Pennsylvania

Michael Wagner
George Washington University

Brian Woodfield
Brigham Young University

Dong Xu
Boise State University

Eva Zurek
University at Buffalo, the State University of New York

Andrew Cooksy

B.A., chemistry and physics, Harvard College, 1984;

Ph.D., chemistry, University of California, Berkeley, 1990;

Postdoctoral Research Associate, Harvard-Smithsonian Center for Astrophysics and Harvard University Department of Chemistry, 1990-1993;

Asst. and Assoc. Professor, University of Mississippi Department of Chemistry, 1993-1999. Asst. and Assoc. Professor, San Diego State University Department of Chemistry, 1999-2010. Professor, San Diego State University Department of Chemistry, 2010-.

Northrop-Grumman Excellence in Teaching Award, 2010

Senate Excellence in Teaching Award, SDSU College of Sciences, 2011

A Rigorous Standard
with a Relaxed Style

"" A course in physical chemistry can describe the physical universe
with uncommon depth, breadth, and clarity. The aim of this book
is to help the reader make the most of the experience. ""

—Andrew Cooksy

PHYSICAL CHEMISTRY is the framework that unites **all** chemistry—providing powerful insight into the discipline as an integrated series of connected concepts.

As an instructor and author, Andrew Cooksy helps students uncover these connections while showing how they can be expressed in mathematical form and demonstrating the power that derives from such expressions.

The text's lively and relaxed narrative illuminates the relationship between the mathematical and the conceptual for students. By formulating the fundamental principles of physical chemistry in a mathematically precise but easily comprehensible way, students are able to acquire deeper insight—and greater mastery—than they ever thought possible.

This innovative approach is supported by several exclusive features:

- Split quantum and thermodynamics volumes can be taught in either order for maximum course flexibility.

- A discrete chapter (Chapter A) included in each volume summarizes the physics and mathematics used in physical chemistry.

- Chapter opening sections orient the students within the larger context of physical chemistry, provide an overview of the chapter, preview the physical and mathematical relationships that will be utilized, and set defined chapter objectives.

- Unique pedagogical features include annotations for key steps in derivations and an innovative use of color to identify recurring elements in equations.

Uncovering connections between foundational concepts

Reflective of the author's popular lecture strategy, chapter opening and closing features ground each topic within the larger framework of physical chemistry and help students stay oriented as they follow the development of chapter concepts.

Learning Objectives outline the skills students should expect to acquire from their study of the chapter.

Visual Roadmaps help students see the relationship between the chapters in each part of the text and the topics in each chapter.

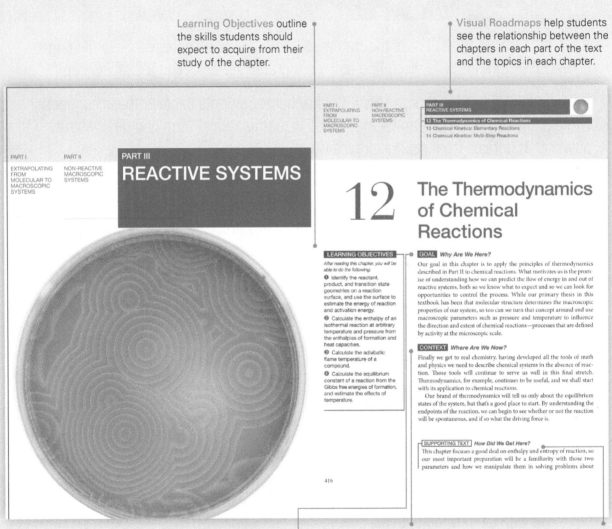

PART I
EXTRAPOLATING
FROM
MOLECULAR TO
MACROSCOPIC
SYSTEMS

PART II
NON-REACTIVE
MACROSCOPIC
SYSTEMS

PART III
REACTIVE SYSTEMS

12 The Thermodynamics of Chemical Reactions
13 Chemical Kinetics: Elementary Reactions
14 Chemical Kinetics: Multi-Step Reactions

PART I
EXTRAPOLATING
FROM
MOLECULAR TO
MACROSCOPIC
SYSTEMS

PART II
NON-REACTIVE
MACROSCOPIC
SYSTEMS

PART III
REACTIVE SYSTEMS

12 The Thermodynamics of Chemical Reactions

LEARNING OBJECTIVES

After reading this chapter, you will be able to do the following:

❶ Identify the reactant, product, and transition state geometries on a reaction surface, and use the surface to estimate the energy of reaction and activation energy.

❷ Calculate the enthalpy of an isothermal reaction at arbitrary temperature and pressure from the enthalpies of formation and heat capacities.

❸ Calculate the adiabatic flame temperature of a compound.

❹ Calculate the equilibrium constant of a reaction from the Gibbs free energies of formation, and estimate the effects of temperature.

GOAL *Why Are We Here?*

Our goal in this chapter is to apply the principles of thermodynamics described in Part II to chemical reactions. What motivates us is the promise of understanding how we can predict the flow of energy in and out of reactive systems, both so we know what to expect and so we can look for opportunities to control the process. While our primary thesis in this textbook has been that molecular structure determines the macroscopic properties of our system, so too can we turn that concept around and use macroscopic parameters such as pressure and temperature to influence the direction and extent of chemical reactions—processes that are defined by activity at the microscopic scale.

CONTEXT *Where Are We Now?*

Finally we get to real chemistry, having developed all the tools of math and physics we need to describe chemical systems in the absence of reaction. Those tools will continue to serve us well in this final stretch. Thermodynamics, for example, continues to be useful, and we shall start with its application to chemical reactions.

Our brand of thermodynamics will tell us only about the equilibrium states of the system, but that's a good place to start. By understanding the endpoints of the reaction, we can begin to see whether or not the reaction will be spontaneous, and if so what the driving force is.

SUPPORTING TEXT *How Did We Get Here?*

This chapter focuses a good deal on enthalpy and entropy of reaction, so our most important preparation will be a familiarity with those two parameters and how we manipulate them in solving problems about

416

Context: Where Do We Go From Here? sections at the end of each chapter afford students a perspective on what they have just learned, and how it provides the foundation for the material explored in the next chapter.

Goal: Why Are We Here? chapter openers prepare students for the work ahead using one to two simple sentences.

Context: Where Are We Now? helps students understand how the chapter they are starting is related to what has come before and its place in the unfolding development of physical chemistry.

Supporting Text: How Did We Get Here? reviews previously introduced concepts, mathematical tools, and topical relationships that the new chapter will draw on.

Active research, tools, and techniques

Through learning about the instruments and methods of modern physical chemistry and meeting researchers at work today, students gain an appreciation for the practical applications of this science to many fields.

TOOLS OF THE TRADE | Bomb Calorimetry

The distinction between heat and temperature was not well established until the end of the 19th century, and consequently there was no consistent theory to describe the release or absorption of heat by chemical processes such as phase changes or chemical reactions. Around 1780, Antoine Lavoisier and Pierre Laplace together developed an instrument for measuring heats of various processes, which they gauged by the amount of ice melted, but the work was ahead of its time. Nearly a century later, Marcellin Berthelot developed the first modern device for measuring the heat flow in a chemical reaction: the bomb calorimeter.

What is a bomb calorimeter? A calorimeter is any device that measures the heat flow during a process. Calorimeters are the chief diagnostic tool in thermodynamics, and we will draw on many results from calorimetry in the chapters ahead. A bomb calorimeter is any calorimeter that operates with the sample at a fixed volume.

Why do we use a bomb calorimeter? Standard benchtop conditions in the laboratory allow us to maintain a constant temperature of the system (using a water bath or

heating mantle) and a constant pressure (by exposure to the atmosphere or—for air-sensitive compounds—by working in a glove-box filled with an inert gas at fixed pressure). Why is fixed pressure important? Keeping the pressure fixed reduces the number of changing variables, which is convenient for record keeping alone, but it also simplifies the thermodynamics whenever we can set one parameter to a constant. By fixing the pressure, we ensure that the enthalpy change during a process is equal to the heat:

$$\Delta H = \int dH = \int (T dS + V dP)$$

$$(\Delta H)_P = \int (T dS + V dP)_P = \int (T dS) = q \ \ if \ dP = 0,$$

where the subscript P indicates that the pressure is kept constant. The enthalpy was *invented* to make this relationship true.

But we have a more general definition—and a more intuitive understanding—of the *energy* E. If we want to measure ΔE instead of ΔH, however, the experiment can be much more challenging. The combustion of sucrose, for example,

$$C_{12}H_{22}O_{11} + 12O_2 \rightarrow 12CO_2 + 11H_2O$$

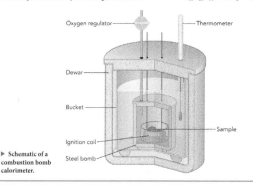

▶ Schematic of a combustion bomb calorimeter.

Labels: Oxygen regulator · Thermometer · Dewar · Bucket · Sample · Ignition coil · Steel bomb

BIOSKETCH | William A. Lester, Jr.

William A. Lester, Jr. is Professor of the Graduate School at the University of California at Berkeley, where he carries out research on the extension of Monte Carlo methods

(see Section 9.5) to quantum mechanical models of chemical reactions. Quantum Monte Carlo (QMC) calculations allow the thermodynamics of the system to be solved at the molecular scale by, for example, randomly sampling possible forms of the electron distribution. This approach allows QMC to address complex problems, because the necessary computer resources—although quite high—grow slowly (relative to other methods) with the system's number of degrees of freedom. Professor Lester and his research group have shown that QMC yields structures and energies of molecules with high accuracy, and they have extended this work to calculating thermochemical properties of reactions. Among recent projects, they have applied QMC to the prediction of bond dissociation energies and heats of formation of hydrocarbons. Theoretical studies such as these are particularly valuable for combustion science, where the several reactions occur simultaneously and often too rapidly to be well characterized by experiment.

Conceptual Insight and Mathematical Precision in a Real World Context

A discrete summary of the prerequisite mathematics and physics adds flexibility and convenience by incorporating the necessary math tools in a single chapter.

TABLE A.5 Solutions to selected integrals. In these equations, a and b are constants, n is a whole number, and C is the constant of integration.

$$\int x^n dx = \frac{1}{n+1}x^{n+1}+C \qquad \int a\,dx = a(x+C)$$

$$\int \frac{1}{x}dx = \ln x + C \qquad \int e^x dx = e^x + C$$

$$\int \ln x\,dx = x\ln x - x + C \qquad \int \frac{dx}{x(a+bx)} = -\frac{1}{a}\ln\left(\frac{a+bx}{x}\right)+C$$

$$\int \sin x\,dx = -\cos x + C \qquad \int \cos x\,dx = \sin x + C$$

$$\int \sin^2(ax)\,dx = \frac{x}{2}-\frac{\sin(2ax)}{4a}+C \qquad \int \cos^2(ax)\,dx = \frac{x}{2}+\frac{\sin(2ax)}{4a}+C$$

$$\int [f(x)+g(x)]dx = \int f(x)dx + \int g(x)dx \qquad \int_a^b dx = x\Big|_a^b = b-a$$

$$\int_0^\infty x^n e^{-ax}dx = \frac{n!}{a^{n+1}} \qquad \int_0^\infty e^{-ax^2}dx = \frac{1}{2}\left(\frac{\pi}{a}\right)^{1/2}$$

$$\int_0^\infty xe^{-ax^2}dx = \frac{1}{2a} \qquad \int_0^\infty x^2 e^{-ax^2}dx = \frac{1}{4}\left(\frac{\pi}{a^3}\right)^{1/2}$$

$$\int_0^\infty x^{2n+1}e^{-ax^2}dx = \frac{n!}{2a^{n+1}} \qquad \int_0^\infty x^{2n}e^{-ax^2}dx = \frac{[1\cdot3\cdot5\dots(2n-1)]\sqrt{\pi}}{2^{n+1}a^{n+(1/2)}}$$

$$\int_0^\infty x^n e^{-ax}dx = \frac{n!}{a^{n+1}}-e^{-ax}\sum_{i=0}^n \frac{n!x^{n-i}}{a^{i+1}(n-i)!}$$

the value of C is lost. When we undo the derivative by taking the integral, we add an unknown constant of integration to the integrated expression. Omit this constant when solving definite integrals, because the limits of integration will determine its value.

3. The function being integrated is the **integrand**, and it is multiplied by the incremental change along the coordinates, called the volume element.

Most of the algebraic solutions to integrals that we need appear in Table A.5.

EXAMPLE A.4 Analytical Integration

PROBLEM Evaluate the numerical value for each of the following expressions.

1. $\int_1^4 \frac{dx}{x}$

2. $\int_0^\infty e^{-2x}dx$

3. $\int_0^{\pi/3}(3\cos\theta^2 - 1)\sin\theta\,d\theta$

SOLUTION These can be solved by substitution of the expressions in Table A.5.

(a) $\int_1^4 \frac{dx}{x} = \ln x\Big|_1^4 = \ln 4 - \ln 1 = 1.386 - 0 = 1.386$

(b) $\int_0^\infty e^{-2x}dx = -\frac{1}{2}e^{-2x}\Big|_0^\infty = -\frac{1}{2}(e^{-\infty}-e^0) = -\frac{1}{2}(0-1) = \frac{1}{2}$

(c) $\int_0^{\pi/3}(3\cos^2\theta - 1)\sin\theta\,d\theta = [-\cos^3\theta + \cos\theta]\Big|_0^{\pi/3}$

$$= \left[-\left(\frac{1}{2}\right)^3 + \left(\frac{1}{2}\right)\right] - [-(1)^3 + (1)] = \frac{3}{8}.$$

By the way, it is possible to apply rules of symmetry to extend some of the analytical solutions in Table A.5. For example, when the integrand is $x^{2n}e^{-ax^2}$, then the function is exactly the same from 0 to $-\infty$ as from 0 to $+\infty$ (Fig. A.3a). Therefore, the integral $\int_{-\infty}^\infty x^{2n}e^{-ax^2}dx$ is equal to 2 times $\int_0^\infty x^{2n}e^{-ax^2}dx$. However, if the power of x is odd, $2n+1$, then the function is negative when $x < 0$ and positive when $x > 0$ (Fig. A.3b). The integral from $-\infty$ to 0 cancels the integral from 0 to $+\infty$, so $\int_{-\infty}^\infty x^{2n+1}e^{-ax^2}dx = 0$.

Numerical Integration

Not all integrals have algebraic solutions, and some have algebraic solutions only between certain limits (such as 0 and ∞). With suitable computers, any integral can be calculated without trying to cram it into some algebraic form. This is accomplished by going back to the definition in calculus,

$$\int_{x_1}^{x_2} f(x)\,dx = \lim_{\delta x \to 0}\left\{\sum_{i=1}^N f[x_1 + i\,\delta x]\right\}\delta x \qquad (A.20)$$

▶ **FIGURE A.3 Symmetry and definite integrals. (a)** If a function $f(x)$ is equal to $f(-x)$ for all values of x, then the integral from $-\infty$ to ∞ is equal to 2 times the integral from 0 to ∞. **(b)** If $f(x)$ is equal to $-f(-x)$, then the integral from $-\infty$ to ∞ is 0.

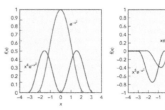

Chapter A provides a comprehensive summary of the physical laws and mathematical tools used to develop the principles of physical chemistry.

The distinctive use of color in the text's mathematical narrative allows students to identify important equation elements (such as the partition function) even as they take on different mathematical forms.

$$= \sum_{v=0}^{\infty} \frac{e^{-\omega_e v/(k_B T)}}{q_{vib}(T)} \omega_e v$$

$$= \left(1 - e^{-\omega_e/(k_B T)}\right) \sum_{v=0}^{\infty} e^{-\omega_e v/(k_B T)} \omega_e v.$$

Thoughtful color-coding in key equations makes it easier for students to follow the development of complex derivations as well as recognize common mathematical elements that appear in the representation of different physical situations.

Derivations Demystified

$$\left(\frac{\partial E}{\partial T}\right)_{V,n} = \left(\frac{T\partial S - P\partial V}{\partial T}\right)_{V,n} \qquad \text{by Eq. 7.12}$$

$$= \left(\frac{T\partial S}{\partial T}\right)_{V,n} \qquad \partial V = 0 \text{ if } V \text{ constant}$$

$$= \left(\frac{dq}{dT}\right)_{V,n} \qquad \text{by Eq. 7.43}$$

$$\equiv C_V(T). \qquad \text{by Eq. 7.39}$$

Derivations are made transparent and comprehensible to students without sacrifice of mathematical rigor. Colored annotations provide crucial help to students by explaining important steps in key derivations.

DERIVATION SUMMARY The Maxwell-Boltzmann Distribution. Maxwell assumed that a Gaussian distribution applied to the velocity components v_X, v_Y, v_Z of all the velocity vectors $\vec{v}$ of the particles in a gas, and we used this assumption to obtain a probability distribution of velocity vectors. To eliminate the dependence of direction, we integrated that distribution over all angles, which introduced a factor of v^2 in addition to the Gaussian exponential e^{-av^2}. We solved for the constant a in the exponential by requiring the mean square speed $\langle v^2 \rangle$ to predict the experimentally determined kinetic energy of a monatomic gas as a function of temperature. We also required the probability to be normalized, so that $\int_0^{\infty} \mathcal{P}_v(v)\,dv = 1$, in order to obtain the value of the coefficient that multiplies the entire function.

Summaries spell out the essential results of difficult derivations, making it easier to accommodate the needs of different courses, the preferences of different instructors, and the study and review habits of different students.

Supporting students' quest for deeper understanding

With numerous worked examples, robust review support, a wealth of end-of-chapter problems, and a solutions manual written by the text's author, students have everything they need to master the basics of physical chemistry.

Worked Examples provide students with context of the problem, clearly describe the parameters of the problem, and walk students step-by-step toward the solution.

EXAMPLE 8.1   Gas Compression in a Pump

CONTEXT In many applications in chemistry and chemical engineering, we have to push gases from one place to another (for example, to get a reactant gas to the inlet of a reactor) or pull gases out of a container (for example, to create a vacuum in which we can manipulate ions in a mass spectrometer). We use gas pumps to do this. A typical configuration is a rotary vane pump, which admits gas at an inlet, traps it, compresses it to a smaller volume and higher pressure, and then releases it through an outlet at its new, higher pressure (Fig. 8.5). Pumps are rated partly by their **compression ratio**, the volume of the gas at the inlet to the gas at the outlet, V_{inlet}/V_{outlet}, which also gives the ratio of the outlet pressure to inlet pressure P_{outlet}/P_{inlet}. To achieve higher compression ratios (and therefore lower inlet pressures), a rotary vane pump may be divided into two stages, with the second stage taking the compressed gas from the first stage and compressing it still more. This can be an energy-intensive process, and we can use our results to estimate the minimum work necessary to compress a gas isothermally.

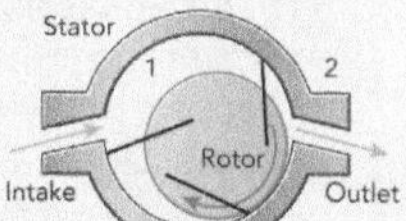

PROBLEM A vacuum chamber in a spectrometer is maintained at an operating pressure of 10.0 mtorr by a two-stage rotary vane pump with an exhaust pressure at the pump outlet of 800. torr. What is the minimum power in watts (J/s) consumed by the pump to keep the chamber at this pressure when there is a flow of 0.22 mmol/s and $T = 300.$ K?

SOLUTION The amount of gas moved in 1 second is 0.22 mol. This gas must be compressed by a factor of $800/0.010 = 80,000$ to achieve the pressure increase from inlet to outlet. The reversible work done per second gives the minimum power needed, because the reversible process wastes none of the work:

$$w_{T,rev} = -nRT \ln\left(\frac{V_2}{V_1}\right)$$

$$\frac{w_{T,rev}}{\Delta t} = -\left(\frac{n}{\Delta t}\right)RT \ln\left(\frac{P_1}{P_2}\right)$$

$$= -(0.22 \cdot 10^{-3} \text{mol/s})(8.3145 \text{ J K}^{-1}\text{mol}^{-1})(300. \text{K}) \ln(1/80,000) = 6.2 \text{ W}$$

▲ **FIGURE 8.5 Schematic of a typical rotary-vane pump.** The rotor spins inside the cavity of the stator, with vanes (shown as black lines) that slide in and out of the rotor as the gap between the rotor and stator changes. Vacuum pump oil makes a gas-tight seal between the regions separated by the vanes. As the rotor spins, gas enters through the intake into a high-volume region 1. As the rotor turns, the gas becomes trapped by the vanes and compressed until it gets pushed through the outlet.

Stator · Rotor · Intake · Outlet · 1 · 2

KEY CONCEPTS AND EQUATIONS

KEY TERMS

OBJECTIVES REVIEW

PROBLEMS

A comprehensive online solutions manual, written by author Andrew Cooksy, is filled with unique solution sets emphasizing qualitative results to help students move beyond the math to a deeper conceptual understanding.

End-of-chapter materials bring students full circle, helping them assess their grasp of current chapter concepts and synthesize information from prior chapters.

MasteringChemistry® for Students

www.masteringchemistry.com

MasteringChemistry provides dynamic, engaging experiences that personalize and activate learning for each student. Research shows that Mastering's immediate feedback and tutorial assistance helps students understand and master concepts and skills—allowing them to retain more knowledge and perform better in this course and beyond.

Student Tutorials

Physical chemistry tutorials reinforce conceptual understanding. Over 460 tutorials are available in MasteringChemistry for Physical Chemistry, including new ones on The Cyclic Rule and Thermodynamic Relation of Proofs.

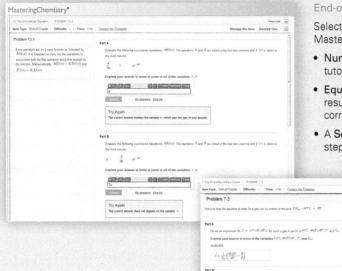

End-of-Chapter Content Available in MasteringChemistry:

Selected end-of-chapter problems are assignable within MasteringChemistry, including:

- **Numerical answer** hints and feedback are only with tutorials in this course
- **Equation and Symbolic answer types** so that the results of a self-derivation can be entered to check for correctness, feedback, and assistance
- A **Solution View** that allows students to see intermediate steps involved in calculations of the final numerical result

MasteringChemistry®

www.masteringchemistry.com

Easy to get started. Easy to use.

MasteringChemistry provides a rich and flexible set of course materials to get you started quickly, including homework, tutorial, and assessment tools that you can use *as is* or customize to fit your needs.

NEW! Calendar Features

The Course Home default page now features a **Calendar View** displaying upcoming assignments and due dates.

- Instructors can schedule assignments by dragging and dropping the assignment onto a date in the calendar. If the due date of an assignment needs to change, instructors can drag the assignment to the new due date and change the "available from and to dates" accordingly.

- The calendar view gives students a syllabus-style overview of due dates, making it easy to see all assignments due in a given month.

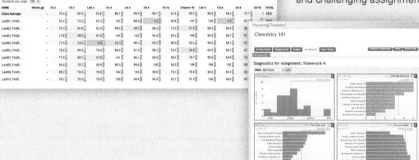

Gradebook

Every assignment is automatically graded. Shades of red highlight struggling students and challenging assignments at a glance.

Gradebook Diagnostics

This screen provides you with your favorite diagnostics. With a single click, charts summarize the most difficult problems, vulnerable students, grade distribution, and even score improvement over the course.

NEW! Learning Outcomes

Let Mastering do the work in tracking student performance against your learning outcomes:

- Add your own or use the publisher-provided learning outcomes.

- View class performance against the specified learning outcomes.

- Export results to a spreadsheet that you can further customize and share with your chair, dean, administrator, or accreditation board.

PART I
EXTRAPOLATING
FROM MOLECULAR
TO MACROSCOPIC
SYSTEMS

PART II
NON-REACTIVE
MACROSCOPIC
SYSTEMS

PART III
REACTIVE
SYSTEMS

A Introduction: Tools from Math and Physics

GOAL *Why Are We Here?*

The goal of this textbook is a concise and elegant exposition of the theoretical framework that forms the basis for all modern chemistry. To accomplish this, we are going to draw regularly on your knowledge of algebra, geometry, calculus, mechanics, electromagnetism, and chemistry. Physical chemistry is both rewarding and challenging in this way.

Mathematics of several varieties is our most valuable tool, and in this text we shall be interested in it only as a tool. It is not necessary, for example, that you remember how to derive the algebraic solution to the integral $\int \ln x \, dx$, but it will help if you know that an algebraic solution exists and how to use it (because with it we will obtain a useful equation for diffusion). This chapter is a summary of the math and physics that serve as our starting point as we explore the theory of chemistry. If you are embarking on this course, you may wish to review any of the following topics that appear alarmingly unfamiliar at first glance.

A.1 Mathematics

Algebra and Units

Basic Formula Manipulations

The use of algebra in this text is similar to its use in introductory physics and chemistry courses. We will routinely encounter the basic manipulations of variables in equations, especially to solve for one unknown in terms of several known constants. A tough example would be to solve for n_B in the equation

$$T_B = T_B' \left[\frac{V_T - V_A}{V_T - V_A'} \right]^{-n_B R / C_B}$$

The key is to see that a solution must be available, because the variable we are solving for appears in only one place, and a series of operations will allow us to isolate it on one side of the equation. Once we recognize that, then we can methodically undo the operations on one side of the equation to leave n_B: divide both sides by T_B', take the logarithm of both sides to bring n_B down to earth from the exponent, and finally divide both sides by the factor that leaves n_B alone on one side of the equation. Those steps eventually bring us to

$$n_B = -\frac{C_B}{R}\frac{\ln\left(\dfrac{T_B}{T_B'}\right)}{\ln\left(\dfrac{V_T - V_A}{V_T - V_A'}\right)}.$$

One issue that makes the algebra something of a challenge is the notation. To put it mildly, we will use a lot of algebraic symbols. In fact, with the exception of "O," which looks too much like a zero, we use the entire Roman alphabet at least twice, and most of the Greek.[1] The symbols have been chosen in hopes of an optimal combination of (a) preventing the same symbol from appearing with different meanings in the same chapter, (b) adherence to the conventional usage in the scientific literature, and (c) clarity of meaning. Unfortunately, these three aims cannot always be satisfied simultaneously. Physical chemistry is a synthesis of work done by pioneers in mathematics, physics, and chemistry, often without any intention that the results would one day become integrated into a general theory of chemistry. We bring together many fields that evolved independently, and the way these fields fit together is one of the joys of this course. Admittedly, the complexity of the notation is not.

The text provides guides to the notation used in long derivations and sample calculations to show how the notation is used. Please be aware, however, that no textbook gimmick can substitute for the reader's understanding of the parameters represented by these symbols. If you recognize the difference between the fundamental charge e and the base of the natural logarithm e, you are in no danger of confusing the two, even though they are both represented by the letter "e," sometimes appearing in the same equation.

Unit Analysis and Reasonable Answers

One of the most helpful tools for checking algebra and for keeping these many symbols under control is unit analysis. If a problem asks you to solve for the value of some variable Y, and you're not certain what units you will get in the end, then it's likely that the meaning of Y has not been made entirely clear. In many cases, including viscosities and wavefunctions, the units are not obvious from the variable's definition in words but are easily determined from an important equation in which the variable appears. Quick: how do you write the units for pressure in terms of mass and distance and time? If you recall the definition of the pressure as force per unit area

$$P = \frac{F}{A}$$

---

[1] If the lower case Greek letter upsilon (υ) didn't look so much like an italic "v" (v), there are at least two places it would have been used. It's bad enough that v and the Greek nu (ν) are so similar and sometimes appear in the same equation.

and know that force has units of mass times acceleration, then pressure must have units of

$$\frac{\text{force}}{\text{distance}^2} = \frac{\text{mass} \times \text{speed/time}}{\text{distance}^2} = \frac{\text{mass} \times \text{distance/time}^2}{\text{distance}^2}$$

$$= \frac{\text{mass}}{\text{distance} \times \text{time}^2} = \text{kg m}^{-1}\,\text{s}^{-2}. \tag{A.1}$$

It will not be worthwhile to attempt a problem before understanding the variables involved.

Unit analysis is also a useful guard against algebraic mistakes. An error in setting up an algebraic solution often changes the units of the answer, and a check of the answer's units will show the mistake. This does not protect against many other mistakes, however, such as dividing instead of multiplying by 10^{10} to convert a length from meters to angstroms. In such cases, there is no replacement for knowing what range of values is appropriate for the quantity. Recognizing a reasonable value for a particular variable is primarily a matter of familiarity with some typical parameters. The values given in Table A.1 are meant only to give common orders of magnitude for various quantities. Answers differing by factors of 10 from these may be possible, but not common.

TABLE A.1 **Some typical values for parameters in chemical problems.** These are meant only as a rough guide to expected values under typical conditions.

Parameter	Value (in typical units)
chemical bond length	1.5 Å
chemical bond energy	400 kJ mol^{-1}
molecular speed	200 m s^{-1}
mass density (solid or liquid)	1 g cm^{-3}

EXAMPLE A.1 Unreasonable Answers

PROBLEM Unit analysis and recognition of a reasonable value can prevent errors such as those that resulted in the following answers. Identify the problem with these results for the requested quantity:

Quantity	Wrong answer
the density of NaCl(s)	$1.3 \cdot 10^{-24}$ g cm^{-3}
the density of NaCl(s)	$3.3 \cdot 10^{7}$ g cm^{-1}
bond length of CsI	12.3 m
speed of a molecule	$4.55 \cdot 10^{11}$ m s^{-1}
momentum of electron	$5 \cdot 10^{-10}$ m s^{-1}

SOLUTION Each of those examples gives an answer of entirely the wrong magnitude (which could arise from using the wrong conversion factor, the wrong units, or both).

Quantity	Wrong answer	Why unreasonable
the density of NaCl(s)	$1.3 \cdot 10^{-24}$ g cm^{-3}	too small
the density of NaCl(s)	$3.3 \cdot 10^{7}$ g cm^{-1}	wrong units
bond length of CsI	12.3 m	too big
speed of a molecule	$4.55 \cdot 10^{11}$ m s^{-1}	too big (greater than speed of light)
momentum of electron	$5 \cdot 10^{-10}$ m s^{-1}	wrong units

In many problems, the units themselves require some algebraic manipulation because several units are products of other units. For example, the unit of pressure, 1 kg m^{-1} s^{-2}, obtained in Eq. A.1, is called the "pascal." We shall also encounter an equation

$$E_n = -\frac{Z^2 m_e e^4}{2(4\pi\varepsilon_0)^2 n^2 \hbar^2},$$

in which E_n has units of energy, Z and n are unitless, m_e has units of mass, e has units of charge, ε_0 has units of charge^2 energy^{-1} distance^{-1}, and $\hbar$ has units of energy $\times$ time. The units on each side of the equation must be identical, and this we can show by substituting in the appropriate units for mass, charge, and energy:

$$1 J = 1 \frac{(\text{kg})(\text{C})^4}{(\text{C}^2 \text{J}^{-1} \text{m}^{-1})^2 (\text{Js})^2}$$

$$= 1 \frac{(\text{kg})(\text{C})^4}{\text{C}^4 \text{s}^2/\text{m}^2}$$

$$= 1 \text{ kg m}^2 \text{ s}^{-2} = 1 \text{ J}. \tag{A.2}$$

This may be a good place to remind you about that bothersome factor of $4\pi\varepsilon_0$ and some other aspects of the SI units convention.

SI Units

The accepted standard for units in the scientific literature is the Système International (SI), based on the meter, kilogram, second, coulomb, kelvin, mole, and candela.[2] It is acceptable SI practice to use combinations of these units and to convert up or down by factors of 1000. So, for example, the SI unit of force should have units of (mass $\times$ acceleration), or kg m s^{-2}, a unit commonly called the newton and abbreviated N. Energy has units of force $\times$ distance, so the SI unit is kg m^2 s^{-2}, also called the joule and abbreviated J. But the joule is inconveniently small for measuring, say, the energy released in a chemical reaction, so one could use the kilojoule (10^3 J) and remain true to the SI standard. We'll give special attention to energy units shortly.

A practical advantage of a single system for all physical units is that—if you're careful—the units take care of themselves. Allowing for the factors of 1000, if all the quantities on one side of an equation are in SI units, the value

---

[2]If you don't recall the candela, that's understandable. It's the unit of luminous intensity, and with that, makes its last appearance in this text.

on the other side will also be in SI units. If an object of mass 2.0 kg rests on a table, subject to the gravitational acceleration of 9.8 m s^{-2}, then I can calculate the force it exerts on the table by multiplying the mass and the acceleration,

$$F = ma = (2.0\text{ kg})(9.8\text{ m s}^{-2}) = 20\text{ N},$$

and I can be certain that the final value is in SI units for force, namely newtons.

Standardization of units takes time, however, and you can be certain that the chemical data you encounter in your career will not adhere to one standard. One formerly common set of units, now widely discouraged, is the **Gaussian** or **CGS system,** similar to SI except that it replaces the meter, kilogram, and coulomb with the centimeter, gram, and electrostatic unit, respectively. Another convention, now on the rise, is the set of atomic units, for which all units are expressed as combinations of fundamental physical constants such as the electron mass m_e and the elementary charge e.

The SI system, while having some features convenient to engineering, suffers from one inconvenience in our applications: elementary calculations that include electric charges or magnetic fields require the use of constants called the permeability μ_0 and permittivity ε_0 of free space. Although these constants originally appeared with a physical meaning attached, for our purposes they are merely conversion factors. In particular, the factor $4\pi\varepsilon_0$ converts SI units of coulomb squared to units of energy times distance, J $\cdot$ m. For example, the energy of repulsion between two electrons at a separation of $d = 1.0 \cdot 10^{-10}$ m is

$$\frac{e^2}{4\pi\varepsilon_0 d} = \frac{(1.602 \cdot 10^{-19}\text{ C})^2}{(1.113 \cdot 10^{-10}\text{ C}^2\text{ J}^{-1}\text{ m}^{-1})(1.0 \cdot 10^{-10}\text{ m})} = 2.306 \cdot 10^{-18}\text{ J. (A.3)}$$

In contrast, the atomic and CGS units fold this conversion into the definition of the charge, and the factor of $4\pi\varepsilon_0$ would *not* appear in the calculation. For all equations in this text involving the forces between charged particles, we conform to the standards of the day and use SI units and the associated factor of $4\pi\varepsilon_0$.

In other cases, however, we will not adhere strictly to the SI standard. Even allowing for factors of 1000, I don't know any chemists who express molecular dipole moments in coulomb meters, a unit too large for its purpose by 30 orders of magnitude (not even prefixes like "micro-" and "nano-" are enough to save it). The conventional unit remains the debye, which is derived from CGS units (adjusted by 18 orders of magnitude, it must be said) and just the right size for measuring typical bond dipoles. The angstrom (Å) also remains in wide use in chemistry because it is a metric unit (1 Å = 10^{-10} m) that falls within a factor of 2 of almost any chemical bond length.

Of all the physical parameters, energy has the greatest diversity in commonly used scientific units. There are several ways to express energy, even after excluding all sorts of nonmetric energy units (such as the British thermal unit, kilowatt-hour, foot-pound, ton of TNT, and—most beloved of chemists—the calorie). Other conventions appear when discussing the interaction of radiation with matter, for which it is common to quantify energy in terms of the frequency (s^{-1}) or reciprocal wavelength (cm^{-1}) of the radiation. Under the proper assumptions, it may also be informative to convert an energy to a corresponding temperature, in units of kelvin. Typical laboratory samples of a compound have numbers of molecules in the range of 10^{20} or more, and molecular energies are therefore often given in terms of the energy per mole of the compound (e.g., $kJ\,mol^{-1}$). These cases will be explained as they appear, and they are summarized in the conversion table for energies on this textbook's back endpapers.

Once these non-SI units are introduced, please make sure you are comfortable with the algebra needed to convert from one set of units to another. This one skill, mundane as it may seem, will likely be demanded of you in any career in science or engineering. Famous and costly accidents have occurred because this routine procedure was not given its due attention.[3]

Complex Numbers

Complex numbers are composed of a real number and an imaginary number added together. For our purposes, a complex number serves as a sort of two-dimensional number; the imaginary part contains data on a measurement distinct from the data given by the real part. For example, a sinusoidal wave that varies in time may be described by a complex number in which the real part gives the shape of the wave at the current time and the imaginary part describes what the wave will look like a short time later.

The imaginary part of any complex number is a real number multiplied by $i \equiv \sqrt{-1}$. (The symbol "$\equiv$" is used throughout this text to indicate a definition, as opposed to the "$=$" symbol, used for equalities that can be proved mathematically.) This relationship between i and -1 allows the imaginary part of a complex number to influence the real-number results of an algebraic operation. For example, if a and b are both real numbers, then $a + ib$ is complex, with a the real part and ib the imaginary part. The **complex conjugate** of $a + ib$, written $(a + ib)^*$, is equal to $a - ib$, and the product of any number with its complex conjugate is a real number:

$$(a + ib)(a - ib) = a^2 - iba + iba - i^2b^2 = a^2 + b^2. \qquad (A.4)$$

Notice that the value of b—even though it was contained entirely in the imaginary parts of the two original complex numbers—contributes to the value of the real number quantity that results from this operation.

Many of the mathematical functions in the text are complex, but multiplication by the complex conjugate yields a real function, which can correspond directly to a measurable property. For that reason, we often judge the validity of the functions by whether we can integrate over the product f^*f. In this text, a well-behaved function f is single-valued, finite at all points, and yields a finite value when f^*f is integrated over all points in space. To be very well-behaved, the function and its derivatives should also be continuous functions, but we will use a few functions that are naughty in this regard.

---

[3]A prominent example is the loss in 1999 of the unmanned Mars Climate Orbiter, a probe that entered the Martian atmosphere too low and burned up because engineers were sending course correction data calculated using forces in pounds to an on-board system that was designed to accept the data in newtons.

EXAMPLE A.2   Complex Conjugates

PROBLEM Write the complex conjugate f^* for each of the following expressions f and show that the value of f^*f is real.

1. $5 + 5i$
2. $-x/i$
3. $\cos x - i\sin x$

SOLUTION

1. $f^* = 5 - 5i$

$$f^*f = (5 + 5i)(5 - 5i) = 25 + 25 = 50$$

2. First we would like to put this in the form $a + ib$, so we multiply by $\frac{i}{i}$ to bring the factor of i into the numerator:

$$f = -\frac{x}{i}\left(\frac{i}{i}\right) = -\frac{ix}{-1}.$$

The real part of this function is zero, but for any complex conjugate, we change the sign on the imaginary term: $f^* = -ix$

$$f^*f = (ix)(-ix) = -i^2x^2 = x^2$$

3. $f^* = \cos x + i\sin x$

$$f^*f = \cos^2 x - i^2\sin^2 x = \cos^2 x + \sin^2 x = 1$$

Trigonometry

Elementary results from trigonometry play an important role in our equations of motion, and therefore you should know the definitions of the sine, cosine, and tangent functions (and their inverses) as signed ratios of the lengths of the sides of a right triangle. Using the triangle drawn in Fig. A.1, with sides of length y, x, and r, we would define these functions as follows:

$$\sin\phi \equiv \frac{y}{r} \qquad \csc\phi \equiv \frac{1}{\sin\phi} = \frac{r}{y}$$

$$\cos\phi \equiv \frac{x}{r} \qquad \sec\phi \equiv \frac{1}{\cos\phi} = \frac{r}{x} \qquad \text{(A.5)}$$

$$\tan\phi \equiv \frac{y}{x} \qquad \cot\phi \equiv \frac{1}{\tan\phi} = \frac{x}{y}$$

The sign is important. If ϕ lies between $90°$ and $270°$, then the x value becomes negative, so $\cos\phi$ and $\sec\phi$ would be less than zero. Similarly, $\sin\phi$ and $\csc\phi$ are negative for ϕ between $180°$ and $360°$.

Please also make sure you are comfortable using the trigonometric identities listed in Table A.2. These are algebraic manipulations that may allow us to simplify equations or to isolate an unknown variable.

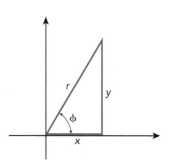

▲ FIGURE A.1 **Right triangle used to define trigonometric functions of the angle ϕ.**

TABLE A.2 **Selected trigonometric identities.**

$\sin^2 x + \cos^2 x = 1$	$\sec^2 x - \tan^2 x = 1$
$\sin(x \pm y) = \sin x \cos y \pm \cos x \sin y$	$\cos(x \pm y) = \cos x \cos y \mp \sin x \sin y$
$\sin x \sin y = [\cos(x-y) - \cos(x+y)]/2$	$\cos x \cos y = [\cos(x+y) + \cos(x-y)]/2$
$\sin x \cos y = [\sin(x+y) + \sin(x-y)]/2$	
$\sin 2x = 2 \sin x \cos x$	$\cos 2x = 2 \cos^2 x - 1$

Coordinate Systems

Mathematical functions are described by their variables, but we have some choice in deciding what those variables are. Rather than defining the function $f(x) = x^2$ as written, we could define it in terms of a new variable $y = 2x$, for which $f(y) = y^2/4$.

For functions that represent distributions in three-dimensional space, there are two common choices of variables: the **Cartesian coordinates,** (x, y, z); and the **spherical polar coordinates,** (r, θ, ϕ). The Cartesian coordinates can each vary from $-\infty$ to $+\infty$. The polar coordinates lie in the ranges

$$0 \leq r < \infty \qquad 0 \leq \theta < \pi \qquad 0 \leq \phi < 2\pi,$$

where π **radians** is equal to $180°$, and the radian is the ratio of a circle's circumference to its diameter. Usually when we move between the two systems, we will take the angle θ as measured in any direction from the positive half of the z axis, and the angle ϕ as the angle measured parallel to the xy plane from the positive x axis towards the positive y axis. The distance r is always measured in any direction from the origin. These definitions are illustrated in Fig. A.2.

The Cartesian and spherical polar coordinate systems satisfy the fundamental requirements for a complete coordinate system in three-dimensional space—namely, that every point in space can be represented by some set of values for these coordinates, and every set of coordinates corresponds to only one point in space. Although the Cartesian coordinate representation of a single point may be easier for us to visualize than the representation in spherical coordinates, functions that have a lot of angular symmetry can be written and manipulated much more easily in spherical coordinates than in Cartesian coordinates.

Converting between Cartesian and spherical coordinates is straightforward but often tedious. The most crucial conversions between Cartesian and spherical coordinates have been done for us by someone else, and we should not be too shy to take advantage of all that hard work. Should it be necessary to convert between the two systems for a particular application, the following equations can be used:

$$x = r \sin\theta \cos\phi \qquad r = (x^2 + y^2 + z^2)^{1/2}$$

$$y = r \sin\theta \sin\phi \qquad \theta = \arccos\left(\frac{z}{r}\right) \qquad \text{(A.6)}$$

$$z = r \cos\theta \qquad \phi = \arctan\left(\frac{y}{x}\right)$$

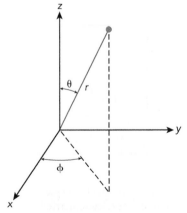

▲ FIGURE A.2 **The relation between spherical polar and Cartesian coordinates.**

The most important conversion we will need is between the **volume elements,** abbreviated $d\tau$, that appear in all integrals. The volume element is so named because its integral, evaluated over some three-dimensional region, is the volume enclosed by that region. For an integral over three-dimensional space, the volume element is

$$d\tau \equiv dx\,dy\,dz = r^2\,dr\sin\theta\,d\theta\,d\phi. \qquad (A.7)$$

Although this equation is not obvious at first glance, we can observe easily that $d\tau$ has units of volume as promised. For the Cartesian volume element, $dx\,dy\,dz$ is the volume of a cube with sides of length dx, dy, and dz and has units of volume. The only spherical coordinate with units of distance, r, appears three times in the spherical volume element: twice in r^2 and once in dr (which has the same units as r), giving units of distance^3 or volume. The remaining terms, $\sin\theta\,d\theta\,d\phi$, are unitless.

EXAMPLE A.3   Cartesian and Polar Coordinates

PROBLEM Convert the following Cartesian expression into spherical coordinates and the spherical polar expression into Cartesian coordinates.

$$f(x,y,z) = z\,e^{-(x^2+y^2+z^2)/a^2}$$
$$g(r,\theta,\phi) = (3\cos^2\theta - 1)\tan\phi$$

SOLUTION We can directly substitute using the expressions in Eqs. A.6:

$$f(r,\theta,\phi) = (r\cos\theta)\,e^{-r^2/a^2}$$
$$g(x,y,z) = \left[3\left(\frac{z}{r}\right)^2 - 1\right]\tan\left[\arctan\frac{y}{x}\right]$$
$$= \left[3\left(\frac{z^2}{x^2+y^2+z^2}\right)-1\right]\frac{y}{x}$$

Linear Algebra

Linear algebra is so named because it grew out of methods for solving systems of linear equations. For our purposes, it is the branch of mathematics that describes how to perform arithmetic and algebra using vectors and matrices.

Vectors

Formally, a **vector** is a set of two or more variable values, but our use of the term will be restricted to *Euclidean* vectors, which are governed by the following definitions and rules:

1. A vector has direction, which can be specified by assuming one of the endpoints to be the origin and giving the coordinates of the other endpoint. As an example, the vector $(1,0,0)$ has one end at the origin and the other end at $x = 1$ on the x axis.

2. A vector $\vec{A} = (A_x, A_y, A_z)$ has a length or **magnitude,** indicated $|\vec{A}|$ or simply A, where

$$|\vec{A}| \equiv A = \sqrt{A_x^2 + A_y^2 + A_z^2}. \qquad (A.8)$$

3. The **dot product** of two vectors $\overrightarrow{A} = (A_x, A_y, A_z)$ and $\overrightarrow{B} = (B_x, B_y, B_z)$ is a scalar quantity (*not* a vector) given by

$$\overrightarrow{A} \cdot \overrightarrow{B} \equiv A_x B_x + A_y B_y + A_z B_z. \tag{A.9}$$

4. The dot product of $\overrightarrow{A}$ and a **unit vector** (vector of length one) parallel to $\overrightarrow{B}$ is called the **projection** of $\overrightarrow{A}$ onto $\overrightarrow{B}$; this is often evaluated with $\overrightarrow{B}$ chosen to be one of the coordinate axes, such as

$$\overrightarrow{A} \cdot \hat{z} = A_z,$$

where $\hat{z} \equiv (0,0,1)$. This quantity gives the extent that the vector $\overrightarrow{A}$ stretches along the z direction, and is often called the z **component** of $\overrightarrow{A}$.

5. The **cross product** of two vectors is also a vector, given by

$$\overrightarrow{A} \times \overrightarrow{B} \equiv (A_y B_z - A_z B_y, A_z B_x - A_x B_z, A_x B_y - A_y B_x). \tag{A.10}$$

The cross product $\overrightarrow{A} \times \overrightarrow{B}$ is always perpendicular to the vectors $\overrightarrow{A}$ and $\overrightarrow{B}$.

6. The vector sum of $\overrightarrow{A}$ and $\overrightarrow{B}$ is given by

$$\overrightarrow{A} + \overrightarrow{B} = (A_x + B_x, A_y + B_y, A_z + B_z) \tag{A.11}$$

and is a vector with maximum magnitude $A + B$ (if the two vectors point in exactly the same direction) and minimum magnitude $|A - B|$ (if they point in exactly opposite directions).

Matrices

Although we will use vectors to represent physical quantities, such as position and angular momentum, to a mathematician a vector is any set of expressions that depend on some index. For example, the position vector $\overrightarrow{r}$ is the set of coordinate values r_i, where $r_1 = x$, $r_2 = y$, and $r_3 = z$. In that example, the index i lets us pick out one part of the vector. A **matrix** is a set of values or functions that depend on at least two different (and usually independent) indices. We will not encounter many matrices in this text, but there are a few places where they allow you to go one step farther in calculating important physical quantities in chemistry.

As an example, we may write the matrix $\mathbf{R}$ of values $r_i r_j$ for each i and j from 1 to 3:

$$\mathbf{R} = \begin{pmatrix} r_1 r_1 & r_1 r_2 & r_1 r_3 \\ r_2 r_1 & r_2 r_2 & r_2 r_3 \\ r_3 r_1 & r_3 r_2 & r_3 r_3 \end{pmatrix} = \begin{pmatrix} x^2 & xy & xz \\ yx & y^2 & yz \\ zx & zy & z^2 \end{pmatrix}.$$

This matrix gives all the possible combinations of x, y, and z with x, y, and z. The matrix $\mathbf{R}$ would be one short way to represent all the terms that would arise from expanding $(x + y + z)^2$:

$$(x + y + z)^2 = x^2 + y^2 + z^2 + 2xy + 2yz + 2xz.$$

It would also represent them in such a way that we could pick out any one of those terms—any single **matrix element** R_{ij}—by itself from the values of the two indices, as for example $R_{13} = r_1 r_3 = xz$.

There is an algebra for matrices. We can multiply a matrix by a constant:

$$c\begin{pmatrix} f \\ g \end{pmatrix} = \begin{pmatrix} cf \\ cg \end{pmatrix}. \tag{A.12}$$

We can also multiply a matrix and a vector $\overrightarrow{A}$, obtaining a new vector $\overrightarrow{B}$ according to the formula $B_i = \sum_j R_{ij} A_j$. For example, the product of any 2×2 matrix and a 2-coordinate vector is given by

$$\begin{pmatrix} r & s \\ t & u \end{pmatrix} \begin{pmatrix} f \\ g \end{pmatrix} = \begin{pmatrix} rf + sg \\ tf + ug \end{pmatrix}. \tag{A.13}$$

The product is a *new* vector. The multiplication just shown forms the basis for one of the most common applications of matrices in physics: changing a vector from one form to another. For example, start with the vector (a, b, c) where a, b, and c are constants giving the length of the vector along the x, y, and z axes, respectively. Now carry out the following multiplication:

$$\begin{pmatrix} 0 & 1 & 0 \\ 0 & 0 & 1 \\ 1 & 0 & 0 \end{pmatrix} \begin{pmatrix} a \\ b \\ c \end{pmatrix} = \begin{pmatrix} b \\ c \\ a \end{pmatrix}.$$

The result is a new vector with the same magnitude but pointing in a different direction, where a is now the length of the vector along the z axis instead of the x axis, and so on. The vector has been rotated by $90°$ around all three coordinate axes. What would be an awkward operation to carry out using trigonometry becomes relatively straightforward when we use matrix algebra. This example also illustrates how we can use a matrix to represent mathematically a real physical process, in this case the rotation of an object in space.

A second common application of matrix algebra is to solve a set of equations of the form

$$\begin{aligned} h_{11}ax + h_{12}by &= cax \\ h_{21}ax + h_{22}by &= cby. \end{aligned} \tag{A.14}$$

Here, the h_{ij}'s can be any coefficients, ax and by together form a vector in the xy plane, and c is some unknown constant that we want to find. Using our rules of matrix multiplication, these equations can be written as a single matrix equation:

$$\begin{pmatrix} h_{11} & h_{12} \\ h_{21} & h_{22} \end{pmatrix} \begin{pmatrix} ax \\ by \end{pmatrix} = c \begin{pmatrix} ax \\ by \end{pmatrix}. \tag{A.15}$$

Equation A.15 is an example of an eigenvalue equation, because after multiplying $\begin{pmatrix} ax \\ by \end{pmatrix}$ by $\begin{pmatrix} h_{11} & h_{12} \\ h_{21} & h_{22} \end{pmatrix}$ on the left, we get $\begin{pmatrix} ax \\ by \end{pmatrix}$ multiplied by a constant c on the right. (The eigenvalue equation is discussed in more detail in Section 2.1 of the *Quantum Mechanics* volume.) We can solve for the values of c that make Eq. A.15 true by a convenient feature of matrix algebra.

Say, for example, that we want to find the values of c that solve the two equations

$$2ax + by = cax$$

$$ax = cby$$

for any given values of a and b. Then, the matrix elements h_{ij} have the values:

$$\begin{pmatrix} h_{11} & h_{12} \\ h_{21} & h_{22} \end{pmatrix} = \begin{pmatrix} 2 & 1 \\ 1 & 0 \end{pmatrix}.$$

Then we find the values of c by **diagonalizing** the matrix. First, subtract the unknown value c from each value h_{ii} (these are the **diagonal elements** of the matrix):

$$\begin{pmatrix} 2-c & 1 \\ 1 & 0-c \end{pmatrix}.$$

Next, take the **determinant** of the matrix and set it equal to zero. The determinant is an algebraic combination of all the elements in a square matrix, with the following formulas for 2×2 and 3×3 matrices:

$$\begin{vmatrix} r & s \\ t & u \end{vmatrix} = ru - st \tag{A.16}$$

$$\begin{vmatrix} r & s & t \\ u & v & w \\ x & y & z \end{vmatrix} = r\begin{vmatrix} v & w \\ y & z \end{vmatrix} + s\begin{vmatrix} w & u \\ z & x \end{vmatrix} + t\begin{vmatrix} u & v \\ x & y \end{vmatrix} = rvz + swx + tuy - rwy - suz - tvx. \tag{A.17}$$

Using the 2×2 case, the determinant we need to set to zero in our example is

$$\begin{vmatrix} 2-c & 1 \\ 1 & -c \end{vmatrix} = (2 - c)(-c) - (1)(1) = c^2 - 2c - 1 = 0.$$

Solving for c with the quadratic formula, we obtain two solutions:

$$c = \frac{1}{2}\left[2 \pm \sqrt{4 + 4}\right] = 1 \pm \sqrt{2}.$$

There are two valid solutions to Eq. A.15, corresponding to the $+$ and $-$ signs. To show that they are solutions, substitute each result for c in Eqs. A.14:

$$2ax + by = (1 \pm \sqrt{2})ax \qquad \qquad h_{11}ax + h_{12}by = cax$$
$$ax + (0)by = (1 \pm \sqrt{2})\,by \qquad \qquad h_{21}ax + h_{22}by = cby$$
$$ax = (1 \pm \sqrt{2})by \qquad \qquad \text{solve for } ax$$
$$2(1 \pm \sqrt{2})by + by = (1 \pm \sqrt{2})^2 by \qquad \qquad \text{replace } ax$$
$$(3 \pm 2\sqrt{2})by = (3 \pm 2\sqrt{2})by.$$

The same method can be used to solve any number of related equations simultaneously, boiling the problem down to a single step: diagonalizing the matrix. Consequently, matrix diagonalization routines comprise a key element in computer programs designed to solve problems and simulate processes in virtually every realm of chemistry and physics.

Differential and Integral Calculus

If, like many of your classmates, you enjoyed everything about organic chemistry except its neglect of your calculus skills, rest assured that we won't make the same mistake in physical chemistry. Much of the problem-solving ahead of us involves taking a process that we understand on a tiny scale and expanding that description to a larger scale. That tiny-scale understanding will often be phrased mathematically using **derivatives,** which are an idealized version of how a property—such as electron position or chemical concentration—changes over a small step. Change makes *everything* interesting: how the colors of the leaves change with time, how the climate changes the closer we get to the coast, and how the taste of ice cream changes with the amount of vanilla added. For another example, we describe the interactions between particles in terms of the forces they exert on one another. Force is proportional to an acceleration, and acceleration is the derivative of the velocity with respect to time. A force describes where a particle is going to move right now. If we want to see a bigger picture, we can undo the derivative with **integration** and extract from the force law an idea of where the particle will be at different times. The force itself is a derivative (with respect to distance) of the energy, and integrating the force over distances can tell us how the energy of a system varies at different locations.

Another form of this extension from small scale to large scale requires us to calculate sums and averages—which are convenient ways to describe huge systems—from functions too detailed to bear patiently. For example, an understanding of the small-scale interaction between molecules and gravity leads us to predict that air is denser near sea level than at high altitudes. A clever equation even tells us how the air density varies with altitude. By integrating this equation over all altitudes, we can find the total amount of air present and drop all the information about the detailed interactions. It is this general approach of extrapolating from small to large that makes a journeyman command of calculus essential for the text.

Derivatives

Solutions to some standard derivatives appear in Table A.3. It does not hurt to know how to obtain derivatives and integrals, but we will be treating these aspects of calculus as just another kind of algebra. In other words, one may replace the derivative or integral expression by the correct algebraic expression, with the appropriate substitutions. This will suffice for almost all the calculus we encounter in the text.

When a function depends on more than one variable, then the derivative of the function with respect to one variable generally depends on the other variables as well. As one example, suppose that we have a variable P that depends on three other variables n, T, and V, and a constant R, such that

$$P = \frac{nRT}{V}.$$

TABLE A.3 Solutions to selected derivatives.

dx^c	$= cx^{c-1}dx \, (c \neq 0)$	$d(cx)$	$= cdx$
$d \ln x$	$= \dfrac{1}{x}dx$	de^x	$= e^x dx$
$d \sin x$	$= \cos x \, dx$	$d \cos x$	$= -\sin x \, dx$
$d[f(x) + g(x)]$	$= d[f(x)] + d[g(x)]$	$d[f(x)g(x)]$	$= f(x)d[g(x)] + g(x)d[f(x)]$
		$d[f(x)/g(x)]$	$= \dfrac{f(x)d[g(x)] - g(x)d[f(x)]}{g(x)^2}$

Then the derivative of P is related to the derivatives of the three variables, because small changes in n, in T, and in V will each contribute to the overall change in P. In general, derivatives of multivariable functions require knowing how all the variables depend on each other. In these instances, we will use the **partial derivative,** represented by the symbol ∂, which is simply the derivative of the function with respect to one variable *treating all the other variables as though they are constants.* The expression

$$\left(\frac{\partial P}{\partial V}\right)_{n,T}$$

represents the partial derivative of P with respect to V, treating n and T as though they were constants, just like R. Using the partial derivative, the total derivative of P may be written as a sum over the derivatives of the variables:

$$dP = \left(\frac{\partial P}{\partial n}\right)_{T,V} dn + \left(\frac{\partial P}{\partial T}\right)_{n,V} dT + \left(\frac{\partial P}{\partial V}\right)_{n,T} dV = \frac{RT}{V}dn + \frac{nR}{V}dT - \frac{nRT}{V^2}dV.$$

(A.18)

In the third partial derivative, for example, the variables n and T are treated as constants and factored out of the derivative. Hence the partial derivative simplifies to

$$\left(\frac{\partial P}{\partial V}\right)_{n,T} = \left(\frac{\partial(nRT/V)}{\partial V}\right)_{n,T} = nRT\left(\frac{\partial(1/V)}{\partial V}\right)_{n,T} = -\frac{nRT}{V^2}. \quad (A.19)$$

Table A.4 contains some useful relations involving partial derivatives.

Analytical Integrals

Please make sure that you understand the following terminology regarding integrals:

1. A **definite integral** is evaluated between **limits,** the quantities a and b in the expression $\int_a^b f(x)\,dx$. The integration of $f(x)$ in this case is only carried out from $x = a$ to $x = b$.

2. When the limits are not specified, the integral is an **indefinite integral.** The derivative of a constant C is zero. Therefore, when we take the derivative of a function $f(x) = g(x) + C$, all the information about

TABLE A.4 Relations involving partial derivatives.

reciprocal rule	$\left(\dfrac{\partial x}{\partial y}\right)_z \left(\dfrac{\partial y}{\partial x}\right)_z$	$= 1$
slope rule	$dz(x, y)$	$= \left(\dfrac{\partial z}{\partial x}\right)_y dx + \left(\dfrac{\partial z}{\partial y}\right)_x dy$
cyclic rule	$\left(\dfrac{\partial x}{\partial y}\right)_z$	$= -\left(\dfrac{\partial x}{\partial z}\right)_y \left(\dfrac{\partial z}{\partial y}\right)_x$
chain rule	$\left(\dfrac{\partial x}{\partial y}\right)_z$	$= \left(\dfrac{\partial x}{\partial w}\right)_z \left(\dfrac{\partial w}{\partial y}\right)_z$
	$\left(\dfrac{\partial x}{\partial y}\right)_z$	$= \left(\dfrac{\partial x}{\partial y}\right)_w + \left(\dfrac{\partial x}{\partial w}\right)_y \left(\dfrac{\partial w}{\partial y}\right)_z$

TABLE A.5 Solutions to selected integrals. In these equations, a and b are constants, n is a whole number, and C is the constant of integration.

$$\int x^n dx = \frac{1}{n+1}x^{n+1} + C \qquad\qquad \int a\,dx = a(x+C)$$

$$\int \frac{1}{x}\,dx = \ln x + C \qquad\qquad \int e^x dx = e^x + C$$

$$\int \ln x\,dx = x\ln x - x + C \qquad\qquad \int \frac{dx}{x(a+bx)} = -\frac{1}{a}\ln\left(\frac{a+bx}{x}\right) + C$$

$$\int \sin x\,dx = -\cos x + C \qquad\qquad \int \cos x\,dx = \sin x + C$$

$$\int \sin^2(ax)\,dx = \frac{x}{2} - \frac{\sin(2ax)}{4a} + C \qquad\qquad \int \cos^2(ax)\,dx = \frac{x}{2} + \frac{\sin(2ax)}{4a} + C$$

$$\int [f(x) + g(x)]dx = \int f(x)\,dx + \int g(x)\,dx \qquad\qquad \int_a^b dx = x\Big|_a^b = b - a$$

$$\int_0^\infty x^n e^{-ax}\,dx = \frac{n!}{a^{n+1}} \qquad\qquad \int_0^\infty e^{-ax^2}\,dx = \frac{1}{2}\left(\frac{\pi}{a}\right)^{1/2}$$

$$\int_0^\infty xe^{-ax^2}\,dx = \frac{1}{2a} \qquad\qquad \int_0^\infty x^2 e^{-ax^2}\,dx = \frac{1}{4}\left(\frac{\pi}{a^3}\right)^{1/2}$$

$$\int_0^\infty x^{2n+1}e^{-ax^2}\,dx = \frac{n!}{2a^{n+1}} \qquad\qquad \int_0^\infty x^{2n}e^{-ax^2}\,dx = \frac{[1\cdot3\cdot5\ldots(2n-1)]\sqrt{\pi}}{2^{n+1}a^{n+(1/2)}}$$

$$\int_0^s x^n e^{-ax}\,dx = \frac{n!}{a^{n+1}} - e^{-as}\sum_{i=0}^n \frac{n!s^{n-i}}{a^{i+1}(n-i)!}$$

the value of C is lost. When we undo the derivative by taking the integral, we add an unknown constant of integration to the integrated expression. Omit this constant when solving definite integrals, because the limits of integration will determine its value.

3. The function being integrated is the **integrand,** and it is multiplied by the incremental change along the coordinates, called the volume element.

Most of the algebraic solutions to integrals that we need appear in Table A.5.

EXAMPLE A.4   Analytical Integration

PROBLEM Evaluate the numerical value for each of the following expressions.

1. $\displaystyle\int_1^4 \frac{dx}{x}$

2. $\displaystyle\int_0^\infty e^{-2x}\,dx$

3. $\displaystyle\int_0^{\pi/3} (3\cos\theta^2 - 1)\sin\theta\,d\theta$

SOLUTION These can be solved by substitution of the expressions in Table A.5.

(a) $\displaystyle\int_1^4 \frac{dx}{x} = \ln x \Big|_1^4 = \ln 4 - \ln 1 = 1.386 - 0 = 1.386$

(b) $\displaystyle\int_0^\infty e^{-2x}\,dx = -\frac{1}{2}e^{-2x}\Big|_0^\infty = -\frac{1}{2}(e^{-\infty} - e^0) = -\frac{1}{2}(0 - 1) = \frac{1}{2}$

(c) $\displaystyle\int_0^{\pi/3} (3\cos^2\theta - 1)\sin\theta\,d\theta = \left[-\cos^3\theta + \cos\theta\right]\Big|_0^{\pi/3}$

$$= \left[-\left(\frac{1}{2}\right)^3 + \left(\frac{1}{2}\right)\right] - [-(1)^3 + (1)] = \frac{3}{8}.$$

By the way, it is possible to apply rules of symmetry to extend some of the analytical solutions in Table A.5. For example, when the integrand is $x^{2n}e^{-ax^2}$, then the function is exactly the same from 0 to $-\infty$ as from 0 to $+\infty$ (Fig. A.3a). Therefore, the integral $\int_{-\infty}^{\infty} x^{2n}e^{-ax^2}\,dx$ is equal to 2 times $\int_0^\infty x^{2n}e^{-ax^2}\,dx$. However, if the power of x is odd, $2n + 1$, then the function is negative when $x < 0$ and positive when $x > 0$ (Fig. A.3b). The integral from $-\infty$ to 0 cancels the integral from 0 to $+\infty$, so $\int_{-\infty}^{\infty} x^{2n+1}e^{-ax^2}\,dx = 0$.

Numerical Integration

Not all integrals have algebraic solutions, and some have algebraic solutions only between certain limits (such as 0 and ∞). With suitable computers, any integral can be calculated without trying to cram it into some algebraic form. This is accomplished by going back to the definition in calculus,

$$\int_{x_1}^{x_2} f(x)\,dx = \lim_{\delta x \to 0}\left\{\sum_{i=1}^{N} f[x_1 + i\,\delta x]\right\}\delta x \tag{A.20}$$

▶ FIGURE A.3 **Symmetry and definite integrals. (a)** If a function $f(x)$ is equal to $f(-x)$ for all values of x, then the integral from $-\infty$ to ∞ is equal to 2 times the integral from 0 to ∞. **(b)** If $f(x)$ is equal to $-f(-x)$, then the integral from $-\infty$ to ∞ is 0.

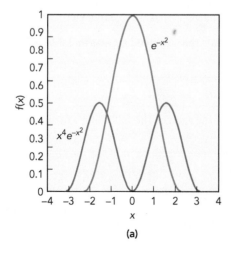

(a)

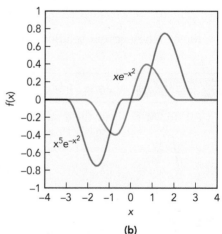

(b)

where

$$\delta x = \frac{x_2 - x_1}{N}.$$

One increment of this sum is illustrated in Fig. A.4. If the computer has $f(x)$ stored not as an algebraic expression but as a list of N points, it can carry out that sum very quickly. This is a simple **numerical** (rather than analytical) method of solving the integral. A thousand points ($N = 1000$) would take much less than a second for any modern computer.

It's straightforward to carry out such an integration with a number of elegant programs. Let's say we need to evaluate the integral

$$\int_0^2 f(r)\, r^2\, dr = \int_0^2 \exp(-\pi^{1/3} r^2) r^2\, dr.$$

The appropriate commands for some selected programs are given in Table A.6. For each program, the result is obtained in less time than it takes to enter the command.

The problem is that in many scientific problems we want to integrate a function of several coordinates, and the number of points we need to sample increases as the *power* of the number of coordinates. As an example, in computational chemistry we often integrate an energy equation over the x, y, and z coordinates of every electron in a molecule. If this has to be done numerically, then we may want to sample 10 points along each coordinate. Even if the molecule has only 10 electrons, then a brute force approach would require that we check the value of the function at 10 points along each of 30 coordinates (x, y, and z for each electron), which would require 10^{30} calculations, clearly too demanding even by today's standards. In practice, we reduce the problem to as few coordinates as possible for what we're trying to study, often by factoring the integral into as many independent, low-dimensional integrals as possible. Clever spacing of the points (where the function changes faster, we want more points per unit x) and other tricks can make this even more efficient. The search for more efficient numerical integrators for many-dimensional functions has fueled research projects in mathematics for decades.

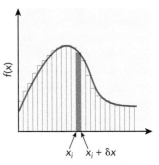

▲ **FIGURE A.4 Numerical integration.** In the simplest numerical integration of a function (solid line), the area of a rectangle of height $f(x_i)$ between two points, x_i and $x_i + \delta x$, is calculated, and the areas of all such adjacent rectangles are summed to obtain an approximate value for the integral.

TABLE A.6 Syntax for numerical integration of the expression $\int_0^2 \exp(-\pi^{1/3} r^2) r^2\, dr$ for common symbolic math programs. The solution is 0.2479.

Program	Command(s)
Maple™	int(exp(-P^(1/3)*r^2)*r^2, r=0..2)
Mathematica®	Integrate[Exp[-Pi^(1/3)*r^2]*r^2, {r,0,2}]
MATLAB®	int(exp(-P^(1/3)*r^2)*r^2, r,0,2)
Octave	function xr = x(r); xr = r^2 * exp(-(pi^(1/3))*r^2); endfunction quad("x",0,2)

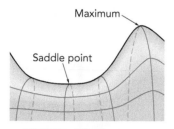

▲ **FIGURE A.5 Stationary points on a function of two variables.**

Volumes and Stationary Points

We are going to consider several different kinds of algebraic functions, including those that describe the energy of the reactants at different stages of a chemical reaction, or the distribution of the electrons in a molecule. These can be so complicated that even the gross features of the function may be difficult to visualize.

One way calculus can help us manage a complicated function is by identifying **stationary points,** regions where the slope of the function is zero. If the function has only one coordinate, such as the time elapsed since a chemical reaction began, then the stationary points are local minima or maxima along the curve. With more dimensions, such as the electron distribution of a molecule in three-dimensional space, the stationary points can also be **saddle points** on the surface of the function, locations where some coordinates reach local maxima while the others are at local minima. Figure A.5 shows a function that could represent the potential energy of a molecule as it changes from one conformation to another, with different stationary points occurring along the surface.

These stationary points are identified by solving for the first derivative of the function and finding the values of the coordinates for which the first derivative— the slope—is zero. For example, $df(x,y,z)/dx$ gives the slope of $f(x,y,z)$ along the x direction. At a minimum, maximum, or saddle point of a three-dimensional function, $df(x,y,z)/dx = df(x,y,z)/dy = df(x,y,z)/dz = 0$.

Another task for which we employ calculus is the computation of areas or volumes contained by functions. These can be obtained by taking the integral of the function between the appropriate limits. Two examples follow.

EXAMPLE A.5   **First Derivatives**

PROBLEM Find the maximum value of the function $3\cos^2\theta - 1$ where $0 \leq \theta < \pi$.

SOLUTION Take the derivative with respect to the variable, in this case θ:

$$\frac{d}{d\theta}(3\cos^2\theta - 1) = -6\cos\theta\sin\theta.$$

This derivative is zero when either $\sin\theta$ or $\cos\theta$ is zero, in other words at $\theta = 0$ or $\pi/2$. To determine which of these values corresponds to the maximum, we can substitute the two values back into the original function, obtaining

$$3\cos^2(0) - 1 = 3\cdot(1)^2 - 1 = 2,$$

$$3\cos^2(\pi/2) - 1 = 3\cdot(0)^2 - 1 = -1.$$

The maximum value of $3\cos^2\theta - 1$ is 2.

EXAMPLE A.6 **Integration**

PROBLEM Find the area of a rectangle with length a along the x axis and width b along the y axis.

SOLUTION This can be obtained from the integral

$$\int_0^a f(x)dx = \int_0^a b\,dx = b\int_0^a dx = b(a-0) = ab.$$

Alternatively, one could evaluate the double integral

$$\int_0^a dx \int_0^b dy = (a-0)(b-0) = ab.$$

EXAMPLE A.7 **Triple Integrals**

PROBLEM Prove the equation for the volume of a sphere of radius R.

SOLUTION In spherical polar coordinates, the volume of a sphere with radius R is found from the triple integral:

$$\int_0^{2\pi} d\phi \int_0^{\pi} \sin\theta\,d\theta \int_0^R r^2\,dr = (\phi)\Big|_0^{2\pi} (-\cos\theta)\Big|_0^{\pi} \frac{1}{3}r^3\Big|_0^R$$

$$= (2\pi - 0)[-(-1)-(-1)]\left(\frac{1}{3}R^3 - 0\right) = \frac{4\pi R^3}{3}.$$

Fourier Transforms

While we're on the subject of integrals, here's a related mathematical manipulation that sees wide application in the sciences, but that you may not have seen before. An integral examines the values of a continuous function at each step along some coordinate, like position or time, and then adds up all the values. The resulting integral doesn't depend on the value of the coordinate any more, because *all* of the coordinate values over the range of integration are used. But if the integrand depends on a second coordinate as well, the integral can also be used to **transform** the function's dependence from one variable to another.

The **Fourier transform** allows us to take any well-behaved function and rewrite it as an integral over sine and cosine functions. If we have a function $f_t(t)$ that varies with (as an example) the time t, then we can rewrite this as a "new" function $f_\omega(\omega)$, where ω is a coordinate with units of $1/t$. The function $f_\omega(\omega)$ is not really a new function but a splitting of the original $f_t(t)$ into many pieces, where each piece is a sine or cosine function with frequency ω.

The simplest case is for a *periodic* function, which starts at $t = 0$, carries on for a period of time τ, and then repeats over and over again. There is *always* a way to write exactly the same function as a sum of sines and cosines:

$$f_t(t) = \frac{a_0}{2} + \sum_{n=1}^{\infty}\left(a_n \cos\frac{2\pi nt}{\tau} + b_n \sin\frac{2\pi nt}{\tau}\right)$$

where n is an integer. The values of the coefficients a_n and b_n are the amplitudes of the cosine and sine functions that we add together to make $f_t(t)$, where each of those functions has an oscillation frequency $\omega = 2\pi n/\tau$. Therefore, we can think of the a_n and b_n coefficients as the values of a function $f_\omega(\omega)$ that tells us how much cosine or sine we must add together to obtain the original $f_t(t)$. The transform is the process of finding the values of those coefficients and, as you were warned, it is an integration:

$$a_n = \frac{2}{\tau} \int_{-\tau/2}^{\tau/2} f_t(t) \cos \frac{2\pi nt}{\tau} \, dt = \frac{2}{\tau} \int_{-\tau/2}^{\tau/2} f_t(t) \cos \omega t \, dt$$

$$b_n = \frac{2}{\tau} \int_{-\tau/2}^{\tau/2} f_t(t) \sin \frac{2\pi nt}{\tau} \, dt = \frac{2}{\tau} \int_{-\tau/2}^{\tau/2} f_t(t) \sin \omega t \, dt.$$

In order to keep track of both the sine and cosine components simultaneously, we can make $f_\omega(\omega)$ a complex function, in which the real part gives the cosine component and the imaginary part gives the sine component:

$$f_\omega(\omega) = a_n + ib_n.$$

For a periodic function, $\omega = 2\pi n/\tau$ is a discrete variable, meaning that it is limited to only certain values. For a non-periodic function, we can still carry out a Fourier transform, but we have to extend the period τ to infinity. As τ becomes infinite, the gaps between values of the frequency ω become smaller and smaller, until ω is a continuous variable, like the time t. In that limit, the Fourier transform is written

$$\mathcal{F}[f_t(t)] \equiv f_\omega(\omega) = \frac{1}{\sqrt{2\pi}} \int_{-\infty}^{\infty} f_t(t) \left[\cos(\omega t) + i \sin(\omega t)\right] dt. \quad \text{(A.21)}$$

Notice that the original function depended only on t, but integrating over all values of t removes the t-dependence, giving a function that depends only on ω. We can reclaim the original function from its Fourier transform by carrying out the inverse Fourier transform:

$$\mathcal{F}^{-1}[f_\omega(\omega)] = f_t(t) = \frac{1}{\sqrt{2\pi}} \int_{-\infty}^{\infty} f_\omega(\omega) \left[\cos(\omega t) - i \sin(\omega t)\right] d\omega. \quad \text{(A.22)}$$

In short, this particular Fourier transform shows us how to rewrite a function of *time* as the sum of many sine waves having different *frequencies*. This procedure can be used, for example, to find the frequencies in an audio recording that carry the data we want to keep, and then design an electronic filter that will remove the high-frequency noise we don't want.

Fourier transforms are not limited to conversions between time and frequency. Clever imaging techniques in astronomy and chemistry employ Fourier transforms involving functions of position, for example.

Differential Equations

Equations in which the variables of interest appear with different orders of their derivatives are called **differential equations.** Examples are $x + (dx/dy) = 0$ and $(d^2x/dy^2) + y = 5(dx/dy)$. The solution to these is essentially a problem in integration, and (like integration) need not have an algebraic solution. Some of

EXAMPLE A.8 Fourier Transform

PROBLEM Find the Fourier transform $f_k(k)$ of a step function $f_x(x)$ (Fig. A.6a), for $f_x(x)$ equal to f_0 in the range $-a \leq x < a$ and zero everywhere else.

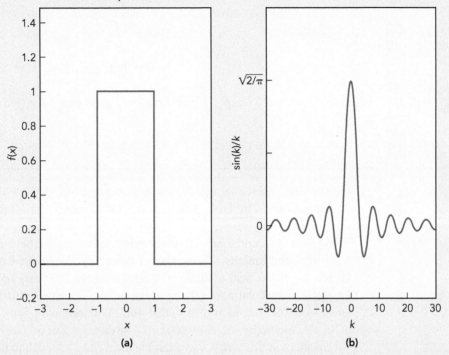

▲ FIGURE A.6 The step function (a) $f_x(x)$ for f_0 and $a = 1$, and (b) its Fourier transform $f_k(k)$.

SOLUTION We'll need to use one of the results from Table A.5 and remember that we can break the integral up into different regions:

$$\mathcal{F}[f_x(x)] = \frac{1}{\sqrt{2\pi}} \int_{-\infty}^{\infty} f_x(x)[\cos(kx) + i\sin(kx)]\,dx$$

$$= \frac{1}{\sqrt{2\pi}} \left\{ \int_{-\infty}^{-a}(0)\,dx + \int_{-a}^{a} f_0[\cos(kx) + i\sin(kx)]\,dx + \int_{a}^{\infty}(0)\,dx \right\}$$

$$= \frac{1}{\sqrt{2\pi}} f_0 \left[\frac{1}{k}\sin(kx) - \frac{i}{k}\cos(kx) \right]_{-a}^{a}$$

$$= \frac{1}{\sqrt{2\pi}} f_0 \left[\frac{2}{k}\sin(ka) - 0 \right]$$

$$= \frac{1}{\sqrt{2\pi}} \frac{2f_0}{k}\sin(ka) \equiv f_k(k).$$

We can plot this new function, $f_k(k)$, versus k and get the graph in Fig. A.6b. This tells us that the step function $f_x(x)$ can be formed by adding together sine waves with wavelengths $1/k$, with the amplitude or height of each sine wave given by the values in Fig. A.6b.

TABLE A.7 Solutions to selected differential equations.

Differential equation	Solution
$\dfrac{dx}{dy} = \dfrac{g(y)}{f(x)}$	$\displaystyle\int f(x)dx = \int g(y)dy + C$
$\dfrac{dx}{dy} = h(w),\ w = x/y$	$\displaystyle\ln y = \int [h(w){-}w]^{-1}\, dw + C, h(w) \neq w$
	$\ln x = \ln y + C, h(w) = w$
$\dfrac{dx}{dy} + g_1(y)x = g_2(y)$	$x\exp\!\left(\displaystyle\int g_1(y)dy\right) = \int g_2(y)\exp\!\left(\int g_1(y)dy\right)dy + C$
$\dfrac{d^2y}{dx^2} = a^2 y$	$y = C_1 e^{ax} + C_2 e^{bx}$

the simplest differential equation solutions are given in Table A.7. As with the integral solutions in Table A.5, the constant of integration C is used when no limits are given.

We will encounter several differential equations, particularly in quantum mechanics and kinetics. There is good news here. Napoleon firmly planted the French flag in the field of differential calculus about 200 years ago. His Académie Française found solutions to the most important differential equations in our field, and it would be rude to neglect all that work. With one exception—the distribution of the electron in the one-electron atom—we do not rigorously solve those differential equations in this text, but we examine them closely enough to find the origins of important physical effects. Some exercises and problems also take advantage of the solutions available in Table A.7 in order to be a little more general or interesting.

In any case, we are not excused from knowing what a differential equation is. If one of our French friends hands us the solution, we should be able to show that it does in fact solve the equation.

EXAMPLE A.9   Differential Equations

PROBLEM Prove that the first three examples in Table A.7 predict the same solution for x at $y = 2$ given the differential equation

$$\frac{dx}{dy} = -\frac{2x}{y},$$

where $x = 2$ at $y = 1$.

SOLUTION Our equation is in the proper form to apply the first solution in Table A.7, with $g(y) = -2/y$ and $f(x) = 1/x$. We rewrite "x" in the integrand as a dummy variable x' to avoid confusing it with x:

$$\int_{2}^{x} f(x')dx' = \int_{1}^{2} g(y)dy$$

$$\int_{2}^{x} \frac{dx'}{x'} = -2\int_{1}^{2} \frac{dy}{y}$$

$$\ln\left(\frac{x}{2}\right) = -2\ln 2$$

$$x = \frac{1}{2}.$$

To apply the second solution in Table A.7, we would set $h(w) = -2w$ and solve:

$$\ln y \Big|_1^2 = \int_2^{x/2} [h(w) - w]^{-1} dw$$

$$\ln 2 = -\frac{1}{3} \int_2^{x/2} \frac{dw}{w}$$

$$\ln 2 = -\frac{1}{3} \ln\left(\frac{x}{4}\right)$$

$$\ln\left(\frac{1}{8}\right) = \ln\left(\frac{x}{4}\right)$$

$$x = \frac{1}{2}.$$

Note that the limits for the integral over w are the limits of $w = x/y$: 2/1 at $x = 2$ and $y = 1$, and $x/2$ at $y = 2$.

For the third solution in Table A.7, set $g_1(y) = 2/y$ and $g_2(y) = 0$:

$$x\exp\left(\int_1^y g_1(y)dy\right) = \int_1^2 g_2(y)\exp\left(\int_1^2 g_1(y)dy\right)dy + C$$

$$x\exp\left(\int_1^y \frac{2dy}{y}\right) = C$$

$$x\exp(2\ln y) = C$$

$$xy^2 = C \qquad \text{for } x = 2, y = 1 \text{ we find } C = 2$$

$$x(2)^2 = 2 \qquad \text{now set } y = 2$$

$$x = \frac{1}{2}.$$

Power Series

On many occasions in the text, we will find it convenient to express a mathematical function as a sum over an infinite number of terms. We shall often employ the **power series** expansion, which has the form

$$\sum_{n=0}^{\infty} a_n x^n = a_0 + a_1 x + a_2 x^2 + a_3 x^3 + \ldots, \qquad (A.23)$$

where the a_i's are constants. This is one way of representing an unknown function in some mathematical form on our way to solving its equation.

Another use appears when the later terms in the expansion are much smaller than the leading terms. If x is near 0, then the terms on the right-hand side decrease precipitously as the exponents get larger, and the series eventually converges to some finite value. The advantage to us is that it allows a sophisticated function to be rewritten in a more approachable, but approximate, form. For example, the function $\sin(2x)^{1/2}$ is not easy to integrate on paper, but when x is small, the power series expansion allows this function to be set roughly equal to $1 - x^2$, which integrates easily to $x - (x^3/3) + C$.

More importantly, approximations such as this make the equations more intuitively meaningful, allowing us to better predict how matter will behave *qualitatively*. An approximate equation may sound like a poor replacement for the exact result, but this course is likely to enhance your appreciation of a simple concept over precision. As satisfying as a ten-significant-digit answer may be, the most highly prized advances in physical chemistry are those that reveal a new insight, and insights are hard to gain from final equations that take up a page or more. (Just wait; you'll see for yourself.) Approximations are essential tools in our work ahead, because they provide some algebraic answers that are unattainable otherwise, and they greatly simplify others that would just be too cumbersome to be informative. Many of our approximations will be based on the power series given next.

Trigonometric and Exponential Series

The power series expansion we will use most frequently is the one for the exponential function:

$$e^x = \sum_{n=0}^{\infty} \frac{x^n}{n!} = 1 + x + \frac{x^2}{2} + \frac{x^3}{6} + \frac{x^4}{24} + \dots \qquad (A.24)$$

This leads to a common approximation when the magnitude of x is small:

$$\text{for } |x| \ll 1: \quad e^x \approx 1 + x. \qquad (A.25)$$

Two other important power series are used for the trigonometric functions $\sin x$ and $\cos x$ (with x in units of radians, not degrees):

$$\sin x = \sum_{n=0}^{\infty} (-1)^{n+1} \frac{x^{2n+1}}{(2n+1)!} = x - \frac{x^3}{6} + \frac{x^5}{120} - \dots \qquad (A.26)$$

$$\cos x = \sum_{n=0}^{\infty} (-1)^n \frac{x^{2n}}{(2n)!} = 1 - \frac{x^2}{2} + \frac{x^4}{24} + \dots \qquad (A.27)$$

Of all the basic math equations that we draw on in the text, these must be two of the most amazing. The functions $\sin x$ and $\cos x$ never return values outside the range -1 to 1, no matter how gargantuan a value of x you plug in. To look at the equation for $\sin x$, summing over x and x^3 and x^5 and beyond, it seems ridiculous that the value should ever converge if we set $x = 1001$, for example, but it always does.

Even if you're not impressed by that, notice that these equations prove the small angle approximations

$$\text{for } |x| \ll 1: \quad \sin x \approx x, \quad \cos x \approx 1.$$

The various trigonometric identities, such as $\cos(-x) = \cos x$, can also be tested using these expansions:

$$\cos(-x) = \sum_{n=0}^{\infty} (-1)^n \frac{(-x)^{2n}}{(2n)!}$$

$$= \sum_{n=0}^{\infty} (-1)^n (-1)^{(2n)} \frac{(x)^{2n}}{(2n)!}$$

$$= \sum_{n=0}^{\infty} (-1)^n \frac{(x)^{2n}}{(2n)!} = \cos x.$$

And better still, we can combine the power series for $\sin x$ and $\cos x$ (see Problem A.9) to obtain the Euler equation,

$$e^{ix} = \cos x + i \sin x \qquad (A.28)$$

which will play a major part in our interpretation of the atomic orbital wavefunctions.

The Taylor Series

The power series given previously for e^x, $\sin x$, and $\cos x$ are special cases of a general power series equation that allows any single function $f(x)$ to be expressed in terms of its derivatives at some point x_0:

$$f(x) = \sum_{n=0}^{\infty} \frac{1}{n!} \left(\frac{d^n f(x)}{dx^n} \right) \bigg|_{x_0} (x - x_0)^n$$

$$= f(x_0) + \left(\frac{df(x)}{dx} \right) \bigg|_{x_0} (x - x_0) + \frac{1}{2} \left(\frac{d^2 f(x)}{dx^2} \right) \bigg|_{x_0} (x - x_0)^2$$

$$+ \frac{1}{6} \left(\frac{d^3 f(x)}{dx^3} \right) \bigg|_{x_0} (x - x_0)^3 + \dots, \qquad (A.29)$$

where the subscript x_0 after the derivative means the derivative is evaluated at the point x_0. This power series is called the **Taylor series,** and its use is confined to those cases when only the leading terms (the lowest order derivatives) are important—in other words, for slowly varying functions. The first derivative of a function is the function's slope, the second derivative is the slope of the slope, and so on. If a function doesn't have rapid oscillations or sharp peaks, its higher derivatives tend to converge toward zero, and the Taylor series becomes a useful approximation.

One common application of the Taylor series expansion is to the natural logarithm $\ln x$. If we choose $x_0 = 1$, since $\ln(1) = 0$, then we get the following:

$$0 < x < 2: \ln x = \ln(1) + \left(\frac{d \ln x}{dx} \right) \bigg|_{x=1} (x - 1) + \frac{1}{2} \left(\frac{d^2 \ln x}{dx^2} \right) \bigg|_{x=1} (x - 1)^2$$

$$+ \frac{1}{6} \left(\frac{d^3 \ln x}{dx^3} \right) \bigg|_{x=1} (x - 1)^3 + \dots$$

$$= 0 + \frac{1}{x} \bigg|_{x=1} (x - 1) + \frac{1}{2} \left(-\frac{1}{x^2} \right) \bigg|_{x=1} (x - 1)^2 + \frac{2}{6} \left(\frac{1}{x^3} \right) \bigg|_{x=1} (x - 1)^3 + \dots$$

$$= (x - 1) - \frac{1}{2} (x - 1)^2 + \frac{1}{3} (x - 1)^3 + \dots = \sum_{n=1}^{\infty} (-1)^{n+1} \frac{1}{n} (x - 1)^n.$$

$$(A.30)$$

The logarithm function is undefined for nonpositive numbers, so $x > 0$. In addition, the Taylor series for the logarithm **diverges** if $x - 1 > 1$, because the denominators are increasing only as n. Consequently, this expansion is useful only over the interval $0 < x < 2$.

In another case, the Taylor expansion is applied to the function $1/(1 + x)$, where $|x| < 1$. Choosing $x_0 = 0$ in the Taylor expansion, we find

$$|x| < 1: \quad \frac{1}{1 + x} = 1 - x + x^2 - x^3 + \ldots = \sum_{n=0}^{\infty} (-1)^n x^n. \tag{A.31}$$

The range on this expansion is also restricted because the series diverges when $|x| \geq 1$.

A third example in which we use the Taylor series is the expansion of the function $(1 + x)^{1/m}$, where $|x| \ll 1$. Again we choose $x_0 = 0$ and expand:

$$|x| < 1: \quad (1 + x)^{1/m} = 1 + \frac{1}{m}(1)^{(1/m)-1}x + \left(\frac{1}{2}\right)\left[\left(\frac{1}{m}\right)\left(\frac{1}{m} - 1\right)\right](1)^{(1/m)-2}x^2 + \ldots$$

$$= 1 + \frac{1}{m}x - \frac{m - 1}{2m^2}x^2 + \ldots. \tag{A.32}$$

This is one form of the **binomial series.** As one example, we will use Eq. A.32 to approximate the square root function, when $m = 2$:

$$|x| < 1: \quad (1 + x)^{1/2} = 1 + \frac{x}{2} - \frac{x^2}{8} + \ldots. \tag{A.33}$$

EXAMPLE A.10   **Taylor Series**

PROBLEM Use the Taylor series expansion to find an approximate equation for $1/x$. Let $x_0 = 1$ and keep terms up to order $(x - x_0)^2$.

SOLUTION Drawing on Eq. A.29,

$$\frac{1}{x} = \frac{1}{x_0} + (x - x_0)(-x_0^{-2}) + \frac{(x - x_0)^2}{2}(2x_0^{-3}) + \frac{(x - x_0)^3}{2 \cdot 3}(-2 \cdot 3x_0^{-4}) + \ldots$$

$$= \frac{1}{x_0} - \frac{(x - x_0)}{x_0^2} + \frac{(x - x_0)^2}{x_0^3} - \frac{(x - x_0)^3}{x_0^4} + \ldots$$

$$= 1 - (x - 1) + (x - 1)^2 - (x - 1)^3 + \ldots$$

$$\approx 2 - x + (x - 1)^2.$$

In the last two steps we set $x_0 = 1$ and then dropped the higher order terms, which is only valid for values of x close to 1. For $x = 1.1$, this equation predicts $1/x = 0.910$, very near the correct value of 0.909. However, for $x = 1.6$, the correct value is 0.625 and this equation predicts 0.760.

There is an important result from this, by the way. The variable in a power series, such as the x in e^x or the θ in $\cos\theta$, must not have physical units, strictly speaking. That is because there is no way to add quantities that represent different physical properties. If we set $x = 2\,\text{cm}$, then e^x is nonsense, because we could then write $e^x = 1 + 2\,\text{cm} + 2\,\text{cm}^2 + \ldots$, and there is no single quantity in the real world that can represent "distance plus area." For this reason, the text also gives most angles in radians (for example, $\pi/4$ instead of $45°$) but leaves the units implied instead of writing them down.[4] Having said all that, this text violates the rule against units in power series in the intermediate steps of some derivations, when it is convenient to split the logarithm function into different pieces. It is okay to write $\ln([A]/[A]_0)$, with both $[A]$ and $[A]_0$ being concentrations, because the units cancel before we take the logarithm. We can replace this, as a mathematical exercise, with $\ln[A] - \ln[A]_0$, but please be aware that the individual terms $\ln[A]$ and $\ln[A]_0$ do not have physically meaningful values until we somehow get rid of the concentration units, perhaps by combining these terms with other concentrations in the problem.

A.2   Classical Physics

Classical or Newtonian physics describes nature on the **macroscopic** scales of time, mass, and energy—measured in seconds, kilograms, and joules—to which we are most accustomed. Quantum mechanics and relativity describe deviations from classical mechanics, but they operate more subtly in our experience because their effects are strongest at energy scales much smaller (quantum) or much larger (relativity) than we normally perceive with our own senses. Our interest in this volume is at the **microscopic** scale, which we will take to mean the scale of individual atoms and molecules: distances of a few nanometers or less, masses less than 1000 atomic mass units, and energies of no more than about 10^{-18} J. Nevertheless, Isaac Newton's laws of motion for macroscopic bodies are often indispensable in visualizing the motions of **microscopic** entities, such as individual electrons, atoms, and molecules, sometimes with no adjustment at all. Therefore, it may be useful to review a few topics from classical physics that will show up in the text.

Force and Energy

In the absence of forces, a moving object with mass m will travel in a straight line with velocity vector $\vec{v}$. The vector has a magnitude (the speed) and a direction (the trajectory). In order to change the object's speed, its trajectory, or both, the object must undergo an **acceleration** $d\vec{v}/dt$. A train leaving a station may accelerate to increase its speed along a straight track (so fixed trajectory), and the moon is subject to an acceleration that continuously changes its trajectory, to maintain a roughly circular orbit around the earth, at roughly constant speed. Acceleration to change a velocity $\vec{v_1}$ to a new velocity $\vec{v_2}$ requires a force F, but the amount of force depends on exactly how the acceleration is applied: all at once, continuously over a long time, or in a series of jerks. The force is minus one times the energy expended per unit distance:

$$-\frac{dE}{dx} \equiv F_x.$$
(A.34)

---

[4]Strictly speaking, the radian is a unit (we measure the size of angles with it) but has no dimensions (because it is effectively the ratio of two distances: a circle's circumference divided by its diameter). Thus, it's still okay to put it in a power series.

Newton's law in turn relates the force to the acceleration:

$$F_x = m\frac{dv_x}{dt}, \text{ or } \overrightarrow{F} = m\frac{d\overrightarrow{v}}{dt}.$$
(A.35)

For motion in a circular orbit of radius r at constant speed v, a **centripetal acceleration** of v^2/r is necessary to keep changing the direction of the velocity vector. Acceleration is caused by a force, and centripetal acceleration is caused by a **centripetal force** of mv^2/r, where m is the mass of the orbiting object.

Conservation of Mass and Energy

At all times before, during, and after any chemical process, the overall mass and overall energy must each remain constant. In chemistry, we can ignore the exceptions to this rule.[5] Whether the process is the collision of two atoms, the combustion of 30 kg of fuel, or the binding of an electron to a proton, we shall enforce both of these constraints—the overall mass is constant, and the overall energy is constant.

In the absence of external forces, the overall linear momentum $\overrightarrow{p} = m\overrightarrow{v}$ is also conserved. This is especially important in mechanics because it vastly simplifies the equations of motion for a system with many colliding bodies. We will use it in our discussions of molecular collisions.

Kinetic, Potential, and Radiant Energy

Energy is a parameter of great importance in chemistry, and one of the parameters that we will be following from beginning to end of the text. It is a convenience to separate energy into different forms, as, for example, the different ways that a molecule can store energy. However, energy can be converted from one form to another, and sometimes the distinction between two forms of energy may suddenly become unclear. There are nevertheless three forms for which we will impose fairly rigid definitions:

1. **Kinetic energy** K: any energy due to the motion of an object with mass m:

$$K = \frac{mv^2}{2},$$
(A.36)

where v is the speed of the object

2. **Potential energy** U: any energy due to the interaction of an object with fields of the fundamental forces:

$$U = -\int F\,ds,$$
(A.37)

where ds is a distance derivative and F is the force arising from gravitational fields, electromagnetic fields, or the fields of the nuclear forces (but don't expect to see that last set often in this text)

3. **Radiant energy**: any energy present in the form of electromagnetic radiation

---

[5]The exceptions include the conversions between mass and energy accompanying reactions in particle physics, the postulated formation of virtual particles, and—for the most part—the scaling of mass and energy at relativistic speeds. This last exception becomes important when treating the motions of core electrons in heavy atoms.

Kinetic energy will be involved when there is any motion of matter: motions of electrons within atoms, atoms within molecules, and molecules within massive solids or gases or liquids. Potential energy in our work will almost always result from the **Coulomb force,** the force on a particle with charge q_1 due to an electric field $\vec{\mathcal{E}}$:

$$\vec{F}_{\text{Coulomb}} = q_1 \vec{\mathcal{E}}. \tag{A.38}$$

A charged particle generates an electric field in the surrounding space,

$$\vec{\mathcal{E}} = \frac{q_1 \vec{r}}{4\pi\varepsilon_0 r^3}, \tag{A.39}$$

where q_1 is the particle's charge and $\vec{r}$ is the vector connecting the particle to the point where the field is measured. For two particles with charges q_1 and q_2, therefore, the Coulomb force acting on each particle is

$$\vec{F}_{\text{Coulomb}} = \frac{q_1 q_2 \vec{r}_{12}}{4\pi\varepsilon_0 r_{12}^3}, \tag{A.40}$$

where $\vec{r}_{12}$ is the vector connecting the two charges. Often we want only the force along the axis connecting the two charges, in which case this force law may be written

$$F_{\text{Coulomb}} = \frac{q_1 q_2}{4\pi\varepsilon_0 r_{12}^2}. \tag{A.41}$$

This force will result in some motion, and therefore some kinetic energy K, unless some canceling force is present.

There is also a potential energy U due to the interaction of the particles with the electric field, whether or not there is any motion. One way of looking at this is as follows. Force is the derivative with respect to position of the energy taken *from the field* and used to accelerate the particle. The potential energy is the energy *still available in the field* to accelerate the particle, before the particle acquires it. The energy consumed would be the integral of the force, and therefore the potential energy is the *negative* integral of the force:

$$U_{\text{Coulomb}} = \int_0^{U(r_{12})} dU = -\int_\infty^{r_{12}} F_{\text{Coulomb}} \, dr = -\int_\infty^{r_{12}} \frac{q_1 q_2}{4\pi\varepsilon_0 r^2} dr = \frac{q_1 q_2}{4\pi\varepsilon_0 r_{12}}. \tag{A.42}$$

The lower limit of the integral in r is chosen to be infinity because that is where U is zero.

Throughout this textbook we will be studying the Coulomb force interactions of various particles: atomic nuclei and electrons, atoms and other atoms, molecules and other molecules. From the potential energy function and the total energy, we can in principle determine *all the possible results* of these interactions. Because the total energy is conserved, the potential energy becomes the key to many central problems throughout physical chemistry. Keep an eye on it.

Both kinetic and potential energy are measured relative to some reference energy. The observable results that we predict cannot depend on which reference points we select, as long as we are careful to stick with them throughout the problem. The kinetic energy depends on the speed v, but this value depends in turn on the observer's own speed. Similarly, the integral $\int F \, ds$ that gives the potential energy has to be integrated from some origin (mathematically, the lower limit of the integral), and we are free to choose whatever origin we like. We usually set zero potential energy to correspond to some convenient physical state, such as an ionized atom or the most stable geometry of a molecule.

We will usually measure positions and speeds relative to the **center of mass** of our system, The center of mass is the location $\vec{r}_{COM}$ such that

$$\sum_{i=1}^{N} m_i(\vec{r}_i - \vec{r}_{COM}) = 0, \tag{A.43}$$

where the sum is over all the particles i, each with mass m_i and position $\vec{r}_i$. The center of mass makes a convenient origin for our coordinate system because most of the motions that we want to study are the *relative* motions of particles, such as the influence of a positively charged nucleus on a nearby electron, rather than the overall motion of the system.

To see how the relative and overall motion can be separated, let's take the case of two particles with masses m_1 and m_2, position vectors $\vec{r}_1$ and $\vec{r}_2$, and velocity vectors $\vec{v}_1$ and $\vec{v}_2$. The center of mass is at position $\vec{r}_{COM}$ such that

$$m_1(\vec{r}_1 - \vec{r}_{COM}) + m_2(\vec{r}_2 - \vec{r}_{COM}) = 0.$$

We convert this to a relationship among the velocities by taking the derivative of both sides with respect to time t:

$$m_1\left(\frac{d\vec{r}_1}{dt} - \frac{d\vec{r}_{COM}}{dt}\right) + m_2\left(\frac{d\vec{r}_2}{dt} - \frac{d\vec{r}_{COM}}{dt}\right)$$

$$= m_1(\vec{v}_1 - \vec{v}_{COM}) + m_2(\vec{v}_2 - \vec{v}_{COM}) = 0.$$

Now we can solve for the center of mass velocity vector,

$$\vec{v}_{COM} = \frac{m_1\vec{v}_1 + m_2\vec{v}_2}{m_1 + m_2}.$$

This velocity tells us how fast and in what direction the center of mass of the system is moving, no matter what the individual motions of the two particles. There is a kinetic energy $mv^2/2$ associated with this motion, where the speed is the magnitude of the $\vec{v}_{COM}$ vector and the mass is the combined mass of both particles:

$$K_{COM} = \frac{1}{2}(m_1 + m_2)\vec{v}_{COM}^2$$

$$= \frac{1}{2}(m_1 + m_2)\frac{(m_1v_1)^2 + 2m_1m_2(\vec{v}_1 \cdot \vec{v}_2) + (m_2v_2)^2}{(m_1 + m_2)^2}$$

$$= \frac{1}{2}\frac{m_1^2v_1^2 + 2m_1m_2(\vec{v}_1 \cdot \vec{v}_2) + m_2^2v_2^2}{m_1 + m_2}. \tag{A.44}$$

This value is not the same as the *total* kinetic energy, however, because it ignores any motions of the particles *relative* to the center of mass, and therefore relative to one another. (By motions "relative to one another," we mean changes in the distance between the particles, or in the direction of one from the other.)

The total kinetic energy K_{tot} is the sum of $mv^2/2$ for both particles. But the number we need to calculate much of the time is the kinetic energy K_{rel} for the relative motion, so we find that by subtracting K_{COM} from K_{tot},

$$
\begin{aligned}
K_{rel} &= K_{tot} - K_{COM} \\
&= \frac{1}{2}(m_1 v_1^2 + m_2 v_2^2) - \frac{1}{2}\frac{m_1^2 v_1^2 + 2m_1 m_2(\vec{v_1}\cdot\vec{v_2}) + m_2^2 v_2^2}{m_1 + m_2} && \text{by Eq. A.44} \\
&= \frac{1}{2}\left[m_1 v_1^2\left(1 - \frac{m_1}{m_1 + m_2}\right) - \frac{2m_1 m_2(\vec{v_1}\cdot\vec{v_2})}{m_1 + m_2} + m_2 v_2^2\left(1 - \frac{m_2}{m_1 + m_2}\right)\right] && \text{combine } mv^2 \text{ terms} \\
&= \frac{1}{2}\left[m_1 v_1^2\left(\frac{(m_1 + m_2) - m_1}{m_1 + m_2}\right) - \frac{2m_1 m_2(\vec{v_1}\cdot\vec{v_2})}{m_1 + m_2} + m_2 v_2^2\left(\frac{(m_1 + m_2) - m_2}{m_1 + m_2}\right)\right] && 1 = (m_1 + m_2)/(m_1 + m_2) \\
&= \frac{1}{2}\left[m_1 v_1^2\left(\frac{m_2}{m_1 + m_2}\right) - \frac{2m_1 m_2(\vec{v_1}\cdot\vec{v_2})}{m_1 + m_2} + m_2 v_2^2\left(\frac{m_1}{m_1 + m_2}\right)\right] \\
&= \frac{1}{2}\frac{m_1 m_2}{m_1 + m_2}\left[v_1^2 - 2(\vec{v_1}\cdot\vec{v_2}) + v_2^2\right] && \text{factor out } m_1 m_2/(m_1 + m_2) \\
&= \frac{1}{2}\frac{m_1 m_2}{m_1 + m_2}(\vec{v_1} - \vec{v_2})^2 \\
&\equiv \frac{1}{2}\mu(\vec{v_1} - \vec{v_2})^2 && (A.45)
\end{aligned}
$$

In the last step, we define the **reduced mass** $\mu = m_1 m_2/(m_1 + m_2)$. The reduced mass is the effective mass that appears in the equation for the relative kinetic energy of the two particles. Notice that the velocity that appears in that expression, $\vec{v_1} - \vec{v_2}$, is the velocity difference between the two particles, so K_{rel} depends only on the relative motion of the two particles.

When we cover the motions of electrons around atomic nuclei and the motions of atoms bound together in a molecule, we will take for granted this breakdown of the kinetic energy into a component for the center of mass and a separate component for the relative motion. In the relative motion term, the masses of the particles will appear in terms of the reduced mass defined in Eq. A.45.

Angular Momentum

Angular momentum is among the most challenging concepts in first-year physics. It is an important concept, however, for we will find that the energies of electrons, atoms, and molecules are often stored in some form of angular motion, and nearly all magnetic properties in chemistry arise from angular motion. The value of angular momentum lies in its being conserved in the absence of external forces, just as energy, mass, and linear momentum are. Angular momentum and all its effects result from changing the *direction* of motion of an object. This requires a force and involves an acceleration. That does not necessarily mean a change in kinetic energy, because circular motion at fixed speed requires a constant acceleration but no change in v nor in $K = mv^2/2$.

The angular momentum $\overrightarrow{L}$ of an object with linear momentum $\overrightarrow{p} = m\overrightarrow{v}$ and position vector $\overrightarrow{r}$ relative to the system's center of mass is defined to be the vector cross-product

$$\overrightarrow{L} = \overrightarrow{r} \times \overrightarrow{p} = m\overrightarrow{r} \times \overrightarrow{v}. \tag{A.46}$$

The angular momentum is also a vector; it has a magnitude $|\overrightarrow{L}| = L$, and it has a direction that is *perpendicular* to the plane of motion. For circular motion, the vectors $\overrightarrow{r}$ and $\overrightarrow{p}$ are always perpendicular, and the magnitude of the cross product simplifies to

$$circular\ motion: \quad L = rp = mvr. \tag{A.47}$$

You may well ask what the angular momentum represents physically. The easiest answer is that it is a purely mathematical expression of great convenience. Its conservation law may suggest that it has a fundamental physical meaning, but the conservation of angular momentum is a direct consequence of conservation of the linear momentum and energy, not a new basic principle. Angular momentum is by no means unique in this way, however, because nearly all the parameters that we study in this text are similar mathematical constructs—including, for example, mass, charge, force, and energy. There is very little that we can claim to measure directly. The astonishing thing is how successfully we can convert those few direct measurements—and a great deal of abstract interpretation—into predictions for how matter will behave. It is that modeling of the behavior of matter, the extrapolation from theory to real observations, that is the subject from this point on.

PROBLEMS

For this chapter, only a few representative and illustrative problems are given, for the benefit of students interested in using some of the skills and concepts that we will draw on in the main text.

Unit Analysis and Reasonable Values

A.1 If the pK_a is $-\log_{10} K_a$, and K_a is equal to $\exp[-\Delta G/(RT)]$, write an equation for the pK_a in terms of $\Delta G/(RT)$.

A.2 Which of the following results is the most reasonable value for the average speed of an electrical signal traveling through a copper wire?
a. $2 \cdot 10^{10}$ m s^{-1} b. $2 \cdot 10^5$ m s^{-1} c. 2 m s^{-1}

A.3 Which of the following results is the most reasonable value for the number of atoms in a sample of crystal with dimensions 5.0 Å × 5.0 Å × 5.0 Å?
a. $8 \cdot 10^{10}$ b. $8 \cdot 10^5$ c. 8

A.4 Are the following values reasonable for the quantities indicated?
a. $25 \cdot 10^{-8}$ m for the C—O bond length in H_2CO.

b. 78 amu for the mass of one 6-carbon hydrocarbon molecule.

A.5 What are the correct units for the constants indicated in the following equations?
a. k in $-d[A]/dt = k[A][B]$, if $[A]$ and $[B]$ are in units of mol L^{-1}, and t is in units of s
b. k_B in $k = Ae^{-E_a/(k_B T)}$, if E_a is in units of J, T is in units of K, and A is in units of s^{-1}
c. K_c in $K_c = [C][D]/([A][B])$, if the concentrations $[X]$ are all in units of mol L^{-1}
d. k in $\omega = \sqrt{k/\mu}$, if ω has units of s^{-1} and μ has units of kg

Mathematics

A.6 Solve for all possible values of x that satisfy the equation

$$-\frac{(2x + 1)^2 e^{-ax^2}}{a^2} = 0,$$

where a is a finite and non-zero constant.

A.7 Find the complex conjugates of the following:

a. $x - iy$
b. ix^2y^2
c. $xy(x + iy + z)$
d. $(x + iy)/z$
e. e^{ix}
f. 54.3

A.8 For the vectors $\vec{A} = (1,0,0)$, $\vec{B} = (1,0,1)$, and $\vec{C} = (0,2,1)$, find the following:

a. $|\vec{C}|$
b. $\vec{A} + \vec{B}$
c. $\vec{A} \cdot \vec{B}$
d. $\vec{A} \cdot \vec{C}$
e. $\vec{A} \times \vec{B}$

A.9 Prove the Euler formula,

$$e^{ix} = \cos x + i \sin x,$$

using the Taylor series expansions for these functions.

A.10 Solve V_m to three significant digits in the following equation using a math program or successive approximation, given the values $P = 1.000$ bar, $a = 3.716$ L mol^{-2}, $b = 0.0408$ L mol^{-1}, $R = 0.083145$ L mol^{-1} K^{-1}, and $T = 298.15$ K:

$$\frac{\left(P - \dfrac{a}{V_m^2}\right)(V_m - b)}{RT} = 1.$$

A.11 Evaluate the derivatives with respect to x for the following functions.

a. $f(x) = (x + 1)^{1/2}$
b. $f(x) = [x/(x + 1)]^{1/2}$
c. $f(x) = \exp[x^{1/2}]$
d. $f(x) = \exp[\cos x^2]$

A.12 Evaluate the following integrals:

a. $\displaystyle\int_0^\infty e^{-ax}dx$

b. $\displaystyle\int_1^5 x^2 dx$

c. $\displaystyle\int_1^5 x^{-3/2}dx$

d. $r^2 \displaystyle\int_0^{2\pi} d\varphi \int_0^\pi \sin\theta d\theta$

Physics

A.13 Calculate the force in N between two electrons separated by a distance of 1.00 Å.

A.14 Express the kinetic energy K of an object in terms of its momentum p and its mass m.

A.15 Let the gravitational force obey the law $F_{gravity} = -mg$, where the constant g, equal to 9.80 m s^{-2}, is the acceleration due to gravity at the earth's surface. What is the gravitational potential energy as a function of the height r and the mass m?

A.16 In dealing with molecular forces, gravity is usually ignored. Show that this is justified by comparing the Coulomb force between the electron and proton at a distance of 0.529 Å to the gravitational force on the hydrogen atom exerted by the earth near the earth's surface (earth's gravitational force $F_{gravity} = -mg$, where $g = 9.80$ m s^{-2}).

A.17 One method for solving Newton's equations of motions for a mechanical system employs a function called the **Lagrangian** L, commonly defined as the kinetic energy K minus the potential energy U:

$$L \equiv K - U.$$

For the mechanics problems we encounter, the potential energy depends *only* on position, and the kinetic energy depends *only* on speed. (For example, if two identical particles pass through the same point in space at different times, they have the same potential energy even if one of them is traveling faster.) Using this, show that the Lagrangian has the feature that

$$\frac{\partial L}{\partial x} = \frac{d}{dt}\frac{\partial L}{\partial v_x}.$$

A.18 Two particles with masses m_1 and m_2 are moving with initial speeds v_1 and v_2, respectively. They collide with each other, and then continue along. None of the energy is lost to other kinds of motion, and there are no external forces working on the particles (i.e., the potential energy can be set to zero). Use conservation of energy and linear momentum to find the speeds v_1' and v_2' of the particles after the collision.

A.19 Calculate the angular momentum L of an electron (assuming that the velocity and position vectors are perpendicular to each other) when the electron is in a state with $K = -U$ at a distance of 1.0 Å from a proton. The electron's kinetic energy is K, and U is its potential energy due to the Coulomb force. Use the proton as the center of mass of the system.

A.20 This is a basic mechanics problem that includes some trigonometry, some vector algebra, and some mechanics. You may solve any or all of the individual pieces; together they add up to a verification of conservation of angular momentum for a very simple collision problem. The collision system is illustrated in the accompanying figure. Two classical, spherical particles (labeled 1 and 2) of equal mass m and equal diameter d approach each other, each with speed v_0 but traveling in opposite directions parallel to the z axis. They initially travel with constant x coordinates, such that $0 \le x_1 < d/2$ and $x_2 = -x_1$. This guarantees that the two particles will eventually collide, changing their velocity vectors. When the particles collide, a line can be drawn between their centers of mass, and we'll designate as θ the angle that this line makes with the z axis, where $0 \le \theta < \pi/2$.

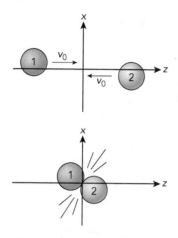

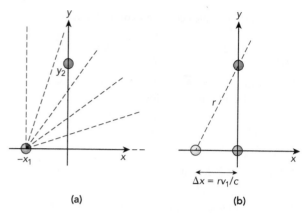

(a)

(b)

a. Show that when the collision occurs, the particle centers of mass are at the positions (relative to the center of mass of the entire system)
$$\vec{r}_1^{(0)} = ((d/2)\sin\theta, 0, -(d/2)\cos\theta) \text{ and }$$
$$\vec{r}_2^{(0)} = (-(d/2)\sin\theta, 0, (d/2)\cos\theta).$$

b. Set the time when the collision occurs to $t = 0$, so t is negative before the collision and positive after the collision. The velocity vectors before the collision are $\vec{v}_1'' = v_0(0,0,1)$ and $\vec{v}_2'' = v_0(0,0,-1)$. Show that the velocity vectors after the collision are $\vec{v}_1' = v_0(\sin 2\theta, 0, -\cos 2\theta)$ and $\vec{v}_2' = v_0(-\sin 2\theta, 0, \cos 2\theta)$.

c. The position vectors are functions of time and may be written $\vec{r}_i = \vec{r}_i^{(0)} + \vec{v}_i t$, where i is 1 or 2, using the velocity vectors $\vec{v}_i''$ or $\vec{v}_i'$ appropriate for t negative or positive, respectively. Show that the angular momentum vector for this system is $\vec{L} = -mdv_0(0, \sin\theta, 0)$ when calculated before *and* after the collision.

A.21 This is a demonstration of how the magnetic field is generated by the motion of a charged particle. Start with two charged particles 1 and 2 at Cartesian coordinates $(-x_1,0,0)$ and $(0,y_2,0)$, respectively. We generally think of these two particles as interacting via their electric fields, which exert forces along field lines emanating from the charged particle as in (a) of the accompanying figure.

We now move particle 1 along the x axis at a constant speed v_1 and check the forces when particle 1 reaches the origin, as shown in (b) of the accompanying figure. The field lines of particle 1 move as well, and this changes the electrical force at particle 2. However, that change is transmitted at the speed of light, not instantaneously. The time required for the field of particle 1 at time t to reach particle 2 is r/c, where r is the separation between the two particles at time t. In the meantime, particle 1 has traveled a distance rv_1/c.

a. Write an equation for the electric field vector $\vec{\mathcal{E}}_1$ generated by particle 1 at the location of particle 2. Write the vector in its Cartesian form, showing the x- and y-components, as for example $\vec{\mathcal{E}}_1 = c(x_0,y_0,0)$. The electric field vector can be defined by its relation to the force: $\vec{F}_{elec} = q_2\vec{\mathcal{E}}_1$.

b. Now let's define the magnetic flux density by the relation $\vec{B} = \frac{1}{c^2}\vec{\mathcal{E}}_1 \times \vec{v}_1$. Write $\vec{B}$ in vector form.

c. Next, find the vector representing the magnetic force exerted on particle 2 in the field generated by particle 1, defining that force to be $\vec{F}_{mag} = q_2 \vec{v}_1 \times \vec{B}$.

d. Finally, we want to show that this magnetic force is the difference between the actual Coulomb force and the classical Coulomb force that would be present if the electric field were transmitted instantaneously. Write the magnitude of the actual Coulomb force F (using the distance r) and the classical force F' (using the distance y_2). Calculate the difference between these, and simplify the result assuming $v_1 \ll c$ to show that you get the same value as predicted by the equations.

EXTRAPOLATING FROM MOLECULAR TO MACROSCOPIC SYSTEMS

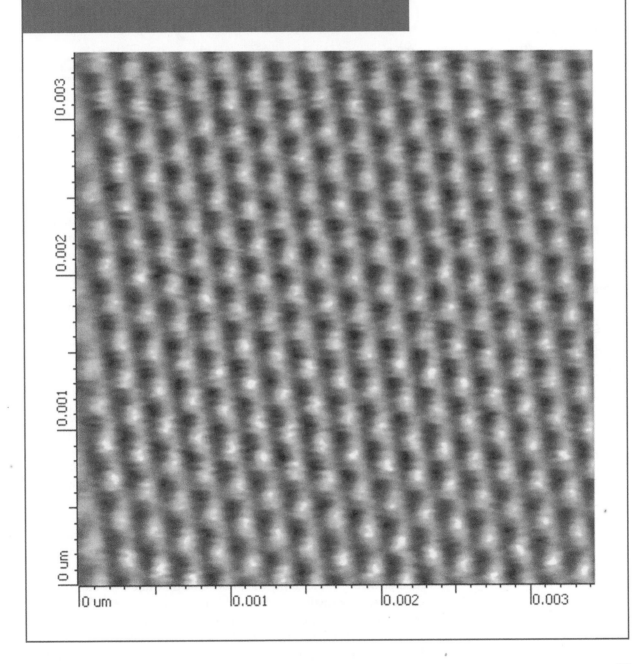

1

Classical Physical Chemistry Sets the Stage

LEARNING OBJECTIVES

After reading this chapter, you will be able to do the following:

❶ Apply the first and second laws of thermodynamics to calculate or place limits on simple thermodynamic processes.

❷ Use the Maxwell-Boltzmann distribution to predict global properties of an ideal gas that depend on the speed.

GOAL   *Why Are We Here?*

Classical chemistry predated an accurate knowledge of basic atomic structure, let alone quantum mechanics. The goal of this chapter is to review those laws of classical chemistry that ultimately led us *beyond* classical chemistry—that forced us to confront the fundamental question of how matter is put together years before the experiments existed that could begin to provide the answer. Theories that described matter at the macroscopic scale pointed increasingly toward a new understanding of matter at the microscopic scale. Several of the principles that grew out of these theories share a common foundation, which will guide our approach to the rest of the material in this volume.

CONTEXT   *Where Are We Now?*

This book has a theme: that chemical properties are determined by atomic and molecular structure. Experiments and calculations support this thesis on a daily basis in laboratories the world over, from high school classrooms to universities and government agencies. Physical chemistry is the study of the fundamental principles that bind chemistry together, and this principle more than any other is the basis of our understanding of chemistry today.

It is all the more remarkable, then, that progress in physical chemistry proceeded at the pace it did—powered by astonishing insights into the nature of matter—well into the 20th century without an accurate theory of the structure of the atom. Scientists of the time knew the atom only as an indiscernibly small mass that could be assigned to one of the elements on the basis of its reactivity or other properties. But the *origin* of those properties—what makes sodium different from magnesium, say, but similar to potassium—and the unifying thread that connects all chemistry together was a mystery.

In this chapter, we briefly review just a few of these basic, shared principles of chemistry that grew out of the arduous work of 19th- and early 20th-century scientists. We'll call these the laws of *classical chemistry,* which guided our understanding of matter prior to the revolutionary theories of quantum mechanics and relativity. General chemistry courses provide ample opportunities to work closely with many of these laws, but amidst the details we rarely have a chance to see these laws together and appreciate how intricately related they are.

At the end of our review, we will still be left with the mystery, the unanswered questions that inspired the work described in the chapters that follow.

1.1 The Classical Properties of Matter

Physical chemistry is the study of the fundamental laws of *ordinary matter,* matter composed of protons, neutrons, and electrons. Those laws are woven together, seamlessly connecting the properties of individual atoms and molecules (the *microscopic scale*) and moles of particles (the *macroscopic scale*). Although the laws of chemistry at the macroscopic scale were developed first, *they are not separate from the laws at the microscopic scale.* Physical chemistry shows how the behavior of billions of molecules can be predicted from the structure of a single molecule.

But we arrived at our modern theoretical framework for chemistry in the opposite direction: experiments at the macroscopic scale showed the need for a more refined theory of microscopic molecular structure. Experiments on energy transfer were at the forefront of the new science of physical chemistry, motivating the invention of properties that had not been conceived by existing theory and painting a dim outline of a new model of the structure of matter.

We begin with a review of measurement and the parameters that could be precisely determined by classical chemistry. These are the parameters that we will seek to understand and the values of which we will seek to predict as we progress through the text.

Rulers and Clocks

We learn about the properties of matter from measurements. But *what* do we measure? This question may sound trivial, but the longer and more sophisticated our experience in the laboratory, the easier it is to forget that there are very few parameters of a system that we measure *directly.* We exist in space and time, and (if we can trust our senses) we can directly measure an extent in space (a distance) by comparison to a ruler or an extent in time (a duration) with a clock.

But distances and durations are not sufficient to describe the physical universe. If we have two clear blocks of the same size and shape, one made of diamond and the other of water ice (Fig 1.1), they'll last indefinitely if the temperature is below freezing, but we can quickly learn that they are not the same thing. It takes more effort to lift the diamond than the ice. If we raise the temperature of the ice, it gradually turns to liquid. Rulers and clocks aren't enough by themselves to tell us how these two materials come to be so fundamentally different. What can we measure instead that will tell us how these two substances differ from one another?

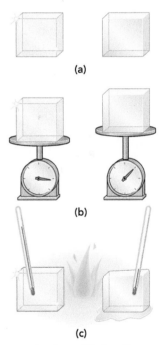

(a)

(b)

(c)

▲ FIGURE 1.1 **Blocks of diamond and ice. (a)** Two transparent blocks have similar appearance. We have invented parameters that explain the observable distinctions between the two. **(b)** We can measure the weight and convert it to a mass. **(c)** We can measure the temperature at which one block melts.

Because it takes more effort to lift the diamond than the ice, everything else being equal, we invent a new parameter: the mass, which is one way of measuring how much matter is present, regardless of how much space it occupies. But how do we quantify mass? One way is to measure the distance that the object, say our diamond block, compresses a spring on a scale as gravity pulls it toward the earth. The diamond block and the ice block compress the spring by different amounts, and from the degree of compression, we can give the mass a number. But see what we're measuring? Not the mass, but a *distance*—the shortening of the spring. The laws of physics are the only connection we have between that distance and the mass of the block. After a year or so of physics class, we know to say that the scale is actually measuring the weight of the block, which is the force of Earth's gravity acting on the object. Force in turn is mass times acceleration, and if the acceleration due to Earth's gravity is a constant $9.8\,\mathrm{m\,s^{-2}}$, we can convert our scale reading to a mass.

The melting of the ice, on the other hand, takes place over time. We recognize that a transformation is occurring from the change in the shape of the material from a rigid block to a puddle of water. Not only do the diamond and the ice differ in mass, they differ in how they evolve. To quantify the change in the ice, we may notice that the ice block is initially cold to the touch, and the water liquid eventually feels as warm as its surroundings. Again, we can invent a parameter—the temperature this time—which helps us quantify the process that converts solid ice to liquid water. How do we measure temperature? Well, yes, you're right: we get a thermometer. But how does the thermometer work? An old-fashioned thermometer works by measuring the expansion of liquid mercury in a tube, measuring a distance again. Modern thermometers are more likely to register temperature by electronic means, such as detecting the change in the voltage across an electrical circuit element. The change in electrical signal generates a force that determines the reading on our thermometer. And force is mass times acceleration—parameters that we measure using rulers and clocks.

This distinction—between directly measured properties and *inferred* properties such as mass and charge—is an important distinction because many of the parameters that we will use in physical chemistry are so familiar to us that we tend to remember only their most superficial meaning. One of the goals of classical physical chemistry was to discover how the properties of different forms of matter were related to each other and how those properties depended on the conditions of the measurement. In many cases, as in our earlier example, parameters were invented in order to explain experimental results that the existing set of parameters did not appear to describe. But whereas the mass and charge are defined by fundamental forces that challenge our deepest understanding of physics, other parameters such as the enthalpy needn't tax our brains so much, being formed from other quantities in combinations that (as we will see later) have a convenient mathematical form for calculations and measurements rather than a basis in the nature of matter.

Let's review a few of the parameters that kept the classical chemists busy as they so capably began building the framework for the modern theory of chemistry.

Classical Parameters

We take our place at the lab bench and introduce ourselves to the thing we're going to study—the **system.** What properties can we measure?

Volume

Everything we study in the laboratory occupies three dimensions in space, and we can measure the distance occupied along each coordinate axis to determine a volume V. Unhappily, the SI unit for volume is the cubic meter, which is a handy unit if you have a metric bulldozer, but not a convenient amount to manage in a typical lab. So you may find that the units for volume are more trouble to handle than the concept itself. In this text, as in the chemistry lab, we will most often report volumes in liters $(1\,\text{L} \equiv 10^{-3}\,\text{m}^3)$ or milliliters (equivalent to cubic centimeters: $1\,\text{mL} = 1\,\text{cm}^3 = 10^{-6}\,\text{m}^3)$.

Mass and Moles

We've already mentioned mass as another parameter of our system that we can quantify. One of the great accomplishments of classical chemistry was the recognition that atoms of the same element shared the same (or nearly the same) mass. The principle had to be refined to account for different isotopes of the same element. A crucial relationship that we learn in general chemistry courses relates the mass of a particular substance to the number of particles—atoms or molecules—that make up the substance. A useful unit for the counting of gram-scale amounts of particles is the mole, defined by Avogadro's number:

$$1\ \text{mol} = 6.022 \cdot 10^{23}. \tag{1.1}$$

If we know the molecular formula of the substance we're studying, we can calculate the molar mass $\mathcal{M}$. The number of particles in a sample of mass M is given in units of moles by

$$n = \frac{M}{\mathcal{M}}. \tag{1.2}$$

If we had just one particle of mass m, then its molar mass would be

$$\mathcal{M} = (6.022 \cdot 10^{23}\,\text{mol}^{-1})m. \tag{1.3}$$

Again, be on the lookout for the units here. The values for atomic masses taken from the usual periodic table are either molar masses $\mathcal{M}$ in $\text{g}\,\text{mol}^{-1}$ or atomic masses m in atomic mass units (amu). Neither value uses SI units, which are $\text{kg}\,\text{mol}^{-1}$ for molar mass and kg for mass. We will routinely convert grams to kilograms to get SI units for masses.

Pressure

Volume lets us measure the space occupied; mass lets us count particles. Neither parameter alone explains how matter can change, however. For change to take place, we need force. The force exerted by a system in a particular direction is generally proportional to the area of the face where the force is operating, so rather than measure the force of our system, we will instead measure its **pressure,** the force F divided by the area A:

$$P = \frac{F}{A}. \tag{1.4}$$

In SI, we measure pressures in pascal (Pa), a unit only marginally more useful than the cubic meter. By a fortunate coincidence, however, if we define the bar to be exactly 10^5 Pa, then the bar turns out to be within 2% of typical atmospheric pressure, so in this text we will normally use the bar for specifying pressures and then convert to pascal when we want to put all quantities into SI units.

Pressure helps us quantify the force exerted by a system, but where does that force come from? Forces come in different forms, after all. Gravity, mentioned earlier, is a relatively weak force in chemistry because the particles that are responsible for the chemical identity of our system—the individual atoms and molecules—have tiny masses. Instead, processes in chemistry are dominated by electromagnetic forces, particularly the Coulomb force that governs the attractions and repulsions of two charged particles q_1 and q_2 across a distance r_{12} (Eq. A.41):

$$F_{\text{Coulomb}} = \frac{q_1 q_2}{4\pi\varepsilon_0 r_{12}^2}.$$

If the force is positive, the two charges are repelled; if it is negative, the charges attract one another. At the distance scales we're accustomed to, greater than a centimeter or so, the Coulomb force in matter is most often evident from the inability of two objects to share the same space. The negatively charged electrons on object A repel the negatively charged electrons on object B, and similarly the nuclei repel each other. Yet the Coulomb force also binds the nuclei and electrons together, maintaining the chemical bonds that give each object its structural integrity.

The particles within our system are always moving, at least slightly. As they move toward the surface of our system, they push against anything that lies at the surface, and that push is what we measure as the pressure. The faster the particles move, the more force they exert, and the higher the pressure. So one of the parameters closely related to the pressure must be the particle speed.

Temperature

The *temperature* can usually be related to the average speed of the particles. If the ice in our earlier example begins at a low temperature compared to the surrounding air, then as it warms to higher temperature the average speed of the molecules increases. Temperature turns out to be a useful parameter for describing the melting of ice because the melting tends to occur when the ice reaches a specific temperature, at which point the average molecular motion in the solid reaches a certain level.

But what *is* temperature? It falls into that category of parameters that we knew how to measure long before we understood what it was we were measuring. We could sense hot and cold and knew that temperature was the parameter that quantified the difference. Temperature is measured as described earlier, by a system's ability to expand the mercury in a thermometer or to alter the voltage of certain circuit components. The units of temperature we use in the laboratory are usually Celsius, originally defined so that liquid water freezes at $0\,°C$ and boils at $100\,°C$ at a pressure of 1.00 bar. In this text, however, we will almost always use the SI unit of temperature, the kelvin (K), where

$$T(\text{K}) = T(°\text{C}) + 273.15, \tag{1.5}$$

because equations in physical chemistry that depend on temperature almost always require that we use an absolute temperature scale.

None of this tells us what the temperature is, however. A precise definition for temperature had to await a theory of how energy is transferred during physical processes: the theory underlying thermodynamics.

1.2 Thermodynamics

Thermodynamics is the study of the conversion of energy among different forms and the transfer of energy between different systems. As chemists, we're interested in molecular structure and how to change it. The two disciplines merge in chemical thermodynamics, which probes the flow of energy during the transformation from reactants to products. Every chemical reaction involves the exchange of energy, as well as the exchange of atoms. And whereas the nature of the reaction determines which atoms we're working with, the principles of chemical thermodynamics apply to *any* chemical reaction. Before we can become familiar with those principles, let's make sure we know what we mean by the *energy*.

Energy

In Section A.2 we divide energy into three classes: kinetic, potential, and radiant energy. Each of these is consistent with a more general definition of energy as the capacity to do work, which means the capacity to apply a force or (equivalently) to oppose an existing force. The kinetic energy of a swinging baseball bat, for example, is a measure of the bat's capacity to knock a baseball out of the park. The potential energy of a rock on the side of a cliff determines how much force it will exert when it hits the ground. And the radiant energy of sunlight can be converted by solar cells into an electrical current that charges a car battery, which then drives the engine. In principle, we can convert any of these forms of energy into any other.

However, in this text we will neglect the conversion of mass into energy, leaving that complication in the hands of our colleagues in the physics department. We will also ignore kinetic energy in the motion of the entire system and potential energy from any external fields that affect the entire system. For our purposes, zero energy is the minimum possible energy at which all the particles in our system are fully assembled and present in whatever container we have constructed for them. This is properly called the **internal energy** of the system, but in this volume we will simply refer to it as the energy. Although the absolute energy E appears in many expressions we encounter,[1] ultimately the experimental properties we measure depend only on the energy *differences* ΔE between the initial and final state energies E_i and E_f. If we shift all of the energies by a constant amount E_0, the value of ΔE is the same:

$$\Delta E = E_f - E_i = (E_f + E_0) - (E_i + E_0). \tag{1.6}$$

Thermodynamics is largely concerned with the nature of the terms that contribute to ΔE.

---

[1] The standard symbol for internal energy is U, but we will use U for potential energy. Since E stands in for the energy in other common symbols (such as E_a for the activation energy in chemical reactions), using E also keeps the notation more internally consistent.

The First Law

The **first law of thermodynamics** divides the change in energy ΔE into two contributions:

- The **work** w is the energy change that drives a *net shift in the distribution of mass in the system.*
- The **heat** q is any other energy change—in particular the energy associated with *changes in the random motions of individual particles in the system.*

The division of energy transfer into heat and work is written mathematically

$$\Delta E \equiv E_{\text{final}} - E_{\text{initial}} = q + w, \tag{1.7}$$

where w is the work done going from the initial state to the final state and q is the heat evolved or absorbed during the change. Equation 1.7 is one form of the **first law of thermodynamics.**

In this textbook, positive w means work is done *to* the sample, increasing the sample's energy—as when, for example, an external pressure is applied that compresses a gas. Negative w is work done *by* the sample, decreasing its energy. With this convention, the signs for heat and work function the same way: positive q or w means an *increase* in the system's energy, and negative q or w means a *decrease* in the system's energy. (This sign convention is not always used, so be careful. Reference books in engineering have adopted the opposite sign convention for work, so the work that the system can do is expressed as a positive number, which has an understandable appeal to a practical mindset.)

Qualitative Definitions of Enthalpy and Free Energy

As the study of chemical thermodynamics expanded, scientists found that the energy was not always the most convenient parameter to incorporate into their mathematical models. For example, an engineer wishing to get work out of the steam in a pipe might want to raise the temperature of the steam by heating the pipe (Fig. 1.2). But some of that energy, instead of raising the temperature of the gas, would be diverted into the work required for the gas to expand against a piston at one end of the pipe. If we calculate the energy change ΔE of the gas, we include both the contributions to the heating of the gas q and the work w done by the expansion. In this example, q is positive because we are adding energy to the system in the form of heat, but w is negative because the work done is energy *lost* by the system. The total change in energy $\Delta E = q + w$ is therefore *less* than the energy needed to heat the steam:

$$\Delta E = q + w < q. \tag{1.8}$$

So instead of calculating ΔE, we calculate the change in a new parameter, invented just for this purpose, which we call the **enthalpy** H:

$$\text{at constant pressure: } \Delta H = q. \tag{1.9}$$

We limit this equality to cases in which the pressure is constant, because if the pressure changes then the energy needed to do the work changes.

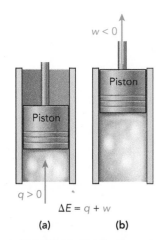

$w < 0$

Piston

Piston

$q > 0$

$\Delta E = q + w$

(a)   (b)

▲ **FIGURE 1.2 The first law of thermodynamics.** In this example, **(a)** heating steam in a pipe adds energy q, which raises the temperature of the gas and **(b)** causes the gas to expand against the surroundings, pushing a piston upward. The expansion requires a loss in energy w (where w is negative) due to the work done by the gas. The total change in energy ΔE is the sum $q + w$.

If we look at the other end of our steam pipe, where we get the steam to do work—by pushing against a piston, for example—we find that not all of the heat that we put into the steam can be converted into work. Some of that energy goes into warming the piston rather than pushing it forward; this is heat lost from the gas to the surroundings. Josiah Willard Gibbs, in 1873, invented a way of looking at physical processes that led to the formulation of another relative of the energy, dubbed the **Gibbs free energy** G. Gibbs and others found that it was possible to define the *maximum amount of work* obtainable other than by expansion (such as by chemical reaction). The Gibbs free energy defines the value for a process at constant temperature and pressure, and G turns out to be a pivotal parameter in chemistry. However, Gibbs found that the mathematical form of the free energy was dependent on a relatively new parameter, one that connected the concepts of energy and temperature.

To understand why we need a new parameter that distinguishes temperature from energy, let's return for a moment to our blocks of diamond and ice, introduced at the beginning of this chapter. As we apply heat to the block of diamond, the temperature rises. (In fact, the temperature will rise faster for diamond than for almost anything else you can find.) We added energy in the form of heat, so $\Delta E > 0$, and we saw a rise in temperature so $\Delta T > 0$. From this we might at first conclude that *temperature* and *energy* are different names for the same phenomenon.

But a marvelous thing happens when we heat the melting block of ice. The ice absorbs the energy as does the diamond, and yet *the temperature stays the same* as long as there is still some ice to melt. This fundamental qualitative observation proves that energy and temperature cannot be the same thing. There is some additional variable, hidden from us until we assemble the experimental evidence and examine it carefully, a variable that connects E and T and provides the basis for at last defining the temperature unambiguously. That hidden variable was formulated in the mid-19th century by Rudolf Clausius. If the epiphanies of quantum mechanics and relativity were earthquakes that rocked the foundations of physics, Clausius' discovery was a more gradual but more profound island drift—such that when it was over we looked up and found ourselves in waters that we scarcely recognized. The workings of the natural world appeared in a new light.

Entropy and the Second Law

Clausius' deduction must number among the most incisive in the science of any century, and he didn't arrive at it instantly. Working over several years in the 1850s and 1860s, he formulated a relationship between heat and temperature based on a new parameter that he eventually christened the **entropy.**

At the time he was working, the Industrial Revolution was pushing the development of steam engines. The goal of these engines was to convert heat (from a burning fuel) into useful work (moving a piston) as efficiently as possible. Any system that accomplishes this is called a **heat engine** (Fig. 1.3). The term *useful work* distinguishes work that we can direct where we want from *turbulence,* which is the random movement of a fluid—technically a form of work but not one that we control. Clausius recognized that the heat used to drive an engine was always greater than the useful work the engine could do. He treated the difference between the input heat and the work done as a heat transfer to the

▶ FIGURE 1.3 **The heat engine.** Heat q is transferred into the engine to be converted into work w, but some of the energy is always dissipated into the surroundings, as heat q_{surr} that cannot be converted into useful work. The entropy S was invented to relate that lost energy to a constant temperature T, such that $q_{surr} = T\Delta S_{surr}$.

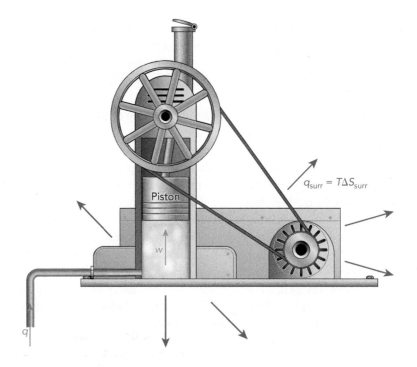

surroundings, q_{surr}, which could not be reclaimed as useful work. Because that lost energy tended to grow in proportion to the operating temperature, Clausius defined the entropy S to relate the heat lost to the temperature:

$$\Delta S_{surr} = \int \frac{dq_{surr}}{T}, \tag{1.10}$$

where ΔS_{surr} is the change in entropy of the surroundings. The engine would complete a cycle in which the piston moved up (doing work) and then returned to its starting position. At the end of a cycle, all the parameters of the *engine* would be the same—the temperature, the position of the piston, and the entropy—but the *surroundings* would have warmed up slightly. Adding the changes in entropy of the engine and of the surroundings to get the total entropy change ΔS_T, Clausius reasoned that running the engine had increased the total entropy.

The concept of entropy would turn out to be central to the theory of thermodynamics. For one, it solved the problem of how temperature and energy were different. In Clausius' work, the entropy was defined in terms of the temperature, and the temperature was an independent parameter. In Chapter 2, we will flip this around, presenting an independent definition of the *entropy* and then defining the temperature to be

$$T \equiv \left(\frac{\partial E}{\partial S} \right)_{V,n}. \tag{1.11}$$

The partial derivative $\left(\frac{\partial E}{\partial S} \right)_{V,n}$ represents the incremental change in energy E of our system with respect to its entropy S, while we keep the volume V and moles of substance n constant. (Notice that the entropy S in this case is the entropy of the system itself, whereas the entropy S_T in Clausius' original definition is the combined entropy of the system and the surroundings.) With this equation defining the temperature, we can begin to understand how T can remain constant as E increases for the heated block of ice. Assume for a moment that the density of liquid water is negligibly different from the density of ice. As we heat the block of ice toward its melting point, the energy of the ice increases at a faster rate than the entropy, causing the temperature to increase. When we reach the melting point, the process changes dramatically: suddenly the energy and entropy are increasing at the same rate, so T remains a constant.

Why is this? That question remained a mystery in Clausius' time, for no answer is possible without an accurate understanding of the molecular structure and interactions that distinguish the solid from the liquid. With the advantage of an intervening century, we can solve that riddle when we address phase transitions in Chapter 10.

However, defining the temperature was not Clausius' goal. He wanted a theory that explained and quantified the inefficiency of engines, and he succeeded beyond his dreams. Not only did Clausius define the change in entropy with Eq. 1.10, he recognized from numerous measurements that for any process,

$$\Delta S_T \geq 0. \tag{1.12}$$

Equation 1.12 states the **second law of thermodynamics:** *the overall entropy never decreases.* The equality, $\Delta S_T = 0$, is reserved for an ideal case that we call a **reversible** process. But for any real process, $\Delta S_T > 0$. There is always some inefficiency, some increase in the overall entropy, and therefore some decrease in the amount of energy in the universe that can be converted to useful work. Eventually, according to the second law, there will be no more work that can be done.

If you're at the start of a new term in a physical chemistry course, with an inkling of the work ahead of us, this conclusion may carry a bit of irony. The immense implications of the second law reach far beyond the field of physical chemistry, but there are few better opportunities than in chemical thermodynamics to ponder the origins and analyze the impact of the second law.

We will revisit the first and second laws—in Chapters 8 and 9, respectively—as well as algebraic definitions for H and G after we have examined the nature of energy and entropy at the microscopic level. For now, we'll continue our review of some of the principal findings of classical chemistry as we prepare to assemble a working model of chemistry at the macroscopic level.

1.3 The Kinetic Theory of Ideal Gases

One of the first great steps toward our modern molecular theory of matter was the development of the **kinetic theory of gases** in the 19th century. The kinetic theory imagined gases as a vast collection of independent, randomly moving particles and showed that it was possible to predict experimental observables from this model. It's difficult to imagine how mysterious the microscopic nature of matter must have seemed to scientists of the age. The very idea of molecules—no matter what they looked like—as distinct particles rather than as a homogeneous soup was controversial. The kinetic theory focused attention on other experiments that might further probe the discrete nature of matter, guiding us toward our present conception of matter as being composed of subatomic particles drawn from a limited set.

Our system is an ideal gas. The ideal gas is a system of point masses that interact with nothing except the walls of their container. The interaction with the container walls is limited to **elastic collisions:** a particle striking any wall continues traveling with the same velocity vector except that the component of the velocity normal to the wall changes sign. In other words, the particle is reflected from the surface as a beam of light would be, with no change in its kinetic energy or its motion parallel to the wall.[2] As such, the ideal gas exerts a pressure on the walls of its container, but the particles never collide with each other and never condense into a liquid no matter how low the temperature. Nevertheless, for all its simplicity the ideal gas serves as an accurate model of real gases under most conditions.

Gases are important components in many reactions—after all, several of the pure nonmetal elements are most stable in gaseous form. However, gases are inconveniently sparse for chemical use. A gas at typical laboratory temperature and pressure occupies a volume a thousand times greater than the same number of moles of a liquid or solid. This makes gas-phase chemistry less popular for a number of reasons, including the long reaction times involved, the inability of gases to dissolve liquids or solids, and the demands on storage space. In this book we will nevertheless begin with systems of gases because, their faults as chemical reagents notwithstanding, gases are by far the easiest phase of matter to describe mathematically. With gases, to a great extent, we can neglect the interaction between molecules without seriously impacting the quantitative conclusions we draw about the substance.

The Maxwell-Boltzmann Distribution

Many experimental studies of gases had established relationships among the pressure, temperature, volume, and moles of gases by the mid-19th century, but James Clerk Maxwell proposed one of the first lasting theories of the behavior of gases in 1860. He began from the idea that molecular motions are random, consistent with the observation by botanist Robert Brown of the rapid, random motions of pollen particles suspended in water under a microscope. Results of random events tend to cluster around some average, with a lower probability of occurrence the further from the average. For example, if we flip a coin 50 times and write down how many times we get heads, and then flip the coin another 50 times and write the number of heads, and repeat the experiment many times, we get the distribution of results shown as individual points in Fig. 1.4. The most likely result is that we get heads 25 times and tails 25 times, but that exact result will be obtained in only about 11% of the experiments we run. Most of the time the number of heads is a little lower or a little higher than 25. The chance of getting heads decreases the further we get from the average value of 25. For example, the chance of getting heads more than 30 times in the 50 flips is only about 6%. In the limit that the number of coin flips is a very large number, the

---

[2]We actually have to allow the collisions to exchange energy with walls just enough so that the walls and the gas stay at the same temperature. Should we decide to heat or cool the gas, the energy transfer to or from the gas can take place through the walls, in which case the collisions are not completely elastic.

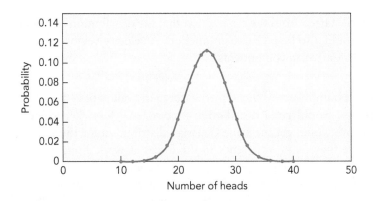

points in Fig. 1.4 become a **Gaussian function,** which is any function of the form e^{-x^2}. The curve in Fig. 1.4 closely matches the function

$$\sqrt{\frac{2}{\pi N}}\, e^{-2(h-\langle h\rangle)^2/N},$$

where h is the number of times we get heads, $\langle h\rangle$ is the mean value of h (in this case, 25), and N is the number of coin flips. (We will show how this extrapolation to large N works in Section 5.2.)

We're going to derive the Maxwell-Boltzmann distribution here, in a manner similar to Maxwell's own original derivation. This takes advantage of empirical data and the assumption of a Gaussian distribution. A very different derivation is carried out in Chapter 3 to take advantage of the methods of statistical mechanics, a field that largely grew out of Maxwell's description of the ideal gas as presented here. If that's one derivation too many for now, you can skip ahead to the **DERIVATION SUMMARY** after Eq. 1.27.

CHECKPOINT Maxwell assumed that speeds were distributed among molecules according to the same law of random distribution that applied to flipping a coin. Although the Gaussian distribution was well-known in statistics at the time, Maxwell's idea of using the Gaussian to describe a physical system was bold. Rather than giving an argument as to why the distribution should be Gaussian, Maxwell showed how the Gaussian distribution correctly predicted trends in the viscosity of gases.

Parameters key

symbol	parameter	SI units
$\mathcal{P}(x)$	probability distribution of any variable x	unitless
$\mathcal{P}_v(v)$	Maxwell-Boltzmann distribution	$\mathrm{s\,m^{-1}}$
$\vec{v}$	velocity	$\mathrm{m\,s^{-1}}$
a	constant in exponential e^{-av^2}	$\mathrm{s^2\,m^{-2}}$
Θ,Φ	angles of orientation	unitless
q	heat	J
n	moles	mol
T	temperature	K
R	gas constant	$\mathrm{J\,K^{-1}\,mol^{-1}}$
K	kinetic energy	J
N	number of particles	unitless
m	mass of one particle	kg
$\mathcal{M}$	molar mass	$\mathrm{kg\,mol^{-1}}$
$\mathcal{N}_A$	Avogadro's number	$\mathrm{mol^{-1}}$

Maxwell started from the assumption that the distribution of velocity vectors in a gas would also be a Gaussian function. A velocity vector can be written in terms of its Cartesian components,

$$\vec{v} = (v_X, v_Y, v_Z).$$

Maxwell's assumption was that if we measured the values of v_X for all the particles in a gas, we would get a distribution proportional to $e^{-av_X^2}$, where a is some constant. We would get the same Gaussian distribution for the values of v_Y and v_Z, because there is no difference in the forces at work along any of the three axes. Unlike the coin flips, the value of v_X is a continuous variable, so we can no longer ask what the probability of finding any particular value of v_X is—the more precise we make our value of v_X, the lower the chances of finding that exact value. Instead, we evaluate the probability of finding a value v_X per unit speed, which we write as $\mathcal{P}_v(v_X)$ (the subscript v indicates that the probability is per unit speed, not a unitless probability as shown in Fig. 1.4). The probability of finding a value of v_X that lies between v_1 and v_2 is then given by

$$\mathcal{P}(v_1 \leq v_X < v_2) = \int_{v_1}^{v_2} \mathcal{P}_v(v_X)dv_X = \int_{v_1}^{v_2} Ae^{-ax^2}dv_X. \tag{1.13}$$

The probability we obtain on the left side of the equation is now a unitless number (so we take away the subscript v).

The Gaussian function $e^{-av_X^2}$ gives the probability the right shape, but we need to **normalize** it. A normalized distribution is multiplied by a coefficient to ensure that when we integrate over a range of values, the result of the integral is equal to a probability between 0 (if that range of values is not found in the distribution) and 1 (if the range of values covers every possible value in the distribution). We normalize $\mathcal{P}_v(v_X)$, for example, by multiplying it by a normalization constant A and then integrating over all possible values of v_X, setting the result equal to 1:

$$\int_{-\infty}^{\infty} \mathcal{P}_v(v_X)dv_X = \int_{-\infty}^{\infty} Ae^{-av_X^2}dv_X$$

$$= 2\int_{0}^{\infty} Ae^{-av_X^2}dv_X. \quad \mathcal{P}_v(v_X) = \mathcal{P}_v(-v_X)$$

This last step puts the integral into the same form as one of our tabulated solutions:

$$\int_{-\infty}^{\infty} \mathcal{P}_v(v_X)dv_X = A\sqrt{\frac{\pi}{a}} \qquad \text{by Table A.5}$$

$$= 1 \qquad \text{if normalized}$$

$$A = \sqrt{\frac{a}{\pi}} \qquad \text{solve for A}$$

$$\mathcal{P}_v(v_X) = \sqrt{\frac{a}{\pi}}e^{-av_X^2} \tag{1.14}$$

We still don't know the value of the constant a, but we'll get to that.

If the value of v_X does not affect the chance of getting any particular value v_Y or v_Z, then the three components of the velocity vector are independent variables. Therefore, the probability of finding a particle with a specific velocity vector $\vec{v}$ is given by the product of the probabilities for a particular v_X, a particular v_Y, and a particular v_Z:

$$\mathcal{P}_{v^3}(\vec{v}) = \mathcal{P}_v(v_X)\mathcal{P}_v(v_Y)\mathcal{P}_v(v_Z)$$

$$= \left(\sqrt{\frac{a}{\pi}}e^{-av_X^2}\right)\left(\sqrt{\frac{a}{\pi}}e^{-av_Y^2}\right)\left(\sqrt{\frac{a}{\pi}}e^{-av_Z^2}\right) \qquad \text{by Eq. 1.14}$$

$$= \left(\frac{a}{\pi}\right)^{3/2}e^{-av_X^2}e^{-av_Y^2}e^{-av_Z^2}$$

$$= \left(\frac{a}{\pi}\right)^{3/2}e^{-a(v_X^2+v_Y^2+v_Z^2)}. \qquad (1.15)$$

The function $\mathcal{P}_{v^3}(\vec{v})$ is a probability per unit speed cubed, because we need to multiply it by $dv_X dv_Y dv_Z$ and integrate over *each* velocity component to get the unitless probability of finding a particular velocity vector $\vec{v}$ within the limits of the integral.

Now to find the probability of a randomly selected molecule having some particular velocity vector, we would integrate

$$\mathcal{P}_{v^3}(\vec{v})\,dv_X dv_Y dv_Z = \left(\frac{a}{\pi}\right)^{3/2}e^{-a(v_X^2+v_Y^2+v_Z^2)}dv_X dv_Y dv_Z$$

over the appropriate values of v_X, v_Y, and v_Z. However, Maxwell didn't care in what direction the molecules were traveling; he cared only about the speed of the molecules, where the speed is the magnitude of the velocity vector. If we had a function $f(x,y,z)$ that depended only on the distance from the origin, we could change coordinates from x,y,z to r,θ,ϕ, and we would find that our function depended only on r, which is the magnitude of the position vector. In the same way, we now want to transform the expression for $\mathcal{P}_{v^3}(\vec{v})$ into an expression that only depends on the magnitude of the velocity vector. In this case, we change from (v_X, v_Y, v_Z) to (v, Θ, Φ), where Θ and Φ are the angles that specify the direction of the velocity vector. In making this change, the volume element becomes (by analogy with Eq. A.7)

$$dv_X dv_Y dv_Z = v^2 dv \sin\Theta\, d\Theta\, d\Phi.$$

To eliminate the dependence on the angles in the volume element, we integrate $\mathcal{P}_{v^3}(\vec{v})$ over all values of the angles. This leaves only the dependence on v, giving us the probability distribution for *speed*:

$$\mathcal{P}_v(v)dv = \int_0^\pi \int_0^{2\pi} \mathcal{P}_{v^3}(\vec{v})v^2 dv \sin\Theta\, d\Theta\, d\Phi$$

$$= \int_0^\pi \int_0^{2\pi} \left(\frac{a}{\pi}\right)^{3/2} e^{-a(v_X^2+v_Y^2+v_Z^2)}v^2 dv \sin\Theta\, d\Theta\, d\Phi \qquad \text{by Eq. 1.15}$$

$$= \int_0^\pi \int_0^{2\pi} \left(\frac{a}{\pi}\right)^{3/2} e^{-av^2}v^2 dv \sin\Theta\, d\Theta\, d\Phi \qquad v_X^2 + v_Y^2 + v_Z^2 = v^2$$

$$= \left(\frac{a}{\pi}\right)^{3/2}\left(\int_0^\pi \sin\Theta\, d\Theta\right)\left(\int_0^{2\pi} d\Phi\right)e^{-av^2}v^2 dv \qquad \text{separate variables}$$

$$= \left(\frac{a}{\pi}\right)^{3/2}(2)(2\pi)e^{-av^2}v^2 dv = 4\pi\left(\frac{a}{\pi}\right)^{3/2}e^{-av^2}v^2 dv. \qquad (1.16)$$

But what is the value of the constant a in the exponential? For that, Maxwell could turn to laboratory data. It was known from experiments on the simplest gases—monatomic gases such as argon—that at constant volume the heat q required to raise the temperature an amount ΔT is given by

$$q = \frac{3}{2} nR\Delta T, \tag{1.17}$$

where R is the **universal gas constant** equal to $8.3145\,\mathrm{J\,K^{-1}\,mol^{-1}}$. If all the energy in the sample is kinetic energy, which is a good approximation for these gases, then q gives the change in the total kinetic energy of the sample K. We can write K as the average kinetic energy per particle multiplied by the total number of particles N:

$$K = N\left\langle \frac{mv^2}{2} \right\rangle = \frac{Nm}{2} \langle v^2 \rangle, \tag{1.18}$$

where m is the mass of *one* particle (for example, the mass of one helium atom or argon atom). The molar mass $\mathcal{M}$ of an atom or molecule is given by Avogadro's number $\mathcal{N}_A$ times the mass of a single particle,

$$\mathcal{M} = \mathcal{N}_A m, \tag{1.19}$$

so we can write the total kinetic energy as

$$K = \frac{Nm}{2} \langle v^2 \rangle = \frac{n\mathcal{N}_A m}{2} \langle v^2 \rangle = \frac{n\mathcal{M}}{2} \langle v^2 \rangle. \tag{1.20}$$

Therefore, the change in kinetic energy ΔK when we heat the gas by an amount q is

$$\Delta K = q = \frac{3}{2} nR\Delta T. \tag{1.21}$$

At this point, we make an unjustified assumption that not only ΔK but K itself is directly proportional to T:

$$K = \frac{3}{2} nRT. \tag{1.22}$$

(It takes a more sophisticated understanding of temperature to justify this assumption, and we'll get to that in Chapter 2.) Combining Eqs. 1.20 and 1.22 allows us to solve for the mean square speed $\langle v^2 \rangle$:

$$K = \frac{n\mathcal{M}}{2} \langle v^2 \rangle = \frac{3}{2} nRT$$

$$\langle v^2 \rangle = \frac{3}{2} nRT \left(\frac{2}{n\mathcal{M}} \right) = \frac{3RT}{\mathcal{M}}. \tag{1.23}$$

We can also solve for $\langle v^2 \rangle$ by applying the **classical average value theorem** to the distribution function $\mathcal{P}_v(v)$. According to this theorem, if $f(x)$ is any function of a coordinate x that has a distribution of values $\mathcal{P}_x(x)$, then the mean value of $f(x)$ is given by the integral

$$\langle f(x) \rangle = \int_{\text{all space}} \mathcal{P}_x(x) f(x)\, dx. \tag{1.24}$$

Effectively this is a weighted average, where each possible value of $f(x)$ is scaled by the likelihood of having the corresponding value of x. Therefore, the mean square speed $\langle v^2 \rangle$ from our probability distribution is

$$\langle v^2 \rangle = \int_0^\infty \mathcal{P}_v(v) v^2\, dv$$

$$= 4\pi\left(\frac{a}{\pi}\right)^{3/2}\int_0^\infty e^{-av^2}v^4\,dv \qquad \text{by Eq. 1.16}$$

$$= 4\pi\left(\frac{a}{\pi}\right)^{3/2}\left(\frac{3\sqrt{\pi}}{2^3 a^{5/2}}\right) \qquad \text{from Table A.5}$$

$$= \frac{12\pi^{3/2}a^{3/2}}{8\pi^{3/2}a^{5/2}} = \frac{3}{2a}. \qquad (1.25)$$

Finally, combining Eqs. 1.23 and 1.25, we can find a,

$$\frac{3}{2a} = \frac{3RT}{\mathcal{M}}$$

$$a = \frac{\mathcal{M}}{2RT}, \qquad (1.26)$$

and use this to arrive at the **Maxwell-Boltzmann distribution** of molecular speeds:

$$\mathcal{P}_v(v) = 4\pi\left(\frac{a}{\pi}\right)^{3/2}e^{-av^2}v^2\,dv = 4\pi\left(\frac{\mathcal{M}}{2\pi RT}\right)^{3/2}v^2e^{-\mathcal{M}v^2/(2RT)}$$

$$\mathcal{P}_v(v) = 4\pi\left(\frac{\mathcal{M}}{2\pi RT}\right)^{3/2}v^2e^{-\mathcal{M}v^2/(2RT)}. \qquad (1.27)$$

The Maxwell-Boltzmann distribution is graphed for two different species in Fig. 1.5, showing how the low mass of He contributes to a broad distribution of speeds. We will examine this distribution in more detail in Chapter 3.

DERIVATION SUMMARY The Maxwell-Boltzmann Distribution. Maxwell assumed that a Gaussian distribution applied to the velocity components v_X, v_Y, v_Z of all the velocity vectors $\vec{v}$ of the particles in a gas, and we used this assumption to obtain a probability distribution of velocity vectors. To eliminate the dependence of direction, we integrated that distribution over all angles, which introduced a factor of v^2 in addition to the Gaussian exponential e^{-av^2}. We solved for the constant a in the exponential by requiring the mean square speed $\langle v^2\rangle$ to predict the experimentally determined kinetic energy of a monatomic gas as a function of temperature. We also required the probability to be normalized, so that $\int_0^\infty \mathcal{P}_v(v)\,dv = 1$, in order to obtain the value of the coefficient that multiplies the entire function.

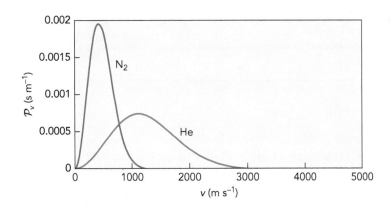

◀ FIGURE 1.5 **The Maxwell-Boltzmann distribution.** Molecular speeds are plotted for N_2 and for He at 300 K.

Maxwell's reasoning had to go through subsequent refinements, particularly modifications by Ludwig Boltzmann, before arriving at the argument presented here. But Maxwell's great innovation was to think about the invisible atoms and molecules as having *distributions* of properties, rather than as a set of particles with identical behavior in all respects. It was 50 years later, in the 1920s, that Otto Stern and coworkers verified the Maxwell-Boltzmann distribution experimentally, and by that time the central notion had taken hold: that macroscopic sets of molecules could only be accurately described in terms of probability distributions. Indeed, this concept came to form the basis of the language of quantum mechanics, which arrived on the scene as the Maxwell-Boltzmann distribution was being demonstrated in the laboratory.

BIOSKETCH | Joseph Francisco

Joseph Francisco is the William E. Moore Distinguished Professor of Earth and Atmospheric Sciences and Chemistry at Purdue University, and he is a recent president of the American Chemical Society. His research has been devoted to experimental and theoretical studies of reactive gas-phase molecules, particularly those that may contribute substantially to the complex chemistry of the atmosphere. These studies bridge the molecular and macroscopic levels of our science, showing how quantum mechanics can be used to predict bulk chemical thermodynamics and kinetics.

For example, he has shown that weak intermolecular interactions—which we ignore for the ideal gases of this chapter—can dramatically enhance the rates of certain reactions. In reactions during which a hydrogen atom changes location, water can attach to a reactant and donate one of its own H atoms while it accepts another as a replacement, allowing the H atom transfer to occur across a greater distance than would be probable otherwise. Professor Francisco and his group have shown how water (as well as other species) can accelerate the conversion of the methoxy radical CH_3O to hydroxymethyl radical CH_2OH. Water accepts the H atom from the carbon as it transfers another to the oxygen, allowing the reaction to take place with a much lower activation energy E_a than if the water were not present. Extrapolating this effect at the molecular level to the bulk limit, effectively by applying Maxwell's theory of gases, predicts the reaction rate in the atmosphere as a function of the concentrations of water and methoxy radical. Properties at the molecular level predict the behavior at the bulk level.

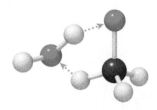

The Ideal Gas Law

Maxwell's understanding of how molecules moved in gases opened the door to a new way of thinking about matter: building up from individual particles to the bulk rather than the other way around. That will also be our approach in this volume, extrapolating from the structure and behavior of single molecules and small groups of molecules up to the properties of mole quantities.

One particularly valuable application of this method predicts the relationships among pressure, volume, temperature, and moles of an ideal gas. Picture N particles of a gas in a rectangular container of volume V at temperature T and pressure P (Fig. 1.6). The container has length a along the X axis, and the area of wall 1 is A, so the volume of the box is equal to $A \times a$. The particles are all of the same composition and so have the same mass m. To find the pressure in this system, we need to know the force per unit area. There are numerous ways of phrasing this, and the one we'll use is based on the force F_1 exerted by a single particle on wall 1 of the container, averaged over time.

When a single particle hits the wall, the velocity component along the X axis v_X is converted to a velocity component $-v_X$ and the particle goes shooting back in the $-X$ direction toward wall 2. This event changes the particle's velocity by an amount $-2v_X$, so there must be an acceleration. When we measure pressure we measure the force averaged over time. In this case, let's average over the time Δt required for a single round-trip from wall 1 to wall 2 and then back again. We start from Newton's $F = ma$, looking only at the acceleration a_X along the X axis, and rewrite this in terms of the change in speed along the X axis, Δv_X. The time-averaged force F_1 exerted on wall 1 by a single particle with mass m traveling along the X axis at speed v_X is balanced (according to Newton's third law) by an equal and opposite force on the particle of $-F_1$, which is then given by

$$-F_1 = ma_X = m\frac{dv_X}{dt} \approx m\frac{\Delta v_X}{\Delta t} = 2m\frac{(-v_X)}{\Delta t}. \qquad (1.28)$$

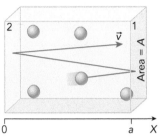

▲ FIGURE 1.6 **The system for the kinetic theory of gases.** Each molecule must travel the length of the box between any pair of opposite walls before it strikes one of those walls twice.

How do we find Δt? In the ideal gas, the particles do not collide with each other. Every particle that collides with wall 1 is repelled back toward wall 2, which bounces it back toward wall 1. The particle travels a total distance of $2a$ along the X axis between collisions with wall 1. The time Δt between these collisions with wall 1 is therefore given by that round-trip distance divided by the speed v_X of the molecule along the X axis:

$$\Delta t = \frac{2a}{v_X}. \qquad (1.29)$$

Now we have from Eq. 1.28

$$F_1 = 2m\frac{v_X}{\Delta t} = 2m\frac{v_X}{2a/v_X} = \frac{mv_X^2}{a} \qquad (1.30)$$

for the force, averaged over time, exerted by one particle on wall 1 as it bounces back and forth within the container. To extend this result to the entire gas, we set the *total* force on wall 1 equal to the average single-particle force times the N particles hitting the wall:

$$\langle F \rangle = N\langle F_1 \rangle = N\left\langle \frac{mv_X^2}{a} \right\rangle = \frac{Nm\langle v_X^2 \rangle}{a}. \qquad (1.31)$$

The parameters N, m, and a are all constants, so we can factor them out of the average.

We have one parameter left to solve: the average square speed along X, $\langle v_X^2 \rangle$. The overall mean square speed $\langle v^2 \rangle$ can be separated into contributions from the three Cartesian coordinates:

$$\langle v^2 \rangle = \langle v_X^2 + v_Y^2 + v_Z^2 \rangle = \langle v_X^2 \rangle + \langle v_Y^2 \rangle + \langle v_Z^2 \rangle. \qquad (1.32)$$

Because nothing makes motion along the X axis different from motion along the Y or Z axes, these averages must all be equal:

$$\langle v_X^2 \rangle = \langle v_Y^2 \rangle = \langle v_Z^2 \rangle, \tag{1.33}$$

and we may rewrite Eq. 1.32 as

$$\langle v^2 \rangle = 3 \langle v_X^2 \rangle. \tag{1.34}$$

We use Eq. 1.23 to find a value for $\langle v_X^2 \rangle$:

$$\langle v^2 \rangle = \frac{3RT}{\mathcal{M}} \qquad \text{by Eq. 1.23}$$

$$= 3 \langle v_X^2 \rangle \qquad \text{by Eq. 1.34}$$

$$\langle v_X^2 \rangle = \frac{RT}{\mathcal{M}}$$

$$\langle F \rangle = \frac{Nm \langle v_X^2 \rangle}{a} = \frac{NmRT}{\mathcal{M}a}$$

$$= \frac{n\mathcal{M}RT}{\mathcal{M}a} = \frac{nRT}{a}. \tag{1.35}$$

Finally, we solve for the pressure on wall 1:

$$P = \frac{\langle F \rangle}{A} = \frac{nRT}{aA} = \frac{nRT}{V}. \tag{1.36}$$

This one equation combines the relationships among four of the parameters discussed in Section 1.1 into one magnificent summation, the **ideal gas law:**

$$PV = nRT. \tag{1.37}$$

Chemical Equilibrium and Chemical Kinetics

Applications of the kinetic theory of gases extend beyond the ideal gas law. Let's briefly examine two examples of particular importance to reaction chemistry.

Chemical reactions change the arrangements of atoms in molecules. This rearrangement requires movement of the atoms, and that kind of movement is governed by the same principles that led to the Maxwell-Boltzmann distribution. The random motions hypothesized by Maxwell not only carry particles in the gas from one part of the container to another, they carry atoms from one molecule to another. In a general sense, which we will refine in later chapters, Maxwell's theory accounts for the motions at the molecular level that transform reactants to products, or—because those motions can go in either direction—from products back to reactants.

The Maxwell-Boltzmann distribution in Eq. 1.27 was written in terms of molecular speeds,

$$\mathcal{P}_v(v) = 4\pi \left(\frac{\mathcal{M}}{2\pi RT} \right)^{3/2} v^2 e^{-\mathcal{M}v^2/(2RT)},$$

but we can also write it in terms of the kinetic energy of the molecules. Call the kinetic energy of one molecule ε, so that $\varepsilon = mv^2/2$. The term $\mathcal{M}v^2/2$ is equal to $\mathcal{N}_A mv^2/2$, which is the kinetic energy per mole of molecules with mass m

and speed v. In the ideal gas, all the energy is kinetic energy, so we may set $\mathcal{N}_A m v^2 / 2$ equal to the molar energy E_m,

$$\frac{\mathcal{M}v^2}{2} = \frac{\mathcal{N}_A m v^2}{2} = \mathcal{N}_A \varepsilon = E_m, \qquad (1.38)$$

the energy per mole of particles. The Maxwell-Boltzmann distribution can then be written as a probability of being at a particular value of E_m:

$$\mathcal{P}_v(v) = 4\pi \left(\frac{\mathcal{M}}{2\pi RT}\right)^{1/2} \left(\frac{\mathcal{M}v^2}{2\pi RT}\right) e^{-\mathcal{M}v^2/(2RT)} = 4\pi \left(\frac{\mathcal{M}}{2\pi RT}\right)^{1/2} \left(\frac{E_m}{\pi RT}\right) e^{-E_m/(RT)}. \qquad (1.39)$$

That exponential term, $e^{-E_m/(RT)}$, plays a starring role in chemistry.

If we allow a chemical reaction to run long enough, the time comes when we measure no more change in the concentrations of the reactants or products. We call this condition *chemical equilibrium*. Even though no change occurs at the macroscopic level, at the molecular scale the reactions are continuing to occur. The balance between the reactant and product concentrations is determined by the rate of motion of the atoms, and those motions are still distributed according to a form of the Maxwell-Boltzmann distribution. If we imagine a probability distribution for the reactants that varies as $e^{-E_{react}/(RT)}$ and for the products that varies as $e^{-E_{prod}/(RT)}$, it makes sense that the ratio of products to reactants should be given approximately by

$$\frac{[\text{products}]}{[\text{reactants}]} \approx \frac{e^{-E_{prod}/(RT)}}{e^{-E_{react}/(RT)}} = e^{-(E_{prod} - E_{react})/(RT)} = e^{-\Delta_{rxn}E^\ominus/(RT)}, \qquad (1.40)$$

where $\Delta_{rxn}E^\ominus$ is the energy change in the system as we transform reactants completely to products. The exact nature of that ratio depends on the conditions under which the reaction is carried out, however. Under typical conditions, allowing the reaction to exchange energy by heat and work with its surroundings, it is the change in the Gibbs free energy $\Delta_{rxn}G^\ominus$ that determines the balance between reactants and products.

For the simplest reaction, $A \rightleftharpoons B$, we express the balance using an *equilibrium constant* K_{eq}, which Chapter 12 will show has the form described earlier:

$$K_{eq} = e^{-\Delta_{rxn}G^\ominus/(RT)} \approx \frac{[\text{B}]}{[\text{A}]}. \qquad (1.41)$$

The exponential term $e^{-E_m/(RT)}$ found in the Maxwell-Boltzmann distribution provides the basis for calculating the equilibrium constant.

Returning to the topic that opened this chapter, how do we measure the properties of the system at chemical equilibrium? The concentrations of the reactants and products can be determined by various means, such as titration, in which the volume of one substance is used to determine the number of moles of another substance. At the point of chemical equilibrium, those concentrations do not change with time. We measure quantities such as volume that can be found using rulers; we don't need clocks.

But what if we can't wait for chemical equilibrium, or don't want to? As the system approaches equilibrium, the concentrations of reactants and products are changing in time. The study of the *time-dependence* of chemical reactions is known as chemical kinetics, and it brings together principles of physical chemistry at every level, from molecular structure to molecular dynamics to thermodynamics. It is a science of clocks as well as rulers, and it adds a dimension both to the complexity of the problem and the depth of our understanding about chemistry.

The rate at which chemical changes occur is given by a *rate law*. For the one-way reaction A → B (such that A converts to B but B does not return to A), the rate law would normally have the form

$$-\frac{d[A]}{dt} = k[A],$$
(1.42)

where k is the *rate constant* for the reaction. In the same way that chemical equilibrium is determined by the distribution of energies of reactants and products, the rate at which reactants transform into products is determined by the energy distribution of the reactants. This time, we ask what is the probability that the reactants have enough energy to overcome an energy E_a, the activation barrier to the reaction. Chapter 13 will show how, once again, the exponential distribution of the energies determines the properties of the system, this time in the form of the *Arrhenius equation*:

$$k = Ae^{-E_a/(RT)}.$$
(1.43)

This one expression for the energy distribution, $e^{-E_m/(RT)}$, appears throughout our study of bulk systems, reminding us that thermodynamics and kinetics and diffusion and other aspects of chemistry are not separate topics at all. They are all linked, and the job of physical chemistry is to discover those links and use them to strengthen our grasp on the nature of matter.

CONTEXT   *Where Do We Go From Here?*

The scientists of the 19th century brought us to the brink of the discoveries about atoms that launched quantum mechanics. Some of those scientists even saw beyond those future discoveries, formulating theories that waited decades for proof or recognition. But fundamental questions remained that only a working model of atomic and molecular structure could answer. While the behavior of bulk systems could be predicted with increasing reliability by the laws of thermodynamics and kinetics, the microscopic origin of those laws was a mystery. Why did the Maxwell-Boltzmann distribution rely on a Gaussian distribution of speeds, rather than, for example, energies? What was the physical meaning of the entropy, and why should it always increase as required by the second law?

Quantum mechanics soon provided the description of matter at the atomic scale that formed the basis for the conjectures of Clausius, Maxwell, Boltzmann, and others, while introducing its own—and even more perplexing—mysteries along the way. For our purposes, quantum mechanics provides a model of matter at the molecular scale that has been rigorously tested by experiments and that we may reasonably trust as the foundation for the macroscopic chemistry that we are concerned with in this volume.

In the rest of this text, we will take advantage of that model, rather than retracing the arduous path of history. Science is the exploration of the mystery of nature, and the mystery will always remain. When we do uncover links among the natural laws, they're worth celebrating and exploring in their own right. We will start with a summary of principles of molecular structure, and use these to explain the meaning of the entropy and the origin of the recurring $e^{-E_m/(RT)}$ term.

KEY CONCEPTS AND EQUATIONS

1.2   Thermodynamics.

a. The first law of thermodynamics breaks the change in energy ΔE of our system into contributions from heat and work:

$$\Delta E \equiv E_{\text{final}} - E_{\text{initial}} = q + w. \qquad (1.7)$$

b. The temperature is given by a partial derivative of the energy with respect to entropy:

$$T = \left(\frac{\partial E}{\partial S}\right)_{V,n}. \qquad (1.11)$$

c. The second law of thermodynamics requires that the overall entropy never decrease during a process:

$$\Delta S_T \geq 0. \qquad (1.12)$$

1.3   Kinetic Theory of Ideal Gases.

a. The Maxwell-Boltzmann distribution gives the probability per unit speed that a molecule in an ideal gas will be found at a particular speed v:

$$\mathcal{P}_v(v) = 4\pi \left(\frac{\mathcal{M}}{2\pi RT}\right)^{3/2} v^2 e^{-\mathcal{M}v^2/(2RT)}. \qquad (1.27)$$

b. The exponential term in the Maxwell-Boltzmann distribution can be written in terms of the molar energy, $E_{\text{m}} = \mathcal{M}v^2/2$, to have the form $e^{-E_{\text{m}}/(RT)}$. This term is the origin of the exponential dependence of the equilibrium constant on the free energy of reaction, $\Delta_{\text{rxn}}G^{\ominus}$:

$$K_{\text{eq}} = e^{-\Delta_{\text{rxn}}G^{\ominus}/(RT)}$$

and of the Arrhenius rate constant on the activation energy E_a:

$$k = A e^{-E_a/(RT)}.$$

KEY TERMS

- **Work** w is a transfer of energy into or out of the system that results in a net movement of mass.
- **Heat** q is any transfer of energy that is not work, including changes in the energy that increase the random microscopic motions of molecules.
- The **internal energy** E is the energy that we can directly control in chemical thermodynamics; it includes energies of molecular motion but not (for example) the conversion of mass into energy or kinetic energy from motion of the entire system.
- The **enthalpy** H is a parameter derived from the energy so that ΔH gives the heat transfer of a process when the pressure is constant.

- The **Gibbs free energy** G gives the useful non-expansion work that can be extracted from a process at constant temperature and pressure.
- The **entropy** S is a measure of the heat that is lost during a process that cannot be converted into useful work.
- A **heat engine** is a device that converts heat into useful work.
- A **normalized** distribution function has been multiplied by a coefficient to ensure that the function, when added or integrated over all of its possible values, sums to 1.
- The **Maxwell-Boltzmann distribution** $\mathcal{P}_v(v)$ gives the probability per unit speed that a particle in an ideal gas will be found at speed v.

OBJECTIVES REVIEW

1. *Apply the first and second laws of thermodynamics to calculate or place limits on simple thermodynamic processes.*

   A gas expands, pushing a piston that does 743 kJ of work. If the total energy of the gas does not change, what is q for this process?

2. *Use the Maxwell-Boltzmann distribution to predict global properties of an ideal gas that depend on the speed.*

   Write an expression for the root mean square momentum $\langle p^2 \rangle^{1/2}$ of nitrogen molecules in air at 298 K, where the momentum is mv.

PROBLEMS

The problems at the end of each chapter are intended to cover most of the chapter's material, ranging from straightforward numerical calculations to more difficult derivations. In later chapters, problems may take the opportunity to combine results from different chapters.

Discussion Problems

1.1 As discussed in Section 1.1, new parameters are often born out of the need to explain inconsistencies between experimental results and prevailing theory. In physics, mass is the property that determines the magnitude of the gravitational force, and charge is the property that determines the magnitude of the electromagnetic force. Particle physicists have ascribed a property they call *color,* which determines the magnitude of the *strong force* that binds protons and neutrons together to form atomic nuclei. What is it about the structure of the atomic nucleus that would lead physicists to believe a new property other than mass and charge was required to explain observations?

1.2 Graph the potential energy for one of the particles in the container used to derive the ideal gas law in Fig. 1.6. What is the force acting on the particle at $x = a/2$? What is the force acting on the particle at $x = a$, where wall 1 is located?

Measurement and Classical Parameters

1.3 If we had only a ruler and a clock as measuring devices, how could we measure the density of water?

1.4 Calculate the height in meters of a column of water that exerts a pressure of 1.00 bar at its base.

1.5 Chemistry buildings built to code require a great deal of ventilation, which necessitates a delicate air-handling system to balance the air flowing into the building and the air flowing out. For example, if the air flow out of the building is too high, the pressure inside the building could drop to 0.990 bar while the pressure outside was 1.000 bar. Calculate the force that would then be necessary to open a door to the outside measuring 78 inches high and 30 inches wide. Neglect the mass of the door itself. (1 inch = 2.54 cm.)

1.6 Estimate the number of moles of butanol on the surface of a circular drop of water with a diameter of 1.0 cm if surface tension experiments show that the effective area per butanol molecule is 33 Å^2.

1.7 Calculate the pressure in bar that would be equivalent to the attractive force between the Na^+ and Cl^- ions in a salt crystal, at a separation of 2.81 Å and approximating the area over which the force works to be 4.00 Å^2.

Thermodynamics and Kinetic Theory of Gases

1.8 A typical engine alternates between compression and expansion of a gas. If during the compression stroke the entropy of the gas is reduced by 0.20 J K^{-1}, what is the minimum change to the entropy of the surroundings?

1.9 An ideal gas is trapped beneath a platform that has an area of 1.00 m^2 and a mass of 1000. kg. The initial volume of the gas is 10.0 L, and the gas supports the platform at a height of 10.0 cm. We heat the gas from 298 K to 475 K, and the platform is raised to a total height of 16.0 cm. The ΔH for this process is 2.92 kJ. Find q, w, and ΔE. The ambient pressure outside the system is 1.00 bar, and the acceleration due to gravity is $g = 9.8 \text{ m s}^{-2}$.

1.10 Heating a sample of neon gas from 298 K to 323 K, we lose heat to the surroundings at a rate of 4.5 J K^{-1}. What is the change in the surrounding entropy, ΔS_{surr}?

1.11 Write the complete integral expression that should be solved to find the fraction of molecules in a sample of helium traveling at speeds between 10^2 m s^{-1} and 10^3 m s^{-1} at 298 K. Your final expression should include *all necessary numerical values* simplified as much as possible. If you have integration software handy, evaluate the integral.

1.12 A distribution function such as the Maxwell-Boltzmann distribution can be described in terms of *moments,* where the nth moment of a distribution $\mathcal{P}(x)$ is written

$$\mu_n \equiv \int_{\text{all space}} \mathcal{P}(x)(x - \langle x \rangle)^n dx.$$

(The moment does not have to be evaluated relative to the mean value $\langle x \rangle$, but it is the convention.) Find the third moment of the Maxwell-Boltzmann distribution.

1.13 Calculate the fraction of N_2 molecules in a sample at 298 K that have speeds less than 10.0 m s^{-1} by integrating the Maxwell-Boltzmann distribution over the appropriate range of speeds (just writing the correct integral is good for a start). Simplify the integral by taking advantage of the fact that

$$\frac{2RT}{\mathcal{M}} \gg v^2.$$

2

Introduction to Statistical Mechanics: Building Up to the Bulk

LEARNING OBJECTIVES

After reading this chapter, you will be able to do the following:

❶ Calculate the entropy of a system from the degeneracy of states using the Boltzmann equation, and from probability expressions using the Gibbs equation.

❷ Use energy and degeneracy expressions to calculate partition functions at some given temperature.

❸ Use the canonical distribution to calculate quantum state populations.

GOAL *Why Are We Here?*

Our goal in this chapter is to predict the bulk properties of a gas-phase sample by integrating over the properties of the individual molecules. Our starting point is a single quantum mechanical particle in a large, three-dimensional box, which we then extend to many, many particles by means of probability mathematics. Along the way, we will craft definitions for the entropy and the temperature, two fundamental parameters of thermodynamics.

CONTEXT *Where Are We Now?*

Over the last century or so, physicists have constructed a theoretical framework that describes the structure, properties, and interactions of molecules on a microscopic scale. This framework provides the foundation for our theory of the behavior of matter in the macroscopic limit, the bulk regime in which we couldn't keep track of the properties of individual particles even if we wanted to. But we need some way to connect the two, to extend what we know of single molecules to huge numbers, sacrificing the details of the individual for a comprehensive understanding of the whole. Our starting point in this chapter is a review of the nature of matter at the atomic scale, which can be described using the laws of quantum mechanics. From a single particle, we obtain results that will guide and support our work in the rest of the text.

2.1 Properties of the Microscopic World

An ageless game of favorites has been this: if you were to be abandoned on an island with only ten songs or ten books of your choice, what would they be? Imagine instead that we have to reboot our entire history of science, and start over with only one equation. What is the one concept that is (or has been) most central to our understanding of nature?

I would argue that it is the second law of thermodynamics. The first law restates the conservation of energy in thermodynamic terms (Section 1.2), and the second law was initially concerned with the extent to which energy can be channeled into useful work. But the second law has come to mean much more. To our knowledge, it applies to nearly all processes, at all distance scales and energy scales, no matter what forces may be at work.

The second law plays a major role in the theoretical basis for physical chemistry, but it was originally introduced as an empirical law, an observation rather than a deduction. One of the goals of this text is to justify the second law, along with other principles of thermodynamics and kinetics, so that not only *how* the law works, but also *why* it works, becomes part of our intuition. Therefore, instead of retracing the steps of history, we take advantage of the hard work of the scientists who have come before us and present a theory for the behavior of molecules that stands together as one whole. The second law, as it applies to chemistry, can be explained on the basis of the structure and motions of individual atoms and molecules, as examined over distances of about 10^{-10} m and energies of 10^{-18} J and less.

In the inaugural section of this chapter, we summarize the major results from that microscopic perspective that will lend us, over the following chapters, a deep appreciation of the essential threads that unite all of classical chemistry. The equations and other reference numbers marked *QM* in this section refer to the companion volume to this text, *Quantum Mechanics and Spectroscopy*, which provides a more comprehensive presentation of all of these topics.

Atomic Structure

The properties of molecules are determined by the properties of the atoms from which they are built, and both atoms and molecules are governed by the rules of quantum mechanics. Let's start there.

Fundamentals of Quantum Mechanics

For the most part, we lose sight of the surprising details inherent in quantum mechanics once our system grows beyond the size of a few molecules. But if we seek a deeper understanding of the chemical sample on our laboratory bench top, if we attempt to ask the questions that take us beyond reading a table of constants and sticking numbers into a memorized equation, then we find that quantum mechanics continues to guide the behavior of molecules, no matter how many molecules you have. As one example, a concept key to our interpretation of the second law of thermodynamics is the limitation *by quantum mechanics* on the number of possible states of our system. There may be no more important lesson in all of physical chemistry than this: the number of quantum states of any real system is a *finite* number, and processes occur so as to make that number grow. So what are these quantum states that determine so much in chemistry?

Quantization of Properties

At the scale of single molecules, particles behave at least partly like waves, and we adapt Newton's laws of motion to take this wave nature into account, obtaining the theory of quantum mechanics. For quantum matter to remain in an unchanging state, analogous to a standing wave, it must satisfy certain boundary conditions. In the same way that a classical standing wave can oscillate at only certain frequencies, the boundary conditions on quantum matter limit parameters such as the energy and angular momentum to certain **quantized values** (Fig. 2.1). A classical system has an infinite number of possible states, with each state differing in some infinitesimal way from another—say, by having slightly more energy or slightly less volume. In contrast, a quantum system has a *finite* number of **quantum states.** For a particle with mass m and speed v, this quantization becomes measurable when the distance to be traveled by the particle is much greater than its **de Broglie wavelength** (*QM* Eq. 1.3),

$$\lambda_{dB} = \frac{h}{mv}, \tag{2.1}$$

where Planck's constant, $h = 6.626 \cdot 10^{-34}\,\text{J s}$, determines the quantum limit.

Our quantum states have well-defined energies, but the wavelike nature of our particles permits other parameters, such as speed, to have a distribution of values at different points in space *at the same time,* because—like a wave but unlike a classical particle—the quantum particle cannot be pinned down to a single location. For one of these parameters $A(\vec{r})$ whose value depends on the position $\vec{r}$, we calculate an *average* value over the distribution using the **average value theorem:**

$$\langle A \rangle = \int_{\text{all space}} \mathcal{P}_{\vec{r}}(\vec{r}) A(\vec{r}) d\vec{r}, \tag{2.2}$$

where $\mathcal{P}_{\vec{r}}(\vec{r})$ is the probability distribution, the fraction of the particle's mass and charge that will be found at position $\vec{r}$ per unit r, and the subscript "all space" means that we integrate over all possible values of $\vec{r}$.

The Particle in a Three-Dimensional Box

The particle in a three-dimensional box is a fundamental problem in quantum mechanics that provides useful lessons for our description of gases. The system consists of a small particle (an atom or molecule, for example) trapped in a rectangular box, free to move anywhere along x, y, or z as long as it stays inside the box. Let the box have sides of length a along x, b along y, and c along z. These dimensions of the box define the *domain* of the system, the distance over which the particle can travel. The possible energy values for the particle are then (*QM* Eqs. 2.40 to 2.42)

$$\varepsilon_{n_x,n_y,n_z} = \frac{\pi^2\hbar^2}{2m}\left(\frac{n_x^2}{a^2} + \frac{n_y^2}{b^2} + \frac{n_z^2}{c^2}\right) \tag{2.3}$$

$$\approx \frac{\pi^2\hbar^2}{2m(abc)^{2/3}}(n_x^2 + n_y^2 + n_z^2)$$

$$= \frac{h^2}{8mV^{2/3}}n^2 \equiv \varepsilon_0 n^2, \tag{2.4}$$

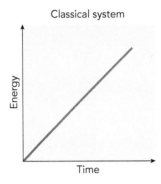

Classical system

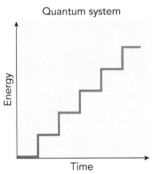

Quantum system

▲ FIGURE 2.1 **Quantized energy.** If we try to add energy gradually to a system, the classical system can absorb any amount because energy is a continuous variable, but the quantum system is allowed to contain only certain energies.

where m is the particle mass, $\hbar = h/(2\pi)$, V is the volume of the box, and n_x, n_y, and n_z are **quantum numbers** restricted to integer values greater than 0. Equation 2.4 assumes that the box is nearly a cube, so a, b, and c can be treated as roughly equal. We have also absorbed the constants into a single parameter ε_0 and combined the sum $(n_x^2 + n_y^2 + n_z^2)$ into a single term n^2. The energy of a single particle is written ε here, to distinguish it from the total energy E when we add more particles. The energy levels of a cubical box, for which the approximation becomes exact, are graphed in Fig. 2.2. The state with the lowest possible energy is always called the **ground state,** and in this system the ground state corresponds to the quantum numbers $(n_x, n_y, n_z) = (1, 1, 1)$. Any other state is called an **excited state.**

When two or more quantum states share the same energy, we call them **degenerate** states, and the number of states with the same energy is the *degeneracy* g of the level. All but a few of the cubical box quantum energy levels correspond to degenerate states, because any combination of integers (n_x, n_y, n_z) such that $n_x^2 + n_y^2 + n_z^2$ is the same value will have the same energy. Figure 2.2 shows, for example, that the lowest energy excited states are a group of three degenerate states $(2, 1, 1)$, $(1, 2, 1)$, $(1, 1, 2)$ at $\varepsilon = 6\varepsilon_0$. On average, the degeneracy increases as the energy increases. When we put more energy into the system, we tend to find more choices in how the energy can be distributed among motions along x and y and z.

Equation 2.4 obeys the **correspondence principle,** which states that

quantum mechanics approaches classical mechanics as we increase the particle's kinetic energy, mass, or domain.

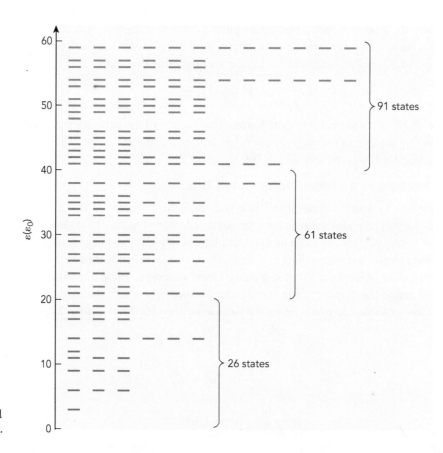

▶ **FIGURE 2.2 The three-dimensional box.** The energy levels of the cubical box graphed in units of ε_0 for the cubical box.

In other words, we use quantum mechanics when λ_{dB} is comparable to the size of the domain, but classical mechanics whenever λ_{dB} is much smaller than the domain. In this case, for example, we find that at very high energies there are so many possible combinations of values for the three quantum numbers n_x, n_y, and n_z that we can no longer distinguish individual quantum states in the laboratory. However, we will see in this chapter that the entropy allows us to indirectly measure the **density of quantum states** $W(\varepsilon)$. For any typical system, as we increase the energy, more quantum states become accessible to the system at that energy. And although the properties of a macroscopic system don't normally depend on any specific quantum state being available, much depends on *how many* quantum states are available. The density of states $W(\varepsilon)$ tells us how many quantum states near energy ε there are per unit of energy.

Quantum States of Atoms

When we can carry out the same analysis to predict the energies of the electron in a one-electron atom, such as H or He$^+$, we find that those energies are given by (QM Eq. 1.15)

$$E_n = -\frac{Z^2 m_e e^4}{2(4\pi\varepsilon_0)^2 n^2 \hbar^2} \equiv -\frac{Z^2}{2n^2} E_h. \tag{2.5}$$

The energy depends on a single quantum number, n, which again may take on only integer values greater than zero. The energy levels of the hydrogen atom, in units of **hartrees** (E_h) are sketched in Fig. 2.3. The hartree is a convenient unit for calculations of electrons in atoms or molecules and is equal to $4.360 \cdot 10^{-18}$ J.

The quantum state of the electron in the atom also depends on the quantum numbers l (which determines the overall angular momentum in the electron's orbital motion), m_l (which determines the orientation of that orbital motion), and m_s (which determines the orientation of the **spin** angular momentum). When we add more electrons to the atom, the energy ceases to obey a simple algebraic expression. For each atom, we can specify the quantum state of the entire set of electrons by giving the electron configuration (which lists the n and

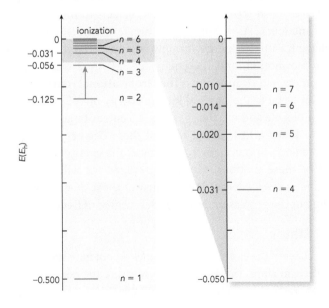

◀ FIGURE 2.3 **Energy levels of the hydrogen atom.**

l values of all the electrons) and the *term symbol* (which is determined from the m_l and m_s values). The ground electronic quantum state of a nitrogen atom, for example, has an electron configuration $1s^2 2s^2 2p^3$ and a term symbol ^4S.

The spin of a fundamental particle, such as the electron, is an intrinsic property of the particle, like its mass or charge. The spin turns out to have a significant impact on the quantum mechanics of systems in which we combine many identical particles. In particular, the set of particles having half-integer spin (including individual electrons, protons, and neutrons) are called **fermions,** and they differ from the behavior of integer spin particles (such as the ^{4}He nucleus), which are called **bosons.** No two indistinguishable fermions—such as two electrons in the same atom—can have the same complete set of quantum numbers at the same time. It is not possible for example to put all the electrons in the atom into the lowest energy $1s$ orbital. This rule is the **Pauli exclusion principle,** and it is an application of the more general **symmetrization principle.** Unlike fermions, several equivalent bosons can occupy the same quantum state simultaneously, giving rise to some interesting effects that we discuss in Chapter 4.

Molecular Structure

We combine atoms to form molecules, using the same forces at work in atoms—the attraction between opposite charges and the repulsion between like charges—to arrive at a much more interesting and more complex structure. When we transfer energy into a molecule, there are now four major forms of motion, or **degrees of freedom,** in which the energy of the molecule can be stored:

1. electronic (motion of the electrons relative to the nuclei)
2. vibrational (motion of the nuclei relative to each other)
3. rotational (motion of the nuclei around the molecular center of mass)
4. translational (motion of the molecular center of mass)

In this order, those degrees of freedom correspond to increasing mass of the moving particles (from electrons, to a few nuclei, to all the nuclei) and increasing distance traveled (from motion within the molecule to motion of the molecule within the container). As listed, these therefore correspond to increasingly classical behavior, with the energy strongly quantized (large gaps between states) in the electronic motion, and the energy essentially a classical, continuously variable parameter in translation. The **translational energy** therefore can normally be set equal to the classical kinetic energy $mv^2/2$, where m is the mass of the whole molecule and v is the speed of its center of mass.

We need to know a little about each of these degrees of freedom, because when we add energy to our laboratory sample, the energy distributes itself into some or all of these motions. Critical questions we will face include *which* motions, and *how much* energy? We can answer those questions only with some understanding of the physics underlying each degree of freedom.

Electronic States

The quantum states of the electrons in molecules do not have the simple form found for the one-electron atom. However, we can measure the relative energies of those states in the laboratory by spectroscopy (*QM* Table 7.1). One way to get a molecule from its ground electronic state into one of its lowest excited electronic states is to apply electromagnetic radiation. We divide a beam of radiation into units called **photons**, where each photon has an energy given by Planck's law (*QM* Eq. 1.2),

$$E_{\text{photon}} = h\nu = \frac{hc}{\lambda}, \tag{2.6}$$

where ν is the frequency of the radiation, λ its wavelength, and c the speed of light, equal to $2.998 \cdot 10^8 \, \text{m s}^{-1}$. The energies needed to excite a molecule into a higher electronic state typically correspond to ultraviolet or visible radiation.

Vibrational States

If we separate out the electronic contribution to the energy, the remaining contributions come from motions of the nuclei. If there are N_{atom} atoms in the molecule, then—because each nucleus can move along the x, y, and z axes—there are $3N_{\text{atom}}$ degrees of freedom. We can assign three of those degrees of freedom to the motion of the center of mass. For a linear molecule, we then need two angles to specify the orientation, and for a non-linear molecule we need three angles (*QM* Section 8.3). That leaves $3N_{\text{atom}}-6$ different **vibrational coordinates** for non-linear molecules and $3N_{\text{atom}}-5$ for linear molecules. For each vibrational coordinate we can define a **vibrational constant** ω_e that depends on the masses of the moving atoms and the rigidity of the chemical bond. According to the **harmonic approximation** (*QM* Eq. 8.18),

$$\omega_e(\text{J}) = \hbar\sqrt{\frac{k}{\mu}}. \tag{2.7}$$

In agreement with the correspondence principle, ω_e decreases with the reduced mass μ of the moving atoms and increases with the force constant k (which measures the rigidity of the bond). For a diatomic molecule AB, the reduced mass is given by

$$\mu = \frac{m_A m_B}{m_A + m_B}.$$

The quantum states for vibrational motion then have **vibrational energies** approximately equal to

$$E_v = \left(v + \frac{1}{2}\right)\omega_e, \qquad v = 0, 1, 2, 3, \ldots \tag{2.8}$$

where the vibrational quantum number v can be any integer 0 or greater. The transition energy from the ground to lowest excited vibrational state is roughly one vibrational constant ω_e and corresponds to the photon energy of infrared radiation. Therefore, most vibrational spectroscopy is carried out at infrared wavelengths.

Rotational States

Non-linear molecules have three **rotational coordinates**, and linear molecules have two. For the simplest case, a linear molecule, the **rotational energies** are given by (*QM* Eq. 9.9)

$$E_J = BJ(J + 1), \qquad J = 0, 1, 2, 3, \ldots \tag{2.9}$$

where B is the **rotational constant** of the molecule and the rotational quantum number J can be any integer 0 or greater. The value of the rotational constant decreases as the molecule gets larger and more massive, again in agreement with the correspondence principle.

There is a degeneracy to rotational states, corresponding to different orientations of the rotational motion, and for linear molecules g_{rot} is equal to $2J + 1$. Why do we need to keep track of the degeneracy of the quantum energy levels? It comes back to the finite number of quantum states of our bench top sample. To determine how that sample can distribute energy that it absorbs, and ultimately to determine which processes can occur and which cannot, the number of available quantum states will be an essential parameter.

Conclusion: the Microscopic World

These individual components of our theory of matter at microscopic scales are sufficient to justify much of what follows. We will extend these few principles toward a conceptual framework that directs our understanding of molecules at laboratory scales. Our ability to magnify our description of one molecule by 20 orders of magnitude to predict the behavior of mole quantities is one of the signal achievements of 20th-century science, and it is the subject of this volume.

2.2   Bulk Properties

To get started on *large* quantities of material, we begin with some definitions. We will need at times to distinguish among the following:

- the **system** (or *sample)*, which is the thing we want to study;
- the **surroundings,** which consist of everything outside the system;
- the **boundary,** which separates the system and surroundings, and
- the **universe,** which is everything: the system, the surroundings, and the boundary in between.

For example, to discuss the properties of the gas in a balloon being inflated, we would consider the gas in the balloon to be the *system*, while the *surroundings* would consist of the atmosphere around the balloon, as well as everything outside of the atmosphere. The *boundary* would be the elastic material of the balloon itself. We need to include the atmosphere in that example, because it has to give way to provide room for the balloon to expand. The boundary defines how the system and surroundings are allowed to interact. For example, the balloon material can stretch, allowing the system to expand into the surroundings. On the other hand, if the gas were placed in a rigid chamber, the boundary would not allow the expansion to take place. We will assume that the boundary itself does not contribute significantly to the thermodynamic parameters such as the volume or energy, so it will not appear explicitly in any calculations.

Extensive and Intensive Parameters

In bench top chemistry, we deal with so many particles that the properties of the group are the same if we take 20 particles away or if we add 40,000 more. We have reached the *bulk*, the limit at which the properties we measure for the entire system no longer change measurably when we add or remove a particle. At what point do we have enough molecules to cross the line into the bulk? That point is never clearly defined, because it depends on what material and what properties are being studied. A cluster of 8 sodium atoms has an ionization energy about 20% higher than the bulk metal value, but the discrepancy drops to less than 10% with the addition of a single atom. Whether or not that difference is significant depends on the precision of our measurements, which vary from one experiment to another. And although the ionization energy is approaching its bulk value, the geometries of the Na_8 and Na_9 clusters still don't resemble the crystal structure of bulk sodium metal, so other properties such as the density are still different from the bulk properties.

In any case, the bulk limit is normally reached well before the sample is visible to the naked eye. A sample size of, say, 10^{10} molecules, is still less than 10^{-13} moles. If water comprised that sample, it would occupy a volume less than 0.01 mm on a side.

Quantum mechanics accurately predicts the properties of individual molecules, and, in principle, quantum mechanics continues to work as the system becomes bigger. But with *moles* of molecules, it is no longer possible to analyze the system using quantum mechanics. Since each new quantum state for any one of the particles corresponds to a new state of the entire system, the number of possible quantum states of the system is staggering when N approaches 10^{10}. We can't write and manipulate equations for so many states. And we don't want to anyway; they yield too much information. Each individual quantum state describes where each molecule can be found and specifies how much energy is stored in the electrons, in vibrations, in rotations, and in translation. When dealing with 10^{10} or more different molecules, we don't care what any single one of them is up to at a particular instant.

Instead, we seek only **macroscopic properties** of the system. We divide these surviving properties into two groups: **extensive** (obtained by *summing* together the contributions from all the molecules in the system) and **intensive** (obtained by *averaging* the contributions from the molecules). Imagine, for example, that we have two subsystems, A and B, in beakers next to each other on a bench top. We won't mix them together or do anything to change what happens inside the beakers. We have the separated subsystems, A and B, and we can also describe the two of them together as a combined system that we'll label A + B. If a parameter X is extensive, then the value of the extensive parameter X_{A+B} for the combined system is given by the sum of its values for the individual subsystems $X_A + X_B$. The volume, the energy, and the number of molecules are such extensive parameters. The value of an intensive parameter for the combined system, by contrast, is an *average* over those for the separate subsystems, not a sum. If A and B are identical—same volume, pressure, and so on—then the value of the intensive parameter Y_{A+B} remains the same as for the individual subsystems $Y_{A+B} = Y_A = Y_B$. If A and B are not identical, as shown in Fig. 2.4, then the value of the intensive parameter Y_{A+B} must be averaged in some fashion (depending on the parameter) over the values of Y_A and Y_B.

Here's a partial list of the macroscopic parameters that will demand our attention:

1. Extensive parameters (sums):

    (a) N, the total number of molecules in the system, or alternatively $n = N/\mathcal{N}_A$, the total number of moles

    (b) V, the total space occupied by the system

    (c) E, the sum of the translational, rotational, vibrational, and electronic energies of the system, measured relative to the system's ground state.

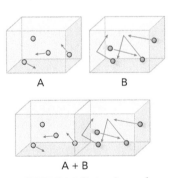

▲ FIGURE 2.4 **Extensive and intensive variables.** When two subsystems A and B are joined side-by-side, the combined system has a total number of molecules $N_A + N_B$ and total volume $V_A + V_B$, but the overall density and the pressure are averages (not sums) of the original values. Therefore, N and V are extensive parameters, while density and P are intensive.

2. Intensive parameters (averages):

   (a) P, the **pressure,** which is the average force per unit area exerted by the molecules on their surroundings

   (b) ρ, the number density, which is the average number of molecules per unit volume N/V, and which is related to the mass density ρ_m by

$$\rho_m = \rho\frac{\mathcal{M}}{\mathcal{N}_A}, \tag{2.10}$$

   where $\mathcal{M}$ is the molar mass and $\mathcal{N}_A$ is Avogadro's number.

Principal Assumptions

In **statistical mechanics** we take advantage of probability theory to describe the properties of macroscopic groups of molecules, based on the common characteristics of the individual molecules. Statistical mechanics extends what we have learned about molecules as quantum-mechanical systems to the behavior of bulk quantities of the substance. Chemistry and statistics fit together splendidly, because (unlike many other applications of statistics) chemical systems are composed of particles that really do behave identically. As noted in Section 2.1, our systems are composed of indistinguishable particles: no experiment allows us to tell the difference between two electrons once we put them in the same atom, or two ^{1}H atoms once we put them in the same H_2 molecule (see *Quantum States of Atoms* in Section 2.1). The things we want to study are *perfect statistical samples*, composed of billions of billions of equivalent particles.

The assumptions common to this field are the following:

1. **Chemically identical molecules share the same physics.** Any molecule is described by exactly the same physics as any other molecule of the same chemical structure (although the molecules may be in different quantum states). Each molecule can occupy any of its quantum-mechanical states within the limits of the available energy. Molecules sharing the same chemical structure and physical environment each have the same chance of occupying a particular quantum state. A single mathematical function can give the probability that any molecule in the system is in a particular quantum state for translation, rotation, vibration, or any other degree of freedom.

2. **Macroscopic variables are continuous variables.** Our system will be such a large group of molecules that its macroscopic properties—such as energy and density—will be completely continuous parameters, even though each individual molecule will continue to have quantized energy levels and other properties.

3. **Measured properties reflect the ensemble average.** If we fix some of the parameters (the volume and number of molecules, for example), we can imagine the very large set of *all* quantum states of that system that correspond to those values. This collection is the **ensemble,** each of the unique quantum states we shall call a **microstate,** and the number of microstates in the ensemble is called the ensemble size Ω, usually an inconceivably large number. If we fix E, N, and V but let the system find its own pressure value, the result will be the average of P over all the microstates—the **ensemble average.** This assumption is called the **ergodic hypothesis.**

How about an example to show what we mean by the ensemble average? Consider the crude system in Fig. 2.5: three F atoms are trapped in a space with only enough room for 4 atoms, and the total energy is sufficient for only 1 atom to be in an excited electronic state (the $^2P_{1/2}$ state); the other 2 atoms are in their ground states (the $^2P_{3/2}$ state). All atoms are in their ground translational states. Because the translational states are quantized in a box this small, there are only 4 translational wavefunctions available to the 3 atoms, corresponding to the 4 columns in Fig. 2.5. For each of these 4 translational states, the energy of the excited electronic state may be added to any of the 3 atoms, resulting in 3 rows for each column and a total of 12 ensemble states. These are 12 distinct microstates, each with unique microscopic properties, but they yield identical values for the macroscopic variables E (the total energy of the system), V (the volume), and N (the total number of particles). If each atom exerts a pressure p_0 on the walls next to it, the average pressure against the top wall in the diagram is $1.5p_0$. That's because half of the ensemble states have one molecule near the top wall (so exerting a pressure of p_0) while the other half of the states put two molecules against the top wall (with a pressure of $2p_0$), and the average over the microstates is $1.5p_0$.

▲ FIGURE 2.5 **A 12-member ensemble.** Each box represents one microstate in the ensemble. The circles represent fluorine atoms in a container with room for only four atoms. The dark circle represents an excited $^2P_{1/2}$ state fluorine atom; the light circles represent the $^2P_{3/2}$ ground state atoms.

The Microcanonical Ensemble

When we approach a problem in statistical mechanics, we must decide which parameters are allowed to vary and which we can fix to specific values throughout. It's a trade-off: the system is more flexible (and often more realistic) if we leave more parameters as variables, but the problem will be easier to solve the more parameters we fix. For example, if we want to predict the behavior of water vapor in a sealed flask, then we may choose to set the number of molecules N to be a constant, whereas if we want to model the behavior of liquid water in an open beaker, free to evaporate into the air, then N may have to be a variable. These choices determine what the ensemble of microstates will be.

BIOSKETCH   Juan J. de Pablo

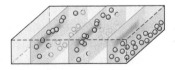

Juan J. de Pablo is the Liew Family Professor in Molecular Theory and Simulations at The Institute for Molecular Engineering at the University of Chicago. Much of Professor de Pablo's work has focused on developing methods in computational statistical mechanics to more accurately predict the behavior of polymers. He has explored how the nature of the ensemble can be adapted to make it easier to predict properties of very large biomolecules or polymers, for example. Recent work in the de Pablo group has included simulations of diblock copolymers—materials made by interweaving strips of two chemically distinct polymers—to test the structures that result when nanoparticles are incorporated into the polymers and to examine methods for encouraging the polymer blocks to assemble themselves over large distances with a minimum of structural defects. Computational studies of these materials, when reliable, can suggest the most promising experiments to carry out, at a great savings in expense and effort.

▶ **FIGURE 2.6 Microcanonical ensemble.** The microstates within the solid line are members of an *ensemble*—they have the same number of particles (N) in the same volume (V), though they differ in their energies (E). In this example, blue arrows indicate particles with $\varepsilon = 1$ (in arbitrary units) while red arrows indicate particles with $\varepsilon = 2$. The microstates enclosed by the dashed line are four members of a single *microcanonical ensemble*—that is, they have the same values for *all* extensive parameters (in this case, N, V, and E). Other microstates with different particle number, volume, or energy, belong to different microcanonical ensembles.

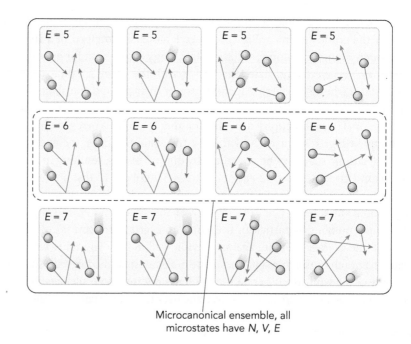

Microcanonical ensemble, all microstates have *N, V, E*

We're going to start off with the **microcanonical ensemble,** which has fixed values for *all of the extensive parameters,* including our current batch: E, V, and N. An illustration of this definition is provided in Fig. 2.6. Because only the extensive variables are fixed, the microcanonical ensemble is particularly convenient for examining changes in the intensive variables (such as the pressure) as functions of the extensive variables.

If we fix all the extensive variables, then any intensive variables that are ratios of the extensive variables, such as the number density $\rho = N/V$, must also be fixed. But then what's left to differ among the microstates? Two things: (1) all the microscopic parameters that we're not even bothering to measure in thermodynamics (such as the energies and positions of individual molecules) and (2) some intensive parameters (such as the pressure) that are not ratios of extensive parameters. For example, one microstate may distribute more energy into the translational motion (as opposed to rotational or vibrational motion) than the next microstate. That will tend to make the pressure in the first microstate higher, because the force exerted against the walls of the container depends more on the speed of the particle than its rotational or vibrational energy. For now, it's okay if the intensive parameters are also fixed, but the microcanonical ensemble does not *require* them to be fixed.

The advantage of the microcanonical ensemble is that we can imagine it being completely isolated from the rest of the universe. Energy and mass are rigorously conserved in the microcanonical ensemble, because E and N are constants. The constant volume means that no mechanical force pushes on the surroundings. This isolation greatly simplifies our first look at macroscopic systems and provides the foundation for more complex and typical chemical systems.

To look at it classically, one microstate differs from the next by changes in the locations and motions of the individual particles. In the simplest case—an ensemble of gas-phase atoms, all in the ground electronic state—we could identify each ensemble state by specifying the X, Y, and Z coordinates *and* the velocity components v_X, v_Y, and v_Z for each of the N atoms.[1] If we picked any two microstates i and j from the ensemble, i and j would differ in their values for at least one velocity or position component out of all these coordinates. Each macroscopic parameter we associate with the system is the average over these states.

Our list of extensive and intensive parameters still lacks a few important contributions, though. We now need to add another extensive parameter, which will also be a constant for our microcanonical ensemble and is arguably the most important parameter in all of science.

2.3 Entropy

The Boltzmann Entropy

Section 1.2 defines the entropy empirically, but with the advantage of a little quantum mechanics we can construct a new definition, one that will let us link the microscopic and macroscopic worlds. We now define the **Boltzmann entropy** S of the system to be proportional to the logarithm of the *total number of microstates* Ω of the system:

$$S \equiv k_B \ln \Omega, \tag{2.11}$$

where the **Boltzmann constant** is given by

$$k_B = 1.381 \cdot 10^{-23} \, \mathrm{J\,K^{-1}}.$$

The units, $\mathrm{J\,K^{-1}}$, reflect the classical definition of the entropy as a ratio of heat to temperature. Equation 2.11, which is *not* based on the temperature, will provide a different view of how temperature and energy are related. The Boltzmann entropy will be our rigorous definition for the entropy, working under any circumstances.

At present, those circumstances are dictated by the conditions of the microcanonical ensemble: fixed values of E, V, and N. In this ensemble, Ω is the total number of microstates that have the same energy, so *under these conditions* the ensemble size Ω is a new name for the degeneracy of states (g in Section 2.1) of our N-particle system. One challenge we face in the notation is to differentiate between the quantum states of the entire ensemble and those of the individual particles within the ensemble. That distinction begins here. Note that we use Ω to represent the total number of microstates in the ensemble (effectively the degeneracy of the entire system), whereas we will continue to use g for the degeneracy of quantum states in some particular individual particle.

A fundamental principle emerges from Eq. 2.11. The entropy measures one property of the system, straightforward in concept:

The entropy S counts the total number of distinct microstates of the system.

That's all it does.

---

[1] We will distinguish between the *velocity* ($\vec{v}$) of a particle, which is a vector and has direction, and the *speed* (v), which is the magnitude of the velocity vector.

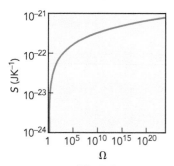

▲ FIGURE 2.7 The Boltzmann entropy. The value of S increases slowly with the number of microstates in the ensemble.

The amazing thing is that, although Ω is enormous, it is *always finite*. The number of arrangements of the system is, in fact, countable. The entropy is the key parameter that links the macroscopic world to the quantum world. In a purely classical system, Ω would be infinity, because there would be infinitely many slightly different ways of repositioning the molecules and redistributing the energy. In defining the entropy in terms of a finite number of microstates, Boltzmann, working in the 1870s, anticipated by 30 years the fundamental concept of quantum mechanics: that microscopic properties of the system must be quantized.

Fine, but why do we bother to introduce S when we already have Ω? One problem with Ω is that if we have two non-interacting subsystems A and B then the degeneracy of the combined system A + B increases as the *product* of the degeneracies Ω_A and Ω_B. For every quantum state of subsystem A, there are Ω_B distinct quantum states of B, so the total number of quantum states of A + B is $\Omega_A\Omega_B$. Whereas the logarithm of Ω, and therefore the entropy, increases as the *sum*:

$$\Omega_{A+B} = \Omega_A\Omega_B$$
$$\ln\Omega_{A+B} = \ln(\Omega_A\Omega_B) = \ln\Omega_A + \ln\Omega_B$$
$$S_{A+B} = S_A + S_B.$$

This assumes that the A and B subsystems are exactly the same before and after we combine them. So, for example, we cannot remove the wall between A and B in Fig. 2.4, because that would increase the volume seen by the A particles. That in turn would increase the degeneracy of the A energy levels, and shift the value of the overall entropy higher.

Another convenience of S over Ω is that the ensemble size of any macroscopic system is huge while its logarithm is more manageable. Figure 2.7 shows that even 10^{23} microstates—as many molecules as are in a mole—gives an immeasurably small entropy. In contrast, the entropy of 1.0 mol of helium gas under typical lab conditions is about $126\,\mathrm{J\,K^{-1}\,mol^{-1}}$. Using Eq. 2.11 to find Ω, we conclude that there are about 10^{95} distinct quantum states. You can see why the logarithm might be useful here.

The Gibbs Entropy

Equation 2.11 ($S = k_B \ln \Omega$) defines the entropy, but this is not always the most convenient form of the definition. We let Ω stand for the total number of states of the ensemble, and in the microcanonical ensemble (constant E, V, N), this number is essentially the degeneracy of the entire system. However, we're about to look at an ensemble in which the energy is not constant, and it becomes less obvious what Ω really means. J. W. Gibbs developed an ingenious general equation for estimating the entropy in any ensemble by using *probabilities* rather than Ω.

Parameters key: the Gibbs entropy

symbol	parameter	SI units
Ω	ensemble size	unitless
N	number of indistinguishable particles in system	unitless
N_i	number of particles in quantum state i	unitless
k_B	Boltzmann constant	$\mathrm{J\,K^{-1}}$
S	entropy	$\mathrm{J\,K^{-1}}$
k	number of quantum states available to each particle	unitless
$\mathcal{P}(i)$	probability of any particle being in quantum state i	unitless

We'll show that Gibbs' definition of the entropy is equivalent to Boltzmann's for a system of N distinguishable particles distributed among k distinct *single-particle* states. The key lies in a bit of *combinatorial mathematics,* the theory of rearrangements of sets. The total number of ways we can rearrange the labels on our N molecules is $N!$ (the first label could be any of the N available labels, the second label any of the remaining $N - 1$ labels, and so on). If the particles are distinguishable, each of these $N!$ sequences is different. For example, imagine that our N particles are six atoms, frozen in place so we can distinguish between them by their positions. Also, let each of $k = 4$ states be a different electronic quantum state. There may be, for example, three atoms in quantum state 1, two atoms in state 2, and one atom in state 3. Call the number of particles in each state N_i, so in this example, $N_1 = 3$, $N_2 = 2$, $N_3 = 1$, and $N_4 = 0$. This means that when we pick one of these particles at random, the probability that the particle is in state 1 is $\mathcal{P}(1) = 3/6$, and similarly $\mathcal{P}(2) = 2/6$, $\mathcal{P}(3) = 1/6$, and $\mathcal{P}(4) = 0$.

Using this example, there are $N! = 6! = 720$ different sequences we could write for the labels 1 through 6. For each sequence, the first three atoms go into state 1, the next two into state 2, and the last atom into state 3. Say that the sequence 123-45-6 means atoms 1, 2, and 3 go into state 1, atoms 4 and 5 into state 2, and atom 6 into state 3. This sequence is different from the sequence 124-35-6, because atom 3 has moved to state 2 and atom 4 to state 1. However, 123-45-6 is the same as 132-45-6, because atoms 2 and 3 are still in state 1 even if we write them in opposite order.

To rephrase this mathematically, there are $N!$ different sequences of all N labels, but within each state there are $N_i!$ ways of rearranging the labels with no effect on the microstate. Therefore, the total number of microstates for this system is

$$\Omega = \frac{N!}{N_1! N_2! \ldots N_k!} = \frac{N!}{\displaystyle\prod_{i=1}^{k} N_i!}. \tag{2.12}$$

For our example system, keeping in mind that $0! = 1$,[2]

$$\Omega = \frac{6!}{(3!)(2!)(1!)(0!)} = 60.$$

That means that there are 60 distinct ways we can assign the 6 labeled atoms such that 3 of them are in state 1, 2 of them are in state 2, and 1 is in state 3. Those microstates are listed in Table 2.1 and a few are shown in Fig. 2.8.

We can now write the entropy in terms of the N values, starting from Boltzmann's law and using Eq. 2.12 for Ω:

$$S = k_B \ln \Omega = k_B \ln \left(\frac{N!}{\displaystyle\prod_{i=1}^{k} N_i!} \right) \qquad \text{by Eq. 2.12}$$

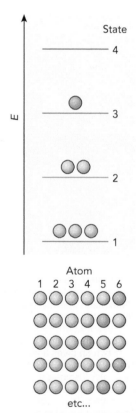

▲ FIGURE 2.8 **Example for calculating the Gibbs entropy.** There are four states available to the six atoms. The set of microstates is all combinations of the six atoms such that three atoms are in quantum state 1, two atoms are in state 2, one atom is in state 3, and no atoms are in state 4.

---

[2]Why is the factorial of 0 equal to 1? If we look at the difference between any two adjacent factorials, we get

$$n! - (n - 1)! = n[(n - 1)!] - (n - 1)! = (n - 1)!(n - 1). \tag{2.13}$$

If you substitute $n = 1$, then you get $1! - 0! = 0!(0)$. If 0! is any defined number, then the right-hand side of that last equation is 0, because we're multiplying by 0. Therefore, $1! - 0! = 0$, so 0! is equal to $1! = 1$. You can use the same argument to show that you can't define a negative number factorial because, using $n = 0$ in Eq. 2.13, you end up with $1 - (-1)! = (-1)!(-1)$. Now adding $(-1)!$ to both sides yields the invalid $1 = 0$.

TABLE 2.1 **Example for calculating the Gibbs entropy.** The $\Omega = 60$ distinct microstates of the ensemble of six distinguishable particles are distributed such that $N_1 = 3$, $N_2 = 2$, and $N_3 = 1$. The microstates are labeled by the three particles in state 1, the two in state 2, and the one in state 3.

1	2	3	1	2	3	1	2	3	1	2	3	1	2	3	1	2	3
123	45	6	123	46	5	123	56	4	124	35	6	124	36	5	124	56	3
125	34	6	125	36	4	125	46	3	126	34	5	126	35	4	126	45	3
134	25	6	134	26	5	134	56	2	135	24	6	135	26	4	135	46	2
136	24	5	136	25	4	136	45	2	234	15	6	234	16	5	234	56	1
235	14	6	235	16	4	235	46	1	236	14	5	236	25	4	236	45	1
145	23	6	145	26	3	145	36	2	146	23	5	146	25	3	146	35	2
156	23	4	156	24	3	156	34	2	245	13	6	245	16	3	245	36	1
246	13	5	246	15	3	246	35	1	256	13	4	256	14	3	256	34	1
345	12	6	345	16	2	345	26	1	346	12	5	346	15	2	346	25	1
356	12	4	356	14	2	356	24	1	456	12	3	456	13	2	456	23	1

$$= k_{\text{B}}\left[\ln N! - \ln \prod_{i=1}^{k} N_i! \right] \qquad \ln(a/b) = \ln a - \ln b$$

$$= k_{\text{B}}\left[\ln N! - \sum_{i=1}^{k} \ln N_i! \right] \qquad \ln(ab) = \ln a + \ln b$$

We simplify this considerably using **Stirling's approximation:**[3]

$$\ln N! \approx N\ln N - N. \qquad (2.14)$$

This expression lets us evaluate $\ln N!$ and all the $\ln N_i$ values:

$$S = k_{\text{B}}\left[N\ln N - N - \left(\sum_{i=1}^{k} N_i \ln N_i - \sum_{i=1}^{k} N_i \right) \right] \qquad \text{by Eq. 2.14}$$

$$= k_{\text{B}}\left[\sum_{i=1}^{k} N_i \ln N - N - \sum_{i=1}^{k} N_i \ln N_i + N \right] \qquad \sum_i N_i = N$$

$$= k_{\text{B}} \sum_{i=1}^{k} \left[N_i \ln N - N_i \ln N_i \right] \qquad \text{cancel } N's, \text{ factor out sum}$$

$$= k_{\text{B}} \sum_{i=1}^{k} N_i \left[\ln N - \ln N_i \right], \qquad \text{factor out } N_i$$

and finally put this in terms of the probabilities, $\mathcal{P}(i) = N_i/N$,

$$= Nk_{\text{B}} \sum_{i=1}^{k} \frac{N_i}{N} \ln \frac{N}{N_i}. \qquad \text{factor out } N \qquad (2.15)$$

This last expression gives us the **Gibbs entropy:**

$$S = -Nk_{\text{B}} \sum_{i=1}^{k} \mathcal{P}(i) \ln \mathcal{P}(i). \qquad (2.16)$$

CHECKPOINT The Gibbs entropy does not look quite like the Boltzmann entropy, but if we look a little longer at Eq. 2.16 we can see that the two definitions are similar. For a very large Ω (which gives a large Boltzmann entropy), we have so many choices among the microstates that we have a very small probability $\mathcal{P}(i)$ of the system being found in any *one* of those microstates. A probability $\mathcal{P}(i)$ close to zero gives a large, negative value of $\ln \mathcal{P}(i)$. To obtain the Gibbs entropy, we use the sum to effectively average all of those large negative logarithms for the different microstates and then multiply by -1, which brings us to a large, positive value for the entropy, consistent with the large Boltzmann entropy.

---

[3]How good is Stirling's approximation? For $N = 6$ it's pretty poor: $\ln 6! = 6.58$ but $6\ln 6 - 6 = 4.75$. By the time we get to 100, however, the approximation is good to within 1%, $\ln 100! = 363.7$ and $100\ln 100 - 100 = 360.5$. The approximate relative error is $\ln(2\pi N)/(2N)$, which decreases roughly as $1/N$. By the time we get to a picomole sample ($10^{-12}\,\text{mol}: N = 6 \cdot 10^{11}$), the approximation is effectively exact. Few calculators or spreadsheets can even calculate 200!.

We will apply the Gibbs entropy to an extension of the example just given. Instead of there being N distinguishable particles distributed among k single-particle quantum states, imagine that we have an ensemble of N quantum states for the entire system distributed among k different energy levels. The Gibbs formula allows us to write the entropy of the system by adding up the probabilities of the system having each possible value of the energy E. This form of the entropy is useful because, for macroscopic systems, the total energy is still an important parameter, but the particular quantum state is not. Using these probabilities allows us to express the entropy in terms of the possible energies of the system without knowing the number or nature of the specific quantum states.

Although we used Boltzmann's equation as the starting point, the Gibbs entropy in Eq. 2.16 is the foundation for a lot of the work we'll be doing in the next few chapters, because it demonstrates convincingly how real chemical questions could be solved by looking at the *statistics* of molecules.

2.4 Temperature and the Partition Function

We're going to narrow our system down to a particular case and use this case to illustrate some general principles of statistical mechanics. Our overall goal for the rest of the text is to predict the properties of bulk materials from the properties of the individual molecules. This is hard. Whenever we run up against a fresh challenge of this magnitude in physical chemistry, we turn first to some imaginary, idealized system and try to smooth out the roughest parts of the math before tackling more realistic problems. Welcome to the ideal gas.

The Ideal Gas

The ideal gas is a collection of particles that don't have any interactions at all, except that they bounce off the walls of their container (Section 1.3). They don't repel each other, so one can pass right through another, and they don't attract each other.

The rest of this chapter is a leisurely application of statistical mechanics to a property of the ideal gas you may know from general chemistry. We want to find an expression for the pressure exerted by an ideal gas on the walls of a container as a function of other macroscopic variables. This exercise allows us to establish some terminology and preliminary conclusions. Several of the tools and results may be familiar to you from previous encounters with classical chemistry.

Temperature

In this and the following section, we will show how the general concept of the ensemble—as vaguely as we've defined it—is sufficient to allows us to define the relationship between energy and entropy, and moreover to derive the pattern by which energy is distributed among the particles in the system. The math is not the most challenging we will come across, but we leave the system so broadly defined that the derivation may seem especially abstract. We want to leave the system in such general terms for now precisely to show how universal the results of this derivation are: in

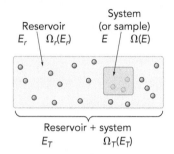

▲ FIGURE 2.9 **The system and reservoir.** The system we are studying will generally sit inside and interact directly with a much larger body, called the reservoir, which is part of the surroundings.

these next few pages lie the origins of *key* concepts in thermodynamics and kinetics. For those eager to get started on the applications rather than reading the math, however, a **DERIVATION SUMMARY** appears after Eq. 2.30.

Our system is now a box of fixed volume V containing a fixed number of molecules N, but this time *we won't keep the energy of the system constant.* We put this system next to a **reservoir.** The reservoir is the part of the surroundings that the system interacts with most strongly. The reservoir has much greater volume and mass than the system, and an energy $E_r \gg E$ (Fig. 2.9). In our example, we could put our sealed ideal gas container (the system) on a lab bench, surrounded by the rest of the lab (the reservoir). The system can exchange energy with the air in the lab through the walls of the container, so its energy is now a variable. The universe consists of the system plus the reservoir, with a total energy E_T (which *is* constant), such that

$$E + E_r = E_T \gg E. \qquad (2.17)$$

The number of microstates available to the reservoir, $\Omega_r(E_r)$, is also much greater than $\Omega(E)$ for the system, because there are many more components to the reservoir. Forgive the "(E)" notation everywhere, but it is crucial for now to remember that the ensemble size is a function of the energy. The ensemble size $\Omega_T(E_T)$ of the universe when the system is at energy E is given by[4]

$$\Omega_T(E_T) = \sum_E \Omega(E)\Omega_r(E_r). \qquad (2.18)$$

There is nothing outside the universe, so its total energy E_T must remain constant; that is, we consider only the states in the microcanonical ensemble of the universe. If the energy of the reservoir rises, the energy of the system decreases by the same amount.

Now we choose a *single* microstate i of the system, with energy E_i. The probability $\mathcal{P}(i)$ of this state occurring among all the microstates of the ensemble is

Parameters key: temperature and the canonical distribution

symbol	Paramter	SI units
E	energy of system	J
E_r	energy of reservoir	J
E_T	energy of combined system plus reservoir	J
Ω	ensemble size of the system	unitless
Ω_r	ensemble size of the reservoir	unitless
Ω_T	ensemble size of combined system plus reservoir	unitless
k_B	Boltzmann constant	J K^{-1}
S	entropy of the system	J K^{-1}
S_r	entropy of the reservoir	J K^{-1}
$\mathcal{P}(i)$	probability that the system is in microstate i	unitless
$\mathcal{P}(E)$	probability that the system has energy E	unitless
T	temperature	K
$Q(T)$	partition function	unitless

[4]The overall degeneracy Ω_T is, in general, a function of both E_T and E. Depending on how the energy is divided between the system and reservoir, Ω_T has more than one value for a given value of E_T. However, since we are about to examine a specific system energy E_i in this derivation, we simplify the notation by writing Ω_T as a function of E_T alone.

equal to the fraction of the total number of microstates for which the system is in state i. If there were nine distinct microstates for which the system was in state i, and 180 microstates altogether, the probability of finding the system in state i would be $9/180 = 0.05$. It follows that, if the *system* is in state i for all nine of those microstates but the nine microstates are somehow distinct, then it must be the state of the *reservoir* that changes in each of those nine microstates. Therefore, we can count the number of microstates for which the system is in a state i with energy E_i by counting the number of states of the reservoir with energy $E_r = E_T - E_i$. An example is described in Fig. 2.10. The probability of the system being in any state i can therefore be written

$$\mathcal{P}(i) = \frac{\Omega_r(E_r)}{\Omega_T(E_T)}. \tag{2.19}$$

Probability distribution functions like this are going to be the focal point of our look at statistical mechanics. We don't want to throw away everything we know about molecular structure, but we need to use it in a way that's manageable for an enormous number of molecules. One way is to keep track of all the values of some molecular property, the vibrational energy for example, without bothering to determine which molecule in the system has which value. That's what the probability distribution functions do for us.

Equation 2.19 is not yet in a convenient form, because it requires us to determine the ensemble size of the reservoir, the number of microstates $\Omega_r(E_r)$ with specific energy E_r. We'd rather make measurements only on the system and not worry about the properties of the reservoir. By introducing a new parameter, we can avoid treating the reservoir explicitly.

We want to replace the $\Omega_r(E_r)$ term in Eq. 2.19 by a function of the energy of our system E_i, if possible. To do this, we will first take advantage of the conservation of energy. If the system and the reservoir are the only bodies exchanging energy, then $E_i + E_r = E_T$, where E_T is a constant. This allows us to replace E_r by $E_T - E_i$. The next step takes a little longer, but introduces the temperature from a completely theoretical perspective that justifies the original empirical definition of temperature.

The real challenge in finding a useful expression for $\mathcal{P}(i)$ is finding a way to write the ensemble size of the reservoir Ω_r in Eq. 2.19 in terms of the ensemble size of the system. Before we get started, it will help to change from Ω_r, which is a huge number

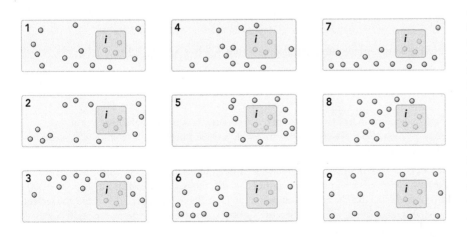

◀ FIGURE 2.10 **The ensemble average.** An example ensemble consists of a total of 180 microstates, of which only the 9 drawn here have the system in a state i. The energies E_i and E_r are fixed for this set. The set of 9 microstates drawn must represent all the different states of the reservoir when it has energy $E_T - E_i$. The probability of the system being in state i is therefore the ratio of $\Omega_r(E_r) = 9$ to $\Omega_T(E_T) = 180$. For this example, $\mathcal{P}(i) = 9/180 = 0.05$.

and very sensitive to the energy value, to the entropy function $S_r \equiv k_B \ln[\Omega_r(E_r)]$. The entropy, because it varies as the logarithm of the ensemble size, is a much more slowly changing function than Ω_r. A function that varies smoothly can be approximated successfully by a Taylor series expansion (Section A.1), which is just what we need here. We can always require that the reservoir energy is much greater than the system energy, and in that case $E_r \approx E_T$. What the Taylor series will let us do is expand S_r around the point where the energy is E_T and the displacement—the difference between E_T and E_r—is given by the system energy E_i. This lets us change from a function of E_r to a function of E_i, which is what we're trying to accomplish. How are the entropies of the reservoir and system related? To find out, we carry out the Taylor series expansion of S_r about E_T:[5]

$$S_r(E_r) = S_r(E_T - E_i) = S_r(E_T) - \left(\frac{\partial S_r}{\partial E_r}\right)_{V,N}\bigg|_{E_T} E_i + \dots . \tag{2.20}$$

The parameters V and N are still held constant, so they appear as constants in the partial derivative $\left(\frac{\partial S_r}{\partial E_r}\right)_{V,N}$. The term $S_r(E_T)$ is the entropy that the reservoir would have if it were at the energy E_T.

To avoid specifying the nature of the reservoir, we can rewrite the partial derivative $\left(\frac{\partial S_r}{\partial E_r}\right)_{V,N}$ in terms of the system entropy by noting that for fixed $E_T = E_r + E_i$ (i.e., conserving the total energy),

$$dE_r = d(E_T - E_i) = -dE_i,$$

and for a given system energy, $\Omega_T(E_T) = \Omega_r(E_r)\Omega(E_i)$ and

$$dS_r = k_B d\ln\Omega_r(E_r) = k_B d\left[\ln\left(\frac{\Omega_T(E_T)}{\Omega(E_i)}\right)\right] = k_B d[\ln\Omega_T(E_T) - \ln\Omega(E_i)]$$
$$= -k_B d\ln\Omega(E_i) = -dS. \tag{2.21}$$

Combining these results, we find

$$\left(\frac{\partial S_r(E_r)}{\partial E_r}\right)_{V,N}\bigg|_{E_T} = \left(\frac{\partial S(E_i)}{\partial E_i}\right)_{V,N}\bigg|_{E_T}. \tag{2.22}$$

Making this substitution and dropping the higher-order terms in the Taylor series,[6] Eq. 2.20 becomes

$$S_r(E_r) = S_r(E_T - E_i) = S_r(E_T) - \left(\frac{\partial S(E_i)}{\partial E_i}\right)_{V,N}\bigg|_{E_T} E_i. \tag{2.23}$$

We now formally introduce the **temperature** T:

$$T \equiv \left(\frac{\partial E}{\partial S}\right)_{V,N}, \tag{2.24}$$

which lets us replace the partial derivative $\left(\frac{\partial S}{\partial E}\right)_{V,N}$ in Eq. 2.23 by $1/T$:

$$S_r(E_r) = S_r(E_T - E_i) = S_r(E_T) - \frac{E_i}{T}. \tag{2.25}$$

---

[5] Forgive a brief reminder here. Be careful not to confuse these two partial derivatives of $f(x)$ with respect to x: $(\partial f(x)/\partial x)_y$ is the derivative with variable y held constant; $(\partial f(x)/\partial x)|_a$ is the derivative evaluated at $x = a$.

[6] Here's a general example to show that the higher-order terms may safely be ignored. If the degeneracy obeys any power law of the form $\Omega = cE^x$ (an assertion we will justify in Section 3.4), then we would find that the first derivative term in Eq. 2.20 is of the order E_i/E_T, while the second derivative term is of order $(E_i/E_T)^2$. Because $E_T \gg E_i$, the higher-order terms are negligible.

To get back to $\Omega_r(E_r)$ in Eq. 2.19, we rewrite this using Boltzmann's law:

$$k_B \ln[\Omega_r(E_r)] = k_B \ln[\Omega_r(E_T)] - \frac{E_i}{T}. \qquad (2.26)$$

Solving for the degeneracy of the reservoir, we obtain

$$\Omega_r(E_r) = \Omega_r(E_T)\, e^{-E_i/(k_B T)}. \qquad (2.27)$$

What is the temperature really? According to our definition, T represents the rate at which the energy of the system rises as the number of microstates increases (Fig. 2.11). There is an odd conservation of complexity in definitions. In classical thermodynamics, temperature is a relatively clear concept: energy flows between two systems in contact until the temperatures are equal. Entropy is then defined in terms of the temperature and becomes rather muddy in the explaining. With our approach, coming from a microscopic perspective, entropy becomes the clearly defined parameter (essentially the logarithm of the number of microstates), but as a consequence we define the temperature as a derivative of the entropy. There are advantages to this mathematical definition of the temperature over the phenomenological definition from classical thermodynamics, but in this form it is not especially satisfying. For now, let it suffice that Eq. 2.24 is our one rigorous definition of the temperature, and we will see more intuitive and qualitative (and more approximate) definitions as we go along, beginning with the very next section.

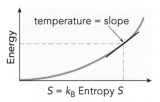

▲ FIGURE 2.11 **Temperature.** The temperature is the change in the system's energy with respect to entropy.

The Canonical Distribution and the Partition Function

We are working now with a different ensemble of the system. Conceptually, it is convenient to fix the extensive variables E, V, and N, but in a typical experiment the system is constantly interacting with its environment (i.e., reservoirs), most commonly by transferring energy into or out of the surroundings. In that case, E is not a fixed quantity. Even though energy is conserved overall, it is not a constant for the system we're studying. However, if the reservoir is big enough, then we can assume that *its* properties are constant, so we fix the temperature T. Under these assumptions, we use the **canonical ensemble,** which is the ensemble of states with fixed T, V, and N, and where the energy of the system is a variable.

Let's return to our effort to describe the probability of the system being in a particular microstate i. We combine Eqs. 2.19 and 2.27:

$$\mathcal{P}(i) = \frac{\Omega_r(E_T)}{\Omega_T(E_T)} e^{-E_i/(k_B T)}, \qquad (2.28)$$

where $\Omega_r(E_T)$ and $\Omega_T(E_T)$ are constants. We eliminate those constants by requiring the probability distribution function to be normalized, so that

$$\sum_{i=1}^{\infty} \mathcal{P}(i) = \frac{\Omega_r(E_T)}{\Omega_T(E_T)} \sum_{i=1}^{\infty} e^{-E_i/(k_B T)} = 1,$$

where the sum is over *all* the microstates of the canonical ensemble (constant V, N, and T). Setting the sum of all the probabilities to one ensures that any single value $\mathcal{P}(i)$ gives the probability as a fraction between 0 and 1 that the system is in state i. We solve for Ω_r/Ω_T and use this solution to define the **partition function** $Q(T)$:

$$\frac{\Omega_r(E_T)}{\Omega_T(E_T)} = \frac{1}{\displaystyle\sum_{i=1}^{\infty} e^{-E_i/(k_B T)}} = \frac{1}{Q(T)},$$

CHECKPOINT If you haven't seen this derivation before, don't let the moment slide by without a modest celebration: few derivations have more impact in chemistry. Let's take a moment to look ahead. The Maxwell-Boltzmann distribution for molecular speeds is the canonical distribution applied to translational states. You may recall that the equilibrium constant for a chemical reaction can be written $\exp[-\Delta_{rxn}G^\circ/(RT)]$. That exponential is again the canonical distribution. The Arrhenius equation $k = Ae^{-E_a/(RT)}$ for the rate constant in chemical kinetics? The canonical distribution again. Chapter 1 ascribes the similarity of these terms to the Maxwell-Boltzmann distribution, but the more fundamental basis is the canonical distribution.

where

$$Q(T) \equiv \sum_{i=1}^{\infty} e^{-E_i/(k_B T)}. \qquad (2.29)$$

Our probability distribution function may now be rewritten more simply as

$$\mathcal{P}(i) = \frac{e^{-E_i/(k_B T)}}{Q(T)}. \qquad (2.30)$$

The probability distribution defined in Eq. 2.30 is called the **canonical distribution** because it is based on the canonical ensemble.

DERIVATION SUMMARY **Temperature and the Canonical Distribution.** With our system free to exchange energy with a much larger reservoir, we used a Taylor series expansion of only two terms to write the reservoir's entropy S_r as a constant plus a term proportional to the energy of the system E. The coefficient of the energy term is the partial derivative (taken from the Taylor series) $\left(\frac{\partial S_r}{\partial E_r}\right)_{V,N}$, which we then *defined* to be $1/(T)$, where T is the temperature. To then find the probability of the system being in state i, we compared the number of ensemble states of the reservoir (based on the equation we obtained for the reservoir entropy) to the number of ensemble states of the combined system plus reservoir. The greater the number of possible ensemble states *overall* for a given state i of the system, the more likely that the system will be found in that state. The ratio of states with the system in state i to the number of possible states overall gives the probability we're looking for, Eq. 2.30, which we call the canonical distribution.

Here we begin to see why the temperature is so important. The probability $\mathcal{P}(i)$ shows us that the likelihood of getting the system into a particular state i drops exponentially with respect to the ratio of the microstate energy E_i to the **thermal energy** $k_B T$. When the system is free to exchange energy with its surroundings, we can state the following:

The temperature establishes how likely the system is to be found at any particular energy.

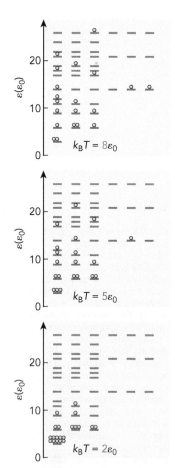

▲ FIGURE 2.12 **Distributions of 20 distinguishable particles in a three-dimensional box.** As the temperature increases, the molecules are spread out over more states and higher energy quantum states.

If microstate i is at an energy E_i much greater than the thermal energy $k_B T$, then it is very unlikely that the sample will find its way to that microstate. Figure 2.12 shows examples of how temperature affects the distributions for 20 particles in a three-dimensional box. We have made these particles distinguishable, so that we can ignore for now the implications of the symmetrization principle (Section 2.1), but we shall investigate those implications in Chapter 4.

Suppose we don't care about the probability of a particular microstate i but want to know the likelihood of finding the system at any particular energy level E. Then we may write an *energy* probability distribution, using $\Omega(E)$ to count all the states i with that energy:

$$\mathcal{P}(E) = \Omega(E)\mathcal{P}(i) = \frac{\Omega(E)e^{-E/(k_B T)}}{Q(T)}, \qquad (2.31)$$

where the sum is carried out over all possible energies of the system. An advantage of the canonical ensemble is that it lets us ignore for now how Ω depends on V and N by fixing those values, allowing us to focus on the energy dependence.

Next we make an important assertion. The probability that we've worked out for the likelihood of the system being at some particular energy E can be extended down to the probability of any particle *within* our system having some particular single-particle energy ε. Molecules within the system exchange energy with each other in the same way that we envision our system exchanging energy with the reservoir. Any molecule we select has a distribution of possible energies and has in effect its own ensemble of possible quantum states, and the chance of that molecule being found at some particular energy ε is also given by the canonical distribution. We can write the probability function in terms of the particular quantum state i as in Eq. 2.30. But in general there are g states with the same energy, and most of the time the energy is what matters to us, not the particular quantum state. In that case, we can solve instead for the probability $\mathcal{P}(\varepsilon)$ that any given molecule has energy ε, regardless of its quantum state, by multiplying $\mathcal{P}(i)$ by the degeneracy $g(\varepsilon)$ of that energy level:

$$\mathcal{P}(\varepsilon) = \frac{g(\varepsilon)e^{-\varepsilon/(k_{\mathrm{B}}T)}}{q(T)}, \qquad (2.32)$$

where the single-particle partition function is

$$q(T) = \sum_{\varepsilon=0}^{\infty} g(\varepsilon)e^{-\varepsilon/(k_{\mathrm{B}}T)}. \qquad (2.33)$$

The partition function normalizes $\mathcal{P}(\varepsilon)$, ensuring that if we examine all possible energy levels, we have a 100% chance of finding the system at one of those energies. A rougher, qualitative way to think of the partition function is this: Q (or q) counts the number of states that the system can easily access.

EXAMPLE 2.1   The Probability Distribution Function

CONTEXT   Ammonia (NH_3) has a vibrational mode for inversion, the "umbrella" motion that flips the three hydrogens from one side of the nitrogen atom to the other. The harmonic approximation does not work well for this vibrational motion, and the lowest excited state, labeled 0^-, is only 0.79 cm^{-1} above the ground state, 0^+. These two quantum states formed the basis of the first working example of the laser principle, demonstrated by Charles Townes and colleagues in 1954. Theirs was a device that generated microwave radiation from the transition of ammonia from the 0^- state to the 0^+. Using such a low-lying excited state helped prevent competing processes from channeling away the energy, and it meant that even at room temperature a significant fraction of the ammonia molecules could be found in the excited state. Calculating what fraction of a system is in a particular state is a crucial step in predicting whether or not a laser can operate (which we will see in Section 6.3).

The probabilities for a molecule being in a particular state differ in the numerator $e^{-\varepsilon/(k_{\mathrm{B}}T)}$ but share the same denominator—the partition function $q(T)$. Therefore, when we evaluate ratios of probabilities, the denominators cancel.

PROBLEM   At 298 K, calculate the ratio of the number of NH_3 molecules in the excited state to the number in the ground state, where the excited state is (a) the 0^- state of the inversion, which lies 0.79 cm^{-1} above the 0^+ ground state; and (b) the 1^+ state, which lies 932.43 cm^{-1} above the 0^+. (The wavenumber unit, cm^{-1}, is conventionally used by spectroscopists as an energy unit, based on the relation between the transition energy in the experiment and the reciprocal wavelength of the photon that induces the transition: $E_{photon} = hc/\lambda$. Because the energy is inversely proportional to the wavelength, the energy is given in units of 1/distance.)

SOLUTION According to the probability distribution function in Eq. 2.30, the ratio will be

$$\frac{\mathcal{P}(v')}{\mathcal{P}(v'')} = \frac{e^{-\varepsilon'/(k_B T)}/q(T)}{e^{-\varepsilon''/(k_B T)}/q(T)} = e^{-\Delta\varepsilon/(k_B T)}.$$

Writing ε/k_B in units of kelvin is a convenient way to get a rough picture of how easily a quantum state can be reached at a particular temperature. In this case, the energy units are written as cm^{-1} (a common convention for vibrational energies), and the Boltzmann constant, which effectively converts units of temperature to units of energy, can be taken from the energy conversion table in the back endpapers: 0.6950 cm^{-1}/K. We then calculate for $v' = 0^{-}$

$$\frac{\Delta\varepsilon}{k_B} = \frac{0.79\,\text{cm}^{-1}}{0.695\,\text{cm}^{-1}/\text{K}} = 1.1\,K,$$

and similarly $\Delta\varepsilon/k_B = 1341.6\,K$ when v' is the 1^{+} state. When $\Delta\varepsilon/k_B$ is low compared to T, the ratio of probabilities will be nearly 1, and where $\Delta\varepsilon/k_B$ is high compared to T, the ratio will be very small. In our example, the $v = 1$ excitation energy is 1341.6 K, compared to 298 K, so we don't expect a lot of molecules to reach the $v = 1$ state:

$$\frac{\mathcal{P}(0^{-})}{\mathcal{P}(0^{+})} = e^{-1.1/298} = 1.0$$

$$\frac{\mathcal{P}(1^{+})}{\mathcal{P}(0^{+})} = e^{-1341.6/298} = 0.0111.$$

The number of molecules in the 0^{-} state is roughly the same as in the ground state. If the Townes group had used the 1^{+} excited state instead, they would have had only about 1% of the number in the ground state.

EXAMPLE 2.2   **The Canonical Distribution: Vibration**

CONTEXT The hydrogen fuel cell combines H_2 and O_2 to form water, releasing energy that can be put to a variety of uses. One of the roadblocks to wider use of these fuel cells is the energy cost in producing H_2 gas in the first place. Molecular hydrogen is formed in chemical reactors by the reaction of methane and water at temperatures as high as 1000 K:[7]

$$CH_4(g) + H_2O(g) \rightarrow CO(g) + 3\,H_2(g).$$

Reaction rates for vibrationally excited $v = 1$ H_2 gas can be orders of magnitude higher than for the ground state $v = 0$ molecule, so potentially we should worry that the H_2 gas will react with other molecules in the mixture as soon as it is formed.

PROBLEM At a temperature of 1000 K, how many vibrational states of H_2 are populated by at least 0.1% of the molecules, given the vibrational constant $\omega_e = 4395\,\text{cm}^{-1}$ and the vibrational energy (relative to the ground state) of approximately $E_{vib} = \omega_e v$?

---

[7]If you recall Le Châtelier's principle, the reaction is endothermic, so high temperatures are necessary to shift the equilibrium to the right. If you've forgotten Le Châtelier's principle, don't worry—we'll get to it in Section 12.5.

SOLUTION The first few vibrational states (relative to the ground state) are at approximately 0, 4395, $2 \times 4395 = 8790$, and $3 \times 4395 = 13185 \, \text{cm}^{-1}$. Solving for the contributions $e^{-\varepsilon/(k_B T)}$ to the partition function, again setting $k_B = 0.6950 \, \text{cm}^{-1}/\text{K}$, we obtain

$$q_{vib} = e^{-\frac{0}{k_B T}} + e^{-\frac{\omega_e}{k_B T}} + e^{-\frac{2\omega_e}{k_B T}} + e^{-\frac{3\omega_e}{k_B T}} + \ldots$$

$$= 1.000 + 0.002 + 3 \cdot 10^{-5} + 5 \cdot 10^{-9} + \ldots,$$

and the values continue to drop by orders of magnitude at a time for higher states. Therefore, at 1000 K, the partition function q_{vib} is approximately 1.002. Each exponential in the sum just given is the $e^{-\varepsilon_i/(k_B T)}$ term we need to find the probability for each state. For example, the ground state has $e^{-0/(k_B T)} = 1$, so the probability of finding a molecule in the ground state is

$$e^{-\frac{0}{k_B T}}/q_{vib} = 1/1.002 = 0.998.$$

In other words, 99.8% of the molecules are in the ground vibrational state. In the same way, we find that $0.002/1.002 = 0.2\%$ in $v = 1$, and less than 0.1% are in the higher levels.

Although 1000 K is much higher than typical reaction temperatures in the laboratory, H_2 has the highest vibrational constant of any molecule, and therefore a *lot* of energy (comparatively) is needed to excite the vibrational states.

EXAMPLE 2.3   Partition Functions

CONTEXT The partition function has uses well outside of the physical chemistry we can cover, including protein dynamics and the behavior of black holes. For example, in protein folding, a long strand of amino acids crumples up to form a blob with the right structure to carry out some biological function. An enzyme, for instance, often takes on a shape that allows it to cradle a substrate molecule as an important reaction is catalyzed. The process of protein folding encompasses an enormous number of possible conformers, and we can predict the outcome of the process partly by estimating the partition function and other properties during the reaction.

As a separate case, the quest for a theory that unifies general relativity and quantum mechanics has led to many efforts to describe the thermodynamics of black holes, where effects of general relativity become dominant. The behavior of the black hole should depend on how many quantum states are available to it, and those states all contribute to the partition function.

In these wide-ranging examples, the partition function can be said to accomplish one task—to obtain an effective count of the states available to the system, weighing in the likelihood of each possible state. The following example uses a simple model to show that—with an extreme energy distribution—the partition function does exactly that.

PROBLEM You discover a molecular system having the energy levels and degeneracies
$$\varepsilon = c(n-1)^6; \quad g = n; \quad \text{for } n = 1,2,3,\ldots \text{ and } k_B T = 400c.$$
Evaluate the partition function, and calculate $\mathcal{P}(\varepsilon)$ for each of the four lowest energy levels.

SOLUTION The four lowest values of the energy are for $n = 1$ through $n = 4$, which yield values of 0, c, $64c$, and $729c$. The partition function (Eq. 2.33) is

$$q(T) = e^{-0/(400c)} + 2e^{-c/(400c)} + 3e^{-64c/(400c)} + 4e^{-729c/(400c)} + 5e^{-4096c/(400c)} + \ldots$$

$$= 1.000 + 1.995 + 2.556 + 0.646 + 2 \cdot 10^{-4} = 6.20,$$

and the probabilities (Eq. 2.32) are

$$\mathcal{P}(n = 1) = \frac{1}{6.20} = 0.16 \qquad \mathcal{P}(n = 2) = \frac{1.995}{6.20} = 0.32$$

$$\mathcal{P}(n = 3) = \frac{2.556}{6.20} = 0.41 \qquad \mathcal{P}(n = 4) = \frac{0.646}{6.20} = 0.10,$$

suggesting that the molecules are divided up among roughly the six lowest energy states, corresponding to $n = 1$ $(g = 1)$ with 16% probability, $n = 2$ $(g = 2)$ with 16% per state, and $n = 3$ $(g = 3)$ with 14%. As the $n = 4$ level has only 2.5%, significantly lower than the others, it contributes little to the partition function. So the value of the partition function, 6.20, tells us that there are roughly six states with significant probability.

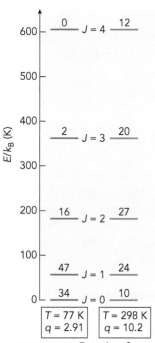

▲ **FIGURE 2.13 Rotational partition functions.** The populations by percentage at each of the lowest rotational energy levels are shown for $^{1}H^{19}F$ at 77 K and at 298 K. The degeneracy is $2J + 1$ for each energy level. The partition function q predicts roughly how many quantum states are occupied: at 77 K, $q = 3$ and the three lowest energy states are found in $J = 0$ and 1, which have 80% of the molecules; at 298 K, $q = 10$ and the ten lowest energy states are found in $J = 0 - 3$, which again account for 80% of the molecules.

For an example, consider the hydrogen nucleus, which has two spin states: spin up $(m_I = 1/2)$ and spin down $(m_I = -1/2)$. In a magnetic field, we can give these states different energies so that, say, the $m_I = -1/2$ state is lower. If we take all the nuclear spin energy out of a system of hydrogen, all the nuclei are in the state $m_I = -1/2$, and the partition function q for nuclear spin will be 1. If we make a lot of energy available so that the nuclei can gain spin energy as easily as they lose it again, both states become equally populated, and q will equal 2. In both cases, q is essentially counting the number of occupied states. In between, where the molecules are split between the two m_I states, q will lie somewhere between 1 and 2. Another example is illustrated in Fig. 2.13.

The canonical distribution (Eq. 2.30) is one of the two most important results we obtain from statistical mechanics,[8] and its significance extends throughout the rest of this text. For now, the canonical distribution gives the qualitative result that we shouldn't ordinarily expect to find molecules in very high energy states, because the $e^{-E/(k_B T)}$ term causes $\mathcal{P}(E)$ to rapidly approach zero as $E/(k_B T)$ becomes large. For a given system and value of T, the denominator—the partition function—is constant, and the numerator decreases exponentially with E. The numerator does increase with Ω, which in turn increases with E. We have seen, however, that for individual molecules, the degeneracy $g(\varepsilon)$ is not a highly sensitive function of the energy. At high energy, $\mathcal{P}(E)$ will be dominated by the exponential decay $e^{-E/(k_B T)}$.

We have a choice as to our zero of energy. If we measure all energies relative to some point E_0, then the probability distribution is (Eqs. 2.32 and 2.33)

$$\mathcal{P}(E) = \frac{g e^{-(E-E_0)/(k_B T)}}{\sum_E g e^{-(E-E_0)/(k_B T)}} \qquad (2.34)$$

$$= \left(\frac{g e^{-E/(k_B T)}}{\sum_E g e^{-E/(k_B T)}} \right) \left(\frac{e^{-E_0/(k_B T)}}{e^{-E_0/(k_B T)}} \right)$$

$$= \frac{g e^{-E/(k_B T)}}{\sum_E g e^{-E/(k_B T)}}. \qquad (2.35)$$

---

[8]The other being the ergodic hypothesis leading directly to the second law of thermodynamics, but we'll get to that in Chapter 9.

TOOLS OF THE TRADE | Pressure Measurement

At the end of this chapter we obtain the ideal gas law, one of the most elegant relationships in chemistry, which relates the pressure of a gas to other parameters. But what *is* pressure?

Among Galileo Galilei's many contributions to science, he was the first (in 1594) to discover that water could not be pulled upward by suction over a distance greater than about 10 meters. Suction pumps for moving water up out of rivers for irrigation or up out of flooding mines were limited to stages of no more than 10 meters of vertical displacement each, no matter how well the pump was engineered. Fifty years after Galileo's discovery, Evangelista Torricelli made the conceptual leap that a vacuum was the absence of matter and made the first measurement of air pressure using a column of mercury. Torricelli resolved the mystery of the water suction pump, showing that it is this force of the atmosphere that pushes water up a pipe toward the pump and that the water stops moving when the force of the atmosphere is equal to the weight of the water in the pipe. Several experiments followed on the heels of Torricelli's work, including work by Blaise Pascal in France and Isaac Newton's mentor Robert Boyle in England, rapidly expanding our understanding of the nature of gases and liquids and how they interact with their surroundings.

Why do we measure pressure? We use pressure measurement chiefly for two purposes: to determine the amount of a gas, and as an indicator of the force exerted by a substance. The ideal gas law tells us that the pressure of a gas is proportional to the amount of gas present, if we keep volume and temperature constant. Whereas we can weigh solids and liquids to determine how much material is present, gases are too buoyant in the air to weigh under typical conditions for their weight to be helpful. Instead, we infer the amount of a gas most often from its gas pressure, which can be measured mechanically by a simple device. Furthermore, because pressure is proportional to force, it also provides a key measurement in predicting the motions of substances. Two interacting subsystems tend to evolve so as to make the pressures equal, advancing toward a point where the forces cancel and the net motion stops. This principle guides the motion of high pressure weather fronts into low pressure regions, the rates of water flow in plumbing and airflow in ventilation systems, and appears in countless other forms throughout nature and engineering.

What can we use to measure pressure? More than a dozen methods exist for measuring pressures, differing in cost, accuracy, measurement range, and sensitivity to environmental factors (such as high temperature or chemical effects). The following table summarizes properties of a few techniques or devices for pressure measurement.

device	operating range (Pa)	operating range (common units)
fluorescence shift	1–100 GPa	10–1000 kbar
bulk modulus cell	0.01–1.4 GPa	0.1–13 kbar
capacitance manometer	0.1–1000 Pa	0.001–1,000 torr
thermocouple gauge	0.1–100 Pa	1–1000 mtorr
ion gauge	10^{-9}–10^{-2} Pa	10^{-11}–10^{-4} torr

How do these devices work? Next we take a quick look at the mechanisms behind a few methods for measuring pressures in the laboratory.

1. *Fluorescence shift.* The highest pressures achieved in the laboratory use tiny volumes trapped between diamond crystals (see *Tools of the Trade*, Chapter 7). At pressures on the order of 10 GPa (10^5 bar), the forces are capable of deforming the structure of other crystals, introducing small but significant alterations in the electron distributions, which in turn shift the fluorescence wavelengths of the crystal. The pressure-dependent shifts in fluorescence of several crystals have been calibrated, allowing pressure measurements in small volumes at very high pressures. Ruby was originally used for this purpose, but other crystals such as samarium-doped yttrium aluminum garnet have since been found to reach higher pressures.

2. *Bulk modulus cell.* This device measures pressures in the range of thousands of bar by completely mechanical means: the high pressure pushes a rod (the probe) into a housing against an elastic packing material (see figure). A narrow stem, connected to the probe, extends out the other side of the cell, and the pressure is measured by the deflection of the stem.

3. *Capacitance manometer.* In electronics, capacitors are components used to allow high-frequency electrical signals to pass while blocking low-frequency or DC signals. The greater the capacitance, the lower the

resistance to a high-frequency signal. A capacitor can be constructed by positioning two conducting plates parallel to one another, separated by a small distance d, and the capacitance is then inversely proportional to d. In a capacitance manometer (see figure), one side of the capacitor is a rigid plate, while the other side is a flexible metal diaphragm, exposed to the system being studied. Higher pressures push the diaphragm toward the plate, decreasing the capacitance. These are reliable devices, and work down to the lowest pressures where mechanical measurements are still usable.

4. *Thermocouple gauge.* A thermocouple gauge does not measure pressure directly. Instead, it measures the thermal conductivity of a gas in a regime where the conductivity increases with gas pressure. Inside a small tube, electrical current runs through a heater coil, which heats the surrounding gas. A short distance away is a thermocouple wire—a wire with a temperature-dependent resistance. The higher the pressure, the higher the temperature recorded at the thermocouple. Thermocouple gauges are much less

expensive than capacitance manometers but must be calibrated for the correct gas. Helium, for example, has a thermal conductivity much higher than air, and a thermocouple gauge calibrated for air will report pressures more than double the correct value.

5. *Ion gauge.* At pressures below about 0.1 Pa, gases exert too little force for mechanical measurements of the pressure to be possible. The ion gauge operates in a fashion similar to the thermocouple gauge, but in place of the heater coil goes a filament. When current runs through the filament, hot electrons are generated and ionize the surrounding gas. The ions are then accelerated by an electric field onto the collector—a solitary wire, which runs to a current meter outside the tube. Because current measurements can be made with high sensitivity, this method works to pressures as low as 10^{-8} Pa. Also like the thermocouple gauge, however, the method suffers from being an indirect measurement. Helium is a notorious culprit again, but this time (because its ionization energy is much higher than most other gases) helium registers a pressure a factor of two or more *lower* than the actual value.

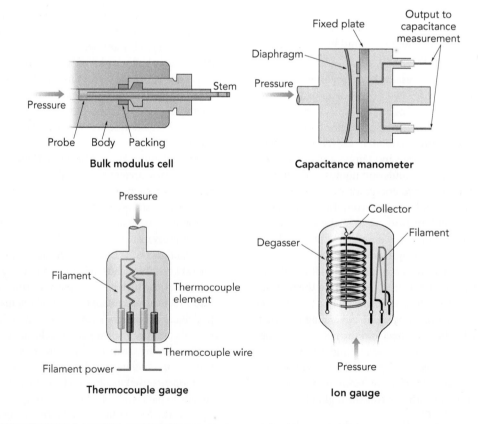

Bulk modulus cell

Capacitance manometer

Thermocouple gauge

Ion gauge

The choice of reference point for the energy, as we would hope, does not alter the probability distribution. For example, our answer in Example 2.1 did not depend on independent quantities E_1 and E_0, but only on $E_1 - E_0$. We have chosen our reference energy to be the *ground state* of the system; in other words, we set the lowest possible energy of the system to zero. This is different from other conventions, such as measuring the energy relative to the separated molecule limit, or relative to the minimum *potential* energy but it differs only by an additive constant.

2.5 The Ideal Gas Law

Let's finally get to the original problem—the pressure of an ideal gas. Let our container be rectangular, as drawn in Fig. 2.14, with one dimension having length s and the faces perpendicular to this dimension having area A, so the volume $V = As$. The pressure P of the gas expresses the force F pushing against the walls of the container per unit area of wall. As long as the container walls remain immovable, there must be a cancellation of forces; that is, there's an equal and opposite normal force per unit area, an equal pressure P. Let us momentarily free one wall from the rest of the box and reduce its force so that it exerts a slightly lower pressure, P_{min}, on the gas.[9]

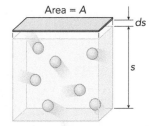

▲ FIGURE 2.14 **Container used for derivation of the ideal gas law.**

The gas, having a greater force per unit area, pushes the wall out an infinitesimal distance ds.

The force F that must be overcome is $F = P_{min}A$, and the energy expended in pushing the wall is

$$dE = -Fds = -P_{min}Ads = -P_{min}dV. \qquad (2.36)$$

Because $V = As$, $dV = Ads$ and

$$dE = -P_{min}dV. \qquad (2.37)$$

If we hold other parameters constant, in particular N and S, then

$$P_{min} = -\left(\frac{\partial E}{\partial V}\right)_{S,N}. \qquad (2.38)$$

In all but a few cases in this text, we will require expansion and compression of the gas to take place under conditions where P and P_{min} are nearly equal in magnitude, allowing us to focus on parameters of the gas itself, rather than its surroundings:

$$\text{if } P = P_{min}: \quad P = -\left(\frac{\partial E}{\partial V}\right)_{S,N}. \qquad (2.39)$$

Although this pressure is determined only by the translational energy of the gas sample, we need not assume that this is the only kind of energy present. There may still be rotational, vibrational, and electronic contributions to the energy which, having no dependence on the volume of the sample, do not contribute to the derivative in Eq. 2.39.

---

[9]We use P_{min}, because the energy needed to move the wall depends only on the force that resists the motion, which is determined by whichever of the two pressures, inside the box or outside the box, is lower. If the pressure outside the box were made slightly higher, then the pressure of the gas would be P_{min}, the wall would move inward a small distance, but Eq. 2.37 would still be valid. Many texts adopt the convention of calling this pressure P_{ext}.

We don't have any statistical equation for $(\partial E/\partial V)_{S,N}$, but we can take advantage of the cyclic rule for partial derivatives (Table A.4),

$$\left(\frac{\partial X}{\partial Y}\right)_Z = -\left(\frac{\partial X}{\partial Z}\right)_Y \left(\frac{\partial Z}{\partial Y}\right)_X, \tag{2.40}$$

to get two derivatives that we *can* evaluate:

$$P = -\left(\frac{\partial E}{\partial V}\right)_{S,N} = \left(\frac{\partial E}{\partial S}\right)_{V,N}\left(\frac{\partial S}{\partial V}\right)_{E,N}. \tag{2.41}$$

We have Eq. 2.24 for the derivative of the energy with respect to *entropy*,

$$T = \left(\frac{\partial E}{\partial S}\right)_{V,N}, $$

but how are we going to evaluate the derivative of S with respect to V?

To accomplish this, we can apply a classical model to our system. Let the translational states of the gas correspond to all the possible ways the gas particles can be arranged inside our box, *regardless of what energy the gas has*. The box is of volume V. Let's break that volume up into units V_0, each one just big enough to hold a single particle. It doesn't matter exactly what size these units are, as long as they are much smaller than the volume overall. Count how many of these units there are in the box and set that equal to M, such that

$$V = MV_0.$$

Next we ask how many possible ways we can put N indistinguishable particles in these M locations. For one particle, there are M options. For two particles, there are M options for the first particle, and for each of those there are $M-1$ options for the second particle, which gives a product of $M(M-1)$ states. But because the particles are indistinguishable, half of these states are the same as the other half except with the identities of particles 1 and 2 reversed. Since we can't tell the difference between particles 1 and 2, there are really only $M(M-1)/2$ distinct states.

If we extend this process, we find that the number of ways Ω of arranging N particles in M slots is $M(M-1)(M-2)\ldots(M-N+1)$, and that we should divide by $N!$ to eliminate all the states that are the same when the particles are indistinguishable. If the system is a gas, then the distances between the particles are, on average, much greater than the sizes of the particles, so we can require that $M \gg N$. We also introduce a constant A to absorb any of the other parameters that affect Ω, such as the energy. The total number of states we arrive at in this way gives us a measure of how the number of translational states available to the gas depends on the volume of the box:

$$\Omega = \lim_{M \gg N} A\frac{1}{N!}M(M-1)(M-2)\ldots(M-N+1) = A\frac{1}{N!}M^N,$$

and because $V = MV_0$, this becomes

$$= A\frac{1}{N!}\left(\frac{V}{V_0}\right)^N = A\frac{1}{V_0^N N!}V^N. \tag{2.42}$$

We will derive a more rigorous expression for Ω in Chapter 3, but Eq. 2.42 has the correct form for our purposes at present. Ignoring everything but the volume-dependence in this equation, we find we can write the entropy as

$$S = k_B \ln \Omega = k_B \ln[\,constant\, V^N] = k_B[\ln constant + N \ln V], \tag{2.43}$$

which gives

$$\left(\frac{\partial S}{\partial V}\right)_{E,N} = k_B N \left(\frac{\partial \ln V}{\partial V}\right)_{E,N} = k_B \frac{N}{V}. \tag{2.44}$$

Substituting the results from Eqs. 2.24 and 2.44 into Eq. 2.41 gives the pressure exerted on the walls of the container:

$$P = \left(\frac{\partial E}{\partial S}\right)_{V,N} \left(\frac{\partial S}{\partial V}\right)_{E,N} = \frac{k_B T N}{V}. \tag{2.45}$$

The Boltzmann constant k_B multiplied by Avogadro's number $\mathcal{N}_A$ gives the gas constant R, so we can also write

$$PV = nRT, \tag{2.46}$$

where n is the number of moles. This is the *ideal gas law*.

As promised, a few new tools and concepts have turned up along the way in this discussion:

- The ensemble is only a convention for describing all the possible configurations of a given system, keeping some set of parameters (e.g., V, N, and E) fixed.
- The distribution function of one of the microscopic parameters (e.g., the energy ε) describes the behavior of individual particles only up to the point necessary to derive macroscopic parameters. All the information about the specific values of those parameters for individual particles is lost.
- The temperature and the distribution function of quantum states in a particular coordinate, such as translation or rotation, give the partition function in that coordinate, which is a measure of how many states are accessible at that temperature.

These concepts are useful in the description of more detailed properties of matter, such as how the molecular structure of a compound results in deviations from the ideal gas behavior described in this chapter.

CONTEXT   *Where Do We Go From Here?*

This introduction to statistical mechanics presents, with the example of the ideal gas, a modern strategy for bridging the microscopic and macroscopic realms of matter:

1. Predict the energies and other properties of individual particles or small groups of particles using the laws of quantum mechanics.
2. Use the resulting energies to obtain the quantum state probability distribution and the partition function.
3. Use the partition function to solve for values of the thermodynamic parameters (such as pressure and entropy) that describe the bulk system.

This is a *remarkably* powerful methodology, enabling us to cover the 23 orders of magnitude difference in mass and volume and energy as we move from the single molecule to the mole.

However, we skipped some of the other rewarding aspects of statistical mechanics in the meantime, including methods for treating more complex particles that have rotational and vibrational degrees of freedom as well as translations. Chapter 3 will show us how we can effectively separate those contributions to the energy, an approach that serves us well when we address heat capacities and other macroscopic parameters in later chapters. Furthermore, in restricting our work to the ideal gas we've applied the strategy outlined earlier only to the simplest form of matter there is. In Chapter 4 we take a look at how to carry out the same extrapolation when there are strong interactions between the particles. Extending the results of quantum mechanics to predict properties of the bulk remains a significant challenge in current research, but the methods we see in these chapters show how the problem is solvable. And this problem—how the microscopic structure of atoms and molecules determines their bulk behavior—is the very essence of chemistry.

KEY CONCEPTS AND EQUATIONS

2.1   **Properties of the microscopic world.** The electronic, vibrational, and rotational energies of individual atoms and molecules are limited to certain values by the laws of quantum mechanics. Only certain quantum states are available to the translational (center of mass) motion of the molecule as well, but there are so many of these and they are so close together in energy that the translational energy appears to be a continuous variable. We will be using the following approximate equations for the vibrational and rotational energies:

$$E_v = \left(v + \frac{1}{2} \right) \omega_e, \qquad (2.8)$$

$$E_J = BJ(J + 1), \qquad (2.9)$$

where the vibrational degeneracy will normally be 1, and the rotational degeneracy for a linear molecule is $2J + 1$.

2.3   **Entropy.** The entropy is a parameter used to measure the number of distinct quantum states Ω in the system. Our rigorous definition of the entropy will be the Boltzmann entropy:

$$S \equiv k_B \ln \Omega, \qquad (2.11)$$

but under conditions where we already know the probabilities of being in the available quantum states,

the entropy may be easier to calculate using the formula for Gibbs entropy:

$$S = -Nk_B \sum_{i=1}^{k} \mathcal{P}(i) \ln \mathcal{P}(i). \qquad (2.16)$$

2.4   **Temperature and the Partition Function.** The temperature is defined by

$$T \equiv \left(\frac{\partial E}{\partial S} \right)_{V,N}, \qquad (2.24)$$

and the ratio $\varepsilon/(k_B T)$ determines the likelihood that any given particle in the system will attain an energy ε when the system is at temperature T. This likelihood is given by the canonical probability distribution

$$\mathcal{P}(\varepsilon) = \frac{g(\varepsilon)e^{-\varepsilon/(k_B T)}}{q(T)}, \qquad (2.32)$$

where the single-particle partition function is

$$q(T) = \sum_{\varepsilon=0}^{\infty} g(\varepsilon)e^{-\varepsilon/(k_B T)}, \qquad (2.33)$$

and where g is the degeneracy of the energy level.

KEY TERMS

- The **system** is whatever we are studying, usually a collection of a large number of particles (atoms or molecules). Depending on how the system has been set up, it may be able to exchange energy or other properties with a **reservoir.**

- An **ensemble** is an imaginary collection of all the possible distinct quantum states of our system. Each of those quantum states is called a **microstate.**
- **Macroscopic properties** describe the entire system at once, rather than individual particles, and are equal

to the average evaluated over all of the ensemble states. Those properties may be divided into **extensive** parameters (such as volume and energy, which are summed over the properties of the individual particles) and **intensive** parameters (such as pressure and density, which are averaged over the particles).

- Ensembles may be constructed assuming that certain macroscopic parameters are held constant. The **microcanonical ensemble** is the set of all microstates with the same values of E, V, and N. The **canonical**

ensemble is the set of all microstates with the same values of T, V, and N.

- The **entropy** is a macroscopic parameter that effectively counts the number of ensemble states.
- The **temperature** T defines the **thermal energy** $k_B T$. The probability that a particle will have a given energy ε rapidly decreases as that energy increases beyond the thermal energy.
- The **ideal gas** is a system of particles that do not interact with each other at all.

OBJECTIVES REVIEW

1. *Calculate the entropy of a system from the degeneracy of states using the Boltzmann equation and from probability expressions using the Gibbs equation.*
   Find the entropy in $J\,K^{-1}$ of a system of five distinguishable molecules in which the total number of microstates is 7776 (using Eq. 2.11 for the Boltzmann entropy), and for the same five molecules if they each have an equal probability of being found in any of six quantum states (using Eq. 2.16 for the Gibbs entropy).

2. *Use energy and degeneracy expressions to calculate partition functions at some given temperature.*
   Calculate the partition function to three significant digits of a system with energy $\varepsilon_n = (100\ K)\,k_B n^2$ and degeneracy $g_n = 3n + 1$ at a temperature of 298 K.

3. *Use the canonical distribution to calculate quantum state populations.*
   What is the probability of finding an N_2 molecule in a non-degenerate quantum state at an energy of $2.2 \cdot 10^{-22}\,J$ at a temperature of 373 K? Set $Q = 1205$.

PROBLEMS

Fundamental Principles of Statistical Mechanics

2.1 The probability of finding a diatomic molecule with vibrational constant $\omega_e = 200\ cm^{-1}$ in the $v = 1$ vibrational state at a temperature of 300 K is which of the following?

a. 0   b. $2.3 \cdot 10^{-8}$ c. 0.24

d. 1   e. 2.6

2.2 Estimate the degeneracy of two argon atoms in a container of volume $1.00 \cdot 10^{-9}\,m^3$ (one cubic millimeter) with a total energy of $2.00 \cdot 10^{-20}\,J$ (roughly liquid nitrogen temperature) and an energy precision of $2.5 \cdot 10^{-26}\,J$ (≈ 2 mK).

2.3 The degeneracy of a system of N_A identical molecules A in a three-dimensional box has the form

$$g = V^{N_A} f(E_A, N_A).$$

If we add N_B more molecules of a different substance B, keeping the volume constant, what is the new equation for the degeneracy? Assume there are no intermolecular interactions.

2.4 If all the states of an ensemble are equally probable, then a 0.05 mol sample of an ideal gas at fixed energy in a 1000 cm^3 container has a chance of being found in a state where a specified 1% region of the volume is not occupied by any molecules. Estimate the probability of finding the

sample in such a state. THINKING AHEAD ▶ [Imagine that the container is a balloon filled with air—what would happen when the balloon encountered this state?]

2.5 Solve the derivative to find an algebraic equation in terms of ε for $\beta = 1/k_B T$ of one particle in a three-dimensional box, letting $d\varepsilon = \varepsilon/n^2$.
THINKING AHEAD ▶ [Should this function increase or decrease with the value of n^2?]

2.6 A system and reservoir are in contact with each other. The volumes and numbers of particles are constant for both, and the overall energy of the universe is conserved. The degeneracy of the system obeys the equation $\Omega = AE^2$, while for the reservoir, $\Omega_r = A_r E_r^{20}$. If the system has an energy of 1.00 J, what is the energy of the reservoir E_r when the system and reservoir are at the same temperature?

2.7 A sample of He^+ ions is prepared at a temperature of 298 K and a density of $2.0 \cdot 10^{10}\,cm^{-3}$. Because these are ions, magnetic fields may be used to trap those ions with kinetic energy less than some threshold value ε_t. The faster-moving ions escape the trap and are lost. Our sample is placed in such a trap with a threshold energy $\varepsilon_t = 1.00 \cdot 10^{-20}\,J$, the fast ions are lost, and the sample is allowed to return to equilibrium (so the temperature is well-defined again). Write an expression for the new temperature of the sample.

Entropy

2.8 The two-particle subsystem A has four possible states, all equally likely. Subsystem B, also with two particles, also has four different possible states, but the probabilities of two of the states are equal to *twice* the probabilities of the other two states; in other words, two states have probability x and two have probability $2x$. Calculate the Gibbs entropy for A and for B.

2.9 Show that in any microcanonical ensemble (having fixed E, V, and N), the Gibbs entropy is identical to the Boltzmann entropy. Do this by writing the probability $\mathcal{P}(i)$ of being in a single state i of the ensemble in terms of Ω.

Partition Functions

2.10 For the partition function Q of *any* typical degree of freedom, find the following values:

a. $\lim_{T \to 0} Q =$

b. $\lim_{T \to \infty} Q =$

2.11 Sodium dimer (Na_2) has a vibrational constant of 159.13 cm^{-1}. Calculate the vibrational partition function of Na_2 to three significant figures at 300.0 K. The degeneracy is 1.

The Canonical Distribution

2.12 The energy level diagram for the lowest energy terms, ^3P and ^1D, of N^+ is drawn in the following figure. The degeneracy at each energy level is $2J + 1$, the number of individual M_J states. For $T = 500$ K, give the term symbol and J value for the energy level that

a. has the highest population *overall*.

b. has the highest population in each of its *individual quantum states*.

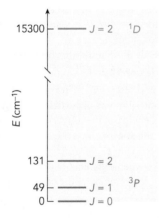

2.13 Atomic carbon has a ground term state ^3P, which is split into three spin-orbit or fine structure levels: the ground state $J = 0$; $J = 1$, which is 16.4 cm^{-1} above the ground state; and $J = 2$, which is 43.4 cm^{-1} above the ground state. The next term state is the $^1D\, J = 2$ state, at an energy 11460 cm^{-1} above the ground state. Recall that the degeneracy of each level is $2J + 1$, the number of M_J states for that level. Atomic carbon is often observed by astronomers in interstellar clouds at temperatures of about 100 K. Calculate the fraction of carbon atoms in each of these four states at $T = 100$ K, assuming that no other states are populated. THINKING AHEAD ▶ [Is the temperature high enough to get many atoms into the $J = 2$ state?]

2.14 In a one-mole sample of H atoms at 1000 K, how many would be in the $2s$ state? THINKING AHEAD ▶ [Does 1000 K seem like a lot of energy for electronic excitation of an atom?]

2.15 In a sample of CO gas ($B_e = 1.9313 \text{ cm}^{-1}$), at what temperature are there half as many molecules in the $J = 0$ rotational level as the $J = 1$ level? THINKING AHEAD ▶ [Qualitatively, how does this ratio vary with temperature?]

2.16 The vibrational partition function for an unknown diatomic molecule at 300 K is 2.33. Find the maximum value of the vibrational constant ω_e in cm^{-1} such that at least 5% of the molecules in a 300 K sample will be found in the $v = 1$ state. THINKING AHEAD ▶ [Does this value of the partition function suggest an approximate value for ω_e?]

2.17 Find the ratio of the number of HF molecules at 300 K in the state $v = 0$, $J = 0$ to the number in the state: (a) $v = 0$, $J = 10$; (b) $v = 1$, $J = 0$; and (c) $v = 10$, $J = 0$, given $\omega_e = 4138.5 \text{ cm}^{-1}$ and $B_e = 20.939 \text{ cm}^{-1}$.

2.18 The ground state $1s^1$ H atom has two hyperfine states, $F = 0$ and $F = 1$, which correspond to different relative orientations of the nuclear spin and electron spin. The transition between these two states has a wavelength of 21 cm and is one of the most important transitions in observational astronomy. The shape of our galaxy and many of its other properties were determined by measurements of this transition. Calculate the fraction of H atoms in each of these states at a temperature of 100 K. The $F = 1$ state is the ground state and there are no other states with significant population.

2.19 If the vibrational partition function of an O_2 sample is 1.045, what percentage of the molecules is in the state $v = 0$?

3

Partitioning the Energy

LEARNING OBJECTIVES

After reading this chapter, you will be able to do the following

❶ Estimate translational, rotational, and vibrational energies as a function of temperature by means of the equipartition principle.

❷ Employ the Taylor series solution to the vibrational partition function to estimate vibrational state populations.

❸ Employ the integral approximation to the rotational partition function to estimate rotational state populations in linear molecules.

❹ Apply the classical average value theorem to determine mean properties of the system related to these degrees of freedom.

GOAL *Why Are We Here?*

The goal of this chapter is to organize our treatment of translations, rotations, and vibrations of molecules so that we can determine the contributions each of these degrees of freedom will make to the macroscopic properties of our system.

CONTEXT *Where Are We Now?*

Chapter 2 presents the ideal gas law, which describes the relationships among the pressure, volume, temperature, and amount of material in a system composed of a large number of non-interacting particles. For such a drastic approximation, the ideal gas provides a surprisingly useful and accurate model of matter. But we're ready to move on to more sophisticated systems, where the molecules not only interact with one another, but store energy in degrees of freedom such as rotation and vibration, as well as the translational motions considered in Chapter 2. In this chapter we will see how we can crudely estimate the energy in each of these degrees of freedom and also determine their partition functions. The effort we spend in this chapter on developing expressions for partition functions is well invested because the partition functions summarize the key details of the molecular structure that we need to carry forward into the macroscopic world. Once we've derived the *general* laws of chemical thermodynamics and kinetics, the partition functions will contain the information that predicts how the behavior of one chemical species differs from another. A lot of chemistry resides in these mathematical expressions.

| SUPPORTING TEXT | *How Did We Get Here?*

- Any two molecules of the same chemical structure and isotopic composition are indistinguishable from one another once we allow them to interact. Each molecule then has an equal chance of being in a particular position or having a particular energy. That indistinguishability makes a large group of molecules a *perfect* sample for the application of statistics. On this basis, we find that the probability of any molecule being at an energy ε in a macroscopic sample with well-defined temperature T is given by the canonical distribution (Eq. 2.32):

$$\mathcal{P}(\varepsilon) = \frac{g(\varepsilon)e^{-\varepsilon/(k_B T)}}{q(T)},$$

where $g(\varepsilon)$ gives the degeneracy of the energy level and the partition function $q(T)$ is (Eq. 2.33):

$$q(T) = \sum_{\varepsilon=0}^{\infty} g(\varepsilon)e^{-\varepsilon/(k_B T)}.$$

- In Section 2.1, we summarize highlights of the quantum mechanics of the particle in a three-dimensional box. Equations 2.3 and 2.4 for the energies are useful:

$$
\begin{aligned}
E_{n_x,n_y,n_z} &= \frac{\pi^2\hbar^2}{2m}\left(\frac{n_x^2}{a^2} + \frac{n_y^2}{b^2} + \frac{n_z^2}{c^2}\right) \\
&\approx \frac{\pi^2\hbar^2}{2m(abc)^{2/3}}(n_x^2 + n_y^2 + n_z^2) \\
&= \frac{h^2}{8mV^{2/3}}n^2 \equiv \varepsilon_0 n^2.
\end{aligned}
$$

These quantum states may be *degenerate,* meaning that more than one state may have the same energy. Counting the number of available quantum states at some energy becomes a critical exercise for predicting the thermodynamics of any system.

- Although the molecules may be indistinguishable from one moment to the next, at any given moment each molecule in the sample will have a unique position and velocity. Across our sample, the positions and speeds and other parameters will have *distributions* of possible values, functions that give the fraction of molecules in our sample with that particular value. For example, $\mathcal{P}_x(x)$ is the algebraic function of x such that $\mathcal{P}_x(5.0 \text{ cm})$ tells us what fraction of molecules per unit x are at the point $x = 5.0$ cm. For a parameter A that we can calculate from x, we use the classical average value theorem, Eq. 2.2, to obtain the mean value of $A(x)$ from the probability distribution $\mathcal{P}_x(x)$:

$$\langle A \rangle = \int_{-\infty}^{\infty} \mathcal{P}_x(x)A(x)dx.$$

- Section 2.1 describes how the vibrations of atoms bound together by a chemical bond may be represented by a harmonic oscillator, which is defined by the potential energy function

$$U(x) = \tfrac{1}{2}kx^2,$$

which results in an energy (Eqs. 2.8 and 2.9)

$$E_v = \hbar\sqrt{\frac{k}{\mu}}\left(v + \frac{1}{2}\right) \equiv \omega_e\left(v + \frac{1}{2}\right),$$

where v is any non-negative integer and where the vibrational constant ω_e is typically between 200 and 3000 cm^{-1}.

- Section 2.1 also describes the rotational energy levels of linear molecules, which are given by (Eq. 2.9, now with an added correction)

$$E_J = BJ(J + 1) - D[J(J + 1)]^2, \tag{3.1}$$

and which have a degeneracy of $2J + 1$, where J is any non-negative integer. The rotational constant B is on the order of 10 cm^{-1} or less, and the distortion constant D is a relatively small correction.

- As we cross the boundary between the microscopic scale of individual molecules and bulk systems, the correspondence principle (Section 2.1) serves to remind us how the density of quantum states must increase—approaching the classical limit—as the mass, energy, and domain of the particles grow.

- We encounter a number of integrals over exponential and Gaussian functions in this chapter. The indefinite integrals do not have analytical solutions, but the definite integrals do, provided that both limits of integration are chosen from the set $-\infty$, 0, and ∞. The analytical solutions we will use are all available in Table A.5, and the ones we need most are the following:

$$\int_0^\infty e^{-ax^2}dx = \frac{1}{2}\left(\frac{\pi}{a}\right)^{1/2} \qquad \int_0^\infty x^2 e^{-ax^2}dx = \frac{1}{4}\left(\frac{\pi}{a^3}\right)^{1/2}.$$

3.1 Separation of Degrees of Freedom

When we analyze the properties of a single molecule in quantum mechanics, we simplify the problem by separating the degrees of freedom as much as we can. Typically, we treat the motions of the electrons as a separate problem from the motions of the nuclei. If we have N_{atom} nuclei, then there must be $3N_{atom}$ total degrees of freedom, because each nucleus can move along X, Y, and Z. However, when we connect the atoms by chemical bonds, we constrain the motions, and it becomes appropriate to break the nuclear motions down into

- translations (motion of the center of mass along X, Y, and Z: 3 coordinates).
- rotations (changes in the overall orientation of the molecule: 2 angles are needed to specify the orientation of a linear molecule, 3 for a nonlinear molecule).
- vibrations (relative motions of the atoms: all the remaining coordinates, so $3N_{atom} - 5$ for a linear molecule, $3N_{atom} - 6$ for nonlinear).

We tend to continue this separation of the degrees of freedom in the macroscopic limit, because some degrees of freedom absorb energy more readily than others, and the amount of energy that each degree of freedom can store varies.

Chemical processes can be seen as largely a matter of energy transfer among different degrees of freedom, so we return to a fundamental question of Chapter 2: how does the energy in a macroscopic sample get distributed among the molecules? The canonical distribution provides an answer: in a system with a well-defined temperature, the distribution of energies ε among individual molecules is given by Eq. 2.32:

$$\mathcal{P}(\varepsilon) = \frac{g(\varepsilon)e^{-\varepsilon/(k_BT)}}{q(T)},$$

where $g(\varepsilon)$ is the degeneracy of that energy level and $q(T)$ is the partition function for that particular degree of freedom (Eq. 2.33):

$$q(T) = \sum_{\varepsilon=0}^{\infty} g(\varepsilon)e^{-\varepsilon/(k_BT)}.$$

The canonical probability function allows us to separate the energies in different degrees of freedom in a single step (although we will take a few steps to show why that works). First, to simplify the notation, let's drop the "(ε)" and just remember that the degeneracy g is a function of the energy. The degeneracy g is the product of the degeneracies in each degree of freedom, so

$$g = g_{\text{vib}}g_{\text{rot}}g_{\text{trans}}. \tag{3.2}$$

Next, if it is true that the energies are separable, so that we may write

$$\varepsilon = \varepsilon_{\text{vib}} + \varepsilon_{\text{rot}} + \varepsilon_{\text{trans}}, \tag{3.3}$$

then the overall partition function is the product of the partition functions for the individual degrees of freedom:

$$\begin{aligned}
q(T) &= \sum_{\varepsilon=0}^{\infty} g(\varepsilon)e^{-\varepsilon/(k_BT)} \\
&= \sum_{\varepsilon_{\text{vib}}=0}^{\infty}\sum_{\varepsilon_{\text{rot}}=0}^{\infty}\sum_{\varepsilon_{\text{trans}}=0}^{\infty} g_{\text{vib}}g_{\text{rot}}g_{\text{trans}}\, e^{-(\varepsilon_{\text{vib}}+\varepsilon_{\text{rot}}+\varepsilon_{\text{trans}})/(k_BT)} \\
&= \sum_{\varepsilon_{\text{vib}}=0}^{\infty}\sum_{\varepsilon_{\text{rot}}=0}^{\infty}\sum_{\varepsilon_{\text{trans}}=0}^{\infty} g_{\text{vib}}g_{\text{rot}}g_{\text{trans}}\, e^{-\varepsilon_{\text{vib}}/(k_BT)}e^{-\varepsilon_{\text{rot}}/(k_BT)}e^{-\varepsilon_{\text{trans}})/(k_BT)} \\
&= \sum_{\varepsilon_{\text{vib}}=0}^{\infty}\left[g_{\text{vib}}e^{-\varepsilon_{\text{vib}}/(k_BT)}\right] \sum_{\varepsilon_{\text{rot}}=0}^{\infty}\left[g_{\text{rot}}e^{-\varepsilon_{\text{rot}}/(k_BT)}\right] \sum_{\varepsilon_{\text{trans}}=0}^{\infty}\left[g_{\text{trans}}e^{-\varepsilon_{\text{trans}}/(k_BT)}\right] \\
&= q_{\text{vib}}q_{\text{rot}}q_{\text{trans}}. \tag{3.4}
\end{aligned}$$

CHECKPOINT Here we re-introduce our color scheme that paints the canonical energy distribution red and the partition function blue. We will shortly see some specific expressions for the energies and partition functions, but the colors will show how they relate back to the fundamental canonical distribution derived in Section 2.4.

This result in turn allows us to break down the probability density for total energy ε into the product of the individual probabilities for vibration, rotation, and translation:

$$\begin{aligned}
\mathcal{P}(\varepsilon) &= \frac{g(\varepsilon)e^{-\varepsilon/(k_BT)}}{q(T)} \\
&= \frac{g_{\text{vib}}g_{\text{rot}}g_{\text{trans}}\, e^{-\varepsilon_{\text{vib}}/(k_BT)}e^{-\varepsilon_{\text{rot}}/(k_BT)}e^{-\varepsilon_{\text{trans}})/(k_BT)}}{q_{\text{vib}}q_{\text{rot}}q_{\text{trans}}} \\
&= \mathcal{P}_{\text{vib}}(\varepsilon_{\text{vib}})\mathcal{P}_{\text{rot}}(\varepsilon_{\text{rot}})\mathcal{P}_{\text{trans}}(\varepsilon_{\text{trans}}). \tag{3.5}
\end{aligned}$$

Each of those probability distribution functions we can evaluate if we know the vibrational, rotational, and translational energy level expressions.

But what good are these probability distributions if we want to find properties of the *entire* system and not just some particular energy? One way that we can summarize much of the information in $\mathcal{P}(\varepsilon)$ with a single value is by applying the classical average value theorem. Say that we measure a parameter $A(\vec{s})$, which is a function of a set of coordinates $\vec{s}$, and we find k different values, some of which we find more than once. The average value of $A(\vec{s})$ is the sum over each possible value A_i multiplied by the number of times N_i that value appears, divided by the total number of values N:

$$\langle A \rangle = \frac{\sum\limits_i N_i A_i}{\sum\limits_i N_i} = \sum_i \frac{N_i}{N} A_i.$$

Each value N_i/N is the probability that the value A_i will show up, so we can also write this average as

$$\langle A \rangle = \sum_i \mathcal{P}(A_i) A_i. \tag{3.6}$$

In the limit that A is a continuous parameter, we can extend the sum to an integral. We must then replace the discrete probability function $\mathcal{P}(A_i)$ with the continuous probability function $\mathcal{P}_{\vec{s}}(\vec{s})$ of being at the point $\vec{s}$. This probability function is no longer unitless, because for a continuous function we determine the probability of being not at a *specific* point $\vec{s}$, but within some range of values ds along each coordinate s. The probability we obtain is a probability *per unit s*. If $\vec{s}$ is a position in three-dimensional space, for example, then $\mathcal{P}_{\vec{s}}(\vec{s})$ is a probability per unit volume. We multiply this probability by the value of A at point $\vec{s}$ and integrate over all possible values of $\vec{s}$ to obtain the classical average value theorem, Eq. 2.2:

$$\langle A \rangle = \int_{\text{all space}} \mathcal{P}_{\vec{s}}(\vec{s}) A(\vec{s}) d\vec{s}.$$

When we use a probability function $\mathcal{P}_{\vec{s}}(\vec{s})$ that has units, we will also want a revised form of the partition function. Let s be a single coordinate. The probability of finding a molecule between the values s_1 and s_2 is still given by the canonical distribution, but now we integrate between those points:

$$\mathcal{P}(s_1 \leq s < s_2) = \int_{s_1}^{s_2} \mathcal{P}_s(s) ds = \int_{s_1}^{s_2} \frac{e^{-\varepsilon_s/(k_B T)}}{q'_s(T)} ds. \tag{3.7}$$

This probability, $\mathcal{P}(s_1 \leq s < s_2)$, *is* still a unitless quantity, but to make it unitless the function $q'_s(T)$ in the denominator needs to have the same units as s, which makes $q'_s(T)$ subtly different from the partition function defined in Chapter 2. We will indicate these derived partition functions with a prime, $q'(T)$. They still serve to normalize the probability distribution, ensuring that

$$\int_{\text{all space}} \mathcal{P}(s) s ds = \int_{\text{all space}} \frac{e^{-\varepsilon_s/(k_B T)}}{q'_s(T)} ds = 1, \tag{3.8}$$

and solving for $q'_s(T)$ we find

$$q'_s(T) = \int_{\text{all space}} e^{-\varepsilon_s/(k_B T)} ds, \tag{3.9}$$

which has units of s.

The probability functions we have in Eq. 3.5 are for the energies in different degrees of freedom, but each degree of freedom uses a different set of coordinates $\vec{s}$. Eventually we will want to focus on each of these degrees of freedom one at a time because they each convey different information about the molecule. But just to get started, an uncommonly convenient approximation takes advantage of a single thread that ties together the translational, rotational, and vibrational energies of the molecule.

3.2 The Equipartition Principle

Let us begin with the translational energy $mv^2/2$ of a single molecule of mass m moving at speed v. The translational energy in a macroscopic system is a continuous variable, which means we may use the classical average value theorem to find its mean value. The set of coordinates $\vec{s}$ that determine the translational energy is the set of velocity components v_X, v_Y, and v_Z. The total translational energy is

$$\frac{mv^2}{2} = \frac{mv_X^2}{2} + \frac{mv_Y^2}{2} + \frac{mv_Z^2}{2}. \tag{3.10}$$

The average translational energy is therefore

$$\left\langle \frac{mv^2}{2} \right\rangle = \left\langle \frac{mv_X^2}{2} + \frac{mv_Y^2}{2} + \frac{mv_Z^2}{2} \right\rangle$$

$$= \frac{m\langle v_X^2\rangle}{2} + \frac{m\langle v_Y^2\rangle}{2} + \frac{m\langle v_Z^2\rangle}{2}. \tag{3.11}$$

In the same way that we separated the energies in Eq. 3.3, these energies have been separated into contributions from motion along each Cartesian coordinate.

Let's first consider motion of a single molecule along only one of these coordinates, v_X, which parametrizes the kinetic energy $\varepsilon_X = mv_X^2/2$ for motion along the X coordinate. The probability for a molecule having a particular value of v_X is given by the canonical distribution, $\mathcal{P}_v(v_X) = e^{-mv_X^2/(2k_BT)}/q'_v(T)$, and we can substitute this expression into the average value theorem to obtain

$$\langle \varepsilon_X \rangle = \int_{-\infty}^{\infty} \mathcal{P}_v(v_X)\frac{mv_X^2}{2}dv_X$$

$$= \frac{m}{2}\frac{\displaystyle\int_{-\infty}^{\infty} e^{-mv_X^2/(2k_BT)}v_X^2 dv_X}{q'_v(T)}, \tag{3.12}$$

where

$$q'_v(T) = \int_{-\infty}^{\infty} e^{-mv_X^2/(2k_BT)}dv_X. \tag{3.13}$$

The integrals have Gaussian integrands of the form $x^{2n}e^{-ax^2}$, and when the limits of the integral are chosen from 0, $-\infty$, and ∞, we can solve each integral using one of the solutions in Table A.5:

$$\int_0^{\infty} e^{-ax^2}dx = \frac{1}{2}\left(\frac{\pi}{a}\right)^{1/2} \qquad \int_0^{\infty} x^2 e^{-ax^2}dx = \frac{1}{4}\left(\frac{\pi}{a^3}\right)^{1/2}.$$

Our integrals actually go from $-\infty$ to ∞, but as long as x^{2n} has an even exponent, the integral from $-\infty$ to 0 has the same value as the integral from 0 to ∞:

$$\int_{-\infty}^{\infty} x^2 e^{-ax^2} dx = \int_{-\infty}^{0} x^2 e^{-ax^2} dx + \int_{0}^{\infty} x^2 e^{-ax^2} dx \qquad \text{break integral at } x = 0$$

$$= \int_{0}^{\infty} (-x)^2 e^{-a(-x)^2} dx + \int_{0}^{\infty} x^2 e^{-ax^2} dx \quad \text{invert 1st integral limits and signs}$$

$$= 2\int_{0}^{\infty} x^2 e^{-ax^2} dx = \frac{1}{2}\left(\frac{\pi}{a^3}\right)^{1/2}.$$

Similarly,

$$\int_{-\infty}^{\infty} e^{-ax^2} dx = 2\int_{0}^{\infty} e^{-ax^2} dx = \left(\frac{\pi}{a}\right)^{1/2}.$$

Setting $x = v_X$ and $a = m/(2k_B T)$, these analytical solutions to the integrals give us the average translational energy for motion along X:

$$\langle \varepsilon_X \rangle = \left(\frac{m}{2}\right) \frac{\int_{-\infty}^{\infty} e^{-mv_X^2/(2k_B T)} v_X^2 dv_X}{\int_{-\infty}^{\infty} e^{-mv_X^2/(2k_B T)} dv_X}$$

$$= \left(\frac{m}{2}\right) \frac{\frac{1}{2}\left(\frac{\pi}{[m/(2k_B T)]^3}\right)^{1/2}}{\left(\frac{\pi}{m/(2k_B T)}\right)^{1/2}} = \left(\frac{m}{2}\right)\left(\frac{1}{2}\right)\left(\frac{2k_B T}{m}\right)$$

$$= \frac{k_B T}{2}. \tag{3.14}$$

The fascinating part of Eq. 3.14 is that the average energy for motion along X *only depends on the temperature.* It does not depend on the mass or the specific coordinate of the motion. In fact, for *any* energy ε_x that varies as the square of its coordinate, so that $\varepsilon_x = ax^2$, the math works out exactly the same way:

$$\langle \varepsilon_x \rangle = \frac{k_B T}{2}, \tag{3.15}$$

regardless of what a and x represent physically. And if we want to know the average value of x^2, we have only to divide this result by a:

$$\langle x^2 \rangle = \frac{1}{a}\langle \varepsilon_x \rangle = \frac{k_B T}{2a}. \tag{3.16}$$

We can apply these results widely because our energies always have a kinetic energy component that varies as some speed squared, v^2. Remember, however, that the integral solution to the partition function is strictly valid only when the probability distribution is a function of a continuous variable like the velocity component v_X, rather than a discrete variable like the vibrational quantum number (which must always be an integer). This integral approximation works if the energy spacing between quantum states is small compared to the thermal energy $k_B T$. Under those conditions, the probability function changes slowly enough from one value of the quantum number to the next that we can treat it like a continuous function.

The other components of the translational energy, for motion along Y and Z, are solved in the same way as for X, giving a total translational energy of

$$\left\langle \frac{mv^2}{2} \right\rangle = \frac{m\langle v_X^2 \rangle}{2} + \frac{m\langle v_Y^2 \rangle}{2} + \frac{m\langle v_Z^2 \rangle}{2} = \frac{3k_B T}{2}. \tag{3.17}$$

We can go further. For rotation, we can approximate $E_{rot} = BJ(J + 1)$ by $E_{rot} \approx BJ^2$ to get a form suitable for applying the same integral solution. The approximation introduces a fractional error of $1/J$, which is small at high temperatures where the rotational quantum number J tends to larger values. Every translational and rotational coordinate of a molecule therefore has average energy $k_B T/2$.

We're still not done. Consider also the vibrational energy in a diatomic under the harmonic approximation *when the thermal energy $k_B T$ is large compared to the energy spacing in that coordinate,* and assuming that the harmonic approximation is valid for the vibrational modes. Under these conditions, the energy depends on the squares of *two* parameters, one for the kinetic energy and one for the potential:

$$\varepsilon_{vib} = K + U = \frac{\mu v_{vib}^2}{2} + \frac{k}{2}(R - R_e)^2, \tag{3.18}$$

where v_{vib} is the relative speed of the two vibrating atoms in the diatomic, and $(R - R_e)$ is the displacement from the equilibrium bond distance. As a result, the average vibrational energy per molecule in a diatomic gas at high temperature T will be

$$\text{for } k_B T \gg \omega_e: \quad \langle \varepsilon_{vib} \rangle = \frac{k_B T}{2} + \frac{k_B T}{2} = k_B T. \tag{3.19}$$

If we can apply the integral solution, *every* vibrational coordinate, having both a kinetic and a potential term, has average energy $k_B T$.

We name this approximation the **equipartition principle,**

$$E = \frac{1}{2}N_{ep}Nk_B T = \frac{1}{2}N_{ep}nRT, \tag{3.20}$$

where N_{ep} is the number of equipartition energy contributions. Equation 1.17 also introduced a new constant R, the **universal gas constant,** which is defined as the Boltzmann constant multiplied by Avogadro's number:

$$R \equiv \mathcal{N}_A k_B = 8.3145 \, \text{J K}^{-1}\text{mol}^{-1}. \tag{3.21}$$

CHECKPOINT The equipartition principle is useful as a conceptual as well as a quantitative guide. It tells us that (to a first approximation) we distribute the total energy of our system equally into every kind of available motion or potential energy (hence the term *equipartition*). Molecules absorb energy based on how many different degrees of freedom are available to store the energy, rather than on the basis of mass or size.

The gas constant R will allow us to shift from thermal energy per molecule ($k_B T$) to thermal energy per mole (RT) as we deal more with the macroscopic properties.

The key to using the equipartition principle is coming up with a reasonable assessment of the number N_{ep}. The following are useful general rules to start from.

At room temperature, all translations and rotations contribute to N_{ep}.

At room temperature, only vibrations with roughly $\omega_e < 300 \text{ cm}^{-1}$ contribute completely to N_{ep}.

Vibrational motions with higher vibrational constants are less likely to be excited at room temperature and therefore do not contribute as much to the total stored energy of the molecule. For those motions that do contribute to N_{ep}, we may write

$$N_{ep} = \text{(number of translational degrees of freedom)}$$
$$+ \text{(number of rotational degrees of freedom)}$$
$$+ 2 \times \text{(number of vibrational degrees of freedom)}. \quad (3.22)$$

Let's calculate the total energies predicted by the equipartition principle per molecule in vibration, rotation, and translation for gases composed of (a) monatomics (such as He or Ar gas), (b) diatomic molecules (such as N_2 or CO gas), (c) nonlinear triatomics (such as SO_2 gas), and (d) linear triatomics (such as CO_2 or HCN gas). Assume for now that all the vibrations contribute to the energy. Each gas has a contribution $3k_BT/2$ from translation. The linear molecules (including the diatomics) each have $3N_{atom} - 5$ (equal to 1 for diatomics and 4 for linear triatomics) vibrational modes; the nonlinear triatomics have $3N_{atom} - 6 = 3$. The linear molecules each have two rotational coordinates; the nonlinear have three. For each vibrational coordinate, there is a contribution of k_BT (counting both the kinetic and potential terms). For each rotational coordinate, there is only a kinetic contribution of $k_BT/2$. Therefore, the predictions are

Gas		Trans	Rotation	Vibration	Total energy
monatomic:	$E = \dfrac{k_BT}{2} \times ($	3	+0	+0	$) = \dfrac{3}{2}k_BT$
diatomic:	$E = \dfrac{k_BT}{2} \times ($	3	+2	$+1 \times 2$	$) = \dfrac{7}{2}k_BT$
linear triatomic:	$E = \dfrac{k_BT}{2} \times ($	3	+2	$+4 \times 2$	$) = \dfrac{13}{2}k_BT$
nonlinear triatomics	$E = \dfrac{k_BT}{2} \times ($	3	+3	$+3 \times 2$	$) = \dfrac{12}{2}k_BT = 6k_BT$

At the same temperature T, a diatomic gas will have more energy than a monatomic gas with the same number of particles, because the diatomic molecule has more degrees of freedom in which to store the energy. The amount of energy predicted by equipartition to go into each degree of freedom for one-, two-, and three-atom molecules is shown in Fig. 3.1. By the time we get to the linear triatomic, vibrations could account for more than half of the energy-carrying capacity of the molecule. If, that is, vibrations were as easily excited as rotations and translations.

However, the equipartition principle relies on the accuracy of the integral approximation for the energy, and therefore the temperature has to be high enough that the integral is a reasonable approximation to the distribution function in that coordinate. At typical lab temperatures, *few vibrational modes contribute to the number of equipartition degrees of freedom,* because vibrational energy spacings often correspond to more than 1000 K.

This critical distinction between vibrations on the one hand and rotations and translations on the other is a manifestation of quantum mechanics in a fundamental macroscopic property. At low temperature, energy can be absorbed only by those degrees of freedom that have small energy gaps—translations and rotations, primarily. The less classical degrees of freedom—vibrations and electronic motions—have energy gaps too great for

CHECKPOINT We are looking for the ways in which each particle in our system will store the energy available to it. There are only a few translational and rotational degrees of freedom available, but they absorb energy easily. In a large molecule, there will be many more possible vibrational degrees of freedom, but they require higher temperatures to absorb energy. Electronic motions typically require even higher temperatures, often beyond what we work with in the lab. For this reason, and because there are no simple equations for the electronic energies and degeneracies, we don't include electronic states in this analysis.

▶ FIGURE 3.1 **Equipartition energy and degrees of freedom.** The number of distinct motions increases with the number of atoms in the molecule, so large molecules are capable of storing more energy (primarily in vibrations) than small molecules.

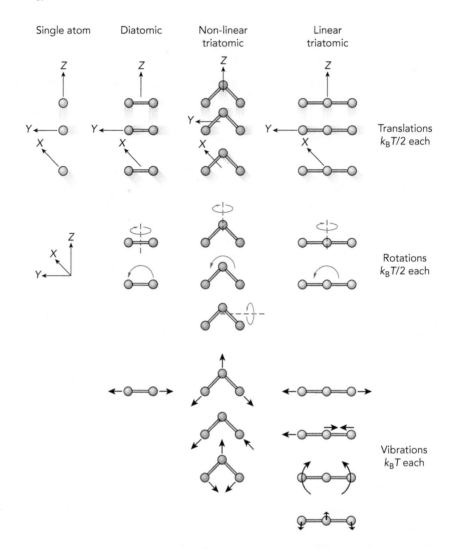

typical molecules to access anything but the ground state at low temperature. Figure 3.2 illustrates this concept by comparing the energy level diagram for some molecule with the corresponding canonical probability distribution. Nearly all of these molecules are in excited translational and rotational states, but very few have enough energy for the energy to be stored in vibration, much less in electronic motion. Consequently, vibrational and electronic excitation may contribute little to the overall energy.

EXAMPLE 3.1   Equipartition and Heating

CONTEXT This example illustrates how we can rapidly estimate heating and cooling requirements without a table of constants but with some awareness of the limitations.

One crucial component in the design of a chemical reactor is the handling of the heat flow. An exothermic reaction on large scale results in enormous amounts of heat that must be channeled away from the vessel to avoid damage. Especially dangerous is the prospect of *thermal runaway*, in which the heat released by a reaction raises the temperature of the mix and accelerates the reaction—potentially leading to explosion.

We can use the equipartition principle to estimate upper and lower limits to the amount of heat that must be removed from a gas-phase sample to reduce the temperature a given amount.

PROBLEM From the equipartition principle, estimate the energies in J that must be carried away to cool (a) 0.500 mol He gas and (b) 0.500 mol CO gas from 1000 K to 900 K at constant volume.

SOLUTION According to the equipartition principle,

$$E = \frac{1}{2}N_{ep}nRT.$$

To change the temperature of the sample from T_1 to T_2, energy must be added in the amount

$$\Delta E \equiv E_2 - E_1 = \frac{1}{2}N_{ep}nR(T_2 - T_1).$$

For a monatomic gas, $N_{ep} = 3$; there are 3 translational degrees of freedom, no rotations, and no vibrations. For a diatomic, $N_{ep} = 7$; 3 translational + 2 rotational + 2(1 vibrational). For He, this result is

$$\frac{1}{2}(3)(0.500)(8.314)(900 - 1000) = -624\,\text{J}.$$

For CO, this result is

$$\frac{1}{2}(7)(0.500)(8.314)(900 - 1000) = -1450\,\text{J}.$$

This result is quite accurate for He, but the correct value for CO is about −1230 J. Including the CO vibration overestimated the change of energy necessary to lower the temperature, because 900 K is still significantly lower in energy than the CO vibrational constant of $2170\,\text{cm}^{-1} = 3120\,\text{K}$. We would obtain a lower limit by omitting vibrations entirely, estimating instead that $N_{ep} = 5$ and obtaining $\Delta E = -1040\,\text{J}$.

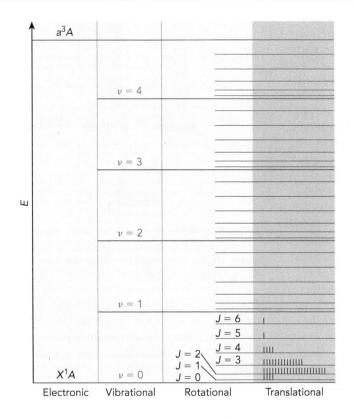

◀ FIGURE 3.2 **Varying amounts of excitation among the molecular degrees of freedom.** Energy levels for a molecule may combine excitations of the electronic, vibrational, rotational, and translational degrees of freedom. However, the corresponding population graph shows that at a typical temperature, very few molecules have enough energy to reach the vibrationally or electronically excited states. In a sample of 100,000 molecules, for example, each vertical bar in the shaded area would represent 2000 molecules. Under these conditions, vibrational and electronic excitation contribute little to the total energy of the sample.

TOOLS OF THE TRADE | Thermometers

Between roughly 1593 and 1693, Galileo Galilei developed some of the first experiments capable of measuring temperature. Among these was a long-necked flask, inverted into a small container of water. By cooling and warming the air in the bulb of the flask, Galileo could raise and lower the level of the water as the gas contracted and expanded. Over the next century, primitive thermometers were built throughout Europe, but no consistent temperature scale was developed until Daniel Gabriel Fahrenheit was able to make the first reliably precise thermometers in 1714, based on the expansion of mercury into a glass tube. Fahrenheit drew on his skills as a glassblower to produce high-quality instruments, which helped to establish his as the first widely reproduced temperature scale.

What are thermometers and why do we use them? A thermometer is any device that measures temperature, most commonly by monitoring a temperature-sensitive property (such as volume or electrical resistance) of a specific substance. In the laboratory, temperature is one of the few parameters that we can control during a chemical reaction, and one that can dramatically impact the speed and outcome of the reaction. Temperature control can be essential to making sure that a reaction has enough energy to go forward, or it can help us to regulate a reaction that risks running out of control. Routine temperature measurements include checking the melting point of a compound after a chemical synthesis to help identify the product. Finally, temperature and pressure are the two intensive parameters commonly established for our system by its interaction with the surroundings. The temperature is often determined by the lab atmosphere, for example, or by a heated water bath. Of T and P, the temperature normally has the most profound effect on the values we measure for thermochemical properties of the system, such as heat capacity or equilibrium constant. For that reason, the temperature is often a critical parameter to record in experimental work.

Fahrenheit's basic design for the thermometer has remained the standard for nearly 300 years. A small bulb of liquid opens into a narrow glass tube. As the liquid absorbs heat, it expands, and this registers as a higher level of liquid in the tube. The expansion is typically a small fraction of the entire volume (less than 0.1% per degree C), so a higher volume of liquid in the bulb helps increase the amount of liquid in the tube, which can improve the visibility and precision of the thermometer. However, the larger the amount of liquid in the bulb, the longer it takes for the temperature of the liquid to adjust to any changes, and the slower the response time of the thermometer.

Fahrenheit's use of mercury was a major advance. Although he and others had found several organic liquids such as alcohols that expanded rapidly on heating—which made them sensitive probes of the temperature—those substances tend to vaporize well below the boiling point of water. Mercury combined a high boiling point with a high rate of expansion, greatly expanding the range of temperatures that could be measured with a single instrument. Alcohol or other organic "spirits" have largely replaced mercury in clinical thermometers, because the toxicity of mercury has made it unpopular and the organic liquids work adequately over the narrow temperature range of the human body. Mercury has long remained in use for laboratory thermometers, but it is gradually being phased out in favor of less toxic alloys with similar properties.

But Fahrenheit's is by no means the only thermometer design. Many applications are not suited to the use of a glass thermometer. If the thermometer cannot be situated where it is readable, if the bulb cannot make good contact with the sample, or simply when an electronic reading is needed, it becomes necessary to find another means of measuring the temperature.

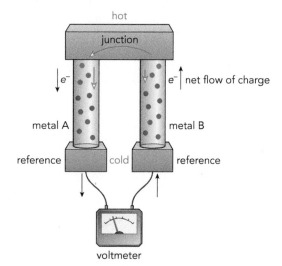

▲ **The thermoelectric effect.** Two wires, one of metal A and one of metal B, meet at the junction. At the junction, the thermoelectric effect creates a voltage, which leads to a net flow of electrons—a detectable electrical current.

How do they work? The most common alternative is the *thermocouple,* a pair of wires made from different conducting materials which meet to form the *thermocouple junction.* Each wire is also in contact with a point at some reference temperature. If the junction is at a higher temperature than the reference, as shown in the figure, then hot electrons will diffuse from the junction toward the cooler reference. However, no two metals will have the same diffusion rates, so a voltage difference builds up across the junction. This is an example of the *thermoelectric effect* or *Seebeck effect*, the generation of a voltage from the temperature difference across two points in contact (and also the generation of a temperature difference by an applied voltage). The voltage is measured using a circuit with high resistance in order to minimize the current across the junction, because current will disrupt the electron thermal distributions which determine the voltage difference.

Thermocouples allow direct conversion of a relative temperature measurement into an electrical signal. Furthermore, the range of materials available to fashion the wires allows thermocouples to cover a temperature range from roughly 4 K to over 2500 K.

3.3 Vibrational and Rotational Partition Functions

The equipartition principle is a powerful tool for estimating the total energy content in each of our system's degrees of freedom, but at times we also need to know in detail how the energy is distributed among the available energy levels. We need to be able to evaluate the probability distributions $\mathcal{P}$ themselves, not just the average energies.

In order to do this, we need to be able to solve the partition functions for these degrees of freedom. The numerator of the probability distribution always has the form $e^{-\varepsilon/(k_B T)}$, which we can evaluate as long as we know the energy expression for ε. But the denominator $q(T)$ or $q'(T)$ depends on a sum or integral over all the possible quantum states. How easy is that to solve?

We begin with the vibrational and rotational degrees of freedom, where the energy levels are sufficiently widely spaced that (unlike the translational states) we routinely measure the gaps between those levels by spectroscopy. The canonical distribution given by Eq. 2.30 is valid for any degree of freedom in any molecule, provided our other assumptions (such as a well-defined temperature) are valid. To determine how much energy is stored in these more quantum mechanical degrees of freedom, we may use Eq. 2.30 together with the partition functions formulated in this section.

Taylor Series Solution to Vibrational Partition Functions

For any single, non-degenerate, harmonic vibrational mode, we know

$$q_{\text{vib}}(T) = \sum_{v=0}^{\infty} e^{-v\omega_e/(k_B T)}, \qquad (3.23)$$

where v is the vibrational quantum number $(0, 1, 2, \ldots)$. If the vibrational constant ω_e is given in cm^{-1}, as is typical, then we can use a value for k_B that correctly converts the units: $k_B = 0.6950\ \text{cm}^{-1}/\text{K}$. Table 3.1 shows that the vibrational constants ω_e of small molecules are usually significantly greater than $300\ \text{cm}^{-1}$, and therefore at room temperature only the lowest vibrational states are

CHECKPOINT The vibrational energy expression that we use here, $\varepsilon_{\text{vib}} = v\omega_e$, is a little simpler than the expression we would use in quantum mechanics because we are measuring the energy from the ground state of the molecule. An additional contribution called the *zero-point energy* becomes important when we allow chemical reactions to occur, but for a non-reacting system it only corresponds to a change in the reference point of our energy, which (as shown in Eqs. 2.34 and 2.35) does not affect our statistics.

usually populated. As a result of the energy levels being quantized, the series in Eq. 3.23 converges rapidly, and only the first and second terms need to be evaluated for 1% accuracy under typical conditions.

However, at high temperature or for very low vibrational constants, there is a useful analytical solution to the power series

$$q_{vib}(T) = 1 + e^{-\omega_e/(k_B T)} + e^{-2\omega_e/(k_B T)} + e^{-3\omega_e/(k_B T)} + \ldots . \quad (3.24)$$

This expression happens to be the same as the Taylor series expansion (Eq. A.29):

$$f(x) = \sum_{n=0}^{\infty} \frac{(x - x_0)^n}{n!} \left(\frac{d^n f(x)}{dx^n} \right) \Bigg|_{x_0},$$

where we set $x = e^{-\omega_e/(k_B T)}$, $x_0 = 0$, and $f(x) = (1 - x)^{-1}$. The nth derivative of $(1 - x)^{-1}$ is $n!(1 - x)^{-n-1} dx$, and the series just shown becomes

$$f(x) = \sum_{n=0}^{\infty} \frac{(x - 0)^n}{n!} \left[n!(1 - 0)^{-n-1} \right] = \sum_{n=0}^{\infty} x^n$$

$$= 1 + x + x^2 + x^3 + \ldots$$

$$= 1 + e^{-\omega_e/(k_B T)} + e^{-2\omega_e/(k_B T)} + e^{-3\omega_e/(k_B T)} + \ldots \quad (3.25)$$

TABLE 3.1 Selected vibrational and rotational constants of diatomic molecules.

Molecule	μ (amu)	R_e (Å)	B (cm^{-1})	α_e (cm^{-1})	D (10^{-6} cm^{-1})	ω_e (cm^{-1})	$\omega_e x_e$ (cm^{-1})
$^1H\,^1H$	0.50	0.742	60.8536	3.0622	46,660	4401.21	121.34
$^1H\,^2D$	0.67	0.742	45.6378	1.9500		3811.92	90.71
$^2D\,^2D$	1.01	0.742	30.442	1.0623		3118.46	117.91
$^1H\,^{19}F$	0.96	0.917	20.9557	0.798	2150	4138.32	89.88
$^1H\,^{35}Cl$	0.98	1.275	10.5934	0.3702	532	2990.95	52.82
$^1H\,^{79}Br$	1.00	1.414	8.3511	0.226	372	2649.67	45.21
$^1H\,^{127}I$	1.00	1.609	3.2535	0.0608	526	2309.60	39.36
$^2D\,^{19}F$	1.82	0.917	11.0000	0.2907	585	2998.19	45.76
$^{12}C\,^{16}O$	6.86	1.128	1.9313	0.0175	6	2169.82	13.29
$^{14}N\,^{14}N$	7.00	1.098	1.9987	0.0171	6	2358.07	14.19
$^{14}N\,^{16}O^+$	7.47	1.063	1.9982	0.0190		2377.48	16.45
$^{14}N\,^{16}O$	7.47	1.151	1.7043	0.0173	-37	1904.41	14.19
$^{14}N\,^{16}O^-$	7.47	1.286	1.427			1372	8
$^{16}O\,^{16}O$	8.00	1.207	1.4457	0.0158	5	1580.36	12.07
$^{19}F\,^{19}F$	9.50	1.418	0.8828			891.2	
$^{35}Cl\,^{35}Cl$	17.48	1.988	0.2441	0.0017	0.2	560.50	2.90
$^{79}Br\,^{79}Br$	39.46	2.67	0.0821	0.0003	0.02	325.29	1.07
$^{127}I\,^{79}Br$	48.66	2.470	0.0559	0.0002	0.008	268.71	0.83
$^{127}I\,^{127}I$	63.45	2.664	0.0374	0.0001	-0.005	214.52	0.61
$^{23}Na\,^{23}Na$	11.49	3.077	0.1548	0.0009	0.7	159.13	0.73
$^{133}Cs\,^{133}Cs$	66.45	4.47	0.0127	0.00003	0.005	42.02	0.08

which is our partition function in Eq. 3.24. Therefore,

$$q_{vib}(T) = f(e^{-\omega_e/(k_B T)}) = \frac{1}{1 - e^{-\omega_e/(k_B T)}}. \tag{3.26}$$

This result is an exact solution to the *harmonic* vibrational energy expression $E_{vib} = v\omega_e$. At high temperatures, omission of the anharmonic terms such as $\omega_e x_e$ may become important, but the approximation usually works well.

Polyatomic molecules have more than one vibrational mode, but as long as each mode is assumed to be independent of the others, Eq. 3.4 shows that the total partition function is the product of the partition functions for the individual contributions:

$$q_{vib}(T) = q_{vib}^{(\nu_1)}(T) \, q_{vib}^{(\nu_2)}(T) \, \ldots \, . \tag{3.27}$$

Degenerate vibrational modes must be treated as distinct motions in this product (see Problem 3.30).

CHECKPOINT Recall that the partition function is roughly a guide to the number of quantum states that will have significant population in the system. Equation 3.26 is consistent with this interpretation: as the temperature decreases, the exponential will approach zero and q_{vib} will approach 1. As the temperature increases, the exponential approaches 1, and q_{vib} begins to grow rapidly.

EXAMPLE 3.2 Vibrational Probability Distribution

CONTEXT Iodine monochloride (ICl) is a useful reagent in synthetic chemistry for the addition of iodine, because the more electronegative Cl atom causes the I atom to have a positive partial charge (making it—if you recall these terms from organic chemistry—*electrophilic* where iodine would normally be *nucleophilic*). Because it has a relatively high mass for a diatomic, ICl also has a low vibrational constant, so a relatively high fraction of its molecules in a reaction mixture may be in vibrationally excited states, which will tend to increase the reaction rate.

PROBLEM Find the abundances of ICl molecules in the ground and lowest excited vibrational states at 300 K, given that the vibrational constant ω_e is 384 cm^{-1}.

SOLUTION The ground state is $v = 0$, and the lowest excited state is $v = 1$. The value

$$\frac{\omega_e}{k_B} = \frac{384 \, \text{cm}^{-1}}{0.6950 \, \text{cm}^{-1}/\text{K}} = 552 \, \text{K},$$

being on the order of the temperature of the sample, alerts us that the vibrational excitation will be significant. The partition function confirms this:

$$q_{vib} = (1 - e^{-\omega_e/(k_B T)})^{-1} = (1 - e^{-552 \, \text{K}/300 \, \text{K}})^{-1} = 1.188.$$

If all the molecules were in the ground vibrational state, q_{vib} would be 1. The value 1.188 is higher than that but still less than 2, indicating that most of the molecules remain in the ground state. With q_{vib} determined, we can find the fraction of the sample in the $v = 0$ and $v = 1$ states from the canonical distribution:

$$\mathcal{P}(v = 0) = \frac{e^{-(0 \times \omega_e)/(k_B T)}}{q_{vib}} = \frac{1}{1.188} = 0.841$$

$$\mathcal{P}(v = 1) = \frac{e^{-(1 \times \omega_e)/(k_B T)}}{q_{vib}} = \frac{e^{-552 \, \text{K}/300 \, \text{K}}}{1.188} = 0.133.$$

Of all the molecules in the sample, we will find that 84.1% are in the ground state and 13.3% are in the $v = 1$ excited state, leaving about 2.5% that will be found in the higher excited states.

Integral Approximation to Rotational Partition Functions

Integrals work best for evaluating the average translational energy, as in Eq. 3.14, because for any macroscopic system the energy spacing between adjacent translational states is too small to measure; the energy becomes a continuous variable. For the internal degrees of freedom of the molecule, however, such as rotation and vibration, the motion is confined to a much smaller scale—the scale of the molecule itself, where quantum effects become important. In this regime, the integral approximation becomes quite poor if the spacing between energy levels approaches the thermal energy $k_B T$.

We now move from vibration to a more classical kind of motion: rotation. Rotational levels have smaller energy gaps than vibrations for two reasons: there is no potential energy curve to constrain the motion, and the effective mass is roughly the mass of the entire molecule. Starting with the simplest case, the rotational energy of a linear molecule is approximated by Eq. 2.9:

$$E_{rot} = BJ(J + 1).$$

If we want to know what fraction of the molecules are in any given state at some temperature T, we need to evaluate the rotational partition function q_{rot} in order to find

$$\mathcal{P}(J) = \frac{g_{rot} e^{-E_{rot}/(k_B T)}}{q_{rot}}. \tag{3.28}$$

To evaluate q_{rot}, we need to recall that $g_{rot} = 2J + 1$, because there are $2J + 1$ different values of M_J. As long as the temperature is high enough that $B \ll k_B T$, the probability function will change gradually as a function of J, and we can transform the sum to the integral with little loss in precision:

$$q_{rot} = \sum_{J=0}^{\infty} g_{rot} e^{-E_{rot}/(k_B T)} \approx \int_0^{\infty} g_{rot} e^{-E_{rot}/(k_B T)} dJ = \int_0^{\infty} (2J + 1) e^{-BJ(J+1)/(k_B T)} dJ. \tag{3.29}$$

This integrand turns out to be convenient because $d[J(J + 1)] = (2J + 1)dJ$, so we have[1]

$$q_{rot} = \int_0^{\infty} e^{-BJ(J+1)/(k_B T)} d[J(J + 1)] = -\frac{k_B T}{B} \left[e^{-BJ(J+1)/(k_B T)} \right]_0^{\infty} = \frac{k_B T}{B}$$

linear molecule: $$q_{rot} = \frac{k_B T}{B}. \tag{3.30}$$

---

[1] As with the common R^{-6} factor in *QM* Section 10.1, one might reasonably wonder if this is mere coincidence, or if some higher symmetry principle is at work here. It's not always the case that g is proportional to dE; for example, for the one-electron atoms $g \approx 2n^2$ (*QM* Eq. 3.57) but $E = -Z^2 E_h/(2n^2)$ (Eq. 2.5). In the case of rotations, we can argue that we're looking at an energy of motion along two angular coordinates (angles Θ and Φ), where the energy is roughly proportional to J^2: $E \approx BJ^2$. We will show, in leading up to Eq. 3.44, that it makes sense that the degeneracy for motion in two dimensions is proportional to the perimeter of a two-dimensional figure of width J, times the increment dJ: $g \approx aJdJ$. It follows that $dE \approx 2BJdJ = (2B/a)g$, so the degeneracy should indeed be directly proportional to the derivative of the energy. This conclusion suggests that the convenient relationship between E and g in this case is a result of our dealing with free motion in two dimensions, so in the end there is a kind of luck involved.

Often this is written $q_{\text{rot}} = T/\theta_{rot}$, where the **characteristic rotational temperature** θ_{rot} is just the rotational constant B in units of K.

In the case of translational energies, the quantum states are so nearly continuous for any macroscopic system that the integral will always be a valid solution to the partition function. For rotations, we have to be just a little more careful. When $B \ll k_B T$, then the probability function $\mathcal{P}(J)$ changes slowly as a function of J, and the integral is accurate. When $B \gg k_B T$, $\mathcal{P}(J)$ drops rapidly, and the integral is not a good approximation. At the same time, however, the sum becomes easy to evaluate because only the first few terms are important. Either way of evaluating the partition function is approximate: the integral neglects the quantization of energy in the coordinate, and the sum has to be truncated at some point. As a yardstick, when $k_B T/B = 3.3$, the error in the integral approximation is just under 10%. To get the error within about 1%, $k_B T/B$ needs to be at least 33. The relationship between the integral and exact solutions is shown in Fig. 3.3.

When we combine rotational and vibrational excitation, we calculate the probability of being in an excited state by the *rovibrational energy,* which to a good approximation is the sum of the rotational and the vibrational excitation energies. In the same way that Eq. 3.27 predicts that we can multiply the partition functions for individual modes to get the overall vibrational partition function, the probability of being in a rotational state J and vibrational state v is given by the product of the probabilities for each excitation:

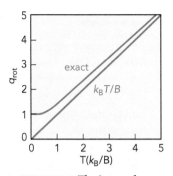

▲ FIGURE 3.3 **The integral approximation to q_{rot}.** The exact q_{rot} and $k_B T/B$ for a linear molecule with rotational constant B are graphed as functions of T. The relative error increases rapidly at low temperature.

$$P(v,J) = \mathcal{P}_{\text{vib}}(v)\mathcal{P}_{\text{rot}}(J) = \left(\frac{g_{\text{vib}} e^{-E_{\text{vib}}/(k_B T)}}{q_{\text{vib}}(T)} \right)\left(\frac{g_{\text{rot}} e^{-E_{\text{rot}}/(k_B T)}}{q_{\text{rot}}(T)} \right) = \frac{g_{\text{vib}} g_{\text{rot}} e^{-(E_{\text{vib}}+E_{\text{rot}})/(k_B T)}}{q_{\text{vib}}(T) q_{\text{rot}}(T)}. \tag{3.31}$$

EXAMPLE 3.3 The Rotational Partition Function

CONTEXT It is rare that most chemists work at temperatures less than the boiling of liquid nitrogen at about 77 K. However, two environments in chemical studies that exist at these temperatures are molecular beams and interstellar molecular clouds, both of which lie typically between about 5 K and 20 K. At these temperatures, special care may be required to compute the partition functions for rotation. An application of particular interest is the measurement of rotationally excited carbon monoxide (CO) in interstellar space by radio telescopes, which can be used to identify regions in our Galaxy where stars may be forming.

PROBLEM Compare the rotational partition functions for CO ($B/k_B = 2.70$ K) obtained by using the integral and by using the direct sum, at 273 K and at 10 K.

SOLUTION The integral yields a partition function of $q_{\text{rot}} = 273/2.70 = 100.72$, at 273 K, and the correct value is 101.06. A comparable value is obtained from the direct sum only by summing from $J = 0$ to 35. However, at 10 K, which is a temperature measured for CO in interstellar space or in some laboratory cooling experiments, the integral predicts a partition function of $10/2.70 = 3.69$, when the correct value is 4.06. The correct value to three digits is predicted by adding the terms from $J = 0$ to 5:

$$\sum_{J=0}^{4} (2J + 1)e^{BJ(J+1)/(k_B T)} = 1 + 3e^{-2(2.70)/10} + 5e^{-6(2.70)/10}$$

$$+ 7e^{-12(2.70)/10} + 9e^{-20(2.70)/10} + 11e^{-30(2.70)/10}$$

$$= 1 + 1.748 + 0.989 + 0.274 + 0.041 + 0.003 = 4.06$$

EXAMPLE 3.4 Rovibrational Partition Function

CONTEXT Gas-phase carbonyl sulfide (OCS) absorbs infrared radiation strongly near 2100 cm^{-1}, making it a convenient reference when spectroscopists need to calibrate their instruments. A large number of rovibrational transitions are usually visible, but the spectrum is relatively simple because the molecule is linear. To calibrate the spectrometer at much higher wavenumbers, it may be necessary to look for transitions from states that have a combination of vibrational motions excited, but those states also tend to have much lower populations. From the temperature and the energy, we can estimate the population and predict whether it is realistic that the molecule can be observed by the spectrometer.

PROBLEM Find the number density of OCS molecules in a 100.0 Pa sample at 298 K that are in the vibrational state $(v_1, v_2, v_3) = (1, 0, 1)$ and rotational state $J = 10$. The constants are CS-stretch $\omega_1 = 859\,\text{cm}^{-1}$, bend $\omega_2 = 527\,\text{cm}^{-1}$, and CO-stretch $\omega_3 = 2080\,\text{cm}^{-1}$, $B = 0.2028\,\text{cm}^{-1}$. The bending mode is a doubly degenerate vibrational motion, so it counts as *two* modes. Give the final value in molecules per cubic centimeter.

SOLUTION At 298 K, we can use the integral approximation for q_{rot} with high precision for any molecule except H_2, as we see here with a quick comparison of B/k_B to the temperature:

$$\frac{B}{k_B} = \frac{0.2028\,\text{cm}^{-1}}{0.6950\,\text{cm}^{-1}/\text{K}} = 0.2918\,\text{K} \ll 298\,\text{K}.$$

We are looking for $\mathcal{P}(v_1 = 1, v_2 = 0, v_3 = 1, J = 10)$, and the state is at energy

$$\varepsilon = v_1\omega_1 + v_2\omega_2 + v_3\omega_3 + BJ(J + 1)$$
$$= \omega_1 + \omega_3 + 110B = 2961\,\text{cm}^{-1}.$$

The fraction of molecules in our state is the product of the probabilities for the excitation in each rotational and vibrational degree of freedom:

$$\mathcal{P}(1,0,1,10) = \mathcal{P}(v_1 = 1)\mathcal{P}(v_2 = 0)\mathcal{P}(v_3 = 1)\mathcal{P}(J = 10)$$

$$= \left(\frac{e^{-v_1\omega_1/(k_BT)}}{q_{\text{vib}(1)}(T)}\right)\left(\frac{e^{-v_2\omega_2/(k_BT)}}{q_{\text{vib}(2)}(T)}\right)\left(\frac{e^{-v_3\omega_3/(k_BT)}}{q_{\text{vib}(3)}(T)}\right)\left(\frac{(2J + 1)e^{-BJ(J+1)/(k_BT)}}{q_{\text{rot}}(T)}\right)$$

$$= \frac{(2J + 1)e^{-\varepsilon/(k_BT)}}{q} = \frac{(21)\exp\left\{(-2961\,\text{cm}^{-1})/[(0.6950\,\text{cm}^{-1}/\text{K})(298\,\text{K})]\right\}}{q} = \frac{1.30 \cdot 10^{-5}}{q}.$$

Now we need to find q. Note that even though we don't consider a state with excitation in the bending mode, we need to factor the bending mode into the partition function, because molecules that are in excited bending states reduce the number that are in the $(1, 0, 1)$ vibrational state. We will count the bending mode twice in evaluating the partition function, because it is a doubly degenerate mode:

$$q_{\text{rovib}}(T) = q_{\text{rot}}(T)\,q_{\text{vib}(1)}(T)\,q_{\text{vib}(2)}(T)\,q_{\text{vib}(3)}(T)$$

$$= \left(\frac{k_BT}{B}\right)\left(\frac{1}{1 - e^{-\omega_1/(k_BT)}}\right)\left(\frac{1}{1 - e^{-\omega_2/(k_BT)}}\right)^2\left(\frac{1}{1 - e^{-\omega_3/(k_BT)}}\right)$$

$$= \left(\frac{0.6950 \cdot 298}{0.2028}\right)\left(\frac{1}{1 - e^{-859/(0.6950 \cdot 298)}}\right)\left(\frac{1}{1 - e^{-527/(0.6950 \cdot 298)}}\right)^2\left(\frac{1}{1 - e^{-2080/(0.6950 \cdot 298)}}\right)$$

$$= (1021)(1.016)(1.085)^2(1.000) = 1222.$$

Now we can calculate the fraction of molecules in our state:

$$\mathcal{P}(1,0,1,10) = \frac{1.30 \cdot 10^{-5}}{1222} = 1.06 \cdot 10^{-8}.$$

Finally, to get the number density in this state, we calculate the number density of the entire OCS sample using the ideal gas law,

$$\rho = \frac{N}{V} = \frac{n\mathcal{N}_A}{V} = \frac{\mathcal{N}_A P}{RT}$$

$$= \frac{(6.022 \cdot 10^{23}\,\mathrm{mol}^{-1})(100.0\,\mathrm{Pa})}{(8.3145\,\mathrm{J\,K}^{-1}\mathrm{mol}^{-1})(298\,\mathrm{K})} = 2.43 \cdot 10^{22}\,\mathrm{m}^{-3} = 2.43 \cdot 10^{16}\,\mathrm{cm}^{-3},$$

and multiply this by $\mathcal{P}(1,0,1,10)$ to get the number density of molecules in the right quantum state:

$$\mathcal{P}(1,0,1,10)\rho = (1.06 \cdot 10^{-8})(2.43 \cdot 10^{16}\,\mathrm{cm}^{-3}) = 2.58 \cdot 10^{8}\,\mathrm{cm}^{-3}.$$

This value is near the detectable limit of a sensitive spectrometer.

The rotational partition functions of non-linear molecules can also be evaluated in terms of the three rotational constants A, B, and C, yielding the approximate result

$$q_{\mathrm{rot(nonlinear)}} \approx \left(\frac{\pi k_B^3 T^3}{ABC}\right)^{1/2}. \tag{3.32}$$

Equation 3.32 works best for the symmetric tops, either $A = B$ (oblate) or $B = C$ (prolate). For asymmetric tops ($A \neq B \neq C$), the approximation suffers in quality but remains useful.

CHECKPOINT In Example 3.4, to find an overall probability for being in a particular rovibrational state, we can separate the rovibrational partition function into individual contributions from the rotational and vibrational partition functions. The probability is the product of the probabilities for the specific vibrational state and for the specific rotational state, and the partition function can be determined separately for each of those degrees of freedom.

Rotation and Nuclear Spin Statistics

Section 1.3 shows how microscopic effects are felt at the macroscopic level by integrating a microscopic term—the speeds of individual molecules—over a macroscopic sample to arrive at the ideal gas law. This effect is not intrinsically a quantum mechanical one, however, since we treated the molecular speed classically, as a continuous parameter. We did not need to introduce purely quantum mechanical features, such as the wavelike nature of the molecules or the quantization of the translational quantum states. There are, however, macroscopic phenomena that result from the purely quantum mechanical principle of spin symmetrization.

Spin Statistics and the Rotational Degeneracy

So far, the rotational partition function hasn't given us much trouble. However, there is an important complication: for molecules in point groups with n-fold proper rotation axes, q_{rot} depends on the molecule's **nuclear spin statistics.**

Many molecules contain atoms that have exactly identical molecular environments, such as the four hydrogen atoms in methane or the two oxygen atoms in O_2. We will call these *equivalent atoms*. If we rotate the molecule HF end over end, we must rotate it a full 2π to return to its original orientation. However, if we rotate an O_2 molecule only π, we get a structure exactly equivalent to the starting structure. We can't tell that the molecule was rotated only part way, because the only difference is that two indistinguishable atoms have changed places.

If a rotation exchanges indistinguishable nuclei, then we return to the symmetrization principle (Section 2.1) to see what symmetry is required of the overall wavefunction. In the same way that the overall wavefunction must be antisymmetric under exchange of any two electron labels, it must also have the proper symmetry if we exchange the labels on any two equivalent nuclei: symmetric if the nuclei are bosons, antisymmetric if they are fermions.

In principle, a molecule has an overall wavefunction Ψ_{total} that specifies a specific quantum state for the combined electronic and nuclear degrees of freedom. We usually write the electronic wavefunction ψ_{elec} as separate from the nuclear wavefunction and then break down the nuclear wavefunction into parts: vibration and (if the molecule is in the gas phase) rotation. If the molecules have symmetry, then we include a third term in the nuclear wavefunction for the nuclear spin, ψ_{ns}. (We don't include translations in the wavefunction because that degree of freedom can normally be treated using classical mechanics.) Altogether, that gives us

$$\Psi_{total} = \psi_{elec}\psi_{vib}\psi_{rot}\psi_{ns}. \tag{3.33}$$

For many simple molecules, including all the homonuclear diatomics, nuclear spin statistics influence the spectra and may also significantly affect some macroscopic properties.

For example, in the molecule $^1H_2$ there are two nuclear spins, one for hydrogen atom A and one for atom B. The spin quantum number I is $\frac{1}{2}$ for the ^1H nucleus, and its projection quantum number m_I may be $-\frac{1}{2}$ or $+\frac{1}{2}$. The total nuclear spin,

$$|I_T| = |\vec{I_A} + \vec{I_B}|,$$

must be 1 or 0. This $I_T = 1$ wavefunction is symmetric with respect to the permutation operator $\hat{P}(A,B)$ that interchanges the two nuclei. The $I_T = 0$ state, on the other hand, is antisymmetric with respect to that exchange.

Because the ^1H nuclei are fermions with $I = \frac{1}{2}$, the overall wavefunction must be antisymmetric when we exchange the two nuclei H_A and H_B. We express their spins in terms of the spin wavefunctions α and β, where α is shorthand for the $m_I = +1/2$ state and β stands for the $m_I = -1/2$ state. The possibilities resolve themselves into two sets. There is the singlet $I_T = 0$ state:

$$\psi_{ns} = \frac{1}{\sqrt{2}}[\alpha(A)\beta(B) - \beta(A)\alpha(B)], \tag{3.34}$$

and the three states of the triplet $I_T = 1$:

$$\psi_{ns} = \alpha(A)\alpha(B)$$

$$\psi_{ns} = \frac{1}{\sqrt{2}}[\alpha(A)\beta(B) + \beta(A)\alpha(B)] \tag{3.35}$$

$$\psi_{ns} = \beta(A)\beta(B).$$

The singlet state is *antisymmetric* with respect to exchange of A and B, while the three states of the triplet are *symmetric*. The overall wavefunction must be *antisymmetric* under this operation. Now we check the symmetries of the other degrees of freedom.

Consider the rotational wavefunctions for $J = 1$ and 2, $M_J = 0$. The corresponding spherical harmonic wavefunctions are $Y_1^0(\Theta) = \cos\Theta$ and $Y_2^0(\Theta) = (3\cos^2\Theta - 1)$. If we define Θ to be the angle from the space-fixed Z axis to the vector $\vec{AB}$ formed by the two atoms, then when the labels A and B are switched, the vector $\overrightarrow{AB}$ reverses direction and the new angle Θ' is equal to $\pi - \Theta$ (Fig. 3.4). Therefore, $\cos\Theta$ changes sign, but $\cos^2\Theta$ does not. The odd J state is antisymmetric with respect to exchange of the nuclei; the even J state is symmetric.

This rule is general, although to prove it for all M_J requires examining the Φ-dependent part of the wavefunction as well. All even J states are symmetric and odd J states antisymmetric with respect to exchange of identical nuclei. In the $^1\Sigma_g^+$ ground state of H_2, the electronic wavefunction is totally symmetric, and for a diatomic the vibrational wavefunction is always totally symmetric, so only the nuclear and rotational states need be tested for our example.

We find, therefore, that the triplet nuclear spin state (symmetric) must occupy only odd J states (antisymmetric) in $^1H_2$, and the singlet nuclear spin state (antisymmetric) may occupy only even J states (symmetric). The degeneracy of the rotational state now depends on the nuclear spins. This issue often requires using group theory at a more advanced level than we have used here, particularly if we also need to account for the electronic state symmetry or (for polyatomics) the symmetry of the vibrational state. For any linear molecule with a center of symmetry, a simple rule may help: g electronic and vibrational wavefunctions are symmetric with respect to exchange of equivalent nuclei; u wavefunctions are antisymmetric.

Although vibrational symmetry affects the rotational partition function when spin statistics are considered, the vibrational partition function is not usually affected by nuclear spin statistics.[2] Although certain combinations of rotational and vibrational quantum numbers may be forbidden altogether by symmetry, no vibrational state is completely forbidden by symmetry. The nuclear spin statistics will always allow certain rotational states to be occupied within any vibrational state.

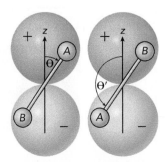

▲ FIGURE 3.4 **Rotational and nuclear spin symmetry.** Exchanging the labels A and B on identical nuclei of a diatomic molecule may change the sign of the rotational wavefunction. The new value Θ' is the complement of the old value, and the wavefunction changes sign if J is odd. The $J = 1$, $M_J = 0$ wavefunction has been outlined in the figure to illustrate how the change of labels reverses the sign of the wavefunction at each nucleus.

Spin Statistics and the Partition Function

Why does this have any impact on macroscopic measurements? We found that the triplet nuclear spin state of $^1H_2$ has an added factor of three in its degeneracy compared to the singlet; there are three values of M_I for $I_T = 1$, and only one value for $I_T = 0$. Nuclear spins are magnetic moments, and very weak ones at that, which means that their interaction with other fields is very weak and changes in I_T are very slow. Different nuclear spin states of a molecule act almost independently of each other and are best described by *separate* partition functions. The energies are even referenced to the lowest energy of the particular spin state. For $^1H_2$, for example,

$$
\begin{aligned}
q_{\text{rot}}(H_2) &= q_{\text{rot}}(I_T = 0) + q_{\text{rot}}(I_T = 1) \\
&= \left[1 + 5e^{-6B/(k_B T)} + 9e^{-20B/(k_B T)} + \ldots \right] \\
&\quad + 3\left[3e^{-(2-2)B/(k_B T)} + 7e^{-(12-2)B/(k_B T)} + 11e^{-(30-2)B/(k_B T)} + \ldots \right],
\end{aligned} \tag{3.36}
$$

---

[2]The reason is that standard vibrational modes do not allow equivalent nuclei to exchange location. An exception is torsional motions, such as the internal rotation of the methyl group in toluene about the C—C bond. This motion corresponds to a vibrational mode, because the positions of the methyl hydrogens change relative to the atoms in the ring, but every 120° of the torsion brings us back to an equivalent geometry. Such a vibrational mode then has to be treated by the same statistical mechanics we would use for the corresponding rotation.

here the rotational energies $BJ(J + 1)$ in the $I_T = 1$ partition function are referenced to the ground state of the $I_T = 1$ levels: $J = 1$ at energy $2B$. (That's why $2B$ is subtracted from all the energy terms in the $I_T = 1$ partition function.)

This effect appears most dramatically in $^1H_2$, because its very high rotational constant (60.9 cm^{-1} $\approx$ 87.5 K) means that only a few levels are populated at room temperature. This quantum effect then becomes important in determining the energy content of a container of hydrogen. The two nuclear spin states of hydrogen are even given different names: **para-hydrogen** for $I_T = 0$ and **ortho-hydrogen** for $I_T = 1$. At low temperatures, the total partition function for H_2 would be the sum of the partition functions for *ortho* (odd J) and *para* (even J) H_2.

The same effect is seen in other linear molecules. For homonuclear boson diatomics, such as O_2, or other symmetric linear systems, such as CO_2, the overall wavefunction must be symmetric. Only even J are populated in the vibrational ground state of CO_2, but only odd J are populated in $^3\Sigma_g^-$ O_2 (because of the antisymmetric electronic state) and in $\nu_3 = 1$ (σ_u antisymmetric stretch) CO_2. As a general (but approximate) rule, rotational partition functions in molecules where rotation exchanges n equivalent $I = 0$ nuclei will be reduced by a factor of $1/n$.

3.4 The Translational Partition Function

The energy in a macroscopic system is stored in many different degrees of freedom. It may be in the form of radiation or stored in molecular translations, rotations, vibrations, or electronic excitation. We largely ignore translations when we study molecular structure, because, of all the molecular degrees of freedom, it is the least dependent on the chemical composition of our sample. Translation is merely motion of the molecular centers of mass.

However, it is also the most easily excited molecular degree of freedom in a gas, and for that reason it is perhaps the most important to take into account now that we are interested in the total energy content of a gas sample. Furthermore, as we explore processes that affect the macroscopic sample, we will encounter several that involve transportation of molecules, such as the flow of a liquid or expansion of a gas, and these processes are carried out along the translational coordinates. To predict the behavior of these systems, we will need expressions for the translational partition function.

There are two factors that make the translational partition function more challenging to derive than the functions for rotation and vibration. First, the degeneracy cannot so easily be spelled out for a continuous variable like translational energy as it can for the discrete energy levels of the quantum mechanical degrees of freedom. Second, whereas we can isolate the effects of vibration and rotation to the individual molecules, one feature of translational motion that interests us most is the opportunity to allow the molecules to interact with one another. As a result, we need to examine not only the translational partition function for a single molecule, but for all N molecules in our sample.

How can we estimate the number of quantum states for a basic macroscopic system? Our sample is a large number N of identical, completely independent atoms, contained within a box of volume V, with a total energy E. The atoms neither attract nor repel one another, nor do they vibrate or rotate. All the energy is in the form of translational motion. How many quantum states does this sample have?

The quantum mechanics we tend to focus on in physical chemistry relates to the electronic, vibrational, and rotational motions, for which we can easily measure the gaps between the quantized energy levels. We don't spend much time on translational states, because the quantized energies of translation are given in Eq. 2.4:

$$\varepsilon = n^2 \frac{h^2}{8mV^{2/3}} = n^2 \varepsilon_0,$$

where m is the mass of one molecule in our gas sample, and where $n^2 = n_X^2 + n_Y^2 + n_Z^2$, with n_X, n_Y, and n_Z the quantum numbers for motion along the X, Y, and Z space-fixed Cartesian coordinates, respectively.[3]

The energies are proportional to $V^{-2/3}$, which is to say, roughly inversely proportional to the surface area of the container. When the volume V gets much larger than a few angstroms, and the molecule is free to travel even one micron, the energy levels are so close together that for experimental purposes they become continuous. Our molecules are rarely confined to a container small enough for us to measure the gaps between the translational states, so usually we treat translation as a classical motion. But to understand the entropy of bulk chemical systems, we need to recognize that the classical limit is an *approximation* of a degree of freedom that remains, at some level, *quantum mechanical*.

---

[3]You may have noticed that this energy should be $h^2(n^2 - 3)/(8mV^{2/3})$, because the ground translational state $(n_x, n_y, n_z = 1, 1, 1)$ has zero-point energy $3h^2/(8mV^{2/3})$, and we're now measuring the energy *relative to the ground state*. Fortunately, this correction is negligible for a macroscopic system. There will be enough math without carrying that added -3 around.

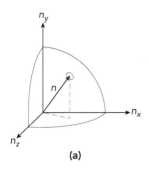

(a)

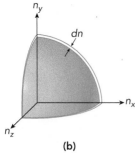

(b)

▲ FIGURE 3.5 **One octant of a sphere in phase space.** The sphere has radius $n = (n_x^2 + n_y^2 + n_z^2)^{1/2}$ in a coordinate system where the axes are n_x, n_y, and n_z. The number of quantum states of the particle in a three-dimensional box at a particular energy $E = n^2\varepsilon_0$ is roughly equal to the number of points with integer values of the coordinates within dn of the surface.

We have the energy expression, but how do we count the number of states available at each energy? Let's start with just one atom. There's a great way to estimate Ω at large values of n^2 using geometry instead of calculus. To estimate the number of quantum states of one atom at some energy ε, we want to know how many values of n_x, n_y, and n_z give the same value of n^2. This quantity can be approximated by graphing n^2 (which is proportional to ε) on a three-dimensional coordinate system with axes n_x, n_y, and n_z, as shown in Fig. 3.5. That function is a sphere of radius $n \equiv \sqrt{n^2} = (n_x^2 + n_x^2 + n_x^2)^{1/2}$. The equation for a sphere of radius r in Cartesian space has the equation $x^2 + y^2 + z^2 = r^2$, and here we have instead $n_x^2 + n_x^2 + n_x^2 = n^2$. In our application, only one-eighth of the sphere contributes to the energy level density, because the quantum numbers n_x, n_y, n_z are all positive.

The degeneracy is a discrete function, being defined only at values of n that are the square root of an integer n^2. The degeneracy will be roughly the number of distinct points (n_x, n_y, n_z) in a shell of width dn that lies between a radius of $\sqrt{n^2}$ and a radius of $\sqrt{n^2 + 1}$. This approach is like filling a hollow ball with sand, and counting all the grains in contact with the inner wall of the ball (Fig. 3.6); if the ball is big enough to hold a lot of grains, the number of grains in contact will increase in proportion to the ball's surface area.[4]

The number of states in this shell is given by the surface area of a sphere, $4\pi n^2$, divided by 8 (to get only positive values of the quantum numbers), times dn. In the limit that n^2 is big, we can estimate the dn by saying that the smallest increment in n^2 is 1, and therefore

$$d(n^2) = 2n\,dn = 1$$

$$dn = \frac{1}{2n}. \tag{3.37}$$

The degeneracy is the same as the ensemble size Ω for the system at this energy, so let's write it as

$$\Omega = \frac{4\pi n^2}{8}dn = \frac{\pi}{2}n^2 dn = \frac{\pi}{2}n^2\frac{1}{2n}$$

$$= \frac{\pi}{4}n = \frac{\pi}{4}\left(\frac{\varepsilon}{\varepsilon_0}\right)^{1/2}. \qquad \text{by Eq. 2.4} \quad (3.38)$$

As we approach the classical limit, we will no longer be able to distinguish between one energy level and the very next energy level, so we replace Ω by a new function: the density of quantum states W, which is averaged over several energy levels. We find W by noting that our equation for Ω is valid over an energy range ε_0, the difference between the energy level $n^2\varepsilon_0$ and the very next possible energy level at $(n^2 + 1)\varepsilon_0$. Hence, the quantum state density as a function of energy is

$$W(\varepsilon) = \frac{\Omega}{\varepsilon_0} = \frac{\pi}{4\varepsilon_0^{3/2}}\varepsilon^{1/2}. \tag{3.39}$$

▲ FIGURE 3.6 **Counting points on a surface.** The number of grains of sand in contact with the surface of their container is proportional to the surface area.

[4]We can get away with saying that the volume is equal to the number of points, because the points we are looking for must be at integer values of the three coordinates n_x, n_y, and n_z. If you imagine a cube 10 units on a side, for example, it will have a volume of 1000 and it will contain 1000 points where the coordinates all have integer values. A sphere is not a cube, of course, but in the limit of large n^2, this argument is also valid for the sphere.

Now we're ready to find the partition function.

To analyze how the translational energy of a macroscopic sample is distributed, we may use the canonical distribution, which requires knowing the energy and the translational partition function. We have an expression for the translational energy of a particle of mass m in a box of volume V (Eq. 2.4), but we still need the partition function. Fortunately, once we have the energy expression, we can always obtain the partition function. Unfortunately, that sometimes takes more than a couple of steps of math.

It's worth it: partition functions don't just normalize the canonical distribution, they also provide a means of deriving almost any important parameter in classical chemistry, including equilibrium constants and rate constants. Presented next are two ways to get the translational partition function for one particle. Method 1 is quicker, but only works for the ideal gas. Method 2 is longer, but can later be applied to samples in which the molecules interact with each other—which is what chemistry is all about. Choose your preferred method, or jump ahead to the **DERIVATION SUMMARY** after Eq. 3.54.

Method 1: Single Energy Integration.

Method 1 parameters key		
Symbol	**Parameter**	**SI units**
h	Planck's constant	J s
k_B	the Boltzmann constant	$J\,K^{-1}$
m	the mass of one particle	kg
V	the volume of the box, equal to abc	m^3
n_X, n_Y, n_Z	the quantum numbers for motion along X, Y, and Z	unitless
ε	the energy of one particle in the box	J
ε_0	$\frac{h^2}{(8mV^{2/3})}$, so that $\varepsilon = n^2\varepsilon_0$	J
g	the degeneracy, the number of quantum states having the same energy	unitless
$W(\varepsilon)$	the density of quantum states	J^{-1}
T	the temperature	K
q_{trans}	the translational partition function for one particle	unitless

The partition function for translation of a single molecule is

$$q_{\text{trans}}(T,V) = \sum_i e^{-\varepsilon_i/(k_B T)} = \sum_{\varepsilon=0}^{\infty} g(\varepsilon)e^{-\varepsilon/(k_B T)}. \qquad (3.40)$$

We use a lower case q to denote this partition function, and ε for the energy, to emphasize that these parameters are for a *single* molecule in the sample.

For macroscopic systems, the translational energy is a continuous variable; the energy spacing between adjacent translational states is too small to measure. As long as the probability distribution function behaves like a function of a continuous variable, we may replace the sum in Eq. 3.40 by an integral:

$$q_{\text{trans}}(T,V) = \int_0^{\infty} g(\varepsilon)e^{-\varepsilon/(k_B T)}. \qquad (3.41)$$

To make the transformation to the integral work, however, we need a volume element. This, it turns out, is provided by the degeneracy term. The density of states $W(\varepsilon)$ is equal to $g(\varepsilon)/d\varepsilon$, where $d\varepsilon$ is the smallest interval between two energy levels. Therefore, we can replace $g(\varepsilon)$ by $W(\varepsilon)d\varepsilon$:

$$q_{trans}(T,V) = \int_0^\infty W(\varepsilon)e^{-\varepsilon/(k_B T)}d\varepsilon \qquad \text{let } g(\varepsilon) = W(\varepsilon)d\varepsilon \ (3.42)$$

$$= \frac{\pi}{4\varepsilon_0^{3/2}}\int_0^\infty \varepsilon^{1/2}e^{-\varepsilon/(k_B T)}d\varepsilon. \qquad \text{by Eq. 3.39} \qquad (3.43)$$

We may do a variable substitution here, letting

$$u = \varepsilon^{1/2}, \quad du = \frac{1}{2}\varepsilon^{-1/2}d\varepsilon, \text{ so } d\varepsilon = 2\varepsilon^{1/2}du = 2u\,du.$$

Inserting these into Eq. 3.43 gives

$$q_{trans}(T,V) = \frac{\pi}{4\varepsilon_0^{3/2}}\int_0^\infty ue^{-u^2/(k_B T)}(2u\,du) = \frac{\pi}{2\varepsilon_0^{3/2}}\int_0^\infty u^2 e^{-u^2/(k_B T)}du$$

$$= \frac{\pi}{2\varepsilon_0^{3/2}}\left(\frac{1}{4}\right)\left[\pi(k_B T)^3\right]^{1/2} = \frac{1}{8}\left(\frac{\pi k_B T}{\varepsilon_0}\right)^{3/2} \qquad \text{by Table A.5}$$

$$= \frac{1}{8}\left(\frac{8\pi mk_B T V^{2/3}}{h^2}\right)^{3/2} \qquad \varepsilon_0 \text{ defined in Eq. 2.48}$$

$$= \left(\frac{2\pi mk_B T}{h^2}\right)^{3/2}V. \qquad (3.44)$$

Method 2: Split Kinetic/Potential Energy Integration.

Method 2 parameters key		
Symbol	**Parameter**	**SI units**
h	Planck's constant	J s
k_B	the Boltzmann constant	J K^{-1}
m	the mass of one particle	kg
a,b,c	the lengths of the sides of the box along X, Y, and Z	m
V	the volume of the box, equal to abc	m^3
n_X,n_Y,n_Z	the quantum numbers for motion along X, Y, and Z	unitless
v_X,v_Y,v_Z	the speeds of motion along X, Y, and Z	m s^{-1}
ε	the energy of one particle in the box	J
ε_0	$\frac{h^2}{(8mV^{2/3})}$, so that $\varepsilon = n^2\varepsilon_0$	J
g	the degeneracy, the number of quantum states having the same energy	unitless
W	the density of quantum states	J^{-1}
T	the temperature	K
q'_K	the integral of $e^{-K/(k_B T)}$ over all velocity components	m^{-3}
q'_U	the configuration integral for one particle, the integral of $e^{-U/(k_B T)}$ over all positions	m^3
q_{trans}	the translational partition function for one particle	unitless

Another way of obtaining Eq. 3.44 takes more effort but yields a form that we can extend to more realistic systems.

To emphasize the quantum mechanical nature of the energy levels when the system is very small, let's begin with the energy expression for the particle in a three-dimensional box (Eq. 2.4):

$$\varepsilon = \varepsilon_0 n^2 = \varepsilon_0(n_X^2 + n_Y^2 + n_Z^2).$$

This energy expression depends on the quantum numbers n_X, n_Y, and n_Z, which are all integers. Therefore, we can express the single-particle translational partition function as

$$q_{\text{trans}}(T,V) = \sum_{n_X=1}^{\infty} \sum_{n_Y=1}^{\infty} \sum_{n_Z=1}^{\infty} e^{-\varepsilon/(k_B T)} = \int_1^{\infty} \int_1^{\infty} \int_1^{\infty} e^{-\varepsilon/(k_B T)} dn_X dn_Y dn_Z. \quad (3.45)$$

In this case, we casually toss in the volume element $dn_X dn_Y dn_Z$ when moving from the sum to the integral, because dn_X is the smallest possible increment in n_X, and that's equal to 1. Similarly, we can set dn_Y and dn_Z equal to 1, so the whole volume element amounts to multiplying by 1.

An odd thing happens, however, when we move from describing the quantum mechanical perspective on our system to the classical. We can describe the quantum system using only *three* quantum numbers. Another way of saying this is that there is a space with coordinate axes n_X, n_Y, and n_Z—we call this the **Hilbert space** of the system—and each quantum state corresponds to a unique point in this space. The strange thing is that when we describe the state of the same particle using classical mechanics we need *six* coordinates: three for the position of the particle (X, Y, Z) and three more for the velocity of the particle (v_X, v_Y, v_Z). The space that has coordinate axes X, Y, Z, v_X, v_Y, v_Z for each particle is called the **phase space** of the system, and in classical mechanics each possible state of the system is a unique point in phase space. The correspondence principle requires that we can move smoothly from the quantum limit to the classical limit as our system gets bigger, which means that at some level the Hilbert space and the phase space of the system must be equivalent to each other. How, then, can the Hilbert space have three dimensions when the phase space has six?

One way to quickly explain this discrepancy is to point out that we've short-changed our quantum mechanical descriptions of systems from the beginning. All of our quantum mechanics has been time-independent, meaning that we've averaged over the motions of our particles and allowed only stationary states, *states that can remain stable forever*, to be among our solutions. We don't demand this constraint of the classical system. Time-dependent quantum mechanics allows the wavefunctions to have a complex phase factor, so now each coordinate n has a real and an imaginary component, which effectively increases the size of the Hilbert space to six dimensions, like the phase space.

Now we wish to change the triple integral in the quantum mechanical expression Eq. 3.45 to an integral over the six classical coordinates. The quantum number n_X corresponds directly to excitation of motion along the Cartesian axis X, so we can replace the integral over n_X by an integral over the velocity component v_X:

$$\varepsilon_X = \frac{n_X^2 \pi^2 \hbar^2}{2ma^2} = \frac{mv_X^2}{2} \qquad \text{by Eq. 2.3}$$

$$n_X^2 = \frac{m^2 a^2}{\pi^2 \hbar^2} v_X^2 \qquad \text{solve for } n_X^2$$

$$n_X = \pm \frac{ma}{\pi \hbar} v_X = \pm \frac{2ma}{h} v_X \qquad \text{solve for } n_X$$

$$dn_X = \frac{2ma}{h} dv_X \tag{3.46}$$

$$\int_1^\infty f(n_X)\, dn_X \approx \int_0^\infty f(v_X) \frac{2ma}{h} dv_X. \qquad \text{by Eq. 3.46}$$

We change the lower limit of the integral from 0 to $-\infty$ because the velocity component v_X should be allowed to have either positive or negative values, and each value of n_X applies to motion in both the $-X$ and $+X$ directions:

$$\int_1^\infty f(n_X)\, dn_X = \frac{1}{2} \int_{-\infty}^\infty f(v_X) \frac{2ma}{h} dv_X$$

$$= \frac{ma}{h} \int_{-\infty}^\infty f(v_X)\, dv_X. \tag{3.47}$$

Extending this result to all three coordinates in Eq. 3.45, we get

$$q_{\text{trans}}(T,V) = \left(\frac{m}{h}\right)^3 abc \int_{-\infty}^\infty \int_{-\infty}^\infty \int_{-\infty}^\infty e^{-\varepsilon/(k_B T)}\, dv_X dv_Y dv_Z. \tag{3.48}$$

To introduce the position coordinates X, Y, and Z, we just multiply this by 1:

$$V = \int_0^a \int_0^b \int_0^c dX dY dZ$$

$$1 = \frac{1}{V} \int_0^a \int_0^b \int_0^c dX dY dZ$$

$$q_{\text{trans}}(T,V) = \left(\frac{m}{h}\right)^3 V \left\{ \int_{-\infty}^\infty \int_{-\infty}^\infty \int_{-\infty}^\infty e^{-\varepsilon/(k_B T)}\, dv_X dv_Y dv_Z \right\} \left\{ \frac{1}{V} \int_0^a \int_0^b \int_0^c dX dY dZ \right\}$$

$$= \frac{1}{V} \left(\frac{m}{h}\right)^3 V \int_{-\infty}^\infty \int_{-\infty}^\infty \int_{-\infty}^\infty \int_0^a \int_0^b \int_0^c e^{-\varepsilon/(k_B T)}\, dv_X dv_Y dv_Z dX dY dZ \tag{3.49}$$

$$= \left(\frac{m}{h}\right)^3 \int_{-\infty}^\infty \int_{-\infty}^\infty \int_{-\infty}^\infty \int_0^a \int_0^b \int_0^c e^{-\varepsilon/(k_B T)}\, dv_X dv_Y dv_Z dX dY dZ \tag{3.50}$$

For now, our energy ε depends only on the speeds of the particles, but the expression in Eq. 3.50 will also be valid when the energy changes with *position*. In other words, this way of writing the partition function lets us include not only the kinetic energy of translational motion, but any potential energy terms that arise as well. We break up the total energy ε into its kinetic and potential contributions K and U:

$$q_{\text{trans}}(T,V) = \left(\frac{m}{h}\right)^3 \int_{-\infty}^\infty \int_{-\infty}^\infty \int_{-\infty}^\infty \int_0^a \int_0^b \int_0^c e^{-(K+U)/(k_B T)}\, dv_X dv_Y dv_Z dX dY dZ$$

$$= \left(\frac{m}{h}\right)^3 \int_{-\infty}^\infty \int_{-\infty}^\infty \int_{-\infty}^\infty e^{-K(v_X, v_Y, v_Z)/(k_B T)}\, dv_X dv_Y dv_Z$$

$$\times \int_0^a \int_0^b \int_0^c e^{-U(X,Y,Z)/(k_B T)}\, dX dY dZ$$

$$\equiv q_K'(T) q_U'(T,V). \tag{3.51}$$

What we've done here is introduce a volume-independent kinetic energy contribution to the partition function:

$$q'_K(T) = \left(\frac{m}{h}\right)^3 \int_{-\infty}^{\infty} \int_{-\infty}^{\infty} \int_{-\infty}^{\infty} e^{-K(v_X,v_Y,v_Z)/(k_B T)} dv_X dv_Y dv_Z \qquad (3.52)$$

$$= \left(\frac{m}{h}\right)^3 \left[\int_{-\infty}^{\infty} e^{-mv_X^2/(2k_B T)} dv_X\right]\left[\int_{-\infty}^{\infty} e^{-mv_Y^2/(2k_B T)} dv_Y\right]\left[\int_{-\infty}^{\infty} e^{-mv_Z^2/(2k_B T)} dv_Z\right]$$

$$= \left(\frac{m}{h}\right)^3 \left[\left(\frac{2\pi k_B T}{m}\right)^{1/2}\right]\left[\left(\frac{2\pi k_B T}{m}\right)^{1/2}\right]\left[\left(\frac{2\pi k_B T}{m}\right)^{1/2}\right]$$

$$= \left(\frac{2\pi m k_B T}{h^2}\right)^{3/2}, \qquad (3.53)$$

and a term q'_U called the **configuration integral** that contains all the dependence on any potential energy terms. If the potential energy is zero inside the box, then the configuration integral reduces quite nicely:

$$q'_U(T,V) = \int_0^a \int_0^b \int_0^c e^{-U(X,Y,Z)/(k_B T)} dX dY dZ$$

$$= \left[\int_0^a e^{-(0)/(k_B T)} dX\right]\left[\int_0^b e^{-(0)/(k_B T)} dY\right]\left[\int_0^c e^{-(0)/(k_B T)} dZ\right]$$

$$= [a][b][c] = V. \qquad (3.54)$$

Combining these two terms gives us the same translation partition function we obtain using Method 1, Eq. 3.44:

$$q_{\text{trans}}(T,V) = q'_K(T)q'_U(T,V) = \left(\frac{2\pi m k_B T}{h^2}\right)^{3/2} V.$$

DERIVATION SUMMARY The Translational Partition Function. In both methods of deriving Eq. 3.44, we begin by summing $e^{-\varepsilon/(k_B T)}$ over all the quantum states, and then we approximate this sum by an integral over the energy. In Method 1, we use the expression for the density of states $W(\varepsilon)$, Eq. 3.39, to account for several states having the same energy. Then the integral turns out to be one we can solve analytically. In Method 2, we divide the integral up into kinetic and potential energy contributions, and then—instead of integrating over the *energy*—we integrate over all possible velocities (in the kinetic energy term) and over all possible positions (in the potential energy term). This way, we account for all the translational states, essentially recalculating $W(\varepsilon)$ on the fly. Both integrals are solved analytically and then combined to give Eq. 3.44.

The advantage of Method 2 is that we will be able to adapt this partition function to systems with more realistic potential energy functions by re-evaluating the configuration integral $q'_U(T,V)$ alone. The integral $q'_K(T)$ will remain unchanged with

$$q'_K(T) = \left(\frac{2\pi m k_B T}{h^2}\right)^{3/2}. \qquad (3.55)$$

The overall partition function here remains unitless, but the configuration integral has units of volume and $q'_K(T)$ has units of 1/volume.

3.5 Temperature and the Maxwell-Boltzmann Distribution

If the gas in our sample is a monatomic ideal gas (no rotations or vibrations), and we assume there is too little energy to affect the electronic states, then energy can be stored only in the translational degrees of freedom. The total energy E is then the sum over the kinetic energies of all the atoms:

$$E = \sum_{j=1}^{N} \left(\frac{1}{2}mv_j^2\right) = \frac{1}{2}mN\langle v^2 \rangle, \tag{3.56}$$

where m is the atomic mass, v_j the velocity of atom j, N the total number of atoms, and $\langle v^2 \rangle$ the squared velocity averaged over all the atoms. If we replace $\langle v^2 \rangle$ with $\langle v_X^2 + v_Y^2 + v_Z^2 \rangle$, this reduces to

$$E = \frac{1}{2}mN\langle v_X^2 + v_Y^2 + v_Z^2 \rangle = \frac{3}{2}mN\langle v_X^2 \rangle. \tag{3.57}$$

The X, Y, and Z directions are all equivalent, so $\langle v_X^2 \rangle = \langle v_Y^2 \rangle = \langle v_Z^2 \rangle$.

To obtain any one of these averages, we just need Eq. 3.16 from our introduction to the equipartition principle: for any degree of freedom such that the energy varies as $\varepsilon = ax^2$, the average of x^2 is given by

$$\langle x^2 \rangle = \frac{k_B T}{2a}.$$

In our case, the x value is v_X, v_Y, or v_Z and $a = m/2$, which leads us to

$$\langle v_X^2 \rangle = \langle v_Y^2 \rangle = \langle v_Z^2 \rangle = \frac{k_B T}{m}. \tag{3.58}$$

Substituting back into the expression for the total energy (Eq. 3.57) gives

$$E = \frac{3}{2}mN\left(\frac{k_B T}{m}\right) = \frac{3}{2}Nk_B T = \frac{3}{2}nRT. \tag{3.59}$$

This is exactly the result we would obtain from the equipartition principle, using Eq. 3.20 and setting $N_{ep} = 3$ for the three translational coordinates.

EXAMPLE 3.5 Translational Probability Distribution

CONTEXT A bolometer is a device that detects heat by monitoring the resistance of a temperature-sensitive material. Although they were developed as detectors for radiation at long wavelengths, bolometers have also been used in the laboratory to intercept molecules. When a molecule under study strikes the bolometer element, it transfers heat to the detector. When a vibrationally excited molecule strikes the element, it transfers even more heat and increases the detector signal. This effect has been used as an energy detector for molecules carried in beams of helium or other noble gases. In this problem, we estimate roughly what the momentum would be for a helium atom traveling in a beam when it strikes a bolometer element, using the root mean square (rms) momentum along the X axis as a representative value.

PROBLEM Calculate the root mean square momentum along the X axis $\langle p_X^2 \rangle^{1/2}$ in kg m s^{-1} of a helium atom in the gas-phase when the temperature is 312 K. (We don't want to calculate $\langle p_X \rangle$ because that should average to zero, motion in the $-X$ and $+X$ directions being equally likely.)

SOLUTION Use the classical average value theorem. We are looking for the average of $p_X^2 = (mv_X)^2$, a function that depends on v_X. The translational energy along the X axis $mv_X^2/2$ is the contribution to the energy that depends on that coordinate, the energy we put into the exponent is $mv_X^2/2$:

$$\langle p_X^2 \rangle = \frac{\displaystyle\int_0^\infty (mv_X)^2 e^{-mv_X^2/(2k_B T)} dv_X}{\displaystyle\int_0^\infty e^{-mv_X^2/(2k_B T)} dv_X} = \frac{\dfrac{m^2\sqrt{\pi}}{4(m/(2k_B T))^{3/2}}}{\dfrac{1}{2}\sqrt{\dfrac{\pi}{m/(2k_B T)}}}$$

$$= mk_B T.$$

The definite integrals were solved using the solutions in Table A.5.

The equipartition principle also predicts this result. Using K_X for the kinetic energy for motion along the X axis,

$$p_X^2 = 2mK_X = 2m(k_B T/2) = mk_B T.$$

Substituting in the mass of one helium atom $m = 6.67 \cdot 10^{-27}$ kg and

$$k_B T = (1.381 \cdot 10^{-23}\ \text{J K}^{-1})(312\ \text{K}) = 4.31 \cdot 10^{-21}\ \text{J, we get}$$

$$\langle p_X^2 \rangle^{1/2} = \sqrt{(6.67 \cdot 10^{-27}\ \text{kg})(4.31 \cdot 10^{-21}\ \text{J})} = 5.36 \cdot 10^{-24}\ \text{kg m s}^{-1}.$$

With Eq. 2.24 we found that the temperature describes how fast the energy of our sample increases relative to the degeneracy of states. A simpler, but less rigorous, concept of the temperature may be taken from our result for the energy of our monatomic ideal gas: *the temperature is roughly the average kinetic energy in the easily excited degrees of freedom in the sample.* This raises two questions:

1. Why does T depend on only the *easily excited* degrees of freedom? Because a degree of freedom that isn't excited doesn't contribute to the degeneracy of quantum states, and therefore it doesn't affect the temperature according to our definition in Eq. 2.24 (see Fig. 3.7).

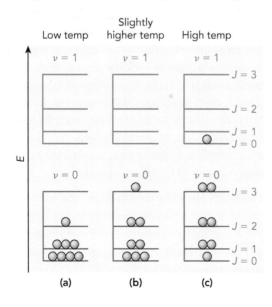

◀ FIGURE 3.7 **Easily excited degrees of freedom.** Temperature is not determined by motions that require much more energy to excite than is easily available. In this case, the vibration represented by quantum number v does not contribute appreciably to the energy of the molecule when the temperature is low. Heating slightly, from **(a)** to **(b)**, is not enough to put a significant number of molecules in the $v = 1$ excited state, so the vibration does not count as an available degree of freedom. If we go to much higher temperatures **(c)**, the vibration becomes an available degree of freedom for storing energy.

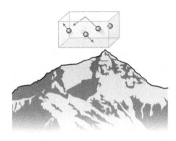

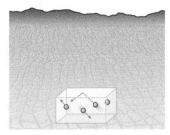

▲ **FIGURE 3.8 Temperature and kinetic energy.** Changing the potential energy of every particle equally—as by moving the sample from low altitude to high altitude—does not affect the motions of the particles or the number of available quantum states, so it does not affect the temperature.

2. Why does T depend on only the *kinetic* energy? Because once we've established the confines of our system and the forces at work, the potential energy (at least in our examples) becomes a fixed parameter. The kinetic energy is what we can adjust to change the degeneracy. The particle in a three-dimensional box, for example, has the same potential energy everywhere, no matter what its quantum state is. Increasing the kinetic energy is what takes us from one set of energy levels to the next set, thereby changing the degeneracy and therefore (according to Eq. 2.24) the temperature. Carrying that box from Death Valley to the top of Mt. Everest increases its potential energy, but if the particle has the same kinetic energy, then the degeneracy of states and the temperature stay the same (see Fig. 3.8).

The Boltzmann constant k_B is then just a conversion factor between the temperature T and the kinetic energy per each accessible degree of freedom. The *average* kinetic energy of the individual atoms is expressed in terms of the temperature by

$$\frac{E}{N} = \langle \varepsilon \rangle = \frac{3k_B T}{2}. \tag{3.60}$$

Because the temperature reflects a parameter averaged over the molecules in the sample, it is an *intensive* parameter, like the pressure.

This picture also points out why temperature is not interchangeable with energy. The temperature gives the average kinetic energy in any easily excited degree of freedom (for example, translation) of a large group of particles, measured from the lowest possible energy for the sample, the ground state.[5] The temperature is insensitive to the exact number of molecules or the number of degrees of freedom available for them to store energy. The system energy E, in contrast, increases with the number of molecules or accessible degrees of freedom, if T is kept constant.

EXAMPLE 3.6   Temperature and Translational Energy

CONTEXT Nitrogen has a remarkable role in environmental biochemistry. Organisms need nitrogen to form the amino part of amino acids, which then form proteins. But where does the nitrogen come from? Nitrogen has a hard time combining with other elements to form rocks, and as a result the earth's crust is *incredibly* nitrogen-poor: nitrogen is the 31st most abundant element (by mole, not by mass) in the earth (after yttrium, for goodness' sake), when it is the 6th most abundant in the universe. Instead, the nitrogen is taken from the air and converted to ammonia, through a process called **nitrogen fixation,** primarily by microorganisms. Iron-based enzymes catalyze this reaction at temperatures near 300 K, where the average kinetic energy of a single nitrogen molecule (as we'll find) is nowhere near enough to overcome the activation barrier of over $500 \, \text{kJ mol}^{-1}$ for the gas-phase reaction to combine N_2 and H_2.

PROBLEM Find the average translational kinetic energy (in kJ mol^{-1}) and the rms speed along the X axis $\langle v_X^2 \rangle^{1/2}$ for a nitrogen molecule at 298 K.

---

[5]Any zero-point energy E_{zp} is immaterial to a determination of the temperature, because it is a constant for the sample and therefore $\partial E_{zp}/\partial S = 0$.

SOLUTION A useful quantity is the thermal energy:

$$k_BT = (1.381 \cdot 10^{-23}\,\text{J}\,\text{K}^{-1})(298\,\text{K}) = 4.115 \cdot 10^{-21}\,\text{J} = 2.478\,\text{kJ}\,\text{mol}^{-1}.$$

(Let's keep one too many significant digits until the end.) This figure is a useful one to remember: the thermal energy at room temperature is about $2.5\,\text{kJ}\,\text{mol}^{-1}$. The average energy is then

$$\varepsilon = \frac{3k_BT}{2} = 3.72\,\text{kJ}\,\text{mol}^{-1}.$$

The rms speed along X is given by

$$\langle v_X^2 \rangle^{1/2} = \left(\frac{k_BT}{m}\right)^{1/2} = \left(\frac{4.115 \cdot 10^{-21}\,\text{J}}{(28.00\,\text{amu})(1.661 \cdot 10^{-27}\,\text{kg}\,\text{amu}^{-1})}\right)^{1/2} = 297\,\text{m}\,\text{s}^{-1}.$$

Here's another good number to remember: typical molecular speeds in a room temperature gas are on the order of $100\,\text{m}\,\text{s}^{-1}$. We'll return to this example of N_2 at 298 K a few more times to help us stay familiar with these benchmark numbers.

We have the translational energies and the translational partition function. What, at last, is the probability distribution for the translational states? The canonical velocity distribution gives the probability that the particle has a velocity vector with specific values of v_X, v_Y, and v_Z, all within some range, so this is a probability per unit speed cubed. We write that distribution as $\mathcal{P}_{\vec{v}}(\vec{v})$, and we evaluate it as follows:

$$\mathcal{P}_{\vec{v}}(\vec{v}) = \frac{e^{-mv^2/(2k_BT)}}{\displaystyle\int_{-\infty}^{\infty}\int_{-\infty}^{\infty}\int_{-\infty}^{\infty} e^{-mv^2/(2k_BT)}\,dv_X\,dv_Y\,dv_Z} = \frac{e^{-mv^2/(2k_BT)}}{h^3 q_K'(T)/m^3}$$

$$= \left(\frac{m}{2\pi k_BT}\right)^{3/2} e^{-mv^2/(2k_BT)}, \tag{3.61}$$

where we have taken advantage of the integral $q_K'(T)$, which appeared in Eq. 3.53 to normalize the probability distribution for velocity.

Usually, we are more interested in the probability distribution for speed than the distribution for the velocity vectors, $\mathcal{P}_{\vec{v}}(\vec{v})$. The function $\mathcal{P}_v(v)$ gives the fractional probability per unit speed of finding a molecule with speed v traveling in *any* direction, and it is obtained by integrating $\mathcal{P}_{\vec{v}}(\vec{v})$ over all angles to obtain the *Maxwell Boltzmann distribution* (Eq. 1.27):

$$\mathcal{P}_v(v)dv = \int_0^{2\pi}\int_0^{\pi} \mathcal{P}_{\vec{v}}(\vec{v})v^2\,dv\,\sin\Theta d\Theta\,d\Phi$$

$$= 4\pi\mathcal{P}_{\vec{v}}(\vec{v})v^2 dv$$

$$\mathcal{P}_v(v) = 4\pi\left(\frac{m}{2\pi k_BT}\right)^{3/2} v^2 e^{-mv^2/(2k_BT)},$$

where we have converted the volume element $dv_X\,dv_Y\,dv_Z$ to $v^2\,dv\sin\Theta\,d\Theta\,d\Phi$ just as one would convert from Cartesian to polar coordinates for an integral over positions. In the same way that r is the magnitude of the position vector, v is the magnitude of the velocity vector, and also like r, v is independent of direction and cannot have negative values.

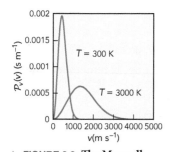

▲ **FIGURE 3.9 The Maxwell-Boltzmann distribution.** Molecular speeds are plotted for N_2 (mass = 28 amu) at 300 K and 3000 K.

This rigorous derivation of the Maxwell-Boltzmann distribution of speeds takes advantage of the translational partition function $q'_K(T)$ that we obtained earlier. The Maxwell-Boltzmann distributions graphed in Fig. 3.9 illustrate the main qualitative features of any typical canonical distribution:

1. There is little population at the lowest speeds, because there the translational degeneracy is small.

2. There is little population at the highest speeds, because there the canonical exponential term $e^{-mv^2/(2k_BT)}$ is small.

3. As the temperature increases, the distribution is shifted toward higher energies.

We see the same effect shaping the Maxwell-Boltzmann distribution for the translational states (Fig. 3.9) that shapes the probability distribution for the rotational states. There are two competing effects: the degeneracy gradually increases with the energy ε, but the energy term $e^{-\varepsilon/(k_BT)}$ drops rapidly once ε gets larger than k_BT.

EXAMPLE 3.7   **The Canonical Distribution: Speed**

CONTEXT A long-running challenge in our ability to probe chemical reactions has been finding the *minimum* collision energy between two reactants necessary for the reaction to take place. A typical sample has molecules spread out over a wide range of kinetic energies, so we normally calculate activation barriers from indirect measurements made over a range of temperatures. An elegant technique to make *direct* measurements of this threshold involves generating narrow beams of the reactant molecules and sending one or both beams through a velocity selector. A typical velocity selector consists of two or more spinning wheels, with holes drilled in each wheel to let the beam pass when the hole lines up with the beam. By timing the rotation rates of the two wheels, it is possible to deflect all the molecules in the beam that are not within a certain velocity window. These experiments are among the most detailed probes of chemical reactions being done, providing extremely sensitive tests of our theoretical understanding of physical chemistry.

PROBLEM In a 1.00 mole sample of N_2 at 298 K, calculate the number of molecules that will have speeds v within 0.005 m s^{-1} of 10.0 m s^{-1}.

SOLUTION The probability function $\mathcal{P}_v(v)$ will give us the fraction of molecules at any given speed. We need to plug the mass of N_2 and the value of v^2 into Eq. 1.27. The mass of N_2 is

$$(28.0 \text{ amu})(1.661 \cdot 10^{-27} \text{ kg amu}^{-1}) = 4.65 \cdot 10^{-26} \text{ kg}.$$

The squared velocity $v^2 = 1.00 \cdot 10^2 \text{ m}^2\text{s}^{-2}$. The thermal energy k_BT is $4.11 \cdot 10^{-21}$ J.

$$\mathcal{P}_v(v) = 4\pi \left(\frac{m}{2\pi k_BT} \right)^{3/2} v^2 e^{-mv^2/(2k_BT)}$$

$$= 4\pi \left[\frac{(4.65 \cdot 10^{-26} \text{ kg})}{2\pi(4.11 \cdot 10^{-21} \text{ J})} \right]^{3/2} (10.0 \text{ m s}^{-1})^2$$

$$\times \exp\left\{ -(4.65 \cdot 10^{-26}\text{kg})(1.00 \cdot 10^2 \text{ m}^2\text{s}^{-2}) / \left[2(4.11 \cdot 10^{-21}\text{ J}) \right] \right\}$$

$$= 3.03 \cdot 10^{-6} \text{s m}^{-1}.$$

This result is the fraction of N_2 molecules at $v = 10.0\,\mathrm{m\,s^{-1}}$ per "volume element" of speed. To estimate the total number of molecules, we multiply $\mathcal{P}_v(v)$ by the range of allowed values, $dv = 0.01\,\mathrm{m\,s^{-1}}$ (plus or minus $0.005\,\mathrm{m\,s^{-1}}$) and by the total number of molecules.

$$
\begin{aligned}
N(v) &= N\,\mathcal{P}_v(v)\,dv \\
&= (6.022 \cdot 10^{23}\,\mathrm{mol^{-1}})(3.03 \cdot 10^{-6}\,\mathrm{s\,m^{-1}})(0.01\,\mathrm{m\,s^{-1}}) \\
&= 1.83 \cdot 10^{16}.
\end{aligned}
$$

The Maxwell-Boltzmann distribution describes the whole set of molecules at once. If we want just a single, representative value, we can distill it from the distribution. For example, let's find an equation for the average molecular speed in our sample, in terms of the temperature T and the molecular mass m. We integrated the velocity components v_X, v_Y, and v_Z between $-\infty$ and ∞, but the speed is the *magnitude* of the velocity vector and cannot have negative values. When integrating over speed, we therefore integrate instead between 0 and ∞:

$$
\begin{aligned}
\langle v \rangle &= \frac{\displaystyle\int_0^\infty \mathcal{P}_v(v)\,v\,dv}{\displaystyle\int_0^\infty \mathcal{P}_v(v)\,dv} = \frac{\displaystyle\int_0^\infty e^{-mv^2/(2k_{\mathrm B}T)}\,v^3\,dv}{\displaystyle\int_0^\infty e^{-mv^2/(2k_{\mathrm B}T)}\,v^2\,dv} \\[2ex]
&= \frac{\dfrac{1}{2}\left(\dfrac{2k_{\mathrm B}T}{m}\right)^2}{\dfrac{1}{4}\left(\dfrac{2\pi k_{\mathrm B}^3 T^3}{m^3}\right)^{1/2}} = \left(\frac{8k_{\mathrm B}T}{\pi m}\right)^{1/2}
\end{aligned}
$$

$$
\langle v \rangle = \sqrt{\frac{8k_{\mathrm B}T}{\pi m}}. \tag{3.62}
$$

CONTEXT *Where Do We Go From Here?*

Extremes can usually be analyzed and understood more easily than what lies in between. In physical chemistry, the extremes are the quantum mechanical limit of individual atoms and molecules and the classical limit of moles and kilograms. The hardest job we have in physical chemistry is connecting the two. But it's also the most important job, because this connection lies at the center of our modern theory of chemistry. Molecular structure determines chemical behavior. The partition functions that we've derived in this chapter provide a means of summarizing key elements of the quantum mechanical properties of the particles in our system, and we call on them several more times through the end of this volume, for example to provide quantitative expressions for the equilibrium constant and rate constant of a chemical reaction.

Before we get to reactions, however, we need to move beyond a gas of non-interacting particles and see what happens when we let molecules bump into each other. This carries us a little from simpler extremes toward the complex realm of real systems. But that's where the excitement in chemistry is, and now we have the tools to explore it.

KEY CONCEPTS AND EQUATIONS

3.2 **Equipartition.** The wavefunction for any state of a system and the energy of that state is a function of the electronic, vibrational, rotational, and translational coordinates. Most molecules will be in the ground electronic state. For the remaining three types of coordinates, the energy contribution is roughly proportional to the square of the coordinate:

$$E_{vib}(v_{vib}, R) = \frac{1}{2}\mu v_{vib}^2 + k(R - R_e)^2$$

$$E_{rot}(J) \approx BJ^2$$

$$E_{trans}(v_X) = \frac{1}{2}mv_X^2.$$

If $k_B T$ is much greater than the energy spacing in that coordinate (B, ω, etc.) then the equipartition principle says that the energy per molecule in that coordinate is $k_B T/2$. The kinetic and potential terms of each vibrational coordinate are counted separately, and rotation about the inertial axes are counted separately (so linear molecules have two rotational coordinates; nonlinear molecules have three).

- The temperature is a measure of the amount of energy available to each coordinate. According to the equipartition principle, that amount of energy is $k_B T/2$ each for kinetic and potential energy in any *single, easily excited* coordinate.

3.3 and **3.4** **Partition Functions.** We are mainly interested in these forms of the partition function:

$$q_{vib}(T) = \sum_{v=0}^{\infty} e^{-v\omega/(k_B T)} = \frac{1}{1 - e^{-\omega/(k_B T)}}. \quad \text{Eq. 3.26, if harmonic}$$

$$q_{rot(linear)}(T) = \sum_{J=0}^{\infty} (2J + 1)e^{-BJ(J+1)/(k_B T)}$$

$$\approx \frac{k_B T}{B} \qquad \text{Eq. 3.30 if } k_B T \gg B$$

$$q_{rot(nonlinear)} \approx \left(\frac{\pi k_B^3 T^3}{ABC}\right)^{1/2} \quad \text{Eq. 3.32 if } k_B T \gg A, B, C$$

$$q_{trans}(T, V) = \left(\frac{2\pi m k_B T}{h^2}\right)^{3/2} V \qquad \text{Eq. 3.44}$$

We treat all vibrations, rotations, and translations as independent coordinates, which means

$$q_{total}(T) = q_{trans}(T)\, q_{rot}(T)\, q_{vib(1)}(T)\, q_{vib(2)}(T) \cdots q_{vib(3N_{atom}-6)}$$

where each of the $3N_{atom} - 6$ (or $3N_{atom} - 5$ for linear molecules) vibrational modes of the molecule is given its own partition function q_{vib}. Each q_{vib} depends only on the vibrational constant ω for that mode.

KEY TERMS

- The **equipartition principle** predicts that each degree of freedom in kinetic energy and each in potential energy tends to absorb about $\frac{1}{2}k_B T$ of energy per molecule, as long as $k_B T$ is much less than the gaps between adjacent quantum energy levels for that degree of freedom.
- The **universal gas constant** R is the molar version of the Boltzmann constant k_B: $R = \mathcal{N}_A k_B$.

- **Nuclear spin statistics** determine how the degeneracies of particular rotational quantum states may be affected by the exchange of equivalent nuclei during the rotation.
- The **configuration integral** q'_U is essentially a partition function for the potential energy of one particle, integrated over all positions.

OBJECTIVES REVIEW

1. *Estimate translational, rotational, and vibrational energies as a function of temperature by means of the equipartition principle.*
   Roughly what would be the contribution to the internal energy of 3.50 mol of O_2 at 355 K from each of the following: translation, rotation, and vibration?

2. *Employ the Taylor series approximation to the vibrational partition function to estimate vibrational state populations.*
   Estimate the fraction of F_2 molecules in the $v = 1$ state at 428 K ($\omega_e = 891 \text{ cm}^{-1}$).

3. *Employ the integral approximation to the rotational partition function to estimate rotational state populations in linear molecules.*
Estimate the fraction of HF molecules in the $J = 4$ state at 428 K ($B = 20.956$ cm^{-1}).

4. *Apply the classical average value theorem to determine mean properties of the system related to these degrees of freedom.*
Show that the average momentum vector $\langle p \rangle$ is zero in the Maxwell-Boltzmann distribution.

PROBLEMS

Discussion Problems

3.1 Begin with a sample of 1.00 mol N_2 gas at 300 K. For each change listed in the following table, estimate the factor by which the energy E_0 of the original sample changes. Assume that the equipartition principle holds, with vibrations *not included* for temperatures below 1000 K and *included* for temperatures above 1000 K.

Changing original sample by	Multiplies E_0 by factor of
raising temperature by 300 K	
raising temperature by 900 K	
removing 0.50 mol N_2 at 300 K	
adding 3.00 mol CO_2 at 300 K	
adding 3.00 mol CO_2 and raising temperature by 900 K	

3.2 Give the value of each probability for a rotating linear molecule:

a. $\lim_{T \to 0} \mathcal{P}(J = 0) =$

b. $\lim_{T \to 0} \mathcal{P}(J = 4) =$

c. $\lim_{T \to \infty} \mathcal{P}(J = 4) =$

Equipartition

3.3 According to the equipartition principle, what is the overall energy in J of one mole of water at 1000 K if translational, rotational, and vibrational degrees of freedom are considered?

3.4 How many equipartition degrees of freedom N_{ep} are present in H_2CO

a. when vibrations are *not* included?

b. when vibrations *are* included?

3.5 Assume we find a system with an energy per molecule $\varepsilon = c(n - 1)^6$ and degeneracies $g = n$, where $n = 1, 2, 3, \ldots$. If the temperature is high enough that this becomes a roughly continuous (classical) distribution of states, write the integral you could use to calculate the average overall energy E in a sample of N molecules at temperature T.

3.6 The F_2 molecule has a vibrational constant of 891 cm^{-1}. Find the minimum temperature at which the actual average vibrational energy in a sample of F_2 gas is within 10% of the value predicted by equipartition.

3.7 Using only the equipartition principle, estimate the difference in energy, including vibrations, in J at 1000 K between (a) 1.000 mol IBr_3 gas and (b) the mixture of 1.500 mol $Br_2 + 0.500$ mol I_2. THINKING AHEAD ▶ [What degrees of freedom are important in each example?]

3.8 As an approximate rule, a particular vibrational mode in a molecule will contribute to the equipartition principle energy at temperatures where $\frac{3}{2} k_B T \geq \omega_e$ for that mode. In the following table are listed all the vibrational constants for several molecules. A sample of 1.00 mol of one of these molecules is heated from 350 K to 360 K, absorbing 698 J of energy in the process. Show which molecule is most likely the one in the sample.

molecule	vibrational constants (cm^{-1})								
AsI_3	71	71	94	219	224	224			
CCl_4	217	217	314	314	314	459	776	776	776
CH_4	1306	1306	1306	1534	1534	2917	3019	3019	3019
Cl_2CO	285	440	567	580	849	1827			
CI_4	90	90	125	125	125	178	555	555	555

3.9 The disinfectant iodoform (CHI_3) has a vapor pressure of $4.5 \cdot 10^{-5}$ bar at 298 K. Assume that all the vibrational modes *except* those involving the H atom contribute to its energy according to the equipartition principle. In a sample of air at 1.00 bar and 298 K containing iodoform at its vapor pressure, estimate the fraction of the total equipartition energy that is present in the iodoform. Assume that air can be treated as a gas of diatomic molecules with no vibrational excitation at this temperature.

Rotational and Vibrational Partition Functions

3.10 The vibrational constant for the (non-degenerate) inversion or "umbrella" motion of NH_3 is 950 cm^{-1}. Calculate the percentage of NH_3 molecules in a sample at 373 K that will be in the $v = 1$ state of the inversion mode.

3.11 Carbonyl sulfide (OCS) has a bending mode with quantum number v_2, degeneracy $g_2 = v_2 + 1$, and a vibrational constant $\omega_2 = 520$ cm^{-1}. If 0.100% of the OCS molecules in a sample can be found in the $v_2 = 2$ vibrational level, then find the temperature T.

3.12 The probability of a linear molecule being in a particular rotational level J depends on the temperature, and for $J > 0$ there is a temperature T_{max} at which the probability for that level reaches a maximum. Derive an expression for T_{max} as a function of J and the rotational constant B, assuming that $k_B T \gg B$. **THINKING AHEAD** ▶ [Once we have an expression for T as a function of J, how do we find T_{max}?]

3.13 Use the equipartition principle to estimate the average relative speed in cm s^{-1} of the two Cl atoms vibrating in the $^{35}Cl_2$ molecule ($\omega_e = 564.9$ cm^{-1}) at 2500 K. Explain whether this value should be higher or lower than the correct value.

3.14 The equipartition principle was applied to rotation in the text by assuming that $J^2 \gg J$ so that E_{rot} could be approximated by BJ^2. Now solve for $\langle E_{rot} \rangle$ explicitly, using the average value theorem but this time using $E_{rot} = BJ(J + 1)$ without the above assumption.

3.15 H_2 has a vibrational constant $\omega_e = 4401.21$ cm^{-1} and a bond energy $D_0 = 435.99$ kJ mol^{-1}. Treating the vibrations as harmonic, estimate how many H_2 molecules in a 10.00 mol sample at 1000 K have enough vibrational energy to spontaneously dissociate. In other words, how many have $E_{vib} > D_0$? Is this an upper or lower limit on the true number of such molecules? **THINKING AHEAD** ▶ [How can we find the probability of being in any state above some vmax without having to add up contributions to $v = \infty$?]

3.16 Infrared studies of atmospheric contaminants in low abundance can be hampered by H_2O and CO_2 absorptions. The strength of these absorptions is roughly proportional to the number of molecules in the initial state of the transition. Find an equation for the J value from which the strongest absorptions will take place for a linear molecule with rotational constant B at temperature T.

3.17 Sodium dimer, Na_2, has a vibrational constant ω_e of only 159 cm^{-1}, whereas for H_2 $\omega_e = 4401$ cm^{-1}. If a sample of Na_2 at 298 K has the *same energy* and number of moles as a sample of H_2, use the equipartition principle to estimate the temperature of the H_2 sample. Both samples are in the gas phase.

3.18 Spectroscopy reveals that 1.0% of the molecules in a sample of $^{79}Br_2$ ($\omega_e = 325.29$ cm^{-1}) at 325 K are in a particular excited vibrational state. Find the value of v for that state.

3.19 Find an expression in the limit of the harmonic oscillator approximation for the probability that a diatomic molecule is in *any* vibrationally excited state.

3.20 Find an equation for the probability that a low-frequency harmonic oscillator with vibrational constant ω_e is in vibrational state v where $v > 0$. Use the first-order Taylor series expansion to simplify the expression in the limit that $k_B T \gg \omega_e$.

3.21 Some reactions of organic molecules with iodine occur best at temperatures of up to 500 °C (773 K), which just happens to be equal to $2.50 \omega_e / k_B$. At that temperature, we may approximate the vibrational energy as a continuous function of the quantum number v. Use that approximation to estimate what fraction of I_2 molecules will be in vibrational states with $v \leq 4$.

3.22 A sample of Cl_2 gas ($\omega_e = 560$ cm^{-1}) is prepared at 240 K, where we can approximate that all the molecules are in the ground vibrational state. The sample is heated as we monitor the number of molecules in the $v = 0$ state. Estimate the temperature when the signal drops to 0.75 of its original value.

3.23 The vibrational constant ω_T for the tritium molecule, $^3H_2$, is lower than the vibrational constant ω_H for $^1H_2$ by a factor of $\sqrt{3}$. The temperature of a sample of $^1H_2$ gas is such that 1.0% of the molecules are in the $v = 1$ state. At the same temperature, what fraction of $^3H_2$ molecules would be in $v = 1$? **THINKING AHEAD** ▶ [Should so decreasing the gap between energy levels increase or decrease the fraction of molecules in the excited state?]

3.24 Find the rotational partition function for chloroacetylene gas (HCCCl) at 500 K. The rotational constant is $B = 0.186$ cm^{-1}.

3.25 The rotational partition function for a sample of carbonyl sulfide (OCS) is found to be 25.0. Above roughly what J value do we expect to see the population (the number of molecules per J level) rapidly decreasing?

3.26 At 298 K, what rotational state in a sample of 1 mol HF ($B = 20.9557\,\text{cm}^{-1}$) will come closest to having the same population as $J = 10$ in a sample of 1 mol IBr ($B = 0.0559\,\text{cm}^{-1}$) at the same temperature?

3.27 The absorption intensity of a single rovibrational transition is proportional to the number of molecules in the initial state of the transition. The intensity is also proportional to the factor $J + J' + 1$, where J and J' are the initial and final state rotational quantum numbers, respectively. (The factor $J + J' + 1$ accounts for the relative orientation of the radiation electric field and the molecular dipole moment.) A portion of the rovibrational absorption spectrum of CO ($B = 3.83\,\text{cm}^{-1}$) taken at 100 K contains a sequence of three adjacent R branch lines with relative intensities (in order of increasing J) 97:100:85. Identify the initial J values of the transitions.

3.28 Find the rotational constant in cm^{-1} of a compound that has 7.0% of its molecules in the ground rotational state when cooled to a temperature of 40 K.

3.29 Assuming the integral approximation is valid for the rotational partition function and that the harmonic approximation is valid for vibrations, write an expression for the fraction of the sample of a gas-phase, diatomic molecule that will be in the rotational *and* vibrational ground state ($J = 0$, $v = 0$), in terms of the temperature T and the constants B and ω_e.

3.30 Find the integral approximation to the vibrational partition function for a degenerate bend in a linear molecule. The energy is $E_{\text{vib}} = \omega_e v$ and the degeneracy is given by $g_{\text{vib}} = v + 1$.

3.31 Calculate the rotational partition function of H_2 at 273 K and at 77 K (the boiling point of N_2). Use the direct sum rather than trying to find an integral expression. The rotational constant is $60.81\,\text{cm}^{-1}$.

3.32 Derive an equation in terms of B and T for the value of J that has the highest population, that is, the value of J that gives the highest value of $\mathcal{P}(J)$. Assume that the molecule is linear and that $k_B T \gg B$.

3.33 The rotational constants of HF and DF are 20.956, cm^{-1} and $11.00\,\text{cm}^{-1}$, respectively. In an isotopically pure sample of hydrogen fluoride at 298 K,

a. what percentage of the HF molecules will be in the $J = 4$ state?

b. what percentage of the DF molecules will be in the $J = 4$ state?

3.34 At 1000 K, what percentage of the molecules in a sample of N_2 gas ($\omega_e = 2359.6\,\text{cm}^{-1}$) are in the $v = 0$ state?

3.35 Calculate the fraction of CO molecules ($B = 1.88\,\text{cm}^{-1}$, $\omega_e = 2170\,\text{cm}^{-1}$) in the rovibrational ground state at $T = 77$ K.

3.36 Calculate the rotational and vibrational partition functions for CO_2 at 300 K. The rotational constant is $B = 0.391\,\text{cm}^{-1}$, and the vibrational constants are $\omega_1 = 1388\,\text{cm}^{-1}$, $\omega_2 = 667\,\text{cm}^{-1}$, and $\omega_3 = 2284\,\text{cm}^{-1}$.

3.37 An isotopically enhanced sample has n_{DI} moles of deuterated hydrogen iodide and n_{HI} moles of common HI at 398 K. Comparison of the infrared intensities of the $v = 1 \rightarrow 2$ transition finds a ratio of 0.100 for DI molecules ($\omega_e = 1640\,\text{cm}^{-1}$) in the $v = 1$ state to HI molecules ($\omega_e = 2310\,\text{cm}^{-1}$) in the $v = 1$ state. What is the ratio $n_{\text{DI}}/n_{\text{HI}}$ in the entire sample?

3.38 HCl has a rotational constant of $10.59\,\text{cm}^{-1}$ and a vibrational constant of $2991\,\text{cm}^{-1}$. Find the temperature, if there is one, at which the $J = 0$ and $J = 1$ rotational states have the same population. Then find the temperature, if there is one, at which the $v = 0$ and $v = 1$ states have the same population.

3.39 Derive the integral approximation to the vibrational partition function under the harmonic approximation and verify that it agrees with the exact solution under the appropriate conditions.

3.40 The linear C_3 molecule has a remarkably low bending vibrational constant, $\omega_2 = 63\,\text{cm}^{-1}$. Recall that the degeneracy of a π bending mode $g_{\text{bend}}(v) = v + 1$. Find the most probable value of v for the bending state at 298 K.

3.41

a. HCO is an asymmetric top molecule formed during the combustion of acetylene. Use appropriate approximations to calculate the partition function for this molecule at $T = 500$ K, given that its rotational constants are $A = 729.366$ GHz, $B = 44.788$ GHz, and $C = 41.930$ GHz, and its vibrational constants are $1868\,\text{cm}^{-1}$ (CO stretch), $1081\,\text{cm}^{-1}$ (bend), and $2483\,\text{cm}^{-1}$ (CH stretch). Include the degeneracy due to rotation, but neglect the hydrogen hyperfine structure, and assume that the hyperfine contribution to the energy is negligible.

b. In calculating the rotational energy, use the equation for a prolate symmetric top, replacing B with $(B + C)/2$. Calculate $\mathcal{P}(v = 0, K_a = 0, J)$ for $J = 0$, 5, 10, 15, 20, 25, and 30 for HCO at $T = 500$ K.

3.42 The possible spin states for one ^{14}N nucleus, which has spin $I = 1$, are $m_I = 1$ (α), $m_I = 0$ (β), and $m_I = -1$ (γ).

a. We saw that the possible spin states for *two* indistinguishable $I = 1/2$ nuclei in a diatomic molecule are $\alpha\alpha$, $\alpha\beta + \beta\alpha$, $\alpha\beta - \beta\alpha$, and $\beta\beta$ (neglecting normalization). List all the distinct *two-particle* nuclear spin states for $^{14}\text{N}_2$.

b. Put an "**e**" next to each of these nuclear spin states that will be associated with even J rotational levels of the N_2 molecule.

3.43 Imagine a set of three N_2 molecules frozen in a crystal so that no rotation or translation can occur. But there is energy in vibration, which totals $E = 3\omega_e$, where the vibrational constant $\omega_e = 2358\,\text{cm}^{-1}$. If we apply the equipartition principle to this sample, what temperature would give this energy?

3.44 Let's examine the sample described in Problem 3.43 more closely.

a. What is the ensemble size Ω of this sample? (Because the N_2 molecules are frozen in space, they are *distinguishable* molecules. Also, neglect any anharmonicity.)

b. What is the entropy (in SI units) of this sample?

c. The ensemble size when $E = 2\omega_e$ is 6. Estimate the temperature for the system between $E = 2\omega_e$ and $E = 3\omega_e$ using the following approximation:

$$T = \left(\frac{\partial E}{\partial S}\right)_{N,V} \approx \Delta E / \Delta S.$$

The Maxwell-Boltzmann Distribution

3.45 Calculate the average magnitude of the momentum for motion in *any* direction in kg m s^{-1} for N_2 molecules in air at 1 bar and 300 K.

3.46 Calculate the average speed of gas-phase UF_6 at 298 K.

3.47 According to the Bohr model of the atom, the $n = 1$ electron in a hydrogen atom orbits the nucleus with a speed of $2.19 \cdot 10^6\,\text{m s}^{-1}$. At what temperature would this be the average speed of a sample of free electrons?

3.48 Calculate the probability of exciting an electron in a one-dimensional box (actually a nanoscale wire) to the $n = 2$ excited state if the box is 10.0 nm long and the temperature is 400.0 K. For the one-dimensional box, $E_n = n^2\pi^2\hbar^2/(2ma^2)$ and the levels are non-degenerate (but remember our convention for measuring energy now). For this example, $\pi^2\hbar^2/(2m_ea^2)$ is equal to $6.02 \cdot 10^{-22}\,\text{J}$ and the partition function is 2.44.

3.49 The dispersion $\langle(\Delta x)^2\rangle$ of a parameter x measured over a large sample is defined as $\langle(x - \langle x\rangle)^2\rangle$. The dispersion gives an idea of how widely spread out the values are in a distribution. Find $\langle(\Delta v)^2\rangle$ for an ideal gas with molecular mass m as a function of temperature T, where v is the speed of the molecule.

3.50 A 2.00 mole sample of N_2 gas is at 298 K. What is the highest speed at which (on average) one molecule out of the entire sample can be found?

3.51 In a gas mixture of 25% He ($m = 4.00\,\text{amu}$) and 75% Ar ($m \approx 40.0\,\text{amu}$) at constant temperature T, the He atoms move on average faster than the Ar atoms. However, with more Ar than He, there is a speed v' at which equal numbers of Ar atoms and He atoms have speeds $v > v'$. Set up an integral equation that you could use to numerically calculate the value of v'. Show what numerical values you would use where necessary, being sure that your units are consistent. If you have an integration tool handy, calculate the value of v'.

3.52 Write (but do not solve) the integral that would tell us how many molecules in one mole of neon gas at 298 K would have speeds greater than $1.00\,\text{km s}^{-1}$. Provide all numerical values needed, including any conversion factors.

3.53 Earth is losing its supply of helium by escape from the atmosphere, because helium has a high median speed and (unlike hydrogen, which has even less mass) it does not chemically react to form heavier molecules as it diffuses into the stratosphere. Write the integral necessary to calculate the fraction of helium atoms at 273 K that have speeds greater than $11\,\text{km s}^{-1}$, roughly the escape velocity. Include any necessary numerical values and unit conversions.

3.54 The function $\mathcal{P}_v(v)$ is graphed in the accompanying figure for a sample of Ne gas at 300 K. Sketch in the graphs of $\mathcal{P}_v(v)$ you would find for (a) Ar gas at 300 K and (b) Ne gas at 150 K.

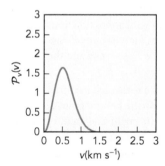

3.55 Calculate the kinetic integral $q'_K(T)$ for N_2 at 300 K. Use this to estimate the fraction of N_2 molecules traveling at speeds within $\pm 500\,\text{cm s}^{-1}$ of 10^3, 10^4, 10^5, and $10^6\,\text{cm s}^{-1}$.

3.56 We obtained an equation for the average speed $\langle v\rangle$ for an ideal gas (Eq. 3.62) using the classical average value theorem. Use the same approach to prove that the average position of one of these gas molecules along the x axis is equal to $a/2$ (i.e., the middle of the box).

PART I
EXTRAPOLATING FROM MOLECULAR TO MACROSCOPIC SYSTEMS

PART II
NON-REACTIVE
MACROSCOPIC
SYSTEMS

PART III
REACTIVE
SYSTEMS

4 Statistical Mechanics and Molecular Interactions

GOAL *Why Are We Here?*

The goal of this chapter is to explore the application of statistical mechanics to systems in which we can no longer neglect the interactions between the molecules. These are relatively advanced topics, but we will be rewarded with a more complete picture of the behavior of gases and with a useful approximation for describing liquids in some of our work beyond this chapter. We will conclude with an examination of how statistical mechanics determines the behavior of certain substances at very low temperatures, focusing particularly on the relatively new field of Bose-Einstein condensates.

CONTEXT *Where Are We Now?*

We've been introduced to several tools for extending molecular properties to large numbers of molecules: the ergodic hypothesis, the canonical distribution of energies, and the equipartition principle, to name three. However, we've kept to idealized conditions while trying these tools out, neglecting the forces between the molecules.

The equipartition principle (Section 3.2) shows that diatomic molecules in the gas phase have roughly half of their energy in the translational motion. If we start turning on the interactions between the molecules, we find that these translational coordinates become associated with the intermolecular forces. Can we relate the intermolecular forces and the bulk properties of this non-ideal gas? Extending our work to a more accurate and sophisticated molecular model is our next task, and it serves as an extended introduction to the central problem of the chemical model: predicting how the microscopic structure of individual molecules determines the behavior of the bulk material. With the ideal gas, we could steer clear of this question. Now we drive straight at it.

SUPPORTING TEXT | *How Did We Get Here?*

The main qualitative preparation we need for the work ahead is a brief survey of how molecular interactions work and the simplifications we commonly use to model them. That survey will form the first section of the chapter. We will also draw on the following equations and sections of text:

- Equation 2.41 gives a general expression for the pressure that uses partial derivatives to relate the pressure to energy, volume, and entropy:

$$P = -\left(\frac{\partial E}{\partial V}\right)_{S,N} = \left(\frac{\partial E}{\partial S}\right)_{V,N}\left(\frac{\partial S}{\partial V}\right)_{E,N}.$$

- The classical average value theorem, Eq. 2.2, calculates the average of a property $A(\overrightarrow{r})$, weighting it by the probability distribution $\mathcal{P}_{\overrightarrow{r}}(\overrightarrow{r})$:

$$\langle A \rangle = \int_{\text{all space}} \mathcal{P}_{\overrightarrow{r}}(\overrightarrow{r})A(\overrightarrow{r})d\overrightarrow{r}.$$

- Section 2.1 explains that the fundamental particles that make up molecules—electrons and atomic nuclei—are classified as either *fermions* (which have half-integer spin) or *bosons* (which have integer spin). Individual electrons, protons, and neutrons all have spins of 1/2 and are fermions. When we combine protons and neutrons, the resulting atomic nucleus is a fermion if the mass number is odd and a boson if the mass number is even. For example, the ^{13}C nucleus (mass number 13) is a fermion, but the ^{14}C nucleus (mass number 14) is a boson. The Pauli exclusion principle applies to fermions: no two indistinguishable fermions may share the same complete set of quantum numbers. This rule does not apply to bosons.

- In Section 3.3, we saw that the spins of equivalent nuclei need to be taken into account when determining the populations of specific rovibrational states in a molecule. In particular, if the equivalent nuclei are fermions, then we must describe them using an antisymmetric wavefunction, which changes sign when we reverse the labels on the nuclei. If the equivalent nuclei are bosons, the wavefunction is symmetric.

4.1 Molecular Interactions

Section 2.1 reviews the properties of individual molecules that we need to justify the ideal gas law and the equipartition of energy in a gas. But we want to be able to predict the properties of real gases, liquids, and solids as well. The distinction between a gas and a liquid, and between a liquid and solid, reflects a balance between the available kinetic energy (which is determined by the temperature) and the potential energy (which is determined by the forces that each particle exerts on its neighbors). In this section we summarize how those forces work and the algebraic expressions that we use to approximate the corresponding potential energies.

Intermolecular Forces

In the **gas phase,** the average distance between molecules is usually so large that intermolecular interactions can be neglected. As we compress the gas, bringing the molecules together, the interactions become stronger and the impact on the properties of the sample become more profound. However, unless the molecules are highly reactive, the intermolecular interactions tend to remain weak compared to the forces that bind the atoms together within each molecule. The physical origins of the mathematical expressions for these interactions are described in *QM* Chapter 10, but here we need only know the expressions themselves and how to use them.

We start by setting the intermolecular potential energy for the sample to zero when there are no interactions between the particles at all. Repulsions between the particles then correspond to positive contributions to the total potential energy, and attractions correspond to negative contributions. Where the total potential energy is positive, for example, molecules will tend to be pushed away from that arrangement.

The most important of these interactions *is* the repulsion between the particles. If we are starting from a non-reactive gas, the electrons and nuclei have usually already organized themselves into molecules with a favorable distribution of charges. When we cram the molecules together, we don't improve the interaction between positive and negative charges but instead begin to press the like charges together—electrons into other electrons and nuclei toward other nuclei—and they push back against one another. The distance at which this repulsion becomes important is small, determined by the atomic radii of the interacting atoms, and it climbs very steeply as the distance between the molecules decreases (*QM* Eq. 10.5):

$$u_{\text{repulsion}} \approx Ae^{-2cR/a_0}. \tag{4.1}$$

If the two particles we push together can chemically react, then the repulsive potential energy curve represents the energy barrier that must be overcome before the reaction can take place. If the two particles do not react, then the repulsive potential corresponds to a much higher barrier that prevents two particles in the sample from occupying the same space. In either case, we will assume for now that the energy available in the sample is not enough to overcome those barriers.

There are also attractive forces, but the attractions between non-reacting molecules are significant over a much smaller range of temperatures than the repulsion. The attractive interactions correspond to negative potential energies, but if the thermal energy $k_{\text{B}}T$ can easily cancel that negative contribution, then most molecules are moving too fast for the attractive forces to strongly influence their motions. Where these attractive forces play a major role is at thermal energies low enough that molecules can be trapped by the attractive forces. Lowering the temperature then tends to bring the molecules of the gas together, first forming small clusters of molecules, then droplets, and eventually leading to formation of a condensed phase—a liquid or solid.

▶ FIGURE 4.1 **Definitions of the angles in the dipole–dipole interaction.** The angles θ_A and θ_B give the orientations of the two dipole moment vectors with respect to the z axis that connects the two molecules, whereas $\phi \equiv \phi_B - \phi_A$ gives the angle of rotation *around* the z axis between the two dipoles.

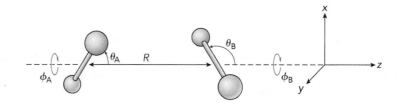

One of the strongest of these attractive intermolecular forces is the **dipole–dipole interaction.** The potential energy expression for the two rigid dipoles μ_A and μ_B interacting over a distance R is (*QM* Eq. 10.18)

$$u_{2-2}(R,\theta_A,\theta_B,\phi) = -\frac{2\mu_A\mu_B}{(4\pi\varepsilon_0)R^3}\left[\cos\theta_A\cos\theta_B - \frac{1}{2}\sin\theta_A\sin\theta_B\cos(\phi_B - \phi_A)\right], \quad (4.2)$$

where the angles θ_A, θ_B, and $\phi \equiv \phi_B - \phi_A$ are defined in Fig. 4.1. To simplify the notation, let's label $u_{2-2}(R,\theta_A,\theta_B,\phi)$ by u for the rest of this section.

When the two dipole moments both lie in the z axis, Eq. 4.2 simplifies to (*QM* Eq. 10.16)

$$u = -\frac{2\mu_A\mu_B}{(4\pi\varepsilon_0)R^3}. \quad (4.3)$$

In the gas and liquid phases, however, the dipole moments are not fixed in place, but rotate constantly. Because the potential energy of the interaction depends on the relative orientation of the two dipole moment vectors, we need to average the potential energy over all those orientations. We now show that the canonical distribution of energies greatly influences this average, leading to a potential energy function with a new R-dependence.

In the gas phase, the molecules rotate through orientations of all types, attractive *and* repulsive. Therefore, we get an accurate picture of this interaction only if we average over the rotational angles of the two molecules. To average u as given by Eq. 4.2 over all angles θ_A, θ_B, and ϕ (as abbreviated using the volume element $d\tau$), we use the classical average value theorem (Eq. 2.2):

$$\langle u \rangle = \int \mathcal{P}(u)\,u\,d\tau,$$

CHECKPOINT What we're seeing here is an application of the canonical distribution, $e^{-\varepsilon/(k_B T)}/q$, to the potential energy, after having applied it primarily to various forms of the kinetic energy in Chapters 2 and 3. Whereas kinetic energy depends on *motion* and is a function of particle velocities, potential energy depends on *position* and is a function of the Cartesian coordinates and orientation angles of our particles. To see these contributions better in the initial math, we again show the exponential energy term in red and the partition function in blue.

where the integral (and all the integrals in this section) is evaluated over all space. The canonical distribution tells us that $\mathcal{P}(u)$ will have the form

$$\mathcal{P}(u) = Ae^{-u/(k_B T)}, \quad (4.4)$$

where A is a normalization constant equal to $1/q$ such that

$$\int \mathcal{P}(u)\,d\tau = A\int e^{-u/(k_B T)}\,d\tau = 1,$$

so

$$A = \frac{1}{\int e^{-u/(k_B T)}\,d\tau}. \quad (4.5)$$

Substituting this integral in for A, we obtain a more explicit expression for the average value of the dipole–dipole potential:

$$\langle u \rangle = A\int e^{-u/(k_B T)}\,u\,d\tau$$

$$= \frac{\int e^{-u/(k_B T)} u \, d\tau}{\int e^{-u/(k_B T)} d\tau}. \qquad (4.6)$$

To carry out the angular averaging while preserving R as a variable, we then integrate over all angles with the volume element $d\tau' = \sin\theta_A \, d\theta_A \sin\theta_B \, d\theta_B \, d\phi$, treating R for now as a constant. We assume that the interaction between the molecules is weak—in other words, that we are at a temperature significantly higher than where the gas would condense. In this limit, we can set $u \ll k_B T$ and therefore use the Taylor series approximation (Eq. A.25) for e^x,

$$\text{for } |x| \ll 1: \quad e^x \approx 1 + x,$$

to replace $e^{-u/(k_B T)}$ by $1 - u/(k_B T)$:

$$\langle u \rangle_{\theta, \phi} = \frac{\int e^{-u/(k_B T)} u \, d\tau'}{\int e^{-u/(k_B T)} d\tau'}$$

$$\approx \frac{\int \left(1 - \dfrac{u}{k_B T}\right) u \, d\tau'}{\int \left(1 - \dfrac{u}{k_B T}\right) d\tau'}$$

$$= \frac{\int u \, d\tau' - \dfrac{1}{k_B T} \int u^2 \, d\tau'}{\int d\tau' - \dfrac{1}{k_B T} \int u \, d\tau'}. \qquad (4.7)$$

CHECKPOINT The volume element $d\tau'$ here is taken from the angular (θ and ϕ) parts of the volume element $r^2 dr \sin\theta \, d\theta \, d\phi$ in spherical coordinates. After carrying out this rotational averaging, we find that the resulting dipole–dipole interaction has the same R−6 dependence shared by two other intermolecular interactions: the dipole-induced dipole and the dispersion force. The derivation here shows that this factor of R−6 stems largely from the symmetry of the dipole–dipole interaction, which causes the $u \, d\tau$ integrals in Eq. 4.7 to vanish.

It is possible to simplify things here considerably by observing that

$$\int u \, d\tau' = \int_0^\pi \int_0^\pi \int_0^{2\pi} \left(-\frac{2\mu_A \mu_B}{R^3}\right) \sin\theta_A d\theta_A \sin\theta_B d\theta_B \left[\cos\theta_A \cos\theta_B - \frac{1}{2}\sin\theta_A \sin\theta_B \cos\phi\right] d\phi$$

$$= -\frac{2\mu_A \mu_B}{R^3} \left\{\left[\int_0^\pi \cos\theta_A \sin\theta_A d\theta_A\right]\left[\int_0^\pi \cos\theta_B \sin\theta_B d\theta_B\right]\left[\int_0^{2\pi} d\phi\right]\right.$$

$$\left. -\frac{1}{2}\left[\int_0^\pi \sin^2\theta_A d\theta_A\right]\left[\int_0^\pi \sin^2\theta_B d\theta_B\right]\left[\int_0^{2\pi} \cos\phi d\phi\right]\right\} = 0,$$

because the integrals $\int_0^\pi \cos\theta \sin\theta \, d\theta$ and $\int_0^{2\pi} \cos\phi \, d\phi$ are both equal to zero. Returning to Eq. 4.7, we get

$$\langle u \rangle_{\theta, \phi} \approx -\frac{\dfrac{1}{k_B T} \int u^2 d\tau'}{\int d\tau'}$$

$$= -\left(\frac{4\mu_A^2 \mu_B^2}{(4\pi\varepsilon_0)^2 R^6 k_B T}\right) \frac{\int_0^\pi \int_0^\pi \int_0^{2\pi} \sin\theta_A d\theta_A \sin\theta_B d\theta_B d\phi \left[\cos\theta_A \cos\theta_B - \frac{1}{2}\sin\theta_A \sin\theta_B \cos\phi\right]^2}{\int_0^\pi \int_0^\pi \int_0^{2\pi} \sin\theta_A d\theta_A \sin\theta_B d\theta_B d\phi}$$

$$= -\frac{4\mu_A^2 \mu_B^2}{(4\pi\varepsilon_0)^2 R^6 k_B T} \frac{\frac{8\pi}{9} + \frac{4\pi}{9}}{8\pi} = -\left(\frac{4\mu_A^2 \mu_B^2}{(4\pi\varepsilon_0)^2 R^6 k_B T}\right) \frac{\frac{4\pi}{3}}{8\pi}$$

$$= -\frac{2\mu_A^2 \mu_B^2}{(4\pi\varepsilon_0)^2 3 R^6 k_B T}.$$

The dipole–dipole force is attractive, on average, because the molecules preferentially find stabilizing orientations. But as the molecules move farther away, the dipole–dipole potential at any orientation drops as R^{-3} *and* the ability of the molecules to find those stabilizing orientations drops as R^{-3} as well. The net effect is that the interaction strength is proportional to R^{-6}.

The dipole–dipole interaction is calculated on the basis of the *permanent* dipole moments of the two molecules. But an external electric field $\mathcal{E}$ will shift the electron distribution of a molecule, creating an **induced dipole moment** μ_{induced} of magnitude (*QM* Eq. 3.36)

$$\mu_{\text{induced}} = \alpha \mathcal{E}, \tag{4.8}$$

where α is the **polarizability** of the molecule. The polarizability is effectively defined by Eq. 4.8, and it measures the looseness of the electron binding to the nuclei: the more tightly bound the electrons are to the nuclei, the lower the polarizability. All molecules are polarizable to some extent, but electron-rich molecules (which often include heavy atoms, π orbitals, or electron lone pairs) tend to have higher values of α.

The dipole moment of a polar molecule A carries an electric field, which generates an induced dipole moment in any neighboring molecule B. The resulting **dipole-induced dipole force** between A and B is always attractive, tending to pull the two molecules toward each other. The potential energy for that interaction is proportional to the dipole moment of A and the polarizability of B (*QM* Eq. 10.23):

$$u_{2-2'}(R) = -\frac{4\mu_A^2 \alpha_B}{(4\pi\varepsilon_0)R^6}. \tag{4.9}$$

The dipole-induced dipole force provides a mechanism for interaction between polar and non-polar molecules, and it can also significantly increase the strength of the attraction between two polar molecules.

Even when two molecules are completely non-polar, there is an attraction between them. The **dispersion force** is a quantum mechanical effect that creates an attraction between all molecules, roughly proportional to the polarizabilities α on each molecule. For the interaction between two molecules of the same chemical composition, we can roughly approximate the potential energy of dispersion using the polarizability and the separation between the ground and lowest excited electronic states, ΔE (*QM* Eq. 10.38):

$$u_{\text{disp}} \equiv E_{\text{disp}} \approx -\frac{\alpha^2 \Delta E}{2R^6}. \tag{4.10}$$

This expression is accurate only to within about a factor of two, but it correctly predicts the R-dependence of the dispersion potential energy.

The three attractive potential energy functions we've considered—the averaged dipole–dipole, the dipole-induced dipole, and the dispersion potential energy—are all proportional to $-1/R^6$. To predict the properties of systems in which these intermolecular interactions are important, we need a single function that accurately mimics the shape of the correct potential energy function: a **model potential.**

We will make use of three model potentials. The most sophisticated combines the $-R^{-6}$ dependence typical of the attractive terms with a steeply rising repulsive wall, resulting in the **Lennard-Jones potential.** The Lennard-Jones potential represents the repulsive wall by a term proportional to R^{-12}, because this makes the entire function easier to manipulate (as we shall see in Section 4.2) than if it used an exponential term for the repulsion. The Lennard-Jones potential may be written (*QM* Eq. 10.44)

$$u_{\text{LJ}}(R) = \varepsilon\left[\left(\frac{R_e}{R}\right)^{12} - 2\left(\frac{R_e}{R}\right)^{6}\right]. \tag{4.11}$$

As shown in Fig. 4.2a, Eq. 4.11 describes a smooth curve that depends on only two adjustable parameters: the well depth ε and the equilibrium distance R_e (the distance at which the potential energy reaches its minimum value). In some cases, the equilibrium distance is not as important to us as the point at which the curve crosses zero, in which case we can also write the potential energy in terms of the Lennard-Jones distance R_{LJ}:

$$u_{\text{LJ}}(R) = 4\varepsilon\left[\left(\frac{R_{\text{LJ}}}{R}\right)^{12} - \left(\frac{R_{\text{LJ}}}{R}\right)^{6}\right]. \tag{4.12}$$

Two other model potentials replace the smooth curve of the Lennard-Jones potential with crudely drawn straight lines, but they have the advantage of greater mathematical simplicity. The **square well** potential (Fig. 4.2b) divides space into an impenetrable region at low R, an attractive region with constant well-depth ε, and a region at large R in which the interactions are turned off completely (*QM* Eq. 10.43):

$$u_{\text{sq}}(R) = \begin{cases} \infty & \text{if } R \leq R_{\text{sq}} \\ -\varepsilon & \text{if } R_{\text{sq}} < R \leq R'_{\text{sq}} \\ 0 & \text{if } R > R'_{\text{sq}} \end{cases} \qquad \text{square well} \tag{4.13}$$

The simplest model potential commonly used is the **hard sphere** (Fig. 4.2c), which eliminates the attractive well completely, retaining only the repulsion between molecules at small distances (*QM* Eq. 10.42):

$$u_{\text{hs}}(R) = \begin{cases} \infty & \text{if } R \leq R_{\text{hs}} \\ 0 & \text{if } R > R_{\text{hs}} \end{cases} \qquad \text{hard sphere} \tag{4.14}$$

The hard sphere potential treats the particles like billiard balls or marbles, bouncing off one another but incapable of sticking together.

The Condensed Phases: Liquids and Solids

The beginning of this section refers to a balance between the intermolecular forces and the kinetic energy of the particles that determines the phase of matter

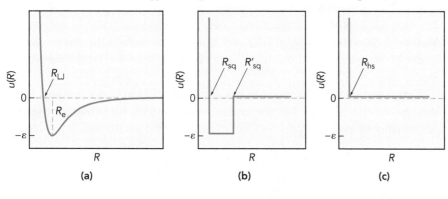

◄ FIGURE 4.2 **Three model potential energy functions.**
(a) The Lennard-Jones potential.
(b) The square well potential.
(c) The hard sphere potential.

of the substance. In a *solid* the strength of the forces holding the atoms together is so great compared to the forces trying to separate them that the atoms do not wander, although they each vibrate about some fixed position. The forces binding the solid may be chemical bonds or even the weaker intermolecular forces described earlier. There are no longer translations or rotations—those motions have been converted into vibrations.

Liquids are the most difficult phase of matter to describe, because the molecules are bound tightly enough to interact strongly but not tightly enough for them to stay fixed in location. As a result, descriptions of liquids usually rely on average properties of the structure. Foremost among these is the **pair correlation function** $\mathcal{G}(R)$, a unitless function that is proportional to the average number of particles at a distance R from any randomly selected particle:

$$\mathcal{G}(R) = \frac{\text{number of molecules at distance } R}{\text{number of molecules at distance } R \text{ in a random distribution}}. \quad (4.15)$$

A formal mathematical definition of $\mathcal{G}(R)$ may be written in which we calculate the average number of molecules $\overline{N}_{R_a,R_b}$ at a distance between R_a and R_b of our random reference molecule:

$$\overline{N}_{R_a,R_b} = 4\pi\rho \int_{R_a}^{R_b} \mathcal{G}(R)\, R^2 \, dR. \quad (4.16)$$

Figure 4.3 illustrates the typical shape of the pair correlation function: it is zero at $R = 0$ (because the reference molecule excludes others from that region), it peaks near one diameter (where one shell of molecules surrounds the reference molecule), and it converges toward one at large R (as the distribution of molecules relative to the reference becomes random). In liquids, additional peaks indicate a layering effect: once one shell has formed, a second shell can form around that. But each additional shell is more weakly influenced by the original reference molecule, and the distribution becomes more random. In a low-temperature compressed gas, we expect to see a single small peak in the $\mathcal{G}(R)$, indicating that the attractive forces such as dispersion make the molecules slightly more likely to be found near each other than separated by large distances. Using the tools we develop in this volume, we will find that we can draw upon the expressions for intermolecular interactions to predict a correct, though approximate, form of $\mathcal{G}(R)$.

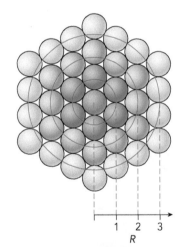

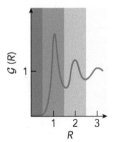

▲ **FIGURE 4.3 The pair correlation function.** The peaks in the pair correlation function $\mathcal{G}(R)$ occur at the distances from the reference particle where other particles are most likely to be found.

4.2 Pressure of a Non-Ideal Gas

With a few expressions for the intermolecular potential energy now available, we can next see how these interactions modify the ideal gas law. We will use our **Method 2** derivation of $q_{\text{trans}}(T,V)$ in Section 3.4 to include the intermolecular forces in our translational partition function, and then solve for the pressure as we did in Section 2.5.

Here is our strategy. We need to find how our solution to the pressure of the ideal gas can be influenced by the intermolecular forces. We solved for the ideal gas pressure by evaluating the product of two partial derivatives in Eq. 2.41:

$$P = -\left(\frac{\partial E}{\partial V}\right)_{S,N} = \left(\frac{\partial E}{\partial S}\right)_{V,N}\left(\frac{\partial S}{\partial V}\right)_{E,N}.$$

The first of these partial derivatives, $\left(\frac{\partial E}{\partial S}\right)_{V,N}$, defines the temperature T so the dependence on the intermolecular interactions must be entirely due to the second factor, a derivative of the entropy S. The entropy is determined by the ensemble size Ω, so we need to find how Ω is affected by these forces.

If two particles bounce off one another, or pull one another together, the degree of freedom that is directly affected is the *translational* motion of the two particles. Chapter 3 shows how the information that governs each degree of freedom is summarized in its partition function. The partition function, in turn, can relate this information to the energy E and the entropy S in our equation for the pressure. In Eq. 2.31,

$$\mathcal{P}(E) = \Omega(E)\mathcal{P}(i) = \frac{\Omega(E)e^{-E/(k_B T)}}{Q(T)},$$

we find that we can relate the partition function $Q(T)$ to the ensemble size $\Omega(E)$ through the canonical distribution. Because we want to look at the influence of each particle on the translational motions of the others, the partition function we need in this case is $Q_{\text{trans}}(T)$, the translational partition function for all N particles in our system.

The following derivation isn't easy, but it's one of the most dramatic examples available of how statistical mechanics lets us extrapolate from the molecular scale—in this case, the interaction between two molecules—up to the bulk. The experience of stepping through this derivation can provide both a *qualitative* understanding of how the large-scale behavior of a gas is affected by molecular structure and a tool for *quantitative* predictions of a laboratory system based only on molecular parameters. But if you're pressed for time, there is a DERIVATION SUMMARY ahead, just after Eq. 4.36.

Parameters key: the non-ideal gas

Symbol	Parameter	SI units
h	Planck's constant	J s
k_B	the Boltzmann constant	J K^{-1}
m	the mass of one particle	kg
a,b,c	the lengths of the sides of the container along X, Y, and Z	m
V	the volume of the container, equal to abc	m^3
T	the temperature	K
P	the pressure	Pa
N	the number of particles in the system	unitless
$\mathcal{N}_A$	Avogadro's number	mol^{-1}
v_X, v_Y, v_Z	the speeds of motion along X, Y, and Z	m s^{-1}
q_{trans}	the translational partition function for one particle	unitless
Q_{trans}	the translational partition function for N particles	unitless
Q'_K	the N-particle integral of $e^{-K/(k_B T)}$ over all velocity components	m^{-3N}
Q'_U	the N-particle configuration integral, the integral of $e^{-U/(k_B T)}$ over all positions	m^{3N}
K	the total kinetic energy for all N particles	J

(continued)

(continued) Parameters key: the non-ideal gas

Symbol	Parameter	SI units
U	the total potential energy of the N-particle system	J
E	the total energy of the N-particle system, equal to $K + U$	J
Ω	the total number of microstates in the system's ensemble	unitless
$u(R)$	the intermolecular potential energy function for two particles	J
R	the separation between the centers of mass of two particles	m
R	the universal gas constant, equal to $\mathcal{N}_A k_B$	$\mathrm{J\,K^{-1}\,mol^{-1}}$

The N-Particle Translational Partition Function

Our immediate goal is to find a suitable expression for the N-particle translational partition function. We can extend the expression for the single-particle function q_{trans} that we used in Eq. 3.51,

$$q_{\text{trans}}(T,V) = \left(\frac{m}{h}\right)^3 \int_{-\infty}^{\infty} \int_{-\infty}^{\infty} \int_{-\infty}^{\infty} e^{-K(v_X, v_Y, v_Z)/(k_B T)} dv_X dv_Y dv_Z$$

$$\times \int_0^a \int_0^b \int_0^c e^{-U(X,Y,Z)/(k_B T)} dX dY dZ,$$

CHECKPOINT We are starting from the basic definition of the partition function: we sum the canonical probability factor $e^{-\varepsilon/(k_B T)}/q$ over every possible state of the system. For the kinetic energy partition contribution, we sum over all possible velocity vectors. For the potential energy contribution, we sum over all possible positions of the particles.

to N particles by carrying out the integral over the velocity components and the positions for *each* of the N molecules. Furthermore, we will assume that the N particles are indistinguishable, so simply switching the labels on any set of particles does not make a new state. To account for the indistinguishability, we therefore divide the integral by $N!$ (as we did in Eq. 2.42) to obtain the many-particle translational partition function

$$Q_{\text{trans}}(T,V) = \frac{1}{N!}\left\{\left(\frac{m}{h}\right)^{3N} \int_{-\infty}^{\infty} \dots \int_{-\infty}^{\infty} e^{-K/(k_B T)} dv_{X1} dv_{Y1} dv_{Z1} \dots dv_{XN} dv_{YN} dv_{ZN}\right\}$$

$$\times \left\{\int_0^a \int_0^b \int_0^c \dots \int_0^a \int_0^b \int_0^c e^{-U/(k_B T)} dX_1 dY_1 dZ_1 \dots dX_N dY_N dZ_N\right\}$$

$$\equiv \frac{1}{N!} Q'_K(T) Q'_U(T,V), \tag{4.17}$$

where Q'_U is the N-particle version of the configuration integral defined in Eq. 3.54:

$$Q'_U(T,V) = \int_0^a \int_0^b \int_0^c \dots \int_0^a \int_0^b \int_0^c e^{-U/(k_B T)} dX_1 dY_1 dZ_1 \dots dX_N dY_N dZ_N. \tag{4.18}$$

The N-particle kinetic energy is the sum over all the single-particle classical kinetic energies,

$$K = \sum_{i=1}^{N} K_i = \sum_{i=1}^{N} \frac{m v_i^2}{2}, \tag{4.19}$$

and so the first term breaks up into a product of N integrals:

$$Q'_K(T) = \left(\frac{m}{h}\right)^{3N} \int_{-\infty}^{\infty} \dots \int_{-\infty}^{\infty} e^{-K/(k_B T)} dv_{X1} dv_{Y1} dv_{Z1} \dots dv_{XN} dv_{YN} dv_{ZN}$$

$$= \left(\frac{m}{h}\right)^{3N} \int_{-\infty}^{\infty} \cdots \int_{-\infty}^{\infty} e^{-(K_1+K_2+\ldots K_N)/(k_B T)}$$

$$\times \; dv_{X1} dv_{Y1} dv_{Z1} \ldots dv_{XN} dv_{YN} dv_{ZN} \qquad\qquad K = \sum_i^N K_i$$

$$= \left(\frac{m}{h}\right)^{3N} \left\{ \int_{-\infty}^{\infty} e^{-K_1/(k_B T)} dv_{X1} dv_{Y1} dv_{Z1} \right\}$$

$$\times \ldots \left\{ \int_{-\infty}^{\infty} e^{-K_N/(k_B T)} dv_{XN} dv_{YN} dv_{ZN} \right\}. \qquad e^{a+b} = e^a e^b$$

To simplify this expression, we recognize that each of these N integrals must have the same value, because each particle behaves in the same way as any other. This lets us rewrite $Q_K'(T)$ in terms of the single-particle function $q_K'(T)$, which we evaluated previously:

$$Q_K'(T) = \left\{ \left(\frac{m}{h}\right)^3 \int_{-\infty}^{\infty} e^{-mv^2/(2k_B T)} dv_X dv_Y dv_Z \right\}^N$$

$$= q_K'(T)^N$$

$$= \left(\frac{2\pi m k_B T}{h^2}\right)^{3N/2}. \qquad\qquad \text{by Eq. 3.53} \quad (4.20)$$

The potential energy contribution cannot be split up this way into single-particle contributions if the intermolecular interactions are important, because in that case the potential energy depends on the distances *between* particles, rather than the positions of individual particles. A completely general partition function for translation may therefore be written as the product of these kinetic and potential energy integrals, Eqs. 4.18 and 4.20:

$$\text{in general:} \quad Q_{\text{trans}}(T,V) = \frac{1}{N!}\left(\frac{2\pi m k_B T}{h^2}\right)^{3N/2} Q_U'(T,V). \qquad (4.21)$$

The kinetic energy contribution is unchanged from that calculated earlier for the ideal gas.[1] Only when the intermolecular potential is zero—when we have an ideal gas—can we separate the configuration integral into a term for each particle and solve:

$$Q_U'(T,V) = \int_0^a \int_0^b \int_0^c \cdots \int_0^a \int_0^b \int_0^c e^{-U/(k_B T)} dX_1 dY_1 dZ_1 \ldots dX_N dY_N dZ_N$$

$$= \int_0^a \int_0^b \int_0^c \cdots \int_0^a \int_0^b \int_0^c e^{-0/(k_B T)} dX_1 dY_1 dZ_1 \ldots dX_N dY_N dZ_N \qquad \text{if } U = 0$$

$$= \left\{ \int_0^a \int_0^b \int_0^c dX_1 dY_1 dZ_1 \right\} \ldots \left\{ \int_0^a \int_0^b \int_0^c dX_N dY_N dZ_N \right\} \qquad e^0 = 1$$

$$= \{V\} \ldots \{V\} = V^N.$$

---

[1] It may seem unrealistic that the kinetic energy partition function is the same as it was for the ideal case, when the intermolecular potential energies clearly affect the motions of the molecules. If we could somehow suddenly increase the attractive forces on all the molecules in a sample, we would expect the kinetic energies to change, but Q_K would keep the same form. The distribution of the kinetic energies is determined entirely by the canonical distribution *for the kinetic energy*. We have engineered this function specifically to be independent of the potential energy.

CHECKPOINT One of the
reasons we want to know the
partition function is because, as
we find in Eq. 4.24, through the
partition function we can find the
ensemble size Ω. The ensemble
size in turn is related through the
Boltzmann equation to the
entropy, a critical parameter to
understand if we are going to be
able to predict how our system
responds to changes. Later on
we'll see that, in fact, all of the
principal thermodynamic
properties of the system can be
calculated from the complete
partition function.

Combining this result with Eq. 4.20 gives the N-particle translational partition function for the ideal gas:

$$\text{ideal gas:} \quad Q_{\text{trans}}(T,V) = \frac{1}{N!}\left(\frac{2\pi m k_B T}{h^2}\right)^{3N/2} V^N. \qquad (4.22)$$

The translational partition function in Eq. 4.21 allows us also to obtain a general equation for Ω, the ensemble size of our system. In the microcanonical ensemble, the probability of the system being in any one translational state i that has energy E is simply the reciprocal of the ensemble size:

$$\mathcal{P}(i) = \frac{e^{-E/(k_B T)}}{Q_{\text{trans}}(T,V)} = \frac{1}{\Omega}. \qquad (4.23)$$

For example, if there are ten states of equal energy ($\Omega = 10$), the sample has a 0.1 chance of being found in state 1, a 0.1 chance of being found in state 2, and so on. Therefore,

$$\Omega = Q_{\text{trans}}(T,V)\,e^{E/(k_B T)} = \frac{1}{N!}\left(\frac{2\pi m k_B T}{h^2}\right)^{3N/2} Q'_U(T,V)e^{E/(k_B T)}, \quad (4.24)$$

and

$$\ln\Omega = \ln Q_{\text{trans}}(T,V) + \frac{E}{k_B T} \qquad (4.25)$$

$$= \ln\left[\frac{1}{N!}\left(\frac{2\pi m k_B T}{h^2}\right)^{3N/2}\right] + \ln Q'_U(T,V) + \frac{E}{k_B T}. \qquad (4.26)$$

The configuration integral $Q'_U(T,V)$ is the link that connects the intermolecular forces to the entropy.

Our next step is to introduce non-ideality into the configuration integral by turning on the intermolecular forces—repulsion between the molecules at small distances and attraction (from such terms as the dipole–dipole or dispersion forces) at intermediate distances.

Intermolecular Potential Energy in the Partition Function

The total potential energy U due to interaction between the component molecules of any system, gas or liquid or solid, is usually calculated by simply adding up the pairwise interaction energies u—the interaction energy between any two molecules j and k at positions (X_j, Y_j, Z_j) and (X_k, Y_k, Z_k):[2]

$$U(X_1,\ldots,Z_N) \approx \sum_{j=1}^{k-1}\sum_{k=2}^{N} u(X_j, Y_j, Z_j, X_k, Y_k, Z_k).$$

We use this idea to simplify the *average* total potential energy:

$$\langle U(X_1,\ldots,Z_N)\rangle = \frac{N(N-1)}{2}u(X_1, Y_1, Z_1, X_2, Y_2, Z_2). \qquad (4.27)$$

---

[2]This expression neglects the way one pair interaction affects the next pair interaction. For example, a polar molecule induces a dipole moment in a second, non-polar molecule. The induced dipole moment then changes the interaction potential energy of the second molecule with a third polar molecule, and so on. It is possible to include these many-body effects by iteration if necessary, but the pair potential approximation is at least a good starting point.

The second step is possible if we assume that the molecules are all indistinguishable.[3] In that case, any pairwise interaction is like any other, depending only on the distance between the two molecules and their relative orientation, not on their labels. If we write all the possible two-particle interactions as a square matrix with rows and columns indexed from 1 to N, as in Table 4.1 for $N = 5$, each matrix element represents an interacting pair and there are N^2 possible combinations. However, the N diagonal elements of the matrix each represent one particle interacting with itself, which we ignore, leaving $N^2 - N = N(N - 1)$. Of those remaining, half are identical to the other half. For example, the interaction pair $(1, 2)$ corresponds to the same contribution as $(2, 1)$, so only one of the pairs is counted. Therefore, the total number of distinct pair interactions is $N(N - 1)/2$.

Solving for the Pressure

We have managed so far to avoid setting any requirements on the intermolecular interactions, except to limit them to the pairwise case. Now we will simplify matters by letting the fluid be a gas, where the molecules interact just enough so that we still need to worry about it. A critical assumption we can make now is that the interaction potential for a given pair of molecules is zero over nearly the entire volume of the sample.

We rewrite the configuration integral by changing variables from X, Y, and Z to unitless variables scaled by the dimensions a, b, and c of the container:

$$m_X = \frac{X}{a} \quad m_Y = \frac{Y}{b} \quad m_Z = \frac{Z}{c}. \tag{4.28}$$

This lets us factor the volume $V = abc$ out of the volume element because

$$dX dY dZ = (a dm_X)(b dm_Y)(c dm_Z) = V dm_X dm_Y dm_Z, \tag{4.29}$$

TABLE 4.1 Pair interaction matrix for $N = 5$. There are $N^2 = 25$ entries, but the $N = 5$ diagonal elements and half of the remaining elements do not represent unique pairwise interactions. This leaves $N(N - 1)/2 = 10$ contributions (for this example) to Eq. 4.27.

	1	2	3	4	5
1	(1,1)	(1,2)	(1,3)	(1,4)	(1,5)
2	(2,1)	(2,2)	(2,3)	(2,4)	(2,5)
3	(3,1)	(3,2)	(3,3)	(3,4)	(3,5)
4	(4,1)	(4,2)	(4,3)	(4,4)	(4,5)
5	(5,1)	(5,2)	(5,3)	(5,4)	(5,5)

---

[3]We ignore for now the distinction between bosons and fermions, which would require us to count states more carefully. The result that we obtain here is valid in the limit that many quantum states are available, a condition that certainly applies to the translational states of a gas at all but the very lowest energies. Cases in which the results depend strongly on the particles being fermions or bosons include translational states of liquid helium (a cold, dense material) and Bose-Einstein condensates (*very* cold substances, but not so dense), which are discussed in Section 4.4, and rotational or vibrational states of symmetric molecules, which are discussed in Section 3.3.

and this also changes the upper limits of the integrals to 1. The configuration integral now has the following form:

$$Q'_U(T,V) \equiv \int_{-\infty}^{\infty} \ldots \int_{-\infty}^{\infty} e^{-U(X_1,\ldots,Z_N)/(k_BT)}\, dX_1 \ldots dZ_N \qquad \text{by Eq. 4.18}$$

$$= V^N \int_0^1 \ldots \int_0^1 e^{-U/(k_BT)} dm_{1X} \ldots dm_{NZ} \qquad \text{by Eq. 4.28}$$

$$= V^N \int_0^1 \ldots \int_0^1 e^{-N(N-1)u(R)/(2k_BT)} dm_{1X} \ldots dm_{NZ}. \quad \text{by Eq. 4.27} \quad (4.30)$$

We can treat this as the product of many identical triple integrals over $e^{-u(R)/(k_BT)}$, one triple integral for each pair of molecules. There are $N(N-1)/2$ pairs, so we set

$$Q'_U(T,V) = V^N \int_0^1 \ldots \int_0^1 \left[e^{-u(R)/(k_BT)} \right]^{N(N-1)/2} dm_{1X} \ldots dm_{NZ}$$

$$\approx V^N \left[\int_0^1 \int_0^1 \int_0^1 e^{-u(R)/(k_BT)} dm_X dm_Y dm_Z \right]^{N(N-1)/2}$$

$$= V^N \left[\int_0^1 \int_0^1 \int_0^1 dm_X dm_Y dm_Z + \int_0^1 \int_0^1 \int_0^1 \left(e^{-u(R)/(k_BT)} - 1 \right) dm_X dm_Y dm_Z \right]^{N(N-1)/2}$$

$$= V^N \left[1 + \frac{\mathcal{I}(T)}{V} \right]^{N(N-1)/2}, \qquad (4.31)$$

where

$$\mathcal{I}(T) \equiv V \int_0^1 \int_0^1 \int_0^1 \left(e^{-u(R)/(k_BT)} - 1 \right) dm_X dm_Y dm_Z. \qquad (4.32)$$

We have broken the integral up into two terms, because the second integrand, $e^{-U/(k_BT)} - 1 \approx -u(R)/(k_BT)$, includes all the *non-ideal* contributions to the solution. Because the non-ideality arises from interactions between the molecules, the integral $\mathcal{I}(T)$ vanishes once R gets to be large, which is over nearly the entire volume of the sample. Indeed, the integral $\mathcal{I}(T)$ is essentially volume-independent, because $u(R)/(k_BT)$ becomes negligible once R is larger than a few angstroms. To simplify this integral, we can change the volume element using Eq. 4.29 to replace $dm_X dm_Y dm_Z$ by $dXdYdZ/V$, and then transform from Cartesian to spherical coordinates, setting $dXdYdZ$ to $4\pi R^2 dR$:

$$\mathcal{I}(T) = V \int_0^1 \int_0^1 \int_0^1 \left(e^{-u(R)/(k_BT)} - 1 \right) dm_X dm_Y dm_Z$$

$$= \int_0^a \int_0^b \int_0^c \left(e^{-u(R)/(k_BT)} - 1 \right) dXdYdZ$$

$$= \int_0^{R_{max}} \int_0^{\pi} \int_0^{2\pi} \left(e^{-u(R)/(k_BT)} - 1 \right) R^2 dR \sin\Theta d\Theta\, d\Phi.$$

$$= 4\pi \int_0^{R_{max}} \left(e^{-u(R)/(k_BT)} - 1 \right) R^2 dR, \qquad (4.33)$$

where R_{max} is the distance to the wall in any direction. Finally, since $u(R)$ approaches zero when R is large, in which case $e^{-u(R)/(k_B T)} - 1$ also approaches zero, we don't have to worry about the actual value of R_{max}; integration to any very large number will give the same result:

$$\mathcal{I}(T) = 4\pi \int_0^\infty \left(e^{-u(R)/(k_B T)} - 1 \right) R^2 dR. \tag{4.34}$$

Finally, we can solve for the pressure:

$$P = \left(\frac{\partial E}{\partial S} \right)_{V,N} \left(\frac{\partial S}{\partial V} \right)_{E,N} \qquad \text{by Eq. 2.41}$$

$$= T \left(\frac{\partial (k_B \ln \Omega)}{\partial V} \right)_{E,N} \qquad \text{by Eqs. 2.11 and 2.24}$$

$$= k_B T \left(\frac{\partial \ln \Omega}{\partial V} \right)_{E,N}. \qquad \text{factor out } k_B \quad (4.35)$$

At this point, we use Eq. 4.26 to rewrite the derivative of $\ln \Omega$ in terms of the configuration integral:

$$\ln \Omega = \ln \left[\frac{1}{N!} \left(\frac{2\pi m k_B T}{h^2} \right)^{3N/2} \right] + \ln Q'_U(T,V) + \frac{E}{k_B T} \qquad \text{by Eq. 4.26}$$

$$\left(\frac{\partial \ln \Omega}{\partial V} \right)_{E,N} = 0 + \left(\frac{\partial \ln Q'_U(T,V)}{\partial V} \right)_{E,N} + 0 \qquad \text{only } Q'_U \text{ depends on } V$$

$$= \frac{1}{Q'_U(T,V)} \left(\frac{\partial Q'_U(T,V)}{\partial V} \right)_{E,N}. \qquad d \ln x = dx/x$$

We replace the partial derivative in Eq. 4.35 with this result, obtaining

$$P = k_B T \left(\frac{1}{Q'_U(T,V)} \right) \left(\frac{\partial Q'_U(T,V)}{\partial V} \right)_{E,N} \qquad \text{by Eq. 4.35}$$

$$= k_B T \left(\frac{1}{Q'_U(T,V)} \right) \left\{ \frac{\partial}{\partial V} \left[V^N \left(1 + \frac{\mathcal{I}(T)}{V} \right)^{N(N-1)/2} \right] \right\}_{E,N} \qquad \text{by Eq. 4.31}$$

$$= \frac{k_B T}{Q'_U(T,V)} \left[N V^{N-1} \left(1 + \frac{\mathcal{I}(T)}{V} \right)^{N(N-1)/2} \right. \qquad d(xy) = x dy + y dx$$

$$\left. + \frac{N(N-1) V^N}{2} \left(1 + \frac{\mathcal{I}(T)}{V} \right)^{[N(N-1)/2]-1} \left(-\frac{\mathcal{I}(T)}{V^2} \right) \right]$$

$$= \frac{k_B T}{Q'_U(T,V)} \left[\frac{N}{V} Q'_U(T,V) - \frac{N(N-1)}{2V^2} \frac{\mathcal{I}(T) Q'_U(T,V)}{1 + \frac{\mathcal{I}(T)}{V}} \right]. \qquad \text{by Eq. 4.31}$$

Letting $N(N-1) = N^2$ (because $N \gg 1$) and $1 + \mathcal{I}(T)/V = 1$ (because $V \gg \mathcal{I}$), we have an equation for the pressure of a *non-ideal* gas:

$$P = \frac{N k_B T}{V} - \frac{N^2 k_B T}{2V^2} \mathcal{I}(T)$$

$$= \frac{nRT}{V} - \frac{n^2 RT}{2V^2} \mathcal{N}_A \mathcal{I}(T). \qquad R = \mathcal{N}_A k_B \quad (4.36)$$

DERIVATION SUMMARY Pressure of a Non-ideal Gas. To obtain Eq. 4.36, we start from an equation that expresses the pressure as the product of two partial derivatives: one equal to the temperature T and the other a derivative of entropy with respect to volume. To find the volume dependence of the entropy, we used a relationship between ensemble size Ω and the configuration integral Q'_U. The ensemble size gives the Boltzmann entropy, and the configuration integral folds together all effects from the forces on the particles, including the repulsion of the walls and the intermolecular interactions. Carefully applying approximations that preserved the effect of the intermolecular forces while simplifying the math, we arrive at Eq. 4.36, which has the form of the ideal gas law with an added correction for non-ideality. The non-ideality is given by an integral (Eq. 4.34) over a function of the intermolecular potential energy.

Deviations from the ideal gas law were measured before we had the statistical theory to arrive at Eq. 4.36. Next we'll look at the two most common forms of the non-ideal gas laws and how they are consistent with Eq. 4.36.

The Virial Expansion

The first method rewrites the pressure as a power series in the density. This approach allows the density-dependence of experimental measurements to be succinctly expressed in tables, without demanding any theoretical interpretation. Equation 4.36 is already consistent with that form. Replacing V/n by the molar volume V_{m}, we get the first two terms in a power series of $1/V_{\mathrm{m}}$:

$$P = RT\left[\frac{1}{V_{\mathrm{m}}} + B_2(T)\frac{1}{V_{\mathrm{m}}^2}\right], \tag{4.37}$$

where

$$B_2(T) \approx -\frac{1}{2}\mathcal{N}_A\mathcal{I}(T) = -2\pi\mathcal{N}_A\int_0^\infty (e^{-u(R)/(k_{\mathrm{B}}T)} - 1)R^2 dR. \tag{4.38}$$

In general there are higher order terms in the power series:

$$P = RT\left[\frac{1}{V_{\mathrm{m}}} + B_2(T)\frac{1}{V_{\mathrm{m}}^2} + B_3(T)\frac{1}{V_{\mathrm{m}}^3} + \ldots\right]. \tag{4.39}$$

This is called the **virial expansion,** and the constant B_i is the ith **virial coefficient.** Our derivation is only valid for the virial expansion up to the B_2 term. Our assumption in Eq. 4.31—that the integral of u over all N molecules amounted to the product of $N(N-1)$ identical integrals—neglects the couplings between three or more molecules that give rise to higher-order terms. Note that the second virial coefficient B_2 has units of volume mol^{-1}. The SI units for B_2 are therefore $\mathrm{m}^3\,\mathrm{mol}^{-1}$, but tabulated values are usually given in $\mathrm{L}\,\mathrm{mol}^{-1}$.

What has happened is this: by including a fairly general term for the intermolecular potential energy, the equation of state has changed from the ideal gas law to an equation in which the pressure P no longer has a simple inverse proportionality to volume at constant temperature. If we ignore terms beyond the second virial coefficient, we obtain a quadratic equation for the molar volume V_{m},

$$P = RT\left[V_{\mathrm{m}}^{-1} + B_2(T)V_{\mathrm{m}}^{-2}\right]$$

$$\left(\frac{P}{RT}\right)V_m^2 - V_m - B_2(T) = 0,$$

which can be solved using the quadratic formula:

$$V_m = \frac{RT}{2P}\left[1 + \sqrt{1 + 4B_2(T)P/(RT)}\right]. \qquad (4.40)$$

For helium, B_2 (273 K) is about 0.0222 L mol^{-1}, and for argon $B_2(273$ K) is -0.0279 L mol^{-1}. Plugging these values into Eq. 4.40, using the value 0.083145 bar L K^{-1} mol^{-1} for the gas constant R, predicts that at 273.15 K and 1 bar, 1 mole of He occupies 22.733 L and 1 mole of argon occupies 22.683 L, in contrast to the ideal gas value 22.711 L. At higher densities, the fractional discrepancy increases, so at 10 bar and 273.15 K, 1 mole helium would occupy 2.293 and 1 mole of argon would occupy 2.243 L when the ideal gas value is 2.271 L.

The van der Waals Coefficients

A second way of expressing the pressure of a non-ideal gas breaks the molecular interactions into two classes, repulsion and attraction, to provide a clearer picture of the trends in behavior. For example, why should the molar volume of helium at 273.15 K be higher than ideal (22.733 L), while argon's is lower (22.683 L)? This can be answered by looking back at the definition of $B_2(T)$ in terms of the intermolecular potential. We found (Eq. 4.38)

$$B_2(T) \approx -2\pi\mathcal{N}_A \int_0^\infty (e^{-u(R)/(k_B T)} - 1)R^2 dR.$$

We divide the intermolecular potential into two regions: one at very low R in which the Coulomb repulsion dominates and $u(R) \gg k_B T$, and a second region in which the weak attractive forces control the interactions. At low R, $u(R)/(k_B T)$ gets very large and $e^{-u(R)/(k_B T)}$ approaches 0. At high R, $|u(R)|$ is small compared to $k_B T$, and $e^{-u(R)/(k_B T)}$ can be approximated by $1 - u(R)/(k_B T)$.

When two identical, spherical molecules get close enough to strongly repel each other, the centers of mass are separated by twice the molecular radius. Let's model the pair potential energy by the Lennard-Jones potential (Eq. 4.12). The two regions of our intermolecular potential are roughly $R \leq R_{LJ}$ (for the small R limit, where $u(R)$ approaches infinity) and $R > R_{LJ}$ (for the large R limit). This lets us break up the integral into two parts:

$$B_2(T) \approx -2\pi\mathcal{N}_A \int_0^{R_{LJ}}\left[(e^{-u(R)/(k_B T)} - 1)\right]R^2 dR$$

$$-2\pi\mathcal{N}_A \int_{R_{LJ}}^\infty\left[(e^{-u(R)/(k_B T)} - 1)\right]R^2 dR$$

$$\approx -2\pi\mathcal{N}_A \int_0^{R_{LJ}}(0 - 1)R^2 dR - 2\pi\mathcal{N}_A \int_{R_{LJ}}^\infty -\frac{u(R)}{k_B T}R^2 dR$$

$$= 2\pi\mathcal{N}_A \int_0^{R_{LJ}}R^2 dR + 2\pi\mathcal{N}_A \int_{R_{LJ}}^\infty \frac{u(R)}{k_B T}R^2 dR. \qquad (4.41)$$

The first term in Eq. 4.41 integrates to $2\pi\mathcal{N}_A R_{LJ}^3/3$. The second term depends on the nature of the intermolecular potential at $R > R_{LJ}$. For the Lennard-Jones potential (Eq. 4.12),

$$u(R) = 4\varepsilon\left[\left(\frac{R_{LJ}}{R}\right)^{12} - \left(\frac{R_{LJ}}{R}\right)^6\right],$$

the second integral in Eq. 4.41 can be solved, giving

$$B_2(T) = \frac{2\pi\mathcal{N}_A}{3}R_{LJ}^3 + \frac{8\pi\mathcal{N}_A\varepsilon}{k_BT}\left(\int_{R_{LJ}}^{\infty}\frac{R_{LJ}^{12}}{R^{10}}dR - \int_{R_{LJ}}^{\infty}\frac{R_{LJ}^6}{R^4}dR\right)$$

$$= \frac{2\pi\mathcal{N}_A}{3}R_{LJ}^3 + \frac{8\pi\mathcal{N}_A\varepsilon}{k_BT}\left(-\frac{R_{LJ}^{12}}{9R^9}\bigg|_{R_{LJ}}^{\infty} + \frac{R_{LJ}^6}{3R^3}\bigg|_{R_{LJ}}^{\infty}\right)$$

$$= \frac{2\pi\mathcal{N}_A}{3}R_{LJ}^3 + \frac{8\pi\mathcal{N}_A\varepsilon}{k_BT}\left(\frac{R_{LJ}^{12}}{9R_{LJ}^9} - \frac{R_{LJ}^6}{3R_{LJ}^3}\right)$$

$$= \frac{2\pi\mathcal{N}_A}{3}R_{LJ}^3 - \frac{16\pi\mathcal{N}_A^2\varepsilon R_{LJ}^3}{9RT} \tag{4.42}$$

$$\equiv b - \frac{a}{RT}, \tag{4.43}$$

where

$$b = \frac{2\pi\mathcal{N}_A R_{LJ}^3}{3} \tag{4.44}$$

is one half the volume of a sphere of radius R_{LJ}, and

$$a = \frac{16\pi\mathcal{N}_A^2\varepsilon R_{LJ}^3}{9} \tag{4.45}$$

is proportional to both the molecular volume and the well depth of the intermolecular potential. Within the limits of our approximations, both constants should always be positive.

If R_{LJ} is the separation between the molecular centers of mass at closest approach, why is b equal to $\mathcal{N}_A$ times half the volume of a sphere with that radius? When we treat our molecules as hard spheres, we find that the center of mass of every molecule can never occupy the volume within a distance R_{LJ} of another molecule; it is excluded from that volume not only by the other molecule's repulsive wall, but by its own (Fig. 4.4).

Therefore, the space forbidden to the molecule's center of mass is the volume $4\pi R_{LJ}^3/3$. Since we can associate this **excluded volume** with every other molecule in the container, the overall excluded volume is this number times the total number of molecules and then divided by two (because the excluded volume applies equally to each molecule), leaving $2N\pi R_{LJ}^3/3$.

Using Eq. 4.42, we can replace the second virial coefficient in Eq. 4.37 with this more detailed expression:

$$P = RT\left[\frac{1}{V_m} + \left(b - \frac{a}{RT}\right)\frac{1}{V_m^2}\right]. \tag{4.46}$$

To put this in a form similar to the ideal gas law, we solve for RT and simplify:

$$P = RT\left(\frac{1}{V_m} + \frac{b}{V_m^2}\right) - \frac{a}{V_m^2}$$

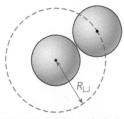

▲ **FIGURE 4.4 The excluded volume.** For every pair of molecules, there is a spherical region of approximate radius R_{LJ} that their centers of mass cannot occupy. The volume of this region is therefore roughly $4\pi R_{LJ}^3/3$.

$$P + \frac{a}{V_m^2} = RT\left(\frac{1}{V_m} + \frac{b}{V_m^2}\right)$$

$$\left(P + \frac{a}{V_m^2}\right)\left(\frac{1}{V_m} + \frac{b}{V_m^2}\right)^{-1} = RT \qquad \text{solve for } RT$$

$$\left(P + \frac{a}{V_m^2}\right)V_m\left(1 + \frac{b}{V_m}\right)^{-1} = RT \qquad \text{factor out } V_m$$

$$\left(P + \frac{a}{V_m^2}\right)V_m\left(1 - \frac{b}{V_m}\right) \approx RT. \qquad (1+x)^{-1} \approx 1 - x \text{ for small } x$$

The last step uses a Taylor series approximation to simplify the expression, and it is valid as long as the excluded volume b is small compared to the molar volume V_m of the substance. (For example, if $b/V_m = 0.01$, the approximation error is only 1%.) This is a good approximation for a gas, but it fails as we approach the density of the liquid. Finally we have the **van der Waals equation in its standard form:**

$$\left(P + \frac{a}{V_m^2}\right)(V_m - b) = RT, \qquad (4.47)$$

where a and b are called the **van der Waals coefficients.** The Lennard-Jones parameters and van der Waals coefficients for several gases are given in Table 4.2. This table also compares experimental values of the second virial coefficients B_2 to values calculated from the Lennard-Jones parameters using Eq. 4.43.

TABLE 4.2 Lennard-Jones parameters ε and R_{LJ}, van der Waals coefficients a and b, and second virial coefficients $B_2(298 \text{ K})$ for selected gases. Values for $B_2(298 \text{ K})$ calculated from Eq. 4.43 are given in the last column for comparison to the experimental values.

Gas	ε/k_B (K)	R_{LJ} (Å)	a (L^2 bar mol^{-2})	b (L mol^{-1})	$B_2(298 \text{ K})$ (L mol^{-1})	$B_2(298 \text{ K})$ (calc) (L mol^{-1})
He	10	2.58	0.0346	0.0238	0.012	0.020
Ne	36	2.95	0.208	0.01672	0.011	0.022
Ar	120	3.44	1.355	0.03201	−0.016	−0.004
Kr	190	3.61	2.325	0.0396	−0.051	−0.042
H_2	33	2.97	0.2453	0.02651	0.015	0.023
N_2	92	3.68	1.370	0.0387	−0.004	0.011
O_2	113	3.43	1.382	0.03186	−0.016	−0.001
CO	110	3.59	1.472	0.03948	−0.008	0.001
CO_2	190	4.00	3.658	0.04286	−0.126	−0.057
CH_4	137	3.82	2.300	0.04301	−0.043	−0.016
C_2H_2	185	4.22	4.516	0.05220	−0.214	−0.062
C_2H_4	205	4.23	4.612	0.05821	−0.139	−0.080
C_2H_6	230	4.42	5.570	0.06499	−0.181	−0.115
C_6H_6	440	5.27	18.82	0.1193	−1.454	−0.542

EXAMPLE 4.1   Van der Waals Coefficients

CONTEXT Helium is the closest of any real gas to an ideal gas, because—as we know from trends in the periodic table—it has the smallest effective radius of any atom (smaller even than a hydrogen atom) and the highest ionization energy. As a result, it has the smallest excluded volume and the lowest polarizability because its electrons are so tightly bound to the nucleus. Argon is larger and more polarizable, which means that both its excluded volume and its inter-particle attractions are greater than for helium. Non-ideality in argon is of particular interest in high-pressure electrical discharges through the gas, because argon does not require high electron energies to ionize and support an electric discharge, while at the same time it is a non-reactive gas and does not rapidly degrade in the discharge the way molecular gases such as nitrogen do. Pressures as high as 10^3 bar are used to generate these plasmas, extending well into the range of densities where non-ideality may be significant.

PROBLEM Estimate the van der Waals coefficients for helium and argon using their Lennard-Jones parameters.

SOLUTION The values for a are obtained by calculating the following:

$$a(\text{L}^2\,\text{bar}\,\text{mol}^{-2}) = \mathcal{N}_A^2 \left(\frac{16\pi\varepsilon R_{LJ}^3}{9} \right) (1.381 \cdot 10^{-23}\,\text{J}\,\text{K}^{-1}) \, (1000\,\text{L}\,\text{m}^{-3})^2$$

$$\times \, (10^{-10}\,\text{m}\,\text{\AA}^{-1})^3 \, (10^{-5}\,\text{bar}\,\text{Pa}^{-1})$$

$$= 2.797 \cdot 10^{-4}\,\varepsilon(\text{K})\,R_{LJ}(\text{\AA})^3,$$

and the values for b similarly:

$$b(\text{L}\,\text{mol}^{-1}) = \mathcal{N}_A \left(\frac{2\pi R_{LJ}^3}{3} \right) (1000\,\text{L}\,\text{m}^{-3}) \, (10^{-10}\,\text{m}\,\text{\AA}^{-1})^3$$

$$= 1.26 \cdot 10^{-3}\,R_{LJ}(\text{\AA})^3.$$

Using $\varepsilon/k_B = 10\,\text{K}$ and $R_{LJ} = 2.58\,\text{\AA}$ for He, we predict $a = 0.048\,\text{L}^2\,\text{bar}\,\text{mol}^{-2}$ and $b = 0.0217\,\text{L}\,\text{mol}^{-1}$. These are comparable to the values $0.0346\,\text{L}^2\,\text{bar}\,\text{mol}^{-2}$ and $0.0238\,\text{L}\,\text{mol}^{-1}$ reported in Table 4.2. Using $\varepsilon/k_B = 120\,\text{K}$ and $R_{LJ} = 3.44\,\text{\AA}$ for Ar, we predict $a = 1.366\,\text{L}^2\,\text{bar}\,\text{mol}^{-2}$ and $b = 0.0513\,\text{L}\,\text{mol}^{-1}$, and the measured values are $1.355\,\text{L}^2\,\text{bar}\,\text{mol}^{-2}$ and $0.03201\,\text{L}\,\text{mol}^{-1}$.

Helium has a very small polarizability and therefore only very weak dispersion forces. Consequently, the well depth ε for the intermolecular potential is negligible, and a for helium is an order of magnitude lower than any other. In contrast, b, which depends only on the effective molecular volume, is small for helium but comparable to other atoms and molecules. Because $B_2(T) = b - a/(RT)$ (and a and b are positive), this implies that $B_2(T)$ should have a small, positive value for helium. Argon, on the other hand, has a much more diffuse and polarizable electron cloud, so ε is comparatively large. Hence, as long as the temperature is low enough that $a/(RT) > b$, then $B_2(T)$ for argon will be negative.

We've accomplished a major feat. Statistical mechanics allowed us to take the microscopic properties of the interaction between just two molecules and extend it to the macroscopic properties of the gas, giving us an equation that not only describes *quantitatively* how the gas will behave, but that also offers an *intuitive* justification for the behavior. Our understanding of the nature of the intermolecular potential allows us to anticipate the sign and magnitude of $B_2(T)$ and how that will vary with temperature. At higher temperatures, meaning greater kinetic energy, the small attractive potential energy term contributes less and less to the total energy, and the gas becomes less sticky. The excluded volume, on the other hand, has negligible dependence on the temperature. As a result, the measured pressure tends to grow, relative to the ideal value, as the temperature increases.

The relationship between the second virial coefficient B_2 and the potential curve parameters R_{LJ} and ε is based on several approximations—the Lennard-Jones potential and neglect of higher order virial coefficients, to name two. Tabulated values of the virial and van der Waals coefficients are obtained from experimental measurements, not from the theoretical argument we used earlier. Values of B_2 calculated from Eq. 4.43 are shown in the last column of Table 4.2, and we see that errors of factors of two or three are common under these assumptions, and some errors are even bigger. Equation 4.43 fares best for examples such as Ne and CH_4 with relatively small a constants. Strong attractive forces increase the average interaction time of each molecule and make the results much more sensitive to the quality of our model for the intermolecular potential. You can see, however, from this derivation how it would be possible to improve the model—and hopefully its predictions—for example, by including anisotropic terms in the potential, and then calculating precise numerical solutions to the integrals.

4.3 The Pair Correlation Function

When the intermolecular attraction or the external pressure becomes so great that the spacing between the molecules is less than the average molecular size, we expect the fluid to behave like a liquid. It is appropriate, therefore, that when we introduce molecular interactions into our pressure law, we keep an eye out for the pair correlation function that we use to describe liquid structure.

The pair correlation function is defined by the qualitative relationship (Eq. 4.15):

$$\mathcal{G}(R) = \frac{\text{number of molecules at distance } R}{\text{number of molecules at distance } R \text{ in a random distribution}}.$$

To obtain numerical values, we need a mathematical expression, which we can write in terms of the number density (Eq. 4.16):

$$\overline{N}_{R_a R_b} = 4\pi\rho \int_{R_a}^{R_b} \mathcal{G}(R)R^2 dR,$$

where $\overline{N}_{R_a R_b}$ is the average number of molecules at distances between R_a and R_b of any selected reference molecule, ρ is the number density (usually expressed as molecules per cm^3), and $\mathcal{G}(R)$ is the pair correlation function. Now that we have some tools from statistics, we can obtain an explicit expression for $\mathcal{G}(R)$.

The pair correlation function gives the probability of finding one molecule at a specific distance from another. In terms of our distribution functions, we can evaluate a similar function, $\mathcal{P}_R(R_{12})$, the probability that we will find two specific molecules 1 and 2 separated by a distance R_{12}.

We begin with the position probability distribution function $\mathcal{P}_{V^N}(X_1,\ldots,Z_N)$, which gives the probability per unit volume of locating a specific molecule 1 in our sample of N molecules at a position (X_1,Y_1,Z_1), molecule 2 at position (X_2,Y_2,Z_2), and so on:

$$\mathcal{P}_{V^N}(X_1,\ldots,Z_N) = \frac{e^{-U(X_1,\ldots,Z_N)/(k_BT)}}{\int_{-\infty}^{\infty}\ldots\int_{-\infty}^{\infty}e^{-U(X_1,\ldots,Z_N)/(k_BT)}dX_1\ldots dZ_N} = \frac{e^{-U(X_1,\ldots,Z_N)/(k_BT)}}{Q'_U(T,V)}. \tag{4.48}$$

This is a function with units of V^{-N}, giving a probability per unit volume about position 1, per unit volume about position 2, and so on. The denominator normalizes the function so that

$$\int_{-\infty}^{\infty}\ldots\int_{-\infty}^{\infty}\mathcal{P}_{V^N}(X_1,\ldots,Z_N)dX_1\ldots dZ_N = 1. \tag{4.49}$$

This function is defined as though the molecules are distinguishable.

However, this general distribution function contains much more information than we intend to retain in $\mathcal{P}_R(R_{12})$. We only want the probability of molecule 2 being at a distance R_{12} from molecule 1. We don't care at all, for example, where the other $N-2$ molecules are. We don't even care exactly where molecule 1 is. To eliminate those unneeded variables from our analysis, we carry out the following steps on our way to obtaining $\mathcal{P}_R(R_{12})$:

1. Integrate over the positions of all the other molecules, 3 through N.

2. Integrate over the direction of molecule 2 from molecule 1.

3. Integrate over the location of molecule 1.

These three steps are what make up our next volley of mathematics, summarized in Fig. 4.5.

1. To remove any dependence on the locations of the other molecules, we integrate $\mathcal{P}_{V^N}(X_1,\ldots,Z_N)$ over all values of the coordinates X_3 through Z_N:

$$\mathcal{P}_{V^2}(X_1,\ldots,Z_2) = \int_{-\infty}^{\infty}\ldots\int_{-\infty}^{\infty}\mathcal{P}_{V^N}(X_1,\ldots,Z_N)dX_3\ldots dZ_N. \tag{4.50}$$

● = 1   ○ = 2

▶ FIGURE 4.5 **A schematic representation of the integrations used in the derivation of** $\mathcal{P}_R(R_{12})$. The probability of every possible arrangement of the five molecules is determined by $\mathcal{P}_{V^N}(X_1,\ldots,Z_N)$. Step 1 adds up all the microstates that have molecules 1 and 2 in the same place, regardless of the positions of molecules 3–5. Step 2 adds in all the additional states where 2 is the right distance from 1, but in any direction. Step 3 adds in all the states where 1 and 2 are the right distance apart, regardless of where 1 is.

(1) Integrate over positions of molecules 3–5.

(2) Integrate over direction of 2 from 1.

(3) Integrate over location of 1.

This function has units of V^{-2}, as it gives the probability per unit volume about position 1 and per unit volume about position 2.

2. To express the location of molecule 2 in terms of its distance from molecule 1, we switch the coordinates of molecule 2 from the Cartesian coordinates X_2, Y_2, Z_2 to the relative polar coordinates R_{12}, Θ_{12}, and Φ_{12}, which are spherical coordinates that use molecule 1 as the origin. Next, we integrate over all values of the angles Θ_{12} and Φ_{12} to find the probability that molecule 1 has a neighbor molecule 2 at distance R_{12} in *any direction*:

$$\int_0^\pi \int_0^{2\pi} \mathcal{P}_{VR}(X_1,Y_1,Z_1,R_{12},\Theta_{12},\Phi_{12})R_{12}^2 \sin\Theta_{12}\, d\Theta_{12}\, d\Phi_{12}$$

$$= 4\pi R_{12}^2\, \mathcal{P}_{VR}(X_1,Y_1,Z_1,R_{12},\Theta_{12},\Phi_{12}) \tag{4.51}$$

$$\equiv \mathcal{P}_{VR}(X_1,Y_1,Z_1,R_{12}).$$

Under the assumption of an isotropic intermolecular potential energy,[4] the probability distribution function does not change as a function of Θ_{12} or Φ_{12}, and the integration merely contributes a coefficient of 4π.

3. Finally, we integrate this new probability function over all possible locations of molecule 1, obtaining the probability that 1 is a distance R_{12} from molecule 2:

$$\int_0^a \int_0^b \int_0^c \mathcal{P}_{VR}(X_1,Y_1,Z_1,R_{12})dX_1 dY_1 dZ_1 = V\mathcal{P}_{VR}(X_1,Y_1,Z_1,R_{12})$$

$$\equiv \mathcal{P}_R(R_{12}). \tag{4.52}$$

In this case, the integration donates a coefficient of V to the expression because all locations in the container are identical.[5]

Combining our three steps, we get

$$\mathcal{P}_R(R_{12}) = \frac{4\pi R_{12}^2 V \int_{-\infty}^\infty \cdots \int_\infty^\infty e^{-U(X_1,\ldots,Z_N)/(k_B T)}\, dX_3 \ldots dZ_N}{Q'_U(T,V)}. \tag{4.53}$$

If there were only two molecules, we could find the fraction of time that the distance between the two molecules is in the range R_a to R_b by integrating $\mathcal{P}_R(R_{12})$ from R_a to R_b. Now we add a third molecule. This changes the potential energy function $U(X_1, \ldots, Z_3)$, but our integration to get $\mathcal{P}_R(R_{12})$ is still valid. So we can now ask what fraction of time *either* molecule 2 or 3 spends at a distance R_a to R_b from molecule 1, and the answer is the integral over $\mathcal{P}_R(R_{12}) + \mathcal{P}(R_{13})$. If molecules 2 and 3 are indistinguishable, then $\mathcal{P}_R(R_{12})$ and

---

[4]Calculating the properties of a macroscopic substance is a much more daunting task when one includes all the orientation-dependence of the intermolecular potentials. As much as possible, we ignore the dependence on the orientations and approximate $u(X_1, \ldots, Z_2)$ using model potentials that depend only on the distance between the two molecules, R. The Lennard-Jones, square well, or hard sphere potentials all serve this purpose, although they can also be mapped onto the shape of the molecule to obtain anisotropic effects. However, several important interactions such as π-π interactions are strongly angle-dependent.

[5]This assumption neglects effects at the walls of our container, where the environment certainly differs from that in the middle of the container. We expect these effects to be negligible. For a container 1 cm on a side, a given molecule will typically spend less than 10^{-6} of the time near the walls.

$\mathcal{P}_R(R_{13})$ are equivalent expressions. Therefore, if we don't mind losing track of which molecule is 2 and which is 3, we may write

$$\mathcal{P}_R(R_{12}) + \mathcal{P}_R(R_{13}) = 2\mathcal{P}_R(R), \tag{4.54}$$

where $\mathcal{P}_R(R)$ is now the probability that molecule 1 is a distance R from *any other single* molecule in the sample.

For a sample of N molecules, this lets us calculate the average number of molecules $\overline{N}_{R_a,R_b}$ at a distance between R_a and R_b from any randomly selected molecule (which we have been calling molecule 1). There are $N - 1$ other molecules in the sample, each with a chance $\mathcal{P}_R(R)$ of being a distance R from molecule 1, so the average number of molecules in the distance range we're looking for is

$$\overline{N}_{R_a,R_b} = (N - 1)\int_{R_a}^{R_b} \mathcal{P}_R(R)\, dR. \tag{4.55}$$

This must be the same as our earlier equation that defined the pair correlation function, Eq. 4.16:

$$\overline{N}_{R_a,R_b} = 4\pi\rho \int_{R_a}^{R_b} \mathcal{G}(R)\, R^2\, dR.$$

At last, by equating the integrands of Eqs. 4.16 and 4.55, we can define the pair correlation function in terms of the intermolecular potential energy function:

$$(N - 1)\,\mathcal{P}_R(R)dR = 4\pi\rho\mathcal{G}(R)R^2 dR \tag{4.56}$$

$$\mathcal{G}(R) = \frac{N - 1}{4\pi\rho\, R^2}\mathcal{P}_R(R)$$

$$= \left(\frac{V(N - 1)}{4N\pi R^2}\right)\mathcal{P}_R(R) \qquad\qquad \rho = N/V$$

$$= \left(\frac{V}{4\pi R^2}\right)\mathcal{P}_R(R) \qquad\qquad N - 1 \approx N$$

$$= \left(\frac{V}{4\pi R^2}\right)\frac{4\pi R^2 V\int_0^a \cdots \int_0^c e^{-U/(k_B T)}dX_3 \ldots dZ_N}{Q'_U(T,V)} \qquad \text{by Eq. 4.53}$$

$$= \frac{V^2 \int_0^a \cdots \int_0^c e^{-U/(k_B T)}dX_3 \ldots dZ_N}{Q'_U(T,V)}. \tag{4.57}$$

CHECKPOINT We integrate over the volume of the container using the dimensions of the three-dimensional box (sides of lengths *a*, *b*, and *c*) as the upper limits of the integrals. At large *R*, *U(R)* approaches zero, and to a first approximation the triple integral over $e^{-U/(kT)}$ is roughly equal to *abc=V*, the volume of the box.

Equation 4.57 will allow us to express several of our results in the subsequent sections in terms of the pair correlation function. The value in this extends from a practical interest (because $\mathcal{G}(R)$ is experimentally measurable in liquids by neutron diffraction) to the theoretical (because this is the function that relates the macroscopic properties of bulk fluids directly to the structure at the microscopic scale).

We may now obtain a very approximate, analytical expression for $\mathcal{G}(R)$ in the dilute fluid limit by starting from Eq. 4.57:

$$\mathcal{G}(R) = V^2\frac{\int_0^a \cdots \int_0^c e^{-U/(k_B T)}dX_3 \ldots dZ_N}{\int_0^a \cdots \int_0^c e^{-U/(k_B T)}dX_1 \ldots dZ_N} \tag{4.58}$$

$$= V^2\frac{V^{N-2}e^{-u/(k_B T)}\left[1 + \dfrac{\mathcal{I}(T)}{V}\right]^{[N(N-1)/2]-1}}{V^N\left[1 + \dfrac{\mathcal{I}(T)}{V}\right]^{N(N-1)/2}} \qquad \text{by Eq. 4.31}$$

$$= e^{-u(R)/(k_B T)} \left[1 + \frac{\mathcal{I}(T)}{V} \right]^{-1}$$

$$\approx e^{-u(R)/(k_B T)}. \qquad \mathcal{I}(T) \ll V \quad (4.59)$$

This formulation of $\mathcal{G}(R)$ is usable for gases but is a terrible approximation for liquids. Equation 4.59 is plotted in Fig. 4.6 for the argon Lennard-Jones potential at various temperatures. Although the curve has the right qualitative shape and temperature dependence (showing stronger structure at lower T), it fails to predict any structure beyond the first solvation shell.

4.4 Bose-Einstein and Fermi-Dirac Statistics

The correction of the rotational partition function for symmetry effects, described in Section 3.3, opens the door to a much wider consideration of particle spin effects in the macroscopic properties of matter. We will just briefly touch on this field.

Introduction

According to the symmetrization principle given in Section 2.1, the wavefunction of a system of indistinguishable particles must be antisymmetric with respect to exchange of the particles if the particles have *half-integer spin* (fermions), and must be symmetric if the particles have *integer spin* (bosons). One manifestation of this rule, the Pauli exclusion principle, applies to all fermions, including electrons, protons, and neutrons, each of which has a spin quantum number equal to $\frac{1}{2}$.

You could reasonably ask, "What other particles am I ever going to run into?" The answer: any larger particles—atomic nuclei or whole atoms—that you can *construct* out of these particles.

Indistinguishability is not limited to what we consider our fundamental building blocks. We cannot distinguish between two electrons in the same helium atom. Once we bring two electrons close enough together that their wavefunctions overlap in space, we've lost hope of knowing which is which. In exactly the same way, we cannot distinguish between two helium atoms if we bring them close enough together that their wavefunctions overlap. Don't be bothered by the fact that the helium atom is composed of several different kinds of particles—electrons, neutrons, and protons. Unless we carry out a measurement that probes certain particles within the helium atom, it may as well be one particle.[6] Consequently, any atom is a boson if its *total count of electrons, neutrons, and protons is an even number*. Bosons, because their wavefunctions must be symmetric overall, are immune to the Pauli exclusion principle.

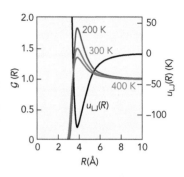

▲ FIGURE 4.6 **Approximate pair correlation functions for the Lennard-Jones potential curve of argon.** These curves are computed using Eq. 4.59.

CHECKPOINT Why must an antisymmetric wavefunction obey the Pauli exclusion principle? Say that particles 1 and 2 share exactly the same wavefunction ψ_a in some total wavefunction $\psi_a(1)\psi_a(2)\psi_b(3)$. When we switch the labels on those two particles, the total wavefunction must remain exactly the same: $\psi_a(2)\psi_a(1)\psi_b(3)$ (the fact we multiply them in opposite order is not significant). The total function *cannot* change sign— cannot be antisymmetric on exchanging the labels—unless every particle is in a unique quantum state.

---

[6]We tacitly employ this simplification in chemistry all the time. When we probe nuclear spins by NMR, we treat the atomic nuclei as single particles, not as collections of independent protons and neutrons. We now know that protons and neutrons themselves are each composed of smaller particles called quarks, but chemistry is rich enough without extending our studies to that scale.

For example, a group of ^{4}He atoms (two protons, two neutrons, two electrons in each atom) is a group of indistinguishable bosons, and all the atoms may share the same quantum state. We're not concerned with the electronic or nuclear spin states here; remember that we're not taking measurements of particles inside the helium atom. These atoms can share the same *translational* state. From our experience with classical particles, that's amazing, because translational states represent position and velocity. Two ^{4}He atoms that occupy the same translational state violate the most fundamental law of classical objects: *they occupy the same space at the same time.* As shown in Fig. 4.7, it is possible (in principle) for two bosons, such as ^{4}He atoms, to approach each other and overlap as though they were waves. By itself, this is not new. We could draw the same picture for two electrons, as long as they had opposite spins. What *is* new is that we can do this with more than two particles. We can keep adding bosons to the same space, and, provided we can overcome the Coulomb repulsion, there is no limit to the number of overlapping particles.

The application of the symmetrization principle to statistical mechanics is identical to its application in the case of rotational state degeneracies in Section 3.3: if we are counting states to evaluate the partition function of the sample, we need to know whether to count antisymmetric or symmetric states. In the case of half-integer spin particles, fermions, this leads to statistical quantities described by **Fermi-Dirac statistics.** For integer spin particles, bosons, the characteristics are those of **Bose-Einstein statistics.**

Take as an example a system of two independent and indistinguishable particles a and b, with the state specified by a single quantum number l. The total wavefunction is $\psi_{l_a,l_b}(1,2) = \psi_{l_a}(1)\psi_{l_b}(2)$. If we lay out a matrix for all possible combinations of states with quantum numbers l_a and l_b, as shown in Table 4.3, limiting the number of states $g(E)$ to 5 for each particle, we find $5^2 = 25$ possible combinations. However, since the particles are indistinguishable, the state with wavefunction $\psi_{l_a,l_b}(1,2)$ is no different from the state with wavefunction $\psi_{l_b,l_a}(1,2)$ and should not be counted separately. That reduces the number of available states for ψ_{l_a,l_b} when $l_a \neq l_b$, but what about the five states where $l_a = l_b$? For fermions, those states are

TABLE 4.3 Two particle states $\psi_{l_a,l_b}(1,2)$ where l_a and l_b are limited to five values.

			l_a		
l_b	**1**	**2**	**3**	**4**	**5**
1	(11)	(12)	(13)	(14)	(15)
2	(21)	(22)	(23)	(24)	(25)
3	(31)	(32)	(33)	(34)	(35)
4	(41)	(42)	(43)	(44)	(45)
5	(51)	(52)	(53)	(54)	(55)

▶ FIGURE 4.7 **Two boson wavefunctions occupying the same space.** The spatial wavefunction for two ^{4}He atoms approach each other (a), overlap (b), and continue. Note that while such a collision is *possible* for a boson, it remains unlikely under normal conditions because of the large Coulomb repulsion between overlapping electrons and nuclei.

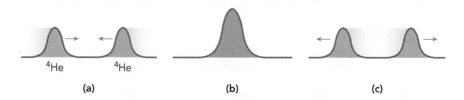

^{4}He $\quad$ ^{4}He

(a) $\qquad\qquad$ (b) $\qquad\qquad$ (c)

not permitted, because the same quantum numbers cannot be shared by identical fermions, but for bosons those states are allowed. Therefore, in this example, fermions would have $(25 - 5)/2 = 10$ states, whereas bosons would have $(25 + 5)/2 = 15$ states. In general, for two fermions with $\Omega(E)$ states accessible to each, the number of two-particle states $\Omega(E)$ at energy E would be $[g(E)^2 - g(E)]/2$, whereas for bosons $\Omega(E)$ would be $[g(E)^2 + g(E)]/2$.

As N approaches infinity, the difference between these two values becomes less significant, $(g^2 \pm g)/2 \approx g^2/2$. The effect becomes even weaker as we include more degrees of freedom. For three particles, one could construct a three-dimensional matrix, and the number of three-particle states $\psi_{l_a, l_b, l_c}(1,2,3)$ accessible for fermions would be $(g^3 - 3g^2 + 2g)/6$, and for bosons would be $(g^3 + 3g^2 + 2g)/6$. In general, for N particles, each with g possible one-particle states accessible, the number of N-particle states accessible to the system Ω for fermions is given by

$$\Omega = \frac{g!}{N!(g - N)!}, \tag{4.60}$$

and for bosons is

$$\Omega = \frac{g^N}{N!}, \tag{4.61}$$

assuming $g > N$. The leading term climbs as the power of the number of particles, with the correction term climbing one power lower.

The number of N-particle states increases with temperature, so at high temperatures the spin statistics become irrelevant. The number of available one-molecule states becomes so large that, even for bosons, no two molecules will be found in the same quantum state. The mathematics of this limit is called **Maxwell-Boltzmann statistics** and is what chemists generally use. However, the effects of the fermion or boson statistics are important in several cases, particularly at low temperature and for low-mass systems, where quantum effects dominate. The statistics of photons (with spin equal to one) must also be dealt with using Bose-Einstein statistics.

Bose-Einstein Condensates

In the laboratory, the difference between fermions and bosons is evident in the disparity between the Fermi-Dirac liquid ^{3}He and the Bose-Einstein liquid ^{4}He. The high degeneracy of states in ^{4}He begins to dominate the properties of the liquid once the temperature is low enough that the number of accessible states is small and the chance of multiple atoms occupying the same quantum state is high. At this boundary, which falls at 2.17 K when the pressure is 1 bar, the liquid expands on cooling and its thermal conductivity becomes discontinuous—both are results of the Bose-Einstein statistics. The exchange repulsion that affects fermions causes the Fermi-Dirac isotope ^{3}He to resist formation of the condensed phases more than ^{4}He. Helium gas at 1 bar condenses at about 3.2 K for ^{3}He, and at 4.22 K for ^{4}He. Although neither isotope solidifies at 1 bar, the solid can be formed by compressing the liquid at temperatures below 2 K. For ^{3}He at a temperature of 1 K, the solid is formed only at pressures in excess of 35 bar; ^{4}He at the same temperature solidifies at about 27 bar.

Because the ^{4}He atoms do not repel each other as strongly as the ^{3}He atoms, liquid ^{4}He becomes a **superfluid** at temperatures below 2.17 K. A superfluid flows without significant resistance, even flowing upward against gravity, along any surface. The capability of the boson wavefunctions to overlap contributes substantially to this effect. Superfluidity also appears in ^{3}He liquid, but only at temperatures below 0.005 K. At this temperature, the atoms pair up, forming dimers with an even number of particles *in total* (four protons, two neutrons, four electrons) and therefore obeying boson statistics. **Superconductivity**—the ability of a material to transmit electrical current with zero resistance—is attributed to a similar manifestation of Bose-Einstein statistics. In this case, the bosons are pairs of electrons with opposite spins, called **Cooper pairs.** At low temperatures, the Cooper pairs may share identical wavefunctions, and the exchange repulsion no longer contributes to the resistance.

The influence of Bose-Einstein statistics on the behavior of liquids and electron Cooper pairs is difficult to isolate, for a familiar reason. In the liquid, the interatomic potential energy is comparable to the thermal energy. The Coulomb and dispersion force interactions between the helium atoms must be accurately modeled before effects from other phenomena, such as the quantum spin statistics, can be understood separately. Similarly, the electrons in a solid are subject to strong interactions with the nuclei.

A theoretically simpler system is the Bose-Einstein gas, cooled to temperatures where the quantum statistical effects can be measured, but still a gas and therefore in a regime where molecular interactions were negligible. The first of these to be studied was a gas of extremely cold ^{87}Rb atoms, and other alkali metal gases have been used subsequently.[7] The alkali metals heavier than lithium have low vaporization temperatures and are easily put into the gas phase in a vacuum-tight chamber. Once this is accomplished, the atoms, which have ^2S ground state terms and are paramagnetic, can be steered using magnetic fields to keep them from hitting the walls of the vacuum chamber. Any hot atoms—atoms with lots of translational energy—are deflected less by the magnetic fields, and they leave this "magnetic trap," eventually reaching the wall of the chamber where they stick and do not re-enter the trap (Fig. 4.8).

By allowing the hot atoms to evaporate from the trap in this fashion, the atoms left in the trap become colder and colder. Further temperature reduction has been induced by laser cooling, canceling some of the atomic momentum with the momentum h/λ of laser photons. Eventually temperatures of less than $2 \cdot 10^{-7}$ K may be achieved. At this point the speeds of the atoms are so slow that the effective size of each atom, as indicated by its de Broglie wavelength $h/(mv)$, extends over distances too large to feel the influence of the inter-particle forces. Being bosons, these wavefunctions may overlap—the atoms may share the

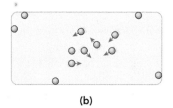

(a)

(b)

▲ **FIGURE 4.8 A schematic of the magnetic evaporation effect.** Gas-phase ions are trapped by magnetic fields in the middle of a vacuum chamber **(a)**. When one ion gains a lot of energy, it can escape the magnetic fields and freeze to the wall of the chamber, leaving behind only the less energetic ions. This reduces the average energy per ion, and therefore the temperature, of the remaining sample **(b)**.

---

[7]These *atoms* are bosons; the nuclei are not. In each of these examples, the nucleus has an odd number of nucleons and has a half-integer spin, and is therefore a fermion. The unpaired electron is necessary so that the magnetic trap can constrain the atoms.

same translational state—without any opposition from the Pauli exclusion principle. A useful parameter is a unitless quantity called the **phase-space density** ρ_{ps} such that

$$\rho_{ps} = \rho \lambda_{dB}^3. \tag{4.62}$$

It has been predicted that Bose-Einstein statistics begin to dominate the distribution of states in such a sample when ρ_{ps} exceeds 2.612.

To verify experimentally that many atoms simultaneously occupy identical translational states, you gently turn off the magnetic fields maintaining the trap. The gas of atoms expands at a rate determined by the thermal speed before the trap was deactivated. The result is that the atoms have speeds too slow and too similar to be achievable in the available space according to quantum mechanics, unless the atoms have been sharing quantum states.

CONTEXT *Where Do We Go From Here?*

Statistical mechanics is a powerful tool, in the same way that a welding torch is a powerful tool: it can be essential for constructing a sturdy framework, but it takes practice and training to use well. From our work in this chapter, we can see the pathway to connecting molecular properties to bulk behavior, and we can also see the challenges that lie along that path. As it happens, our remaining topics in this field do not require quite the same thoroughness to see the rewards. Having investigated the fundamental distribution of energy in *static* macroscopic systems, we next consider what happens when we let the system evolve. We consider two kinds of change in the system: mass transfer and energy transfer. With help from the statistical mechanics we've already done, these topics will help us understand the origins of the effects seen in all macroscopic processes as we lead up to our study of thermodynamics.

KEY CONCEPTS AND EQUATIONS

4.2 **Pressure of a Non-ideal Gas.** For the ideal gas, the intermolecular potential energy $u(R)$ is zero everywhere, and

$$P = -\frac{\partial E}{\partial V} = \left(\frac{\partial E}{\partial S}\right)\left(\frac{\partial S}{\partial V}\right).$$

We use the fact that for a single state i, $\Omega(E_i) \propto V^N e^{-E_i/(k_B T)}$, to obtain:

$$P = (k_B T)\left(N\frac{\partial \ln V}{\partial V}\right) = \frac{k_B T N}{V}.$$

To get a simple non-ideal gas law, we include a general intermolecular potential $u(R)$, with the constraints that it is isotropic (depends only on distance R) and weak compared to the thermal energy $k_B T$ except near the repulsive wall. Non-ideal gas laws can be written

any of several ways. Two of the most popular are the virial expansion,

$$P = RT\left[V_m^{-1} + B_2(T)V_m^{-2} + B_3(T)V_m^{-3} + \ldots\right], \tag{4.39}$$

and the van der Waals equation,

$$\left(P + \frac{a}{V_m^2}\right)(V_m - b) = RT. \tag{4.47}$$

The second virial coefficient $B_2(T)$ is given by

$$B_2(T) = -2\pi\mathcal{N}_A \int_0^\infty (e^{-u(R)/(k_B T)} - 1)R^2 dR.$$

For most forms of $u(R)$, $B_2(T)$ can be written in terms of the van der Waals constants a and b:

$$B_2(T) = b - \frac{a}{RT},$$

where all the temperature dependence is in the $a/(RT)$ term, which depends on the strength of the intermolecular attractive forces. The b term is essentially a correction to the volume of the container involving the non-zero size of the molecules.

4.3   The Pair Correlation Function. The pair correlation function $\mathcal{G}(R)$ describes the average structure of a substance by mapping the probability of finding any neighboring particles at any specific distance from a randomly selected reference particle. For a gas, $\mathcal{G}(R)$ may be crudely approximated by

$$\mathcal{G}(R) \approx e^{-u(R)/(k_B T)}, \qquad (4.59)$$

where $u(R)$ gives the intermolecular potential energy.

4.4   Bose-Einstein and Fermi-Dirac Statistics. Particles in chemistry are composed of three fundamental particles: protons, neutrons, and electrons. Any composite particle, such as an atomic nucleus or a whole atom, which contains an *even* number of these fundamental particles, is a boson, and any with an *odd* number of fundamental particles is a fermion. Bosons do not obey the Pauli exclusion principle, and at low temperatures many such particles may occupy the same quantum state. This effect is observed in the creation of Bose-Einstein condensates, characterized by temperatures lower than those that can be attained by a fermionic substance.

KEY TERMS

- Major contributors to the attractions between molecules are the **dipole–dipole** force between two polar molecules, the **dipole-induced dipole** force in which a polar molecule distorts the electron distribution of a second molecule, and the **dispersion** force, which is an attractive force shared by all molecules.

- The **polarizability** of a particle is a measure of how easily its electron distribution is shifted in the presence of an external electric field.

- The **virial expansion** expresses the pressure of a non-ideal gas as a power series in the reciprocal of the molar volume, $1/V_m$.

- The **van der Waals equation** for non-ideal gases breaks the non-ideality into two corrections: a correction to the pressure for the attractions between the molecules proportional to the coefficient a, and a correction for the repulsions proportional to the **excluded volume** b. The value of the van der Waals b coefficient is the volume of a container per mole of substance that is effectively occupied by particles of that substance.

- **Fermi-Dirac statistics** describe the probabilities of finding fermions in particular states, and **Bose-Einstein statistics** do the same for bosons. In particular, Fermi-Dirac statistics obey the Pauli exclusion principle, and Bose-Einstein statistics do not.

- A **superfluid** is a fluid with zero viscosity, and a **superconductor** is a material with zero electrical resistance. Both effects may be ascribed to the quantum properties of bosons.

OBJECTIVES REVIEW

1. *Outline the steps necessary to extrapolate from properties at the molecular scale to properties at the bulk scale.*
   Draw a rough flow chart for the steps to be carried out if we wanted to predict the mean Z value of particles of mass m in our system under the influence of gravity, where the gravitational potential energy is equal to mgZ with g a constant and Z the height.

2. *Estimate properties of non-ideal gases using the van der Waals and virial equations.*
   Using Table 4.2, predict the temperature at which ethane has a molar volume of 1.00 L and a pressure of 24.0 bar. Compare this to the value for the ideal gas.

3. *Approximate the pair correlation function for a gas from its Lennard-Jones parameters.*
   Write an approximate expression for the pair correlation function of ethane, using the values in Table 4.2.

4. *Predict the qualitative behavior of low-temperature fluids based on their spin statistics.*
   Identify any of the following gases that could form Bose-Einstein condensates: (a) ^{1}H (b) ^{19}F$^-$ (c) ^{20}Ne.

PROBLEMS

Discussion Problems

4.1 The electronic energy for the one-electron atom is $E_{\text{elec}} = -Z^2 E_{\text{h}}/(2n^2)$, where Z is the atomic number and n is the principal quantum number, with $n = 1, 2, 3, \ldots$. The degeneracy is equal to $2n^2$. We can apply the integral approximation to the partition function for this system, but it isn't helpful. What do you expect for the partition function of the one-electron atom in the limit that the integral expression is valid?

4.2 The a and b van der Waals coefficients for selected molecules are given in the following table. Rank the molecules in order of increasing size (R_{LJ}) and intermolecular attraction (ε).

Molecule	a (L^2 bar mol^{-2})	b (L mol^{-1})
A	11.29	0.088
B	13.93	0.117
C	16.02	0.112

a. increasing R_{LJ}:
b. increasing ε:

4.3 Simplify

$$\mathcal{I}(T) = 4\pi \int_0^\infty \left(e^{-u(R)/(k_{\text{B}}T)} - 1 \right) R^2 dR$$

in the limit of the ideal gas.

4.4 Given the following values for the van der Waals coefficients, which one of these gases should have the lowest molar volume at 0.20 bar and 298 K?

Molecule	a (L^2 bar mol^{-2})	b (L mol^{-1})
Fluoromethane	4.692	0.05264
Methanol	9.649	0.06702
Ethane	5.562	0.0638
Cyanogen	7.769	0.06901
Propane	8.779	0.08445
Sulfur dioxide	6.803	0.05636
Silane	4.377	0.05786

Non-ideal Gases

4.5 Helium is listed as having an atomic radius of only 0.31 Å (the smallest of any element) and the atomic radius

of neon is given as 0.71 Å. Then how is it that helium comes have a *larger* van der Waals coefficient b (0.0238 L mol^{-1}) than neon (0.01672 L mol^{-1}) in Table 4.2?

4.6 In the simplest of our intermolecular model potentials, the hard sphere potential, the potential energy is infinite when $R < R_{\text{hs}}$ and zero everywhere else. In the limit of the hard sphere potential, write the numerical value or simplest algebraic expression for the following:

a. the van der Waals coefficient a

b. the van der Waals coefficient b

c. the configuration integral Q'_U

4.7 The correct expression for the average potential energy in a sample of N molecules at temperature T is

a. $\displaystyle\int_0^\infty \ldots \int_0^\infty e^{-U/(k_{\text{B}}T)} du_1 \ldots du_N / Q'_U$

b. $\displaystyle\int_0^\infty \ldots \int_0^\infty e^{-U/(k_{\text{B}}T)} U du_1 \ldots du_N / Q'_U$

c. $\displaystyle\int_0^\infty \ldots \int_0^\infty e^{-U/(k_{\text{B}}T)} U dX_1 \ldots dZ_N / Q'_U$

d. $\displaystyle\int_{-\infty}^\infty \ldots \int_{-\infty}^\infty e^{-U/(k_{\text{B}}T)} U dX_1 \ldots dZ_N / Q'_U$

4.8 In terms of the van der Waals coefficients, what are the predicted values of the second virial coefficient $B_2(T)$ in the limits of (a) high temperature and (b) low temperature?

4.9 Calculate the van der Waals coefficients a (in L^2 bar mol^{-2}) and b (in L mol^{-1}) from the Lennard-Jones potential for a molecule with $\varepsilon = 200\,\text{cm}^{-1}$ and $R_{\text{LJ}} = 3\,\text{Å}$.

4.10 Write the integral (including volume element and limits) that we would need to find the average distance between any two molecules in a gas phase sample of N molecules at temperature T and volume V.

4.11 Write the value (if available from the information) or the simplest expression for the following terms in the *ideal gas limit*.

a. $\mathcal{I}(T)$

b. $Q'_U(T,V)$

c. $u(R)$

d. $\mathcal{G}(R)$

e. P

4.12 The van der Waals constants for selected molecules are listed in the following table.

Molecule	a (L^2 bar mol^{-2})	b (L mol^{-1})
A	15.52	0.0904
B	52.88	0.3051
C	12.74	0.0926
D	45.03	0.2439

a. Rank these in order of *increasing* molecular size.
b. Rank these in order of *increasing* intermolecular bond strength ε.

4.13 A gas sample consists of N molecules with molecular diameter R_{LJ}. Write an integral, as explicitly as possible, that would calculate the total number of *pairs* of molecules such that $R_{12} < 2R_{LJ}$.

4.14 Consider a typical gas-phase sample in a three-dimensional box of dimensions a, b, and c. If the intermolecular interactions are important only over small distances, show whether the integral

$$\int_0^a \int_0^b \int_0^c e^{-u(X,Y,Z)/(k_B T)}dXdYdZ$$

can be approximated by

$$\int_{-\infty}^\infty \int_{-\infty}^\infty \int_{-\infty}^\infty e^{-u(X,Y,Z)/(k_B T)}dXdYdZ.$$

4.15 Using slightly different approximations from those we used in our derivation of the virial expansion, we can obtain the **pressure equation:**

$$P = \frac{Nk_B T}{V} - \frac{2\pi N^2}{3V^2}\int_0^{V^{1/3}}\left(\frac{\partial u}{\partial R}\right)_{E,N}\mathcal{G}(R)R^3 dR. \quad (4.63)$$

Show whether Eqs. 4.63 and 4.36 agree when all of the following conditions are satisfied:

a. The gas is dilute, so $\mathcal{G}(R) \approx e^{-u/(k_B T)}$.
b. The thermal energy $k_B T$ is large compared to the intermolecular potential energy $u(R)$.
c. The intermolecular potential energy is dominated by R^{-6} attractive terms.

4.16 Write the form you would expect the pressure equation (Eq. 4.63) to take if the pair potential energy u were also a function of the angles Θ_1, Θ_2, and Φ.

4.17 The non-ideality of a gas may be expressed as a **compressibility factor,** z:

$$z \equiv \frac{PV_m}{RT}.$$

a. Find the value of z for the ideal gas.
b. Given the van der Waals coefficients listed in the following table, put an X next to the name of

the molecule that will have the *greatest* value of z at a pressure of 1.00 bar and 298 K.

Molecule	a (L^2 bar mol^{-2})	b (L mol^{-1})
carbon disulfide	11.3	0.073
HCN	11.3	0.088
dichloromethane	12.4	0.087
xenon difluoride	12.4	0.070

4.18 If thermal properties reveal a sample of N_2 gas to have a one-molecule translational partition function q_{trans} of $1.45 \cdot 10^{26}$, and a second sample of a pure, unknown gas is found to have a q_{trans} of $2.76 \cdot 10^{24}$ at the same temperature, what is the second gas?

4.19 The kinetic energy partition function $Q_K(T,V)$ is often written in the form

$$Q_K(T,V) = \frac{1}{N!}\left(\frac{1}{\Lambda}\right)^{3N}V^N,$$

where Λ is called the **thermal wavelength.** Find an equation for the de Broglie wavelength of the molecules in our sample traveling at the root mean square momentum, and compare this to the thermal wavelength. Calculate the value for the de Broglie wavelength when $T = 298\,K$ and $m = 100.0\,amu$.

4.20 Derive a general expression for the partition function $q(T,a)$ of a single particle in a one-dimensional box when $k_B T \gg E_1$, with the energy levels given by $E_n = \frac{n^2 h^2}{8ma^2}$. Extend this to the N-particle partition function, $Q(T,a)$. THINKING AHEAD ▶ [How should the partition function depend on m, on a, and on n?]

4.21 Prove the following equation for the non-ideal gas using the methods that yielded Eq. 4.59:

$$\langle U \rangle \approx -\frac{N^2}{2V}\frac{\partial \mathcal{I}(T)}{\partial \beta}, \quad \text{where } \beta = 1/(k_B T).$$

4.22 Using the Lennard-Jones 6-12 potential, with $\varepsilon = 200\,cm^{-1}$ and $R_{LJ} = 2.0$ Å, calculate the average potential energy $\langle U \rangle$ in J for 1 mole of N_2 gas at 300 K in a 20 L container. The average potential energy can be found using the equation in Problem 4.21.

4.23 Calculate the average kinetic energy $\langle K \rangle$ for the system in Problem 4.22.

4.24 The average intermolecular potential energy for one pair of molecules is $\langle u \rangle = -\frac{\partial \ln q}{\partial \beta}$, where $\beta = 1/(k_B T)$. In terms of q, k_B, and T, what is $\frac{\partial \langle u \rangle}{\partial T}$?

4.25 Derive the equivalent form of the equipartition principle for vibration in a molecule of mass m with a square well potential,

$$u(R \leq R_{sq}) = \infty \quad u(R_{sq} < R \leq R'_{sq}) = -u_0$$
$$u(R > R'_{sq}) = 0.$$

In other words, derive an equation for $\langle E_{vib} \rangle$ assuming $k_B T$ is large. Estimate the zero-point energy using the particle in a box.

4.26 For the case in which $u(R)$ is given by the square well potential with $u_0 = -200\,\text{cm}^{-1}$, $R_{sq} = 1.50\,\text{Å}$, and $R'_{sq} = 3.00\,\text{Å}$, solve for the parameters a and b in the van der Waals equation and the second virial coefficient B_2 at 298 K.

4.27 Find an expression for the second virial coefficient $B_2(T)$ of the square well potential that will work at *low* temperatures $T \leq \varepsilon / k_B$.

4.28 The pair potential energy for two non-rotating dipoles may be written as

$$u_{2-2}(R) = cR^{-3}.$$

By replacing the R^{-6} attractive term in the Lennard-Jones potential with this term, show that this interaction *cannot* be responsible for the attractive contribution to the pair potential energy in our analysis of the second virial coefficient $B'_2(T)$.

4.29 For methane, the van der Waals coefficients a and b are $2.30\,\text{L}^2\,\text{bar}\,\text{mol}^{-2}$ and $0.0430\,\text{L}\,\text{mol}^{-1}$, respectively. At what temperature will methane act most like an ideal gas? This temperature is called the **Boyle temperature.**

4.30 Derive expressions for the van der Waals coefficients if the intermolecular pair potential is given by

$$u(R \leq R_0) = \infty$$
$$u(R > R_0) = -\zeta e^{-R/R_0}$$

instead of the Lennard-Jones 6-12 potential. Calculate a and b for the case $R_0 = 2.0\,\text{Å}$, $\zeta = 200\,\text{cm}^{-1}$.

4.31 In our derivation of the van der Waals equation, we assumed $|u(R)| \ll k_B T$. What will be the qualitative effect on a and b as the temperature is reduced to a value $T < |u(R)|/k_B$?

4.32 Calculate the pressure in bar of 1210 mol krypton in a 150.0 L container at 298 K, if $a = 2.325\,\text{L}^2\,\text{bar}\,\text{mol}^{-2}$ and $b = 0.0396\,\text{L}\,\text{mol}^{-1}$.

4.33 The van der Waals b coefficient for CCl_2F_2 is $0.0998\,\text{L}\,\text{mol}^{-1}$. Estimate R_{LJ} in Å for this compound.

4.34 Most molecules have some angular dependence to their intermolecular potential functions. Find an approximate expression for $B_2(T)$ for the following intermolecular potential:

$$u(R,\Theta,\Phi) = 4\varepsilon\left[\left(\frac{R_{LJ}}{R}\right)^{12} - \left(\frac{R_{LJ}}{R}\right)^{6}\right]\left[\alpha + \beta\cos\Theta\right].$$

4.35 In the virial expansion to two terms,

$$P = RT\left[V_m^{-1} + B_2(T)V_m^{-2}\right],$$

there appears to be a point at which adding gas to a container at constant V and at low, constant T actually

causes the pressure to *decrease.* Find an expression for the number of moles n_P at this point in terms of the van der Waals coefficients.

4.36 A rotating electric dipole moment μ in a constant electric field $\mathcal{E}$ has a potential energy

$$u(\theta) = -\mathcal{E}\mu\cos\theta,$$

where θ is the angle between the directions of the dipole moment and the electric field.

a. Calculate the average of this potential energy as a function of T, $\mathcal{E}$, and μ by the same method and approximations used to average the dipole–dipole potential energy.

b. Write the integral needed to find the average angle $\langle \theta \rangle$ in this sample.

c. Solve this integral in the limit that the potential energy well depth is much less than the thermal energy.

4.37 In Section 4.1 we describe the dipole-induced dipole force while ignoring the angle dependence of the electric field. The electric field of the dipole actually has the equation

$$\vec{\mathcal{E}} = \frac{\mu}{R^3}(3\sin\theta\cos\theta, 3\sin\theta\cos\theta, 3\cos^2\theta - 1).$$

The potential energy for the dipole-induced dipole interaction is

$$u(R,\theta,\phi) = -\frac{\alpha\vec{\mathcal{E}} \cdot \vec{\mu}}{R^3}.$$

Show whether or not this potential energy still varies strictly as R^{-6} after averaging over all the angles (it is not necessary to solve any integrals).

4.38 Our sample consists of N gas-phase ions of mass m at temperature T, confined in a chamber that measures from $-a/2$ to $+a/2$ along each of the X, Y, and Z axes. An electric field is applied such that the potential energy seen by each ion is $u(X,Y,Z) = \mathcal{E}_0 Z^2$, where $\mathcal{E}_0$ is a constant. The gas is sufficiently diffuse that we may neglect any potential energy for interactions between the ions.

a. Write an expression for the *total* potential energy of the sample, in terms of the parameters given earlier.

b. Find the *total* translational partition function of the sample, and simplify the expression as much as possible in the limit that $\mathcal{E}_0 a^2 \ll k_B T$.

4.39 Sketch the pair correlation function for a hard sphere potential, where $u(R < R_{hs}) = \infty$, $u(R \geq R_{hs}) = 0$.

Bose-Einstein and Fermi-Dirac Statistics

4.40 Identify each of the following particles as a fermion or boson.

a. electron

b. ^{235}U nucleus

c. ^{235}U neutral atom

d. ^{19}F$^-$ ion

e. ^{1}H$_2$ molecule

4.41 Identify each of the following as either a fermion or a boson:

a. a pair of electrons

b. the ^{23}Na atom

c. the ^{24}Mg atom

d. the photon

4.42 The ^{87}Rb Bose-Einstein condensate described in Section 4.4 achieved a temperature of $2 \cdot 10^{-7}$ K. Let's say that if a fermion gas of ^{88}Rb atoms were used instead, it would need to maintain an average de Broglie wavelength *less* than 100 Å to avoid overlap between neighboring particles. Estimate the minimum temperature the fermion sample could attain. Assume that $\langle \lambda_{dB} \rangle \approx \frac{h}{m\langle v \rangle}$.

4.43 Consider an atom with $g_s = 3$ available states: $l = 0$, 1, and 2. How many two-atom states $\psi_{l_a,l_b}(1,2)$ are available if the atoms are indistinguishable and obey (a) Maxwell-Boltzmann statistics, (b) Fermi-Dirac statistics, and (c) Bose-Einstein statistics?

4.44 Repeat Problem 4.43, replacing $N = 2$ atoms with $N = 3$ and counting the number of three-atom states $\psi_{l_a,l_b,l_c}(1,2,3)$ instead of two-particle states. How will Fermi-Dirac and Bose-Einstein statistics compare to Maxwell-Boltzmann in the limit of a large number of particles N?

4.45 A set of three helium atoms in liquid helium have the possible translational states $(n_1,n_2,n_3) = (9, 2, 2)$, $(4, 7, 3)$, $(3, 3, 5)$, $(1, 5, 9)$, $(5, 5, 5)$, $(3, 5, 3)$, and $(9, 5, 1)$ where n_1 identifies the quantum state for atom number 1, and so on. (a) How many three-atom states from this set are there if these are ^{4}He atoms (bosons)? (b) How many three-atom states from this set are there if these are ^{3}He atoms (fermions)? (c) If the values of n were suddenly limited to 0 and 1 only for every atom, how many three-atom states would be possible for three ^{3}He atoms?

4.46 Calculate the de Broglie wavelength of ^{87}Rb atoms at a temperature of $170 \cdot 10^{-9}$ K, and use this to estimate the maximum average separation between atoms in a gas-phase ^{87}Rb Bose-Einstein condensate.

5

Mass Transport: Collisions and Diffusion

LEARNING OBJECTIVES

After reading this chapter, you will be able to do the following:

❶ Calculate representative values for collision properties such as mean free path and average collision frequency from system parameters such as pressure and temperature and molecular parameters such as mass and collision cross-section.

❷ Estimate diffusion constants from molecular properties, and use the diffusion constant to predict the rate of diffusion using the Einstein diffusion equation.

❸ Apply Fick's laws to describe the diffusion of one fluid through another.

GOAL *Why Are We Here?*

Our purpose in this chapter is to apply principles and results from statistical mechanics, particularly the Maxwell-Boltzmann distribution, to the motions of atoms and molecules in our system. These motions will be relevant to forming solutions, to the appearance of certain features in spectroscopy, and to the mixing of reactants in a chemical reaction.

CONTEXT *Where Are We Now?*

Previous chapters use statistical mechanics to describe a few macroscopic properties of matter, such as gas pressure and temperature. Ultimately, however, we want to understand not only the static properties of matter, but also the *dynamics*. Dynamic systems depend on time, as well as on all the coordinates of the static system, so by nature they pose a greater challenge to our intuition about chemistry.

In this chapter, we will look for ways to predict the behavior of a chemical system when there is some net motion of either matter or energy through our system. The study of **transport processes** in chemistry probes the mechanisms for these motions, in which the average density of one or more substances changes as a function of location *and* as a function of time.

SUPPORTING TEXT *How Did We Get Here?*

For matter to move from one location to another in our systems, the matter must be a *fluid*: generally meaning a liquid or a gas. Our work in this chapter will draw on the general idea that the molecules in a fluid will have a distribution of positions, speeds, and trajectories, and that we can often reduce the properties of the distribution to a few representative numerical values. More detailed concepts to support the ideas developed in this chapter can be found in the following equations and ideas:

- Equation 1.27 gives the Maxwell-Boltzmann distribution of molecular speeds v in a gas:

$$\mathcal{P}_v(v) = 4\pi \left(\frac{\mathcal{M}}{2\pi RT} \right)^{3/2} v^2 e^{-\mathcal{M}v^2/(2RT)},$$

where $\mathcal{M}$ is the molar mass, R the universal gas constant, and T the temperature. By applying the average value theorem to this distribution, we then found the average or mean molecular speed (Eq. 3.62):

$$\langle v \rangle = \sqrt{\frac{8k_B T}{\pi m}} = \sqrt{\frac{8RT}{\pi \mathcal{M}}},$$

where we may use either values appropriate to the molecular scale (m is the mass of one molecule, k_B the Boltzmann constant) or to the bulk scale (the molar mass $\mathcal{M} = \mathcal{N}_A m$ and $R = \mathcal{N}_A k_B$).

- Section 3.2 derives the equipartition principle, which predicts that if there are N_{ep} easily excited degrees of freedom in each molecule, then the internal energy of the system will be given approximately by Eq. 3.20:

$$E = \frac{1}{2} N_{ep} N k_B T = \frac{1}{2} N_{ep} n R T.$$

In this chapter we will be primarily concerned with translational motion— the motion of whole molecules from one place to another—and for translation in a three-dimensional system, $N_{ep} = 3$.

- Section 4.2 shows how the excluded volume term b in the van der Waals equation arises from the repulsive part of the intermolecular potential energy function, offering one method of mathematically defining the size of a particle in our system.

5.1 Collision Parameters

We study chemistry in environments that range from solids and liquids, with densities of over 10^{22} particles per cubic centimeter, out to interstellar nebula, which may have fewer than a thousand particles in the same volume. Our task in this section is to identify some key parameters that we can use to describe the most basic motions and interactions of the atoms and molecules in *any* of these systems.

Consider molecular motion when there are no interactions at all between the molecules; $u(R)$ is 0 everywhere. Not only are there no attractive forces, but without a hard-sphere repulsion, the molecules have no volume; they never even bounce against each other. This case is the simplest starting point, but it's not very interesting.

So next we add an interaction term, making the system more realistic and, of course, more complicated. The most significant feature on any intermolecular potential energy curve is the repulsive wall at low separation distance R, so we include that term first. Our intermolecular interactions are now represented by the hard sphere potential, which allows for purely elastic collisions. There can be no coupling between the center of mass motion and the internal degrees of freedom of the particles because there *are* no internal degrees of freedom (no rotations and no vibrations), and there is no conversion between kinetic and potential energy because the potential energy at all locations is either zero or infinity. So what turns out to be the most critical feature when we first look at molecular dynamics is that molecules *hit* each other. We can go a long way with just that fact, so for now, let's look for the collision properties of a fluid made up of hard spheres.

We first consider a few characteristic parameters of molecular collisions in a statistical sample. The most common parameters used to describe collisions between two molecules are

- various average molecular velocities.
- the average collision energy
- the effective cross-sectional area of the collision partners, called the **collision cross section**
- the average collision frequency
- the average distance a molecule travels between collisions, or **mean free path**

We will limit ourselves to two-particle collisions, except when stated otherwise. Most of the results we obtain in this section assume the substance is a gas, because the collisions in liquids are not so easily counted (adjacent molecules are in constant contact, but should we call that colliding?). Nevertheless, the concepts, if not the algebraic formulas, are valid for either liquids or gases.

Collision parameters key

symbol	parameter	SI units
m	particle mass	kg
v	speed	m s^{-1}
k_B	Boltzmann constant	J K^{-1}
T	temperature	K
μ	reduced mass $= (m_A m_B)/(m_A + m_B)$ for two particles	kg
$\langle v_{AA} \rangle$	average relative speed	m s^{-1}
ρ	number density	m^{-3}
σ	collision cross section	m^2
d	minimum distance between two colliding particles (roughly the average of the two diameters)	m
γ	average collision frequency	s^{-1}
λ	mean free path $=$ average distance travelled between collisions	m

TABLE 5.1 Definite integrals of use in calculating properties of the canonical distribution. These and other integrals are found in Table A.5.

$$\int_0^\infty e^{-ax^2}dx = \frac{1}{2}\left(\frac{\pi}{a}\right)^{1/2}$$

$$\int_0^\infty xe^{-ax^2}dx = \frac{1}{2a}$$

$$\int_0^\infty x^2 e^{-ax^2}dx = \frac{1}{4}\left(\frac{\pi}{a^3}\right)^{1/2}$$

$$\int_0^\infty x^3 e^{-ax^2}dx = \frac{1}{2a^2}$$

$$\int_0^\infty x^4 e^{-ax^2}dx = \frac{3}{8}\left(\frac{\pi}{a^5}\right)^{1/2}$$

$$\int_0^\infty x^{2n} e^{-ax^2}dx =$$

$$\left(\frac{1\cdot 3\cdot 5\ldots(2n-1)}{2^{n+1}a^n}\right)\sqrt{\frac{\pi}{a}}$$

$$\int_0^\infty x^{2n+1} e^{-ax^2}dx = \left(\frac{n!}{2a^{n+1}}\right)$$

The Velocity Distribution and Characteristic Speeds

The distribution of molecular speeds in our canonical sample obeys the Maxwell-Boltzmann distribution. In any statistical distribution, it is common to define three characteristic values: the mean (average), the **mode** (having the highest probability), and the **median** (such that half the population is below and half above). In Chapter 3, we obtained an expression for the mean molecular *speed* of a sample of molecules with molecular mass m and temperature T (Eq. 3.62):

$$\langle v \rangle = \sqrt{\frac{8k_BT}{\pi m}}.$$

We obtained this solution using the Maxwell-Boltzmann distribution and some analytical integral solutions of the type in Table 5.1.

Another informative way we could characterize the molecular motion uses the **mode speed**, v_{mode}, given by the maximum of $\mathcal{P}_v(v)$. We can find this expression by setting the first derivative of $\mathcal{P}_v(v)$ to zero and solving for v:

$$\frac{d\mathcal{P}_v(v)}{dv} = \left(\frac{2m^3}{\pi k_B^3 T^3}\right)^{1/2}\left[2ve^{-mv^2/(2k_BT)} + v^2 e^{-mv^2/(2k_BT)}\left(-\frac{m}{2k_BT}\right)(2v)\right] \quad (5.1)$$

Set this to zero when $v = v_{mode}$ and solve for v_{mode}:

$$0 = 1 - v_{mode}^2\left(\frac{m}{2k_BT}\right) \quad (5.2)$$

$$v_{mode} = \sqrt{\frac{2k_BT}{m}}. \quad (5.3)$$

By definition of the median speed v_{med}, half the molecules travel at speeds greater than v_{med} and half at lower speeds. This value can be determined by numerical integration (Problem 5.11) and yields $1.538\sqrt{\frac{k_BT}{m}}$.

The average *relative* speed $\langle v_{AB} \rangle$ is the crucial quantity when examining molecular collisions, however, and it may be obtained from Eq. 3.62 by the following qualitative argument. The classical mechanics of the interaction between two moving bodies is identical to the case of one fixed body of infinite mass and a second body of mass μ, where μ is the reduced mass of the original two bodies. This removes the center of mass motion from the problem, leaving only those effects dependent on the relative motion. Therefore, if the average speed depends on the mass m, the average relative speed is identical but for substitution of μ for m:

$$\langle v_{AB} \rangle = \sqrt{\frac{8k_BT}{\pi \mu}}, \quad (5.4)$$

where μ is the reduced mass $(m_A m_B)/(m_A + m_B)$. For the special case where the collision partners A and B are molecules of the same compound (A = B) with mass m, this expression simplifies to

$$\langle v_{AA} \rangle = \sqrt{\frac{16k_BT}{\pi m}} = \sqrt{2}\langle v \rangle. \quad (5.5)$$

For these evaluations, we have defined the speed v to be positive. The average *velocity* $\langle \vec{v} \rangle$, on the other hand, is zero, because molecules have an equal probability of moving in any direction, and vectors in opposite directions cancel. For this reason, it is also common to define the **root mean square** or **rms velocity**

$$v_{rms} \equiv (\langle v^2 \rangle)^{1/2} = (3\langle v_X^2 \rangle)^{1/2} \tag{5.6}$$

$$= \sqrt{\frac{3k_B T}{m}}. \qquad \text{by Eq. 3.17, } \langle v_X^2 \rangle = k_B T/m$$

These results are all summarized in Table 5.2.

Average Collision Energy

The collision energy is the energy of a pair of colliding particles measured relative to the potential energy when the two particles are separated by a large distance. In essence, we are interested in the *relative* kinetic energy in translation of the colliding particles. Speed is a relative quantity—its value depends on the inertial frame of reference chosen. Therefore, the kinetic energy also depends on the frame of reference. For example, in the system drawn in Fig. 5.1a, two atoms approach each other with equal speed v. Each appears to have the same kinetic energy, $mv^2/2$. In Fig. 5.1b, the same event occurs, but at the same time the box containing the atoms is moving at a speed v. The kinetic energy in our new frame of reference has become $m(2v)^2/2 = 2mv^2$ for particle A and 0 for particle B. Nevertheless, the *relative speed* of the two atoms is still $2v$, and the collision, when it takes place, is just as energetic as when the container was at rest.

The kinetic energy relevant to collision strength is measured in the **center of mass frame,** the reference frame in which the center of mass of the collision pair is at rest. If two helium atoms approach each other at equal speed v, measured relative to the laboratory, we measure the kinetic energy with respect to the laboratory frame, because in that frame the center of mass of the two helium atoms is at rest. The relative speed is $v_{AA} = 2v$, and the collision energy is $2m_{He}v^2/2 = m_{He}v^2$, which is the same as $\mu v_{AA}^2/2$. If, on the other hand, the atoms move at the same speed in parallel trajectories, the collision energy is measured with reference to the frame that moves at the same speed in the same direction. The collision energy in that case is zero, because in the center of mass frame the two atoms are at rest; they will not collide. For two particles A and B, the collision energy is $\mu v_{AB}^2/2$ and the average collision energy is

$$\langle E_{AB} \rangle = \frac{\mu \langle v_{AB}^2 \rangle}{2}. \tag{5.7}$$

By analogy with $\langle v^2 \rangle$, we could correctly predict that

$$\langle v_{AB}^2 \rangle = \frac{3k_B T}{\mu}. \tag{5.8}$$

The average collision energy turns out to have the same value as the equipartition value for the translational kinetic energy, $3k_B T/2$:

$$\langle E_{AB} \rangle = \frac{\mu(3k_B T)}{2\mu} = \frac{3k_B T}{2}. \tag{5.9}$$

TABLE 5.2 Various representative speeds for the Maxwell-Boltzmann distribution.

$\langle v \rangle$	$\sqrt{\dfrac{8k_B T}{\pi m}}$	$= 1.596\sqrt{\dfrac{k_B T}{m}}$
v_{rms}	$\sqrt{\dfrac{3k_B T}{m}}$	$= 1.732\sqrt{\dfrac{k_B T}{m}}$
v_{mode}	$\sqrt{\dfrac{2k_B T}{m}}$	$= 1.414\sqrt{\dfrac{k_B T}{m}}$
$\langle v_{AA} \rangle$	$\sqrt{\dfrac{16k_B T}{\pi m}}$	$= 2.257\sqrt{\dfrac{k_B T}{m}}$
v_{med}		$= 1.538\sqrt{\dfrac{k_B T}{m}}$

CHECKPOINT As Table 5.2 shows, the values $\langle v \rangle$, v_{mode}, v_{med}, $\langle v_{AA} \rangle$, and v_{rms} differ by less than a factor of two. If we want only a rough idea of the rate at which molecules are flying around the sample, it hardly matters which of the characteristic speeds we choose. For precise results related to molecular motions, we should integrate over the velocity distribution of the molecules anyway. However, the other equations in this section depend on the speed, and certain values are more accurate than others. We will generally use $\langle v_{AB} \rangle$ or $\langle v_{AA} \rangle$ to calculate parameters related specifically to *collisions* between particles.

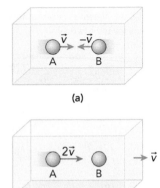

(a)

(b)

▲ **FIGURE 5.1 Frame of reference.** Identical two-atom collisions in a container when **(a)** the container is at rest and the atoms move with equal and opposite speeds: $K = 2(mv^2/2) = mv^2$; and **(b)** the container is moving at speed v, one atom is at rest ($v = 0$) and the second atom moves toward the first at speed $2v$: $K = m(2v)^2/2 = 2mv^2$. The speeds for the atoms and the container are all given relative to the same laboratory reference frame. To an observer inside the container, events **(a)** and **(b)** are identical.

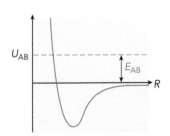

▲ **FIGURE 5.2 Intermolecular potential energy surface for two molecules with collision energy E_{AB}.**

The average collision energy can be drawn on the diagram for the potential energy between the collision partners along that coordinate, as shown in Fig. 5.2.

It is the energy gap between the dissociation limit of the two molecules, where we usually put our zero of energy, and the total energy of the **collision state**— our pair of colliding particles. In a classical elastic collision, these molecules will approach each other until the energy of the collision state is equal to the potential energy, after which they repel other. This is the same collision energy we will use in describing chemical reactions on the microscopic scale, where E_{AB} will be matched against some potential energy barrier to determine if the reaction goes forward or not.

Collision Cross Section

The collision cross section σ expresses the maximum separation between interacting molecules in the form of an area. We can think of this area as roughly πd^2, where d is some average *apparent* molecular diameter of the collision partners. A big thing is easier to hit, so size should influence the chance of two molecules colliding, but why an area? The principle behind defining a cross-sectional area for collision is illustrated in Fig. 5.3. A single reference molecule enters a container of other molecules. The likelihood of a collision depends not on the occupied *volume*, but on the occupied cross-sectional *area* of the container, and the corresponding area of our reference molecule. Our reference molecule is just as likely to collide with a sphere of radius r as with a flat disk of radius r if the disk is oriented face-on to the reference molecule.

For small molecules and many large ones, the orientation is not critical, if all we care about is how many times molecules hit each other. The bond lengths of most diatomics are small compared to their effective radius for intermolecular repulsion, and the shape of the molecule for purposes of collisions is nearly spherical. For reactive collisions, however, the relative orientation of the collision partners becomes especially important because the reaction often requires a specific matching of reactive sites on the two molecules.

Why is σ given by the area πd^2 and not the area of a single molecule, $\pi(d/2)^2$? For the same reason the excluded volume b in the van der Waals equation depends on diameter rather than radius (Section 4.2). As shown in Fig. 4.4, the center of mass of a spherical molecule is excluded from the volume of radius d (two molecular radii) around the center of mass of an equivalent molecule.

Any molecule that lies within a distance d of the trajectory of our reference molecule will be hit. The cross-sectional area πd^2 reflects not only the size of the target molecules, but also the size of the reference molecule.

Typical values for the collision cross section are given in Table 5.3. These values are obtained for collisions between molecules or atoms of the same composition, Ar colliding with Ar or SO_2 with SO_2. These values are roughly equal to the area of a circle with radius equal to the Lennard-Jones distance R_{LJ}. For example, πR_{LJ}^2 for CH_4 is equal to 46 Å^2, and σ is 52 Å^2. In fact,

TABLE 5.3 Selected collision cross sections, σ (Å^2).

He	22	Ne	25
H_2	27	Ar	36
O_2	36	N_2	37
NH_3	43	CO_2	45
CH_4	52	H_2O	62
C_2H_4	71	C_2H_2	72
C_2H_6	84	SO_2	90
$n\text{-}C_4H_{10}$	148	$CHCl_3$	156

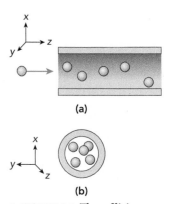

▲ FIGURE 5.3 **The collision cross section.** (a) A reference molecule enters a chamber containing other molecules. (b) The probability of a collision occurring depends on the apparent *area* presented by the target molecules perpendicular to the axis of motion.

these values fluctuate by several percent from one reference to another, because different experimental and analytical methods will yield slightly different results. Real molecules, after all, do not have a strictly defined boundary; the electron density drops smoothly with distance, and so do the forces between molecules. The apparent size of the molecule therefore depends on, for example, how sensitive our measurements are, whether we are testing at one specific collision energy or taking an average at some particular temperature, and even the rovibrational quantum states of the colliding molecules.[1]

When it becomes necessary to describe collisions between dissimilar molecules, we can approximate the heterogeneous collision cross section by the following argument. If the collision cross section σ_A is the area of a circle with radius d_A, then the cross section for collisions between two molecules A and B should be given by the average of the two radii:

$$\sigma_{AB} = \pi d_{AB}^2 = \pi \left(\frac{d_A + d_B}{2} \right)^2$$

$$= \frac{\pi}{4} (d_A^2 + 2d_A d_B + d_B^2)$$

$$= \frac{\pi}{4} \left(\frac{\sigma_A}{\pi} + \frac{2\sqrt{\sigma_A \sigma_B}}{\pi} + \frac{\sigma_B}{\pi} \right)$$

$$= \frac{1}{4} (\sigma_A + 2\sqrt{\sigma_A \sigma_B} + \sigma_B). \qquad (5.10)$$

For some cases, particularly when ions are involved, strong long-range attractive forces can make the collision cross section much larger than we would predict from the standard atomic radii.

---

[1]Although we will assume our sample to be in a thermal distribution of speeds and collision energies, more detailed cross-section measurements are available, for example, from *crossed molecular beam* experiments. These measure how molecules are deflected by collisions at the intersection of two jets of gas in a vacuum. Specific molecular velocities can be selected so that the collisions occur with a well-defined collision energy, and in many cases it is also possible to select specific quantum states of the collision partners. From the density of molecules in the two jets, σ can be calculated.

Average Collision Frequency

In a statistical sampling of a gas of molecules with collision cross section σ, number density ρ, and average relative speed $\langle v_{AA} \rangle$, how many collisions would be suffered by one reference molecule traveling a small distance l among the other molecules? Assume the probabilities are the same in all directions, so the average time after each collision before the next collision is independent of the new trajectory. The number of molecules N_{coll} in the path of our reference particle (and the number of collisions it will have) is

$$N_{coll} = \rho \sigma l, \tag{5.11}$$

where σl is the effective volume of the test particle's path as it travels the distance l. The rate at which these collisions occur is determined by the relative speed of the colliding particles and the collision frequency γ of our test particle is the number of collisions N_{coll} divided by the time Δt_{coll} required to travel the distance l, which is l/v_{AA}. Averaged over all the relative speeds, this gives

$$\gamma = \frac{N_{coll}}{\Delta t_{coll}} = \frac{\rho \sigma l}{l/\langle v_{AA} \rangle} = \rho \sigma \langle v_{AA} \rangle.$$

$$\gamma = \rho \sigma \langle v_{AA} \rangle. \tag{5.12}$$

Don't be bothered that the reference molecule's trajectory will change after every collision. Provided that the fluid is evenly distributed, changes in direction will not alter the likelihood of hitting another molecule.

Mean Free Path

One last parameter useful for fluid transport is the mean free path λ. The average distance a particle travels between collisions is given by the average speed divided by the average collision frequency. In a gas, this gives

$$\lambda = \frac{\langle v \rangle}{\gamma} = \frac{1}{\sqrt{2} \rho \sigma}. \tag{5.13}$$

EXAMPLE 5.1   Collision Parameters in Air

CONTEXT The collision parameters we discuss here have many applications. We will use them to predict rate constants for chemical reactions (Section 13.1), and they provide important reference quantities for the design of experimental apparatus (for example, ensuring a low enough pressure in the mass spectrometer described in the *Tools of the Trade*). An additional application is to the dynamics of electrical discharges. Paschen's law, named for German physicist Friedrich Paschen,[2] predicts the voltage at which electrical current will start to flow across a gap between two electrodes by ionization of the gas in between.

---

[2] Paschen's name also appears in *QM* Section 1.4, where it is used to label the $n'' = 3$ transitions in the hydrogen atom spectrum.

This *breakdown voltage* is a function of the mean free path of an ionizing electron through the gas, which is determined by the number density and collision cross section of the gas. In the Paschen equation, these are represented by the pressure of the gas and a pair of parameters that are adjusted to each gas (and which also account for the ionization properties of the gas). In air at 1.0 bar and 298 K, the minimum breakdown voltage is 330 V over a gap of 7.5 μm.

Collisional properties in air are dominated by those of N_2 and O_2, which account for roughly 98% of the particles in air at sea level. In this example, we shall find the collisional properties of the atmosphere under the assumption that O_2 is similar enough to N_2 in these respects that we may just use N_2.

PROBLEM What are ρ, λ, $\langle v_{AA} \rangle$, and γ for N_2 at 1.00 bar and 298 K? What is the average collision energy in units of cm^{-1}?

SOLUTION The collision cross section for N_2 in Table 5.3 is 37 $Å^2$. The number density ρ can be found from the ideal gas law:

$$\rho = \frac{n\mathcal{N}_A}{V} = \frac{P\mathcal{N}_A}{RT} = 2.43 \cdot 10^{19} \, cm^{-3}. \tag{5.14}$$

Therefore,

$$\lambda = \frac{1}{\sqrt{2}(2.43 \cdot 10^{19} \, cm^{-3})(37 \, Å^2)(10^{-8} \, cm \, Å^{-1})^3} = 7.9 \cdot 10^2 \, Å. \tag{5.15}$$

Under typical conditions, the molecules in air travel about 1000 Å between collisions.

The average relative speed, $\langle v_{AA} \rangle$, can be calculated once we know the mass of the N_2 molecule, 28 amu = $4.65 \cdot 10^{-26}$ kg:

$$\langle v_{AA} \rangle = \sqrt{\frac{16k_B T}{\pi m}} = \left(\frac{16(1.381 \cdot 10^{-23} \, J \, K^{-1})(300 \, K)}{\pi(4.65 \cdot 10^{-26} \, kg)} \right)^{1/2} \tag{5.16}$$
$$= 6.71 \cdot 10^2 \, m \, s^{-1}.$$

The collision frequency $\gamma = \rho\sigma \langle v_{AA} \rangle$ can then be calculated:

$$\gamma = (2.43 \cdot 10^{19} \, cm^{-3})(100 \, cm/m)^3(37 \cdot 10^{-20} \, m^2)(6.71 \cdot 10^2 \, m \, s^{-1}) = 6.0 \cdot 10^9 \, s^{-1}. \tag{5.17}$$

Each molecule in the air collides with other molecules at a rate of roughly six billion times per second.

The average collision energy is

$$\langle E_{AA} \rangle = \frac{3k_B T}{2} = 6.17 \cdot 10^{-21} \, J = 311 \, cm^{-1}. \tag{5.18}$$

This is too little energy to excite even the lowest vibrational transitions in N_2 ($\omega_e = 2358 \, cm^{-1}$) or O_2 ($\omega_e = 1580 \, cm^{-1}$). Under these conditions, collision energies are very weak compared to typical chemical bond strengths.

TOOLS OF THE TRADE | **Ion Cyclotron Resonance Mass Spectrometry (ICR-MS)**

J. J. Thomson, following his groundbreaking measurement of the mass of the electron in 1897, adapted the instrument used in that work to detect charged atoms. His coworker Francis Aston and others continued to develop methods of determining atomic and molecular masses through the 1920s and 1930s, during which they verified the existence of different isotopes of the same element. With the need to analyze and separate isotopes of uranium during the development of the first atomic weapons in World War II, mass spectrometry suddenly became a technique of intense interest.

Two general methods for mass selection—*quadrupole* and *time-of-flight*—are described briefly in *QM Section 11.2*. Here we take a look at a more elaborate method, the Cadillac of mass spectrometry techniques, capable of extraordinarily high resolution and sensitivity.

What is ICR-MS? An **ion cyclotron resonance mass spectrometer** or *ion cyclotron plasma mass spectrometer* (ICP-MS) is a high-sensitivity, high-resolution method of mass spectrometry that takes advantage of Fourier transform analysis to simultaneously analyze all the ions in the sample. For this reason, ICR-MS is also known as Fourier transform mass spectrometry (FTMS).

Why do we use ICR-MS? The goal of any mass spectrometer is to measure the relative amounts of chemical species present in a sample as a function of their atomic and molecular masses. The sensitivity of ICR-MS is higher than most other techniques because the Fourier transform technique allows the gathering of data simultaneously for all the ions in the system, rather than the monitoring of the ion signal one mass value at a time. Furthermore, the resolution is the highest of any commercially available method, more than an order of magnitude better than most other methods, allowing for isotopes of different elements that *have the same mass number* to be distinguished in some cases. For example, a quadrupole mass spectrometer cannot distinguish between $^{12}C\,^{16}O$ and $^{14}N_2$, because both molecules have a total mass number of 28. However, an ICR-MS can resolve the difference between the mass of 27.99 amu for $^{12}C\,^{16}O$ and 28.00 amu for $^{14}N_2$, making it capable of much more discriminating chemical analysis than other techniques.

How does it work? A mass spectrometer ionizes the sample in the gas phase, acts on the ions with electric and/or magnetic fields to discriminate among ions with different mass-to-charge ratios m/z (where z is the charge on the ion), and finally measures the number of ions at each m/z value with a detector sensitive to the current generated by the ions. In practice, the masses are identified by the relative acceleration due to forces acting on ionized forms of the species. The accelerations arising from an applied electric field $\mathcal{E}$ or magnetic field B acting on a particle with mass m and charge $q = ze$ are

$$\text{electric:}\quad \overrightarrow{a} = \frac{\overrightarrow{F}_{\text{Coulomb}}}{m} = \frac{q\overrightarrow{\mathcal{E}}}{m} = \frac{ze\overrightarrow{\mathcal{E}}}{m}$$

$$\text{magnetic:}\quad \overrightarrow{a} = \frac{\overrightarrow{F}_{\text{magnetic}}}{m} = \frac{q(\overrightarrow{v} \times \overrightarrow{B})}{m} = \frac{ze(\overrightarrow{v} \times \overrightarrow{B})}{m}.$$

In either case, the acceleration is proportional to z/m, and mass spectra are often printed as functions of the ratio m/z, rather than the mass m. In a typical mass spectrum, the $z = +1$ ion is more abundant than any other, but in cases where, for example, the dication $z = +2$ can be formed, it is important to remember that it will respond in the same fashion as a $z = +1$ ion with half the mass.

Conventional mass spectrometers separate the ions in space prior to detection, so the signal at each value of m/z arrives at the detector at a different time. The ICR-MS relies instead on the continuous circulation of ions within a vacuum chamber called the FTMS cavity. The system of circulating ions is the *cyclotron*. A schematic is shown in the figure. The oscillation of the ions within the FTMS cavity generates a parallel oscillating *image current* of electrons in the electrodes of the FTMS cell, and the detector is a current meter monitoring this image current.

Once the sample is ionized and injected into the FTMS cavity, the circulation of the ions is excited by a pulse of radiofrequency (rf) radiation. The rf pulse consists of intense oscillating electric and magnetic fields, which accelerate the circulation rate for each ion by an amount dependent on its m/z value. Because the

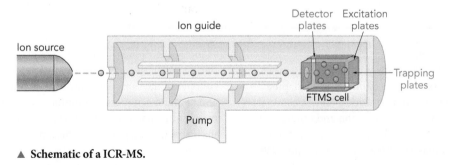

▲ **Schematic of a ICR-MS.**

frequency of the circulation for each m/z value is unique, the current measured at the detector is the sum of many currents oscillating at distinct frequencies. Carrying out a Fourier transform on the signal (Section A.1), after acquiring data over a period of about a tenth of a second, we decompose the current measure as a function of time into signals at different frequencies, which correspond to the different values of m/z. We have obtained our mass spectrum.

However, the high resolution of the technique is dependent on a relatively long data sampling period, during which the ions *must not collide with each other or with other particles in the chamber.* In order to attain an average time between collisions greater than 0.1 s, we need a collision frequency of only $\gamma \approx 10 \text{ s}^{-1}$. For air at $T = 298$ K, this requirement corresponds to a pressure of less than $3 \cdot 10^{-9}$ bar or $2 \cdot 10^{-6}$ torr and a mean free path of nearly 50 m.

5.2 The Random Walk

All the parameters we just evaluated are independent of the change in the molecule's trajectory after each collision. They're useful for describing just how dynamic the system is, but they don't directly tell us how far we will be able to transport one substance through or across another. We need to know how far, *averaged over all these collisions,* the molecules will travel. Even in the simple case of elastic collisions, the path of the molecule undergoing a large number of collisions is a difficult calculation. Rather than trying to trace this path, we fall back on statistical mechanics and ask instead how far in *any* direction a given molecule is likely to travel after N_{coll} collisions if each collision results in purely random redirection of the molecule's trajectory but no change in its speed.

Randomness, paradoxically, makes the problem manageable by allowing us to use many-particle distribution functions to describe the system rather than deterministic, one-particle-at-a-time equations. In this case, we have reduced our system to a classic **random walk** problem.

The common feature to all random walk problems is a series of steps, each of which result in some random change in one or more of the parameters describing the system. Let's state our problem as follows:

1. An atom travels along a single line, suffering a collision at regular intervals.
2. For each collision, the atom either continues in the same direction or exactly reverses its direction.
3. After N collisions, how far has the atom traveled?

A derivation that shows how we can solve this problem is waiting in the wings. It starts from an equation that exactly solves an ancient coin-flipping problem and then applies some clever approximations and extensions to find itself at the distribution of a molecule bouncing randomly off of other molecules. If you'd just as soon skip the details, a DERIVATION SUMMARY appears just after Eq. 5.30.

If we let the flip of a penny determine the random choice at each step, heads meaning one step forward and tails meaning one step back, then our progress k can be measured by the number of times i that we get heads minus the number of times j that we get tails:

$$k = i - j, \quad \text{where} \quad N = i + j. \tag{5.19}$$

Parameters key: the random walk

symbol	parameter	SI units
N	total number of events (steps taken, coins flips, etc.)	unitless
i	number of positive event outcomes (forward steps or heads on coin flip)	unitless
j	number of negative event outcomes (backward steps or tails on coin flip)	unitless
k	net number of positive outcomes $= i - j$	unitless
$\mathcal{P}(k)$	probability of any particular value of k after N steps	unitless
λ	mean free path	m
L	$= \sqrt{N}\lambda$, a characteristic distance of travel after N steps	m
$\mathcal{P}_X(X)$	probability per unit distance of particle ending at X	m^{-1}
$\mathcal{P}_V(X, Y, Z)$	probability per unit volume of particle ending at (X, Y, Z)	m^{-3}
$\mathcal{P}_r(r)$	probability per unit distance of particle ending at distance r from origin	m^{-1}

Because the coin flip is random, we will get many different values for k, so we can't ask where to find any one molecule. We can only ask what is the probability of obtaining a particular value of this parameter. There's some real work ahead, so let's pull that answer out of any decent applied math textbook:

$$\mathcal{P}(k) = \frac{N!}{2^N i! j!}$$

$$i = N - j = N - (i - k) \qquad\qquad j = N - i = N - (j + k)$$
$$\;\;= N - i + k \qquad\qquad\qquad\quad\;\; = N - j - k$$
$$i = \frac{N + k}{2} \qquad\qquad\qquad\qquad j = \frac{N - k}{2}$$

$$\mathcal{P}(k) = \frac{N!}{2^N \left(\dfrac{N + k}{2}\right)! \left(\dfrac{N - k}{2}\right)!}. \qquad (5.20)$$

Although we will not prove this equation, its validity is easily tested. Consider the case in which we flip the coin three times ($N = 3$). There are eight possible results: HHH, HHT, HTH, THH, HTT, THT, TTH, and TTT (with H indicating "heads" and T indicating "tails"). These eight states constitute our ensemble. Each result is equally possible, and the results correspond to values of k of $3, 1, 1, 1, -1, -1, -1, -3$, respectively, with a 3/8 chance each of getting $k = +1$ or -1, and a 1/8 chance each of getting $k = +3$ or -3. We get the same distribution from Eq. 5.20:

$$\mathcal{P}(-3) = \frac{3!}{2^3 3! 0!} = \frac{6}{8 \cdot 6 \cdot 1} = \frac{1}{8}$$

$$\mathcal{P}(-1) = \frac{3!}{2^3 2! 1!} = \frac{6}{8 \cdot 2 \cdot 1} = \frac{3}{8}.$$

Equation 5.20 will prove unwieldy when we look at molecular systems, however, because we will need large values of N and then the factorials will become enormous numbers. For that reason, we need to rewrite Eq. 5.20, using two common approximations from the calculus of infinite series. Instead of solving all the factorials, we evaluate the natural logarithm of $\mathcal{P}(k)$:

$$\ln \mathcal{P}(k) = \ln N! - N\ln 2 - \ln\left(\frac{N+k}{2}\right)! - \ln\left(\frac{N-k}{2}\right)!. \quad (5.21)$$

We simplify this using Stirling's approximation (Eq. 2.14):

$$\ln N! \approx N\ln N - N.$$

This gives us

$$\ln \mathcal{P}(k) \approx (N\ln N - N) - N\ln 2 - \left[\frac{1}{2}(N+k)\left[\ln(N+k) - \ln 2\right] - \frac{1}{2}(N+k)\right]$$

$$- \left[\frac{1}{2}(N-k)\left[\ln(N-k) - \ln 2\right] - \frac{1}{2}(N-k)\right]$$

$$= N\ln N - \frac{1}{2}\left[(N+k)\ln(N+k) + (N-k)\ln(N-k)\right]. \quad (5.22)$$

Next, we assume that for a large number of steps N, $k \ll N$ for the region of significant probability. This assumption will be supported by our results, but we already know that the maximum value of k is N, which is obtained only if all N steps are in the same direction. The probability of all the steps being in the same direction is $(2/2^N)$, which becomes vanishingly small as N climbs.[3]

Therefore, our assumption is a reasonable one, and this allows us to use the power series

$$\ln x = (x-1) - \frac{1}{2}(x-1)^2 + \frac{1}{3}(x-1)^3 - \frac{1}{4}(x-1)^4 + \dots,$$

which holds as long as $|x-1| < 1$. We rewrite the last two terms in Eq. 5.22 as

$$(N+k)\ln(N+k) + (N-k)\ln(N-k)$$

$$= (N+k)\ln\left[N\left(1 + \frac{k}{N}\right)\right] + (N-k)\ln\left[N\left(1 - \frac{k}{N}\right)\right]$$

$$= (N+k)\left[\ln N + \frac{k}{N} - \frac{1}{2}\left(\frac{k}{N}\right)^2 + \frac{1}{3}\left(\frac{k}{N}\right)^3 - \frac{1}{4}\left(\frac{k}{N}\right)^4 + \dots\right]$$

$$+ (N-k)\left[\ln N - \frac{k}{N} - \frac{1}{2}\left(\frac{k}{N}\right)^2 - \frac{1}{3}\left(\frac{k}{N}\right)^3 - \frac{1}{4}\left(\frac{k}{N}\right)^4 - \dots\right]$$

$$= (N+k)\ln N + k + \frac{k^2}{N} - \frac{1}{2}\frac{k^2}{N} - \frac{1}{2}\frac{k^3}{N^2}$$

$$+ \frac{1}{3}\frac{k^3}{N^2} + \frac{1}{3}\frac{k^4}{N^3} - \frac{1}{4}\frac{k^4}{N^3} - \frac{1}{4}\frac{k^5}{N^4} + \dots$$

$$+ (N-k)\ln N - k + \frac{k^2}{N} - \frac{1}{2}\frac{k^2}{N} + \frac{1}{2}\frac{k^3}{N^2}$$

$$- \frac{1}{3}\frac{k^3}{N^2} + \frac{1}{3}\frac{k^4}{N^3} - \frac{1}{4}\frac{k^4}{N^3} + \frac{1}{4}\frac{k^5}{N^4} - \dots$$

$$= 2N\ln N + 2\frac{k^2}{N} - \frac{k^2}{N} + \frac{2}{3}\frac{k^4}{N^3} - \frac{1}{2}\frac{k^4}{N^3} + \dots \quad (5.23)$$

---

[3]As an illustration, the probability of 33 steps all being taken in the same direction is $2 \cdot 10^{-10}$, or one chance in 5 billion.

We are left with

$$(N + k)\ln(N + k) + (N - k)\ln(N - k) = 2N\ln N + \frac{k^2}{N} + \frac{k^4}{6N^3} + \dots \quad (5.24)$$

where the remaining terms are of order N^{-5} and smaller. This series converges very quickly, since each order increases as only the odd powers of $1/N$. Substituting this expression back into Eq. 5.22 gives

$$\ln \mathcal{P}(k) \approx N\ln N - \frac{1}{2}\left[2N\ln N + \frac{k^2}{N}\right] = -\frac{k^2}{2N}, \quad (5.25)$$

where we have truncated terms of order N^{-3} and beyond.

We evaluated $\ln \mathcal{P}(k)$ rather than $\mathcal{P}(k)$ directly because the slower variation of the logarithm as a function of N makes simplifications like Stirling's approximation possible. All the approximations we have used are valid in the limit of large N, and so in that limit we may rewrite our expression for the probability as

$$\mathcal{P}(k) = \sqrt{\frac{2}{\pi N}}e^{-k^2/(2N)}, \quad (5.26)$$

where the constant in front of the exponent is just a normalization constant so that

$$\sum_{k=-N}^{N} \mathcal{P}(k) = 1. \quad (5.27)$$

CHECKPOINT There are a few steps that may come across as sleight-of-hand in this derivation the first time you see it. The key step is using Stirling's approximation to replace $\ln N!$ by $N\ln N - N$. This step converts the expression of factorials into a continuous function that we can treat using normal algebraic techniques. As a result of that step, we see explicitly that the distribution of final positions $\mathcal{P}(k)$ approaches a Gaussian distribution $e^{-k^2/(2N)}$ in the limit of many steps.

This function is a Gaussian distribution, and the derivation is useful to a wide number of statistical problems.[4]

The function is shown in Fig. 5.4, where the curve given by Eq. 5.26 is graphed for $N = 100$ and $N = 10,000$. As expected, the greater the number of steps, the wider the range of possible final states. The peak probability remains at the starting point, but that peak probability decreases as N increases.

Figure 5.4 also illustrates the meaning of Eq. 5.26 in relation to colliding molecules. The probability that the molecule will be found any great distance from its starting point is small, but its chances increase as the square root of the number of steps:

$$\mathcal{P}_X(X) = \frac{1}{\sqrt{2\pi}L}e^{-X^2/(2L^2)}, \quad (5.28)$$

where the distance traveled from the starting point along one dimension is X, and L gives the width of the probability curve. In our simplified case, the collisions took place at regular intervals, and the speed of the atom was invariant, so the atom traveled the same distance between each collision. If we set this distance equal to the mean free path λ, then $X = k\lambda$, and $L = \sqrt{N}\lambda$ makes Eq. 5.28 the same as Eq. 5.26 to within the normalization constant.[5]

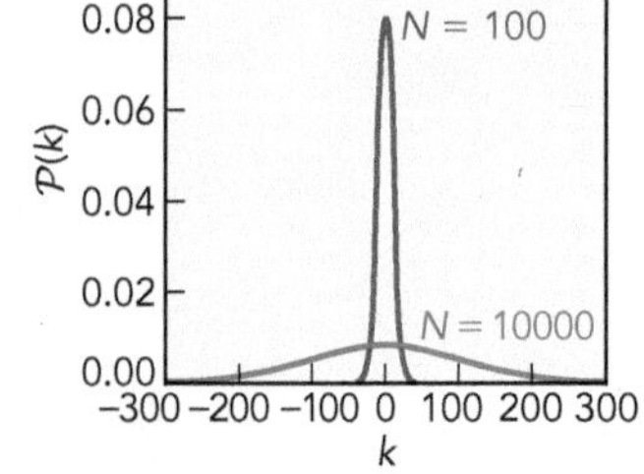

FIGURE 5.4 $\mathcal{P}(k)$ versus k for $N = 100$ and $N = 10000$.

---

[4] The most familiar example may be the **standard error distribution,** which predicts that in a series of N measurements subject to random sources of error, the number of measured values x_i obtained when the true value is x_0 is distributed as $e^{-(x_i - x_0)^2/(2s^2)}$, where s is the **standard deviation.** Comparison to Eq. 5.26 shows that s^2 is proportional to N, and therefore s is proportional to $\sqrt{N}$. The mean value of x, the parameter that's being measured, is obtained by adding all N values and dividing by N. The standard deviation in the mean is then s/N, which is proportional to $\sqrt{N}/N$. By repeating a measurement N times, we improve the precision of our mean value by a factor of $1/\sqrt{N}$.

[5] The change in normalization corrects for an artifact in our derivation of Eq. 5.26, namely that for even N, only even k was possible, and likewise, for odd N only odd k.

Extending this result to three dimensions is a simple matter as long as we keep the translational coordinates independent of each other:

$$\mathcal{P}_V(X, Y, Z) = \mathcal{P}_X(X)\mathcal{P}_Y(Y)\mathcal{P}_Z(Z) = \frac{1}{\sqrt{8\pi^3 L^3}} e^{-(X^2 + Y^2 + Z^2)/(2L^2)}. \quad (5.29)$$

It is more convenient to write this in terms of r, the distance in any direction from the starting point, remembering that the volume element in that case changes from $dX\,dY\,dZ$ to $4\pi r^2 dr$:

$$\mathcal{P}_r(r) = \frac{4\pi}{\sqrt{8\pi^3 L^3}} e^{-r^2/(2L^2)} r^2. \quad (5.30)$$

DERIVATION SUMMARY The Random Walk. We derived in succession several different probability functions, moving gradually from the simplest random walk model to the motions of molecules. We started with Eq. 5.20, a formula from combinatorial mathematics for $\mathcal{P}(k)$ of the discrete, one-dimensional random walk. Equation 5.20 is an exact expression, but factorials make it difficult to evaluate for a large number of steps N, so we applied Stirling's approximation for the logarithm of a large factorial to obtain Eq. 5.25, which shows that $\mathcal{P}(k)$ gives a Gaussian distribution in the limit of large N. We next converted the discrete variable k to the continuous variable X by assuming that each step traveled on average a distance equal to the mean free path, in order to obtain $\mathcal{P}_X(X)$. To get the *three*-dimensional probability distribution $\mathcal{P}_V(X, Y, Z)$, we multiplied $\mathcal{P}_X(X)$ by equivalent expressions for the distribution along Y and Z. Finally, we set $X^2 + Y^2 + Z^2$ equal to r^2 to get the distribution as a function of distance r from the origin.

5.3 Transport without External Forces

By themselves, the results from Section 5.2 are useful for describing the motion of matter in liquids, gases, and even solids. However, they say nothing about a net change in the average density of any particular substance with time, which is our working definition of mass transport. To apply these results to transport problems, we need to specify the parameters of the system more exactly. We begin with examples in which there are no forces working on the system other than the intermolecular repulsions.

Diffusion

One important ramification of Eq. 5.30 is that the volume in which the reference molecule is most likely to be found increases with the number of collisions and, therefore, with the time that the molecule wanders in a bath of other molecules. Let's now take a more realistic case, in which we treat not just one but a large number of reference molecules that are of a different composition from the bath.

We begin with the reference compound isolated on one side of a container, initially in a distinct compartment separated by a wall, and then allow the reference substance to move through the bath by removing the wall between them. According to Eq. 5.30, if we wait long enough, L becomes large and the exponential approaches 1, leaving the probability $\mathcal{P}_r(r)$ proportional to r^2. That's what we would find for an evenly mixed solution, where the chance of finding molecules at distance r from some origin is proportional to the surface of area of the sphere with radius r. We've proven a result that you could have guessed, but an important one nevertheless: *in the limit of very long time, two fluids will eventually mix homogeneously due to random elastic collisions.* When two fluids are combined, elastic collisions will tend to cause a net flow to even out the mixture. This process, wherein the density of the reference compound increases in the bath but decreases at its starting point due to random motion, is called **diffusion.**

Diffusion in Gases and Liquids

Assume the bath is a liquid or gas. We want to discuss the density as a function of time, so we change Eq. 5.30 to a function of time t using the collision parameters derived earlier. The width L of the distribution is $\sqrt{N_{coll}}\,\lambda$ in each direction, and N_{coll} is equal to γt, where γ is the collision frequency, so

$$L^2 = \lambda^2 \gamma t. \tag{5.31}$$

Combining this with Eq. 5.30 gives

$$\mathcal{P}_r(r,t) = \frac{4\pi}{\sqrt{8\pi^3 \gamma^3 t^3 \lambda^3}} e^{-r^2/(2\lambda^2 \gamma t)} r^2,$$

which we usually write in terms of the following **diffusion equation:**

$$\mathcal{P}_r(r,t) = \frac{\pi}{2(\pi D t)^{3/2}} e^{-r^2/(4Dt)} r^2, \tag{5.32}$$

where we have introduced a theoretical expression for the **diffusion constant**

$$D = \frac{\lambda^2 \gamma}{2} = \frac{\langle v_{AA} \rangle}{4\rho\sigma} = \frac{L^2}{2t}. \tag{5.33}$$

For the case of molecule B diffusing through a medium of molecule A (as when a solute diffuses through a solvent), we use instead the relative collision cross section σ_{AB} and the average relative speed $\langle v_{AB} \rangle$:

$$D_{B:A} \equiv \frac{\langle v_{AB} \rangle}{4\rho_A \sigma_{AB}} = \frac{1}{2}\lambda\langle v_B \rangle. \tag{5.34}$$

A more careful derivation replaces the factors of 1/2 in Eqs. 5.33 and 5.34 by a factor of 1/3, and yields better agreement with experimental values.

Our equations for γ and λ depend on the number density ρ, the temperature T, and the mass and collision cross section of the molecules. The number density and temperature are not intrinsic to the substance, so you might suspect that the diffusion constant D is not much of a constant. You'd be right: diffusion rates vary substantially with temperature and density. For liquids, the number density is nearly constant, which helps a bit, but D will also vary with molarity once the solute concentration becomes large enough for B:B collisions to contribute to the dynamics.

Equation 5.32 is often reduced to a convenient average, the root mean square distance traveled by a diffusing molecule $\langle r^2 \rangle^{1/2}$. Evaluate the integral

$$\langle r^2 \rangle = \int_0^\infty r^2 \mathcal{P}_r(r,t)dr = 6Dt, \tag{5.35}$$

and you find the **Einstein equation for diffusion:**

$$r_{rms} = \langle r^2 \rangle^{1/2} = \sqrt{6Dt}. \tag{5.36}$$

CHECKPOINT In this derivation, we see the distinction between the probability distribution (which describes all the molecules using a mathematical *function*) and averaged parameters such as the diffusion constant (which reduce the properties of the system to single *numerical values*). Our expressions in Eqs. 5.33 and 5.34 are only approximate, because the actual diffusion rate depends not just on the mean free path and average collision frequency or speed, but on the functional form of the distribution of positions and speeds of the molecules. In fact, different values of the diffusion constant are used depending on (for example) whether the flow of particles is in all three dimensions or down a channel. The value of D is a useful reference point, however, especially for comparison of relative diffusion rates among different molecules.

The distance traveled increases as the square root of time, in agreement with our qualitative description of how the Gaussian width depends on L. Typical values of D at 298 K for solutes moving through liquids are given in Table 5.4. They generally range from 10^{-6} to 10^{-4} cm^2 s^{-1}, with slower diffusion corresponding to greater molecular mass (which reduces the molecular speed at given temperature) and to greater polarity or hydrogen-bonding (which magnifies the intermolecular attractions). This corresponds to values of r_{rms} of about 0.01 cm after 1 second, 0.05 cm after 1 minute, and 0.5 cm after 1 hour. Diffusion in a solution, in the absence of stirring or other disturbances, tends to be a very slow way to mix things.

Fick's Laws

Many solution-phase experiments operate under conditions in which the concentration of one or more substances changes with position and/or time within the system. One important example is the flow of ions back and forth across the membrane of a cell, signaling the cell to carry out a task. To understand how these systems work, we need to know the effect of the density gradient on the diffusion rate. From the random walk theory, we now derive *Fick's laws of diffusion*, which govern this relationship.

We take as our sample a container with our fluid A diffusing along the Z axis through a homogeneous medium B (Fig. 5.5). The medium is unchanged by the diffusion of A. The flux $J(Z)$ is defined as the rate at which mass flows through a given cross-sectional area. For our example, we count the molecules of A that move from our container at $Z \leq Z_0$ to $Z > Z_0$ per unit time through a rectangular region of area A lying in the XY plane at $Z = Z_0$. The total number of molecules between $Z = 0$ and $Z = Z_0$ at time t is given by the integral

$$N(Z \leq Z_0) = N_T \int_0^{Z_0} \mathcal{P}_Z(Z,t)dZ, \qquad (5.37)$$

where N_T is the total number of molecules A in the sample.

The flux $J(Z_0)$ can then be obtained through the following steps:

$$J(Z_0) \equiv \frac{1}{A}\left(\frac{dN}{dt}\right)\bigg|_{Z_0} = -\frac{N_T}{A}\frac{d}{dt}\left[\int_0^{Z_0} \mathcal{P}_Z(Z,t)dZ\right]$$

$$= -\frac{N_T}{A}\frac{d}{dt}\left[\int_0^{Z_0} \frac{1}{\sqrt{2\pi}L}e^{-Z^2/(2L^2)}\right]$$

$$= -\frac{N_T}{A}\frac{d}{dt}\left[\int_0^{Z_0} (4\pi Dt)^{-1/2}e^{-Z^2/(4Dt)}dZ\right]$$

TABLE 5.4 **Selected diffusion coefficients at 298 K, (10^{-5} cm^2 s^{-1}).**

acetone in water	1.28	H_2 in CCl_4 (l)	9.75
1-butanol in water	0.56	N_2 in CCl_4 (l)	3.42
ethanol in water	1.24	O_2 in CCl_4 (l)	3.82
sucrose in water	0.52	water in acetone	4.56
ethanol in benzene	3.02		

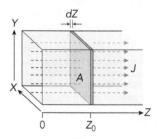

▲ FIGURE 5.5 **The system considered for the derivation of Fick's laws.** The flux at point Z_0 is the number of molecules passing through the area A per unit time.

The negative sign in the first step appears because the number of molecules at $Z < Z_0$ is dropping with time as diffusion takes place, so dN/dt is negative.

This integral cannot be solved analytically, but its derivative can be. We start from an identity (borrowed from calculus) that lets us write the definite integral of a Gaussian function as a power series:

$$\int_0^u e^{-w^2} dw = \sum_{k=0}^{\infty} \frac{(-1)^k u^{2k+1}}{k!(2k+1)}. \tag{5.38}$$

Then we set $w = Z/(4Dt)^{1/2}$, $dw = dZ/(4Dt)^{1/2}$, and $u = Z_0/(4Dt)^{1/2}$. Here's what happens:

$$
\begin{aligned}
J(Z_0) &= -\frac{N_T}{A} \frac{d}{dt}\left[\pi^{-1/2} \sum_{k=0}^{\infty} \frac{(-1)^k [Z_0/(4Dt)^{1/2}]^{2k+1}}{k!(2k+1)} \right] \\
&= -\frac{N_T}{A}\left[\pi^{-1/2} \sum_{k=0}^{\infty} \frac{(-1)^k}{k!(2k+1)} \frac{d}{dt}[Z_0/(4Dt)^{1/2}]^{2k+1} \right] \\
&= -\frac{N_T}{A}\left\{ \pi^{-1/2} \sum_{k=0}^{\infty} \frac{(-1)^k}{k!(2k+1)}\left[-\frac{(2k+1)}{2} t^{-(2k+3)/2} \left(\frac{Z_0}{\sqrt{4D}} \right)^{2k+1} \right] \right\} \\
&= -\frac{N_T}{A}\left[\pi^{-1/2} \sum_{k=0}^{\infty} \frac{(-1)^k(2k+1)}{k!(2k+1)} \left(\frac{Z_0}{\sqrt{4Dt}} \right)^{2k+1} \left(-\frac{1}{2t} \right) \right] \\
&= \frac{N_T}{2\sqrt{\pi A t}} \left(\frac{Z_0}{\sqrt{4Dt}} \right) \sum_{k=0}^{\infty} \frac{(-1)^k}{k!} \left(\frac{Z_0}{\sqrt{4Dt}} \right)^{2k} \\
&= \frac{N_T}{2\sqrt{\pi A t}} \left(\frac{Z_0}{\sqrt{4Dt}} \right) e^{-Z_0^2/(4Dt)} \tag{5.39}
\end{aligned}
$$

CHECKPOINT This derivation of Fick's first law treads through several lines of calculus but can be described in one step. We write the flux J in terms of the density distribution $\mathcal{P}_z(Z,t)$, and then work over the right-hand side of the equation until we have written it in terms of the number density ρ.

We may obtain another expression for the flux using the number density ρ. The number density at Z_0 is the number of molecules $N\mathcal{P}_Z(Z_0,t)dZ$ divided by the volume AdZ:

$$\rho(Z_0, t) = \frac{N_T}{A} \mathcal{P}_Z(Z_0, t). \tag{5.40}$$

CHECKPOINT Fick's laws were formulated in the 1850s by Adolf Fick, to analyze data obtained from his experiments on the diffusion of salts through water. The process was poorly understood, as there was no working model of matter to rely on, and Fick's laws provided some of the evidence for the subsequent development of the particle model. Fick's laws have subsequently been used to describe processes in medicine (Fick himself was a physiologist) as well as chemistry and engineering.

We find that

$$
\begin{aligned}
-D\left(\frac{d\rho}{dZ} \right)\Big|_{Z_0} &= -D\frac{N_T}{A}\left[\frac{d}{dZ} \mathcal{P}_Z(Z,t) \right]\Big|_{Z_0} \\
&= -DN_T A\left[\frac{d}{dZ}[(4\pi Dt)^{-1/2} e^{-Z^2/(4Dt)}] \right]\Big|_{Z_0} \\
&= -DN_T A(4\pi Dt)^{-1/2}\left(\frac{-2Z_0}{4Dt} \right) e^{-Z_0^2/(4Dt)} \\
&= \frac{N_T}{2\sqrt{\pi A t}}\left(\frac{Z_0}{\sqrt{4Dt}} \right) e^{-Z_0^2/(4Dt)}. \tag{5.41}
\end{aligned}
$$

Equating the results of Eq. 5.39 and 5.41,

$$J(Z_0) = -D\left(\frac{d\rho}{dZ} \right)\Big|_{Z_0}. \tag{5.42}$$

This is **Fick's first law,** and it quantifies the logical conclusion that molecular mass shifts more rapidly as the density change becomes steeper. A closetful of junk will spill out rapidly onto a clean floor when the closet door is opened, but the junk doesn't shift as fast if the floor outside the closet is also filled with junk.

If we want to know the change in number density with time instead, we can rewrite ρ as $N/V = N/(A\,dZ)$ where we measure the flux:

$$\frac{d\rho}{dt} = \frac{d}{dt}\left[\frac{N(Z_0)}{A\,dZ}\right] = \frac{d}{dZ}\left[\frac{-N(Z_0)}{A\,dt}\right] = -\frac{dJ(Z_0)}{dZ}. \tag{5.43}$$

Substituting in the value for $J(Z_0)$ from Fick's first law, we obtain **Fick's second law:**

$$\frac{d\rho}{dt} = D\frac{d^2\rho}{dZ^2}. \tag{5.44}$$

EXAMPLE 5.2   **Fick's Laws for Linear Concentration Gradient**

CONTEXT If we introduce a substance A into a medium B at a single point and allow it to diffuse, our random walk theory predicts that the concentration will follow a Gaussian profile centered on the point of origin. However, there are many ways to bring A and B together, including the use of external forces (as described in Section 5.4). The simplest profile for a concentration that changes with position is a linear gradient, in which the concentration increases or decreases at a constant rate as we move through the system. This linear gradient is unusually stable, according to Fick's laws, which makes it desirable for systems in which the concentration of A should remain fixed. For example, linear concentration gradients of sucrose have been used to thicken the medium used in experiments to separate compounds by centrifuge. In more recent work, bacteria have been induced to move in particular directions down micrometer-diameter channels on *microfluidic devices* by establishing a linear concentration gradient of a particular *chemical attractant* along the channel. The attractant is a chemical compound known to attract the bacteria, but the bacteria will continue moving only if the attractant concentration continues to increase. Directing the movement of the bacteria allows the microfluidic device to be used as a sensor for the effect of different compounds on the bacteria. A linear gradient helps to maintain a consistent concentration profile during the experiment.

PROBLEM Find the flux $J(Z)$ and the flow rate $d\rho/dt$ if the concentration is given by the linear gradient

$$\rho(Z) = \rho_0 - cZ.$$

SOLUTION From Fick's first law, we predict

$$J(Z) = -D\left(\frac{d\rho}{dZ}\right) = cD.$$

The flux is the same at all values of Z. As a result, when we look at the flow rate, we find from Fick's second law that

$$\frac{d\rho}{dt} = D\frac{d^2\rho}{dZ^2} = 0.$$

The concentration at point Z does not change with time, because mass is flowing away to larger Z at exactly the same rate it is replenished by flow in from smaller Z.

Diffusion and Solids

Molecules can diffuse through and across the surfaces of solids, and these processes are described by the same equations already derived, with one important qualification: diffusion across a surface involves fewer dimensions than diffusion through space. As a consequence, the proper phrasing of the Einstein equation for diffusion involves extension from the one-dimensional random walk to only two dimensions, giving

$$\mathcal{P}_r(r,t) \;=\; \frac{1}{2Dt}e^{-r^2/(4Dt)}r, \tag{5.45}$$

where r is the distance along the surface. This sort of action is important in surface catalysis of reactions, as for example the hydrogenation of alkenes and alkynes on platinum.

Diffusion through solids, although described by the same formulas, takes place more slowly than through fluids, being characterized by diffusion constants of the order 10^{-6} cm^2 s^{-1} and lower. Hydrogen and helium are important exceptions, since their very compact electron shells make for small effective volumes, allowing them to slip easily through gaps in crystal structures.

Viscosity

When a macroscopic quantity of liquid or gas moves from one location to another, it moves at a rate much less than the average speed of the free molecule $\langle v \rangle$. The average speed of a single H_2O molecule at 300 K is about 1000 m s^{-1}, which might imply water would flow across a tabletop 1 meter long in 1 millisecond. The actual flow speed is much slower, partly because the attractive forces between the molecules are large and also because the repulsive forces allow the molecules to exchange momentum. The effect of momentum exchange is briefly illustrated in Fig. 5.6a for motion of a group of molecules along just one dimension. Even with some net flow in a single direction, caused for example by gravity or by an external pressure, there is a canonical distribution of motions in all directions. At any given time, many molecules move—if only for a short time—against or perpendicular to the direction of flow. With collision rates of 10^9 s^{-1} in a typical gas and faster in the liquid, the exchange of momentum occurs on sufficiently rapid time scales to slow the motion of the molecules. Generally some combination of attractive and repulsive interactions contributes to the overall **viscosity,** or resistance to flow.

If we take the simpler case of the hard sphere gas, there are only the repulsive forces. When these molecules collide, they exchange momentum, but the kinetic energy of each partner remains unchanged in the center of mass frame of the collision. If two molecules collide while traveling in the same direction, the fast molecule is always slowed by the collision (Fig. 5.6b, Problem A.18). The viscosity can be accurately estimated in this scenario in a few steps.

We want an expression that describes how the viscosity depends on the parameters of our system. Consider the case of a gas flowing along the Z axis through a shaft with rectangular walls (Fig. 5.7). The exchange of momentum with the walls will depend on the number of collisions between the gas

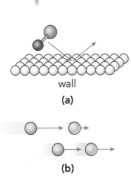

wall

(a)

(b)

▲ FIGURE 5.6 **Momentum transfer.** **(a)** Molecules collide with surfaces across which they flow, imparting momentum to the surface and reducing the momentum of the molecular flow. **(b)** Subsequent momentum exchange between flowing molecules always slows the faster molecules.

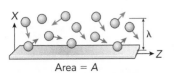

Area = A

▲ FIGURE 5.7 **The system used for deriving the elementary viscosity law.** There is a net motion of the molecules parallel to a wall of area A, but there are also collisions with the wall that transfer momentum from the flow, slowing the flow.

molecules and the wall, and we have all the tools to determine that. We choose a region on one of the walls, perpendicular to the X axis, with area A, and all the molecules within the mean free path λ from the wall within that region. The volume we are considering is therefore $A\lambda$, and the number of molecules in that volume is $\rho A\lambda$. Of those, half will on average have a velocity component toward the wall along the X axis. The rest will be moving away from the wall and therefore will not collide with it. The $\rho A\lambda/2$ molecules that will collide with the wall are moving with average speed $\langle v_X \rangle$ toward the wall, which means that on average it takes each molecule time $\lambda/\langle v_X \rangle$ to collide with the wall. The number of collisions N_{coll} that take place in that area of the wall in a time Δt is

$$
\begin{aligned}
N_{\text{coll}} &= \frac{(\text{number of colliding molecules in volume})}{(\text{time per molecule per collision})}\Delta t \\
&= \frac{\rho A\lambda/2}{\lambda/\langle v_X \rangle}\Delta t = \frac{\rho A\langle v_X \rangle \Delta t}{2}.
\end{aligned} \tag{5.46}
$$

How much does the momentum change for each collision, on average? Let's call this change Δp_i and set it equal to $m\Delta v_i$, where m is the molecular mass and Δv_i is the change in speed of molecule i during the collision. We can rewrite Δv_i as follows:

$$
\Delta v_i = \Delta X \frac{\Delta v_i}{\Delta X} \approx \Delta X \frac{dv_i}{dX}. \tag{5.47}
$$

We want to consider speed changes only within one mean free path λ from the wall, because beyond that distance, collisions with other molecules will be important. That means we may set ΔX equal to λ:

$$
\Delta v_i = \lambda \frac{dv}{dX}. \tag{5.48}
$$

This leaves us with $\Delta p_i = m\lambda dv_i/dX$. The total change in momentum during interval Δt is just Δp_i multiplied by the number of collisions, yielding

$$
\Delta p = N_{\text{coll}}\Delta p_i = \left(\frac{\rho A\langle v_X \rangle \Delta t}{2}\right)\left(m\lambda \frac{dv}{dX}\right). \tag{5.49}
$$

Viscosity is generally expressed in terms of a restraining force. Newton's law can be written several ways:

$$
F = ma = m\frac{dv}{dt} = \frac{d(mv)}{dt} = \frac{dp}{dt} \approx \frac{\Delta p}{\Delta t}, \tag{5.50}
$$

where the approximation in the last step is valid as long as the density of molecules is high, so that a short time interval Δt still involves a large number of collisions. Our restraining force due to viscosity is therefore

$$
F_{\text{viscosity}} \approx \frac{\Delta p}{\Delta t} = \left(\frac{\rho A\langle v_X \rangle}{2}\right)\left(m\lambda \frac{dv}{dX}\right) \equiv \eta A \frac{dv}{dX}. \tag{5.51}
$$

This is the expression that defines the viscosity η. To push a fluid some distance along a stationary wall of area A, it is necessary to overcome this viscous force, which increases in proportion to the area of the surface and the flow differential dv/dX. The flow differential increases as we push the liquid at a faster flow rate v, because a greater drop in the forward speed of the molecules is then required as they approach the surface of the stationary wall. The viscosity is the proportionality constant for this relation and measures the efficiency with which the forward momentum of the flowing liquid is transferred along the X axis to the wall.

Finally, we solve for the viscosity, solved here for flow only in one dimension:

$$\eta = \frac{1}{2}\rho\langle v_X\rangle m\lambda = \frac{1}{2}\rho\sqrt{\frac{8k_BT}{\pi m}}m\left(\frac{1}{\rho\sigma}\right) = \sqrt{\frac{2mk_BT}{\pi\sigma^2}}. \qquad (5.52)$$

The viscosity increases with molecular speed and decreases with cross section because small, fast molecules will reach the wall most quickly, allowing the most efficient momentum transfer from the liquid to the wall. The temperature dependence in gases is straightforward, as shown in Eq. 5.52. However, this result is not valid when the gas density drops to where the mean free path is greater than the size of the container, a condition often required in laboratory studies of reactive molecules and solid surfaces.

This derivation is invalid for liquids because there the intermolecular attraction is too important. Although the viscosity of a liquid does depend on the temperature and on the compound, the temperature dependence is more complicated, and the constant is not given by Eq. 5.52. Table 5.5 illustrates this point: while D and η are predicted to within a factor of 2 for non-polar gases and to within a factor of 3 for H_2O vapor, the strong interactions within liquid water cause errors by factors of more than 10 for D and more than 100 for η.

There are many more specific laws governing the motion of and through fluids, and their derivations are generously housed in huge and handsome textbooks on engineering. Let's be content to point out a few aspects of two such laws that should make some qualitative sense to us by now.

1. **Stokes' law** states that for a fluid solvent with viscosity η, the speed v of a sphere of radius r, large compared to the molecular dimensions of the solvent, meets a resisting force

$$F_{\text{viscosity}} = -6\pi\eta rv. \qquad (5.53)$$

TABLE 5.5 Observed (obs) and calculated (calc) values for the diffusion constant D $(cm^2\,s^{-1})$ and the viscosity η $(g\,cm^{-1}\,s^{-1})$ for selected samples and conditions. Calculated values of D are obtained using Eq. 5.34; calculated values of η are obtained using Eq. 5.52.

sample	D_{obs}	D_{calc}	sample	η_{obs}	η_{calc}
H_2 in N_2 293.15 K, 1.01 bar	0.772	1.06	H_2 300 K, 0.01 bar	$9.0\cdot10^{-5}$	$10.8\cdot10^{-5}$
Ar in N_2 293.15 K, 1.01 bar	0.190	0.312	Ar 300 K, 1.00 bar	$22.9\cdot10^{-5}$	$36.3\cdot10^{-5}$
Ar in N_2 573.15 K, 1.01 bar	0.615	0.853	Ar 600 K, 1.00 bar	$39.0\cdot10^{-5}$	$50.8\cdot10^{-5}$
H_2O in N_2 293.15 K, 1.01 bar	0.242	0.290	H_2O 300 K, 0.10 bar	$10.0\cdot10^{-5}$	$14.3\cdot10^{-5}$
H_2O in acetone (liq), 298 K	$4.56\cdot10^{-5}$	$4.7\cdot10^{-4}$	H_2O (liq), 298 K	$8.90\cdot10^{-3}$	$14.3\cdot10^{-5}$

Although the coefficient and the power law are not obvious, it makes sense that the force working against the motion of the sphere should increase as the interaction between the sphere and the medium increases (r increases) and as the sphere has more momentum to exchange with the medium (v increases). In terms of our collision parameters, we expect the number of collisions between the solvent and the sphere to increase with size (σ) and with speed. This form of the force law shows that for large molecules or particles moving through a fluid, the viscosity acts as a form of **friction,** a force that grows with speed to resist an increase in speed.

2. **Poiseuille's formula** states that for a viscous fluid flowing through a tube of radius r with pressure drop ΔP over a distance l, the volume flow rate dV/dt is

$$\frac{dV}{dt} = \frac{\pi r^4 \Delta P}{8 \eta l}. \tag{5.54}$$

The flow rate is faster

- for lower viscosity,

- for less interaction with the walls of the tube (a larger radius or shorter tube), and

- for more force pushing the liquid through (larger ΔP).

Poiseuille's formula is useful for calculating how large the diameter of a pipe needs to be (in a laboratory pumping line, or a public park's sprinkler system) to get the minimum necessary flow through the system.

BIOSKETCH | Facundo M. Fernandez

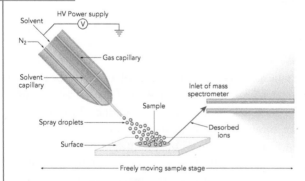

Facundo M. Fernandez is a professor of chemistry at Georgia Tech University, where he heads a research group that investigates novel approaches to the mass spectrometry of biomolecules. For decades, while mass spectrometers routinely analyzed samples of relatively simple organic and inorganic molecules, they could not be effectively used for large biomolecules because the ionization step was too destructive. Ionization required enough energy that not only was the electron removed, but dozens of chemical bonds would also be broken, resulting in a complex mass spectrum of one hundred or more distinct fragment ions formed from the original biomolecule. In recent years, more gentle ionization techniques have been developed. Professor Fernandez's group has been working, for example, on the development of *desorption electrospray ionization,* a method for analyzing the chemical components on a surface without having to subject the surface to special sample preparation. In this approach, the surface is sprayed with droplets of a partly ionized solvent. The solvent picks up the chemical analytes and provides the charge needed for the mass discrimination without the violent ionization step of older techniques. Professor Fernandez is also studying how these methods can be combined with liquid chromatography, another technique that relies on differences in mass transport to accomplish chemical analysis.

5.4 Transport with External Forces

We can now apply these results to a couple of cases where the motion of a compound through a medium is used as a diagnostic tool. These examples use an external field to bias the results toward a particular feature of the molecules under analysis.

Sedimentation

Early in the text we emphasized that gravity is not a significant factor in the physical chemistry of small molecules and atoms. However, chemistry has never been restricted exclusively to such manageable systems, and techniques have long been in use to study extremely large molecules, including organic oligomers and biomolecules. One such technique is **sedimentation,** which uses gravity or angular acceleration to characterize compounds of different masses in solution when the masses are 1000 amu or more. The earth's gravity is normally much too small to be effective over a reasonably short time scale, so sedimentation studies are conducted using **ultracentrifuges,** which are capable of effective forces in excess of 10^6 gravities (Fig. 5.8).

These molecules are much larger than the solvent molecules in which they are suspended. If spherical, they are therefore subject to Stokes' law, Eq. 5.53. Often, however, we can only write the more general friction force law

$$F_{\text{viscosity}} = -fv, \tag{5.55}$$

where f is the **frictional coefficient** of the molecule and depends on the molecule's size, shape, and attractive forces—in short, on the entire intermolecular potential, which is nightmarishly complicated for such large systems.

This frictional force is working against the effective force due to angular acceleration in the ultracentrifuge, which accelerates the molecule away from the center of rotation with a force

$$F_{\text{cent}} = (m - m_0)\omega^2 r \equiv m\omega^2 rb, \quad b \equiv \left(1 - \frac{m_0}{m}\right), \tag{5.56}$$

where m is the mass of the macromolecule, m_0 is the mass of the solvent displaced by the macromolecule as it moves through the solvent, ω is the angular velocity (2π times the number of rotations per second) of the ultracentrifuge, and r is the distance of the molecule from the center of the ultracentrifuge. The constant b is called the **buoyancy correction** and is a function of both the solvent and solute, with typical values in the neighborhood of 0.25 ($m_0/m \approx 0.75$). The difference between these masses is clearly important; if the solvent is more dense than the macromolecule, the macromolecule is forced upward because the gravitational force on the solvent is stronger. It is precisely this effect that makes centrifuges useful for separating compounds with different densities in biological samples; fractions of the sample that are less dense than the solvent will be forced to the top; the rest will be forced to the bottom.

A constant or **steady state speed** v_{ss} is attained when these two forces exactly cancel:

$$fv_{ss} = m\omega^2 rb, \tag{5.57}$$

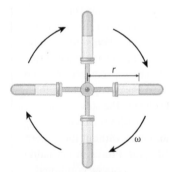

▲ **FIGURE 5.8 Schematic of the ultracentrifuge.** The system is rotating clockwise with an angular frequency ω and radius r.

which occurs at the speed:

$$v_{ss} = \frac{m\omega^2 rb}{f}.$$ (5.58)

This equation is useless without some value for f, and this is usually obtained from another form of the Einstein equation for diffusion,

$$D = \frac{k_B T}{f},$$ (5.59)

which allows the sedimentation condition to be expressed as a function of the diffusion coefficient D:

$$v_{ss} = \frac{m\omega^2 rbD}{k_B T}.$$ (5.60)

Electrophoresis

The same exploitation of transport properties is used in **electrophoresis,** except that the frictional force due to viscosity opposes the electrostatic force acting on a charged macromolecule moving through an electric field. In solution, many of these very large molecules will be charged—proteins, for example, have carboxylic acid groups (COOH), which are easily deprotonated (leaving COO^-). This means that when placed in an electric field, they will be accelerated along the field lines, and the rate of acceleration will depend again on the mass.

The mobility of charged macromolecules through a medium in electrophoresis is much more complicated experimentally than sedimentation, however, because there are multiple charge sites available in some of these molecules and because charged clusters may easily form. These problems are somewhat diminished by conducting the experiment across an amorphous polymer gel rather than in a solution. In gel electrophoresis, the mobility v of a macromolecule is usually proportional to the logarithm of its molecular weight:

$$v \propto \ln m.$$ (5.61)

This process has the advantage over sedimentation that it not only determines approximate masses, but it allows many different components of a mixture of macromolecules to be separated on the basis of molecular weight. DNA strands are often separated in this way, allowing particular sequences to be isolated from others and used in subsequent experiments.

Both electrophoresis and sedimentation are somewhat fallible in that the mobility of the macromolecule is also determined by its shape, and this is neglected in the results quoted earlier. Therefore, particular care must be taken in using the approximate formulas to describe molecules with large aspect ratios—that is, non-spherical shapes.

Convection and Chromatography

Convection is the net flow of a gas or liquid sample, usually through some other fluid medium. The medium must be distinguishable from the sample by some macroscopic parameter, such as the chemical composition, energy content, or density. This very broad definition includes diffusion of one substance through another; for example, the steady mixing of sugar dissolved in water.

For convenience, let us further narrow the definition of convection in this section to motion of our sample through another fluid *without mixing.*[6] For a sample moving down a tube lying parallel to the Z axis, we can then ascribe to our sample a well-defined average convection speed through the tube:

$$\langle v_{conv} \rangle \equiv \frac{dZ}{dt}. \tag{5.62}$$

The average is carried out over all the molecules in the sample to cancel those random components of the motion that don't contribute to net flow of the sample down the tube. This kind of convection requires the influence of forces beyond those between the molecules—a reactant being mechanically pushed through a reactor by the forced flow of solvent, or the unequal gravitational pull on two fluids of different densities (which causes hot air to rise through cool air, for example).

Any real fluid will diffuse while convection takes place, however. We can use Fick's law to find the distribution of molecules in our fluid at any given time as a sample simultaneously diffuses and convects. The rate at which our sample accumulates at a particular location Z_0 at time t_0 is determined both by the diffusion rate (using Fick's second law, Eq. 5.44)

$$\left(\frac{d\rho}{dt} \right)_{diff} = D \frac{d^2\rho}{dZ^2},$$

and by the convection rate:

$$\left(\frac{d\rho}{dt} \right)_{conv} = \frac{d\rho}{dZ/\langle v_{conv} \rangle} = \langle v_{conv} \rangle \frac{d\rho}{dZ}. \tag{5.63}$$

The combined accumulation rate is the sum of these two:

$$\frac{d\rho}{dt} = \left(\frac{d\rho}{dt} \right)_{diff} + \left(\frac{d\rho}{dt} \right)_{conv}$$

$$= D \frac{d^2\rho}{dZ^2} + \langle v_{conv} \rangle \frac{d\rho}{dZ}. \tag{5.64}$$

This is a second-order differential equation, which may be solved according to standard techniques. The result, however, is simply that the Z in our one-dimensional diffusion equation

$$\mathcal{P}_Z(Z,t)_{diff} = (4\pi Dt)^{-1/2} e^{-Z^2/(4Dt)}$$

is corrected for convectional motion:

$$\mathcal{P}_Z(Z,t) = (4\pi Dt)^{-1/2} e^{-(Z-\langle v_{conv} \rangle t)^2/(4Dt)}. \tag{5.65}$$

This final equation describes the distribution of sample in a chromatography column as a function of time and position, and it can be used to demonstrate the balance between convection speed and diffusion rate required for the effective separation of two substances.

---

[6]This is a common definition of convection when the topic is mass transfer, whereas discussions of heat transfer tend to use a broader definition of convection (incorporating diffusion).

CONTEXT *Where Do We Go From Here?*

Although chemistry is devoted to the study of atoms and molecules, you can see that it's impossible to separate the behavior of matter from the energies that control the structure and dynamics of chemical species. Having just examined the way matter moves through our system, we next consider the transport of *energy*. Energy transport has particular relevance to methods in spectroscopy, perhaps the chief analytical technique in chemistry, and also to systems with temperature variations, in which heat flows within the system to evenly distribute the energy among the system's degrees of freedom.

KEY CONCEPTS AND EQUATIONS

5.1 **Collision Parameters.** Collisions between particles in our system occur at rates that depend on the temperature T, the number density $\rho = N/V$, and the collision cross section σ. From these values, we can calculate several parameters useful for characterizing the collisions in our system:

average relative speed: $\langle v_{AB} \rangle = \sqrt{\dfrac{8k_B T}{\pi \mu}}$ (5.4)

average collision energy: $\langle E_{AB} \rangle = \dfrac{3k_B T}{2}$ (5.9)

average collision frequency: $\gamma = \rho \sigma \langle v_{AA} \rangle$ (5.12)

mean free path: $\lambda = \dfrac{1}{\sqrt{2}\rho\sigma}$. (5.13)

5.3 **Transport without External Forces.**

a. A large number of particles moving randomly in all three dimensions from a point of origin leads to a Gaussian distribution

$$\mathcal{P}_r(r,t) = \frac{\pi}{2(\pi Dt)^{3/2}} e^{-r^2/(4Dt)} r^2,$$ (5.32)

where r is the distance from the origin, t is the duration of the travel, and D is the diffusion constant.

b. The **Einstein equation for diffusion** provides a convenient simplification of the dynamics of diffusion, expressing the root mean square distance traveled in time t by particles diffusing in three dimensions:

$$r_{rms} = \sqrt{6Dt}.$$ (5.36)

c. Fick's laws describe how the concentration of one chemical species changes due to diffusion through a medium:

Fick's 1st law: $J(Z_0) = -D\left(\dfrac{d\rho}{dZ}\right)\Big|_{Z_0}$ (5.42)

Fick's 2nd law: $\dfrac{d\rho}{dt} = D\dfrac{d^2\rho}{dZ^2}$, (5.44)

where $J(Z_0)$ is the flux (the net number of particles flowing per unit area).

5.4 **Transport with External Forces.** In a centrifuge, large molecules of diverse masses can be separated because the steady state speed v_{ss} of each molecule toward the outside of the centrifuge is proportional to the mass:

$$v_{ss} = \frac{m\omega^2 rbD}{k_B T}.$$ (5.60)

KEY TERMS

- The **mean free path** λ is the average distance a molecule travels between collisions.
- The **collision cross section** σ is an effective cross-sectional area quantifying how the size of the particle affects its collisions with other particles.
- The **random walk** is a model for the motion of a particle through a medium, assuming that the direction of motion of the particle is redirected at regular intervals in random directions. This model explains the observed properties of particles subject to diffusion.
- **Diffusion** is the net motion of one chemical component through another component or across a boundary, in the absence of external forces.
- **Viscosity** is a fluid's resistance to flow when a force is applied, which occurs because particles in the fluid lose momentum by collisions and transfer that loss in momentum to neighboring particles.

OBJECTIVES REVIEW

1. *Calculate representative values for collision properties such as mean free path and average collision frequency from system parameters such as pressure and temperature and molecular parameters such as mass and collision cross section.*

   Calculate the mean free path of ozone (O_3, $\sigma = 121$ Å^2) in the stratosphere, assuming a temperature of 220 K and a pressure of 0.23 bar.

2. *Estimate diffusion constants from molecular properties, and use the diffusion constant to predict the rate of diffusion using the Einstein diffusion equation.*

   Predict the value of the diffusion constant for gas-phase acetylene in N_2 and estimate how long it would

   take for the rms diffusion distance to reach 1.0 m if $T = 298$ K and $P = 1.0$ bar.

3. *Apply Fick's laws to describe the diffusion of one fluid through a medium.*

   If the concentration of sodium ions drops by $1.0 \cdot 10^{-4}$ M across a 10 nm thick membrane, and the diffusion constant is $1.0 \cdot 10^{-11}$ cm^2 s^{-1} through the membrane, what will be the flux of ions through the membrane in mol s^{-1} m^{-2}? Approximate the derivative $d\rho/dZ$ by $\Delta\rho/\Delta Z$.

PROBLEMS

Discussion Problems

5.1 A closed gas cylinder of constant volume and initial temperature of 300 K is heated to 600 K. By what factor do the following parameters change? (a) the most probable speed; (b) the average speed; (c) the mean free path; and (d) the average collision energy

5.2 For the following compounds in the gas phase at 298 K and 1 bar, put a σ next to the one with the largest collision cross section, put a v next to the one with the lowest average speed, and put a λ next to the one with the largest mean free path.

a. He

b. C_6H_6

c. Cl_2

5.3 When discussing the average kinetic energy, we noted that kinetic energy is a relative quantity, dependent on the relative velocities of the system and the observer. Resolve the following paradox. We consider two molecules, each with mass m, with a mutual attractive interaction.

a. In the center of mass frame, the molecules move toward each other with equal speed. At a distance R_1 between the molecules, each has speed v_1, and at a closer distance R_0, the speed of each molecule has doubled to $v_0 = 2v_1$.

b. Now we follow exactly the same process, but making our measurements from the standpoint of one of the molecules, which we call the reference. The reference molecule appears to be still, with no kinetic energy, and the other molecule moves toward it at speed $2v_0$ when at a distance R_0, and at a distance $2v_1 = 4v_0$ when at a distance R_0.

If energy is conserved in frame (a), the center of mass frame, the potential energy must have decreased by an amount equal to the rise in the kinetic energy:

$$E_1 = U_1 + \frac{mv_1^2}{2} + \frac{mv_1^2}{2} = U_1 + mv_1^2$$

$$= E_0 = U_0 + \frac{mv_0^2}{2} + \frac{mv_0^2}{2} = U_0 + 4mv_1^2;$$

$$U_1 - U_0 = 3mv_1^2.$$

Applying the same logic to our measurements in frame (b), however, we find

$$E_1 = U_1 + \frac{m(2v_1)^2}{2} = U_1 + 2mv_1^2$$

$$= E_0 = U_0 + \frac{m(2v_0)^2}{2} = U_0 + 4mv_1^2;$$

$$U_1 - U_0 = 2mv_1^2.$$

Does the shape of the potential energy curve also depend on the choice of reference frame?

5.4 Section 5.3 finds that a fluid of hard spheres, given enough time, will evenly distribute itself within another fluid of hard spheres. This happens even though the combined fluid after diffusion is at exactly the same energy as before. Identify the driving force for diffusion in this case, from the standpoint of the canonical ensemble of states.

5.5 Which of the following conditions *best* describes the necessary conditions for separation of two compounds A and B in solution by column chromatography? Let the convection speeds v be such that $v_A > v_B$, let t_A be the time required for compound A to traverse the column (the retention time), and d be the length of the column.

a. $v_A \gg v_B$

b. $v_A D_A \gg v_B D_B$

c. $\dfrac{v_A}{D_A} \gg \dfrac{v_B}{D_B}$

d. $v_A - v_B \gg \sqrt{\dfrac{6}{t_A}}(\sqrt{D_A} + \sqrt{D_B})$

e. $d\left(1 - \dfrac{v_B}{v_A}\right) \gg \dfrac{6t_A}{v_A}(D_A + D_B)$

f. $t_A\left(1 - \dfrac{v_B}{v_A}\right) \gg \sqrt{\dfrac{6d}{v_A}}(\sqrt{D_A} + \sqrt{D_B})$

g. $\displaystyle\int_0^{t_A} v_A D_A \mathcal{P}_r(r,t)\,dt \gg \int_0^{v_A t_A} v_B D_B \mathcal{P}_r(r,t)\,dr$

h. $\displaystyle\int_0^{t_A}\left(\dfrac{v_A}{D_A}\right)\mathcal{P}_r(r,t)\,dt \gg \int_0^{v_A t_A}\left(\dfrac{v_B}{D_B}\right)\mathcal{P}_r(r,t)\,dr$

Collision Parameters

5.6 A sample of H_2 gas is at 298 K and 1.00 bar. Deuterium (^2H) appears in the sample at its natural abundance of 0.02%. Calculate the collision frequency γ in s^{-1} with which any given D_2 molecule in the sample collides with another D_2 molecule.

5.7 We heat a sample of O_2 gas to get an average collision energy of 12.0 kJ mol^{-1}. Calculate the temperature and the average relative speed of the O_2 molecules under these conditions.

5.8 Write an equation that could be used to estimate the collision cross section of a molecule from its b van der Waals coefficient. THINKING AHEAD ▶ [What does dimensional analysis suggest about the relationship between σ and b?]

5.9 Calculate the mean free path for H atoms in warm interstellar atomic clouds, where the temperature is roughly 1000 K and the number density is about $10^3\,cm^{-3}$. Use 25 $Å^2$ for the collision cross section.

5.10 In an equimolar gas mixture of ^4He ($\sigma = 22$ $Å^2$) and ^{40}Ar ($\sigma = 36$ $Å^2$) at 298 K, out of 1000 collisions suffered by a single typical helium atom, how many of those collisions are with other helium atoms?

5.11 The median speed v_{med} is the speed such that there are equal numbers of molecules with speed $v > v_{med}$ and with speed $v < v_{med}$. To find the median speed of molecules with mass m at temperature T, it is necessary to evaluate a numerical integral. Find an equation that could be solved to find v_{med}; don't bother to solve the numerical integral (it is already solved in Table 5.2).

5.12 Express the average time between molecular collisions Δt for a gas-phase molecule in terms of the mean free path λ and the average relative velocity $\langle v_{AB} \rangle$.

5.13 The total number of collisions N_{coll} in 1 L of N_2 at 273 K and 1.1 bar over a period of 1 s is $1.66 \cdot 10^{35}$. Find N_{coll} under the following conditions: (a) 100 cm^3, 273 K, 1.1 bar; (b) 1 L, 100 K, 1.1 bar; and (c) 1 L, 273 K, 2.2 bar.

5.14 For benzene, the collision cross section σ is 88 $Å^2$. Estimate the average collision frequency γ for benzene gas at 0.1 bar, 300 K.

5.15 For 300 K, at what number density ρ and mean free path λ will benzene have an average collision frequency γ comparable to typical intermolecular vibrational frequencies, roughly 10^{12} s^{-1}? What does this suggest about benzene at this density?

5.16 At what temperature will the collision frequency γ be $1.00 \cdot 10^9$ s^{-1} per atom in a sample of Ar ($\sigma = 36$ $Å^2$) at 1 bar?

5.17 In a sample of 0.10 mole of N_2 gas at 300 K and 1.00 bar, how many molecular collisions *total* take place each second?

5.18 At what pressure (in bar) does the mean free path in N_2 ($\sigma = 37$ $Å^2$) at 298 K drop to 100 $Å$?

5.19 From the rotational spectrum of a pure sample of molecule X, it is determined that the mean free path is $9.6 \cdot 10^{-7}$ m, at a pressure of 0.100 bar and temperature of 298.2 K. Find the collision cross section of X in SI units.

5.20 What will be the average speed (*not* relative speed) of $^{19}F_2$ molecules in a sample where the average collision energy is 15.0 kJ mol^{-1}?

5.21 Assume that the value of σ for Ar in Table 5.3 is measured at 298 K. Calculate from this value the effective collision diameter d for Ar. Then, using the parameters in Table 5.2, calculate the classical turning points $R_{class}(T)$ for the Lennard-Jones potential of Ar at energies corresponding to 298 K and at 998 K above the dissociation limit at $U(R) = 0$. Finally, assuming that d is proportional to $R_{class}(T)$, estimate the value of σ if it is measured at 998 K. THINKING AHEAD ▶ [How should the turning points compare to a typical chemical bond length?]

5.22 Two gas-phase samples of C_2H_2 ($\sigma = 72$ $Å^2$, $m = 26$ amu), have the *same volume and number of moles* but different temperatures. Given the values at 300.0 K in the following table, write the corresponding values for 600.0 K, in the same units.

parameter	300.0 K	600.0 K
$\rho\,(m^{-3})$	$6.8 \cdot 10^{24}$	
$\lambda\,(m)$	$5.7 \cdot 10^{-7}$	
$\langle v_{AA} \rangle\,(m\,s^{-1})$	699	
$\gamma\,(s^{-1})$	$1.2 \cdot 10^6$	

5.23 Two gas-phase samples, Ar ($\sigma = 36\,\text{Å}^2$, $m = 40$ amu) and C_2H_2 ($\sigma = 72\,\text{Å}^2$, $m = 26$ amu), have the same pressure and temperature. Given the values for Ar in the following table, write the corresponding values for C_2H_2 in the same units.

parameter	Ar	C_2H_2
$\rho\,(\text{m}^{-3})$	$4.3 \cdot 10^{23}$	
$\lambda\,(\text{m})$	$6.5 \cdot 10^{-6}$	
$\langle v_{AA} \rangle\,(\text{m s}^{-1})$	696	
$\gamma\,(\text{s}^{-1})$	$1.1 \cdot 10^8$	

5.24 In experiments on combustion intermediates, highly reactive molecules are formed in the gas phase in a 2.0 m long tube with diameter 75 mm, at total pressures around 500 mtorr ($6.6 \cdot 10^{-4}$ bar) and temperatures around 400 K. Assuming the reactive molecules start in the center of the tube and have the mass and collision cross section of CO_2 ($\sigma = 45\,\text{Å}^2$), calculate the time it takes for the rms travel distance of the molecules to equal the distance to the wall of the tube. This allows an estimate of the lifetime of the intermediates in the system.

5.25 In an ideal gas sample, PV is one measure of the energy content of the sample (PV equals the energy needed to fill a vessel to a final volume V at a constant pressure P and at constant temperature). Set $PV_0 = \langle E_{AB} \rangle$, and find an expression for the characteristic volume V_0 in terms of only the mean free path and collision cross section.

5.26 Prove that the average collision energy $\langle E_{AA} \rangle$ for like molecules still satisfies the equipartition principle. This involves integrals that can be separated into an integral over the relative speed of two molecules v_{12} and an integral over the center of mass speed v, where

$$(v_1^2\,dv_1)(v_2^2\,dv_2) = (v^2 dv)\,(v_{12}^2\,dv_{12})$$

and

$$v_1^2 + v_2^2 = 2v^2 + \frac{1}{2}v_{12}^2.$$

5.27 Some reactions require the simultaneous collision of at least three molecules in order to occur. Assume that for a collision to be characterized as three-body, the collision with the third molecule must occur while the other two molecules are within 4 Å of each other (4 Å is the effective radius of an N_2 molecule). Calculate an approximate ratio of three-body to two-body collisions in N_2 gas at 300 K and 1 bar, given that the collision cross section is 43 Å^2.

5.28 Find the average collision energy in kJ mol^{-1} of a 0.100 mol sample of an ideal gas at 0.831 bar and a volume of 4.00 L.

5.29 Experiments in surface science are normally carried out in vacuum chambers at extremely low pressures to prevent the surface from becoming quickly contaminated by stray molecules. Estimate how low the pressure (in bar) must be if a square patch of surface 10.0 nm on a side is to suffer on average fewer than one collision per hour when the temperature is 298 K. Assume the chief contaminant is air and that its molecules have an average speed $\langle v \rangle$ of 650 m s^{-1} and collision cross section of 37 Å^2. (The surface is stationary, so you should use $\langle v \rangle$ instead of $\langle v_{AA} \rangle$.)

5.30 A typical tank of oxygen gas ($\sigma = 36\,\text{Å}^2$) is transported from the seller at 298 K and a pressure of about 170 bar. Find the mean free path in Å.

5.31 To initiate a reaction in chlorine gas, we estimate that we need at least $1.0 \cdot 10^7$ collisions of the chlorine per second, with an average collision energy of at least $7.0 \cdot 10^{-21}$ J. What are the minimum temperature and pressure (in bar) of chlorine that we should use? Assume the collision cross section is 63 Å^2.

Random Walks

5.32 If we flip a coin an even number of times N, there's a chance that we will get an equal number of heads and tails.

a. Find a general expression for this probability in terms of N.

b. Find the minimum number of flips so that this probability is less than 1/3.

5.33 The text discusses the probabilities $\mathcal{P}(k)$ of the outcomes for three flips of a coin where

$$k = (\text{number of heads}) - (\text{number of tails}).$$

Graph the probabilities for six coin flips, and label the axes.

5.34 Calculate $\mathcal{P}(k = 16)$ for $N = 20$ using Eq. 5.26 and compare to the exact solution to show that the approximation is within a factor of 2 of the exact value.

5.35 Using the integral solution to the one-dimensional random walk, calculate the probability that after 100 collisions, a molecule would be (a) where it started, (b) 10 mean free paths from where it started, and (c) 100 mean free paths from where it started.

5.36 In repeated trials, it is found that molecules starting at the origin in a one-dimensional random walk system have a 2.9% chance of being found back at the origin after $1.0 \cdot 10^{-3}$ seconds. Estimate the collision frequency γ.

5.37 Calculate the probability that a molecule in a three-dimensional medium will be found between 0.99 and 1.01 cm from its starting point after 1.00 s if the number density is $1.00 \cdot 10^{19}$ cm^{-3}, the average relative velocity is $1.00 \cdot 10^5$ cm s^{-1}, and the collision cross section is 100 Å^2.

Diffusion, Viscosity, Convection

5.38 The diffusion constant for SF_6 in air is 0.150 cm^2 s^{-1} at 373 K. What is the rms distance in cm traveled by SF_6 after 1 hour?

5.39 The root mean square distance traveled by a radioactive tracer molecule in a sample of nitrogen gas is found to be 65 cm after 2.00 hours. We then double the temperature (in K) and cut the pressure in half. Estimate the new rms distance covered in 2.00 hours.

5.40 Prove that for a mixture of two gases A and B with constant temperature and constant total pressure everywhere in the sample, the constant D_A for diffusion of A through the mixture must be equal to the constant D_B for diffusion of B through the mixture.

5.41 In Table 5.5, D_{calc} overestimates the liquid H_2O/acetone diffusion constant by a factor of 26. Based on this, state a quantitative conclusion about the interaction between H_2O and acetone in the liquid phase.

5.42 Write, but do not evaluate, the integral (with all appropriate numerical values) necessary to calculate the fraction of a glycine sample that has diffused in water a distance of at least 5 cm at 20 s ($D = 1.055 \cdot 10^{-5}$ cm^2 s^{-1}).

5.43 Calculate the diffusion constant of H_2O in liquid H_2O at 298 K, based on our idealized random walk model.

5.44 Write the expression necessary to calculate the average *inverse time* $1/t$ for diffusion in three dimensions of the molecules in a sample to a distance r_0 from their origin. Take into account that you are probably not starting from a correctly normalized function. (By the way, it is necessary to average $1/t$ because the average diffusion time is infinite. At infinite time, the molecules are evenly distributed, so—considering only the net motion—some molecules *never* diffuse away from the origin.)

5.45 Find $\langle r^2 \rangle^{1/2}$ in terms of D and t for the case of diffusion across a surface.

5.46 A long tube is filled with N_2 gas at constant pressure. A small sample of a compound B is introduced at one end of the tube. Write the integral in terms of D and t that you would need to evaluate to find the number of molecules of B flowing past point Z_0 midway down the tube between times t_1 and t_2.

5.47 For flow in one direction, draw a curve for $\rho(Z)$ near the point Z_0 such that

a. the concentration is decreasing as Z increases, and

b. the concentration at Z_0 will *drop* with time.

5.48 A liquid is added to solvent with an initial (normalized) distribution at $t = 0$ of $\mathcal{P}_Z(Z) = Ae^{-(Z/a)^6}$, where $A = 0.53896$ cm^{-1}. This is a nearly constant value from $Z = 0$ to $Z = \pm a$, where it rapidly drops to zero, as shown in the graph. Find an expression for the flux as a function of Z, and sketch that function on a graph.

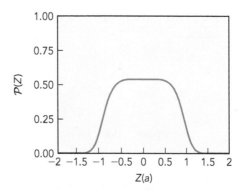

5.49 The collision cross section for H_2 is 27 Å^2 and for benzene is 88 Å^2. Which is less viscous at 300 K, H_2 gas or benzene gas?

5.50 In deriving our equation for the viscosity of a gas, we used a number $N_{coll} = \rho A \langle v_X \rangle \Delta t/2$ for the number of collisions with a certain area A of the wall over time Δt. If that number is initially $1.00 \cdot 10^8$ for a particular gas, what does it become if we do the following?

a. double the pressure but keep the same temperature

b. double the temperature but keep the same pressure

c. change the molecule to double the collision cross section

5.51 For a dense gas, Poiseuille's formula states that $\Delta V/\Delta t$ of a tube with diameter r, length $l \gg r$, and pressure drop ΔP has the form

$$\frac{\Delta V}{\Delta t} \propto \frac{(r^4 \Delta P)}{(\eta l)}.$$

Estimate how $\Delta V/\Delta t$ depends on each of these parameters $(r, \Delta P, \eta, l)$ in the diffuse gas limit where $l \gg \lambda$, the mean free path.

5.52 The Stokes equation, $F_{visc} = -6\pi\eta rv$, predicts the viscous force opposing the motion of a spherical particle with radius r at speed v through a medium with viscosity η. See if you can use the definition of the viscosity to set up a differential equation in the velocity that integrates over the surface of the spherical particle to predict the *general* form of the Stokes equation (ignoring that factor of 6).

5.53 In a common technique in chemical synthesis, helium is used to carry a volatile mixture out of a reactant flask through two "cold traps" connected in series, as drawn in the following figure. The first trap is cooled by dry ice (195 K) and the second by liquid nitrogen (77 K). The first trap removes water and solvent from the gas, and the isolated product condenses and freezes in the second trap. Indicate by 1, 2, 3, 4, and 5 the order of increasing gas flow rate (in cm s^{-1}) through the tubing in the five different regions (a) through (e) of the apparatus shown, if the pressures change as shown. Assume that the measurements are made in regions of equal cross sectional area A and that the pressures are steady.

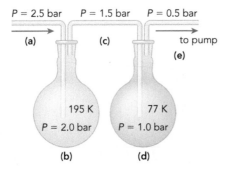

5.54 Find the steady state sedimentation speed v_{ss} for sucrose ($C_{12}H_{22}O_{11}$) in water ($D = 5.216 \cdot 10^{-6}$ cm^2 s^{-1}) at 298 K if there is no centrifuge and the only external force working on the molecule is gravity ($g = 980.7$ cm s^{-2}). Use any of the conditions assumed when a centrifuge is used.

5.55 Sedimentation is a useful way to measure molecular weight only as long as the sedimentation rate is fast compared to the rate of diffusion. Assume that this can be determined simply by comparing the steady state speed for sedimentation to the diffusion rate $d\langle R^2 \rangle^{1/2}/dt$. For what range of diffusion time t is sedimentation useful for a molecule at $T = 300$ K with buoyancy correction $b = 0.25$, molecular mass $m = 25000$ amu, and diffusion constant $D = 1.1$ cm^2 s^{-1}, when placed in an ultracentrifuge 3 cm from the center of rotation and spun at 20000 rpm?

5.56 Two plots are shown of signal versus distance for the chromatographic separation of two substances at different times. Sketch what the true chromatogram will look like as a function of **time** if the column is 20 cm long.

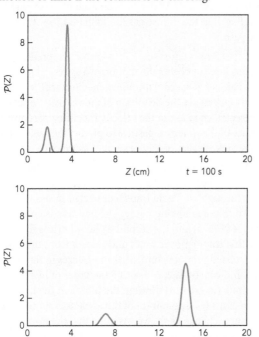

5.57 In a series of experiments, a gullible professor sits at one end of a long tank of water, tasting the water for evidence of chemicals that are being added at the opposite end. Assume that the professor is equally sensitive to the taste of each chemical. An initial study using sugar finds that it takes 2500 seconds for the professor to detect its presence. In the following table, estimate the amount of time that will elapse before the chemical is detected when the initial experiment is altered in the specified manner. Exact answers may not be possible in all cases. Each experiment is the same as the initial experiment except for the change described.

experiment	time
initial expt	2500 s
change chemical to double the diffusion constant	
double the length of the tank	
double the concentration of sugar added initially	

5.58 The diffusion constant of a lysozyme in water is $1.11 \cdot 10^{-6}$ cm^2 s^{-1}, and the diffusion constant of propane in water at the same temperature is $1.21 \cdot 10^{-5}$ cm^2 s^{-1}. If we let propane diffuse in water for $3.60 \cdot 10^3$ s and measure the rms diffusion distance, how long would it take the lysozyme to reach the *same* rms diffusion distance?

5.59 We add a drop of dye to a solvent in a tube such that the initial distribution is Gaussian along the axis of the tube,

$$\rho(Z) = Ae^{-Z^2/a^2},$$

where A and a are constants. Find the flux $J(Z)$ and the rate of change in the number density $d\rho/dt$ both as functions of Z. Graph the results.

PART I
EXTRAPOLATING FROM MOLECULAR TO MACROSCOPIC SYSTEMS

PART II
NON-REACTIVE
MACROSCOPIC
SYSTEMS

PART III
REACTIVE
SYSTEMS

6

Energy Transport: Radiation and Matter

LEARNING OBJECTIVES

After reading this chapter, you will be able to do the following:

❶ Identify which heat transfer mechanisms are operating in a given system.

❷ Graph the radiation density of a perfect blackbody at a given temperature.

❸ Calculate absorbances using the Beer-Lambert law.

❹ Estimate spectroscopic linewidths due to Doppler and collision broadening.

GOAL *Why Are We Here?*

The purpose of this chapter is to inform us about the mechanisms by which energy moves from one location to another in a system. We will pay particular attention to energy in the form of radiation and its interaction with matter because this form of energy transfer has taught us most of what we know about molecular structure and the fundamental nature of chemistry.

CONTEXT *Where Are We Now?*

In Chapter 5, we investigated the way that matter moves through space, including its motion through other matter. Now we explore how *energy* moves through space, and again we will want to consider its motion through matter. In chemistry, we tend to focus on how the molecules get around in the system, often assuming that the energy we need is *invested in* the molecules, so any required energy transport will be taken care of by the mass transport. However, we will see that there are other mechanisms available for transferring energy in our system. And even when we do use the molecules to carry energy around the system, we need to consider some issues apart from the mass transport.

We deal with energy transfer now because, as Part I draws to a close, we take the opportunity to look at the microscopic dynamics of our system in detail before tackling thermodynamics. In thermodynamics, we will increasingly fold these properties into empirical parameters so that we can enlarge our perspective to take in the macroscopic behavior of the system. This is also an opportunity to examine in some depth a few specific examples of the interaction of radiation with bulk matter that have particular significance to chemistry: blackbody radiation, spectroscopy, and lasers. We will use statistical mechanics to justify the basic laws underlying these processes, which provide among our clearest examples of macroscopic properties of a system that can only be explained from the molecular perspective. We won't be done for good with partition functions by the end of this chapter, but we should have a better idea how useful they are.

From here we will venture next into chapters on thermodynamics, which continue to deal with energy transfer but in a more formal fashion, without dwelling quite so carefully on the details. This does not diminish the power of thermodynamics at all, but illustrates the tremendous range and adaptability of our chemical model. We can adjust the scope of our vision to the magnitude of the system before us. As the system grows in size and complexity, we are increasingly able to formulate the problems we want to solve in ways that make them tractable. There are many rewards in studying science, but experiencing this seamless fabric of the theory of chemistry is surely one of the most gratifying.

| SUPPORTING TEXT | *How Did We Get Here?*

The main qualitative preparation we need for the work ahead is an understanding that energy can be interconverted among three forms: kinetic energy (the motion of masses), potential energy (the potential for motion due to a net force acting on a particle with mass), and electromagnetic radiation (no mass, just electric and magnetic fields propagating through space). For a more detailed background, we will draw on the following equations and information to support the ideas developed in this chapter:

- We are reminded in Section 2.1 that electromagnetic radiation can be broken down into indivisible units of photons, each carrying an energy (Eq. 2.6)

$$E_{\text{photon}} = h\nu = \frac{hc}{\lambda},$$

  where h is Planck's constant, ν is the radiation frequency, c is the speed of light, and λ is the radiation wavelength.

- Section 3.2 presents the equipartition principle, a crude but effective way to divide the energy of our system among its degrees of freedom (Eq. 3.20):

$$E = \frac{1}{2}N_{\text{ep}}Nk_{B}T = \frac{1}{2}N_{\text{ep}}nRT,$$

  where N_{ep} is the number of equipartition energy contributions, typically one for every translational and rotational motion, and two (one kinetic and one potential energy contribution) for each vibrational motion with a vibrational constant less than or comparable to the thermal energy $k_{B}T$. This equation also offers one of our clearest pictures of the relationship between energy and temperature: the temperature is roughly the average energy per degree of freedom that the system can easily tap into.

- Chapter 5 introduces several parameters for describing the collisions between molecules, including the collision frequency γ (Eq. 5.12),

$$\gamma = \rho\sigma\langle v_{\text{AA}}\rangle,$$

  where ρ is the number density, σ is the collision cross-section, and $\langle v_{\text{AA}}\rangle$ is the average relative speed of the particles. Chapter 5 also includes a derivation of Fick's laws, which relate a change in concentration in our system to how the system will evolve in time. Fick's second law, in particular, gives the concentration change as a function of time in terms of the diffusion coefficient D (Eq. 5.44):

$$\frac{d\rho}{dt} = D\frac{d^2\rho}{dZ^2}.$$

6.1 Conduction, Convection, and Radiation

Chemical systems are dynamical systems, full of motion, and our look at mass transport addresses only a fraction of the dynamics. If we keep the forces at work in our system fixed, then the potential energy function for our system stays roughly constant. We can then regard the flow of energy as a redistribution of the molecular kinetic energies, perhaps involving radiant energy as well. The first law of thermodynamics divides the energy change ΔE of our system into energy transfer among molecular degrees of freedom in the form of heat q and the flow of energy that results from net translation of matter in the form of work w. Of these two types of energy transfer, heat tends to be the most important to us as chemists for the very reason that it is intricately tied to transformations at the molecular scale. Here we will briefly review some basic points related to heat in general.

We divided energy transfer up into heat and work, and now we divide heat itself up into three sub-categories: convection, conduction, and radiation (Fig. 6.1).

Convection

We first encountered convection in Section 5.4, defining it there as any net motion of a fluid, usually through another fluid. While this is a way of transporting mass, convection also carries the energy of the molecules wherever they go. If the fluid on the move has a different energy content than the medium it moves through, convection becomes a form of heat, a way of transferring energy through the microscopic degrees of freedom of the sample.

Recall that energy content is not the same as *temperature*; a substance with more equipartition degrees of freedom N_{ep} carries more energy at the same temperature. Take an air-tight tank divided into two sections, the bottom filled with 1.00 bar of He gas at 298 K, and the top filled with 0.100 bar SF_6 gas also at 298 K. When we remove the barrier between the two, the SF_6 and the He will gradually mix. (The distribution of the heavier SF_6 sags towards the bottom slightly, but the difference in concentration from top to bottom is only about 0.03% per meter of height.) The SF_6 has a greater internal energy than the He, because each SF_6 molecule has 15 vibrational modes and 3 rotations to store energy in addition to 3 translations, whereas He has only the 3 translational coordinates. Even though the two gases are at the same temperature, there is a net transfer of energy from the top of the chamber to the bottom as a result of this convection.

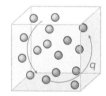

(a) Convection

$q \longrightarrow$

(b) Conduction

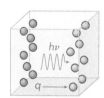

(c) Radiation

▲ FIGURE 6.1 **Three types of heat transfer.** (a) In convection, the heat is carried by *moving* matter. (b) In conduction, the heat is carried by matter that is on average, *stationary*, with collisions transferring the energy from one particle to the next. (c) In radiation, the heat is carried by photons. In each case, the net direction of heat transfer within the sample is indicated by q.

Convection is the trickiest of the three processes to describe with a general mathematical formula. The amount of energy transferred and the rate of transfer typically depend on the properties of two fluids and the effects of external forces (such as gravity) that influence the convection. Even the convection of hot air through cold air is a complex process, involving the relative diffusion rates of the two gases, the exchange of heat between the two gases (which balances convection against conduction), and the buoyancy of the hot air in the cold air (an effect of gravity). Furthermore, in many chemical applications, the mathematical treatment depends highly on the geometry of the system.

Let's do the best we can in short order, however. We have raised two fundamental questions:

1. How much energy is transferred by convection?

2. How fast is the energy transferred?

How much energy is transferred? For simplicity, we will assume that the energy is carried by two substances A and B, which remain distinguishable throughout the process, and that there is no energy exchange between A and B. The only energy transfer that takes place is based on the motion of A and B through the system. Since the total energy within the system is a constant, the total ΔE for the system is zero, but we're interested at the moment in how energy moves from one part of the system to another. So we'll divide the system into two regions, 1 and 2, such that at the outset region 1 contains all n_A moles of substance A, and region 2 contains all n_B moles of substance B. We then let A and B move around in the system until we have n_{A1}' moles of A and n_{B1}' moles of B in region 1 and n_{A2}' moles of A and n_{B2}' moles of B in region 2. Call the initial energy of region 1 E_1 and the final energy E_1'. We would then calculate the energy change in region 1 as follows:

$$E_1 = E_A$$

$$E_1' = E_A \frac{n_{A1}'}{n_A} + E_B \frac{n_{B1}'}{n_B}$$

$$\Delta E_1 = E_1' - E_1$$

$$= E_A\left(\frac{n_{A1}'}{n_A} - 1\right) + E_B\frac{n_{B1}'}{n_B} \tag{6.1}$$

Similarly,

$$\Delta E_2 = E_B\left(\frac{n_{B2}'}{n_B} - 1\right) + E_A\frac{n_{A2}'}{n_A}. \tag{6.2}$$

These equations tell us how much energy moves from one region to the adjoining region. The energy content of each *substance* (A and B) stays constant in this example, but the energy content in each *region* (1 and 2) of the system changes. If we use the equipartition principle to approximate the energies as C_VT, where C_V is a constant for each substance, we can express the energy change in terms of the temperatures of the two substances:

$$\Delta E_1 = C_{VA}T_A\left(\frac{n_{A1}'}{n_A} - 1\right) + C_{VB}T_B\frac{n_{B1}'}{n_B}.$$

To show that the total energy is conserved, we can compare the energy changes in regions 1 and 2:

$$\Delta E_1 + \Delta E_2 = \left[E_A\left(\frac{n_{A1}'}{n_A} - 1\right) + E_B\frac{n_{B1}'}{n_B} \right] + \left[E_B\left(\frac{n_{B2}'}{n_B} - 1\right) + E_A\frac{n_{A2}'}{n_A} \right] \quad \text{by Eqs. 6.1, 6.2}$$

$$= E_A\left[\left(\frac{n_{A1}'}{n_A} - 1\right) + \frac{n_{A2}'}{n_A} \right] + E_B\left[\left(\frac{n_{B2}'}{n_B} - 1\right) + \frac{n_{B1}'}{n_B} \right] \quad \text{factor out } E_A, E_B$$

$$= E_A\left[\frac{n_{A1}' + n_{A2}'}{n_A} - 1 \right] + E_B\left[\frac{n_{B1}' + n_{B2}'}{n_B} - 1 \right] \quad \text{combine fractions}$$

$$= E_A[1 - 1] + E_B[1 - 1] = 0 \qquad n_{A1}' + n_{A2}' = n_A \quad (6.3)$$

From Section 5.3, we know that diffusion tends to distribute a substance evenly throughout its environment. Therefore, in the absence of external forces, the gases in our example will eventually become completely mixed, and the energy per unit volume will be the same anywhere in our system.

How fast is the energy transferred? In the limit that no energy is transferred between substances A and B, the rate of heat transfer by convection is equal to the relative rates of *mass* transfer of A and B. The rate at which it transfers heat from one region to another dq_A/dt through an area A is

$$\frac{dq_A}{dt} = \frac{E_A}{n_A}\frac{dn_A}{dt},$$

where dn_A/dt is the number of moles of substance A crossing our boundary per unit time. With a few substitutions we can relate this back to the convection speed v_{conv} of substance A through B:

$$\frac{dq_A}{dt} = \frac{E_A}{n_A}\frac{d(\mathcal{N}_A\rho_A V)}{dt} \qquad n_A = \mathcal{N}_A N_A = \mathcal{N}_A \rho_A V$$

$$= \frac{E_A}{n_A}\mathcal{N}_A\rho_A\frac{dV}{dt} \qquad \text{factor out constants}$$

$$= \frac{E_A}{n_A}\mathcal{N}_A\rho_A\frac{A\,dZ}{dt} \qquad dV = A\,dZ$$

$$= \frac{E_A}{n_A}\frac{n_A}{V}\frac{A\,dZ}{dt} = \frac{AE_A}{V}\frac{dZ}{dt} = \frac{AE_A}{V}v_{conv}. \qquad (6.4)$$

If the motion is simply diffusion of A through B, then we can use Fick's second law (Eq. 5.44) to determine how fast the diffusion occurs and base the heat transfer rate on that.

CHECKPOINT The energy carried by convection is determined by how much energy one substance carries away as compared to the energy carried back in. If these balance, then the convective motion has not resulted in any flow of heat. Often, however, the convection is *driven* by an imbalance in the energy content, as, for example, when the lower density of hot air causes it to rise through cold air.

Conduction

Conduction is the transfer of energy from one particle to another by direct contact. In a solid, for example, the only molecular degrees of freedom may be vibration and electronic motion. Heat is conducted in the solid when the repeated vibrational excitation of one molecule causes the next molecule to start vibrating more, and then the next molecule, and the next. Conduction also occurs when electrical energy promotes electrons into the conduction band of a metal, allowing them to carry the excitation energy rapidly from one end of the solid to the other. In liquids and gases, where the electronic and vibrational degrees of freedom of two molecules are only weakly connected, energy is more easily transferred by intermolecular vibration (for liquids) or translations (for gases). Conductance doesn't *require* any net motion of the molecules, although we know that in the liquid and gas the molecules will always be moving anyway.

For a simple picture of heat conduction, let's position two samples A and B, at different temperatures, on opposite sides of a thermal conductor, a wall of thickness l and area A. To a first approximation, conduction of heat through the wall may be assumed to take place in the same manner as the conduction of a fluid through a tube as described by Poiseuille's formula (Eq. 5.54),

$$\frac{dV}{dt} = \frac{\pi r^4 \Delta P}{8 \eta l},$$

where the flow rate is proportional to the pressure difference ΔP between opposite ends of the tube and inversely proportional to the distance l over which the fluid has to travel. Here we will assume that the total rate of energy flow due to conduction dq_{cond}/dt (in watts) is proportional to the temperature difference $T_A - T_B$ across the wall and inversely proportional to the length of the wall l:

$$\frac{dq_{cond}}{dt} = \frac{kA(T_A - T_B)}{l}, \tag{6.5}$$

where k is the thermal conductivity of the wall material in $W\,K^{-1}\,m^{-1}$. (The watt, W, is the SI unit for power, where $1\,W = 1\,J\,s^{-1}$.) Equation 6.5 is known as **Fourier's law of cooling.** Unlike fluid flow down Poiseuille's tube, we don't have viscosity to make the energy flow different in different regions of the tube. Instead, the energy transport increases in efficiency in direct proportion to the area of the wall A.

If we consider the microscopic system more carefully, we will realize that k may be a highly temperature-sensitive parameter of many materials. Metals will be able to conduct heat efficiently through the electrons beginning at very low temperatures, but covalent crystals may require higher temperatures before the molecular vibrations are sufficiently excited to effectively transmit energy. In solids, high temperatures can reduce thermal conductivity as the increased random vibrational motion throughout the solid interferes with the directed transportation of energy from hot to cold sides of the material. Table 6.1 presents values of the thermal conductivity for selected materials, illustrating some of this temperature dependence.

The rest of this section studies the process of radiative energy transfer in greater detail. We'll find that a general description of energy transfer by radiation is much more feasible than general descriptions of conduction and convection, which depend heavily on the specific properties of individual chemical systems. Moreover, a better understanding of radiative energy transfer lets us examine a few points of particular interest in spectroscopy and related applications.

TABLE 6.1 **Thermal conductivities at representative temperatures for selected materials.** Gases are at a pressure of 1 bar.

Substance	$T\,(K)$	$k\,(W\,K^{-1}\,m^{-1})$	Substance	$T\,(K)$	$k\,(W\,K^{-1}\,m^{-1})$
Al (sol)	1	41.1	Pyrex (sol)	173	0.90
	10	235.0		373	1.25
	100	302	Ar (gas)	100	0.0062
	300	237		300	0.0179
NaCl (sol)	20	300	air (gas)	100	0.0094
	300	6.0		300	0.0262
diamond (sol)	5	20.6	H_2O (liq)	273	0.5610
	10	140.		373	0.6791
	100	1410	Hg (liq)	273	7.25
	300	895		373	9.43
graphite $\parallel$ (sol)	10	81.1	graphite $\perp$ (sol)	10	1.16
	100	4970		100	16.8
	300	1950		300	5.70

Radiation

Radiation is an important companion of matter in our exploration of quantum mechanics, although (as chemists) our chief interest remains with the properties of the matter. Radiation has the convenient feature of allowing us to transfer energy into and out of matter while avoiding, if we choose, the complications of molecule–molecule interactions. Whether or not the intermolecular forces are negligible, radiation provides in some cases a dominant source of energy transfer within a chemical system.

The interaction of matter and radiation is closely tied to the molecular perspective of matter, and so in the remaining sections of this chapter, we examine first a seemingly classical problem in radiative energy transfer and then some additional, related mechanisms that show how quantum mechanics manifests itself in everyday laboratory work.

6.2 Blackbody Radiation

A problem of great historical significance appears when we try to understand the light released by a heated object. When a bar of metal gets extremely hot, between 500 and 1000 K, it glows red. As it gets hotter, the glow becomes white, and eventually bluish white. This is the result of the thermal energy exciting the vibrations in the solid, which relax to lower energy states by emitting light. This much was appreciated before the development of quantum mechanics, but efforts to quantitatively explain the dependence of the emission spectrum on the temperature T failed.

BIOSKETCH | Stephen J. Klippenstein

Stephen J. Klippenstein is a senior scientist in the Chemical Sciences and Engineering Division of Argonne National Laboratory in Illinois. Dr. Klippenstein's work focuses on modeling the complex energetics of hydrocarbon combustion. Our ability to test the design of a new engine by simulating its performance on the computer is limited partly by our understanding of the interdependence of the hundreds of individual chemical reaction steps involved in burning fuel. Dr. Klippenstein and his colleagues at Argonne use an arsenal of computational tools from the fields of quantum mechanics, statistical mechanics, and chemical kinetics to unravel the dynamics of energy transfer within these systems. For example, by solving the electronic Schrödinger equation using high-quality approximations, he can predict the potential energy surface for the vibrational motions and use that surface in turn to deduce the energies of the vibrational quantum states. These are then combined to obtain the temperature-dependent partition function for the molecule, which (as we will see in Sections 12.5 and 13.3) allow us to estimate the equilibrium constants and rate constants for reactions that occur too quickly or at concentrations too low to be probed directly by experiment. A major challenge in this work is the modeling of the energy transfer between different regions of the system, because 10% changes in temperature can correspond to factors of two or more in reaction rate.

Figure 6.2 shows a few examples of emission spectra from a classical object heated to different temperatures. Neglecting sharp features from specific quantum state transitions, all solids at a temperature T identically emit light with these broad frequency profiles, once T is high enough to get away from molecule-specific effects in the vibrational energies. Prior to quantum mechanics, it was commonly accepted that solids were made up of atoms locked into place by attractive forces. This picture suggested a *classical* vibrational energy model that treated the vibrating molecules as weights joined by perfect springs. In such a classical system, any vibrational frequency would be possible. Such a material would be an ideal **blackbody,** a substance that absorbs and emits light at any frequency, and its emission is called **blackbody radiation.**

The classical theory treated the vibrational motions generating classical radiation waves in a cavity within the solid. In order for the vibrations to constructively interact with each other inside the solid, they had to correspond to standing waves, some integer number n of half-wavelengths $\lambda/2$ fit inside the solid of length a:

$$n\frac{\lambda}{2} = a, \tag{6.6}$$

so the possible values of λ are $2a/n$. These wavelengths correspond to possible emission frequencies of the solid:

$$\nu = \frac{c}{\lambda} = \frac{nc}{2a}. \tag{6.7}$$

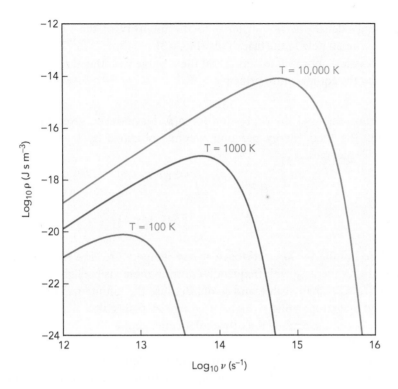

◄ FIGURE 6.2 Emitted radiation
density as a function of
frequency for an ideal blackbody.

The classical density of states emitting at frequencies between ν and $\nu + d\nu$ can be determined by the same means that give the ensemble size for the particle in a three-dimensional box (Eq. 3.38). The radiation carries an additional degree of freedom not available to the particle in a box: the plane of polarization for the electric field vector. For radiation propagating in the z direction, there are two orthogonal polarization planes, which we can set equal to the xz and yz planes. Therefore, we include an extra factor of two in Eq. 3.38, set the volume V of the solid equal to a^3 to simplify matters, and get

$$\Omega = \pi n^2 dn \qquad \text{by Eq. 3.38}$$

$$= \pi \left(\frac{2a\nu}{c} \right)^2 \left(\frac{2a}{c} \right) d\nu \qquad \text{by Eq. 6.7} \quad (6.8)$$

$$= \frac{8\pi V}{c^3} \nu^2 d\nu. \qquad a^3 = V \quad (6.9)$$

This is the number of standing wave oscillations with frequency between ν and $\nu + d\nu$. The equation tells us, as we might expect, that there are more ways to fit short wavelength (high ν) waves into a given volume than long wavelength waves. The spectral radiation density $\rho(\nu)d\nu$ inside the blackbody is the radiant energy per unit volume, measured over a small frequency range $d\nu$ centered at the frequency ν. We use $\rho(\nu)$ to eliminate the dependence on the size of the blackbody and multiply the vibrational energy by the number of modes Ω with this energy. We are left with

$$\rho(\nu)d\nu = \frac{\Omega \langle \varepsilon_{\text{vib}} \rangle}{V} = \frac{8\pi \nu^2 \langle \varepsilon_{\text{vib}} \rangle d\nu}{c^3}. \qquad (6.10)$$

The energy density is proportional to the intensity of the radiation emitted through a small hole in the blackbody (Fig. 6.3).

The classical approach assumed that the average vibrational energy was given exactly by the equipartition principle:

$$\langle \varepsilon_{\text{vib}} \rangle = k_B T.$$

Substituting this into the expression for $\rho(\nu)d\nu$, however, gives an impossible solution. The total energy per unit volume, obtained by integrating $\rho(\nu)d\nu$ over all frequencies,

$$\int_0^\infty \rho(\nu)d\nu = \int_0^\infty \frac{8\pi\nu^2 k_B T d\nu}{c^3}$$

$$= \left(\frac{8\pi k_B T}{c^3}\right)\left(\frac{\nu^3}{3}\right)\Big|_0^\infty, \tag{6.11}$$

would be infinite for *any* substance at *any* temperature. The radiation density would always increase with frequency, emitting more visible light than infrared, more ultraviolet than visible, and so on. Because the infinite radiation density is predicted by extrapolating to higher frequencies than visible light, the failure of this derivation was dubbed the **ultraviolet catastrophe.**

What is the fallacy? We derived the equipartition principle in the classical limit by assuming that the energy gap between adjacent quantum states was much less than the thermal energy $k_B T$, allowing us to treat the energy as a continuous variable. That's an approximation, but where is it likely to cause a problem? The number of available energy levels increases dramatically as the energy increases, so we might have suspected a problem at high temperatures. But the integral in Eq. 6.11 diverges no matter what T is; the ultraviolet catastrophe doesn't stop being a catastrophe at low temperature. Surprisingly, the quantum nature of the vibrational states influences the degeneracy at all temperatures. To see this, we need to use the quantum mechanical solution to the average vibrational energy.

In our derivation of the van der Waals equation, we obtained the expression for the ensemble size Ω of translational states (Eq. 4.25):

$$\ln \Omega = \ln Q(T, V) + \frac{E}{k_B T} = \ln Q(\beta, V) + \beta E,$$

where for convenience we use β to replace $1/(k_B T)$.

This equation relates the energy to temperature, as we want, and it holds for *any* degree of freedom, including vibrations in solids as well as translations in gases. Let's solve Eq. 4.25 for βE to isolate the energy on one side of the equation:

$$\beta E = \ln \Omega - \ln Q(\beta, V).$$

CHECKPOINT The ultraviolet catastrophe was mystifying to researchers working just before the dawn of quantum mechanics because the equipartition principle seemed so successful at describing other aspects of the energetics of substances, and there was not yet any idea that energy was quantized. We arrived at the equipartition principle by integrating over energies, assuming we could treat the energy as a continuous rather than quantized parameter. It is a fundamentally classical approach to thinking about the energy, and the blackbody problem showed that a purely classical approach was not sufficient.

▶ **FIGURE 6.3 The blackbody.** A blackbody is a perfect black box with a small hole in it. The walls radiate energy from vibrational emission with a frequency distribution and intensity determined by their temperature. The radiation density inside the blackbody is $\rho(\nu)$ and is proportional to the intensity of the radiation emitted through the hole.

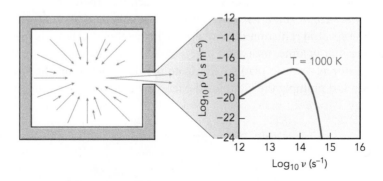

Next we take the derivative of both sides with respect to β, holding the other canonical parameters V and N constant. The value of Ω is not a function of β, so it disappears from the derivative:

$$E = -\left(\frac{\partial \ln Q(\beta, V)}{\partial \beta}\right)_{V,N} = -\frac{1}{Q(\beta, V)}\left(\frac{\partial Q(\beta, V)}{\partial \beta}\right)_{V,N}, \qquad (6.12)$$

where Q is the N-particle partition function.

To get our energy expression in terms of the temperature, we take the derivative of β:

$$\frac{d\beta}{dT} = \frac{1}{k_B}\frac{(dT^{-1})}{dT} = -\frac{1}{k_B T^2}. \qquad (6.13)$$

Therefore,

$$\partial\beta = -\frac{\partial T}{k_B T^2}, \qquad (6.14)$$

and substituting this into Eq. 6.12 gives:

$$E = \frac{k_B T^2}{Q(T, V)}\left(\frac{\partial Q(T, V)}{\partial T}\right)_{V,N} = k_B T^2\left(\frac{\partial \ln Q(T, V)}{\partial T}\right)_{V,N}. \qquad (6.15)$$

This equation is useful because it allows us to find the energy once we know the partition function Q and how Q depends on temperature. Chapter 4 shows how we can obtain Q from the energy and degeneracy equations for the translational, rotational, and vibrational quantum states. Equation 6.15 then becomes one of the bridging equations that allows us to predict the properties of macroscopic samples based on the quantum mechanics of individual molecules. For example, we will use this bridge to cross from the vibrational properties of atoms in a crystal to the bulk crystal's temperature change on heating in Chapter 7.

Equation 6.15 is written for an N-particle system and is also true for the ensemble average energy of a system with only one particle. For each degree of freedom, we may therefore write

$$\langle \varepsilon \rangle = \frac{k_B T^2}{q(T, V)}\left(\frac{\partial q(T, V)}{\partial T}\right)_{V,N}. \qquad (6.16)$$

(Please note that the q in this equation is the single-particle partition function, *not* the heat.)

We now have the energy expressed as a function of temperature, as required. Next we need an explicit expression for the partition function. For vibrations, we've found the partition function (Eq. 3.26)

$$q_{vib}(T, V) = \frac{1}{1 - e^{-\omega_e/(k_B T)}}. \qquad (6.17)$$

To find the average energy, we take the derivative of q_{vib} with respect to T:

$$\frac{\partial q_{vib}(T, V)}{\partial T} = \frac{\omega_e e^{-\omega_e/(k_B T)}}{k_B T^2(1 - e^{-\omega_e/(k_B T)})^2}. \qquad (6.18)$$

Multiplying this by $k_B T^2/Q$ to get $\langle \varepsilon \rangle_{vib}$ (the average molecular energy per vibrational mode) gives

$$\langle \varepsilon_{vib} \rangle = \frac{\omega_e e^{-\omega_e/(k_B T)}}{1 - e^{-\omega_e/(k_B T)}} = \frac{\omega_e}{e^{\omega_e/(k_B T)} - 1}. \qquad (6.19)$$

Next, to find the power curve of the blackbody, we take Eq. 6.19 and use it to find out how much energy the blackbody is losing:

$$\langle \varepsilon_{\text{vib}} \rangle = \frac{\omega_e}{e^{\omega_e/(k_B T)} - 1}$$

$$= \frac{h\nu}{e^{h\nu/(k_B T)} - 1} \qquad \text{set } \omega_e = h\nu$$

$$\rho(\nu)d\nu = \frac{8\pi h\nu^3 d\nu}{c^3(e^{h\nu/(k_B T)} - 1)} \qquad \text{plug into Eq. 6.10}$$

$$\rho(\nu)d\nu = \frac{8\pi h\nu^3 d\nu}{c^3(e^{h\nu/(k_B T)} - 1)}. \qquad (6.20)$$

Equation 6.20 predicts the ideal blackbody radiation curve exactly, giving the curves shown in Fig. 6.2.

What a small correction this is! In the limit that $k_B T \gg h\nu$, we may approximate (Eq. A.24)

$$e^{h\nu/(k_B T)} = 1 + \frac{h\nu}{k_B T} + \frac{1}{2}\left(\frac{h\nu}{k_B T}\right)^2 + \frac{1}{6}\left(\frac{h\nu}{k_B T}\right)^3 + \dots .$$

Dropping all but the first two terms, $1 + h\nu/(k_B T)$ gives the classical solution. Only the higher-order terms $[h\nu/(k_B T)]^2/2 + \dots$ keep the energy density from diverging at high frequency.

6.3 Spectroscopic Intensities

Spectroscopy is the study of the frequency-dependence of the interaction between electromagnetic radiation and matter. A typical spectrum measures the intensity I (of absorption or emission or scattering) as a function of the radiation frequency ν, which is proportional to the photon energy, $E_{\text{photon}} = h\nu$. Both the transition energies and the intensities carry information about the molecular structure, but historically we have paid more attention to the energies than to the intensities. In experiments, we can usually measure radiation frequency or wavelength much more precisely than we can measure the amount of absorbed or emitted energy. Energy has to travel from the system we're studying to a detector, and a surprising number of things can go wrong on the way (light can be scattered or absorbed by surfaces, stray light from other sources can interfere), and more things can go wrong after the energy reaches the detector (detectors are not 100% efficient, and the electronic circuit of the detector adds random noise to the signal). All of these complications increase the uncertainty of the measurement.

In other words, for any typical spectroscopy experiment, the photon energy is more precisely determined than the number of photons. Furthermore, the photon energy of a spectroscopic transition immediately tells us the energy difference between two quantum levels of the system.[1] Those energy differences are largely determined by the properties of the individual molecules. (*QM* Chapters 7, 8, and 9 describe the general principles of molecular energy levels and show how a measured transition energy can be used to infer features of the molecular structure.)

---

[1] We'll refer to quantum "levels" for now, with the understanding that a single level may be degenerate, consisting perhaps of several individual quantum states.

Although the transition intensity carries useful information about other parameters of the quantum states involved in the transition, it is often more difficult to extract from the experimental data. For one thing, to interpret the transition intensity, we need an understanding of the behavior of *all* the molecules in the system, not just any individual molecule. In this section, we examine the factors that determine the intensity of a spectroscopic transition, zeroing in on the terms that we can now treat using statistical mechanics.

The Beer-Lambert Law

The intensity or **flux density** of a beam of radiation equals the energy per unit time per unit area passing a specific point. We use photodetectors to convert photon energy into electrical signals, and, if we keep everything else constant, the total signal will be proportional to the time spent in the beam and to the fraction of the beam's cross-sectional area that hits the detector. Intensity is a practical parameter, taking the total energy seen at the photodetector and dividing out the time and area dependencies:

$$I_{rad} = \frac{E_{rad}}{A(\Delta t)}. \tag{6.21}$$

When we measure an emission spectrum, the energy E_{rad} originates from a transition between quantum levels in an energized molecule. An absorption spectrum, on the other hand, measures the decrease in intensity of radiation from a lamp or laser when the molecule borrows energy from the beam to get to a higher energy quantum level (Fig. 6.4). We will first consider the **absorption intensity** I_{abs}, the intensity lost from the external beam of radiation to the molecular transitions in an absorption spectrum:

$$I_{abs} = I_0 - I_l, \tag{6.22}$$

where I_0 is the intensity of the radiation entering the sample, and I_l is the intensity when the radiation exits the sample of length l (Fig. 6.5).

If we look over an incremental length of the beam, dz, while it travels through the sample, the incremental absorption intensity for the jump from quantum level i to level j can be divided into four contributions:

1. The transition energy, $h\nu_{ij}$, which the beam deposits into the sample for each transition

2. The number of molecules per unit area that lie in this small volume, dN/A

3. The fraction of those molecules that are in the right initial quantum level, the population factor $\mathcal{P}(i)$

4. The fraction of *those* molecules per second that will actually undergo the transition from initial level i to final level j, the transition probability $\mathcal{P}(i \rightarrow j)$

Combining these terms, we obtain this equation for the *incremental* absorption intensity:

$$dI_{abs} = h\nu_{ij}\frac{dN}{A}\mathcal{P}(i)\,\mathcal{P}(i \rightarrow j). \tag{6.23}$$

Remember that this equation is over a vanishingly small distance dz through the sample.

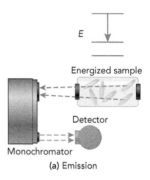

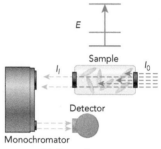

▲ FIGURE 6.4 **Schematic of the apparatus for a traditional spectrometer.** In (a) emission, the molecules are the light source; for (b) absorption spectroscopy, an external light source is provided, and the spectrum is recorded as a decrease in power incident on the detector. In both cases, the photon energy is measured, often by selecting specific energies with a monochrometer, as shown.

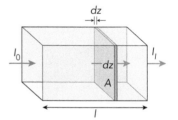

▲ FIGURE 6.5 **Parameters used for Beer-Lambert law absorbance calculation.** Radiation of intensity I_0 enters a sample cell of length l along the z axis and cross-sectional area A, exiting with a lower intensity I_l following absorption by the sample.

The transition energies come directly from the energies of the two quantum levels,

$$h\nu_{ij} = E_j - E_i. \tag{6.24}$$

Assuming that the molecules in our sample are evenly distributed, the number of molecules N in any incremental volume $A dz$ can be written as the number density ρ multiplied by the volume:

$$\frac{dN}{A} = \frac{\rho A \, dz}{A} = \rho \, dz. \tag{6.25}$$

The population factor we now know to calculate from the canonical distribution:

$$\mathcal{P}(i) = \frac{g_i e^{-E_i/(k_B T)}}{Q(T)},$$

where the expressions for degeneracy g_i, energy E_i, and partition function $Q(T)$, all depend on the degrees of freedom that characterize the initial level.

The transition probability, $\mathcal{P}(i \rightarrow j)$, is the toughest nut to crack in this equation. This quantity determines the likelihood that a molecule will meet a photon with energy $h\nu_{ij}$ *and* that the photon gets absorbed. The probability that the molecule encounters the photon is proportional to the **spectral radiation density,** $\rho_{rad}(\nu_{ij})$, the total energy in the beam per unit volume, at a specific frequency ν_{ij}. Effectively, $\rho_{rad}(\nu_{ij})$ counts the number of photons of a particular frequency per unit volume, a convenient factor for determining how often a molecule runs into one of these photons. Fortunately, the radiation density is related to the radiation intensity by a simple factor. Imagine that we count the number $N_{photon}(\nu_{ij})$ of photons at frequency ν_{ij} flying at the speed of light c through a volume of our sample with area A and incremental length dz:

$$\begin{aligned}
\rho_{rad}(\nu_{ij}) &\equiv \frac{N_{photon}(\nu_{ij})(h\nu_{ij})}{V} \\
&= \frac{N_{photon}(\nu_{ij})(h\nu_{ij})}{A \, dz} \\
&= \frac{N_{photon}(\nu_{ij})(h\nu_{ij})}{A(c \, dt)} \\
&= \frac{I(\nu_{ij})}{c}
\end{aligned} \tag{6.26}$$

because the intensity counts the energy per unit area per unit time.

The chance that the molecule then absorbs the photon lies in the transition strength integral (QM Eq. 6.10), which calculates how strongly the transition moment operator $\hat{\mu}_t$ is able to connect the initial and final quantum states, described by the functions ψ_i and ψ_j:

$$\mathcal{P}(i \rightarrow j) \propto \left| \int \psi_j^* \hat{\mu}_t \psi_i d\tau \right|^2.$$

The transition moment operator $\hat{\mu}_t$ is determined by the polarization of the radiation and type of interaction (e.g., electric dipole, magnetic dipole) that couples the photon to the molecule. We can calculate an approximate numerical value for the integral by applying the principles of quantum mechanics to the particles in the individual molecules, but here we will treat that value as an empirical parameter. It is enough to know that if the transition is forbidden, then this integral approaches zero; if the transition is strong, this number approaches one. We combine the transition moment integral and the factor for the radiation density to get

$$\mathcal{P}(i \to j) \propto \frac{I(\nu_{ij})}{c} \left| \int \psi_j^* \hat{\mu}_t \psi_i d\tau \right|^2.$$

The constant of proportionality depends on the type of transition. Using the appropriate coefficients for an electric dipole transition, we obtain

$$\mathcal{P}(i \to j) = \frac{\pi I(\nu_{ij})}{3c\hbar^2 \varepsilon_0} \left| \int \psi_j^* \hat{\mu}_t \psi_i d\tau \right|^2. \tag{6.27}$$

Bringing together all four factors in Eq. 6.23, we have a fairly complete picture of the incremental absorption intensity of an electric dipole transition:

$$dI_{abs} = (h\nu_{ij})(\rho\, dz) \left(\frac{g_i e^{-E_i/(k_B T)}}{Q(T)} \right) \left(\frac{\pi I(\nu_{ij})}{3c\hbar^2 \varepsilon_0} \right) \left| \int \psi_j^* \hat{\mu}_t \psi_i d\tau \right|^2. \tag{6.28}$$

To put this equation in terms of typical experimental measurements, we replace the number density by the concentration C in molar units,

$$\rho \equiv \frac{N}{V} = \frac{n\mathcal{N}_A}{V} = \mathcal{N}_A C, \tag{6.29}$$

and we fold the fundamental constants and transition-specific parameters into a single value called the **molar absorption coefficient** ε:[2]

$$\ln 10\, \varepsilon \equiv (h\nu_{ij})\mathcal{N}_A \left(\frac{g_i e^{-E_i/(k_B T)}}{Q(T)} \right) \left(\frac{\pi}{3c\hbar^2 \varepsilon_0} \right) \left| \int \psi_j^* \hat{\mu}_t \psi_i d\tau \right|^2. \tag{6.30}$$

Now we write the incremental absorption intensity (Eq. 6.28) in the satisfyingly concise form

$$dI_{abs} = (\ln 10)\, \varepsilon\, C\, I(\nu_{ij})\, dz, \tag{6.31}$$

with the factor of $\ln 10$ to be justified in a moment.

An absorption spectrum is commonly recorded as a graph of absorbance A versus radiation wavelength, where

$$A = -\log_{10}\left(\frac{I_l}{I_0} \right). \tag{6.32}$$

The *total* absorption intensity I_{abs} comes from integrating Eq. 6.31 over the length of the sample, from $z = 0$ to $z = l$:

$$dI_{abs} = -dI(\nu_{ij}) \qquad\qquad \text{by Eq. 6.22}$$

$$-dI(\nu_{ij}) = (\ln 10)\, \varepsilon\, C\, I(\nu_{ij})\, dz \qquad\qquad \text{by Eq. 6.31}$$

$$\int_{I_0}^{I_l} \frac{dI(\nu_{ij})}{I(\nu_{ij})} = -(\ln 10) \int_0^l \varepsilon\, C\, dz \quad \text{separate variables and integrate}$$

$$\ln \frac{I_l}{I_0} = -(\ln 10)\, \varepsilon\, C\, l \qquad\qquad\qquad \int_a^b \frac{dx}{x} = \ln \frac{b}{a}$$

<hr>

[2]This popular parameter goes by several other names, including the **extinction coefficient,** the **absorbency index,** and the **molar absorptivity.**

$$\frac{I_l}{I_0} = 10^{-\varepsilon C l} \qquad\qquad \text{take exponential}$$

$$A = -\log_{10}\left(\frac{I_l}{I_0}\right) = \varepsilon\, C\, l. \qquad\qquad (6.33)$$

Equation 6.33, one version of the **Beer-Lambert law,** tells us that the absorbance grows in proportion to the absorption coefficient of the transition, to the concentration of the sample, and to the pathlength through the sample. Using the Beer-Lambert law, analytical chemists can determine the concentration of a particular molecule in a sample from its absorbance, assuming that the absorption coefficient ε has been determined from samples of known concentration.

Three warnings: (1) Eq. 6.30 tells us that the value of ε changes with temperature and with each transition in the molecule, so values can vary considerably for a single molecule, (2) at high concentrations, scattering of the light can become competitive with absorption, and (3) at high values of ε and at high radiation densities, Eq. 6.33 breaks down. The Beer-Lambert law assumes that the number of molecules in the upper state of the transition is negligible compared to the number in the lower state. At high values of ε or I_0, so many molecules are promoted to the upper state that the number of molecules in each state can become comparable. At that point, we say that the transition is becoming *saturated*, and the net number of photons absorbed increases more slowly than we assumed in our derivation. The maximum absorbance depends partly on the number of molecules in the path of the radiation but also on a more detailed understanding of the available radiative processes, which we pursue next.

The Einstein Coefficients

If we shoot a photon at the matter in our system, we can expect one of three broadly defined results: (1) the photon is *transmitted*, passing through the matter without any change in trajectory or energy, (2) the photon is *scattered*, exiting the matter but with altered trajectory and/or energy, and (3) the photon is *absorbed*, increasing the energy content of the matter. To this list we can add a fourth interaction that doesn't require an initial photon: (4) if the system has any energy at all, a photon can be *emitted* from the system. An absorption or emission takes the system from one stationary quantum state to another. Stationary states are quantum states of the system that can remain stable indefinitely. They are called stationary because their properties do not vary with time. Scattering generally involves non-stationary states of the system, and we will not consider scattering further in this section.

Instead, the spectroscopic absorption and emission transitions that interest us are *resonances*, meaning that they occur between two states of well-defined energy. Coming to the rescue once again, it was Albert Einstein who described in detail these resonant interactions of radiation with matter, grouping them into three classes.

1. In **spontaneous emission,** the excited energy level 2 of an atom or molecule releases a photon to arrive at a lower energy level 1. The rate of spontaneous emission is given by the **Einstein A coefficient,** and this is one possible contributor to our transition probability $\mathcal{P}(i \to j)$:

$$\mathcal{P}(2 \to 1)_{\text{spont}} = A_{21}. \tag{6.34}$$

2. In **induced absorption,** the molecule is in any energy level 1, not necessarily the ground state. To push the molecule to a higher energy level, we put the molecule in the oscillating field of a beam of radiation, and the resulting interaction can accelerate an electron, a vibration, or the molecular rotation into a higher energy level. Because absorption, unlike spontaneous emission, requires an energy field to be present before the transition can take place, this transition rate is proportional to the intensity of the beam of radiation. When writing the transition rate, however, we more often express that dependence using the radiation density $\rho_{\text{rad}}(\nu_{12})$. The induced absorption rate is then given by

$$\mathcal{P}(1 \to 2)_{\text{induced}} = B_{12}\rho_{\text{rad}}(\nu_{12}), \tag{6.35}$$

where B_{12} is the **Einstein B coefficient** for induced absorption. This is the same process that the Beer-Lambert law describes, and comparison of Eq. 6.35 to Eq. 6.27 gives this expression for the Einstein induced absorption coefficient:

$$B_{12} = \frac{\pi}{3\hbar^2\varepsilon_0} \left| \int \psi_j^* \hat{\mu}_t \psi_i d\tau \right|^2. \tag{6.36}$$

3. Finally, there is **induced** or **stimulated emission.** Whereas much about spontaneous emission and induced absorption was understood before Einstein took an interest, this process was not recognized until he introduced the concept in 1916. As with spontaneous emission, we start with the molecule in excited level 2. But this time the photon, instead of leaving spontaneously, is given a push. The interaction between the radiation field and the molecule is symmetric, in that it has an equal chance of slowing the molecular motion as it has of speeding things up. This can be justified classically by looking at a relevant trigonometric relation. For example, a radiation field has an oscillating electric field $\mathcal{E}_{\text{rad}}\cos(\nu_{\text{rad}}t)$ and interacts with a polar molecule with a rotational wavefunction $A\cos(\nu_{\text{rot}}t)$. The force acting on the molecule is proportional to the dot product of the oscillating field and the oscillating molecular charge distribution:

$$[\mathcal{E}_{\text{rad}}\cos(\nu_{\text{rad}}t)][A\cos(\nu_{\text{rot}}t)] = \frac{A\mathcal{E}_{\text{rad}}}{2}\left[\cos(\nu_{\text{rad}} + \nu_{\text{rot}})t + \cos(\nu_{\text{rad}} - \nu_{\text{rot}})t\right]. \tag{6.37}$$

The interaction can result in a higher *or a lower* frequency motion, with equal probability. The result is that our induced emission rate is

$$\mathcal{P}(2 \to 1)_{\text{induced}} = B_{21}\rho_{\text{rad}}(\nu_{12}), \tag{6.38}$$

which has exactly the same form as the induced absorption rate, and even the same symbol for the Einstein B coefficient.

CHECKPOINT Imagine that the particles in one molecule are a set of stiff springs strung together between two stiff walls (the boundaries set by the Coulomb attractions among the particles). Each quantum state is then like a particular stable motion among these springs. By pushing back and forth on one wall at the correct frequency, we can drive the system into a different state, a different kind of motion. But once the system is already moving, we are just as likely to take energy away from the springs by pushing on the wall as we are to put energy into the springs. Which way the energy flows depends on the frequencies of the stable motions and the frequency with which we drive the wall, based on the cosine identity used in Eq. 6.37 (and given in Table A.2).

Why did Einstein use the label B for *both* induced emission and induced absorption? Let's start with a sample of molecules, some of them in initial level i. If level i is excited, the molecules will spontaneously emit radiation at a rate

$$\frac{dN_i}{dt} = -N_i \sum_{k<i} \mathcal{P}(i \rightarrow k)_{\text{spont}} = -N_i \sum_{k<i} A_{ik}, \qquad (6.39)$$

where the sum is only over levels k that are lower energy than i.

CHECKPOINT In Eq. 6.40, the sum over A_{ik} is over all states k with energy lower than the initial state i, because spontaneous emission can occur to *any* lower energy state (although not with equal probability). There is no sum for the B_{ij} and B_{ji} terms because the stimulated absorption and emission occur only for the transitions $i \rightarrow j$ and $j \rightarrow i$ that have transition energy ΔE equal to the photon energy $h\nu$ of the incident radiation.

Next, let a beam of radiation pass through the same sample. To simplify the interaction, let's say that the radiation is **monochromatic,** that it contains photons all of roughly the same frequency ν. In that case, induced emission and induced absorption can occur only for transitions that are resonant with the radiation frequency. In the gas phase, energies for two different transitions rarely overlap, so we can normally assume that only one pair of levels is resonant with the external radiation field, say level i and the final energy level j of our resonant transition $i \rightarrow j$. Meanwhile, spontaneous emission can still occur to any lower energy level:

$$\frac{dN_i}{dt} = -N_i \sum_{k<i} A_{ik} - N_i \rho_{\text{rad}}(\nu_{ij})B_{ij} + N_j \rho_{\text{rad}}(\nu_{ij})B_{ji}. \qquad (6.40)$$

The radiation field cuts both ways, so if we can drive molecules from level i to level j with our beam, we should also recognize that we can drive them back again. This explains why the term *resonance* is sometimes more appropriate than *transition*: induced emission and induced absorption coincide, unless one of the two energy levels is completely unoccupied.

We can relate the three Einstein coefficients in Eq. 6.40. Say that we look at the connection between only the two lowest energy levels in the sample, the ground energy level 1 and the lowest excited level 2. Spontaneous emission can only occur from 2 to 1. We introduce radiation at the resonant frequency for levels 1 and 2, and then we wait for the system to reach a steady state, so the energy of the radiation field remains constant. At that point, the number of molecules in levels 1 and 2 must also be constants. If we also require the system to obey the canonical distribution (which may require keeping the radiation density suitably low), then we can relate N_1 and N_2:

$$\frac{dN_2}{dt} = -N_2 A_{21} - N_2 \rho_{\text{rad}}(\nu_{12})B_{21} + N_1 \rho_{\text{rad}}(\nu_{12})B_{12} = 0 \qquad \text{by Eq. 6.40}$$

$$N_1 \rho_{\text{rad}}(\nu_{12})B_{12} = N_2(A_{21} + \rho_{\text{rad}}(\nu_{12})B_{21}) \qquad \text{rearrange}$$

$$N\frac{g_1 e^{-E_1/(k_B T)}}{Q(T)}\rho_{\text{rad}}(\nu_{12})B_{12} = N\frac{g_2 e^{-E_2/(k_B T)}}{Q(T)}(A_{21} + \rho_{\text{rad}}(\nu_{12})B_{21}) \qquad \text{by Eq. 2.30}$$

$$g_1 e^{-E_1/(k_B T)}\rho_{\text{rad}}(\nu_{12})B_{12} = g_2 e^{-E_2/(k_B T)}(A_{21} + \rho_{\text{rad}}(\nu_{12})B_{21}) \qquad \text{divide by } N/Q(T)$$

$$\rho_{\text{rad}}(\nu_{12}) = \frac{g_2 e^{-E_2/(k_B T)}A_{21}}{g_1 e^{-E_1/(k_B T)}B_{12} - g_2 e^{-E_2/(k_B T)}B_{21}} \qquad \text{solve for } \rho_{\text{rad}}(\nu_{12})$$

$$= \frac{g_2 A_{21}}{g_1 e^{(E_2 - E_1)/(k_B T)}B_{12} - g_2 B_{21}}$$

$$= \frac{A_{21}/B_{21}}{\dfrac{g_1 B_{12}}{g_2 B_{21}} e^{h\nu/(k_B T)} - 1}. \qquad E_2 - E_1 = \Delta E = h\nu \ (6.41)$$

Planck found a similar equation for the spectral radiation density of a thermal system (from Eq. 6.20):

$$\rho_{rad} = \frac{8h\pi\nu^3/c^3}{e^{h\nu/(k_BT)} - 1}.$$ (6.42)

Equations 6.41 and 6.42 both hold true at any frequency, and therefore the numerators in the two equations must be equal:

$$A_{21}/B_{21} = \frac{8h\pi\nu^3}{c^3}.$$ (6.43)

The equations are also true at any temperature, so the two denominators are also equal:

$$\frac{g_1 B_{12}}{g_2 B_{21}} = 1.$$ (6.44)

From these two relations, we discern two facts about the interaction of molecules with radiation:

1. If we turn up the power on an external radiation source, we eventually reach a threshold radiation density $\rho_{rad,0}(\nu_{12})$ where the induced emission rate is equal to the spontaneous emission rate:

$$-N_2 B_{21}\rho_{rad,0}(\nu_{12}) = -N_2 A_{21}$$

$$= -N_2 \frac{8h\pi\nu^3}{c^3} B_{21}$$

$$\rho_{rad,0}(\nu_{12}) = \frac{8h\pi\nu^3}{c^3}.$$ (6.45)

This threshold increases *rapidly* with frequency; for example, if we want to control quantum state populations by induced emission, it takes roughly 10,000 times more power to match the spontaneous emission rate in the visible than in the infrared.

2. By induced absorption, we can move molecules from one quantum state to a higher energy state, and the higher the density $\rho_{rad}(\nu_{12})$ of the external radiation, the more molecules we can push upward. However, we never come close to emptying the lower energy state. Assume that the radiation density is high enough that spontaneous emission is negligible. Then, as we approach the limit that $N_1/g_1 = N_2/g_2$, the induced emission rate becomes equal to the induced absorption rate, and the populations settle into a steady state:

$$\frac{dN_1}{dt} = -N_1 B_{12}\rho_{rad}(\nu_{12}) + N_2 B_{21}\rho_{rad}(\nu_{12})$$

$$= -N_1 B_{12}\rho_{rad}(\nu_{12}) + \left(\frac{g_2}{g_1}\right)N_1\left(\frac{g_1}{g_2}\right)B_{12}\rho_{rad}(\nu_{12})$$

$$= 0.$$ (6.46)

This limits the maximum possible absorbance of the transition, and can cause the Beer-Lambert law to fail for strong transitions and high-intensity radiation.

EXAMPLE 6.1	Population Differences in NMR Spectroscopy

CONTEXT Proton spins dominate the use of nuclear magnetic resonance (NMR) for several reasons: (1) hydrogens are abundant and provide a readily available probe of every distinct region of a typical organic molecule; (2) the most abundant isotope of hydrogen has a spin of 1/2, making it NMR active (unlike ^{12}C or ^{16}O) without the added complexity of a higher nuclear spin (such as ^{14}N, with a nuclear spin of 1). A third major reason for the importance of ^1H, especially in the early days of the technology when magnetic fields of 2 T were about the maximum, is its high sensitivity to the field, as measured by its gyromagnetic factor $g_I = 5.586$. This is the highest of any commonly used nucleus in NMR, and it is much higher than ^{13}C ($g_I = 1.405$) and ^{15}N ($g_I = -0.283$), probably the next two most-studied atomic nuclei. The greater sensitivity of the proton means that at these very low transition energies, it is possible to get a greater population difference in ^1H NMR spectra than in most other nuclei.

PROBLEM Calculate the relative population differences $(N_1 - N_2)/N_1$ in ^1H and in ^{13}C spectra at 298 K and a magnetic field of $B = 2.0$ T (the maximum field easily available in the 1970s). The ^1H nucleus has a spin of 1/2 and $g_I = 5.586$, the ^{13}C nucleus has a spin of 1/2 and $g_I = 1.405$, and the energies of the nuclear spin quantum states are given approximately by

$$E_{mag,I} = g_I \mu_N B m_I,$$

where $\mu_N = 5.051 \cdot 10^{-27}$ J T^{-1} and where $m_I = -1/2$ is the ground state 1 and $m_I = +1/2$ is the excited state 2. Both states are non-degenerate ($g = 1$).

SOLUTION The probability that radiation is absorbed is proportional to the difference between the number of molecules in the lower and upper states, and the canonical distribution lets us estimate that at a given temperature

$$\frac{N_1 - N_2}{N_1} = \frac{e^{-E_1/(k_B T)} - e^{-E_2/k_B T}}{e^{-E_1/(k_B T)}}$$

$$= 1 - e^{(E_1 - E_2)/(k_B T)}$$

$$E_1 - E_2 = g_I \mu_N B \left[-\frac{1}{2} - \left(\frac{1}{2}\right) \right] = -g_I \mu_N B$$

$$= -g_I (5.051 \cdot 10^{-27} \text{ J T}^{-1})(2.0 \text{ T}) = -g_I (1.01 \cdot 10^{-26} \text{ J})$$

$$^1H: \frac{E_1 - E_2}{k_B T} = -\frac{(5.586)(1.01 \cdot 10^{-26} \text{ J})}{(1.381 \cdot 10^{-23} \text{ J K}^{-1})(298 \text{ K})} = -1.37 \cdot 10^{-5}$$

$$\frac{N_1 - N_2}{N_1} = 1 - e^{(E_1 - E_2)/(k_B T)} = 1 - \exp(-1.37 \cdot 10^{-5}) = \boxed{1.4 \cdot 10^{-5}}$$

$$^{13}C: \frac{E_1 - E_2}{k_B T} = -\frac{(1.405)(1.01 \cdot 10^{-26} \text{ J})}{(1.381 \cdot 10^{-23} \text{ J K}^{-1})(298 \text{ K})} = -3.45 \cdot 10^{-6}$$

$$\frac{N_1 - N_2}{N_1} = 1 - e^{(E_1 - E_2)/(k_B T)} = 1 - \exp(-3.45 \cdot 10^{-6}) = \boxed{3.4 \cdot 10^{-6}}.$$

Notice that the energy differences are so small compared to the thermal energy $k_B T$ that the values $1 - e^{(E_1 - E_2)/(k_B T)}$ can be approximated by $1 - e^{-x} \approx 1 - (1 - x) = x$, using the Taylor series expansion for e^{-x} when $|x| \ll 1$. These results show that the transition intensity of proton NMR is about a factor of $1.4 \cdot 10^{-5}/3.4 \cdot 10^{-6} = 4.0$ stronger than ^{13}C NMR on the basis of the population differences.

6.4   Laser Dynamics

Equation 6.46 tells us that we can't drive molecules into the excited state beyond a ratio $N_1/g_1 = N_2/g_2$. As a result, with just these two states and using only radiative transitions, we can never prepare the system so that the induced emission rate $B_{21}\rho_{\text{rad}}(\nu)$ exceeds the induced absorption rate $B_{12}\rho_{\text{rad}}(\nu)$. In order to accomplish that, we would need somehow to generate a **population inversion,** the condition that

$$\frac{N_2}{N_1} > \frac{g_2}{g_1}. \tag{6.47}$$

If we introduce radiation at frequency ν_{12} into a system with this population inversion, then

$$N_2 B_{21}\rho_{\text{rad}}(\nu_{12}) = N_2\left(\frac{g_1}{g_2}\right)B_{12}\rho_{\text{rad}}(\nu_{12})$$

$$> \left(\frac{g_2}{g_1}\right)N_1\left(\frac{g_1}{g_2}\right)B_{12}\rho_{\text{rad}}(\nu_{12}) = N_1 B_{12}\rho_{\text{rad}}(\nu_{12}), \tag{6.48}$$

and the induced emission rate will be greater than the induced absorption rate.

Why would we want this? Normally when we send a molecule into an excited state, it releases the energy in all kinds of ways. Spontaneous emission from an excited electronic state often appears as fluorescence to any of a huge number of vibrational states in the electronic ground state. The remarkable thing about stimulated emission is that it allows photons to be expelled from an energized system with complete uniformity. If the electric field of the *stimulating* radiation field can be represented by a sine wave with a single phase ϕ and a single frequency ω,

$$\mathcal{E}(t) = \mathcal{E}_0\cos(\omega t + \phi), \tag{6.49}$$

then the *emitted* radiation will have the same phase, the same frequency, and the same polarization and trajectory as the stimulating beam. But it will have more power. The acronym **laser** is derived from **l**ight, obeying Eq. 6.49, **a**ppearing to have been **a**mplified by the **s**timulated **e**mission of **r**adiation. Different applications of lasers may take advantage of any of the properties just listed: the concentration of light into a beam with well-defined trajectory makes many lasers function as intense sources of light that can travel far with little loss in power, a feature exploited in laser pointers and communications devices; the constancy of phase and frequency allows lasers to generate holograms; and all of these qualities make them ideal radiation sources for spectroscopy.

The substance that will be stimulated into releasing photons is called the **laser medium.** To make a laser, we need the laser medium to undergo a population inversion. Therein lies the challenge because we found in Section 6.3 that we can never accomplish a population inversion between states 1 and 2 simply by induced absorption from 1 to 2. No matter how intense we make the radiation, it can't increase the population in state 2 once the induced absorption and induced emission rates are equal. A well-defined temperature (i.e., a canonical distribution) will *never* result in a population inversion.

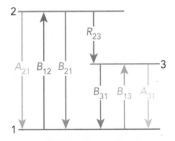

▲ **FIGURE 6.6 Energy level diagram for a three-state laser system.** The crucial steps are the pumping transition $1 \to 2$, the fast relaxation $2 \to 3$, and the lasing transition $3 \to 1$ by stimulated emission. All other rates are assumed to be relatively slow.

Two approaches are used to get around this problem. One simply uses a non-radiative method to transfer a lot of energy into the laser medium, creating an excess of excited state atoms or molecules. If the distribution of states is not too great, perhaps enough molecules will be in a single excited state that it will have greater population (taking into account any degeneracies) than the ground state, and lasing becomes possible. However, because the molecules will rapidly collide with each other and return to a canonical distribution, this does not create a sustainable laser. The second solution allows the laser to operate continuously by working from a system of at least three energy levels instead of two (Fig. 6.6).

In the continuously operating three-level system, we provide a pump transition from state 1 to excited state 2, generally by induced absorption. We cannot generate a population inversion between states 1 and 2, but we put molecules into state 2 as fast as we can. The next step requires a non-radiative transition from energized state 2 to a lower energy state 3 at a rate R_{23}. This non-radiative process often involves collisions between molecules to remove the energy, and the key is that *this* transition can be quite fast. (In thermodynamic terms, we are not letting the system achieve equilibrium, a topic we discuss in the next chapter.) We are not allowed to build up a population inversion between states 1 and 2, but if the molecules move immediately from state 2 to 3, this is no problem. If the lasing transition is then relatively slow, the molecules have a chance to build up in state 3 fast enough to maintain the required population inversion.

We can back up that qualitative picture with the following math. For a steady-state solution to the system, the numbers of molecules in states 1, 2, and 3 all stay constant, so the derivatives dN_i/dt are zero for each state:

$$\frac{dN_1}{dt} = -N_1\big[B_{12}\rho_{\text{rad}}(\nu_{12}) + B_{13}\rho_{\text{rad}}(\nu_{13})\big] + N_2\big[B_{21}\rho_{\text{rad}}(\nu_{12}) + A_{21}\big] + N_3\big[B_{31}\rho_{\text{rad}}(\nu_{13}) + A_{31}\big]$$
$$= 0$$

$$\frac{dN_2}{dt} = N_1\big[B_{12}\rho_{\text{rad}}(\nu_{12})\big] - N_2\big[B_{21}\rho_{\text{rad}}(\nu_{12}) + A_{21} + R_{23}\big] = 0$$

$$\frac{dN_3}{dt} = N_1\big[B_{13}\rho_{\text{rad}}(\nu_{13})\big] + N_2\big[R_{23}\big] - N_3\big[B_{31}\rho_{\text{rad}}(\nu_{13}) + A_{31}\big] = 0$$

$$N_2 = N_1\left[\frac{B_{12}\rho_{\text{rad}}(\nu_{12})}{B_{21}\rho_{\text{rad}}(\nu_{12}) + A_{21} + R_{23}}\right]$$

$$N_3 = \frac{N_1\big[B_{13}\rho_{\text{rad}}(\nu_{13})\big] + N_2\big[R_{23}\big]}{\big[B_{31}\rho_{\text{rad}}(\nu_{13}) + A_{31}\big]}$$

$$= N_1\frac{B_{13}\rho_{\text{rad}}(\nu_{13}) + \dfrac{B_{12}\rho_{\text{rad}}(\nu_{12})R_{23}}{B_{21}\rho_{\text{rad}}(\nu_{12}) + A_{21} + R_{23}}}{B_{31}\rho_{\text{rad}}(\nu_{13}) + A_{31}}$$

$$\frac{N_3}{N_1} = \frac{B_{13}\rho_{\text{rad}}(\nu_{13}) + \dfrac{B_{12}\rho_{\text{rad}}(\nu_{12})R_{23}}{B_{21}\rho_{\text{rad}}(\nu_{12}) + A_{21} + R_{23}}}{B_{31}\rho_{\text{rad}}(\nu_{13}) + A_{31}}. \tag{6.50}$$

In the limit that the spontaneous emission rates are negligible, Eq. 6.50 allows a population inversion to develop as long as

$$\frac{B_{12}\rho_{\text{rad}}(\nu_{12})R_{23}}{B_{21}\rho_{\text{rad}}(\nu_{12}) + A_{21} + R_{23}} \gg A_{31} \tag{6.51}$$

because this condition allows N_3/N_1 to be greater than B_{13}/B_{31}. The more efficient our pump transition is, as determined by B_{12}/B_{21}, and the more efficient our relaxation rate R_{23}, the greater the population inversion we can generate.

TOOLS OF THE TRADE | Lasers

The fundamental operating principle of a laser is described in detail in Section 6.4, but the laser has proven so useful and takes so many different forms that some notes on the technology are appropriate.

Why do we use lasers? Lasers can offer the following advantages over conventional light sources:

1. The *bandwidth* of the laser—the range of wavelengths it emits—is unusually narrow because only wavelengths that satisfy the boundary conditions of the laser optics can be amplified. This wavelength precision has made lasers important radiation sources in spectroscopy.
2. Not only is the wavelength well-defined, but the *phase* of the radiation field is well-defined, making the laser a *coherent* light source, in which the peaks and troughs in the field are clear and evenly spaced. This property makes lasers ideal radiation sources in applications that rely on interference patterns, such as holography and high-resolution distance measurement.
3. Because the amplification must occur along a straight path between the front and rear optics of the laser, the light can be highly *collimated,* meaning that the emitted rays of light are nearly parallel. As a result, the power of the radiation can remain concentrated in a relatively small area over large distances. This feature has led to applications from laser pointers to sophisticated *optical traps,* which employ the momentum of photons to keep tiny particles confined to a small volume.

How do lasers work? The schematic of a typical laser is shown in the accompanying figure. A typical laser requires three elements:

1. An *excitation mechanism,* such as a powerful incandescent lamp or electrical discharge, which provides the initial infusion of energy
2. The *laser medium,* which is the chemical substance that emits the radiation
3. An *optical cavity,* established by a reflective surface at each end of the laser medium

To get the laser to work, we activate the excitation mechanism, which pumps energy into the laser medium in such a way as to generate the required population inversion. The excited state particles begin to emit radiation, and a fraction of that initial emission is intercepted by the optics and reflected back into the optical cavity. These photons trigger the emission of more photons at the same wavelength *and trajectory,* so a build-up of power occurs along the axis of the optical cavity. One of the reflective surfaces, called the *output coupler,* is partly transparent, so a fraction (typically 10%–30%) of the radiation is allowed to leave the laser cavity.

That said, lasers are built in a wide variety of configurations, adapted to reach certain wavelengths or powers or to optimize certain characteristics such as bandwidth. The following table is just a partial list of lasers and their common parameters and properties (many variations are possible even within the set given here).

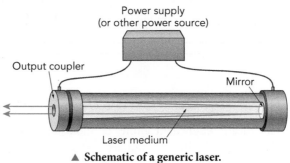

Power supply
(or other power source)

Output coupler

Mirror

Laser medium

▲ **Schematic of a generic laser.**

laser	wavelengths	power	laser medium and transition	excitation source	applications
HCOOH	250—750 μm	5 mW	rovibrational transitions	CO_2 laser	wavelength calibration, spectroscopy
CO_2	9.4—10.6 μm	150 W	CO_2 vibrations	electric discharge, energy transfer by N_2	machining, surgery
HF chemical	2.7—2.9 μm	$\gg$ 1 kW	HF vibrations	chemical reaction $F + H_2 \rightarrow HF + H$	weapons research
Ti:sapphire	650—1150 nm	1 W	electronic transitions in Ti^{3+} ions	doubled Nd:YAG or other laser	spectroscopy, including generation of short laser pulses
Nd:YAG	1064 nm	500 mJ/pulse	electronic transitions in Nd^{3+} ions in a garnet crystal	flashlamp	powering other lasers, surgery
dye laser	400—900 nm	0.5 W	electronic transitions in organic dyes	doubled Nd:YAG or other laser	spectroscopy
doubled Nd:YAG	532 nm	150 mJ/pulse	Nd:YAG frequency-doubled by a crystal	flashlamp	powering other lasers, eye surgery
helium-neon	632.8 nm	0.5 W	electronic transitions in neon gas,	electric discharge, energy transfer by helium	pointers, laser alignment, wavelength calibration
AlGaAs,GaN	405,650, 780 nm	300 mW	semiconductor p-n junction	electrical current	Blu-ray, CD, DVD readers, barcode readers
excimer	193,248,308, 353 nm	300 mJ/pulse	electronic transitions in dissociating ArF, KrF, XeCl, XeF	electric discharge	laser surgery, vaporizing solids
capillary plasma-discharge Ar^{8+}	47 nm	140 μJ/pulse	core electronic transitions in Ar^{8+}	high-current pulsed electrical discharge	coherent x-ray source

6.5   Spectroscopic Linewidths

Chemically equivalent atoms or molecules are indistinguishable entities, and the energies of their stationary states could, in principle, be determined with infinite precision if we were willing to spend eternity making the measurement.

However, when we observe the spectrum of a molecule, each transition appears over a *range* of photon energy. Graphing the absorption or emission intensity as a function of photon energy, each transition has some nonzero **spectroscopic linewidth.**

Many different phenomena contribute to the linewidth of spectroscopic transitions, and they break down into two classes: **inhomogeneous broadening,** which arises when molecules in the sample do not all share the same transition energy, and **homogeneous broadening,** which is an intrinsic broadening that occurs equally for all the molecules in the sample. The chief examples from each of these classes we can now assess, taking advantage of our look at molecular dynamics.

Doppler Broadening

Molecules move, but the speed of light (in a given medium) remains constant. We set up the absorption spectrometer in Fig. 6.7, with a sample cell of length l. Imagine a classical beam of light passing through our sample along the z axis, with an electric field vector $\vec{\mathcal{E}}(z,t)$ oscillating at frequency ν. Let's make this a gas sample so that we can assume the speed of the photon is essentially the speed of light in a vacuum c. Consider molecule A, which starts at one end of the sample cell $z = l$ and moves at speed v_z toward the radiation source, eventually arriving at the opposite end of the cell $z = 0$. Molecule A's travel time through the cell is $\Delta t = l/v_z$, and during that time it has seen the electric field of the radiation oscillate many times.

How many times? The number of $\vec{\mathcal{E}}(z,t)$ oscillations at a particular *position* in the cell is given by the product of the radiation frequency and the time. But the electric field also oscillates with position in the cell, and the number of those oscillations at a particular *time* in the cell is equal to the number of wavelengths that fit inside the cell, l/λ. With the radiation propagating in the $+z$ direction, molecule A during its trip passed through all of the oscillations that reached $z = l$ between times $t = 0$ and $t = \Delta t$,

$$N_{z=l, t=0-\Delta t} = \nu \Delta t = \nu(l/v_z),$$

and through all of the oscillations between $z = 0$ and $z = l$ that haven't yet gotten down to the end of the cell at time $t = \Delta t$:

$$N_{z=0-l, t=\Delta t} = (l/\lambda) = \nu(l/c).$$

CHECKPOINT The Doppler effect appears in many forms. One example analogous to the system described here is a canoe on the water, with waves coming in at a regular frequency from some point we'll call z=0. As we paddle the canoe towards z=0, the waves hit the boat more frequently. The frequency of the waves has not really changed, but the effect on the canoe is exactly as if the frequency had increased in proportion to the speed at which we paddle. By the way, although the symbols for speed and frequency look a lot alike in this section, the speed can be recognized by its subscripts, as it is always v_z or v_{mode} in these expressions.

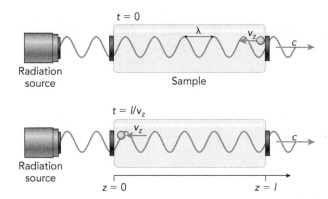

◀ FIGURE 6.7 **The Doppler effect.** In an absorption spectrometer, molecule A, which is moving toward the radiation source, observes a radiation frequency ν_A higher than the frequency ν measured in the rest frame.

So altogether, in time Δt, molecule A saw the electric field of the radiation go up and down $N_{z=l,t=0-\Delta t} + N_{z=0-l,t=\Delta t}$ times. Therefore, as far as molecule A is concerned, the radiation is at frequency

$$\nu_A = \frac{N_{z=l,t=0-\Delta t} + N_{z=0-l,t=\Delta t}}{\Delta t} = \frac{\nu(l/\nu_z) + \nu(l/c)}{l/\nu_z}$$

$$= \left(1 + \frac{\nu_z}{c}\right)\nu. \tag{6.52}$$

This is an example of the **Doppler effect,** which causes observers moving in different directions to experience the same wave as though it had different frequencies. Molecules moving toward the light source experience an effective increase in photon frequency, or (equivalently) shortening of the wavelength. To look at it another way, we can take a particle traveling at speed ν_z towards the radiation source (Fig. 6.8a) and imagine a change in the frame of reference, so that the particle was at rest and the spectrometer moving at speed ν_z towards it (Fig. 6.8b). Again, the light appears to oscillate at a higher frequency than if the particle were stationary. Similarly, molecules moving away from the source will see the radiation frequency effectively decrease. Molecules that move perpendicular to the beam of the radiation absorb at the **rest frequency** of the transition ν_0, which falls at the center of the spectral line. Therefore, as we increase the frequency of our radiation source past the rest frequency ν_0 of a transition, the molecules moving toward the source absorb first, at a frequency slightly below ν_0 (because they experience a frequency that is slightly higher than that), and those moving away from the source only absorb when we get to frequencies higher than ν_0. Therefore, molecules will absorb the radiation when they see an apparent frequency of ν_0, so we set $\nu_A = \nu_0$ and solve for the z-component of the speed at which absorption can occur:

$$\nu_A = \nu_0 = \left(1 + \frac{\nu_z}{c}\right)\nu$$

$$\nu_z = c\left(\frac{\nu_0}{\nu} - 1\right) = c\left(\frac{\nu_0 - \nu}{\nu}\right). \tag{6.53}$$

Different molecules in the sample absorb at different frequencies, so Doppler broadening belongs to the class of inhomogeneous broadening. The amount of radiation absorbed as a function of ν as we scan across this transition is proportional to the number of molecules traveling at the right speed to bring the transition into resonance with the photons, and this depends only on the

▶ **FIGURE 6.8 The Doppler effect, using the frame of reference.** If **(a)** molecule A is moving toward the radiation source at speed ν_z, then **(b)** in its rest frame, the radiation source is moving towards it at the same speed, ν_z. This causes the particle to see the electric field oscillate more times over the same period, meaning that the radiation frequency is higher than the frequency ν measured in the rest frame.

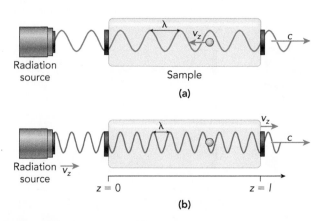

Maxwell-Boltzmann distribution for speeds along the axis of the radiation beam (applying Eq. 3.61 to one dimension),

$$\mathcal{P}_v(v_z) = \left(\frac{m}{2\pi k_B T}\right)^{1/2} e^{-mv_z^2/(2k_B T)} = \left(\frac{m}{2\pi k_B T}\right)^{1/2} e^{-(v_z/v_{mode})^2}. \quad (6.54)$$

This is a Gaussian distribution, with an exponent equal to $-(v_z/v_{mode})^2$, where v_{mode} is the most probable speed of the molecules (Table 5.2). To find the equation for the absorption intensity of a Doppler broadened absorption signal, therefore, we rewrite Eq. 6.54 in terms of the frequency using Eq. 6.53:

$$I(\nu) = I_0 \exp\left[-\left(\frac{v_z}{v_{mode}}\right)^2\right] \quad (6.55)$$

$$= I_0 \exp\left[-\left(\frac{c(\nu_0 - \nu)}{v_{mode}\nu}\right)^2\right], \quad (6.56)$$

which is a Gaussian function with a linewidth of

$$\delta\nu_{Doppler} = \frac{4\nu_0}{c}\sqrt{\frac{2k_B T \ln 2}{m}}, \quad (6.57)$$

measured as the width in frequency ν of the Gaussian at amplitude $I(\nu) = I_0/2$, and therefore called the full width at half-maximum, or **FWHM linewidth.** The Doppler linewidth varies with the rest frequency of the transition because the Doppler effect is more pronounced when the oscillations are faster. As a result, Doppler broadening is usually not important in microwave spectroscopy, but it is the likeliest contributor to line broadening in the UV or visible spectra of gas phase samples.

Collision Broadening

A second major contributor to line broadening occurs when one of our cherished assumptions frays around the edges. We have long taken for granted that the energy levels in our systems are stationary states, with no time dependence. However, in most samples molecules are constantly running into each other, and these collisions affect the local potential energy of each molecule, shifting the energy levels as a function of time, if only slightly. As long as the density and temperature and other macroscopic variables in our sample cell are the same everywhere, this **collision broadening** affects all the molecules in the sample equally, and is a form of homogeneous broadening.

If we allow that these collisions shift the energies, the effect of the collisions on the transition frequency can be analyzed using a Fourier transform of the energy time-dependence. To summarize, the collisions add a time-dependence to the energies, which now vary over a time scale of roughly $1/\gamma$, where γ is the collision frequency from Section 5.1. The Fourier transform of the energy as a function of time yields transition frequencies that have been broadened by roughly γ. The collision broadened lineshapes are given by **Lorentzian** curves, with the equation

$$I(\nu) = I_0\left(\frac{\gamma^2}{(\nu - \nu_0)^2 - \gamma^2}\right), \quad (6.58)$$

which has a FWHM linewidth of

$$\delta\nu_{collision} = 4\gamma. \quad (6.59)$$

CONTEXT *Where Do We Go From Here?*

Notice that as we move to the macroscopic realm, we still focus on how much energy is in the system and how it moves around. Differences between one bulk material and another now depend not only on the structure of individual molecules, but also on the variation in interactions *between* molecules. For that reason, we've begun to pay more attention to the intermolecular potential U, taking for granted that the potential energy terms of the individual molecules can be rolled up into their individual properties. And the measure of how many states have the same energy has taken on a new form, the entropy that will come to control much of the discussion of bulk matter.

The most significant change, however, is in the number of degrees of freedom we use to describe the system. Through statistics we have reduced the number of degrees of freedom in our sample from $3N$, where N is the number of particles, to a handful. Exactly what these coordinates measure, however, now depends on the type of sample, the desired level of precision, and other choices in how the averaging is carried out. Rather than identifying the quantum numbers of each degree of freedom of each molecule in the sample, we use the temperature to describe the distribution of energy among the molecular states. Rather than locating each molecule in space, we use the volume to find the extent of space that is occupied by the sample molecules and the ensemble size or entropy to find how many different ways the sample molecules could be arranged to fill that same volume.

If we wanted to predict the behavior of a non-ideal gas or solid, however, it is insufficient to know only T, P, V, and E; we must also have an expression for the intermolecular potential. From the potential energy function we obtain the van der Waals a and b constants or the virial coefficients for the gas, or the pair correlation function for the liquid. The more precise the results we want, the more we must know about the microscopic properties of the sample.

In demonstrating some applications of statistical mechanics, we have already ventured into our next topic: the thermodynamics of chemical systems. Having seen these examples, we shall generally take the statistical results for granted in the following chapters. Remember, however, that the equations and molecular constants we use in the remainder of the text are all justifiable in terms of the chemical model at the molecular scale.

KEY CONCEPTS AND EQUATIONS

6.1 **Conduction, Convection, and Radiation.** Heat transfer processes are generally divided into the following categories: **convection,** in which heat is transferred by the motion of matter; **conduction,** in which heat is transferred from particle to particle without net motion of the matter itself; and **radiation,** in which the energy is transferred in the form of photons.

6.2 **Blackbody Radiation.** Any macroscopic substance at a nonzero temperature releases radiation as higher energy quantum states relax to lower energy states. In liquids and solids, where the interactions between particles are so strong that the quantum states blur into a continuum, the emitted radiation approaches a continuous distribution called **blackbody radiation.** Using the partition function for a vibrating molecule, we can show that the radiation density $\rho(\nu)$ obeys the equation

$$\rho(\nu)d\nu = \frac{8\pi h\nu^3 d\nu}{c^3(e^{h\nu/(k_B T)} - 1)}. \tag{6.20}$$

6.3 **Spectroscopic Intensities.** In typical absorption spectroscopy, the absorbance A is given by the **Beer-Lambert law** as

$$A = -\log_{10}\left(\frac{I_l}{I_0}\right) = \varepsilon C l. \qquad (6.33)$$

6.4 **Laser Dynamics.** The amplification of emitted light in a laser requires a **population inversion** in which there are more particles in an excited state than in a lower energy state. The population inversion ensures that as photons of energy $h\nu$ pass through the laser medium, they encounter more particles that will emit at the same frequency than particles that will absorb the energy.

6.5 **Spectroscopic Linewidths.** Two common sources of line-broadening in spectroscopy are **Doppler broadening,** caused by the motion of the particles relative to the light source and giving a linewidth (full width at half maximum) of

$$\delta\nu_{\text{Doppler}} = \frac{4\nu_0}{c}\sqrt{\frac{2k_{\text{B}}T\ln 2}{m}}, \qquad (6.57)$$

and **collision broadening,** caused by the scrambling of quantum energy levels when particles collide and giving a width of

$$\delta\nu_{\text{collision}} = 4\gamma. \qquad (6.59)$$

KEY TERMS

- The **Einstein coefficients** quantify the rates at which a sample will release a photon due to **spontaneous emission** (A_{21}) or **stimulated emission** (B_{21}) or will absorb a photon by **induced absorption** (B_{12}).
- The **molar absorption coefficient** is the factor in the Beer-Lambert law (Eq. 6.33) that absorbs all of the molecule- and temperature-dependence, including the intrinsic transition strength as determined by the quantum mechanics and the relative populations of the initial and final states.

- A **laser** is a device that converts energy into a beam of coherent radiation by the stimulated emission of large numbers of photons all with the same phase and frequency.
- The **linewidth** of a spectroscopic signal is the range of photon energies over which a given transition between two quantum states will actually absorb energy.

OBJECTIVES REVIEW

1. *Identify which heat transfer mechanisms are operating in a given system.*
   When we heat a reaction to 373 K in a water bath, is the heat transferred to the reaction mix by conduction, convection, radiation, or some combination of these?

2. *Graph the radiation density of a perfect blackbody at a given temperature.*
   Find the wavelength at which a 373 K blackbody emits its maximum power.

3. *Calculate absorbances using the Beer-Lambert law.*
   Find the molar absorption coefficient ε of dicarbocyanine dye if a $3.0 \cdot 10^{-6}M$ sample in a 1.0 cm pathlength cuvette has an absorbance of 0.85.

4. *Estimate spectroscopic linewidths due to Doppler and collision broadening.*
   For the $v = 0 \rightarrow 1$ vibrational transition of CO in air at roughly 2150 cm^{-1} at 1.00 bar and 298 K, find the Doppler-broadened and collision-broadened linewidths.

PROBLEMS

Discussion Problems

6.1 Circle the letter for each of the following statements that is true for a perfect blackbody.

a. As T increases, more energy is radiated at low frequency.

b. As T increases, more energy is radiated at high frequency.

c. As T increases, the most intense radiation shifts to longer wavelength.

d. As T increases, the number of photons *and* the average photon energy both increase.

Blackbody Radiation

6.2 We used the following expression to get the average energy per molecule for several heat capacity and blackbody equations:

$$\langle \varepsilon \rangle = \frac{k_B T^2}{Q} \frac{\partial Q}{\partial T}.$$

What does this predict to be the average rotational energy per molecule in the usual limit that $k_B T \gg B$? Simplify your answer as much as possible.

6.3 Find the total energy per unit volume E/V in J cm^{-3} emitted at all frequencies by a blackbody at 300 K using the radiation density $\rho(\nu)$.

6.4 Find an equation in terms of $\rho(\nu)$ for the rate dT/dt at which a blackbody, with volume V and heat capacity C_V, will cool off when the heating is discontinued. Assume that blackbody radiation is the only cooling mechanism and that V and C_V remain constant.

6.5 Find the wavelength of the most intense radiation for a blackbody at temperature 300 K.

6.6 The maximum emission from a blackbody at 770 K (about 500 °C) is at a wavelength of about 6600 nm. What will be the wavelength for the most intense emission when we increase the temperature to 1540 K?

6.7 The radiation density of a blackbody is shown in Problem 6.5 to peak at a wavelength
$\lambda_{max}(cm) = 0.510/T(K)$.
Call the radiation density at this wavelength $\rho_{max}(T)$. Find the ratio $\rho_{max}(T_2)/\rho_{max}(T_1)$ for a blackbody cooling from $T_1 = 700$ K to $T_2 = 350$ K.

6.8 The solution to the blackbody radiation problem was found by treating the energy modes of the sample as quantum harmonic oscillators. Find an equation for $\rho(\nu)$, assuming that the energy is stored in quantum *rotations* of a large number of linear molecules, each with rotational energy $\varepsilon_{rot} = BJ(J + 1)$ and degeneracy $g(J) = 2J + 1$.

Radiation Transport

6.9 Find the value of the absorption coefficient (including units) of a 0.100 M solution of pyrene that absorbs 12% of the radiation intensity when the pathlength is 1.00 cm.

6.10 The Fe(III)-ferrozine complex has an extinction coefficient or molar absorptivity of $2.8 \cdot 10^4$ L mol^{-1} cm^{-1}.

a. What concentration would give an absorbance of 0.01 in a 1.0 cm pathlength cuvette?

b. At this concentration, what percent of the incident light is *absorbed*?

6.11 In our derivation of the Beer-Lambert law, we assumed that the concentration was constant. Now let the concentration change according to the exponential decay we expect for diffusion, such that $\rho(z) = \rho_0 e^{-z/l}$,

where ρ_0 is the number density at $z = 0$ and l is the pathlength. What will be the ratio of the absorbance A_2 of this sample to the absorbance A_1 of our original sample, with a fixed number density ρ_0 throughout the sample?

6.12 An electronic transition in argon ions is observed at a wavelength of 488 nm. Find the intensity in W/mm^2 required for a laser beam with a frequency width of 10^3 MHz to exceed the threshold radiation density $\rho_0(\nu)$, where induced and spontaneous emission rates are equal for this transition.

6.13 The absorbed power for the spectroscopic transition from $1 \rightarrow 2$ is proportional to the difference in the number of molecules in the upper and lower states of the transition, $D \equiv N_1 - N_2$. Find an expression for D in terms of the Einstein coefficients, the total number of molecules N, and the radiation density $\rho(\nu_{12})$. Assume this is a two-state system in steady state, and that both states are non-degenerate.

6.14 Determine if the three-level system drawn in the following figure could function as a laser, and if so what criteria must be satisfied. Assume the degeneracies in all states are equal.

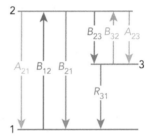

6.15 The infrared spectroscopy of molecular ions was at one time greatly impeded by overlapping transitions of much more abundant neutral molecules. A clever way of addressing this is *velocity modulation*. The sample cell is a long tube with an electrode at each end, and the radiation propagates down the length of the tube. The electrodes are energized to create an electric discharge through the gas in the sample cell, which generates the ions, although the neutral molecules are still much more abundant. The trick is to make the direction of the electric discharge flip back and forth at a rate of 25 kHz. The ions are Doppler shifted in one direction, then the next, while the neutral molecules remain stationary on average. Subtracting the signal obtained with the electric field in the $+z$ direction from the signal with the field in the $-z$ direction cancels the signals from the neutral molecules, leaving only the ion signals. Assume that the ion is accelerated by an electric field of 10 V cm^{-1} in the cell over a typical travel distance of 4 mm before the electric field changes direction. Estimate the Doppler shift of the HN$_2^+$ ion at a wavelength of 3230 cm^{-1}. Compare this to the Doppler width of the transition at 400 K.

PART II

NON-REACTIVE MACROSCOPIC SYSTEMS

PART I
EXTRAPOLATING
FROM MOLECULAR
TO MACROSCOPIC
SYSTEMS

PART II
NON-REACTIVE MACROSCOPIC SYSTEMS

PART III
REACTIVE
SYSTEMS

7 **Introduction to Thermodynamics: Heat Capacity**

7 Introduction to Thermodynamics: Heat Capacity

LEARNING OBJECTIVES

After reading this chapter, you will be able to do the following:

❶ Find relations among thermodynamic parameters using derivatives of the thermodynamic potentials, Maxwell relations, and partial derivative identities.

❷ Estimate the temperature-dependent heat capacity of a solid, liquid, or gas, based on the system's molecular structure and properties.

❸ Use the heat capacity to convert between the changes in energy and temperature of the system.

GOAL *Why Are We Here?*

The goal of this chapter is to introduce the terms and relationships that will guide our application of thermodynamics to chemical systems and to show by means of the heat capacity how the principles of statistical mechanics covered in Part I can be used to justify the numerical values needed to make thermodynamics quantitative.

CONTEXT *Where Are We Now?*

If you're tired of squinting to see everything from the molecular perspective, things are about to get bigger.

Now we discuss macroscopic systems themselves. When working to understand any complex chemical system—a crystal structure or a chemical reactor, for example—it is necessary to study the components of the system. That is what we've done up to this point. We have seen that the laws of chemistry at the macroscopic scale follow from the application of statistics to molecular properties. Although many laws of macroscopic chemistry can be explained on a purely classical basis, the molecular constants these laws rely on (such as collision cross sections, dipole moments, and state degeneracies) are predictable only with some understanding of microscopic molecular structure. Examples of this appear throughout the remainder of the text.

In the next few chapters, we examine processes of bulk substances, leaving out chemical reactions so we can concentrate on those in the concluding chapters. To advance from our statistical analysis to the laws governing *changes* in bulk materials, we will gather several of our recent results into a simplified set of parameters and equations. What we are really doing is redefining some of our terms to conform to the definitions and equations that evolved from the empirical study of chemistry in the macroscopic limit. This is where we unite 19th-century classical chemistry with our more recent understanding of the nature of molecular structure.

SUPPORTING TEXT | *How Did We Get Here?*

The main qualitative idea we need for the work in this chapter is from Chapter 3, that molecules can store energy by channeling it into different degrees of freedom: vibrations and—for a gas—rotations and translations. For a more detailed background, we will draw on the following equations and sections of text to support the ideas developed in this chapter:

- Section 2.2 divides the universe up into the *system* (the thing we want to study) and its *surroundings*. A *reservoir* is any part of the surroundings that the system interacts with strongly. We also learn in this section that we may broadly categorize the thermodynamic parameters that describe our system into *extensive* parameters (such as energy E and volume V, which increase with the size of the system) and *intensive* parameters (such as temperature T and pressure P, which represent averages and do not depend explicitly on the size of the system).

- One of these thermodynamic parameters has special significance: the entropy, which we define using Boltzmann's law (Eq. 2.11),

$$S \equiv k_B \ln \Omega,$$

where k_B is Boltzmann's constant, equal to $1.381 \cdot 10^{-23}\,\mathrm{J\,K^{-1}}$, and where Ω is the number of microstates in the ensemble of all possible microstates of the system. With the entropy defined, we then define the temperature to be (Eq. 2.24)

$$\left(\frac{\partial E}{\partial S} \right)_{V,N} \equiv T.$$

We have a similar expression, also a partial derivative of the energy, for the pressure (Eq. 2.39):

$$P = -\left(\frac{\partial E}{\partial V} \right)_{S,N}.$$

- Thermodynamics deals with changes in parameters of our system, and mathematically we write those changes as derivatives. For that reason, a review of the partial derivative identities in Table A.4 may be helpful.

- The first law of thermodynamics has already made an appearance in Section 1.2, where it expresses any change ΔE in the energy of our system as the sum of *heat q* and *work w* (Eq. 1.7):

$$\Delta E \equiv E_{\mathrm{final}} - E_{\mathrm{initial}} = q + w. \tag{7.1}$$

- Diving deeper into the origins of some thermodynamic parameters, we will rely on the equipartition principle (Eq. 3.20),

$$E = \frac{1}{2} N_{\mathrm{ep}} nRT,$$

for as long it is useful. What constitutes the number of equipartition degrees of freedom, N_{ep}, is one question we have to re-examine in this chapter.

- For approximating the behavior of liquids, we will lean on the Lennard-Jones potential (Eq. 4.12),

$$u_{\mathrm{LJ}}(R) = 4\varepsilon \left[\left(\frac{R_{\mathrm{LJ}}}{R} \right)^{12} - \left(\frac{R_{\mathrm{LJ}}}{R} \right)^6 \right],$$

which allows us to simplify $u(R)$ in the pair correlation function (Eqs. 4.58 and 4.59),

$$\mathcal{G}(R) = \frac{V^2 e^{-U(X_1, \ldots, Z_N)/(k_B T)}}{Q'_U(T,V)} \approx e^{-u(R)/(k_B T)}.$$

- Section 2.1 describes the harmonic model of vibrational motion, and from this we obtain the partition function for vibrations, Eq. 3.26:

$$q_{\mathrm{vib}}(T) = f(e^{-\omega_e/(k_B T)}) = \frac{1}{1 - e^{-\omega_e/(k_B T)}}.$$

- We can draw on tables elsewhere in the text (or in other sources) to find values for the molecular properties that influence the thermodynamic quantities we will need in this chapter. In particular:
  - Table 3.1 gives rotational and vibrational constants for several molecules,
  - Table 4.2 lists values for the van der Waals coefficients and other measures of the intermolecular forces in certain gases, and
  - QM Table 4.2 gives electronic excitation energies of selected atoms.

7.1 The First Law of Thermodynamics

The desire to build better engines motivated the development of thermodynamics, founding the laws that govern how one form of energy can be converted to another. We subsequently learned that the same principles can be applied to the conversion of *chemical* energy, and with this realization, the field of *chemical* thermodynamics was born. We begin our study of chemical thermodynamics with a look at one of the fundamental concepts in general thermodynamics and how we can express it using properties relevant to chemical systems.

Macroscopic Parameters

Statistical mechanics reduces the number of coordinates in our immensely complex macroscopic system to some manageable set. In Chapter 4, after integrating over the motions of all the molecules of a non-ideal gas and over all space, we were able to express the pressure of the gas in terms of the volume of the container, the temperature, the number of molecules, and the intermolecular potential surface, yielding the van der Waals equation. Macroscopic chemistry is ideal for the application of statistics because molecules of the same composition and structure behave identically and because the number of molecules involved is huge.

We have done the difficult work and can focus on the results. We will study the properties of a *system* or *sample,* some collection of molecules, which is part of the much larger *universe.* The *surroundings* are whatever is not part of the system, and the part of the surroundings that interacts strongly with the system is called the *reservoir.* The reservoir is always much larger than the system, enough so that its properties are not measurably affected by the interaction. Within the combination of the system and any connected reservoirs, we conserve energy and mass.

Just what parameters *do* survive the statistical averaging process? The answer depends on how many microscopic parameters we integrate over during the statistical analysis and is partly up to us. Because some microscopic parameters are more significant than others, we will usually choose to leave out relatively small contributions to the macroscopic behavior of our sample. For example, in Section 4.2 the volume V of the container turned out to be a much more significant parameter in the behavior of gases than the excluded volume b of the molecules. We only included b in our analysis to derive a more precise prediction of gas properties.

Given what we've already done, the most obvious choices are those parameters mentioned at the beginning of Chapter 2:

- Extensive parameters: energy E, volume V, number of molecules N, entropy S

- Intensive parameters: pressure P, temperature T, number density ρ

There are some minor changes we need to make now to adopt the language conventionally used for macroscopic systems. For example, because the numbers have gotten so large, we will usually count the number of moles $n = N/\mathcal{N}_A$, rather than the number of molecules. The number of ensemble states Ω is an even more impossibly large number, and we will rarely refer to it from now on, preferring to use the entropy. The definition of the entropy (Boltzmann's law, Eq. 2.11) is important enough to repeat often:

$$S \equiv k_B \ln \Omega.$$

Like the other extensive parameters, the total entropy of two subsystems A and B is the sum of the individual entropies $S_A + S_B$ (Fig. 7.1).

A single **thermodynamic state** consists of the entire set of quantum states that are consistent with those values of the macroscopic parameters T, S, P, V, n, E, and so on. Whenever we measure one of these properties, such as the pressure, enough time elapses so that we average the measurement over a large number of the available quantum states. Microscopic fluctuations in the pressure could occur because the number and force of the molecules hitting the walls of the container change from one instant to the next; now we average over any such changes. If the system begins in some given thermodynamic state A—with parameters T_A, S_A, V_A, and so on—then whenever the system returns to state A we will measure the *same* values of *all* of those parameters. The thermodynamic state is defined by that set of values for T_A, S_A, V_A, and all the other macroscopic parameters. And because the thermodynamic state specifies the values of all those parameters, the parameters themselves are known as **state functions.** Why do we need a special name for parameters of our system? As a counterexample, the term *state function* would not apply to the speed of a particular molecule in the system because that parameter does not describe the *entire* system.

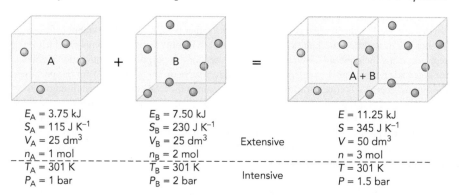

$E_A = 3.75$ kJ
$S_A = 115$ J K^{-1}
$V_A = 25$ dm^3
$n_A = 1$ mol
$T_A = 301$ K
$P_A = 1$ bar

$E_B = 7.50$ kJ
$S_B = 230$ J K^{-1}
$V_B = 25$ dm^3
$n_B = 2$ mol
$T_B = 301$ K
$P_B = 2$ bar

Extensive

Intensive

$E = 11.25$ kJ
$S = 345$ J K^{-1}
$V = 50$ dm^3
$n = 3$ mol
$T = 301$ K
$P = 1.5$ bar

◀ FIGURE 7.1 **Extensive and intensive parameters.** When two ideal gas samples A and B are combined (but not allowed to mix), the extensive parameters E, S, V, and n of the combined system A+B are the sums of the original values; the intensive parameters P and T are averages of the original values. The pressure in this case would be the average force per unit area on (for example) the shared bottom wall.

Many possible values of that one speed would still correspond to the same thermodynamic state. The state functions are all sums or averages over the microscopic properties of the system. Because they are determined only by the particular thermodynamic state, the state functions cannot depend on how we arrived at that state. The heat q and work w are kinds of energy *flow* and their values vary depending on how we change from one thermodynamic state function to another. Therefore, they also are *not* state functions.

An equation that expresses one of the state functions in terms of the others is an **equation of state.** When we derived the ideal gas law, we obtained an equation of state for ideal gas systems. This equation of state and others like it often arise from combining aspects of classical mechanics, such as our definition of the pressure (Eq. 2.39)

$$P = -\left(\frac{\partial E}{\partial V}\right)_{S,N},$$

with results from statistical mechanics, such as our expression (Eq. 2.24)

$$\left(\frac{\partial E}{\partial S}\right)_{V,N} \equiv T.$$

The two equations then yielded an expression for the temperature in terms of the pressure, volume, and number of moles—an equation of state (Eq. 1.37):

$$PV = nRT.$$

Energy was at the heart of the Schrödinger equation and our treatment of quantum mechanics, and it is also the center of attention in thermodynamics. We may not always measure the energy directly, but any macroscopic change in the system requires the transfer of energy between different degrees of freedom. Our plan is to find the laws that govern the flow of energy within a system and then apply these laws to learn how other parameters of the system will change.

The Internal Energy

The ensemble average of a chemical system's total energy is called the internal energy in classical chemistry. The internal energy accounts for all the kinetic and potential energies in electronic motions, vibrations, rotations, and translations of the molecules.[1] Because a system cannot be cooled below its zero-point energy, the internal energy is measured relative to the zero-point energy E_{zp}:

$$E_{internal} \equiv E - E_{zp}, \tag{7.2}$$

which is a convention we've been using since Eq. 2.34.

---

[1] In this text, we give the internal energy the symbol E rather than the accepted symbol U for two reasons. (1) In statistical mechanics it is still important to identify the potential energy, which we represent using the symbol U, which is commonly used for this purpose. The conventional symbol for potential energy in physics is V, which we need to reserve for the volume. (2) For our purposes, the internal energy is identical to a many-particle version of the total energy E used throughout the QM volume of this textbook, except for a shift in the reference point. The change in reference point is irrelevant as long as we study only ΔE values. With the exception of the zero-point energy, the terms we typically exclude from the energy calculation in quantum chemistry (the relativistic mass-energy equivalence, translational or potential energy of the *entire* system) are the same terms we exclude in calculating the thermodynamic internal energy.

So what is really changing here anyway? For one thing, unlike our previous look at the energies of microscopic systems, we will now be more attentive to the possible *paths* we can take from the initial to the final state. We previously defined two classes of energy transfer: work and heat (Section 1.2). Work is the energy required to move an object a certain distance, to change the volume or number of molecules. For work to be done, there must be some net motion of the system's mass. When we derived the ideal gas law through statistical mechanics, we considered the energy required to push one wall of the container a distance ds; that energy transfer was work. Other manifestations of work include diffusion of a solute through a solvent, the motion of a liquid up a tube through capillary action, and the flow of electrons in an electrical circuit.

The second form of energy transfer is heat, associated with the microscopic parameters of the system—electronic, vibrational, or rotational energy, the energy in chemical bonds—and with any *random* translational energy. (In our examples, work will usually transfer energy by means of *non-random* translational motion, causing a net change in the mass distribution of the system.) The glow of a red-hot stove burner, for example, is the loss of vibrational energy from an object by the emission of photons and is a form of heat. There is plenty of motion but no shift in the distribution of the masses *on average*. Heat, because it involves the transfer of energy in and out of the quantum states that describe matter at the microscopic scale, is always accompanied by a change in entropy of the system. In fact, we can use this to define the heat unambiguously: any infinitesimal change dS in the entropy of our system at a temperature T is accompanied by an infinitesimal transfer of energy in the form of heat $đq$ such that

$$đq \leq TdS. \tag{7.3}$$

The division of energy transfer into heat and work was introduced in Chapter 1 as the first law of thermodynamics (Eq. 1.7):

$$\Delta E \equiv E_{\text{final}} - E_{\text{initial}} = q + w, \tag{7.4}$$

where w is the work done going from the initial state to the final state and q is the heat evolved or absorbed during the change. Our convention is that w or q are positive for contributions that increase the energy of the system. When looking at the path along which the sample changes from initial state to final state, we may examine the incremental change in energy dE over a small section of that path, as in Figure 7.2, and ask how much of dE is heat and how much is work:

$$dE = đq + đw. \tag{7.5}$$

The fraction of dE that is heat and the fraction that is work need not remain constant throughout a process. For example, heat can be converted to work during an explosion, and work is converted into heat by friction.

In Eq. 7.5, the symbol $đ$ indicates an **inexact differential.** The inexact differential is a convention adopted in thermodynamics to remind us that q and w are not state functions. We can think of the system energy E as having a defined value at every point in time during any process that we will consider. The differential dE is then defined as the difference in E between two of these points separated by an infinitesimally small time. In contrast, *there is no defined*

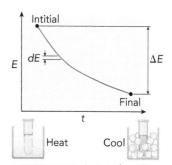

▲ FIGURE 7.2 **A state function in an infinitely slow process.** A sample is heated to some energy E_{initial} and is then cooled with ice down to energy E_{final}. The total energy difference is $\Delta E = E_{\text{final}} - E_{\text{initial}}$. At any point during the cooling process, we may measure the incremental change in the energy dE over an infinitesimally small period of time.

value of the heat q at each of these points, because q is defined by the process itself, not by the state of the sample at any given time. Because q does not have a value for any individual point, its derivative cannot be defined the same way as we defined dE, by taking the difference between two points. Instead, $đq$ is the tiny heat flow that takes place between two points with energy difference dE. In practice, we will treat $đq$ and $đw$ in much the same way we would treat any other derivative, but with this one rule:

To evaluate q or w, we cannot in principle integrate $đq$ or $đw$ directly.

Instead, we must always replace the inexact differential with an exact differential, and that substitution will depend on the nature of the process that takes us between the initial and final points.

We will frequently return to this form of the first law (Eq. 7.5), because it reminds us that we can divide even the smallest incremental energy change for any process into these two contributions. Before that becomes useful, however, we will have to pick up a few more tools and look at some idealized systems to see how we can relate the heat and work to other properties of matter.

7.2 Approximations and Assumptions

Our look at thermodynamics deals largely with limiting cases in order to simplify the problem of calculating the behavior of macroscopic systems. This is a good time to identify a few of the cases, conventions, and terminology we will use.

Equilibrium

The energy of the sample will be measured for states where there is no net change in any of the macroscopic parameters over the time of our measurement. These are called **equilibrium states,** and our brand of thermodynamics has difficulty dealing with anything else. If the system is at equilibrium, then from a few macroscopic parameters (such as n, T, and P) all the other macroscopic parameters (say, V and E) are obtainable from the equation of state. More parameters are necessary to specify the state of a system that has multiple chemical components or that is influenced by external fields such as electromagnetic radiation, static electric fields, or gravity.

Types of Systems

Thermodynamic analysis is most convenient when we can place our systems under certain constraints. Unless otherwise indicated, assume that each of our samples is a **closed system** for which the number of molecules and their composition is fixed, so n is no longer a variable. Many of our equations will be derived under this assumption.[2]

---

[2]This definition of "closed" lets us exclude chemical changes from the system, but more often the term is used to mean only that mass cannot be exchanged with the surroundings. Other references sometimes use "closed system" to refer to a system that, in addition to the n_i values, also fix the total energy and volume; this text will stipulate separately if E or V is fixed.

If all the molecules have the same chemical structure, the sample is **pure,** or single-component. A distinct question is whether the sample is **homogeneous,** with all the chemical components in the same phase of matter, or **heterogeneous,** with more than one phase present. For example, water ice floating in liquid water is a pure and heterogeneous sample, and ethanol mixed with liquid water makes a homogeneous, two-component sample.

An important component of systems that we have encountered before is the reservoir, an immense source of something—for example, pressure, temperature, or quantum states—big enough so that any interaction between our sample and the reservoir leaves the reservoir essentially unchanged. In chemical thermodynamics, the reservoir is usually used to provide a constant value of temperature or pressure or both.

Types of Processes

One parameter that our equilibrium states do *not* depend on is time. Therefore, when we change our sample—say, by doing some work on it or adding some heat—we will only make measurements after the sample has had enough time to adjust to those changes and has no memory of how the change occurred. This requirement is sometimes awkward, implying we cannot make any measurements while the system is changing. Therefore we introduce the concept of a **quasistatic** process, one that occurs so slowly that the sample is always in thermodynamic equilibrium, to within the precision of our measurements. We will often restrict our processes to an idealized quasistatic path along which the *total* entropy of the system S_T (including contributions from reservoirs) never changes. This path is called a **reversible process.**[3] A reversible process must also be quasistatic. In an **irreversible process,** the total entropy increases. We will discuss the significance of the total entropy in Chapter 9, after we have had a chance to get used to processes where we focus on the system alone.

The terms we've just defined limit how both the system and its surroundings behave during a given process. The remaining terms describe restrictions placed specifically on the system. For example, we may confine the process to a path along which no heat flows between the system and its surroundings: $q = 0$. Such a process is called **adiabatic,** and *adiabatic* is used to describe a container that does not permit the transfer of heat to or from systems outside the container. Containers that are permeable to heat flow are called **diathermal.**

If we carry out an adiabatic process reversibly, then $đq = 0$ as the process proceeds, and Eq. 7.3 requires that $dS = 0$ as well. Therefore, the system entropy S remains constant, and we have an **isentropic** process. Other processes keep other parameters constant: **isoenergetic** processes occur at constant energy, **isothermal** processes at constant temperature, **isobaric** at constant pressure, and **isochoric** at constant volume. All of these definitions are summarized in Table 7.1.

---

[3]These definitions of "quasistatic" and "reversible" are also not universal, and are sometimes switched, which can make a study of entropy using different sources quite bewildering. Our work in thermodynamics will be primarily limited to processes that are both quasistatic and reversible, with the exception of some irreversible expansions in Chapter 8.

TABLE 7.1 **Definitions of terms used for thermodynamics.** The *system* is a collection of molecules on which measurements are made; the *universe* consists of the system and any surroundings or reservoirs that interact with the sample.

systems	
open	may exchange mass and energy with surroundings
closed	all n_i fixed (no mass exchange and no reactions)
pure	one chemical component (implied unless stated otherwise)
isolated	no mass or energy exchange with surroundings
homogeneous	one phase of matter
heterogeneous	more than one phase of matter
processes	
quasistatic	system and surroundings always in equilibrium
reversible	system and surroundings at constant S_T ($\Delta S_T = 0$)
adiabatic	no heat flow ($q = 0$)
isentropic	system at constant S ($\Delta S = 0$)
isothermal	system at constant T ($\Delta T = 0$)
isobaric	system at constant P ($\Delta P = 0$)
isochoric	system at constant V ($\Delta V = 0$)
isoenergetic	system at constant E ($\Delta E = 0$)
containers	
adiabatic	walls prevent heat flow
diathermal	walls allow heat flow

The Standard State

As in quantum mechanics, we rarely measure absolute energies, and occasionally will want some reference state of the sample with which to compare our measurements. The workhorse for this application is the **standard state,** labeled by a superscript $\ominus$, for which the pressure is exactly 1 bar (equal to 10^5 Pa or 0.987 atm), and any mixture or solution is separated into its pure components. Unlike STP (standard temperature and pressure), the standard state does not have a specified temperature. The energy of the standard state, $E^{\ominus}$, changes with temperature, for example. We shall have to revisit this when we get to more sophisticated systems.

7.3 Mathematical Tools

Thermodynamics involves several similar parameters. It's easy to get confused over how they are related and how they differ and why so many different forms are used. Let's just throw them all out on the table at once so that we can see the mathematical distinctions between them.

Partial Derivatives of the Energy

We can express our intensive parameters in terms of extensive parameters. Since the values of extensive parameters are proportional to the amount of material, the ratio of one extensive parameter to another extensive parameter must be an intensive parameter. For example, the number density ρ is the ratio of two extensive parameters: N/V. The factors that determine the total amount of material cancel, and the result is an intensive parameter.

The same cancellation takes place when we take the derivative of one extensive parameter with respect to another. For example, we took the derivative of energy with respect to volume to get an intensive parameter, the pressure (Eq. 2.39):

$$P = -\left(\frac{\partial E}{\partial V}\right)_{S,n},$$

(where we have moved to our macroscopic notation by fixing the number of moles n in the partial derivative rather than the number of particles N).

One of our other pivotal intensive parameters, the temperature T, has also been expressed as a partial derivative of the energy with respect to the entropy, another extensive parameter (Eq. 2.24)

$$T = \left(\frac{\partial E}{\partial S}\right)_{V,n}.$$

One more partial derivative remains in this series—what happens when we take the partial derivative of the energy with respect to our other extensive parameter, the number of moles n? This derivative introduces a new intensive parameter, the **chemical potential** μ:

$$\mu \equiv \left(\frac{\partial E}{\partial n}\right)_{S,V}. \tag{7.6}$$

The chemical potential gives the rate at which the energy increases as each mole of molecules is added to the sample, while S and V are constant. Under those conditions, in other words, μ is the molar energy of the sample. We will often neglect the chemical potential while we get comfortable with some simpler, non-chemical processes, but it becomes a central player as we approach real chemistry in Part III.

The Thermodynamic Potentials

Courses in algebra condition us to worry when we have too many variables: we don't want three variables when we have only two equations available to solve them, for example. In thermodynamics, however, the problem is more often finding how to relate the variables we want to know to the ones that we can measure directly. We actually invent new variables in thermodynamics in order to express some of these relationships, chief among these being parameters related to the energy.

The Fundamental Equation

If we have a system composed of k different chemical components, having numbers of moles $n_1, n_2, \ldots, n_k$, then our original series of equations for the intensive parameters becomes

$$T = \left(\frac{\partial E}{\partial S}\right)_{V,n_1,\ldots n_k} \tag{7.7}$$

$$P = -\left(\frac{\partial E}{\partial V}\right)_{S,n_1,\ldots n_k} \tag{7.8}$$

$$\mu_i = \left(\frac{\partial E}{\partial n_i}\right)_{S,V,\text{ all } n_{j\neq i}}. \tag{7.9}$$

The chemical potential must be evaluated separately for each component.

Equations 7.7, 7.8, and 7.9 relate pairs of extensive and intensive **conjugate variables;** S and T are one pair of conjugate variables, V and P another pair, and each set n_i and μ_i is another. Our experience with the relation between P and V in deriving the ideal gas law is indicative of why these variables appear in pairs. For two samples A and B in contact across a movable wall, the pressures P_A and P_B dictate which way the wall will move, changing the volumes V_A and V_B. If $P_A > P_B$, then V_A will increase and V_B will decrease until the two pressures are equal. Similarly, two samples in contact that have different temperatures will exchange entropy to make the temperatures equal, and samples with different chemical potentials will exchange mass to make the chemical potentials equal.

We have written the intensive parameters as derivatives of the energy with respect to their conjugate extensive parameters. This approach implies that the energy E is a function of the extensive parameters: the entropy S, the volume V, and the mole numbers n_i of the sample. We will call this the *explicit variable dependence* of the energy.[4] This rule is sometimes stated as the seemingly redundant **fundamental equation of thermodynamics:**

$$E = E(S, V, n_1, \ldots n_k). \tag{7.10}$$

You won't see much more of the fundamental equation, but its derivatives are going to be with us for the rest of our look at thermodynamics. What thermodynamics follows is the energy *flow* in a variety of processes, meaning that we will be looking not at E but at ΔE and its derivative dE. By applying the slope rule for derivatives to the fundamental equation, Eq. 7.10, we get all the functional dependence:

$$dE = \left(\frac{\partial E}{\partial S}\right)_{V, n_1, \ldots, n_k} dS + \left(\frac{\partial E}{\partial V}\right)_{S, n_1, \ldots, n_k} dV +$$

$$\left(\frac{\partial E}{\partial n_1}\right)_{S, V, n_2, \ldots, n_k} dn_1 + \ldots + \left(\frac{\partial E}{\partial n_k}\right)_{S, V, n, \ldots, n_{k-1}} dn_k$$

$$= TdS - PdV + \mu_1 dn_1 + \ldots + \mu_k dn_k. \tag{7.11}$$

The change in energy can be expressed in terms of the intensive parameters (T, P, μ_i) and differentials of their conjugate extensive parameters (dS, dV, dn_i).

Legendre Transformations

A differential expression such as Eq. 7.11 is most convenient when a number of the derivatives on the right-hand side can be set to zero. For a closed system, we may say that the mole numbers of all the components are constant so $dn_i = 0$. However, this still leaves

$$\text{closed system} : dE = TdS - PdV. \tag{7.12}$$

---

[4]This point is a subtle one. After all, we already have an expression from the equipartition principle that writes the energy in terms of n and T (Eq. 3.20), but now we say that T is not one of the variables that explicitly determine the energy. One of the tough jobs of thermodynamics is to organize a large number of mutually related parameters. Here we have *chosen* a set of parameters that we will say provide the explicit variable dependence for the energy. In Section 9.1 we will show that the translational entropy S can be derived as a function of only E, V, and n, so we can use that as a justification for defining E in terms of S, V, and n.

Neither of the derivatives on the right-hand side are convenient in most chemical applications. Entropy is not under our direct control at all, since it is defined in terms of the number of ensemble states. Fixing the volume in an experiment, while conceptually simple, can be disastrous if the experiment has other ideas: extremely high pressures may build up when constant-volume systems are subjected to small stresses. We would normally prefer an equation that lets us find the change in energy in terms of derivatives of the intensive parameters T and P. These are the stereotypical lab bench-top conditions; a sample in a beaker in equilibrium with the atmosphere is at the ambient pressure and temperature. Under those conditions, T and P are constants, so dT and dP are zero.

The energy is not explicitly a function of the intensive parameters T and P, but we can invent a *new* function, related to the energy by a simple equation, that explicitly depends on T and P. We will use the **Legendre transformation** to make this variable adjustment, replacing the dependence on V by dependence on the conjugate variable P. The principle of the Legendre transformation is a matter of straightforward geometry and differential calculus. We will just cite the results initially, and then will show how the transformation works to our advantage. For simplicity, we assume the energy to depend only on two extensive variables X_1 and X_2 such that

$$\left(\frac{\partial E}{\partial X_1}\right)_{X_2} = Y_1 \quad \left(\frac{\partial E}{\partial X_2}\right)_{X_1} = Y_2. \tag{7.13}$$

The partial Legendre transform of the energy with respect to X_1 switches the functional dependence from one variable (X_1) to its conjugate variable (Y_1), yielding a new extensive parameter $Z(Y_1, X_2)$ such that

$$Z(Y_1, X_2) = E(X_1, X_2) - X_1 Y_1, \tag{7.14}$$

where

$$\left(\frac{\partial Z}{\partial Y_1}\right)_{X_2} = -X_1 \quad \left(\frac{\partial Z}{\partial X_2}\right)_{Y_1} = Y_2, \tag{7.15}$$

and

$$dZ = dE - d(X_1 Y_1) = \left(\frac{\partial E}{\partial X_1}\right)_{X_2} dX_1 + \left(\frac{\partial E}{\partial X_2}\right)_{X_1} dX_2 - Y_1 dX_1 - X_1 dY_1$$

$$= Y_1 dX_1 + Y_2 dX_2 - Y_1 dX_1 - X_1 dY_1 \qquad \text{by Eq. 7.13}$$

$$= -X_1 dY_1 + Y_2 dX_2. \tag{7.16}$$

The new function varies as Y_1 and X_2, rather than as X_1 and X_2, so we have accomplished our goal of altering the explicit variables of E.

We apply these results directly to the energy to get three other functions with the other possible parameter dependencies. Relying on Eq. 7.14, we can obtain the following:

- the **enthalpy** H; the Legendre transform of E with respect to V:
$$H(S, P, n_1, \ldots n_k) = E + PV \tag{7.17}$$

- the **Helmholtz free energy** F; the Legendre transform of E with respect to S:
$$F(T, V, n_1, \ldots n_k) = E - TS \tag{7.18}$$

- the **Gibbs free energy** G; the Legendre transform of E with respect to V and S (also equal to the transform of H with respect to S):

$$G(T,P,n_1, \ldots n_k) = E - TS + PV = H - TS \qquad (7.19)$$

These functions and the energy are called the **thermodynamic potentials.** These give derivative equations, showing the different situations under which each of the thermodynamic potentials might be useful:

$$dE = TdS - PdV + \mu_1 dn_1 + \ldots + \mu_k dn_k \qquad (7.20)$$

$$dH = TdS + VdP + \mu_1 dn_1 + \ldots + \mu_k dn_k \qquad (7.21)$$

$$dF = -SdT - PdV + \mu_1 dn_1 + \ldots + \mu_k dn_k \qquad (7.22)$$

$$dG = -SdT + VdP + \mu_1 dn_1 + \ldots + \mu_k dn_k. \qquad (7.23)$$

If you appreciate symmetry when it appears, this little derivation may greatly clarify the origin of the relationships between these easily confused parameters.

For example, we put a closed system in contact with a temperature and pressure reservoir, fixing T and P and setting the corresponding derivatives dT and dP to zero. The entropy change dS during a process in this sample is related to the energy change by the equation

$$\text{constant } T,P,n_i : dS = \frac{1}{T}(dE + PdV). \qquad (7.24)$$

The changes in two parameters, the energy and volume, need to be measured in order to find dS. On the other hand, if we use the enthalpy, then under the same conditions

$$\text{constant } T,P,n_i : dS = \frac{dH}{T}. \qquad (7.25)$$

By using the enthalpy instead of the energy, we simplify the mathematics.

For a similar reason, the Gibbs free energy is a central parameter in chemistry. Under constant temperature and pressure, the change in G is

$$\text{constant } T,P : dG = \sum_{i=1}^{k} \mu_i dn_i. \qquad (7.26)$$

Because μ_i represents the molar energy of chemical component i, dG monitors changes in the *chemical energy* of the sample, omitting contributions to dE from thermal energy and expansion.

Equations 7.17, 7.18, and 7.19 define H and F and G. Keep this in mind because after one works with nothing but ΔH values (for example) for a long time, it is easy to forget just why we started using H in the first place and just how it differs from the energy. Here's the exact difference: the enthalpy is the energy plus PV.

Still, you have a right to be suspicious of H. Later on, it is tempting to calculate enthalpy changes and discuss them in terms of the properties or activity of a single molecule. But whereas we could always calculate E for a single molecule, how do we determine P if we want to calculate ΔH from $H = E + PV$ for one molecule? If we adhere to our rigorous definitions, then it is meaningless to discuss the H, F, or G values of a microscopic sample. Because the enthalpy and the free energies are defined in terms of *ensemble averages* such as pressure P and temperature T, we need an ensemble of states to average over, so we will treat H, F, and G as strictly macroscopic parameters. For the same reason, we wouldn't

Martin Gruebele holds the James R. Eiszner Chair in Chemistry at the University of Illinois at Urbana-Champaign, and he also serves as a professor of physics. His work covers a broad range of projects, including one of the most challenging questions in biochemistry: how do proteins fold? Although proteins are (for the most part) long, single strands of amino acids joined by peptide bonds, any protein normally exists in a folded state that stabilizes the molecule by optimizing favorable hydrogen-bonding and other weak attractions. The size of these molecules makes detailed studies of all but the smallest peptides impossible, so approximations are used to model the thermodynamics of the process as a *denatured* (unfolded) protein gathers itself into the folded form. The process normally takes proteins through several relatively stable, partially folded thermodynamic states, which can be probed by spectroscopically monitoring a light-sensitive part of the molecule called the *chromophore*. However, the chromophore's response may change at different stages of the folding, complicating the interpretation of the data. Professor Gruebele and his coworkers have shown that the rate of response of the chromophore can be used to effectively separate the variability of the chromophore from the folding of the protein in the analysis. Our text is largely concerned with obtaining bulk thermodynamics by extrapolation from the molecular scale, but this work shows how thermodynamics can itself be applied *at the molecular scale* to better understand structure and dynamics.

ordinarily use temperature, pressure, or chemical potential as characteristics of one molecule. In contrast, the energy E, entropy S, and volume V *can* be defined for single molecules.

We'll call G the Gibbs energy and F the Helmholtz energy from now on. The **free energy** we will use as a general term for any thermodynamic potential (such as G or F) that depends on T and n rather than S and n.

The Maxwell Relations

One immediate application of the thermodynamic potentials is to relationships among the derivatives of our many parameters. There are *many* parameters available in thermodynamics, and they all depend on one another. If P, V, and n are known respectively to be 1.00 bar, 25.0 L, and 1.00 mol in an ideal gas sample, for example, then T must be 301 K. The temperature cannot vary independently. If the parameters themselves are related, then their derivatives must also be related.

To illustrate, we take the partial derivatives of the energy E for a one-component, closed sample ($dn = 0$) with respect to S at constant V, and then with respect to V at constant S, yielding

$$\left[\frac{\partial}{\partial V}\left(\frac{\partial E}{\partial S}\right)_{V,n}\right]_{S,n} = \left(\frac{\partial T}{\partial V}\right)_{S,n}. \qquad \text{by Eq. 7.7}$$

For any continuous function, the value of this derivative must be the same no matter what order we take the derivatives. This expression must therefore be the same as the result if we reverse the order of differentiation:

$$\left[\frac{\partial}{\partial V}\left(\frac{\partial E}{\partial S}\right)_{V,n}\right]_{S,n} = \left[\frac{\partial}{\partial S}\left(\frac{\partial E}{\partial V}\right)_{S,n}\right]_{V,n}.$$

However,

$$\left[\frac{\partial}{\partial S}\left(\frac{\partial E}{\partial V}\right)_{S,n}\right]_{V,n} = -\left(\frac{\partial P}{\partial S}\right)_{V,n}, \qquad \text{by Eq. 7.8}$$

so we have

$$\left(\frac{\partial T}{\partial V}\right)_{S,n} = -\left(\frac{\partial P}{\partial S}\right)_{V,n}. \qquad (7.27)$$

Three more identical derivations from each of the other thermodynamic potentials result in similar expressions:

$$\left(\frac{\partial T}{\partial P}\right)_{S,n} = \left(\frac{\partial V}{\partial S}\right)_{P,n} \qquad (7.28)$$

$$\left(\frac{\partial S}{\partial V}\right)_{T,n} = \left(\frac{\partial P}{\partial T}\right)_{V,n} \qquad (7.29)$$

$$\left(\frac{\partial S}{\partial P}\right)_{T,n} = -\left(\frac{\partial V}{\partial T}\right)_{P,n}. \qquad (7.30)$$

These relations are known collectively as **Maxwell relations.** Notice what they do: they exchange one pair of variables in the derivative (for example S and P on the left side of Eq. 7.30) for the two conjugate variables (V and T on the right side of Eq. 7.30). How are they useful? In Eq. 7.30, for example, the derivative of entropy with respect to pressure can be replaced by the derivative of volume with respect to temperature. Isothermal changes in the elusive entropy with pressure can be studied by more straightforward isobaric measurements of the volume and temperature.

The partial derivatives that appear in Eqs. 7.27–7.30 are common features of thermodynamics. Because thermodynamics is concerned with understanding energy flow, our attention is drawn as often to the *change* in the values of state functions (such as S, T, P, and V) as to the values themselves. As examples, certain volume-scaled derivatives correspond to tabulated material properties and are often cited in thermodynamic expressions:

1. The isobaric **coefficient of thermal expansion:**

$$\alpha \equiv \frac{1}{V}\left(\frac{\partial V}{\partial T}\right)_{P,n} \qquad (7.31)$$

2. The **isothermal compressibility:**

$$\kappa_T \equiv -\frac{1}{V}\left(\frac{\partial V}{\partial P}\right)_{T,n} \qquad (7.32)$$

3. The **adiabatic compressibility:**

$$\kappa_S \equiv -\frac{1}{V}\left(\frac{\partial V}{\partial P}\right)_{S,n} \qquad (7.33)$$

Whenever we can express a result in terms of these parameters, we may be able to obtain a numerical solution by referring to standard tables of material properties.

Thermodynamic Parameters and the Partition Function

With this glossary of terms, most of the discussion that follows can rely on the language of classical chemistry, using parameters like the free energies and the entropy. It becomes easy, maybe even tempting, to forget the connection between the chemical laws of bulk materials and the molecular structure that determines it. Part of my job is to keep you from forgetting.

Having just introduced some of these parameters, let's examine how they are related not just to each other, but to the fundamental features of the molecular structure. Throughout our tour of statistical mechanics, the partition function maintained a link between the bulk properties we were after and the molecular parameters we already had. In the simplest examples, the rotational and vibrational partition functions of a diatomic molecule were functions of the rotational and vibrational constants (Eqs. 3.30 and 3.26):

$$q_{\text{rot}} = \frac{k_B T}{B} \qquad q_{\text{vib}} = \frac{1}{1 - e^{-\omega_e/(k_B T)}},$$

which in turn depend on the moment of inertia μR^2 and the stretching force constant k, respectively.

Our new thermodynamic parameters can also be written in terms of the partition functions. For a pure sample, we need be concerned with only one kind of molecule. That molecule has an overall partition function determined by its structure and the temperature of its environment:

$$Q(T) = q_{\text{elec}}(T)\, q_{\text{vib}}(T)\, q_{\text{rot}}(T)\, Q_{\text{trans}}(T). \qquad (7.34)$$

We have been ignoring the electronic partition function, assuming that it is equal to one if we keep the temperature reasonably low, but it should be included here for completeness's sake. There may be other contributions to the partition function as well (various magnetic terms, for example) in certain circumstances. We have also seen that the temperature may not be enough to specify the partition function for a molecule—for example, the volume was also needed to get Q_{trans} of the ideal gas. But once we have the partition function identified, we may use it to show how these molecular properties influence the thermodynamic parameters.

We start by leaving a few of our thermodynamic parameters alone so that they can define the environment of the molecules. We initially kept E, V, and N (now the number of moles n) constant by using the microcanonical ensemble in Chapters 2 and 4. In switching to the canonical ensemble, we traded E in for T, because T is an easier parameter to keep constant for our sample in experiments.[5] So sticking with the canonical ensemble—meaning that we know T, V, and n—we can find the other parameters.

- **The energy** E. Our hard work (well, actually Max Planck's) is paying off: we already had to find this relation to explain the blackbody emission spectrum.

---

[5]In typical bench-top experiments, the sample and its surroundings can change energy by heat transfer. Keeping the temperature constant is often much simpler.

Equation 6.15 puts the energy in terms of the temperature and partition function:

$$E = k_{B}T^{2}\left(\frac{\partial \ln Q(T)}{\partial T}\right)_{V,n}.$$

- **The entropy** S. Looking this time at our derivation of the translational partition function, the equation we used for the energy (Eq. 4.25) already has Ω in it, and we need that to get S:

$$\frac{E}{k_{B}T} = \ln \Omega - \ln Q(T)$$

$$k_{B}\ln \Omega = k_{B}\ln Q(T) + \frac{E}{T}$$

$$S = k_{B}\ln Q(T) + k_{B}T\left(\frac{\partial \ln Q(T)}{\partial T}\right)_{V,n}. \tag{7.35}$$

- **The Helmholtz energy** F. The Helmholtz free energy is a natural thermodynamic potential to relate to our partition function because it is defined in terms of the same variables we have used for describing the molecular environment. Sure enough, the resulting equation is quite compact:

$$F = E - TS$$

$$= k_{B}T^{2}\left(\frac{\partial \ln Q(T)}{\partial T}\right)_{V,n} - T\left[k_{B}\ln Q(T) + k_{B}T\left(\frac{\partial \ln Q(T)}{\partial T}\right)_{V,n}\right]$$

$$F = -k_{B}T\ln Q(T). \tag{7.36}$$

This equation opens the door to our other two principal intensive parameters: the pressure and chemical potential.

- **The pressure** P. We start from the Helmholtz energy because its equation is so simple:

$$P = -\left(\frac{\partial F}{\partial V}\right)_{T,n} = k_{B}T\left(\frac{\partial \ln Q}{\partial V}\right)_{T,n}. \tag{7.37}$$

- **The chemical potential** μ. In the same way,

$$\mu = \left(\frac{\partial F}{\partial n}\right)_{T,V} = -k_{B}T\left(\frac{\partial \ln Q}{\partial n}\right)_{T,V}. \tag{7.38}$$

The point of these last few equations is to make clear that all of these thermodynamic parameters ultimately depend on molecular properties, as represented here by the partition function. We want to remember that *always*.

Our next step is to show this relationship between the molecular and bulk properties by applying the tools of statistical mechanics to one of the central problems in thermodynamics: how much energy does it take to raise the temperature of any given substance?

7.4 Heat Capacities

Lord Kelvin first defined thermodynamics as the study of the relationship between heat and force, and force is a derivative of the potential energy. What better place to begin our study of thermodynamics, then, than with the parameter that defines the relationship between heat and energy?

Heat Capacities at Constant Volume or Pressure

Statistical mechanics gives us a substantial toolkit for evaluating bulk properties. One such bulk property is the **heat capacity,** which quantifies the distinction between how much energy we inject into a sample and how much the temperature rises. The heat capacity is a critical parameter in—among many other instances—large-scale chemical reactors, where it determines how much energy has to be carried into or carried away from the reaction mixture to control the temperature. The heat capacity is the *flow of heat* per unit change in temperature:

$$C(T) \equiv \frac{\bar{d}q}{dT}. \qquad (7.39)$$

Because heat capacity is the derivative of an extensive parameter with respect to an intensive property, it is itself an extensive property, directly proportional to the number of moles n. To get an intensive parameter, we can divide by n to get the molar heat capacity:

$$C_{\mathrm{m}}(T) \equiv \frac{1}{n}C(T) = \frac{1}{n}\frac{\bar{d}q}{dT}. \qquad (7.40)$$

These are formal definitions, however. Measured heat capacities are evaluated as *partial* derivatives because the heat capacity is a function of several variables, and the heat flow will change depending on which parameters are held constant. We measure C at either constant volume (C_V), which is most convenient for gases and theoreticians, or at constant pressure (C_P), which is the most common tabulated form.

How is the heat capacity related to our other thermodynamic parameters? We can find this by seeing that C measures an incremental change in one form of transferred energy. We have already shown that the incremental change in energy of a closed system may be written as (Eq. 7.12)

$$dE = TdS - PdV = \bar{d}q + \bar{d}w.$$

Changes in volume correspond to a net motion of mass in the system and are therefore usually associated with work.[6] Although work can be carried out in many ways (Fig. 7.3), this is perhaps the most straightforward kind of work, called **pressure-volume** or PV work—*the mechanical energy expended or absorbed as a sample expands or contracts.*

In Chapter 2, we obtained Equation 2.39,

$$\text{if } P = P_{\min}: \quad P = -\left(\frac{\partial E}{\partial V}\right)_{S,N},$$

under the condition that the pressure used in determining the energy loss during an expansion is the pressure that *resists* the motion—in other words, $P_{\min}$, the lower of the two pressures on either side of the moving wall. For example, an ideal gas expanding into a vacuum does no work because it is not pushing *against* anything, regardless of its initial pressure. If PV work is the only kind of work allowed, then the incremental work $\bar{d}w$ is the change in energy due to the incremental expansion or contraction:

$$\bar{d}w = -P_{\min}dV. \qquad (7.41)$$

[6]An exception is the expansion into a vacuum, when no force opposes the expansion.

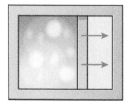

PdV

P = pressure
V = volume

(a)

γdA

γ = surface tension
A = surface area

(b)

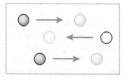

μdn

μ = chemical potential
n = number of moles

(c)

▲ FIGURE 7.3 **Different types of work.** The corresponding contribution to the incremental change in energy dE and the intensive and extensive parameters are shown for **(a)** *PdV* work due to expansion or contraction; **(b)** *γdA* work for changes in surface area, such as the flattening of a spherical droplet when it lands on a surface; **(c)** chemical work for changes in concentration.

Equation 7.41 is essentially the same as Eq. 2.37, $dE = -P_{min}dV$, which we use to obtain the ideal gas law under conditions where the system is completely isolated, letting us keep the entropy constant. Therefore, we could set $đq = 0$ and dE equal to $đw$. Here we look at the more general case, where both $đw$ and $đq$ may be nonzero.

To change the volume, we need to shift the position of some boundary between the sample and its surroundings. In the case of a *reversible* process, the change in volume is carried out slowly enough that the pressures on both sides of this boundary are always equal to within our ability to measure the difference. In other words, we can set P equal to P_{min} for the reversible expansion,

$$đw_{rev} = -PdV. \tag{7.42}$$

Combining this with Eq. 7.12, $dE = TdS - PdV = đq + đw$, we discover that the incremental heat for a reversible process must therefore be

$$đq_{rev} = dE - đw_{rev} = TdS - PdV - (-PdV) = TdS,$$

or simply,

$$đq_{rev} = TdS. \tag{7.43}$$

This relationship between $đq_{rev}$ and TdS makes sense because T is defined to be $\partial E/\partial S$ at constant V and n. In other words, TdS gives the change in energy *when no work is allowed,* as in this case when we don't let the system expand. The product TdS is the incremental change in energy that arises from microscopic changes to the system, and that energy change is what we mean by heat.

Ah, but if no work is allowed, then the only contribution to the incremental energy change must come from the flow of heat, right? If we evaluate the partial derivative of the energy with respect to T, keeping the volume constant (no PV work) and also keeping n constant (for a closed sample), then we find

$$\left(\frac{\partial E}{\partial T}\right)_{V,n} = \left(\frac{TdS - PdV}{\partial T}\right)_{V,n} \qquad \text{by Eq. 7.12}$$

$$= \left(\frac{TdS}{\partial T}\right)_{V,n} \qquad \partial V = 0 \text{ if } V \text{ constant}$$

$$= \left(\frac{đq}{dT}\right)_{V,n} \qquad \text{by Eq. 7.43}$$

$$\equiv C_V(T). \qquad \text{by Eq. 7.39}$$

We have found a new expression for the heat capacity at constant volume:

$$C_V(T) = \left(\frac{\partial E}{\partial T}\right)_V, \tag{7.44}$$

where the subscript V implies that the volume is kept fixed.

We can predict the value of C_V for ideal gases. We begin with the equipartition principle (Eq. 3.20),

$$E = \frac{1}{2}N_{ep}Nk_BT = \frac{1}{2}N_{ep}nRT,$$

CHECKPOINT As long as it's understood that we're working with a closed system of one chemical component, we can drop the "*n*" subscript on our partial derivatives to simplify the notation. We'll have to bring it back when we start looking at how phase changes and mixing other transformations can alter the number of moles of a particular substance.

where N_{ep} is the number of *equipartition* degrees of freedom in translation, rotation, and vibration for each molecule, counting the kinetic and potential energy contributions to each vibrational coordinate separately. The heat capacity for the ideal gas at constant volume is

$$C_V = \left(\frac{\partial E}{\partial T}\right)_V = \frac{1}{2}N_{ep}nR\left(\frac{\partial T}{\partial T}\right)_V = \frac{1}{2}N_{ep}nR. \qquad (7.45)$$

However, the number of available equipartition degrees of freedom effectively increases with temperature, so real heat capacities increase with temperature.

For molecules, the equipartition principle must be applied with caution because the vibrations normally do not contribute much to the overall energy until the temperature approaches ω_e/k_B (i.e., until the thermal energy is comparable to the vibrational transition energy). The trustworthiness of Eq. 7.45, therefore, depends strongly on the temperature and the compound in question. But for a monatomic gas such as helium or argon, there are no rotations or vibrations; there are only the three translational coordinates for each atom, so $N_{ep} = 3$:

$$\text{monatomic ideal gas}: C_V = \frac{3}{2}nR. \qquad (7.46)$$

EXAMPLE 7.1 **Heat Capacity of Mercury Gas**

CONTEXT The heat capacity, like all the thermodynamic parameters that depend on the temperature or entropy, is determined by the microscopic degrees of freedom of the substance. In the Einstein equation for heat capacity, as we'll see, it is determined by the effective vibrational frequencies of the particles. Classical thermodynamics does not attempt to predict the value of the heat capacity from these fundamental considerations, but seeks instead the relationships between this macroscopic parameter and the macroscopic behavior of the sample as heat flows in or out. By understanding the origin of the numerical values, however, we can use macroscopic measurements to learn about microscopic structure. Like bromine, mercury in liquid form is usually written as a diatomic molecule: $Hg_2(liq)$. When bromine evaporates, it forms a gas of diatomic molecules. It is reasonable to ask whether the same is true of mercury.

PROBLEM You have a constant-volume sample with a known quantity of mercury gas, a thermometer, and a calorimeter that allows you to add or remove measured amounts of energy from the sample. How could you use this apparatus to determine whether gas-phase mercury was in monatomic or diatomic form, that is, Hg(gas) or Hg_2(gas)? What numerical results would you look for in your measurement?

SOLUTION The apparatus allows you to measure the heat capacity of the gas at constant V by dividing small changes in the energy added to the sample by the temperature change. The monatomic gas would have a heat capacity C_{Vm} of roughly $3R/2 = 12.5\,\text{J K}^{-1}\,\text{mol}^{-1}$. The diatomic will also have contributions from rotations and vibrations (Hg is so massive that the vibration

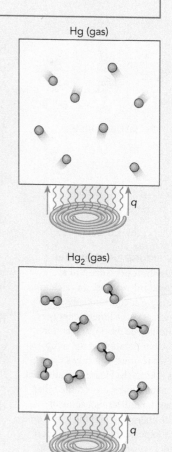

Hg (gas)

Hg_2 (gas)

of Hg_2 would be easily excited at room temperature). These degrees of freedom would predict a much higher heat capacity, about $7R/2 = 29.1 \text{ J K}^{-1} \text{ mol}^{-1}$. In fact, the molar heat capacity C_{Vm} is $12.5 \text{ J K}^{-1} \text{ mol}^{-1}$, which means the gas is monatomic.

EXTEND This very question was asked in the 1890s, when there was a huge effort to fill in the gaps in Mendeleev's periodic table. Argon and helium had recently been isolated, but there were questions about their atomic masses. The molar mass $\mathcal{M}$ of a sample could be determined by using the ideal gas law to find n and dividing that into the mass of the sample m: $\mathcal{M} = mRT/(PV)$. For argon, the value was found to be roughly 40 g/mol. However, it wasn't known if this was the mass of one mole of argon atoms (each with a mass of 40 amu) or of one mole of diatomic argon molecules Ar_2 (each with a *molecular mass* of 40 amu, meaning that argon would have an *atomic mass* of 20 amu). By comparison to mercury, among other elements, it was clear from the heat capacity of argon that it had to be a monatomic gas.

SAMPLE CALCULATION Heat Capacity at Constant Volume of an Ideal Monatomic Gas. *Molar* heat capacity is an intensive parameter but heat capacity is extensive. The value of C_V for 2.00 mol of neon gas is predicted by Eq. 7.46 to be

$$C_V = \frac{3}{2}nR = \frac{3}{2}(2.00 \text{ mol})(8.3145 \text{ J K}^{-1} \text{ mol}^{-1}) = 24.9 \text{ J K}^{-1}.$$

The heat capacity at constant pressure is not easily evaluated from the energy, but it is easier to measure experimentally. We want a thermodynamic potential that gives us TdS when the pressure and number of moles are constant. The enthalpy is well-suited for this task, because its derivative varies as dP and dn,

$$dH = TdS + VdP + \mu dn. \tag{7.47}$$

For a closed sample at constant pressure, we set dP and dn to zero. Recalling that $H = E + PV$, for a closed sample ($dn = 0$), we find that

$$dH = dE + PdV + VdP = đq + đw + PdV + VdP, \tag{7.48}$$

Therefore, at constant pressure and with $dw = -PdV$, $dH = đq$, and

$$C_P \equiv \left(\frac{\partial q_{rev}}{\partial T}\right)_P = \left(\frac{\partial H}{\partial T}\right)_P. \tag{7.49}$$

Why should C_P be different from C_V? Briefly, because if the volume can change, some of the heat added to the system is used to do work by expansion, leaving less to raise the temperature of the sample.

The distinction between C_V and C_P illuminates the thread that ties thermodynamic parameters to molecular structure, so let's go over this point again, a little more carefully. At constant *volume*, the added energy cannot go into work, so it all goes into increasing the energy of the atoms in the sample, therefore raising the temperature and the pressure of the sample (Fig. 7.4a). If instead we keep the *pressure* constant, then we have to increase the volume of the sample (Fig. 7.4b) so that the increasing force of motion in the sample is distributed over a greater area (which allows the pressure = force/area to remain constant). The increase in volume is accomplished by expansion against the external pressure of the sample. At constant pressure, to get the same change in temperature it takes some energy to raise the kinetic energy of the sample, but also additional energy to expand the sample.

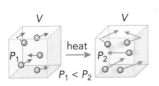

(a) Constant V

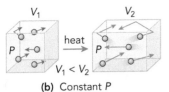

(b) Constant P

▲ FIGURE 7.4 **Heating at constant pressure and at constant volume.** (a) Heating a sample at constant volume only increases the kinetic energy of the atomic motions; therefore, the temperature of the sample rises relatively quickly as energy is added. (b) Heating a sample at constant pressure, on the other hand, requires that some of the energy go into the work done by expansion. As a result, more energy is needed to raise T at constant P than at constant V, so C_P is greater than C_V.

Then how are C_P and C_V related? Since we defined the enthalpy H as (Eq. 7.17)

$$H = E + PV,$$

we can expand our expression for C_P as follows:

$$C_P = \left(\frac{\partial(E + PV)}{\partial T} \right)_P = \left(\frac{\partial E}{\partial T} \right)_P + P \left(\frac{\partial V}{\partial T} \right)_P. \tag{7.50}$$

The first partial derivative, $\left(\frac{\partial E}{\partial T} \right)_P$, can be converted to a term at constant volume using a calculus identity (Table A.4):

$$\left(\frac{\partial X}{\partial Y} \right)_Z = \left(\frac{\partial X}{\partial Y} \right)_W + \left(\frac{\partial X}{\partial W} \right)_Y \left(\frac{\partial W}{\partial Y} \right)_Z.$$

Replacing W, X, Y, Z by V, E, T, P (respectively), our calculus identity becomes

$$\left(\frac{\partial E}{\partial T} \right)_P = \left(\frac{\partial E}{\partial T} \right)_V + \left(\frac{\partial E}{\partial V} \right)_T \left(\frac{\partial V}{\partial T} \right)_P,$$

which we insert into Eq. 7.50 to get

$$C_P = \left(\frac{\partial E}{\partial T} \right)_V + \left(\frac{\partial E}{\partial V} \right)_T \left(\frac{\partial V}{\partial T} \right)_P + P \left(\frac{\partial V}{\partial T} \right)_P. \tag{7.51}$$

The first term, $\left(\frac{\partial E}{\partial T} \right)_V$, is the C_V we've been looking for. The partial derivative $\left(\frac{\partial E}{\partial V} \right)_T$ in the second term is called the **internal pressure** because it resembles the term $-\frac{\partial E}{\partial V}$ defined as the gas pressure in Eq. 2.39.[7] It represents the shift in energy as the volume changes at fixed temperature. This energy shift is determined by the intermolecular forces, and in general the internal pressure is positive in the condensed phase. Compressing a solid or liquid reduces the average distance between the particles, raising the vibrational energies, so $\left(\frac{\partial E}{\partial V} \right)_T > 0$. Typical values range from 5 to 100 bar.

The change in volume with temperature at constant pressure $\left(\frac{\partial V}{\partial T} \right)_P$ is a straightforward physical property. Using the definition for the coefficient of thermal expansion α (Eq. 7.31),

$$\alpha \equiv \frac{1}{V} \left(\frac{\partial V}{\partial T} \right)_{P,n},$$

we have

$$C_P = C_V + V\alpha \left[\left(\frac{\partial E}{\partial V} \right)_T + P \right], \tag{7.52}$$

or, dividing both sides by n to put the values in molar quantities,

$$C_{Pm} = C_{Vm} + V_m \alpha \left[\left(\frac{\partial E}{\partial V} \right)_T + P \right]. \tag{7.53}$$

The heat capacity at constant pressure is normally larger than at constant volume, by an amount proportional to the energy lost in thermal expansion against a pressure P, with a correction for the change in the energy as the system expands.

In the ideal gas limit, the internal pressure is zero because we neglect the intermolecular forces, and the coefficient of thermal expansion becomes

$$\text{ideal gas} : \alpha = \frac{1}{V} \left(\frac{\partial V}{\partial T} \right)_P = \frac{1}{V} \left[\frac{\partial}{\partial T} \left(\frac{nRT}{P} \right) \right]_P = \frac{nR}{PV} = \frac{1}{T}. \tag{7.54}$$

---

[7] In some references, the internal pressure is defined to include the minus sign, so look carefully.

The ideal gas heat capacity at constant pressure is therefore

$$\text{ideal gas} : C_P = C_V + V\alpha P = C_V + nR. \tag{7.55}$$

SAMPLE CALCULATION **Heat Capacity at Constant Pressure of a Monatomic Gas.** The value of C_P for 2.00 mol of neon gas is predicted by Eq. 7.55 to be

$$C_P = C_V + nR = \frac{3}{2}nR + nR = \frac{5}{2}nR$$

$$= \frac{5}{2}(2.00 \text{ mol})(8.3145 \text{ J K}^{-1} \text{ mol}^{-1}) = 41.6 \text{ J K}^{-1}.$$

EXAMPLE 7.2   **Heat Capacity at Maximum Density**

CONTEXT We normally think of hydrogen bonding as attracting adjacent molecules. However, hydrogen bonds are strongest when the atoms in the bond fall roughly in a line, so hydrogen bonding in liquid water (for example) tends to keep the average distance between molecules *greater* than if they were simply packed as tightly as possible. As the temperature of liquid water rises, starting from about 0 °C, there are two competing effects on the density of the water: (1) the added energy starts to break hydrogen bonds, which *decreases* the average distance between molecules, and (2) the added energy allows the hydrogen bonds to stretch more, which *increases* the average distance. This balance between the two effects is one sensitive test of computer simulations of liquid water, because it determines a temperature of maximum density (TMD). The experimental value is 4 °C, whereas simulations of water often predict higher values. At 4 °C, we find that the heat capacities have a simple relationship.

PROBLEM Use the definition of the coefficient of thermal expansion to find the relationship between C_P and C_V at the TMD for any substance.

SOLUTION At the TMD, the density $\rho = m/V$ reaches its maximum value as we change T, and (with m being constant) that means that V reaches a *minimum* value. At a critical point, such as a maximum or minimum, the derivative of a function becomes zero. Therefore, when we vary the temperature of our sample to attain the maximum density (minimum volume) of our sample, the derivative of V with respect to T must go to zero, and that derivative defines the coefficient of thermal expansion:

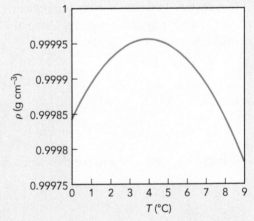

$$\alpha = \frac{1}{V}\left(\frac{\partial V}{\partial T}\right)_P\bigg|_{\text{TMD}} = 0 \qquad \text{because at a minimum}$$

$$C_P = C_V + V\alpha\left[\left(\frac{\partial E}{\partial V}\right)_T + P\right] = C_V.$$

At the TMD, the substance does not expand when heated, so no work is done by expansion and the heat capacities C_P and C_V are equal.

Although the heat capacity of the monatomic ideal gas does not depend on temperature, heat capacities for real substances do. The general relationship between the heat q and the temperature change in the substance is

$$q = \int_{T_1}^{T_2} đq = \int_{T_1}^{T_2} \frac{đq}{dT} dT = \int_{T_1}^{T_2} C(T) dT. \tag{7.56}$$

If we can assume that the heat capacity of a substance does not change with temperature, then we can factor out C, and the relation simplifies to

$$q = \int_{T_1}^{T_2} C dT = C \int_{T_1}^{T_2} dT = C\Delta T. \tag{7.57}$$

Heat capacities do increase with temperature—especially at low temperatures, as we shall see shortly, but near room temperature the temperature dependence is often weak, as shown in Fig. 7.5 for a number of substances. Over temperature ranges of a few tens of degrees, this approximation typically leads to errors of less than 10% and in some cases is much better.

SAMPLE CALCULATION Heat for Temperature-Independent Heat Capacity. If C_{Pm} for neon is $20.786 \text{ J K}^{-1} \text{ mol}^{-1}$, calculate the energy necessary to reversibly heat 2.00 mol of neon gas from 25.0 °C to 35.0 °C at constant pressure. The temperature change is 35.0 °C − 25.0 °C = 10.0 °C or 10.0 K. The temperature *difference* ΔT has the same numerical value whether in centigrade or kelvin, so there is no need to convert temperature units here.[8] The heat is therefore

$$q = nC_{Pm}\Delta T = (2.00 \text{ mol})(20.786 \text{ J K}^{-1} \text{ mol}^{-1})(10.0 \text{ K}) = 415 \text{ J}.$$

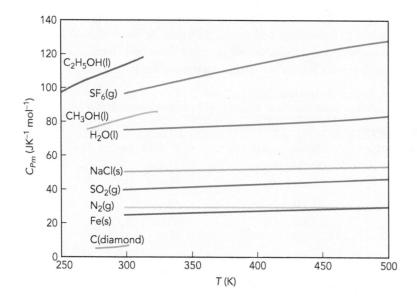

◄ FIGURE 7.5 **Heat capacities for several substances.** These values are measured at a pressure of 1.00 bar.

---

[8]But any time you *multiply or divide* by T, the temperature must be in kelvin. If you *add* temperatures, for example, to calculate an average, then you may use °C or K, but the answer is valid only in the units you used initially. It is only when subtracting temperatures that the result is the same in either set of units.

Heat Capacities for Real Gases

We can now examine the accuracy of the equipartition principle in some detail. In Table 7.2, heat capacities at constant pressure for several real gases are compared to the ideal gas values at 298 K. Let's take a look at some of the trends.

1. *Heat capacities for gas-phase atoms tend to lie very close to the equipartition value* because the principal degrees of freedom are translational, which equipartition handles effectively, and electronic, which is often negligible at temperatures as low as 298 K. The heat capacities of F and Cl atoms, for example, are slightly higher than 20.786 J K^{-1} mol^{-1} (the equipartition value) because some of the energy is used to excite a small fraction of the atoms from the ground $^2P_{3/2}$ electronic state to the excited $^2P_{1/2}$ state. That excitation energy (*QM* Table 4.2) is 404 cm^{-1} (581 K) for F, 882 cm^{-1} (1270 K) for Cl, 3685 cm^{-1} (5302 K) for Br, and 7603 cm^{-1} (10,940 K) for I. At 298 K, the excited states of Br and I are not significantly populated; essentially all the energy is in translation, and the heat capacity is accurately predicted by the equipartition principle.[9]

2. *The more atoms there are in the molecule, the more the equipartition heat capacity overestimates the experimental heat capacity* if vibrations are included (designated "full equipartition" in Table 7.2). Monatomic chlorine gas has a heat capacity within about 5% of that value, whereas Cl_2 and CCl_4 have heat capacities 10% and 23% lower than the full equipartition values, respectively. That's because full equipartition assumes that each vibrational mode absorbs its share of the energy, even if ω_e is too great for there to be efficient vibrational excitation. Treating the vibrational energy classically, as a continuous variable, is a mistake—the quantization of the vibrational energy is too large to ignore. Because each vibration is assumed to take twice as much energy as each translational and rotational degree of freedom, and because each atom added to a molecule brings with it three new vibrational modes,[10] full equipartition is usually a terrible approximation for large molecules.

TABLE 7.2 Molar heat capacities C_{Pm} (J K^{-1} mol^{-1}) of selected gases and corresponding equipartition values (without vibrations, and "full" with vibrations), evaluated at 298.15 K and 1 bar.

monatomics		diatomics		nonlinear 5-atom	
equipartition	20.786	equipartition	29.101	equipartition	33.26
H	20.786	H_2	28.824	CH_4	35.31
N	20.786	N_2	29.088	SiH_4	42.84
F	22.744	F_2	31.302	CF_4	61.09
Cl	21.840	Cl_2	33.907	N_2O_3	65.61
Br	20.786	Br_2	36.048	CCl_4	83.30
I	20.786	I_2	36.887	$SiCl_4$	90.25
full equipartition	20.786	full equipartition	37.415	full equipartition	108.09

[9] The lowest energy excited electronic states for the other atoms in Table 7.2 are at even higher energies, corresponding to over 110,000 K for H ($^2S_{1/2} \rightarrow {}^2S_{1/2}$) and 8300 K for N ($^4S_{3/2} \rightarrow {}^2D_{5/2}$).

[10] There is an exception to the rule that one atom adds three vibrations: an atom added to a linear molecule that makes the molecule nonlinear adds two new vibrational modes and one new rotational coordinate.

3. *The heat capacities for diatomic and polyatomic gases are more accurately predicted by full equipartition when the atoms are heavier.* H_2 has a heat capacity only 77% of the full equipartition value, compared to 98% for I_2. This result is again attributable to the influence of vibrations. As the atoms become heavier, the vibrational constants become lower ($\omega_e = \hbar\sqrt{k/\mu}$), and the heat capacity approaches the estimate based on full equipartition. Vibrations involving H atoms tend to have high vibrational constants, over 3000 cm^{-1}, because the reduced mass μ remains near the atomic mass of hydrogen. On the other hand, I_2 has a vibrational constant of only 214.52 cm^{-1} = 309 K, and at 298 K the I_2 stretching vibration can absorb energy almost as efficiently as the other degrees of freedom.

4. *A lower limit to the heat capacity can therefore be estimated by neglecting all the vibrations* but retaining contributions from rotation and translation— the ordinary equipartition values in Table 7.2. For linear molecules, the lower limit is $7R/2 = 29.10$ J K^{-1} mol^{-1}, and for nonlinear the lower limit is $8R/2 = 33.26$ J K^{-1} mol^{-1}, regardless of how many atoms are in the molecule. These lower limits are fairly successful at predicting the heat capacities of compounds such as H_2 and CH_4 where the vibrations all involve hydrogens and therefore have fairly high excitation energies.

For gas-phase molecules of just a few atoms, the vibrational contribution can dominate the heat capacity. To estimate this contribution more precisely, we need to find how fast energy is added to the vibrational degrees of freedom as the

---

EXAMPLE 7.3 Equipartition and Heat Capacity in Polyatomics

CONTEXT Carbon tetrachloride has found many uses: as a refrigerant, fire retardant, and solvent, to name a few applications. One technique that chemical plants use to produce CCl_4 is to pass Cl_2 gas over heated solid carbon disulfide, which forms the gas. The gas is then condensed and shipped as a liquid. As the gas is cooled before condensation, heat is released and must be carried away. To calculate that heat release, we need the heat capacity of the CCl_4, and this is determined by how much energy is stored in each of its degrees of freedom.

PROBLEM The molar heat capacity at constant pressure of CCl_4 is 83.30 J K^{-1} mol^{-1} at 298 K. If we heated a 1.00 mol sample of CCl_4 gas with 1.00 J, starting near 298 K, roughly what percent of the heat would go into rotation, and what percent into vibration? Assume that the temperature change is small enough that the heat capacity may be treated as constant.

SOLUTION If the heat capacity is constant, then

$$\Delta H = \int_{T_i}^{T_f} C_P \, dT = C_P \, \Delta T = nC_{Pm} \, \Delta T.$$

We expect the rotations and translations in a nonlinear polyatomic molecule to each demand $3nR\Delta T/2$ of the heat and an amount $nR\Delta T$ to be expended as work in the expansion of the gas (Eq. 7.55), with any remaining heat going into partial excitation of the vibrational modes. The fraction of the heat in rotations should therefore be roughly

$$\frac{q(\text{rot})}{q} = \frac{3nR\Delta T/2}{nC_{Pm}\Delta T} = \frac{3R}{2C_{Pm}} = 0.15,$$

and the fraction in vibrational motion will be approximately

$$\frac{q(\text{vib})}{q} = \frac{(C_{Pm} - R)n\Delta T - q(\text{rot}) - q(\text{trans})}{nC_{Pm}\Delta T}$$

$$= \frac{(C_{Pm} - 4R)}{C_{Pm}} = 0.62.$$

So roughly 15% of the heating goes into exciting rotations, and 62% into exciting the vibrations.

temperature increases. In other words, we need $\partial E_{\text{vib}}/\partial T$. To get this, we want a more accurate expression for the vibrational energy E_{vib} than the equipartition value. To find E_{vib} as a function of temperature, we first treat each vibrational mode of the molecule as an independent degree of freedom. We assess how much energy goes into each mode and then take the derivative with respect to T, combining the results for each mode to find the total vibrational contribution to the heat capacity only at the end. The total vibrational energy is

$$E_{\text{vib}} = N\langle \varepsilon_{\text{vib}} \rangle = N\sum_{i=1}^{N_{\text{Vib}}} \langle \varepsilon_i \rangle, \tag{7.58}$$

where $\langle \varepsilon_{\text{vib}} \rangle$ is the average *total* vibrational energy per molecule, $\langle \varepsilon_i \rangle$ is the average energy in vibrational mode i per molecule, and $N_{\text{vib}} = 3N_{\text{atom}} - 5$ for a linear molecule and $3N_{\text{atom}} - 6$ for a nonlinear molecule. The values for $\langle \varepsilon_i \rangle$ we find by using the classical average value theorem and the harmonic approximation to the vibrational partition function (Eq. 3.26):

$$\langle \varepsilon_i \rangle = \sum_{v=0}^{\infty} \mathcal{P}_v(v)\varepsilon_v \tag{7.59}$$

$$= \sum_{v=0}^{\infty} \frac{g_v e^{-\varepsilon_v/(k_B T)}}{q_{\text{vib}}(T)}\varepsilon_v.$$

CHECKPOINT In Eq. 7.59, we briefly revisit the color-coding to distinguish between the contributions of the canonical energy distribution and the partition function to the average vibrational energy. We can see how this corresponds to a classical average, $\bar{x} = \Sigma x/N$, where dividing by the partition function (in blue) is roughly equivalent to dividing by the number of measurements. Recall that the partition function effectively counts the number of populated quantum states, so we are averaging the energy values, weighted by the canonical distribution, and divided by the number of quantum states available to the molecule.

Set $\varepsilon_v = \omega_e v$ and set $g_v = 1$ if the mode is non-degenerate:

$$= \sum_{v=0}^{\infty} \frac{e^{-\omega_e v/(k_B T)}}{q_{\text{vib}}(T)}\omega_e v$$

$$= (1 - e^{-\omega_e/(k_B T)})\sum_{v=0}^{\infty} e^{-\omega_e v/(k_B T)}\omega_e v. \tag{7.60}$$

Now take the temperature derivative at constant volume:

$$\left(\frac{\partial \langle \varepsilon_i \rangle}{\partial T}\right)_V = -\frac{\omega_e}{k_B T^2}e^{-\omega_e/(k_B T)}\sum_{v=0}^{\infty}\left[e^{-\omega_e v/(k_B T)}\omega_e v\right]$$

$$+ (1 - e^{-\omega_e/(k_B T)})\sum_{v=0}^{\infty}\left[\frac{\omega_e v}{k_B T^2}e^{-\omega_e v/(k_B T)}\omega_e v\right]$$

$$= \sum_{v=0}^{\infty}\left[-\frac{\omega_e^2 v}{k_B T^2}e^{-\omega_e/(k_B T)}e^{-\omega_e v/(k_B T)} + \frac{\omega_e^2 v^2}{k_B T^2}e^{-\omega_e v/(k_B T)} - \frac{\omega_e^2 v^2}{k_B T^2}e^{-\omega_e/(k_B T)}e^{-\omega_e v/(k_B T)}\right]$$

$$= \frac{\omega_e^2}{k_B T^2}\sum_{v=0}^{\infty}\left\{e^{-\omega_e v/(k_B T)}\left[v^2 - v(v + 1)e^{-\omega_e/(k_B T)}\right]\right\}. \tag{7.61}$$

This expression yields the effective heat capacity per vibrational mode per molecule. To make this a molar quantity, we multiply by Avogadro's number, and to get the result in units of R (which is convenient for comparing to the other contributions), we divide by R:

$$C_{m,vib}(i) = \frac{\mathcal{N}_A}{R}\left(\frac{\partial\langle\varepsilon_i\rangle}{\partial T}\right)_V R$$

$$= \frac{1}{k_B}\left(\frac{\partial\langle\varepsilon_i\rangle}{\partial T}\right)_V R$$

$$= \left(\frac{\omega_e}{k_B T}\right)^2 \sum_{v=0}^{\infty}\left\{e^{-\omega_e v/(k_B T)}\left[v^2 - v(v+1)e^{-\omega_e/(k_B T)}\right]\right\}R. \quad (7.62)$$

We can evaluate this expression for each vibrational mode i and then add each of these vibrational contributions to the expression for C_{Pm} of the ideal gas, with translations and rotations treated according to equipartition:

$$C_{Pm} = C_{Vm} + R = \left(\frac{3}{2}\,(\text{trans}) + \frac{N_{rot}}{2}\,(\text{rot})\right)R + \sum_{i=1}^{N_{Vib}} C_{m,vib}(i) + R$$

$$= \frac{5 + N_{rot}}{2}R + \sum_{i=1}^{N_{Vib}} C_{m,vib}(i), \quad (7.63)$$

where $N_{rot} = 2$ for a linear molecule and 3 for nonlinear.

For a diatomic molecule (which must be linear and must have a single vibrational mode), the heat capacity can be estimated as

$$C_{Pm} = \frac{7}{2}R + C_{m,vib}(1)$$

$$= \frac{7}{2}R + \left(\frac{\omega_e}{k_B T}\right)^2\sum_{v=0}^{\infty}\left\{e^{-\omega_e v/(k_B T)}\left[v^2 - v(v+1)e^{-\omega_e/(k_B T)}\right]\right\}R$$

$$= \frac{7}{2}R + \left(\frac{\omega_e}{k_B T}\right)^2\left\{e^{-\omega_e/(k_B T)}\left[1 - 2e^{-\omega_e/(k_B T)}\right] + e^{-2\omega_e/(k_B T)}\left[4 - 6e^{-\omega_e/(k_B T)}\right] + \dots\right\}R. \quad (7.64)$$

EXAMPLE 7.4 **Heat Capacity of a Vibrating Diatomic**

CONTEXT Many processes in chemical manufacturing take place at temperatures of a few hundred °C, where the compounds involved may have not been carefully characterized. The heat capacity is a critical parameter for such processes, determining how much energy the reactor must be able to add or remove if the temperature of the reaction is to be controlled, and in that temperature range the heat capacity of a compound may be more than double its room temperature value. Rather than trying to measure the heat capacity directly at high temperature, we often take advantage of infrared spectra, obtained at room temperature and in solvents where the molecules may be easily handled, to find the vibrational constants of the molecule, and use these to calculate the high-temperature heat capacity. Although not exact, these calculations are typically within a few percent of the actual heat capacities. We will test this out on a relatively simple case at low temperature (so that the sum does not have to include many vibrational states).

PROBLEM The vibrational constant of $^{35}Cl_2$ is 560.5 cm^{-1}. Estimate the value of C_{Pm} for $^{35}Cl_2$ at 298.15 K.

SOLUTION It is once again convenient to evaluate ω_e/k_B, which in this case is 806.5 K. Using Eq. 7.64, and keeping only the terms up to $v = 2$, we then predict

$$
\begin{aligned}
C_{Pm} &= \frac{7}{2}R + \left(\frac{\omega_e}{k_B T}\right)^2 \left\{ e^{-\omega_e/(k_B T)}\left[1 - 2e^{-\omega_e/(k_B T)}\right] + e^{-2\omega_e/(k_B T)}\left[4 - 6e^{-\omega_e/(k_B T)}\right] \right\} R \\
&= \frac{7}{2}R + \left(\frac{806.5}{298.15}\right)^2 \left\{ e^{-806.5/298.15}\left[1 - 2e^{-806.5/298.15}\right] + e^{-2\cdot806.5/298.15}\left[4 - 6e^{-806.5/298.15}\right] \right\} R \\
&= (3.500 + 0.579 + 0.161)R = 4.042R = 33.60 \text{ J K}^{-1}\text{ mol}^{-1}.
\end{aligned}
$$

The actual value of C_{Pm} for Cl_2 in Table 7.2 is 33.907 J K^{-1} mol^{-1}.

EXTEND Some of the discrepancy with the tabulated value arises from truncating the sum at $v = 2$, but most of it comes from the contribution of the fairly abundant ^{37}Cl isotope. Nearly half of natural Cl_2 occurs in the heavier $^{35}Cl\,^{37}Cl$, and $^{37}Cl_2$ forms. Because these have slightly lower vibrational constants, they can absorb heat more easily into vibrational excitation than $^{35}Cl_2$, raising the heat capacity.

When we look at the condensed phases, the idea remains the same, but the math steps up a level because we don't have so simple an equation for the energy of the sample. Crystals are not as simple as gases, but they are much simpler than liquids, so we look at them next.

Heat Capacities of Crystals

To find a useful equation for the heat capacity of crystals, we again need an expression for the energy content, this time of a solid, in terms of its temperature T. The equipartition principle gave us an appropriate energy equation for gases, and a similar derivation is useful here. As before, the total energy is just the sum of the average energy per particle over all the particles:

$$E = N\langle\varepsilon\rangle, \tag{7.65}$$

so we first want $\langle\varepsilon\rangle$ as a function of temperature. But what kind of energy does ε represent?

We can answer this question more easily for solids than for liquids because each atom in a solid stays near its equilibrium position instead of wandering all over as it would in a liquid. Crystals, in particular, are convenient because the ordered structure lets us generalize the energy expression for one lattice point in the crystal to all the others. The internal energy in solids is *primarily a function of the vibrational degrees of freedom*, because translation and rotation are both quenched by the strong binding forces (in other words, the particles are locked into place and can only shift and twist back and forth about this position).

Albert Einstein applied Eq. 6.19 to derive the heat capacity for an idealized crystal, called the **Einstein solid.** Einstein's model for the heat capacity of a crystal therefore treats each particle as an independent harmonic oscillator (Fig. 7.6), drawing on the approximation of vibrating chemical bonds as harmonic oscillators (Section 2.1).

Einstein postulated that he could find a value for ε by assuming some single effective vibrational constant ω_E (the **Einstein frequency**) for any given solid. The Einstein frequency is determined empirically by fitting the value to

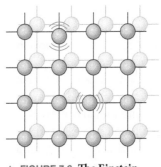

▲ **FIGURE 7.6 The Einstein solid.** This idealized crystal is a three dimensional lattice of N independent harmonic oscillators.

experimental data on temperature-dependent heat capacities. We can write out a partition function for vibrations (Eq. 3.26) in terms of ω_E,

$$q_{\text{vib}}(T,V) = \frac{1}{1 - e^{-\omega_E/(k_B T)}}. \tag{7.66}$$

Here we have written q_{vib} as a function of the volume (as well as temperature) because the volume determines the average spacing R between the particles, which (we will assume) influences the effective vibrational frequency ω_E. As long as we evaluate the heat capacity at constant volume, we may treat ω_E as a constant.

TOOLS OF THE TRADE | Diamond Anvil Cell (DAC)

In the mid-20th century, the quest to study matter under extreme conditions was led not by chemists but by geologists. The temperatures and pressures at which thermodynamic data had been gathered in chemistry laboratories rarely approached the scales that were relevant to studies of the earth's interior. Percy Bridgman at Harvard developed a method of generating high pressures between two flat steel faces (the *anvils*) pushed together by means of a lever or screw. At about the same time, researchers at the University of Chicago realized that the enormous strength of diamond made it an ideal material for containing high pressure samples. Furthermore, diamond is largely transparent to x-rays, and therefore allows crystal structures to be analyzed at high pressure by x-ray diffraction. At the then National Bureau of Standards (NBS, now NIST—the National Institute of Standards and Technology), staff scientist (and former Bridgman student) Charles Weir combined these two approaches and built the first diamond anvil cell in 1958. His colleague, Alvin Van Valkenburg, discovered that the cell was completely transparent when viewed *through* the diamonds, parallel to the axis of compression, thereby allowing a host of analytical measurements to be made at higher pressures than ever before.

What is a DAC? A DAC is a device for generating pressures beyond $5 \cdot 10^{11}$ Pa (5 Mbar) over tiny volumes by forcing together the flattened tips of two diamonds. A sample caught in the volume between the two faces can then be probed by spectroscopy (IR, UV/vis, Raman), microscopy, x-ray diffraction, and other techniques. Diamond also remains structurally sound at both low and high temperatures (from 0.01 K up to the melting point of the diamond), permitting the sample to be studied over an extraordinary range of conditions.

Why do we use DACs? A DAC is capable of replicating the pressures found below the earth's crust, beneath mountains and glaciers, and deep within the other planets of our solar system. The DAC has particular relevance to chemical thermodynamics because many of the properties that interest us as chemists (such as melting points or equilibrium constants in solution) often are not very sensitive to pressure. With a DAC, we can achieve pressures high enough to precisely measure the effect on parameters such as heat capacity and compressibility. In addition to characterizing the specific sample, these measurements provide an important check on our understanding of the laws of thermodynamics. For example, the DAC has allowed water to be pressurized to such an extent that it retains the density of the liquid even at temperatures at which its blackbody radiation makes it glow white-hot. At lower temperatures, the DAC has been used to explore the crystalline structure of water, finding ten different stable crystal structures at pressures over 1000 bar (see Fig. 10.10).

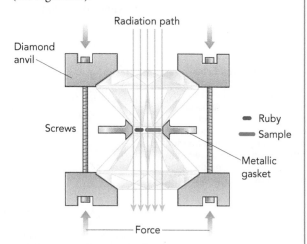

▲ **Schematic of an anvil cell.**

How does it work? The key concept at work in a DAC is that a large pressure P can be generated by expressing a force F over a very small area A, because $P = F/A$.

The design of a typical diamond anvil cell setup is sketched in the figure, illustrating the following components:

1. *Diamond anvils.* Two diamonds with flattened bottom faces, called *culets,* are placed so that the culets can be pressed together, making contact over an area of less than 1 mm². The diamonds need to be of high quality, because structural imperfections can cause the anvil to fracture at high pressure. (The cost of the diamonds was not an issue for the group at NBS, who were entitled to use a supply of contraband diamonds that had been confiscated by the U.S. Customs Office.)
2. *Vise.* The anvils are held in a compressible device, which is adjusted to push the two anvils together with the desired force. Because the force applied to the vise is extended over a much larger area than the culets, the vise can be made from steel, which (strong as it is) lacks the strength of diamond.
3. *Gasket.* The culets do not touch each other, but are kept separated by a gasket, usually made of metal, although other materials (such as boron nitride or an epoxy-diamond compound) may also be used if, for example, an electrically insulating material is required. The gasket also contains the sample (although in the original experiments the compressed sample itself would solidify around the edges, providing an ad hoc gasket).
4. *Pressure sensor.* The operating pressure of the cell is too great and the sample volume too small for mechanical pressure measurement. Instead, a small chip of ruby in the sample cavity typically serves that role. A laser beam excites the ruby, which then fluoresces at a wavelength that shifts with pressure, according to a curve that was calibrated against the compression of NaCl crystals.

The DAC has proven to be a remarkably versatile sample support, extending the range of thermodynamic and other studies to pressures six orders of magnitude beyond typical laboratory conditions, while making the sample available to many of the same analytical techniques we would use in the lab.

The Einstein Heat Capacity

The function we want, ε, is the average energy per particle per vibrational coordinate. Assume each particle is a single atom to simplify the picture. Because the solid is a three-dimensional lattice, there are three vibrational coordinates for each atom. Another way to think of this is that if the solid is treated as one very large molecule, it has $3N_{atom} - 6$ vibrational modes (Section 2.1), where N_{atom} is so large now that $3N_{atom} - 6 \approx 3N_{atom}$. *If we assume that each atom vibrates as an independent harmonic oscillator,* the total energy per mole (when $N_{atom} = \mathcal{N}_A$) is therefore

$$E_{m} = 3\mathcal{N}_A \langle \varepsilon \rangle_{vib} = \frac{3\mathcal{N}_A \omega_E}{e^{\omega_E/(k_B T)} - 1}, \tag{7.67}$$

where the average vibrational energy is given by Eq. 6.19. Finally we get the molar heat capacity[11] from the derivative with respect to temperature:

$$
\begin{aligned}
C_{Vm} &= \left(\frac{\partial E_m}{\partial T} \right)_V \\
&= 3\mathcal{N}_A \omega_E (-1)(e^{\omega_E/(k_B T)} - 1)^{-2} e^{\omega_E/(k_B T)} \left(-\frac{\omega_E}{k_B T^2} \right) \\
&= \frac{3\mathcal{N}_A \omega_E^2 \, e^{\omega_E/(k_B T)}}{k_B T^2 (e^{\omega_E/(k_B T)} - 1)^2}.
\end{aligned}
\tag{7.68}
$$

CHECKPOINT The Einstein heat capacity is remarkably successful, given the simplicity of its assumptions. Although we will extend this to a more accurate expression, the Einstein heat capacity accounts for most of the discrepancy between the real heat capacity and the value predicted by the equipartition principle.

---

[11]But is this really an equation for C_V only, or can we fit it to C_P data as well? The Einstein heat capacity is calculated assuming an ideal crystal made up of independent harmonic oscillators. One of the characteristics of a single harmonically oscillating bond is that the *average* bond distance stays the same, no matter how much energy is put into the motion. Therefore, the Einstein solid *does not expand upon heating,* so the heat capacities at constant pressure and at constant volume are the same. Nevertheless, the Einstein heat capacity is often described as an estimate of C_V, because the model is based on only the internal energy absorbed, with no explicit treatment of work that might be done by expansion. Paradoxically, data fitted to the Einstein heat capacity are usually C_P values, because C_V is rarely measured for crystals. The bottom line is that the Einstein model is simply too primitive to distinguish between C_V and C_P.

At high temperature, this fit is usually successful, but a flaw in the energy expression causes significant discrepancies at low temperature. Equation 7.67 oversimplifies the vibrational motions because it neglects long-range interactions between the molecules, treating all the motions as high-frequency diatomic stretches. When the solid is cooled below the point that those stretches can be excited, the energy and heat capacity become dominated by low-frequency, collective motions of the atoms called **phonon modes.**

The Debye Heat Capacity

The **Debye heat capacity** improves on Einstein's model by taking into account the collective motions of the particles in a solid, the phonon modes (Fig. 7.7). Debye assumed first that the low frequency vibrations could be treated as a set of motions in three dimensions, characterized by a *range* of vibrational constants, rather than a single vibrational constant (as in Einstein's approximation). Debye argued that the density of quantum states W for these oscillations is proportional to the volume V and to the square of the vibrational constant ω, using the same logic that we apply to the states of the three-dimensional box in Section 3.4 and the radiating blackbody in Section 6.2:

$$W(\omega)d\omega = AV\omega^2 \, d\omega,$$

where A is a normalization constant for this distribution of phonon modes. But this time, the vibrational modes are restricted to vibrational constants less than or equal to an upper limit, the **Debye frequency.** To find A, we require that the total number of vibrational modes is equal to three times the number of atoms that can vibrate:

$$3N = \int_0^{\omega_D} W(\omega) \, d\omega = AV \int_0^{\omega_D} \omega^2 \, d\omega = AV\frac{\omega_D^3}{3}$$

$$A = \frac{9N}{\omega_D^3 V}.$$

This revises Eq. 6.19 for the average vibrational energy per particle by introducing a distribution function for the vibrational constants. We apply that distribution function, leaving it in terms of an integral over the vibrational constant:

$$E_m = \int_0^{\omega_D} \frac{\omega W(\omega)d\omega}{e^{\omega/(k_B T)} - 1} \qquad \text{from Eq. 6.19}$$

$$= \frac{9\mathcal{N}_A}{\omega_D^3} \int_0^{\omega_D} \frac{\omega^3 d\omega}{e^{\omega/(k_B T)} - 1} \qquad W(\omega) = 9N\omega^2 d\omega/\omega_D^3$$

$$= \frac{9\mathcal{N}_A(k_B T)^4}{\omega_D^3} \int_0^{\omega_D/(k_B T)} \frac{\left(\dfrac{\omega}{k_B T}\right)^3 d\left(\dfrac{\omega}{k_B T}\right)}{e^{\omega/(k_B T)} - 1} \qquad \text{factor out } k_B T$$

$$= \frac{9\mathcal{N}_A(k_B T)^4}{\omega_D^3} \int_0^{\omega_D/(k_B T)} \frac{x^3 dx}{e^x - 1}. \qquad \text{set } x = \omega/(k_B T) \quad (7.69)$$

Again we take the derivative with respect to T to find the heat capacity:

$$C_{Vm} = \frac{9\mathcal{N}_A k_B^4 T^3}{\omega_D^3} \int_0^{\omega_D/(k_B T)} \frac{x^4 e^x dx}{(e^x - 1)^2}, \qquad (7.70)$$

where the characteristic frequency ω_D is again an empirical parameter of the solid.

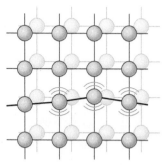

▲ FIGURE 7.7 **Phonon modes in crystals.** The Debye model of the solid allows for low-frequency vibrations that couple the simultaneous motions of many atoms.

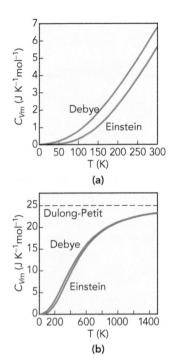

▲ FIGURE 7.8 **The Einstein and Debye heat capacities for diamond.** The curves shown are calculated using $\omega_E = 967\,\text{cm}^{-1}$ and $\omega_D = 1300\,\text{cm}^{-1}$. The Einstein heat capacity converges much too quickly to zero at low temperature **(a)**, differing from the experimental and Debye heat capacities by a factor of 3 at 100 K. However, at high temperature **(b)**, the two curves converge toward the equipartition value of $3R = 24.9\,\text{J K}^{-1}\,\text{mol}^{-1}$, consistent with the law of Dulong and Petit.

The integrals in these equations do not have analytical solutions, and they must be evaluated numerically. However, in the low-temperature limit, ω_D is much greater than the thermal energy $k_B T$, so the upper limit of the energy integral can be replaced by ∞. The solution is then

$$\int_0^\infty \frac{x^3 dx}{e^x - 1} = \frac{\pi^4}{15}. \tag{7.71}$$

Substituting this into the expression for the energy (Eq. 7.69) gives

$$E_m \approx \frac{3\pi^4 \mathcal{N}_A k_B^4 T^4}{5\omega_D^3}, \tag{7.72}$$

and taking the derivative with respect to temperature then yields an expression for the heat capacity in the low-temperature limit:

$$C_{Vm} = \left(\frac{\partial E_m}{\partial T}\right)_V \approx \frac{12\pi^4 \mathcal{N}_A k_B^4 T^3}{5\omega_D^3}. \tag{7.73}$$

The Einstein and Debye heat capacities of diamond are graphed in Fig. 7.8 as functions of T. Although the two functions appear similar from a distance, the Debye heat capacity is higher in the low temperature limit because it includes the influence of the low-energy phonon modes. In the high temperature limit, the equipartition principle predicts that vibration along the three Cartesian coordinates results in a heat capacity of $3R$. This is the **Dulong-Petit heat capacity,** and both the Einstein and Debye heat capacity curves asymptotically approach that value at high temperature (Problem 7.59).

Specific Heat Capacities

The low densities of gases make it difficult to measure their masses by weight, so heat capacities of gases are often given as *molar* heat capacities at constant pressure (since the ideal gas law directly relates the number of moles of any dilute gas to P, V, and T). For liquids and solids, however, it is often easiest to measure the weight of the sample in the lab or in the field, so heat capacities may be given per unit mass instead of per mole. These values are called **specific heat capacities,** and they are symbolized by a lowercase c. Specific heat capacities also apply to materials that do not have a well-defined molecular formula (and therefore no clear molar mass), as in many natural mixtures (such as seawater or soils) or complex structures (such as various woods or rocks). When we do know the molecular formula, we can convert between C_{Pm} and c using the molar mass.

SAMPLE CALCULATION Specific Heat Capacity from Molar Heat Capacity. The molar heat capacity C_{Pm} of calcium chloride solid is $72.6\,\text{J K}^{-1}\,\text{mol}^{-1}$ To calculate the specific heat capacity (at constant pressure) of calcium chloride, we convert moles to grams by *dividing* by the molar mass:

$$c_P = \frac{C_{Pm}}{\mathcal{M}} = \frac{72.6\,\text{J K}^{-1}\,\text{mol}^{-1}}{110.98\,\text{g/mol}} = 0.654\,\text{J K}^{-1}\,\text{g}^{-1}.$$

Table 7.3 gives the heat capacities of several solids. The pure elements in the table are arranged to show the general trend toward greater heat capacity with greater mass of the atoms.

TABLE 7.3 **Heat capacities of selected solids.**

solid	$\mathcal{M}$	C_{Pm}	c_P	solid	c_P
	g mol^{-1}	$\text{J K}^{-1}\text{mol}^{-1}$	$\text{J K}^{-1}\text{g}^{-1}$		$\text{J K}^{-1}\text{g}^{-1}$
lithium	6.94	24.6	3.55	hematite	0.67
beryllium	9.01	16.4	1.71	Pyrex glass	0.84
carbon (graphite)	12.01	8.23	0.69	gypsum	1.09
carbon (diamond)	12.01	6.12	0.51	loose wool	1.26
sodium	22.99	28.2	1.22	paper	1.34
silicon	28.09	20.0	0.71	white pine	2.5
sulfur	32.07	22.7	0.71	paraffin wax	2.9
arsenic	74.92	24.6	0.33		
tantalum	180.9	25.3	0.14		
uranium	238.0	27.7	0.12		

Heat Capacities of Monatomic Liquids

Liquids, having strongly interacting particles but no long-range order, are not described well by simple equations. A common strategy is to express the property we're interested in as an unsolved integral involving the pair correlation function. That's because the very existence of the liquid phase is a delicate balance between the kinetic energy of the molecules and the intermolecular potential energy. In contrast, we can approximate gases as having motion without attractions or repulsions (kinetic, but no potential energy), and solids having a nearly rigid structure with little motion (structure dominated by the potential energy). As a result, we can only predict the properties of the liquid by taking both the motions *and* the structure into account simultaneously. The pair correlation function $\mathcal{G}(R)$ ties these two together.

For the heat capacity, we will initially follow this approach, using $\mathcal{G}(R)$ to describe the liquid structure, and simplifying by considering only the monatomic liquid. For the monatomic liquid, the intermolecular potential energy depends only on R, and we can ignore internal rotational or vibrational degrees of freedom. A **DERIVATION SUMMARY** appears after Eq. 7.76.

We evaluate the heat capacity by taking the derivative of the energy. Therefore, we need an equation for the energy of the liquid. Although we haven't seen a formula for that energy yet, all of the information about the energy is couched in the translational partition function that we used for the non-ideal gas.[12] But now, the intermolecular potential energy function, described by the configuration integral $Q'_U(T,V)$ of Eq. 4.18, cannot be assumed to be small.

---

[12]Here we have taken for granted that we should treat the liquid as a pathological version of the gas. In fact, scientists have studied effective ways to treat the theory of liquids for decades, and one continuing debate is whether it is more rewarding to approach the liquid from the gas-phase limit or from the solid. One approach starts from the solid and models the liquid by effectively decreasing the time it takes for the structure to fracture and slip, until the structure is changing constantly. We will use the present approach only because the starting point (a gas) is simpler, and does successfully predict the major qualitative distinction between heat capacities of liquids and the other phases.

Equation 4.26 connects the ensemble size with the energy:

$$\ln\Omega = \ln\left[\frac{1}{N!}\left(\frac{2\pi mk_BT}{h^2}\right)^{3N/2}\right] + \ln Q'_U(T,V) + \frac{E}{k_BT}.$$

We begin with Eq. 4.26, but this time set $1/(k_BT)$ equal to β:

$$\ln\Omega = \ln\left[\frac{1}{N!}\left(\frac{2\pi m}{h^2\beta}\right)^{3N/2}\right] + \ln Q'_U(\beta,V) + \beta E,$$

and, after a deep breath, take the derivative of both sides with respect to β to isolate and solve for E:

$$0 = \frac{\partial}{\partial\beta}\ln\left[\frac{1}{N!}\left(\frac{2\pi m}{h^2\beta}\right)^{3N/2}\right] + \frac{\partial\ln Q'_U(\beta,V)}{\partial\beta} + E \quad \text{differentiate Eq. 4.26}$$

$$E = -\frac{\partial}{\partial\beta}\ln\left[\frac{1}{N!}\left(\frac{2\pi m}{h^2\beta}\right)^{3N/2}\right] - \frac{\partial\ln Q'_U(\beta,V)}{\partial\beta} \quad \text{solve for } E$$

$$= \frac{\partial}{\partial\beta}\left[\ln\left(\frac{1}{N!}\right) - \frac{3N}{2}\ln\left(\frac{2\pi m}{h^2\beta}\right)\right] - \frac{1}{Q'_U(\beta,V)}\frac{\partial Q'_U(\beta,V)}{\partial\beta}$$

$$\ln(ab) = \ln a + \ln b, \, d\ln x = dx/x$$

$$= -\frac{3N}{2}\frac{\partial}{\partial\beta}\ln\left[\left(\frac{2\pi m}{h^2\beta}\right)\right]$$

$$- \frac{1}{Q'_U(\beta,V)}\frac{\partial}{\partial\beta}\int_0^a\cdots\int_0^c e^{-\beta U}dX_1\ldots dZ_N \quad \text{by Eq. 4.18}$$

$$= -\frac{3N}{2}\left[\left(\frac{2\pi m}{h^2\beta}\right)\right]^{-1}\frac{\partial}{\partial\beta}\left[\left(\frac{2\pi m}{h^2\beta}\right)\right]$$

$$+ \frac{1}{Q'_U(\beta,V)}\int_0^a\cdots\int_0^c e^{-\beta U}UdX_1\ldots dZ_N$$

$$= -\frac{3N}{2}\left[\left(\frac{2\pi m}{h^2\beta}\right)\right]^{-1}\left[\left(\frac{2\pi m}{h^2}\right)\right](-\beta^{-2})$$

$$+ \frac{N(N-1)}{2Q'_U(\beta,V)}\int_0^a\cdots\int_0^c e^{-\beta U}u(R)dX_1\ldots dZ_N \quad \text{by Eq. 4.27}$$

$$= \frac{3N}{2\beta} + \frac{N(N-1)}{2Q'_U(\beta,V)}$$

$$\times \int_0^a\cdots\int_0^c u(R)\left[\int_0^a\cdots\int_0^c e^{-\beta U}dX_3\ldots dZ_N\right]dX_1\ldots dZ_2$$

$$= \frac{3N}{2\beta} + \frac{N^2}{2V^2}\int_0^a\cdots\int_0^c u(R)\mathcal{G}(R)dX_1\ldots dZ_2 \quad \text{by Eq. 4.57}$$

$$= \frac{3N}{2\beta} + \frac{N^2}{2V}\int_0^\infty u(R)\mathcal{G}(R)(4\pi R^2 dR) \quad \text{switch to spherical coords}$$

$$= \frac{3Nk_BT}{2} + \frac{2\pi N^2}{V}\int_0^\infty u(R)\mathcal{G}(R)R^2 dR. \quad \beta = k_BT \quad (7.74)$$

The first term in Eq. 7.74 is the kinetic energy for translation, as predicted for gases by the equipartition principle. The second term, which depends on the pair correlation function, is the translational potential energy arising from the intermolecular forces.

We obtain the heat capacity C_V by taking the derivative of Eq. 7.74 with respect to temperature at constant volume:

$$C_V = \left\{ \frac{\partial}{\partial T} \left[\frac{3Nk_BT}{2} + \frac{2\pi N^2}{V} \int_0^\infty u(R)\mathcal{G}(R)R^2 dR \right] \right\}_V$$

$$= \frac{3Nk_B}{2} + \frac{2\pi N^2}{V} \int_0^\infty u(R) \left(\frac{\partial \mathcal{G}(R)}{\partial T} \right)_V R^2 dR$$

$$= \frac{3Nk_B}{2} + 2\pi N\rho \int_0^\infty u(R) \left(\frac{\partial \mathcal{G}(R)}{\partial T} \right)_V R^2 dR. \qquad (7.75)$$

Tabulated heat capacities are usually evaluated per mole of the substance. The molar heat capacity at constant volume C_{Vm} of our liquid is found by setting N equal to Avogadro's number $\mathcal{N}_A$ and recalling that $\mathcal{N}_A k_B = R$:

$$C_{Vm} = \frac{3R}{2} + 2\pi \mathcal{N}_A^2 \left(\frac{\rho_m}{\mathcal{M}} \right) \int_0^\infty u(R) \left(\frac{\partial \mathcal{G}(R)}{\partial T} \right)_V R^2 dR, \qquad (7.76)$$

where ρ_m is the mass density and $\mathcal{M}$ the molar mass. One factor of $\mathcal{N}_A$ converts the pair potential energy $u(R)$ to a molar energy, the other partly cancels the units of the molar mass.

DERIVATION SUMMARY **The Heat Capacity of a Monatomic Liquid.** We started from a general relationship between the potential energy partition function (actually, the configuration integral Q'_U) and the total energy E, taken from statistical mechanics. This allows us to find an approximate expression for E in terms of the intermolecular potential (which gives the molecule-by-molecule interaction) and the pair correlation function (which tells us how many interactions we need to count). From there we go back to more classical thermodynamics, deriving C_V as the change in energy with temperature, $\partial E/\partial T$. Because we introduced intermolecular interactions as a small correction to non-interacting molecules, the expression we get for C_V leads off with the ideal gas heat capacity, and the intermolecular potential comes in as an integral over the potential energy and the temperature-dependence of the pair correlation function.

The first term in Eq. 7.76, $3R/2$, is the heat capacity of the monatomic ideal gas, and it arises solely from the kinetic contribution to the energy. The second term, containing the potential energy dependence, tends to be larger and can be qualitatively explained as arising from the heat absorbed as intermolecular interactions weaken, leading to more random motion (and loss of order) in the liquid, as sketched for a triatomic liquid in Fig. 7.9. The approximations we have used overestimate the heat capacity, particularly at low temperature. However, Eq. 7.76 successfully predicts that the heat capacity of the liquid is greater than that of the gas, and by roughly the correct magnitude at typical temperatures.

As a crude demonstration, let's first assume that the liquid occupies that middle region of the intermolecular kinetic energy scale where $k_BT \approx \varepsilon$, a reasonable assumption as it turns out. Next, let's play fast and loose with some of our available approximations for the pair correlation function and pair potential energy to see what emerges:

$$C_{Vm} = \frac{3R}{2} + 2\pi \mathcal{N}_A^2 \left(\frac{\rho_m}{\mathcal{M}} \right) \int_0^\infty u(R) \left(\frac{\partial \mathcal{G}(R)}{\partial T} \right)_V R^2 dR \qquad \text{by Eq. 7.76}$$

CHECKPOINT In Eq. 7.76, the R in the first term is the universal gas constant, but in the second term each value of R is the distance between the particles.

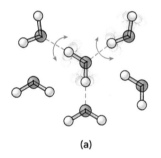

(a)

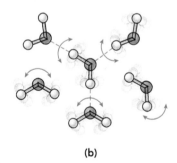

(b)

▲ **FIGURE 7.9 Heating in liquids.** As the temperature of a liquid rises, the structure of the liquid tends to decrease because the increasing motion gradually randomizes the distribution of the molecules. Effectively, the increased energy begins to break intermolecular bonds. This causes changes to the pair correlation function $\mathcal{G}(R)$ with temperature, $\partial \mathcal{G}(R)/\partial T$, which appears as the non-ideal correction in Eq. 7.76.

$$\approx \frac{3R}{2} + 2\pi\mathcal{N}_A^2 \left(\frac{\rho_m}{\mathcal{M}}\right) \int_0^\infty u(R) \left(\frac{\partial e^{-u(R)/(k_B T)}}{\partial T}\right)_V R^2 dR \qquad \text{by Eq. 4.59}$$

$$= \frac{3R}{2} + 2\pi\mathcal{N}_A^2 \left(\frac{\rho_m}{\mathcal{M}}\right) \int_0^\infty u(R) \, e^{-u(R)/(k_B T)} \left(\frac{u(R)}{k_B T^2}\right) R^2 dR$$

$$\approx \frac{3R}{2} + 2\pi\mathcal{N}_A^2 \left(\frac{\rho_m}{\mathcal{M}}\right) \int_0^\infty \left(\frac{u(R)^2}{k_B T^2}\right) R^2 dR \qquad \text{let } e^{-u(R)/(k_B T)} \approx 1$$

$$\approx \frac{3R}{2} + \left(\frac{2\pi\mathcal{N}_A^2}{k_B T^2}\right) \left(\frac{\rho_m}{\mathcal{M}}\right) \int_{R_{LJ}}^\infty \left\{ 4\varepsilon \left[\left(\frac{R_{LJ}}{R}\right)^{12} - \left(\frac{R_{LJ}}{R}\right)^6\right]\right\}^2 R^2 dR \quad \text{by Eq. 4.12}$$

$$= \frac{3R}{2} + \left(\frac{2\pi\mathcal{N}_A^2}{k_B T^2}\right) \left(\frac{\rho_m}{\mathcal{M}}\right) \left\{\frac{304\varepsilon^2 R_{LJ}^3}{315}\right\}$$

$$\approx \frac{3R}{2} + (2\pi\mathcal{N}_A^2 k_B R_{LJ}^3) \left(\frac{\rho_m}{\mathcal{M}}\right) \qquad \text{let } \frac{\varepsilon}{k_B T} \text{ and } \frac{304}{315} \approx 1$$

$$\approx \frac{3R}{2} + (\mathcal{N}_A k_B)(3b) \left(\frac{\rho_m}{\mathcal{M}}\right) \qquad \text{by Eq. 4.44}$$

$$= \frac{3R}{2} + \left(\frac{3Rb\rho_m}{\mathcal{M}}\right) = 3R\left(\frac{1}{2} + \frac{b\rho_m}{\mathcal{M}}\right). \qquad \text{by Eq. 3.21} \quad (7.77)$$

SAMPLE CALCULATION   **Heat Capacity of Liquid Argon.** For liquid argon near its boiling point of 87 K, $b = 0.03201 \, \text{L mol}^{-1}$, $\rho_m = 1.4 \, \text{g cm}^{-3}$, and $\mathcal{M} = 39.948 \, \text{g mol}^{-1}$. Estimate the molar heat capacity at constant volume.

$$C_{Vm} \approx 3R\left(\frac{1}{2} + \frac{b\rho_m}{\mathcal{M}}\right)$$

$$= 3R\left(\frac{1}{2} + \frac{(0.03201 \, \text{L mol}^{-1})(1.4 \, \text{g cm}^{-3})(10^3 \, \text{cm}^3/\text{L})}{39.948 \, \text{g mol}^{-1}}\right)$$

$$= 3(8.3145 \, \text{J K}^{-1} \, \text{mol}^{-1})(1.622) = 40. \, \text{J K}^{-1} \, \text{mol}^{-1}.$$

For liquids of the noble gases and simple molecules, such as those listed in Table 7.4, Eq. 7.77 predicts C_{Vm} values of about 41 J K^{-1} mol^{-1}, equal to about $5R$. The polyatomics have higher heat capacities because they have additional vibrational/rotational motions available to store the energy. (The intramolecular vibrations in N_2 and CH_4 are not appreciably excited at the temperatures in Table 7.4.)

There's a surprise lurking in that value of $5R$. Start with a nearly ideal monatomic gas, such as argon, with a heat capacity of about $3R/2$. Compress it to the density of the liquid. If there were no intermolecular interactions, the heat

TABLE 7.4 Molar heat capacities of simple liquids and gases near their boiling points.

substance	T_b (K)	C_{Vm}(liq) (J K^{-1} mol^{-1})	C_{Vm}(gas) (J K^{-1} mol^{-1})	C_{Vm}(liq) $-$ C_{Vm}(gas) (J K^{-1} mol^{-1})
N_2	77.4	54.5	29.3	25.2
Ar	87.3	40.6	14.5	26.1
CH_4	111.7	51.6	27.3	24.3
Xe	165.1	37.3	12.4	24.9

capacity would stay $3R/2$. If we then turn up the intermolecular vibrations in the liquid so that they became perfect harmonic oscillators, that would double the number of equipartition degrees of freedom (because we add the vibrational potential energy). We might therefore expect the heat capacity of the liquid to approach a maximum value twice the ideal gas value, or $3R$. But we're predicting that the heat capacity of the liquid is *more* than $3R$.

For each of the molecules in Table 7.4, the heat capacity decreases from liquid to gas, because the potential energy of the intermolecular vibrations no longer contributes substantially to the heat capacity. On closer examination, we find that at the boiling point, the differences between $C_{Vm}(\text{liq})$ and $C_{Vm}(\text{gas})$ for each of these liquids cluster around $3R = 24.9\,\text{J}\,\text{K}^{-1}\,\text{mol}^{-1}$. In other words, on releasing the particles from the liquid into the gas phase, we appear to lose both the potential *and* kinetic energy contributions of the harmonic oscillator. The heat capacity of the liquid behaves as though there is *one* set of coordinates for translation, allowing the particles to move around within the liquid, and a *separate* set of coordinates for the intramolecular vibrations.

This occurs because the particles in a liquid specifically populate the flat regions of the intermolecular potential energy curve, where the degeneracy of states rapidly climbs. What may seem odd is that the liquid should be able to absorb more energy for a given change in T than the gas, which is unbounded and has a much greater density of energy levels. That's because heat capacity does not depend on how many energy levels there *are*; it depends on how fast the number of those energy levels *changes* as we go up in energy. As we approach the dissociation limit of the intermolecular bond, the number of available energy levels shoots through the roof, so it takes much more energy to raise the temperature; C_V becomes *greater* than the equipartition value. As shown schematically in Fig. 7.10, the liquid, unlike the gas and the solid, sees the greatest change in the shape of the potential energy function as the energy rises, as purely vibrational degrees of freedom give way to translational motions. The liquid's unique balance of freedom of motion and constraining potential energy wells gives it a higher heat capacity than either the gas or the solid.

Now that we have these basic tools and relations of thermodynamics, we will spend the next few chapters applying them to chemical systems.

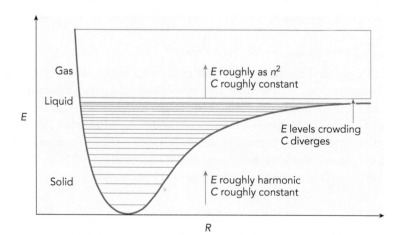

◀ FIGURE 7.10 **Energy level populations in different phases of matter.** In this schematic representation of the populations relative to the intermolecular potential energy curves, the liquid is defined by its density to stay near the top of the potential well, where the molecules are only loosely bound together. In this region, $(\partial E/\partial T)$ becomes large.

EXAMPLE 7.5   **Heat Capacity of Water**

CONTEXT Heat capacities play a crucial role in the climate of coastal lands. If we treat the air as transparent to sunlight, then the temperature of the air is determined by heating from the surface of the land or water beneath it. But the land and the sea have vastly different heat capacities, and they warm and cool at different rates. As a result, when the sun rises or sets, temperature differences rapidly develop between the marine and land air, with the warm air expanding and rising. This generates breezes that circulate the air in coastal areas and keep overall temperature changes relatively mild.

PROBLEM The specific heat capacities of rock, sandy soils, and clays range between roughly $0.75 \text{ J g}^{-1} \text{ K}^{-1}$ and $0.90 \text{ J g}^{-1} \text{ K}^{-1}$, and the densities are roughly $1.0 - 1.8 \text{ g cm}^{-3}$. When sunlight falls on soil, the heat generated penetrates to a depth of roughly 10. cm, whereas the same sunlight warms seawater to a depth of about 10 m. Calculate the heat capacity *per square centimeter* of surface for the land and for the water that are warmed by the sun. Estimate the specific heat capacity and density of seawater to be the same as pure water ($c = 4.184 \text{ J g}^{-1} \text{ K}^{-1}$ and $\rho_{m,\text{water}} = 1.00 \text{ g cm}^{-3}$), and for soil use $c_{\text{soil}} = 0.80 \text{ J g}^{-1} \text{ K}^{-1}$ and a density $\rho_{m,\text{soil}} = 1.5 \text{ g cm}^{-3}$.

SOLUTION We multiply the specific heat capacities by the density to get a heat capacity per unit volume, and then multiply this by the depth to get the heat capacity per unit area. Unit analysis is helpful here to make sure that you are multiplying properly:

$$\frac{C_{\text{water}}}{A} = (4.184 \text{ J g}^{-1} \text{ K}^{-1})(1.00 \text{ g cm}^{-3})(1000 \text{ cm}) = 4.2 \cdot 10^3 \text{ J K}^{-1} \text{ cm}^{-2}$$

$$\frac{C_{\text{land}}}{A} = (0.80 \text{ J g}^{-1} \text{ K}^{-1})(1.5 \text{ g cm}^{-3})(10 \text{ cm}) = 16 \text{ J K}^{-1} \text{ m}^{-2}.$$

The heat capacity per unit area of water is greater than that for the land by a factor of roughly $4.2 \cdot 10^3/16 = 260$.

EXTEND Land warms and cools much more quickly than water, both because it has a much lower heat capacity and because the sunlight warms a much thinner layer of the soil than seawater. As a result, at sunrise the land warms more quickly than the water, and the air above the land expands and rises, creating a breeze from the denser and colder air over the water. Over the course of the day, the water gradually warms up so the air temperatures over land and water become more similar. At sunset, the land cools quickly as the sunlight decreases, and the air above may become cooler than the sea air, causing a **land breeze** to blow out to sea. However, the land breeze is usually weaker than the morning **sea breeze,** partly because sea air at the same temperature is denser than the land air, since it extends to a lower altitude. These morning and evening breezes account for much of the distinctive ambiance of coastal regions.[13]

CONTEXT *Where Do We Go From Here?*

The field of physical chemistry grew out of the efforts of scientist-engineers to design better engines for the industrial revolution. The fundamental problem that faced these pioneers was, "How can we get the most work out of a given amount of fuel?" The methods and parameters developed to solve this problem became the formalism of thermodynamics. Perceptive minds soon realized that the same fundamental laws governing the transformation of heat into work could also be applied to the transformation of chemical energy.

---

[13]The author can vouch for the dependability of these breezes, which fight his morning bicycle ride to work and then often reverse direction in the evening to try and keep him from getting home.

KEY CONCEPTS AND EQUATIONS

7.3 **Mathematical tools.** The **fundamental equation** of thermodynamics, Eq. 7.10, states that the energy E is an explicit function of the extensive parameters of the system, in particular S, V, and n. The conjugate intensive parameters T and P, and the **chemical potential** μ may each be written as a derivative of E with respect to one of these extensive parameters.

- The **Legendre transformation** of E with respect to different variables allows us to create new **thermodynamic potentials**, related to the energy but having different parameter dependencies (Eqs. 7.17, 7.18, 7.19): the **enthalpy** $H(S,P,n) = E + PV$, the **Helmholtz free energy** $F(T,V,n) = E - TS$, and the **Gibbs free energy** $G(T,P,n) = E - TS + PV$.

- The **Maxwell relations** are a set of relationships among partial derivatives of the thermodynamic parameters, obtained by taking mixed second derivatives of the thermodynamic potentials.

7.4 **Heat capacities.** In the simplest, reversible processes, the incremental energy transfer as work $đw$ occurs solely by a volume change dV in the sample, an expansion or contraction, at pressure P:

$$đw_{\text{rev}} = -PdV. \qquad (7.42)$$

The incremental heat flow $đq$ can similarly be related to the temperature T and the change in the entropy of the sample dS:

$$đq_{\text{rev}} = TdS. \qquad (7.43)$$

The transfer of heat into the sample will increase the sample's energy content, and (in general) the temperature as well. The relationship between the energy increase and the change in temperature depends on the number of molecular degrees of freedom available to absorb the energy. That relationship is defined by the heat capacity $đq/dT$, which at constant volume can be expressed as

$$C_V = \left(\frac{\partial E}{\partial T}\right)_V, \qquad (7.44)$$

and at constant pressure as

$$C_P = \left(\frac{\partial H}{\partial T}\right)_P. \qquad (7.49)$$

These two forms of the heat capacity can be interconverted by the equation

$$C_P = C_V + V\alpha\left[\left(\frac{\partial E}{\partial V}\right)_T + P\right]. \qquad (7.52)$$

- For an ideal gas, these expressions simplify to the following:

$$C_V = \frac{1}{2}N_{\text{ep}}nR \quad \text{and} \quad C_P = C_V + nR \quad (7.45)$$

- For a crystal, a harmonic approximation to the vibrational motions leads to the Einstein heat capacity,

$$C_{V\text{m}} = \frac{3\mathcal{N}_A\omega_E^2 e^{\omega_E/(k_B T)}}{k_B T^2 (e^{\omega_E/(k_B T)} - 1)^2}, \qquad (7.68)$$

but this underestimates the heat capacity at low temperatures. Debye carried out a more accurate analysis, incorporating low-frequency phonon modes, and predicted at low temperature

$$C_{V\text{m}} = \left(\frac{\partial E_m}{\partial T}\right)_V \approx \frac{12\pi^4 \mathcal{N}_A k_B^4 T^3}{5\omega_D^3}. \qquad (7.73)$$

- A much more drastic series of approximations allows us to crudely estimate the heat capacity of a monatomic liquid by

$$C_{V\text{m}} \approx 3R\left(\frac{1}{2} + \frac{b\rho_m}{\mathcal{M}}\right), \qquad (7.77)$$

which at least correctly predicts that the molar heat capacity of a liquid is always higher than the gas form of the same substance.

KEY TERMS

- A **state function** is a thermodynamic parameter, a quantity that characterizes the thermodynamic state of a system. The value of a state function does not depend on how we got to that state. In contrast, heat q and work w are not state functions, because they are only defined for changes in the state of the system, and their values depend on the path taken by the process.

- An **equation of state** is any equation that defines a relationship among two or more state functions, usually for a particular type of system. The ideal gas law, for example, is an equation of state that relates the state functions P, V, n, and T to each other for the particular case of the ideal gas.

- Several terms for various types of systems and processes are summarized in Table 7.1.

- The **heat capacity** of a substance is the incremental ratio of the heat transfer of the sample to the sample's change in temperature.

- A **specific heat capacity** is the heat capacity per unit mass of the substance.

OBJECTIVES REVIEW

1. *Find relations among thermodynamic parameters using derivatives of the thermodynamic potentials, Maxwell relations, and partial derivative identities.*

   Show that $\left(\dfrac{\partial V}{\partial S}\right)_{T,n} = \kappa_T/\alpha$.

2. *Estimate the temperature-dependent heat capacity of a solid, liquid, or gas, based on the sample's molecular structure and properties.*

   Estimate the molar heat capacity at constant pressure of gas-phase iron pentacarbonyl, $Fe(CO)_5$ at 5000 K.

3. *Use the heat capacity to convert between the changes in energy and temperature of the sample.*

   Find the energy in joules needed to heat 32.0 g of sulfur ($C_{Pm} = 0.71\ \mathrm{J\,K^{-1}\,mol^{-1}}$) at constant pressure from 5.0 °C to 15.0 °C.

PROBLEMS

Discussion Problems

7.1 Water expands when it freezes at a pressure of 1 bar. If we fill a glass bottle with liquid water and then put the bottle in a freezer, the expansion of the water upon freezing into ice breaks the bottle. What are the signs of q, w, and ΔE for the water during this process?

7.2 Change a single parameter in the following incorrect Maxwell relation to make the relation true:

$$\left(\frac{\partial n_1}{\partial \mu_2}\right)_{S,V,n_2} = \left(\frac{\partial n_2}{\partial \mu_2}\right)_{S,V,n_1}$$

7.3 For each of the following parameters, write "+" if the parameter increases, "−" if it decreases, and "0" if it stays the same as the *temperature increases*.

parameter	as *T* increases
C_{Vm} for a real, polyatomic gas	
maximum value of $\mathcal{G}(R)$ for a real liquid	
q_{vib} for a real solid	
wavelength of maximum output power for a blackbody	
total output power for a blackbody	

7.4 Which of the following gases has the largest heat capacity at 298 K?

a. Cl_2 b. HBr

c. CCl_4 d. C_2H_6

7.5 $H_2(gas)$($\omega_e = 4401\ \mathrm{cm^{-1}}$, $B_e = 60.9\ \mathrm{cm^{-1}}$) has a heat capacity C_{Pm} of 28.824 J K^{-1} mol^{-1}, which is *lower* than the value $7R/2 = 29.1$ predicted from the equipartition principle *without* vibrations. Explain why. (Problem 7.75 is a quantitative version of this one.)

7.6 If substance A has a greater Einstein frequency ω_E than substance B, which substance do you expect to be *harder* (i.e., capable of scratching the surface of the other)?

7.7 Answer the following with E (electronic), V (vibrational), R (rotational), and/or T (translational). The best answer may include more than one of these options.

a. Which degree(s) of freedom are likely to contribute the most to the heat capacity of ethene gas at room temperature?

b. Which degree(s) of freedom are likely to contribute the most to the heat capacity of 1,2-dibromoethene gas at room temperature?

State Functions and Equations of State

7.8 Circle the letter for *each* of the following that is equal to the pressure P of the sample.

a. $k_B T\left(\dfrac{\partial \ln Q}{\partial V}\right)_{T,n}$ d. $\dfrac{1}{\kappa_T}\dfrac{dV}{V}$

b. $T\left(\dfrac{\partial S}{\partial E}\right)_{T,n}$ e. $\dfrac{k_B T}{\rho}$ for an ideal gas

c. $-\left(\dfrac{\partial F}{\partial V}\right)_{T,n}$ f. $-\dfrac{\bar{d}w_{rev}}{dV}$

7.9 Find the equations of state for P, T, and μ if the ensemble size in a system is governed by the equation

$$\Omega = (a_0 NE)^2 V^N,$$

where N is the number of molecules in the sample, E is the energy, V the volume, and a_0 a constant.

7.10 For N indistinguishable, non-interacting bosons of mass m in a three-dimensional box of volume V, find expressions for the pressure and chemical potential.

7.11 Express the enthalpy H for a monatomic ideal gas in terms of n, P, and V.

Derivatives and Maxwell Relations

7.12 Find the value of $(\partial S / \partial n)_{T,V}$ for the monatomic ideal gas.

7.13 Which of the following is equal to $\left(\frac{\partial T}{\partial V}\right)_P$?

a. $\left(\dfrac{\partial V}{\partial S}\right)_P$ b. $-\left(\dfrac{\partial P}{\partial S}\right)_T$

c. $\left(\dfrac{\partial E}{\partial S}\right)_V$ d. $-\left(\dfrac{\partial P}{\partial S}\right)_P$

7.14 Find an expression for $\left(\frac{\partial F}{\partial T}\right)_P$ in a one-component closed system ($dn = 0$) in terms of S, P, and T.

7.15 The grand canonical potential $U_C(T,V,\mu)$ is the Legendre transform of the energy $E(S,V,n)$ with respect to S and n. Find the expression for dU_C for a one-component system. Derive a Maxwell relation from the grand canonical potential $U_C(T,V,\mu)$.

7.16 Find the Maxwell relation, assuming a one-component system, for $\left(\frac{\partial \mu}{\partial T}\right)_{n,P}$.

7.17 Find the Maxwell relation, assuming a one-component system, for $\left(\frac{\partial P}{\partial \mu}\right)_{V,S}$.

7.18 Find a Maxwell relation for $\left(\frac{\partial \mu_1}{\partial n_2}\right)_{n_1,P,T}$ in a two-component system.

7.19 For a liquid sample, we may add to the equation for dE a term σdA, where σ is the surface tension and A the area of the exposed surface of the liquid:

$$dE = T\, dS - P\, dV + \mu\, dn + \sigma\, dA.$$

Find the Maxwell relation obtained by taking the second derivative of the *enthalpy* H with respect to P and A. Don't forget to show what variables are held constant.

Manipulating the Thermodynamic Parameters

7.20 In substances with strong molecular interactions, the heat capacity depends strongly on the pressure. From an equation of state for the volume, this dependence can be determined using the general expression

$$\left(\frac{\partial C_P}{\partial P}\right)_T = -T\left(\frac{\partial^2 V}{\partial T^2}\right)_P.$$

Prove this relation.

7.21 A process is carried out such that $\Delta H = -6.2\ \mathrm{kJ\ mol^{-1}}$, $\Delta G = +23.7\ \mathrm{kJ\ mol^{-1}}$, and $\Delta F = +27.6\ \mathrm{kJ\ mol^{-1}}$. Find ΔE.

7.22 For each of the following, write a single thermodynamic parameter equal to the expression.

a. $\dfrac{nRT}{V}$ for ideal gas   b. $\left(\dfrac{\partial H}{\partial P}\right)_{S,n}$

c. $\left(\dfrac{\partial H}{\partial T}\right)_{P,n}$ d. $F + TS$

7.23 Rewrite the following expressions in terms of S,T,P, V, μ, n, α, C_{Pm}, and/or C_{Vm}, in their simplest form so that no derivatives appear.

a. $\left(\dfrac{\partial H}{\partial P}\right)_{S,n}$ b. $\left(\dfrac{\partial H}{\partial T}\right)_{P,n}$

7.24 The Lagrangian L is a useful function in mechanics, commonly defined as

$$L \equiv K - U,$$

where K and U are the kinetic and potential energies, respectively. Find the Legendre transform of the Lagrangian with respect to all of the velocities v_i of the N particles in the system. THINKING AHEAD ▶ [What units will the transformed function have?]

7.25 Find the Legendre transform of $V(E,S,n)$ with respect to E.

7.26 If we use a thermodynamic potential $U_2(S,P,n_1,\mu_2)$ to describe a two-component system, find an equation relating U_2 to E.

7.27 If the chemical potential μ is defined by the equation

$$\mu \equiv \left(\frac{\partial E}{\partial n}\right)_{S,V},$$

where n is the number of moles of substance, use the equipartition principle to estimate the chemical potential of CH_4 gas at 298 K.

7.28 To see how dangerous an experiment at constant volume can be, find an expression for $\left(\frac{\partial P}{\partial T}\right)_{V,n}$ in terms of α and κ_T, and calculate the value for liquid mercury ($\alpha = 1811\ \mathrm{K^{-1}}$, $\kappa_T = 401\ \mathrm{bar^{-1}}$ at 293 K).

7.29 Simplify the thermodynamic expression $(dE - dF)_T$ to a single term.

7.30 Find the simplest expression for

$$-\left(\frac{\partial V}{\partial T}\right)_{G,n}\left(\frac{\partial T}{\partial G}\right)_{V,n}$$

in terms of any of (or any combination of) α, κ_T, κ_S, C_{Pm}, and/or C_{Vm}. THINKING AHEAD ▶ [None of the thermodynamic parameters listed so far involve a thermodynamic potential (such as G) as the constant in a partial derivative. How can we replace G as the variable held constant?]

7.31 The enthalpy can be a convenient parameter for monitoring the heat flow at constant pressure because $\left(\frac{\partial H}{\partial T}\right)_{P,n} = \left(\frac{\partial q}{\partial T}\right)_{P,n}$. Find the thermodynamic potential X that can similarly be used to monitor the work flow at constant temperature. In other words find X such that $\left(\frac{\partial X}{\partial P}\right)_{T,n} = \left(\frac{\partial w}{\partial P}\right)_{T,n}$.

7.32 Consider an impermeable ($dn = 0$) container in contact with a temperature reservoir at temperature T_r and composed of two systems separated by a movable wall, such that the total volume $V_1 + V_2 = V$ is constant. Using the Helmholtz free energy F, show that the pressures P_1 and P_2 on either side of the wall are equal at equilibrium, where $dF = 0$.

7.33 Equation 2.41 in our derivation of the ideal gas law can be rewritten as

$$P = k_B T \left(\frac{\partial \ln \Omega}{\partial V}\right)_{E,N}.$$

Prove that this equation is consistent with conservation of energy in a closed system (i.e. that $\Delta E = 0$).

7.34 Prove that

$$\kappa_T = \kappa_S + \alpha \left(\frac{\partial T}{\partial P}\right)_S.$$

7.35 Calculate the isothermal compressibility κ_T for 1 mole of an ideal gas at 298 K and 1 bar, where

$$\kappa_T = -\frac{1}{V}\left(\frac{\partial V}{\partial P}\right)_{T,n}.$$

7.36 Find the isothermal compressibility κ_T for a gas obeying the virial expansion in terms of $P, T, V,$ and B_2.

7.37 Express

$$-P^2\left(\frac{\partial (G/P)}{\partial P}\right)_{T,n}$$

in terms of the thermodynamic potentials only.

7.38 Prove that

$$\left(\frac{\partial \mu}{\partial q_{rev}}\right)_P = \frac{1}{nT}\left[T - \left(\frac{\partial X}{\partial S}\right)_P\right],$$

where $X = H - n\mu$ and $đq_{rev} = TdS$.

7.39 Find what thermodynamic parameters are represented by X, Y, and Z in the equation

$$\kappa_S = \frac{C_V}{VT}\left(\frac{\partial T}{\partial X}\right)_Y\left(\frac{\partial T}{\partial X}\right)_Z.$$

7.40 Simplify the following derivatives for a one-component sample:

a. $\left(\frac{\partial E}{\partial S}\right)_{V,n}$ b. $\left(\frac{\partial (H - E)}{\partial P}\right)_{V,n}$ c. $\left(\frac{\partial G}{\partial n}\right)_{P,T}$

Heat Capacities

7.41 Show that the non-ideal term in our equation for the heat capacity of the liquid, Eq. 7.75, vanishes as the temperature approaches absolute zero if we use Eq. 4.59 to approximate the pair correlation function.

7.42 For crystalline SiO_2, the heat capacity is 44 J K^{-1} mol^{-1} at 298 K. Find the Einstein frequency ω_E for $SiO_2(s)$.

7.43 Find an expression for the heat capacity of a solid when the vibrational partition function is given by $q_{vib}(T) = aT + bT^2$.

7.44 Calculate the molar heat capacity C_{Vm} in J K^{-1} mol^{-1} for gas-phase I_2, using the equipartition principle and assuming that translations, rotations, and vibrations all contribute.

7.45 Use the equipartition principle excluding vibrations to estimate the number of moles of CO_2 gas that would have a total heat capacity C_V of 100 J K^{-1}.

7.46 In a nanometer-scale electrical circuit of the near future, a linear chain of 20 atoms is a molecular wire connecting two points. The atom at each end is fixed, but the atoms in between are free to vibrate in all directions. What does the equipartition principle predict for the molar heat capacity C_{Vm} of the wire? Assume that vibrations contribute to the maximum extent possible.

7.47 Consider the atmosphere between sea level and an altitude of 1 km, equal to approximately $2 \cdot 10^{19}$ moles of air (primarily N_2 and O_2). According to the equipartition principle, how much energy in J is released by this region of the atmosphere as it cools from 300 K to 280 K? Do not include vibrations.

7.48 The derivation of C_{Vm} for a fluid shows that the heat capacity is larger for a liquid than for a gas after evaluating the expression

$$C_{Vm} = \frac{3R}{2} + 2\pi \mathcal{N}_A^2\left(\frac{\rho_m}{\mathcal{M}}\right)\int_0^\infty u(R)\, e^{-u(R)/(k_B T)}\left(\frac{u(R)}{k_B T^2}\right) R^2 dR$$

using the Lennard-Jones potential for $u(R)$. Find an equation for C_{Vm} if we instead assume the material is a solid, with a potential $u(R) = -\varepsilon$ if $R_{LJ} < R < 2R_{LJ}$ and $u(R) = \infty$ everywhere else.

7.49 Assume that the pair correlation function for a liquid is roughly given by

$$\mathcal{G}(R) \approx e^{-u(R)/(k_B T)}.$$

Use a square well intermolecular potential with $R_{sq} = 2.8$ Å, $R_{sq}' = 3.2$ Å, and $\varepsilon = 300$ cm^{-1} to estimate the heat capacity of a liquid with a molar mass of 18 g mol^{-1} and a density of 1.0 g cm^{-3} at a temperature of 300 K.

7.50 Find an expression for the molar heat capacity if the thermal energy in a solid obeyed the equipartition principle for vibration exactly.

7.51 Find C_{Vm} in J K^{-1}mol^{-1} at 273 K for a solid with Einstein frequency of 200 cm^{-1}, using the Einstein equation, Eq. 7.68.

7.52 Express the heat capacity at constant volume in terms of the partition function; simplify the equation as much as possible, but leave it general for any phase of matter (solid, liquid, gas).

7.53 Find the heat capacity C_V in terms of n and V for a system of N particles with average energy

$$\langle \varepsilon \rangle = \frac{k_B T^2}{q}\left(\frac{\partial q}{\partial T}\right)_{V,n}, \quad \text{where } q = \left(\frac{2\pi m k_B T}{h^2}\right)^{3/2} V.$$

7.54 Find the average molar energy in kJ mol^{-1} of a solid with Einstein frequency of 200 cm^{-1} at 273 K, using the same approximations used to obtain the Einstein heat capacity.

7.55 Find an expression for the pressure exerted by an expanding solid in terms of N, T, ω_E, and $\partial \omega_E/\partial V$, within the approximations of the Einstein equation.

7.56 If $\varepsilon = \varepsilon_\infty e^{-\varepsilon_0/(k_B T)}$ is the energy per molecule of a sample, find the molar heat capacity at constant volume.

7.57 Imagine a crystal where each vibrational mode can absorb a finite amount of energy, given by the equation

$$\varepsilon(T) = \frac{\varepsilon_0 k_B T}{\varepsilon_0 + k_B T}.$$

This expression gives the energy *per mode* in the crystal.

- Find an equation for C_{Vm}.
- Find $\lim_{T \to 0} C_{Vm}$.
- Find $\lim_{T \to \infty} C_{Vm}$.

7.58 The Debye equation for the heat capacity of a solid is based on the degeneracy of the vibrational modes (or phonon modes) of the solid at a given frequency ν. Find an equation for this degeneracy $g(\nu)$ as a function of the volume and the speed of sound c_s in the solid, where

$$\nu = \frac{c_s}{\lambda}.$$

7.59 Verify that the Debye equation for the heat capacity of a solid predicts the Einstein equation (Eq. 7.68) in the high temperature limit, with $\omega_E = \omega_D$. Do this by evaluating the integral in Eq. 7.70 in the limit that $\omega_D/(k_B T) \ll 1$. (An easier alternative is to show instead that both equations give the same result at high temperature.)

7.60 If we increase the average distance between the atoms in the cubic lattice we used to derive the Einstein heat capacity, as from (a) to (b) in the drawing, does the heat capacity increase or decrease? To simplify the scenario, find the effect on the vibrational potential energy for a single atom vibrating along the x axis between two

fixed walls and held in place by two bonds each with force constant k and equilibrium separation R_e. Let the two walls initially be separated by $2R_e$, and calculate the potential energy for motion of the atom. Then increase the spacing between the two walls by $2R'$ and calculate the new potential energy. What is the effect of the change on the heat capacity?

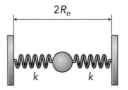

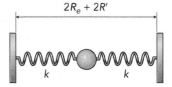

7.61 One neglected term in our equations for the heat capacity is the electronic contribution. At typical temperatures, this is not an important term for most covalent solids because the thermal energy at 1000 K is still less than 0.1 eV (typical electronic excitation energies for covalent molecules are greater than 1 eV). However, metals have their conduction and valence bands extremely close in energy, and the valence band excited states are easily accessible. Therefore metals have a contribution to the heat capacity additional to the Debye contribution for vibrational energy. This electronic term turns out to be roughly linear in temperature, and the low-temperature molar heat capacity may be approximated by the **Debye-Sommerfeld equation:**

$$C_{Vm} \approx \frac{12\pi^4 \mathcal{N}_A k_B^3 T^3}{5\omega_D^3} + \gamma T.$$

Use this equation to find an equation for E_{elec}, the electronic energy of the solid, as a function of T.

7.62 Rank the following in order of increasing heat capacity C_{Pm} at 298 K.

a. $^{13}C_4H_{10}$(gas)   b. $^{12}C_4H_8$(gas)

c. $^{12}C_4H_{10}$(liq)   d. $^{12}C_4H_{10}$(gas)

7.63 Rank the following gas-phase compounds in order of increasing molar heat capacity C_{Pm}:

a. C_3H_8 b. HCCH

c. C_3H_7I d. H_2CO

7.64 How much energy (in J) is necessary to heat 10 L of liquid water ($C_{Pm} = 75.3$ J K^{-1} mol^{-1}) from 300 K to 373 K at 1 bar?

7.65 Find the energy in J necessary to heat one mole of a substance by 10 K at constant pressure and 298 K if C_{Pm} is given by the expression

$$C_{Pm}(J\ K^{-1}) = 10.0\ T^{1/2} - 2 \cdot 10^6\ T^{-2}.$$

7.66 The heat capacity gives the heat transfer per unit temperature change, $\partial q/\partial T$, and we define these at constant V or at constant P. In the same way, we can define "work capacities" at constant T or S, B_T and B_S, by

$$B_T = \left(\frac{\partial w}{\partial P}\right)_T \quad B_S = \left(\frac{\partial w}{\partial P}\right)_S.$$

Find an expression for B_T in terms of B_S, replacing any remaining partial derivatives by common thermodynamic parameters.

7.67 For solid magnesium, the heat capacity at constant pressure roughly obeys the following equation:

$$C_{Pm}\ (J\ K^{-1}\ mol^{-1}) = 22.30 + 0.01025 \cdot T(K)$$
$$- 43100 \cdot T(K)^{-2}.$$

Calculate ΔH and ΔS for the cooling of 24.31 g of Mg from 600 K to 298 K.

7.68 The heat capacity C_{Pm} for 1,2-dibromoethane gas is 96.8 J K^{-1} mol^{-1} at 383 K and 1.00 bar. As we heat the substance from 382 K to 384 K, how much heat (in J mol^{-1}) is transferred into each of these three degrees of freedom: translation, rotation, and vibration? Based on your results, roughly how many vibrational modes appear to be absorbing energy at this temperature?

7.69 From the following data evaluated at 298 K and 1.00 bar, calculate the internal pressure $\left(\frac{\partial E}{\partial V}\right)_{T,n}$ in SI units for liquid water.

C_{Pm}	75.32 J K^{-1} mol^{-1}	C_{Vm}	74.53 J K^{-1} mol^{-1}
α	$2.57 \cdot 10^{-4}$ K^{-1}	density	0.99704 g cm^{-3}
κ_T	$4.63 \cdot 10^{-5}$ bar^{-1}	κ_S	$4.48 \cdot 10^{-5}$ bar^{-1}

7.70 Calculate the enthalpy change in kJ for heating 2 moles of N_2 gas from 77 K to 300 K at constant pressure, using the data in Table 7.2.

7.71 Use the equipartition principle carefully to estimate the molar heat capacity C_{Pm} at 298 K of CHI_3.

7.72 A simple alkane C_nH_{2n+2} has a molar heat capacity C_{Pm} of 205.9 J $K^{-1}mol^{-1}$ at 1500 K. What is the alkane?

7.73 For atomic fluorine, transitions from the ground $^2P_{3/2}$ state (degeneracy $g = 4$) to the lowest excited $^2P_{1/2}$ state ($g = 2$) at 404 cm^{-1} absorb some of the thermal energy.

Find the average electronic energy per atom of atomic fluorine at 298 K, and use this to predict the heat capacity C_{Pm} with greater accuracy than with the equipartition principle alone.

7.74 For the non-linear triatomic molecule FNO, the average vibrational energy per molecule between 280 K and 320 K is given roughly by the equation

$$\langle \varepsilon_{vib} \rangle \approx k_B[a + b(T - T_0) + c(T - T_0)^2],$$

where $a = 33.71$ K, $b = 0.630$, $c = 0.00232$ K^{-1}, and $T_0 = 222.2$ K. This includes all of the vibrational modes. Find the *total* molar heat capacity C_{Pm} of FNO at 298 K, assuming the sample otherwise behaves like an ideal gas.

7.75 In some cases, we need to go back to treating certain degrees of freedom as individual quantized energy levels. This problem steps through one example.

a. Show that a general equation for the heat capacity C_V may be written as

$$C_V = \frac{1}{k_BT^2}\left[\frac{1}{Q}\sum_i \left(g_iE_i^2e^{-E_i/(k_BT)}\right) - \langle E \rangle^2\right].$$

b. Use this equation to calculate the contribution to the molar heat capacity from rotations in H_2(gas) at 298 K. An accurate result will take the H_2 nuclear spin statistics into account (Section 4.4).

c. Add in the other expected contributions to predict the molar heat capacity at constant pressure for H_2(gas). How does it compare to the equipartition value of $7R/2$ and to the actual value of 28.824 J K^{-1} mol^{-1}?

7.76 Estimate as accurately as possible the molar heat capacity C_{Pm} for gas-phase BrF ($\omega_e = 646$ cm^{-1}) at 298 K.

7.77 Find an expression for the molar heat capacity C_{Vm} of a solid when the total partition function is given by $Q(T) = (aT^3)^N$.

7.78 Find the total (not molar) heat capacity at constant volume in J K^{-1} of one L of a solid at 300 K with energy given by the equation

$$E = (200\ J\ dm^{-3}\ K^{-3})\ VT^3.$$

7.79 Find whether C_P is greater or less than $C_V + \alpha PV$ for H_2O(liq) at 273.15 K, the freezing point. At 273.15 K, the liquid has a density of 1.000 g cm^{-3} and the solid has a density of 0.917 g cm^{-3}.

PART III
REACTIVE
SYSTEMS

8

The First Law: Expansion and Engines

GOAL *Why Are We Here?*

The goal of this chapter is to describe how we use thermodynamic quantities to predict the properties of fundamental processes involving expansion and also to demonstrate how crucial features of these processes depend on the molecular properties of the system. The examples in this chapter focus on how the energy evolves during expansion and contraction and on the conversion of heat into work by means of an engine. The treatment of these processes allows us to demonstrate several fundamental points in classical thermodynamics, as well as prepare the groundwork for our later discussion of phase transitions and chemical reactions.

CONTEXT *Where Are We Now?*

Chapter 7 assembled a small toolkit for dealing with problems in thermodynamics, and demonstrated how statistical mechanics can predict some of the numerical values—in that case, heat capacities—that those tools use in finding a solution. We saw how to use the heat capacity to predict the relationship between the changes in energy and temperature for heating and cooling.

Now, to extend our use of those tools, we're going to follow the energy as it moves around during different fundamental processes: expansion and contraction. Whereas the processes in Chapter 7 focus primarily on the heat q, much of our attention in this chapter will be invested in the work w. There are at least two reasons these processes are of interest in chemistry. First, temperature changes are often accompanied by work arising from changes in the volume of the system as it expands or contracts, and so a proper accounting of the energy requires that we understand the work component as well as the heat. Second, a primary motivating factor for the study of chemical thermodynamics has long been the conversion of heat released by combustion into work by means of an engine. We will examine

277

the energy flow during one-step expansions first and then combine steps to make an engine cycle at the end of the chapter. Our toolkit from Chapter 7 is all we will need analyze the basic (and remarkable) operation of engines.

SUPPORTING TEXT   *How Did We Get Here?*

The main *qualitative* result we will draw on for the work in this chapter is the notion from Chapter 2 that the entropy drives a sample to expand when it can and to exchange energy with another sample when it can to equalize their temperatures. For a more detailed background, we will draw on the following equations and sections of text to support the ideas developed in this chapter:

- From Chapter 7, we will need the first law of thermodynamics,

$$dE = đq + đw, \tag{7.5}$$

  which divides the change in energy into incremental components that we will use to obtain the contributions to ΔE from the heat q and the work w during an expansion.

- For processes that involve a temperature change, we will also need the relationship between energy and temperature as expressed using the heat capacity at constant volume:

$$C_V = \left(\frac{\partial E}{\partial T}\right)_{V,n} \tag{7.44}$$

8.1   Expansion of Gases

Although our primary interest will be in the transfer of energy as heat, this is often inseparable—or inconvenient to separate—from energy transfer in the form of work, such as thermal expansion. Therefore, we will take a look at the thermodynamics of systems in which work is also done. To take advantage of the ideal gas law, we will stick for now to PV work.

For problems involving gases in closed systems, it is often informative to plot the process on a P versus V (or "P-V") graph. The temperature is specified by the pressure and volume according to the ideal gas law or one of its non-ideal cousins (e.g., the van der Waals equation), and the energy by the temperature; therefore, every point on a P-V graph corresponds to a unique thermodynamic state. Any quasistatic process may then be drawn on the graph as a curve, because each point along the path of a quasistatic process must correspond to a thermodynamic state. For example, the quasistatic heating of a monatomic ideal gas at constant pressure would correspond to a single horizontal line (Fig. 8.1), indicating that the pressure was fixed but the volume was increasing linearly with T. Such plots, and their relatives, are especially useful for more complex problems, as we will see shortly.

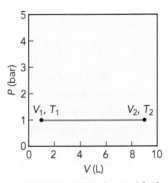

▲ **FIGURE 8.1 Heating an ideal gas.** The process in this case is carried out isobarically (at constant pressure), increasing the volume of the gas from 1 L to 9 L at a constant pressure of 1 bar.

Isothermal Expansion

Consider the work done in the reversible expansion between two thermodynamic states of a gas with volumes V_1 and V_2, where $V_2 > V_1$. Going back to our version of the first law (Eq. 7.5) and combining it with Eq. 7.11:

$$dE = đq + đw = TdS - PdV + \mu\,dn,$$

and restricting ourselves to a closed system (so that $dn = 0$) and a reversible process, our finding that $đq_{rev} = TdS$ (Eq. 7.43) means that our incremental work is what's left over:

$$đw_{rev} = -PdV.$$

We choose the system illustrated in Fig. 8.2: a cylindrical chamber containing 1.00 mole of a compressed gas at an initial volume of 2.478 L, initial pressure of 10.0 bar, and initial temperature of 298 K. Outside the chamber, the pressure is 1.00 bar and the temperature is 298 K. One wall of the chamber consists of a piston, which can be pushed in (reducing the volume and compressing the gas further) or can be released (allowing the gas to expand). For the following expansion problems, we will permit the gas to expand until the pressure inside the chamber is equal to the external pressure of 1.00 bar.

To simplify matters further, we shall assume that the temperature stays constant, making this an *isothermal* expansion. In the terminology developed earlier, we perform the expansion with the gas in contact with a temperature reservoir through diathermal walls so that heat can flow freely between the sample and the reservoir. Even so, there are always several ways of proceeding from one thermodynamic state to another, and although the energy and other parameters are fixed for each state, the heat and work during the transition depend on the *path* between the states as well as the states themselves.

A *reversible* expansion (Fig. 8.3a) is possible only if the gas is allowed to expand slowly, maintaining equal pressure on both sides of the piston at all times. If the gas pressure differed significantly from the pressure exerted by the walls, then the sample would expand or contract suddenly, not quasistatically. An *irreversible* expansion (Fig. 8.3b) would take place, pushing rapidly on the surrounding gas and therefore heating it up. For irreversible processes, we must acknowledge that these pressures are not necessarily equal and return to the more general form of (Eq. 7.41) $đw = -P_{min} dV$. For a reversible process, the change in entropy of the sample dS need not be zero. The change in the *total*

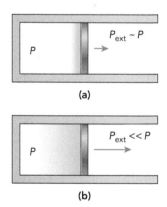

▲ FIGURE 8.3 **Reversible and irreversible expansion.** (a) The reversible expansion occurs when the pressure P of the sample and the external pressure P_{ext} differ by an infinitesimal amount ($P = P_{ext} = P_{min}$). (b) If P is much greater than P_{ext}, then the expansion occurs rapidly and irreversibly ($P > P_{ext} = P_{min}$).

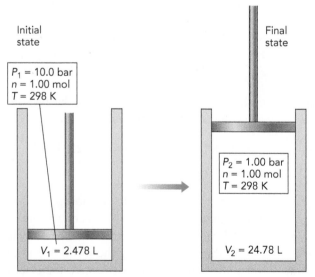

▲ FIGURE 8.2 **The expansion apparatus.** The walls of the chamber are diathermal for the isothermal expansions.

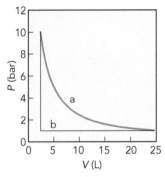

▲ **FIGURE 8.4** *P-V* **plots for two isothermal expansion processes.** The curves are for the **(a)** reversible isothermal expansion and **(b)** an irreversible and instantaneous expansion. The work done by the process is equal in each case to the area under the curve, so the maximum work is done by the reversible process (a).

CHECKPOINT When solving problems that relate gas pressures and volumes to the temperature, we use the form of the gas constant R in units of bar L K^{-1} mol^{-1}. When the expression instead relates energy to temperature, we use $R = 8.3145$ J K^{-1} mol^{-1}.

entropy must be zero, and that's no easy trick. The entropy of the sample may increase during a reversible process only if it is balanced by a decrease in the entropy of a reservoir or some other component of the system.

For the reversible isothermal expansion, we plot the initial and final states on a P-V graph and draw the reversible isothermal expansion as the curve $P = nRT/V$ joining the two points. This is done in curve a of Fig. 8.4. Non-quasistatic processes may involve states in which pressure and volume are poorly defined.

For the reversible isothermal process, we will write the work as $w_{T,\text{rev}}$, where the T indicates that T is held constant. The work can be evaluated as

$$w_{T,\text{rev}} = \int_{V_1}^{V_2} đw_{T,\text{rev}} = -\int_{V_1}^{V_2} P dV, \tag{8.1}$$

where the pressure P is a state function determined by the other parameters, such as T and V. The magnitude of this integral is the area under the curve P between the two points representing the initial and final states. The sign of w is determined by which state is the upper limit and which is the lower limit of the integral; expansion requires work to be done *by* the system, whereas compression requires work be done *to* the system. We can solve the integral by recognizing that

$$V_2 = \frac{nRT}{P_2}$$

$$= \frac{(1.00\,\text{mol})(0.083145\,\text{bar L K}^{-1}\text{mol}^{-1})(298\,\text{K})}{1.00\,\text{bar}}$$

$$= 24.8\,\text{L},$$

and by expressing the pressure in terms of the parameter V and the constants n and T:

$$P = \frac{nRT}{V}.$$

For the reversible isothermal expansion,

$$w_{T,\text{rev}} = -\int_{V_1}^{V_2} P dV = -nRT \int_{V_1}^{V_2} \frac{dV}{V}$$

$$w_{T,\text{rev}} = -nRT \ln\left(\frac{V_2}{V_1}\right). \tag{8.2}$$

This is a general result for the work obtained from the reversible isothermal expansion of an ideal gas. Substituting in the values from Fig. 8.2 gives

$$w_{T,\text{rev}} = -(1.00\,\text{mol})(8.3145\,\text{J K}^{-1}\text{mol}^{-1})(298\,\text{K})\ln\left(\frac{24.78}{2.478}\right)$$

$$= -5.71 \cdot 10^3\,\text{J} = -5.71\,\text{kJ}.$$

The integral in Eq. 8.1 cannot be used for irreversible processes, for which the pressure need not be a continuous function at all. When we introduced Eq. 7.41, we used P_{min} instead of P for the sample to express the work, because the amount of work done depends on the force opposing the work. In any case, this is the convenient way to write the equation for cases where the external pressure is well-known but the pressure of the gas itself is not a smooth function of T and V. For example, if the tension on our piston is released completely, the expansion of

EXAMPLE 8.1 Gas Compression in a Pump

CONTEXT In many applications in chemistry and chemical engineering, we have to push gases from one place to another (for example, to get a reactant gas to the inlet of a reactor) or pull gases out of a container (for example, to create a vacuum in which we can manipulate ions in a mass spectrometer). We use gas pumps to do this. A typical configuration is a rotary vane pump, which admits gas at an inlet, traps it, compresses it to a smaller volume and higher pressure, and then releases it through an outlet at its new, higher pressure (Fig. 8.5). Pumps are rated partly by their **compression ratio,** the volume of the gas at the inlet to the gas at the outlet, $V_{\text{inlet}}/V_{\text{outlet}}$, which also gives the ratio of the outlet pressure to inlet pressure $P_{\text{outlet}}/P_{\text{inlet}}$. To achieve higher compression ratios (and therefore lower inlet pressures), a rotary vane pump may be divided into two stages, with the second stage taking the compressed gas from the first stage and compressing it still more. This can be an energy-intensive process, and we can use our results to estimate the minimum work necessary to compress a gas isothermally.

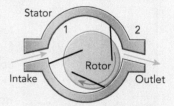

▲ FIGURE 8.5 **Schematic of a typical rotary-vane pump.** The rotor spins inside the cavity of the stator, with vanes (shown as black lines) that slide in and out of the rotor as the gap between the rotor and stator changes. Vacuum pump oil makes a gas-tight seal between the regions separated by the vanes. As the rotor spins, gas enters through the intake into a high-volume region 1. As the rotor turns, the gas becomes trapped by the vanes and compressed until it gets pushed through the outlet.

PROBLEM A vacuum chamber in a spectrometer is maintained at an operating pressure of 10.0 mtorr by a two-stage rotary vane pump with an exhaust pressure at the pump outlet of 800. torr. What is the minimum power in watts (J/s) consumed by the pump to keep the chamber at this pressure when there is a flow of 0.22 mmol/s and $T = 300.$ K?

SOLUTION The amount of gas moved in 1 second is 0.22 mmol. This gas must be compressed by a factor of $800/0.010 = 80,000$ to achieve the pressure increase from inlet to outlet. The reversible work done per second gives the minimum power needed, because the reversible process wastes none of the work:

$$w_{T,\text{rev}} = -nRT \ln\left(\frac{V_2}{V_1}\right)$$

$$\frac{w_{T,\text{rev}}}{\Delta t} = -\left(\frac{n}{\Delta t}\right)RT \ln\left(\frac{P_1}{P_2}\right)$$

$$= -(0.22 \cdot 10^{-3}\,\text{mol/s})(8.3145\,\text{J K}^{-1}\text{mol}^{-1})(300.\,\text{K}) \ln(1/80,000) = 6.2\,\text{W}$$

the compressed gas begins with a 9.0 bar pressure difference across the piston. We find the work done by the system when we integrate Eq. 7.41, where $P_{\min}$ in this case is the constant external pressure of 1.00 bar:

$$w_{T,\text{irr}} = \int_{V_1}^{V_2} đ w_{\text{irr}}$$

$$= -\int_{V_1}^{V_2} P_{\min} dV$$

$$= -P_{\min}(V_2 - V_1), \tag{8.3}$$

and for our case that gives

$$w_{T,\text{irr}} = -(1.00\,\text{bar})(24.78\,\text{L} - 2.478\,\text{L})$$
$$= -22.3\,\text{bar L} = -2.23\,\text{kJ}.$$

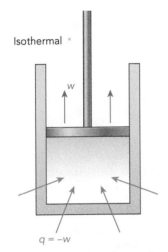

Isothermal

w

$q = -w$

▲ **FIGURE 8.6 The balance of energy in an isothermal expansion.** As work is done by an isothermally expanding ideal gas, energy must be transferred into the gas as heat in order for the energy to remain constant.

CHECKPOINT It may seem bizarre that q is not zero when ∆T is zero, but this is an example of when we must not confuse *heat* with *temperature*. In this case, the distinction arises because heat quantifies the *transport* of energy, whereas the temperature represents (roughly) the average over a quantity of *stored* energy.

This process is plotted as curve b in Fig. 8.4. Again, acknowledging the sign convention, w gives the energy lost through work done *by* the sample, so it is negative.

Of these examples, the reversible expansion does the most work, and this result holds for all processes. This is because in the reversible expansion, the forces that oppose each other must be as evenly matched as possible at every point during the process. In our irreversible expansions, P_{min} is significantly lower than P, and therefore the integral of $P_{min}dV$ is lower. The system does less work because it doesn't have to push as hard.

The energy E for a monatomic ideal gas is just $\frac{3}{2}nRT$ (Eq. 3.59), a function only of the number of moles and the temperature. Therefore, in the isothermal expansion of an ideal gas, $\Delta E = q + w$ is zero because the temperature of the gas doesn't change. Since the sample does work during the expansion, meaning w is negative, then *heat must be absorbed from the reservoir in order to keep the sample temperature the same*, with the total heat influx equaling the amount of work done (Fig. 8.6). So, for example, in our reversible isothermal expansion, the heat flow is

$$q_{T,\text{rev}} = nRT \ln\left(\frac{V_2}{V_1}\right), \tag{8.4}$$

and for the irreversible isothermal expansion as just given,

$$q_{T,\text{irr}} = P_{min}(V_2 - V_1). \tag{8.5}$$

In either case, heat flows from the reservoir while the temperature doesn't change. The reservoir is so vast that heat can flow constantly in and out of it, without measurably affecting the properties of the reservoir itself.

Also worth noticing is that even though this is a reversible process, the entropy of the sample *does* change: $đq_{\text{rev}} \neq 0$ and $dS = đq_{\text{rev}}/T$. This requires the reservoir to respond with its own entropy change that exactly cancels the sample's change in entropy.

Adiabatic Expansion

So can we keep the total entropy *and* the system entropy constant? Consider now the case when the expanding gas is contained within adiabatic walls, which forbid the transfer of heat from the outside. Again we carry out the reversible expansion, from the same initial conditions until the final pressure equals the external pressure of 1.00 bar. For such an adiabatic expansion, $dS = 0$ because $đq_{\text{rev}} = TdS = 0$, so the change in energy comes from work alone:

$$\text{reversible adiabatic process:} \quad dE = đw_{S,\text{rev}} = -PdV, \tag{8.6}$$

where now the subscript S indicates that this is the work for the adiabatic process, where S of the sample is constant. We are continuing to assume a closed sample (meaning $dn = 0$), so the only work done is PV work.

We assume furthermore that the ideal gas law still applies:

$$dE = -PdV = -\frac{nRTdV}{V}. \tag{8.7}$$

If we want only to solve for the work during this process, Eq. 8.7 may look sufficient; we could integrate it from V_1 to V_2 to get ΔE, and that would be the work. There are two problems with that: (*i*) because the sample is thermally insulated from the surroundings, its temperature no longer needs to be constant and therefore *T* cannot be factored out of the integral; and (*ii*) with the temperature changing, we no longer know what the final volume V_2 is. Somehow we ended up with not enough equations for the number of unknowns we have.

TOOLS OF THE TRADE | **Bomb Calorimetry**

The distinction between heat and temperature was not well established until the end of the 19th century, and consequently there was no consistent theory to describe the release or absorption of heat by chemical processes such as phase changes or chemical reactions. Around 1780, Antoine Lavoisier and Pierre Laplace together developed an instrument for measuring heats of various processes, which they gauged by the amount of ice melted, but the work was ahead of its time. Nearly a century later, Marcellin Berthelot developed the first modern device for measuring the heat flow in a chemical reaction: the bomb calorimeter.

What is a bomb calorimeter? A calorimeter is any device that measures the heat flow during a process. Calorimeters are the chief diagnostic tool in thermodynamics, and we will draw on many results from calorimetry in the chapters ahead. A bomb calorimeter is any calorimeter that operates with the sample at a fixed volume.

Why do we use a bomb calorimeter? Standard benchtop conditions in the laboratory allow us to maintain a constant temperature of the system (using a water bath or heating mantle) and a constant pressure (by exposure to the atmosphere or—for air-sensitive compounds—by working in a glove-box filled with an inert gas at fixed pressure). Why is fixed pressure important? Keeping the pressure fixed reduces the number of changing variables, which is convenient for record keeping alone, but it also simplifies the thermodynamics whenever we can set one parameter to a constant. By fixing the pressure, we ensure that the enthalpy change during a process is equal to the heat:

$$\Delta H = \int dH = \int (T dS + V dP)$$

$$(\Delta H)_P = \int (T dS + V dP)_P = \int (T dS) = q \ \ if \ dP = 0,$$

where the subscript *P* indicates that the pressure is kept constant. The enthalpy was *invented* to make this relationship true.

But we have a more general definition—and a more intuitive understanding—of the *energy E*. If we want to measure ΔE instead of ΔH, however, the experiment can be much more challenging. The combustion of sucrose, for example,

$$C_{12}H_{22}O_{11} + 12O_2 \rightarrow 12CO_2 + 11H_2O$$

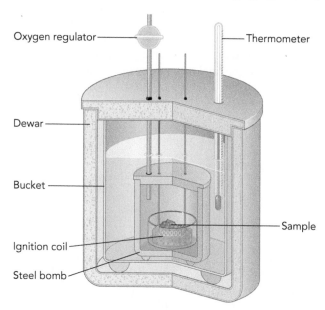

Oxygen regulator

Thermometer

Dewar

Bucket

Sample

Ignition coil

Steel bomb

▶ **Schematic of a combustion bomb calorimeter.**

liberates roughly 6000 kJ of heat per mole of sucrose and forms 23 product molecules for every 13 reactant molecules. In order to directly measure the ΔE of the reaction, we would choose to carry it out at constant *volume* rather than constant pressure. In that case, the heat measured by the calorimeter is equal to the change in energy:

$$\Delta E = \int dE = \int (TdS - PdV)$$

$$(\Delta E)_V = \int (TdS - PdV)_V = \int (TdS) = q.$$

A bomb calorimeter yields the energy change during a reaction directly, and as such provides a more direct link between the heat released or absorbed by a reaction and our understanding of chemical bond energies. What's not to like?

The challenge is that the reaction takes place quickly, and therefore the 6000 kJ/mol released by the reaction initially heats the products, raising the temperature and pressure of the system to many times its initial pressure. The calorimeter risks exploding like a bomb.

The distinction between enthalpy and energy in solution-phase chemistry is usually very small, so

constant-pressure calorimeters are more common. Bomb calorimeters today are primarily to characterize gas-phase reactions and combustion processes, which may occur so rapidly or violently that the sample needs to be contained. Related processes studied by bomb calorimetry include the incineration of toxic wastes and the metabolism of foods.

How do they work? A typical bomb calorimeter consists of a small steel drum (the bomb) submerged in 2 L of water, with a thermometer monitoring the water temperature to high precision. The sample is in a rigid container, so it can do no work, but heat can be transferred between the sample and the surrounding water. By conservation of energy, we can write the energy change of the system as

$$\Delta E_{sys} = -\Delta E_{water} = -q_{water} = -C_V \Delta T_{water}.$$

From the measured change in the temperature of the water and the known heat capacity of the water, the energy change of the system can be calculated. Bomb calorimeters designed specifically for combustion reactions, such as the one shown in the figure, also include a regulated oxygen supply and a nickel heating coil that ignites the fuel.

We can figure out what that missing equation is. Whenever we want to find ΔE for a sample while T varies, *we will need the heat capacity* because that defines the relationship between ΔE and ΔT. So the equation we are missing must involve the heat capacity. For an ideal gas, the energy from the equipartition principle is (Eq. 3.20)

$$E = \frac{1}{2} N_{ep} nRT.$$

For a closed sample of an ideal gas, E is a function only of the temperature. Now we can find the final thermodynamic state of our reversible adiabatic expansion:

$$dE = \left(\frac{\partial E}{\partial T}\right)_V dT = C_V dT \qquad (8.8)$$

$$C_V dT = -\frac{nRTdV}{V}. \qquad \text{combine Eqs. 8.7 and 8.8}$$

$$\int_{T_1}^{T_2} \frac{C_V dT}{T} = -\int_{V_1}^{V_2} \frac{nRdV}{V} \qquad \text{isolate like variables and integrate}$$

$$C_V \ln \frac{T_2}{T_1} = -nR \ln \frac{V_2}{V_1} \qquad \int (dx/x) = \ln x + C \quad (8.9)$$

$$C_V \ln \frac{P_2 V_2/(nR)}{P_1 V_1/(nR)} = C_V \ln \frac{P_2 V_2}{P_1 V_1} = -nR \ln \frac{V_2}{V_1} \qquad PV = nRT$$

$$C_V \left[\ln \frac{P_2}{P_1} + \ln \frac{V_2}{V_1}\right] = -nR \ln \frac{V_2}{V_1} \qquad \ln xy = \ln x + \ln y$$

$$C_V \ln \frac{P_2}{P_1} = -(C_V + nR)\ln \frac{V_2}{V_1} = -(C_P)\ln \frac{V_2}{V_1} \qquad \text{combine terms, Eq. 7.55}$$

$$V_2 = V_1 \left(\frac{P_2}{P_1}\right)^{-C_V/C_P}. \qquad \text{solve for } V_2 \quad (8.10)$$

Here the heat capacity is assumed constant over the temperature range in the integral, an assumption with limitations illustrated by Fig. 7.5. The accuracy of the approximation is improved by averaging the heat capacity over the temperature range. The Appendix gives heat capacities and other data for various species.

Let's use again our expansion apparatus, redrawn in Fig. 8.7, but this time we cannot calculate the final volume or the work done until we have a value for the heat capacity C_V. Assume the gas to be an ideal monatomic gas with $C_{Vm} = 3R/2$. When the pressure of the system is lessened reversibly, and the ambient pressure is 1.00 bar, the final volume V_2 is not 24.78 L but

$$V_2 = V_1 \left(\frac{P_2}{P_1}\right)^{-C_V/(C_V+nR)} = (2.478\,\text{L})\left(\frac{1}{10.0}\right)^{-3/5} = 9.87\,\text{L}.$$

This time there is no flow of heat to maintain the same energy, and E decreases during the expansion because $\Delta E = w$. With E directly proportional to the temperature, the temperature of the gas drops also. This in turn affects the product PV of the gas. In this adiabatic expansion, the gas cannot expand to the final volume of 24.78 L reached in the isothermal expansion, because now the gas *cools* as it expands, and so it reaches the stipulated final pressure of 1.00 bar at a lower volume. The work done during this expansion is equal to $\Delta E = C_V \Delta T$, and T_2 can be found using the ideal gas law:

$$T_2 = \frac{P_2 V_2}{nR} = \frac{(1.00\,\text{bar})(9.87\,\text{L})}{(1\,\text{mol})(0.083145\,\text{bar L K}^{-1}\,\text{mol}^{-1})} = 119\,\text{K}.$$

The work done is

$$w_{S,\text{rev}} = nC_{Vm}(T_2 - T_1)$$

$$= (1.00\,\text{mol})\frac{3R}{2}(119\,\text{K} - 298\,\text{K}) = -2.23\,\text{kJ}.$$

CHECKPOINT Why does the math surrounding the adiabatic expansion seem so much more involved than the math for the isothermal expansion? For the isothermal expansion, T and n are fixed and the ideal gas law can then relate P and V directly. The adiabatic expansion fixes S and n, but S does not appear in the ideal gas law, so we still have three parameters—P, V, and T—all varying. That is why we need to introduce an additional equation before we can solve for all the variables.

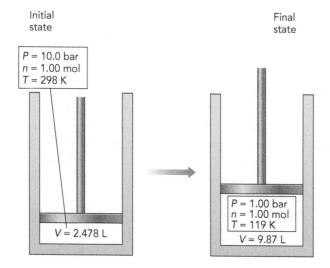

Initial state

Final state

$P = 10.0\ \text{bar}$
$n = 1.00\ \text{mol}$
$T = 298\ \text{K}$

$V = 2.478\ \text{L}$

$P = 1.00\ \text{bar}$
$n = 1.00\ \text{mol}$
$T = 119\ \text{K}$
$V = 9.87\ \text{L}$

◀ FIGURE 8.7 **The adiabatic expansion.** The walls do not allow heat to flow in, so work done by the expanding gas channels energy out of the sample, lowering the gas temperature.

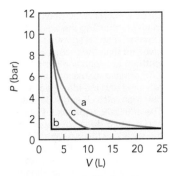

▲ **FIGURE 8.8** *P-V* plots for three expansion processes described in the text. The curves show the following expansions: **(a)** reversible and isothermal, **(b)** irreversible and isothermal, and **(c)** reversible and adiabatic. The work done by the process is equal in each case to the area under the curve.

The *P-V* graph for this expansion is given in curve c of Fig. 8.8.

Rapid, uncontrolled expansions are often approximated to take place under adiabatic conditions because there simply may not be enough time for heat to be transferred into the sample. Combining the expressions we have for ΔE and for V_2/V_1, we can get a common expression for the energy released by our expansion in terms of the initial pressure and volume and the final pressure:

$$w_{S,\text{rev}} = C_V(T_2 - T_1) \tag{8.11}$$

$$= C_V\left[\frac{P_2 V_2}{nR} - \frac{P_1 V_1}{nR}\right] \qquad PV = nRT$$

$$= \frac{C_V}{nR}\left[P_2 V_1\left(\frac{P_2}{P_1}\right)^{-C_V/C_P} - P_1 V_1\right] \qquad \text{by Eq. 8.10}$$

$$= \frac{C_V}{nR}(P_1 V_1)\left[\frac{P_2}{P_1}\left(\frac{P_2}{P_1}\right)^{-C_V/C_P} - 1\right] \qquad \text{factor out } P_1 V_1$$

$$= \frac{C_V}{C_P - C_V}(P_1 V_1)\left[\left(\frac{P_2}{P_1}\right)^{1-(C_V/C_P)} - 1\right] \qquad \begin{array}{c}\text{combine } P_2/P_1 \text{ terms,}\\ \text{Eq. 7.55}\end{array}$$

$$= \frac{1}{(C_P/C_V)-1}(P_1 V_1)\left[\left(\frac{P_2}{P_1}\right)^{1-(C_V/C_P)} - 1\right]$$

$$= \frac{P_1 V_1}{\gamma - 1}\left[\left(\frac{P_2}{P_1}\right)^{1-(1/\gamma)} - 1\right]. \qquad \text{set } \gamma = C_P/C_V \tag{8.12}$$

The **heat capacity ratio** $\gamma = C_P/C_V$ is at its maximum value of 5/3 for the ideal monatomic gas. As the complexity and mass of the gas molecule increases, the number of available equipartition degrees of freedom N_{ep} also rises, so the ratio of $C_P = C_V + nR$ to C_V approaches 1. For gas molecules, however, the value is rarely less than 1.1.

Equation 8.12 shows—as we might expect—that the work done depends primarily on (*i*) $P_1 V_1$, proportional both to the amount of material n available at the beginning of the expansion and to its temperature T_1; and on (*ii*) the compression ratio P_2/P_1 between the final and initial pressures. If we divide w by $n = P_1 V_1/(RT_1)$ to get roughly the work per unit of material, then we find that the work done does not vary enormously over the range of likely γ values (Fig. 8.9). At low compression ratios, a gas with high heat capacity (smaller γ) can do more work because it contains more energy at a given temperature than, for example, a monatomic gas. At very high compression ratios, however, a high heat capacity becomes a liability for doing work because the gas loses more energy upon cooling, and it rapidly reaches too low a temperature to continue expanding.

However, the work increases steeply with the compression ratio P_2/P_1, so a highly compressed gas can do a lot of work—or a lot of damage—as it expands.

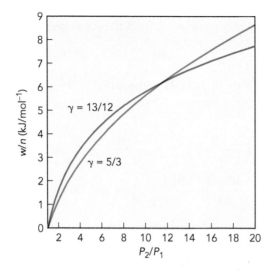

◄ FIGURE 8.9 **Compression ratios and work.** The work per mole $w_{S,rev}/n$ is graphed as a function of the compression ratio P_2/P_1, based on Eq. 8.12 and assuming an initial temperature of $T_1 = 300\,K$. The curve for $\gamma = 5/3$ corresponds to the monatomic gas, and $\gamma = 13/12$ corresponds to a 5-atom molecule with all vibrations included (full equipartition in Table 7.2). The two cross at high compression ratio.

EXAMPLE 8.2   **Explosions as Adiabatic Expansions**

CONTEXT In the latter part of the 19th century, one rapid area of development in chemistry was the synthesis of new explosives, which were needed for mining coal to fuel the industrial revolution. From this period, we learned how to make a number of organic peroxides, such as HMTD (hexamethylene triperoxide diamine). These are compounds which, following a general rule of explosives, combine combustible elements such as carbon and hydrogen with oxygen. This strategy allows the combustion reaction to occur on the time scale of molecular motion, rather than diffusion, and the release of energy propagates through the material faster than the speed of sound. However, the organic peroxides combine this with extremely fragile OO bonds, making compounds that may blow up at the slightest shock. A temperamental explosive is a liability in most settings, and HMTD is no longer used for industrial applications. The energy of such an explosion can be channeled into: (a) the energy required to break the container, (b) the energy of the pressure front at the head of the expansion, and (c) the kinetic energy of any projectiles propelled by the expanding gas. If the container is fragile, as much as 80% of the energy released by the explosion is likely to be in the pressure wave.

PROBLEM If 1.0 L of hexamethylene triperoxide diamine (HMTD) explodes by suddenly decomposing to gases at a pressure of 1.0 kbar at the ambient temperature, calculate the grams of TNT that would release a pressure wave of equal energy, assuming $4680\,J/g\,TNT$, and assuming the explosion is reversible.[1] Assume an average value for γ of 1.4, an ambient pressure of 1.0 bar, and that 70% of the energy is in the pressure wave.

---

[1] It has been common practice to model explosions as reversible adiabatic expansions, as shown in this example, although the fast time scale and generation of a shock wave clearly indicate that the process is irreversible. This approximation has the advantage of establishing an upper limit to the power of the explosive so that a reasonable margin of safety can be set, but irreversible models do tend to agree better with experiment. Problem 8.10 offers a simple example of an irreversible adiabatic expansion.

SOLUTION Equation 8.12 allows us to predict the maximum work from the expansion:

$$w = \frac{P_1 V_1}{\gamma - 1}\left[\left(\frac{P_2}{P_1}\right)^{1-(1/\gamma)} - 1\right].$$

P_1 is the initial pressure of the expanding gases ($1.0 \cdot 10^3$ bar), V_1 is the initial volume (1.0 L), and P_2 is the ambient pressure (1.0 bar, the pressure at which the expansion will stop). We substitute these numbers and multiply by the factor of 70% to find

$$w = (0.70)\frac{(1.0 \cdot 10^3 \text{bar})(1.0\text{L})}{1.4 - 1}\left[\left(\frac{1.0}{1.0 \cdot 10^3}\right)^{1-(1/1.4)} - 1\right] = -1.5 \cdot 10^3 \text{bar L}$$

$$m_{\text{TNT}} = (1.5 \cdot 10^3 \text{bar L})\left(\frac{100\text{J}}{1\text{bar L}}\right)\left(\frac{1.0\text{g TNT}}{4680\text{J}}\right) = 32\text{g TNT}.$$

EXTEND Notice that the initial temperature of the gases was assumed to be ambient, rather than extremely hot. Oddly enough, although we often associate explosions with highly exothermic reactions, the destructive force in the most common explosions is primarily *mechanical energy*—not chemical energy—generated by the rapid expansion of gases that suddenly find themselves at much higher pressure than their surroundings. It is difficult to study the chemical mechanisms at work in explosions, because the dynamics are so rapid, but HMTD is one example where the molecule may fall apart so quickly that relatively little combustion actually occurs. Because energy is needed for the initial bond breaking, the net reaction is sometimes only weakly exothermic and may even be endothermic. Nonetheless, the mechanical energy released when one solid-phase molecule becomes four or five gas-phase molecules—driven to occupy thousands of times their current volume—can have devastating results.

Joule-Thomson Expansion

There is a special case of the adiabatic expansion in which the enthalpy is held constant, even while P, V, and T are all permitted to change. A schematic apparatus for the experiment is shown in Fig. 8.10. A gas is initially at pressure P_1, volume V_1, and temperature T_1, and is separated by a permeable plug from a second container. The plug allows gas to flow, but it *prevents the temperatures and pressures of the two containers from equilibrating*. We press a piston in container 1 to push the gas into container 2, where a second piston is pulled out to make a volume for the gas to expand into. The critical feature is that the expansion from one container into the other is carried out at constant pressure P_1 in the first container and a constant pressure P_2 in the second container. In order for the gas to flow from 1 to 2, P_2 is always less than P_1. The experiment ends when all the gas has been pushed from container 1 to container 2. No heat is permitted to flow into the system or between the two containers.

Because no heat could flow, the change in energy in each container is just equal to the work w in each container, and ΔE overall is equal to the sum of the work in container 1 and the work in container 2:

$$\Delta E = w_2 + w_1. \tag{8.13}$$

Furthermore, because we kept the pressure in each container constant, the work in each container is just $-P\Delta V$. Therefore,

$$\Delta E = -P_2(V_2) - P_1(-V_1) = P_1 V_1 - P_2 V_2. \tag{8.14}$$

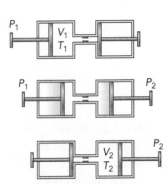

▲ FIGURE 8.10 **The Joule-Thomson experiment.** Gas is forced through a permeable plug at a constant pressure P_1 into a lower pressure P_2.

This difference doesn't have to be zero, because we can pick V_1 and V_2 to be almost anything we want, so we can get the energy to change during this process. However, it turns out that these conditions force the *enthalpy* to be constant, because the change in energy is exactly balanced by the change in PV:

$$\Delta H = \Delta(E + PV) \qquad\qquad H = E + PV$$
$$= \Delta E + \Delta(PV)$$
$$= (P_1V_1 - P_2V_2) + (P_2V_2 - P_1V_1) \qquad \text{by Eq. 8.14}$$
$$= 0. \tag{8.15}$$

This is called the **Joule-Thomson expansion,** after James Joule and William Thomson (also known as Lord Kelvin, of absolute temperature fame).

To calculate the actual work and ΔE values is a task we'll leave for an exercise, but a simpler question we can answer for now is this: at the plug that separates the two containers, how much does the temperature change as the pressure drops? Mathematically, we are looking for the derivative of T with respect to P at constant enthalpy; this number is called the **Joule-Thomson coefficient,** $\left(\frac{\partial T}{\partial P}\right)_H$. It can be expressed in terms of the heat capacity of the gas by use of the chain rule (Table A.4):

$$\left(\frac{\partial T}{\partial P}\right)_H = -\left(\frac{\partial T}{\partial H}\right)_P\left(\frac{\partial H}{\partial P}\right)_T = -\frac{1}{C_P}\left(\frac{\partial H}{\partial P}\right)_T. \tag{8.16}$$

The remaining partial derivative in Eq. 8.16 may be written in terms of the coefficient of thermal expansion α as follows:

$$\left(\frac{\partial H}{\partial P}\right)_T = \left(\frac{T\partial S + V\partial P}{dP}\right)_T \qquad\qquad dH = TdS + VdP$$
$$= T\left(\frac{\partial S}{\partial P}\right)_T + V \qquad\qquad \partial P/\partial P = 1$$
$$= -T\left(\frac{\partial V}{\partial T}\right)_P + V = -TV\alpha + V. \tag{8.17}$$

The third step uses one of our Maxwell relations (Eq. 7.30).

The Joule-Thomson coefficient is

$$\left(\frac{\partial T}{\partial P}\right)_H = \frac{TV\alpha - V}{C_P}. \tag{8.18}$$

This formula is valid for any gas in a closed system, ideal or not. For the ideal gas, we have already shown (Eq. 7.54) that $\alpha = 1/T$, which sets the Joule-Thomson coefficient to zero. For the ideal gas, the temperature of the gas does not change during this expansion, so $P_1V_1 = nRT = P_2V_2$, and therefore $\Delta E = 0$ as well.

To estimate how a real gas will deviate from the ideal gas, we add the simplest non-ideal term to the ideal gas law—namely, the second virial coefficient $B_2(T)$ from Eq. 4.39:

$$P = RT[V_m^{-1} + B_2(T)V_m^{-2}]$$

The Joule-Thomson coefficient depends on V, T, α, and C_P (Eq. 8.18). We leave the temperature in the expression, temperature being one of the parameters we measure, and we treat the heat capacity C_P as an empirical parameter for the non-ideal gas. The non-ideality appears in V and in α.

To evaluate V and α, we want a simple formula for the volume that includes the correction from the virial expansion:

$$V = \frac{nRT}{P}\left(1 + \frac{n}{V}B_2(T)\right). \qquad (8.19)$$

Taking the derivative of this equation to get $V\alpha$ is straightforward, keeping in mind that V is a function of T:

$$V\alpha = \left(\frac{\partial V}{\partial T}\right)_P \qquad \text{differentiate Eq. 8.19}$$

$$= \frac{nR}{P}\left(1 + \frac{n}{V}B_2(T)\right) + \frac{nRT}{P}\left[-\frac{n}{V^2}B_2(T)\left(\frac{\partial V}{\partial T}\right)_P + \frac{n}{V}\left(\frac{\partial B_2}{\partial T}\right)_P\right]$$

$$= \frac{V}{T} + \frac{nRT}{P}\left[-\frac{n}{V^2}B_2(T)\left(\frac{\partial V}{\partial T}\right)_P + \frac{n}{V}\left(\frac{\partial B_2}{\partial T}\right)_P\right]. \qquad (8.20)$$

We have neglected $nB_2(T)/V$ compared to 1 in the first term. Because $\left(\frac{\partial V}{\partial T}\right)_P$ appears on both sides of the equation, we combine both sides to solve for $V\alpha$:

$$V\alpha = \left(\frac{\partial V}{\partial T}\right)_P = \frac{\dfrac{V}{T} + \dfrac{nRT}{P}\dfrac{n}{V}\left(\dfrac{\partial B_2}{\partial T}\right)_P}{1 + \dfrac{nRT}{P}\dfrac{n}{V^2}B_2(T)}. \qquad (8.21)$$

Finally, we go back to the Joule-Thomson coefficient,

$$\left(\frac{\partial T}{\partial P}\right)_H = \frac{TV\alpha - V}{C_P}$$

$$= \frac{1}{C_P}\left[\frac{V + \dfrac{nRT^2}{P}\dfrac{n}{V}\left(\dfrac{\partial B_2}{\partial T}\right)_P}{1 + \dfrac{nRT}{P}\dfrac{n}{V^2}B_2(T)} - V\right] \qquad \text{by Eq. 8.21}$$

$$= \frac{1}{C_P}\left[\frac{\dfrac{nRT^2}{P}\dfrac{n}{V}\left(\dfrac{\partial B_2}{\partial T}\right)_P - \dfrac{nRT}{P}\dfrac{n}{V}B_2(T)}{1 + \dfrac{nRT}{P}\dfrac{n}{V^2}B_2(T)}\right] \qquad \text{combine terms}$$

$$= \frac{1}{C_P}\left\{\frac{\dfrac{n^2RT}{PV}\left[T\left(\dfrac{\partial B_2}{\partial T}\right)_P - B_2(T)\right]}{1 + \dfrac{nRT}{P}\dfrac{n}{V^2}B_2(T)}\right\} \qquad \text{factor out } n^2RT/(PV)$$

$$= \frac{1}{C_P}\left\{\frac{T\left(\dfrac{\partial B_2}{\partial T}\right)_P - B_2(T)}{\dfrac{PV}{n^2RT} + \dfrac{B_2(T)}{V}}\right\} \qquad \text{cancel like factors}$$

$$= \frac{1}{C_P}\left\{\frac{T\left(\dfrac{\partial B_2}{\partial T}\right)_P - B_2(T)}{\dfrac{1}{n} + \dfrac{B_2(T)}{V}}\right\} \qquad \text{set } PV/(nRT) = 1$$

$$\approx \frac{1}{C_P}\left\{\frac{T\left(\dfrac{\partial B_2}{\partial T}\right)_P - B_2(T)}{\dfrac{1}{n}}\right\} \qquad V_m \ll B_2(T)$$
$$\text{so } 1/n = V_m/V \ll B_2(T)/V$$

$$= \frac{T\left(\dfrac{\partial B_2}{\partial T}\right)_P - B_2(T)}{C_{Pm}}. \tag{8.22}$$

The final expression, Eq. 8.22, is simple, but not terribly informative yet. We need only one more step to pull out predictions, however, and that is to rewrite $B_2(T)$ and $\left(\frac{\partial B_2}{\partial T}\right)_P$ in terms of the van der Waals coefficients a and b (Table 4.2):

$$B_2(T) = b - \frac{a}{RT} \qquad \left(\frac{\partial B_2}{\partial T}\right)_P = \frac{a}{RT^2}. \tag{8.23}$$

Now we can write

$$\left(\frac{\partial T}{\partial P}\right)_H = \frac{\dfrac{a}{RT} - \left(b - \dfrac{a}{RT}\right)}{C_{Pm}}$$

$$\left(\frac{\partial T}{\partial P}\right)_H = \frac{\dfrac{2a}{RT} - b}{C_{Pm}}. \tag{8.24}$$

Recall that b represents the excluded volume—roughly the volume of the container occupied by the molecules themselves—and a represents the strength of the intermolecular attraction. Both are positive. We consider two limits:

at high T: $$\left(\frac{\partial T}{\partial P}\right)_H \approx -\frac{b}{C_{Pm}} \tag{8.25}$$

at low T: $$\left(\frac{\partial T}{\partial P}\right)_H \approx \frac{2a}{RTC_{Pm}}. \tag{8.26}$$

Although we have not considered the exact form of C_{Pm}, it's safe to say that it is always positive. When the temperature is high, therefore, the Joule-Thomson coefficient is negative—which means that the temperature decreases as the pressure increases. For our adiabatic expansion, the temperature would *rise* as the gas expands and the pressure drops. At low temperature, on the other hand, the a term dominates, and the coefficient is positive—meaning that the gas cools off as it expands.

Consider the following qualitative pictures of what happens. Gas-phase molecules that share a large a coefficient tend to form weakly bonded complexes, easily dissociated by the impact of another molecule. When this sample expands rapidly into a vacuum, the density decreases. Many of the weakly bonded complexes are dissociated by collisions with other molecules and are not replaced, because the formation of weakly bonded complexes is unlikely at low densities. Therefore, the sample goes from a state in which many of the molecules are weakly bound, at low potential energy, to a state in which nearly all the molecules are unbound, and therefore at higher potential energy. This requires a conversion of kinetic energy into potential energy (collisions lose energy to the breaking of the van der Waals bonds), which lowers the temperature (Fig. 8.11a).

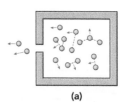
(a)

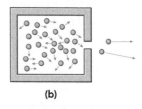
(b)

▲ FIGURE 8.11 The Joule-Thomson expansion in different limits of the intermolecular forces. (a) When the attractive forces dominate the interaction, the molecules are more stable clustered together in the container than sparsely separated in the surroundings; the potential energy increases, the kinetic energy drops, and the temperature therefore drops. (b) When the attractive forces are insignificant, the molecules repel each other, and the potential energy is *lower* when the molecules are separated: less potential energy means more kinetic energy, and therefore *higher* temperature.

On the other hand, molecules with very low a coefficients, or at very high temperatures, form no significant number of complexes. Instead, the chief interaction between the molecules involves only the repulsive wall of the intermolecular potential. At high densities, the molecules spend much of their time in regions of high potential energy next to other molecules. When such a sample expands, it is like spilling a jar of marbles; the potential energy required to keep the molecules stored next to each other is converted to kinetic energy as the molecules roll away from each other (Fig. 8.11b). The sample heats up.

A compressed gas cylinder of helium or methane typically gets filled to a pressure of over 100 bar. Compared to this, 1 bar is negligible, and the liberation of a gas from one of these high-pressure cylinders—to fill a balloon or fuel a methane torch, for example—is approximately a Joule-Thomson expansion. Most gases at room temperature are still in the low temperature limit, and the gas cools as it expands. For H_2 and He gas, however, the molecules are so non-polarizable and the intermolecular forces so weak that the a constant is extremely small, and even at room temperature the Joule-Thomson coefficient is negative. These gases warm up as they expand against a low pressure. The temperature at which the Joule-Thomson coefficient changes sign is called the **Joule-Thomson inversion temperature,** and it tells us where the gas crosses the boundary from the dominance of attractive intermolecular forces to dominance by the repulsions.

EXAMPLE 8.3 **Joule-Thomson Coefficients**

CONTEXT Joule-Thomson cooling occurs in many applications, and understanding the process may be critical to the success of some ventures. For example, one proposal for the reduction of carbon dioxide in the atmosphere is **CO_2 sequestration:** pumping the CO_2 into gas-tight pockets deep underground, left vacant by the extraction of petroleum.[2] One concern, however, is that the nearly enthalpy-free expansion of CO_2 from a transfer tube into the pocket might lead to substantial Joule-Thomson cooling, potentially leading the CO_2 to freeze and clog the line. The first step toward understanding these dynamics is calculation of the Joule-Thomson coefficient.

PROBLEM Use the van der Waals constants to estimate the Joule-Thomson coefficients for He, H_2, N_2, and CO_2 at 300. K.

	a(L^2 bar mol^{-2})	b(L mol^{-1})	C_{Pm}(J K^{-1} mol^{-1})
He	0.034	0.0237	20.79
H_2	0.247	0.0266	28.82
N_2	1.37	0.0387	29.13
CO_2	3.66	0.0429	37.11

SOLUTION The Joule-Thomson coefficient is approximately equal to

$$\left(\frac{\partial T}{\partial P}\right)_H \approx \frac{\frac{2a}{RT} - b}{C_{Pm}},$$

---

[2]In fact, one idea is that pumping carbon dioxide *into* these pockets will enhance the removal of methane *from* the pockets, and the methane can then be sold in order to help finance the sequestration. You might reasonably wonder just what's supposed to happen with the methane that we would sell to finance the removal of greenhouse gases from the atmosphere.

where $R = 0.083145\,\text{bar}\,\text{L}\,\text{K}^{-1}\,\text{mol}^{-1}$ and $T = 300\,\text{K}$. This equation gives the Joule-Thomson coefficient $\left(\frac{\partial T}{\partial P}\right)_H$ in the awkward units L K J^{-1}, so we convert L J^{-1} to bar^{-1} by multiplying by $(10^{-3}\ \text{m}^3/\text{L})^3$ $(10^5\ \text{Pa/bar})$, which increases the value by a factor of 100.

	$\frac{2a}{RT}$ (L mol^{-1})	$\left(\frac{\partial T}{\partial P}\right)_H$ (from Eq. 8.24) (K bar^{-1})	$\left(\frac{\partial T}{\partial P}\right)_H$ (experiment) (K bar^{-1})
He	$2.8 \cdot 10^{-3}$	-0.100	-0.062
H_2	$2.01 \cdot 10^{-2}$	-0.024	-0.03
N_2	0.110	0.24	0.27
CO_2	0.293	0.675	1.11

EXAMPLE 8.4   The Joule-Thomson Inversion Temperature

CONTEXT Gases in industrial applications suffer from one inconvenient characteristic: at STP, a gas occupies a thousand times the volume of the same substance in liquid form. Numerous chemicals that are gases under normal conditions are liquefied to save shipping costs (since the liquid weighs just as much as the gas, but takes up a lot less space), or to take advantage of the cryogenic properties of liquids with very low boiling points (especially liquid N_2, which boils at 77 K, and liquid helium, which boils at 4 K). One standard method for liquefying a gas is to compress it to a high pressure and then let it cool by Joule-Thomson expansion down to its boiling point, where some of the gas will condense. But this backfires if the gas *heats up* as it expands, if the Joule-Thomson coefficient is negative. A basic liquefier design compresses the gas at ambient temperature, lowers the temperature using a heat exchanger, carries out a Joule-Thomson expansion to convert some of the gas into liquid, and then sends the remaining gas back through the heat exchanger (to cool the incoming gas) whereupon it will cycle through the compressor again. For gases where the Joule-Thomson inversion temperature is very low, it is necessary to pre-cool the gas *below* its Joule-Thomson inversion temperature. Otherwise, the compressed gas heats up when expanding and doesn't liquefy.

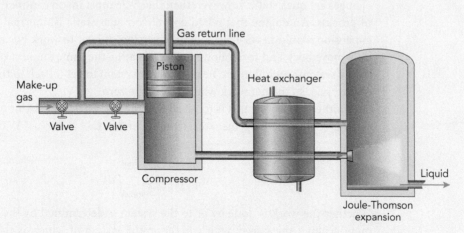

PROBLEM Given $a = 0.034$ L^2 bar mol^{-2} and $b = 0.0237$ L mol^{-1} for He, and 2.25 L^2 bar mol^{-2} and 0.0428 L mol^{-1} (respectively) for CH_4, calculate the Joule-Thomson inversion temperatures for these gases.

SOLUTION Helium, because its attractive forces are so weak, has a very low inversion temperature and requires significant cooling before liquefying:

$$\left(\frac{\partial T}{\partial P}\right)_H = \frac{\dfrac{2a}{RT} - b}{C_{Pm}} = 0$$

$$\frac{2a}{RT} = b$$

$$T = \frac{2a}{Rb} = 35 \text{ K for He, } 1260 \text{ K for } CH_4.$$

On the other hand, even a relatively light molecule such as methane has an inversion temperature of over 1000 K. A glance at the polarizabilities explains the discrepancy: helium, the smallest of all atoms, has a polarizability of only $0.20\,\text{Å}^3$, an order of magnitude smaller than the polarizability of methane, $2.59\,\text{Å}^3$. The higher the polarizability, the greater the dispersion force pulling the particles together.

8.2 Engines

We have looked at isolated types of mechanical processes—adiabatic and isothermal expansions in particular. For many applications, however, the interesting cases are when different types of processes are combined.

We call a device that continuously converts heat into work a **heat engine**. Since the heat source can deliver only a finite amount of energy at any given time, practical engines operate by repeating a single cyclic process that performs this conversion, relying on a stable or reproducible heat source.

Let's consider a couple of general points about these cycles. We'll keep the assumption that the processes in the engine cycle are quasistatic, because in the gas expansion problems we found that the maximum work was done by the reversible process, and a reversible process must be quasistatic. If all the changes are quasistatic, however, then the cycle must involve more than one type of process. An engine that relied entirely on quasistatic isothermal or adiabatic expansion would never have a net conversion of heat to work because it would only move back and forth along the same path—the same amount of work done by the system would have to be done *to* the system to get it back to the beginning of the cycle, so the net work would always be zero.

Another way to see this is to recall that the PV work done by or to the system can be represented by the area integral under the P versus V curve of the process:

$$w_{\text{rev}} = -\int_{initial}^{final} P\,dV. \tag{8.27}$$

Whether the work is done by or to the system is determined by the direction of motion along the curve: which is the initial state and which is the final. This determines which point is the upper limit of the area integral and which is

---

BIOSKETCH | Richard B. Peterson

Richard B. Peterson is professor of mechanical engineering and director of the Advanced Tactical Energy Systems Program at Oregon State University. One of many projects in his research group involving basic thermodynamics is the development of new methods to transfer heat, for example to cool vehicle cabins or protective clothing. It is an ancient problem: how to remove energy from a system that may already be cooler than its surroundings. Opposing the natural flow of heat toward the cooler body requires energy and relies on a series of processes similar to the heat engine cycles described in this chapter. Professor Peterson is investigating mechanical designs based on microchannel components that allow the energy released by a burning hydrocarbon fuel to efficiently pump heat from one location to another, drawing on his expertise in the use of microscopic channels for heat and mass transfer.

---

the lower limit, and it therefore determines the sign of w_{rev}. In the case of a cyclic process, the net work is given by the sum of the area integrals for the steps of the process. For example, if the cycle moves among four states in the order $A \rightarrow B \rightarrow C \rightarrow D \rightarrow A$ and then repeats,

$$w = -\int_A^B P\,dV - \int_B^C P\,dV - \int_C^D P\,dV - \int_D^A P\,dV. \qquad (8.28)$$

If the cycle simply runs back and forth along the same path, the area integrals sum to zero. So to get net work done by the engine, more than one kind of process needs to be used.

The Carnot Cycle

One of the simplest such cycles is the **Carnot cycle,** named for Nicolas Carnot, who developed this first model of the heat engine in 1824. (It was Carnot's model that later inspired Clausius's studies, eventually leading him to formulate the concept of entropy.) Our system is an idealized steam engine (Fig. 8.12), consisting of an ideal gas in a cylindrical chamber of diathermal walls, trapped by a piston that moves when work is done by the engine. We can heat the gas to some high temperature T_{hot} by filling a jacket around the chamber with steam, or we can cool the gas down to a much lower temperature T_{cold} using cold water. The piston is hooked up to a *compressor,* a smaller engine that we can use to compress the gas.

We can choose to begin with the piston down (the gas is compressed) and the diathermal wall heating the system to temperature T_{hot}. Call this state A of the engine. (We will label the states by letter and will number the steps that take us from one state to the next.) The cycle then consists of the following four steps (Fig. 8.13):

1. **Isothermal expansion.** The gas expands isothermally at temperature T_{hot} from pressure P_A and volume V_A to P_B and V_B.

2. **Adiabatic expansion.** We expel the steam heating the diathermal wall. Now the system begins to cool off, expanding adiabatically until it reaches temperature T_{cold} at pressure P_C and volume V_C.

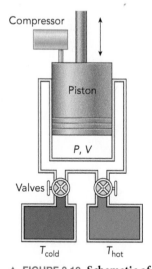

▲ **FIGURE 8.12 Schematic of the Carnot engine.** The walls of the chamber may be heated or cooled by use of the two temperature reservoirs, or they may be thermally insulated from the surroundings for adiabatic expansion and compression.

3. **Isothermal compression.** Now we compress the cooled gas isothermally, using the little engine to push the piston back in, until we get to the conditions P_D and V_D, at T_{cold} (which is maintained by flowing cold water against the wall).

4. **Adiabatic compression.** Finally, we purge the cold water and return to the starting conditions in step 1 by adiabatically compressing the gas until it reaches T_{hot} as a result of the *compression*, rather than by heating the diathermal wall again. From here the cycle can begin again by disengaging the compressor and letting the steam back in behind the diathermal wall.

The important point to see in this cycle is that while the gas cycles back and forth between pressure limits of P_A and P_C, less work is done on the system in the return half of the cycle (steps 3 and 4) than the system does during the power stroke (steps 1 and 2). From a careful look at Fig. 8.13, we can see why this is so: the return stroke involves compressing the *cooled* gas, the power stroke is expansion of the *heated* gas. Heated gas exerts more pressure than cooled gas at the same volume, so it does more work as it expands and requires more work to compress it. The secret of the Carnot cycle is to compress the gas while it is cold and exerts less pressure.

Now let us find the heat and work transferred during each of these stages in terms of the state functions V and T. The volumes are all unequal, with $V_A < V_D < V_B < V_C$ and $T_{cold} < T_{hot}$.

1. **Isothermal expansion.** The work for step 1 is:

$$w_1 = -\int_{V_A}^{V_B} P(V)dV = -\int_{V_A}^{V_B} \frac{nRT_{hot}}{V} dV = -nRT_{hot} \ln \frac{V_B}{V_A}. \quad (8.29)$$

Since the process is isothermal and we are assuming an ideal gas, the energy E is constant. Therefore, $\Delta E = q + w = 0$, and

$$q_1 = -w_1 = nRT_{hot} \ln \frac{V_B}{V_A}. \quad (8.30)$$

Work is done by the system ($đw_1 < 0$) as it expands against the surroundings, and heat must flow into the system ($đq_1 > 0$) to keep it at constant temperature.

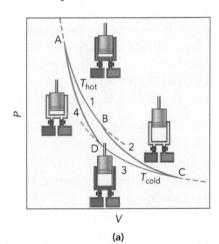

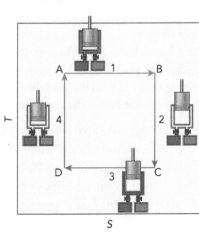

▶ **FIGURE 8.13 The Carnot cycle.** (a) The P-V and (b) T-S graphs are shown for the Carnot engine. Labels indicate the stages of the Carnot cycle as given in the text. The horizontal lines in (b) show the isotherms (constant temperature curves) for the two temperatures.

(a)

(b)

2. **Adiabatic expansion.** For the ideal gas, $E = C_V T$, so $\Delta E = C_V \Delta T$. Furthermore, for an adiabatic process $q = 0$, so $\Delta E = w$, giving

$$w_2 = C_V(T_{cold} - T_{hot}) \qquad q_2 = 0. \qquad (8.31)$$

Work is still being done by the system, but no heat can flow in to replace the energy lost as work, so the system cools off.

3. **Isothermal compression.** This step is analogous to step 1, but in the opposite direction:

$$w_3 = -\int_{V_C}^{V_D} P(V)dV = -nRT_{cold}\ln\frac{V_D}{V_C} \qquad q_3 = nRT_{cold}\ln\frac{V_D}{V_C}. \quad (8.32)$$

Now work is being done to compress the system ($đw > 0$), which would heat it up if it were not for the cold temperature reservoir, so heat flows out ($đq < 0$).

4. **Adiabatic compression.** And this step is analogous to step 2, but in the opposite direction:

$$w_4 = -C_V(T_{cold} - T_{hot}) \qquad q_4 = 0. \qquad (8.33)$$

Work is still being done to the system, but now the system heats up since no heat can flow out of the system.

The work done *by* the system in step 2 exactly cancels the work done *to* the system in step 4, so the total work over the cycle is

$$w = w_1 + w_2 + w_3 + w_4$$
$$= -nR\left(T_{hot}\ln\frac{V_B}{V_A} + T_{cold}\ln\frac{V_D}{V_C}\right). \qquad (8.34)$$

We can simplify this because we learned while working out the properties of the adiabatic expansion that the ratio of the initial and final volumes depends only on the ratio of the initial and final temperatures (Eq. 8.9):

$$C_V\ln\frac{T_2}{T_1} = -nR\ln\frac{V_2}{V_1}.$$

Our steps 2 and 4 are adiabatic expansions at different volumes and pressures, but between the same two temperatures, T_{hot} and T_{cold}. Rewriting Eq. 8.9 for the states in the Carnot cycle, we obtain

$$C_V\ln\frac{T_{hot}}{T_{cold}} = -nR\ln\frac{V_B}{V_C} = -nR\ln\frac{V_A}{V_D}.$$

Therefore, we can relate the volume ratios V_B/V_A and V_D/V_C that appear in Eq. 8.34:

$$\frac{V_D}{V_A} = \frac{V_C}{V_B}$$
$$\frac{V_D}{V_C} = \frac{V_A}{V_B}.$$

The total work done can then be rewritten:

$$w = -nR\left(T_{hot}\ln\frac{V_B}{V_A} + T_{cold}\ln\frac{V_D}{V_C}\right) = -nR\left(T_{hot}\ln\frac{V_B}{V_A} + T_{cold}\ln\frac{V_A}{V_B}\right)$$
$$= -nR(T_{hot} - T_{cold})\ln\frac{V_B}{V_A}. \qquad \ln\frac{V_A}{V_B} = -\ln\left(\frac{V_B}{V_A}\right) \quad (8.35)$$

This quantity is negative (because $V_B > V_A$ and $T_{hot} > T_{cold}$), which implies that over one complete cycle *work is done by the system.*

The total heat absorbed by the system is the sum of the heats from steps 1 and 3:

$$q = nR\left(T_{hot} \ln \frac{V_B}{V_A} + T_{cold} \ln \frac{V_D}{V_C} \right) = nR(T_{hot} - T_{cold})\ln \frac{V_B}{V_A}. \quad (8.36)$$

This is -1 times the work, which is a requirement because we have returned to the original thermodynamic state in step 1 by the end of the cycle, so from beginning to end of the cycle $\Delta E = q + w$ must be zero, and $q = -w$.

The feature that measures an engine's usefulness is its ability to convert fuel, which provides the source of the high temperature reservoir at T_{hot}, into work. The cooling temperature reservoir is usually available as some recyclable commodity (unlike the fuel), such as water or air. The **efficiency** ε of the engine is therefore given by the ratio of the work obtained to the heat provided by the fuel, q_1:

CHECKPOINT If we reverse the direction of the Carnot cycle, then we have a device that carries out an isothermal expansion at T_{cold} (which cools the cold water) and an isothermal compression at T_{hot} (which warms the hot water). In other words, the device moves energy from cold to hot, in the direction opposite to the natural flow of heat. This is accomplished by putting work *into* the engine. This is the operating principle of a refrigerator or a heat pump.

$$\varepsilon = \frac{-w}{q_1} = \frac{nR(T_{hot} - T_{cold})\ln \dfrac{V_B}{V_A}}{nRT_{hot} \ln \dfrac{V_B}{V_A}} = \frac{T_{hot} - T_{cold}}{T_{hot}}. \quad (8.37)$$

The efficiency cannot be greater than 1, which is required by the conservation of energy. Its highest value is for a very large temperature difference $T_{hot} - T_{cold}$, or in other words, for a very hot fuel source.

Note that the P versus V graph is not always the easiest way to draw these cyclic processes. For the Carnot cycle, it would actually be easier to draw the T versus S graph, since the isothermal processes are at constant T and the adiabatic processes are at constant S. This curve is useful for representing the heat evolved or absorbed at each stage of the cycle, just as the P versus V graph is illustrative of the work.

The Otto Cycle

We can apply these same methods to a device that is more common today than Carnot's steam engine: the internal combustion engine. An idealized version of the operation of an internal combustion engine is given by the **Otto cycle**. The system in this case is an adiabatic cylinder with a piston, again connected to a compression engine. This time instead of two temperature reservoirs we need two material reservoirs: a fuel supply line and an exhaust line. Finally, we need a spark plug, which will ignite the fuel.

The steps in the cycle are as follows:

1. Starting this time *before* the power stroke at P_A and T_A, we ignite the compressed fuel at volume V_A. The combustion is much more rapid than any mechanical motion, so the pressure and temperature increase to P_B and T_B *isochorically* (at constant volume).

2. Now comes the power stroke. The pressurized combustion products expand adiabatically to pressure P_C and volume V_C.

3. The piston stops, and the gas cools isochorically, with pressure dropping to P_D. This is a considerable simplification over real devices.

4. Finally, the compressor returns the system adiabatically to volume V_A, pressurizing it in the process. In an internal combustion engine, this coincides with replacement of the exhaust gases with fresh fuel.

This process is more easily drawn on an S-V graph, consisting only of horizontal adiabats and vertical isochors. The efficiency of this engine, found the same way as for the Carnot cycle, is given by the equation (Problem 8.28):

$$\varepsilon = \frac{(T_{hot} - T_{cold})}{T_{hot}} = 1 - \left(\frac{V_A}{V_B}\right)^{(C_P - C_V)/C_V}. \quad (8.38)$$

The ratio V_A/V_B is the compression ratio, and it appears with an exponent that is positive (since $C_P > C_V$). The bigger the difference in volumes, therefore, the closer the efficiency comes to the ideal value of 1. Greater compression ratios imply greater fuel efficiency.

Since the maximum work is accomplished when the process is reversible, and a reversible process must be quasistatic, uneven or too rapid combustion reduces the engine efficiency. The combustion rate is a function of the fuel as much as the igniter, and control over this parameter has long been a target of petroleum engineering. Lead anti-knocking agents were promoted in the 1920s to improve engine efficiency but were subsequently outlawed as the highly toxic lead found its way into the environment.

CONTEXT *Where Do We Go from Here?*

In the low-temperature Joule-Thomson expansion, the gas is at lower potential energy before the expansion than after. The property that compels the gas to climb up out of the cylinder, expanding and breaking intermolecular bonds to arrive at a *less energetically favored state,* is the subject of our next chapter: the entropy.

With this introduction to fundamental processes in thermodynamics, we can now look more closely at the entropy and how it controls the evolution of our systems. In particular, we want to find ways of predicting changes in entropy that take advantage of measurements that are straightforward to make, such as temperature, volume, and pressure measurements. Once we have a feel for how entropy directs processes, we will be able to extend these rather formal case studies in thermodynamics to real chemical problems, such as phase transitions and solvation.

CHECKPOINT The Carnot cycle and the Otto cycle, although the specific paths differ, perform the same job: converting fuel to work, and in the process absorbing heat from T_{hot} and releasing it again to T_{cold}. For this reason, the efficiencies of the Carnot and Otto cycles are both given by Eq. 8.37. All heat engines with the same T_{hot} and T_{cold} have the same theoretical efficiency.

KEY CONCEPTS AND EQUATIONS

8.1 Expansion of Gases. During the reversible **isothermal expansion** of an ideal gas from volume V_1 to V_2 at a constant temperature, the change in energy due to work is

$$w_{T,rev} = -nRT \ln\left(\frac{V_2}{V_1}\right). \quad (8.2)$$

For the temperature to remain constant, the energy lost as the sample does work must be replaced by an equal energy transferred as heat from the surroundings:

$$q_{T,rev} = nRT \ln\left(\frac{V_2}{V_1}\right). \quad (8.4)$$

The irreversible expansion always does less work. If the expansion opposes a constant pressure P_{min}, the work is

$$w_{T,irr} = -P_{min}(V_2 - V_1). \quad (8.3)$$

In an **adiabatic expansion,** heat cannot flow to replace the energy lost to the PV work, so the temperature drops. The work done can be calculated from the initial and final temperatures or pressures:

$$w_{S,rev} = C_V(T_2 - T_1) = \frac{P_1 V_1}{\gamma - 1}\left[\left(\frac{P_2}{P_1}\right)^{1-(1/\gamma)} - 1\right]. \quad (8.12)$$

The Joule-Thomson expansion is a special case of the adiabatic expansion, carried out at constant enthalpy.

The van der Waals equation of gases allows us to predict that the Joule-Thomson coefficient,

$$\left(\frac{\partial T}{\partial P}\right)_H = \frac{\frac{2a}{RT} - b}{C_{Pm}}, \qquad (8.24)$$

will be positive for most real gases, meaning that the gas will cool down as the expansion pulls against the intermolecular attractions. In hydrogen, helium, and other gases with very weak intermolecular attractions, the gas may instead heat up as it expands.

KEY TERMS

- A **Joule-Thomson expansion** allows a gas to expand at constant enthalpy by maintaining a constant pressure inside and outside the original container. Under these conditions it is possible to observe the balance between attractive and repulsive intermolecular forces.
- At the **Joule-Thomson inversion temperature** of a gas, the **Joule-Thomson coefficient** $\left(\frac{\partial T}{\partial P}\right)_H$ goes from a positive value (gas cools upon expansion) to a negative value (gas heats up upon expansion).
- A **heat engine** is a device that converts heat into useful work.

- The **Carnot cycle** is a model for a heat engine that operates on a series of reversible isothermal and adiabatic expansions and compressions, loosely based on the design of a steam engine.
- The **Otto cycle** for a heat engine uses a series of reversible isochoric and adiabatic expansions and compressions. The Otto cycle better represents the operation of an internal combustion engine.
- The **heat capacity ratio** γ for a substance is equal to C_P/C_V.

OBJECTIVES REVIEW

1. *Predict the transfer of energy as work and as heat in reversible and irreversible expansions and compressions. Calculate w for 0.100 mol of an ideal gas compressed reversibly from 2.00 L to 1.00 L at a constant $T = 298$ K.*

2. *Predict the direction of a temperature change in a sample undergoing a Joule-Thomson expansion, based on the*

relative importance of the intermolecular forces. Estimate the Joule-Thomson coefficient of neon at 100 K. (See Table 4.2 for values.)

3. *Calculate the maximum efficiency for a Carnot heat engine and calculate the heat and work at each stage of the cycle. If a Carnot engine has a maximum efficiency of 0.30 and $T_{cold} = 298$ K, what is the value of T_{hot}?*

PROBLEMS

Discussion Problems

8.1 Indicate whether each of the following parameters would be positive ("+"), negative ("−"), or zero ("0") during an irreversible, adiabatic compression of an ideal gas from volume V_1 to a smaller volume V_2.

Parameter	+, −, or 0		Parameter	+, −, or 0
T_2			q	
ΔT			ΔS	
ΔP			ΔE	
w			ΔH	

8.2 A leak in a container allows an ideal gas to escape our sample irreversibly and isothermally at constant pressure. For each of the following parameters, identify any that stay unchanged (for the sample) during this process, any that decrease, and any that increase.

$$\begin{array}{cccc} S & n & V & E \\ S_{tot} & T & H & \mu \end{array}$$

8.3 For the isobaric heating of an ideal gas, indicate whether each of the following would be positive ("+"), negative ("−"), or zero ("0"):

a. ΔE b. q c. ΔP d. ΔT e. ΔV

Expansions

8.4 The reversible isothermal expansion of an ideal gas does 1.00 kJ of work at 325 K, ending at a final volume $V_2 = 10.0$ L and pressure $P_2 = 1.00$ bar. Find the initial volume V_1.

8.5 A 2.5 L volume of a monatomic ideal gas at 298 K is contained by a piston at a constant pressure of 1.00 bar. The piston is lifted by reversible, isothermal, and isobaric addition of 1.00 mole of the same gas through a port in the container. Calculate the work w in J for this process.

8.6 Calculate ΔG for the reversible isothermal expansion of 2.00 mol of an ideal gas from 1.00 L to 20.0 L at 298 K.

8.7 We repeat the isothermal expansion of 1 mol of an ideal gas from 2.478 L to 24.78 L at 298 K, but this time we have a slowly responding piston, such that the pressure inside the sample P is always equal to $2P_{ex} - (1\ \text{bar})$. Calculate the work, w, for the expansion in kJ.

8.8 An irreversible isothermal expansion is carried out in the following system at a temperature of $T = 298.15\,\text{K}$. An ideal gas is initially at a pressure $P_1 = 10.00\ \text{bar}$ in the chamber. One wall of the chamber is movable, but expansion of the gas is resisted by a spring on the other side of the wall. The spring has a force constant of 504.0 N/cm, the diameter of the tube is 10 cm, and the initial length z_1 of the gas chamber is 3.16 cm. The spring exerts a force $F = k(z - z_1)$. Find the values P_2, V_2, and z_2 for the final state, and q, w, ΔE, and ΔH for the entire process.

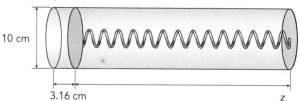

8.9 A chamber has a piston with a cross-sectional area of 10.0 cm^2. How far does the piston move during an isothermal expansion of 0.0020 mole of air from 10.0 bar to 1.0 bar at 298 K?

8.10 Calculate the work in J done during the irreversible process when 0.001 moles of product gases (use $C_{Vm} = 3R$) at 4000 K in a 10 cm^3 volume expand adiabatically during an explosion in air, when the air is at 300 K and 1 bar.

8.11 Find an equation for the rate of change in temperature during an adiabatic expansion, $\left(\frac{\partial T}{\partial V}\right)_{S,n}$, in terms of V, α, κ_T, and κ_S.

8.12 Two thermally insulated gas chambers, A and B, are separated by a freely moving adiabatic wall. The chambers contain ideal gases, both initially at 300 K, 10^5 Pa, and 10 L. Combustion occurs in chamber A, causing the temperature to increase instantaneously to 1000 K and the number of molecules to double. The gases in both chambers now obey the equation $E = 6nRT$. Calculate the *overall* ΔE and ΔH for the expansion (do not include the combustion), and the final values of V_A, V_B, P_A, P_B, T_A, and T_B.

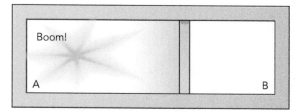

8.13 Start from 0.100 mol of a monatomic ideal gas at a volume of 5.00 L at 298 K. Find the final pressure if we change the volume adiabatically to bring the gas to a final temperature of 410 K.

8.14 Give the final values of P, V, and T for a reversible, adiabatic expansion of 1.00 mol of an ideal gas, starting from 2.00 L and 415 K, if 2.0 kJ of work is done by the expansion. The molar heat capacity at constant volume of the gas is 2.5 R.

8.15 For the reversible isobaric expansion of n moles of an ideal gas at pressure P from an initial volume V_i to a final volume V_f,

a. find an equation for the work w.

b. find an equation for the heat q.

Use the molar heat capacity of the gas C_{Pm} where necessary.

8.16 This chapter describes a reversible, adiabatic expansion of 1.00 mol of a monatomic ideal gas with the following initial and final conditions:

	Initial	Final
P	10.00 bar	1.00 bar
V	2.478 L	9.87 L
T	298 K	119 K

We found that $\Delta E = w = -2.24\,\text{kJ}$. Find ΔH for this process.

8.17 Calculate q and w for the reversible, isothermal compression of 0.100 mol of air from 2.50 L to 0.25 L at 298 K.

8.18 In this chapter, we described the reversible, isothermal expansion of 1.00 mol of an ideal gas from 2.48 L at 10.00 bar and 298 K to a final pressure of 1.00 bar. Repeat the process from the same starting point, but this time set the final pressure P_2 to 0.100 bar. Calculate the following parameters:

a. V_2 b. ΔE c. w d. ΔS e. ΔF

8.19 Two 10.0 L samples A and B each of 1.00 mole helium are placed in a rigid 20.0 L container on either side of a movable wall. Initially, sample A is at a temperature of 400 K while sample B is at a temperature of 250 K. A thin wire in the movable wall allows heat to transfer reversibly between the two regions, and the samples come to equilibrium. Find the volumes V_A and V_B at equilibrium and the heat transfer q_A for sample A for the process. For helium, $C_{Pm} = 20.88\,\text{J K}^{-1}\text{mol}^{-1}$.

8.20 Sketch a $P-V$ graph of the curve for the following cyclical process for an ideal gas:

a. The sample gas expands adiabatically from P_1, V_1 to volume V_2.

b. The sample is compressed isothermally to volume V_3.

c. The sample warms isobarically to the initial state.

This process is graphed while the gas refrigerates a container that is in thermal contact with our sample for only one step of this cycle. Label that step "refrig" on your graph.

8.21 For a monatomic ideal gas, set $\mu = \frac{3}{2}RT$. We reversibly add Δn moles of a monatomic ideal gas to a sample at constant pressure P and temperature T. Find an equation for the work w during this process in terms of P, T, and/or Δn. Because adding the gas involves net motion of the molecular mass, the $\mu\, dn$ term contributes to the work in this problem.

8.22

a. Find a general expression for $\left(\frac{\partial P}{\partial T}\right)_{F,n} + \left(\frac{S}{P}\right)\left(\frac{\partial P}{\partial V}\right)_T$ in terms of any of α, C_P, C_V, or the compressibilities κ, where

$$\kappa_T \equiv -\frac{1}{V}\left(\frac{\partial V}{\partial P}\right)_T \quad \kappa_S \equiv -\frac{1}{V}\left(\frac{\partial V}{\partial P}\right)_S.$$

b. Simplify this expression when the material is an ideal gas.

8.23 Prove that the Helmholtz free energy F decreases during a spontaneous expansion at constant P and T.

8.24 The system sketched in the following figure consists of two rigid (fixed volume) chambers, A and B, which are connected by a porous, thermally insulated membrane. Initially, chamber A has n moles of a gas and B is empty. Chamber A is heated at a constant temperature T_A, causing the gas to migrate through the membrane into chamber B, which is maintained at a much lower constant temperature T_B. Prove that one of the thermodynamic potentials remains a constant during this process.

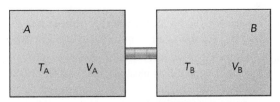

8.25 Compress an ideal gas *adiabatically* from pressure P_1, volume V_1, and temperature T_1 to one-half the volume. What is the final temperature T_2 in terms of the initial temperature T_1 and the molar heat capacities C_{Vm} and C_{Pm}?

Engines

8.26 Find an expression for the heat and work done during steps 1 and 2 of the Otto cycle.

8.27 Calculate the net heat in J consumed by the Carnot cycle graphed in the following figure:

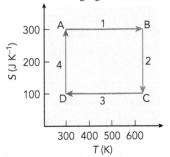

8.28 Prove that the efficiency ε of the Otto cycle is given by

$$\varepsilon = 1 - \left(\frac{V_1}{V_2}\right)^{(C_P - C_V)/C_V}.$$

8.29 Consider two engines, one running on the Carnot cycle and one on the Otto cycle. The adiabatic steps of the power strokes are identical for the two engines. Find which engine is more efficient.

8.30 Find the ratio $\varepsilon_C/\varepsilon_X$, where ε_C is the efficiency of a Carnot engine and ε_X is the efficiency of an engine with the following cycle.

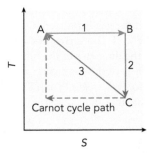

8.31 Consider a closed (constant n) heat engine operating on four steps: (a) isobaric expansion from volume V_1 to V_2, (b) isochoric cooling from temperature T_2 to T_3, (c) isobaric compression from volume V_3 to V_4, (d) isochoric heating from temperature T_4 to T_1. Find expressions for the energy transferred as work and heat at each step and the total work per cycle, in terms of V_1, V_2, T_1, T_3, and n.

8.32 For the process drawn in the following figure, calculate the net work done in one cycle.

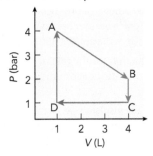

PART I
EXTRAPOLATING
FROM
MOLECULAR TO
MACROSCOPIC
SYSTEMS

PART II
NON-REACTIVE MACROSCOPIC SYSTEMS

7 Introduction to Thermodynamics: Heat Capacity
8 The First Law: Expansion and Engines
9 The Second and Third Laws: Entropy
10 Phase Transitions and Phase Equilibrium
11 Solutions

PART III
REACTIVE
SYSTEMS

9

The Second and Third Laws: Entropy

GOAL *Why Are We Here?*

We're here because the most important parameter in all of thermodynamics is the entropy, and it's time we give it a closer look from the standpoint of bulk systems. Although the absolute entropy is not hard to define, it is *very* difficult to measure in comparison to many other thermodynamic quantities such as T, P, n, V, q, and w. With no convenient yardstick for the entropy, it can be challenging to acquire an intuitive grasp of what entropy means physically. The goal of this chapter is to show how the concept of entropy as we've already defined it leads naturally to the second and third laws of thermodynamics and in fact can be used to obtain reasonably accurate numerical values for the entropy.

CONTEXT *Where Are We Now?*

We have focused so far on macroscopic parameters such as volume, temperature, pressure, and energy that are more familiar to our practical experience than the entropy. With the advantage of a molecular perspective that was not available to our 19th-century thermodynamic trailblazers, we can now approach entropy with a clear and quantitative definition and with an appreciation of the enormous scope of its significance. Indeed, we can hardly make any further progress in our study of thermodynamics until we devote this time to the entropy. Entropy directs the flow of every macroscopic process, including phase changes and solvation—the subjects of the next two chapters.

As part of this discussion, we will closely examine the second law of thermodynamics, a law of literally cosmic importance, governing the dynamics of the universe at nearly every scale. We will also see how the assumptions that we used as the basis for statistical mechanics lead directly to the classical law that heat in a thermodynamic system always flows from one body to another in the direction that will equalize the two temperatures.

> ⌐**SUPPORTING TEXT**   *How Did We Get Here?*
>
> The main preparation we need for the work ahead is Boltzmann's equation (Eq. 2.11),
>
> $$S \equiv k_{\mathrm{B}} \ln \Omega,$$
>
> which expresses the entropy in terms of the number of microstates Ω of the ensemble. From this equation, we will obtain all the principal results of this chapter. In addition, we will take advantage of the following equations and sections of text as we explore the material in this chapter:
>
> - Section 3.4 obtains an expression for the translational partition function by extrapolating from the properties of one quantum mechanical particle in a cubical box. There we find that the possible energies ε of the particle obey Eq. 2.4,
>
> $$\varepsilon_{n_X, n_Y, n_Z} = \frac{h^2}{8mV^{2/3}}(n_X^2 + n_Y^2 + n_Z^2) \equiv \varepsilon_0(n_X^2 + n_Y^2 + n_Z^2) \equiv \varepsilon_0 n^2.$$
>
>   We can count the number of these quantum states per unit energy—the *energy level density W*—in the limit of high energy, and we find that the value is given approximately by (Eq. 3.39)
>
> $$W = \frac{\Omega}{\varepsilon_0} = \frac{\pi}{4\varepsilon_0^{3/2}}\varepsilon^{1/2}.$$
>
> - Stirling's approximation (Eq. 2.14) comes in handy when we need to deal with factorials of large numbers, because it lets us cast the logarithm of the factorial in non-factorial terms:
>
> $$\ln N! \approx N \ln N - N.$$
>
> - In Section 2.4, we introduced the concept of the *temperature reservoir*, a body in contact with the system that can absorb or deliver any amount of heat. In Section 9.2, we will re-examine the flow of energy between the system and the temperature reservoir more closely.

9.1 Entropy of a Dilute Gas

The results from Chapter 2 allow us to calculate the total entropy of a gas at constant energy and volume. We will assume that the gas is dilute so that we can still ignore the intermolecular forces that give rise to the van der Waals coefficients and second virial coefficient in Chapter 4, but we will allow the molecules to rotate and vibrate. Or you can skip to the **DERIVATION SUMMARY** after Eq. 9.6.

If we start from the particle in a three-dimensional box, we have energy levels given by (Eq. 2.4)

$$\varepsilon_{n_X, n_Y, n_Z} = \frac{h^2}{8mV^{2/3}}(n_X^2 + n_Y^2 + n_Z^2) \equiv \varepsilon_0(n_X^2 + n_Y^2 + n_Z^2) \equiv \varepsilon_0 n^2.$$

We showed in Section 3.4 that the number of quantum states per unit energy is given by (Eq. 3.39)

$$W = \frac{\Omega}{\varepsilon_0} = \frac{\pi}{4\varepsilon_0^{3/2}}\varepsilon^{1/2}.$$

Let's replace ε_0 in this equation for W with the expression in Eq. 2.4 to see the mass and volume dependence:

$$W = \frac{\pi}{4}\left(\frac{8mV^{2/3}}{h^2}\right)^{3/2}\varepsilon^{1/2}$$

$$= \frac{4\sqrt{2}\pi m^{3/2}V\varepsilon^{1/2}}{h^3}. \tag{9.1}$$

So we find that at large values of n, the density of states W increases as the volume V increases, as the mass of the particle m increases, and as the energy ε increases. Classical systems have relatively large values of V, m, and ε and so will have very large quantum state densities. As an example, a single hydrogen atom ($m = 1.66 \cdot 10^{-27}$ kg) at an energy corresponding to about 100 K ($\varepsilon = 1.38 \cdot 10^{-21}$ J) in a box of volume 1 cm^3 = 10^{-6} m^3 has an ensemble size of roughly $5.1 \cdot 10^4$ and an energy density of about $1.5 \cdot 10^{44}$ J^{-1}.

Now we check just how quickly the density of quantum states climbs when we go to N particles, with N some gigantic number in a macroscopic system. Because N particles in a three-dimensional box will have $3N$ quantum numbers, we extend the trick we used for three dimensions to $3N$ dimensions. Imagine that we could assemble a box in more than three dimensions; how much space would it contain? For example, a square with sides of length a encloses an area equal to a^2. A cube with sides the same length encloses a volume equal to a^3. We can extrapolate these geometric properties to solids that occupy more than the three dimensions that we perceive. For example, the cube can be made by assembling six two-dimensional squares in the right pattern (Fig. 9.1). Similarly, eight cubes can be assembled in four dimensions to obtain a four-dimensional solid called a *tesseract,* and it contains a four-dimensional volume equal to a^4. By extension, a rectangular solid of D dimensions, made from sides of length a, encloses a space equal to a^D.

However, to count the number of quantum states of a particular energy, we count points on the *surface* of the solid, rather than the points contained within. That evaluation is still straightforward for a rectangular solid: the surface of a square is the perimeter and is equal to $4a$; the surface area of a cube is the sum of the area of 6 squares, or $6a^2$. Extrapolating these to any D-dimensional solid, we find that the size of the surface is always $2Da^{D-1}$ (Table 9.1).

CHECKPOINT In Section 3.4, we construct a sphere that has a radius n, in order to use the surface area of that sphere to count how many combinations of the quantum numbers n_X, n_Y, and n_Z have the same value of n^2. Our goal here is the same– to find how many different sets of quantum numbers give the same total energy–and we will use the same approach. But now, with N particles, there are $3N$ quantum numbers (three for each particle) rather than just three, so instead of a *three*-dimensional sphere (with one axis each for n_X, n_Y, and n_Z), we need a $3N$-dimensional sphere (with one axis each for n_{X1}, n_{Y1}, ..., n_{ZN}).

TABLE 9.1 **Properties of rectangular shapes, as a function of the dimensionality.** All sides are of equal length a, and all edges that meet at the same corner are at right angles to each other.

dimensions	name	made from	vertices	size of boundary	contained space
1	line segment		2	—	a (length)
2	square	4 line segments	4	$4a$ (perimeter)	a^2 (area)
3	cube	6 squares	8	$6a^2$ (surface area)	a^3 (volume)
4	tesseract	8 cubes	16	$8a^3$ (surface volume)	a^4 (hypervolume)
⋮			⋮	⋮	⋮
D	hypercube		2^D	$2Da^{D-1}$	a^D

▶ **FIGURE 9.1** *N*-dimensional **solids.** We can visualize the properties of solids in many dimensions by extrapolating from related shapes in fewer dimensions.

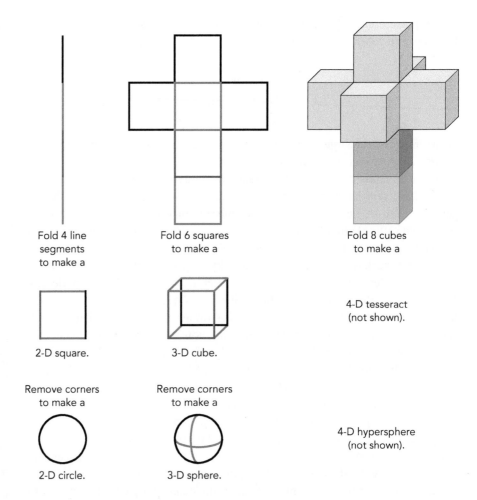

Fold 4 line segments to make a

2-D square.

Fold 6 squares to make a

3-D cube.

Fold 8 cubes to make a

4-D tesseract (not shown).

Remove corners to make a

2-D circle.

Remove corners to make a

3-D sphere.

4-D hypersphere (not shown).

To count the states of equal energy, we need the number of points on the surface of a multidimensional *sphere*, rather than a cube. The surface of a *hypersphere* of D dimensions and radius a occupies a space equal to $C_D D a^{D-1}$, following the same trend as the hypercube but with the constant C_D roughly equal to $\pi^{D/2}/(D/2)!$.[1] We want the number of points on this surface when $D = 3N$. Furthermore, we look only along the positive half of each coordinate axis, because we want only positive values for the quantum numbers, so we take the total surface and divide it by $2^D = 2^{3N}$. The number of points on this surface gives us the ensemble size of N particles in a three-dimensional box. Using the same procedure that we used to find the ensemble size Ω for one particle in a three-dimensional box, Eq. 3.38, we would find that for N *distinguishable* particles, the total number of microstates available is

$$\Omega_{\text{dist}} = \frac{C_D D n^{D-1}}{2^D} dn$$

$$= \frac{\pi^{D/2}}{(D/2)!} \frac{D n^{D-1}}{2^D} dn \qquad \text{plug in } C_D \text{ value}$$

---

[1]This expression for C_D is valid for even values of D. If D is odd, then $C_D = 2^{D+1} \pi^{(D-1)/2}$ $[(D + 1)/2]!/(D + 1)!$. Both expressions yield similar values in the limit of large D, so we use the simpler expression for even D here.

$$= \frac{\pi^{3N/2}}{(3N/2)!} \frac{3Nn^{3N-1}}{2^{3N}} dn. \qquad D = 3N \qquad (9.2)$$

The problem with Eq. 9.2 is that, in our system, we can't tell the difference between, say, particle 1 and particle 2. So if we switch the locations of particles 1 and 2, but leave everything else the same, that's still the same microstate. To divide out all those ways of rearranging the particle labels without changing the microstate, we divide Ω_{dist} by $N!$:

$$\Omega = \frac{\pi^{3N/2}}{N!(3N/2)!} \frac{3Nn^{3N-1}}{2^{3N}} dn. \qquad (9.3)$$

Now we rewrite this expression to get Ω in terms of the N-particle energy E, where Eq. 2.4 still holds, but the value of n^2 now includes contributions from all N molecules:

$$E = \varepsilon_0 n^2, \; n^2 = n_{X1}{}^2 + n_{Y1}{}^2 + n_{Z1}{}^2 + n_{2X}{}^2 + \ldots + n_{NZ}{}^2 \quad \text{Eq. 2.4}$$

$$n = \left(\frac{E}{\varepsilon_0}\right)^{1/2}$$

$$dn = \frac{1}{2n} = \frac{1}{2}\left(\frac{\varepsilon_0}{E}\right)^{1/2}. \qquad \text{by Eq. 3.37} \quad (9.4)$$

Folding Eq. 9.4 into Eq. 9.3, we get

$$\Omega = \frac{\pi^{3N/2}}{N!(3N/2)!} \frac{3N(E/\varepsilon_0)^{(3N-1)/2}}{2^{3N}} \frac{1}{2}\left(\frac{\varepsilon_0}{E}\right)^{1/2} \qquad (9.5)$$

$$= \frac{1}{N!(3N/2)!}\left(\frac{3N}{2}\right)\left(\frac{\pi E}{4\varepsilon_0}\right)^{(3N/2)-1}\frac{\pi}{4} \qquad \text{combine terms}$$

$$= \frac{1}{N![(3N/2)-1]!}\left(\frac{\pi}{4}\right)\left(\frac{2\pi m E V^{2/3}}{h^2}\right)^{(3N/2)-1} \qquad \text{replace } \varepsilon_0 \quad (9.6)$$

By the way, because the entropy S depends only on Ω, this expression lets us know that the entropy S, like Ω, can also be written as a function of E, V, and N. If we can write S as a function of E, V, and N only, then it stands to reason that E can be written as a function of only S, V, and N, which is the assertion made by the fundamental equation of thermodynamics, Eq. 7.10.

DERIVATION SUMMARY The Ensemble Size of the Three-Dimensional Box. In quantum mechanics, unlike classical mechanics, there is a *large but finite* number of possible states of the sample. We constructed a model in which the energy was proportional to the radius of a sphere. For one particle moving in three dimensions, this was the ordinary three-dimensional sphere, but for N particles we allowed the sphere to have $3N$ dimensions. We used geometry to count how many points on the surface of this hypersphere corresponded to the allowed integer values of the quantum numbers, ultimately obtaining Eq. 9.6:

$$\Omega = \frac{1}{N![(3N/2)-1]!}\left(\frac{\pi}{4}\right)\left(\frac{2\pi m E V^{2/3}}{h^2}\right)^{(3N/2)-1}.$$

CHECKPOINT Our main goal through this derivation is to show how the ensemble size Ω (and by extension, the entropy S) of a gas depends on volume and number of particles. Qualitatively, all we are showing is that there are more ways to arrange the particles if the box gets bigger or if we add more particles. However, by being careful here, we will also be able to extend this result to actually predict numerical values for the entropy of real gases.

As the correspondence principle predicts, the ensemble size Ω becomes huge as the mass m, energy E, or volume V approaches a bulk value. We can tell because the fraction in the middle of Eq. 9.6 is roughly $(mEV^{2/3}/h^2)$ raised to some positive power. If we keep the mass small (say, around $m_e \approx 10^{-30}$ kg) and the energy small (say, 10^{-23} J, which is equivalent to about 1 K), the factor of h^2 in the denominator is so small that even a volume of 1 cubic millimeter (10^{-9} m^3) has roughly 10^5 microstates per particle. And the number of states increases as the *power* of the number of particles, so four electrons in this 1 mm^3 box—with the same overall energy—would have nearly 10^{20} microstates.

If we rearrange Eq. 9.6 to emphasize the volume dependence, we see that expanding the box increases the ensemble size approximately as V^N:

$$\Omega = \frac{1}{N!\left[(3N/2) - 1\right]!}\left(\frac{\pi}{4}\right)\left(\frac{2\pi mE}{h^2}\right)^{3N/2}\left(\frac{h^2}{2\pi mEV^{2/3}}\right)V^N. \qquad (9.7)$$

Values of Ω at small values of N and E have been calculated for Fig. 9.2. In these drawings, the box's volume increases by factors of 8, and the ensemble size increases in multiples of (roughly) 8 when there is one particle, (roughly) $8^2 = 64$ when there are 2 particles, and (very roughly) $8^3 = 512$ when there are 3 particles. Keep in mind that we obtained Eq. 9.7 by assuming E was

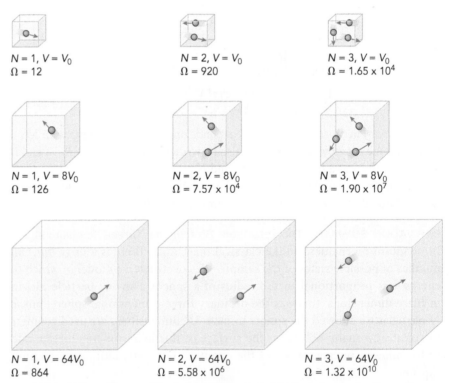

$N = 1, V = V_0$
$\Omega = 12$

$N = 2, V = V_0$
$\Omega = 920$

$N = 3, V = V_0$
$\Omega = 1.65 \times 10^4$

$N = 1, V = 8V_0$
$\Omega = 126$

$N = 2, V = 8V_0$
$\Omega = 7.57 \times 10^4$

$N = 3, V = 8V_0$
$\Omega = 1.90 \times 10^7$

$N = 1, V = 64V_0$
$\Omega = 864$

$N = 2, V = 64V_0$
$\Omega = 5.58 \times 10^6$

$N = 3, V = 64V_0$
$\Omega = 1.32 \times 10^{10}$

▲ FIGURE 9.2 **Example ensemble sizes for N identical particles in boxes of various volumes.** The energy range selected is 50 to 52.5 in units of $h^2/(8mV_0^{2/3})$.

large (so that we could use the surface area of the N-dimensional solid to estimate Ω in Eq. 9.2).[2]

The energy in a macroscopic system is stored in many different degrees of freedom. It may be in the form of radiation or stored in molecular translations, rotations, vibrations, or electronic excitation. We largely ignore translations when we study molecular structure, because, of all the molecular degrees of freedom, it is the least dependent on the chemical composition of our sample. Translation is merely motion of the molecular centers of mass.

However, it is also the most easily excited molecular degree of freedom in a gas, and for that reason it is perhaps the most important to take into account now that we are interested in the total energy content of a gas sample. Furthermore, as we explore processes that affect the macroscopic sample, we will encounter several that involve transportation of molecules, such as the flow of a liquid or expansion of a gas, and these processes are carried out along the translational coordinates. To predict the behavior of these systems, we will need expressions for the translational partition function.

Let's go back to Eq. 9.6 for the ensemble size Ω,

$$\Omega = \frac{1}{N!\left[(3N/2)-1\right]!}\frac{\pi}{4}\left(\frac{2\pi m E V^{2/3}}{h^2}\right)^{(3N/2)-1}.$$

We can calculate the entropy by applying Boltzmann's equation and letting N be a very large number:

$$S = k_B \ln \Omega \qquad \text{by Eq. 2.11 (9.8)}$$

$$= k_B\left[-\ln N! - \ln\left(\frac{3N}{2}-1\right)! + \ln\left(\frac{\pi}{4}\right) + \left(\frac{3N}{2}-1\right)\ln\left(\frac{2\pi m E V^{2/3}}{h^2}\right)\right] \quad \ln\left(x^a\,y\right) = a\ln x + \ln y \;(9.9)$$

$$= k_B\left\{-N\ln N + N - \left(\frac{3N}{2}-1\right)\ln\left(\frac{3N}{2}-1\right) + \left(\frac{3N}{2}-1\right)\right.$$

$$\left. + \ln\left(\frac{\pi}{4}\right) + \left(\frac{3N}{2}-1\right)\ln\left(\frac{2\pi m E V^{2/3}}{h^2}\right)\right\} \qquad \text{by Stirling's approx. (Eq. 2.14)}$$

$$\approx k_B\left\{-N\ln N + N - \left(\frac{3N}{2}\right)\ln\left(\frac{3N}{2}\right) + \left(\frac{3N}{2}\right)\right.$$

$$\left. + \ln\left(\frac{\pi}{4}\right) + \left(\frac{3N}{2}\right)\ln\left(\frac{2\pi m E V^{2/3}}{h^2}\right)\right\} \qquad \text{let } 3N/2 \text{ be } \gg 1 \text{ and } \gg \ln\left(\frac{\pi}{4}\right)$$

$$= k_B\left\{N\left[-\ln N + 1 - \frac{3}{2}\ln\frac{3N}{2} + \frac{3}{2}\right.\right.$$

$$\left.\left. + \frac{3}{2}\ln\left(\frac{2\pi m E V^{2/3}}{h^2}\right)\right]\right\} \qquad \text{factor out } N$$

---

[2]This point marks a convenient division between our microscopic and macroscopic perspectives. When the energy gets too small to use this approximation, or N gets too small to use Stirling's approximation (Eq. 2.14), then we have a microscopic system again—even if the volume is quite large. Although the boundary between microscopic and macroscopic is artificial, it has a great practical importance because many of our results from here on will be useful only in the macroscopic limit. As a quick example, if we set E to zero for the ground state in Eq. 9.7, we get $\Omega = 0$. And if $\Omega = 0$, then the entropy $S = k_B\ln\Omega$ becomes undefined. So Eq. 9.7 is useless for predicting properties of the system in its lowest energy states.

$$= k_B \left\{ N \left[\frac{5}{2} - \ln N + \ln \left(\frac{2}{3N} \right)^{3/2} + \ln \left(\frac{2\pi m E V^{2/3}}{h^2} \right)^{3/2} \right] \right\} \qquad \text{combine terms}$$

$$= Nk_B \left\{ \frac{5}{2} + \ln \left[\left(\frac{4\pi m E}{3Nh^2} \right)^{3/2} \frac{V}{N} \right] \right\}. \qquad \text{rearrange terms (9.10)}$$

CHECKPOINT The Sackur-Tetrode equation shows that the molar entropy of the gas should increase with temperature (because more energy allows us to access more translational quantum states) and decreases with pressure (because higher pressure implies more crowding of the molecules, so less space per mole for molecules to move).

Setting N equal to Avogadro's number $\mathcal{N}_A$ will then yield the molar entropy, where $\mathcal{N}_A k_B$ can be replaced by R. Furthermore, for a monatomic ideal gas, we can write the entropy in terms of the temperature by setting the energy equal to the equipartition value of $3Nk_BT/2$. Making these two substitutions gives one form of the **Sackur-Tetrode equation:**

$$S_m = R \left\{ \frac{5}{2} + \ln \left[\left(\frac{2\pi m k_B T}{h^2} \right)^{3/2} \frac{V}{N} \right] \right\} = R \left\{ \frac{5}{2} + \ln \left[\left(\frac{2\pi m k_B T}{h^2} \right)^{3/2} \frac{RT}{\mathcal{N}_A P} \right] \right\}. \quad (9.11)$$

The last expression uses the ideal gas law to replace V/N by $RT/(\mathcal{N}_A P)$. This predicts, for example, absolute molar entropies of 126.2 J K^{-1} mol^{-1} for helium and 169.7 J K^{-1} mol^{-1} for Xe at 1.00 bar, 298 K. The actual values are 126.0 J K^{-1} mol^{-1} and 169.6 J K^{-1} mol^{-1}, respectively, so the agreement is quite good.

The Sackur-Tetrode equation deals with only the translational degree of freedom in the gas, so the values just given are limited to monatomic gases, which don't have rotations or vibrations. Although the application is limited, Eq. 9.11 gives us an expression for the entropy that accounts for the most fundamental form of motion, which is convenient for the next section. We will return to consider the rotational and vibrational contributions to the entropy in Section 9.4.

9.2 The Second Law of Thermodynamics

The assumption that led us to describe chemical systems by statistical mechanics is this: given several possible microstates of the ensemble, quantum states that all satisfy the fixed parameters of the system, the system is equally likely to be found in any one of those microstates at any time.

It's worth stating that again. The system is equally likely to be found in any of the available microstates at any time.

If we accept this assumption, we must also accept the following implication. Consider our expanding gas under a piston from Chapter 8 to be many particles in a three-dimensional box, where the box can expand along one dimension. Our initial volume is 2.445 L, and the total energy is fixed. Many quantum states are consistent with these values of V and E; the total number of those microstates is Ω. Starting again from Eq. 9.6, rearranged to pull out factors of V:

$$\Omega(E, V, N) = \frac{1}{N! \left[(3N/2) - 1 \right]!} \left(\frac{2\pi m E}{4\pi^2 \hbar^2} \right)^{(3N/2)-1} \frac{\pi}{4V^{2/3}} V^N.$$

With a fixed, large number of particles N (1 mole for our example) and fixed energy, there is always a *hugely* greater number of states available at larger volume. If the volume changes by an amount dV, Ω changes at a rate of

$$\left(\frac{\partial \Omega}{\partial V} \right)_{E,N} \approx \frac{1}{N! \left[(3N/2) - 1 \right]!} \left(\frac{2\pi m E}{4\pi^2 \hbar^2} \right)^{(3N/2)-1} \frac{\pi}{4V^{2/3}} NV^{N-1} = \frac{N}{V} \Omega. \quad (9.12)$$

For example, if the volume increases by a fraction $dV/V = 10^{-10}$, the number of microstates increases by a factor of

$$\frac{d\Omega}{\Omega} = N\frac{dV}{V} = (6.02 \cdot 10^{23})(10^{-10}) = 6.02 \cdot 10^{13}.$$

The chance the system will be found at the larger volume than the initial volume is, according to our statistical assumption, vastly greater because the number of states has increased by a *factor* of 60,200,000,000,000. If the original height of our piston was, say, 10.0 cm, this is the increase in Ω that comes from raising the piston by less than an angstrom. As soon as we enlarge the volume of the sample by the slightest amount, we make this huge number of new quantum states available. The sample can stay at its original volume, with ensemble size Ω, or it can expand and increase the ensemble size by orders of magnitude. Say that the sample changes quantum state every time any two molecules collide, roughly 10^{30} times per second in this 1.00 mol sample. The probability at *each* of these collisions that the system will happen to remain at the original volume is only one in $6.02 \cdot 10^{13}$. In other words, the sample will expand immediately.

This is intended to give you an intuitive feel for what might be the most important principle in this whole book. When possible, systems *always* move toward a greater number of quantum states, toward greater entropy. This holds true for any system, no matter how simple or complicated, and no matter what the nature of the quantum states. The **second law of thermodynamics** can be stated as follows: for any process in a system totally isolated from its surroundings, the macroscopic entropy measured over the entire system S_T can never decrease:

$$\Delta S_T \geq 0. \tag{9.13}$$

There are a few qualifiers in this statement, and they should be examined. Why must the system be totally isolated, and why does the entropy have to be measured over the whole system? The entropy *can* decrease in a process, but only in a local region or (for a microscopic system) over a short time (see this chapter's *Biosketch*). If we allowed the system to communicate with its surroundings, the entropy change in the system could decrease, but it would have to be matched by an increase in the entropy of the surroundings.

The Second Law and Temperature

Remember the system in thermal contact with a reservoir in Section 2.4? The system had fixed volume and number of particles, but it could exchange energy with the surrounding reservoir provided that the total energy E_T was conserved. What does the second law predict about this interaction?

For the simple systems we've looked at, the density of microstates increases with the energy. As energy is transferred from the reservoir into the sample, the energy E of the sample rises and so does the number of available microstates $\Omega(E)$. However, at the same time, the energy of the reservoir E_r is decreasing, and so is the number of reservoir microstates $\Omega_r(E_r)$. Figure 9.3 shows this schematically. As long as the sample and reservoir only exchange energy and don't have any other interactions, the ensemble size of the entire system, $\Omega_T(E_T)$, is the product of $\Omega(E)$ and $\Omega_r(E_r)$. If either number gets too

▶ **FIGURE 9.3 Schematic of the trend in number of quantum states Ω for the system and reservoir.** With the system and reservoir in thermal contact, the total energy E_T is conserved, so if the energy of the sample E climbs, the energy of the reservoir E_r drops an equal amount. The total number of microstates of the system, $\Omega_T(E_T)$, reaches a maximum for intermediate values of E and E_r.

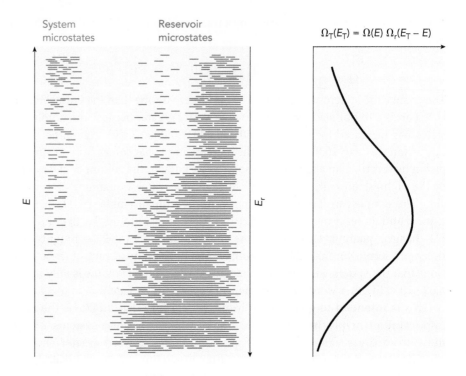

small, then $\Omega_T(E_T)$ decreases. So the largest ensemble, the configuration with the most microstates, is somewhere in the middle, with the sample and reservoir both sharing energy.[3]

We can isolate the dependence of Ω on the energy because we will keep V and n constant for both the sample and the reservoir. We want to find the conditions under which we obtain the most ensemble states possible for the combined system and reservoir, in other words, when the value of Ω_T reaches a maximum as a function of the system energy E. This maximum is a critical point that we can identify by taking the derivative of Ω_T with respect to E and setting the derivative equal to zero:

$$\left(\frac{\partial \ln \Omega_T}{\partial E}\right)_{V,n}\bigg|_{max} = 0 = \left(\frac{\partial \ln(\Omega \Omega_r)}{\partial E}\right)_{V,n}\bigg|_{max} \qquad \text{critical point}$$

$$= \left(\frac{\partial (\ln \Omega + \ln \Omega_r)}{\partial E}\right)_{V,n}\bigg|_{max} \qquad \ln(xy) = \ln x + \ln y$$

$$= \left(\frac{\partial \ln \Omega}{\partial E}\right)_{V,n}\bigg|_{max} + \left(\frac{\partial \ln \Omega_r}{\partial E}\right)_{V,n}\bigg|_{max}$$

Now we take advantage of the conservation of energy, requiring that $E + E_r = E_T$ is a constant, and therefore $dE = -dE_r$:

$$0 = \left(\frac{\partial \ln \Omega}{\partial E}\right)_{V,n}\bigg|_{max} - \left(\frac{\partial \ln \Omega_r}{\partial E_r}\right)_{V,n}\bigg|_{max} \qquad dE = -dE_r$$

---

[3]This is the same as finding that for any pair of numbers that add up to some constant (i.e., a conserved energy), the greatest product of those numbers occurs when the two are equal. For example, if $x + y = 6$, then the biggest value of xy is for $x = y = 3$, which gives $xy = 9$. The other possibilities if x and y are integers give smaller values: 2 and 4 ($xy = 8$), 1 and 5 ($xy = 5$), or 0 and 6 ($xy = 0$).

$$\left(\frac{\partial \ln \Omega}{\partial E}\right)_{V,n}\bigg|_{\max} = \left(\frac{\partial \ln \Omega_r}{\partial E_r}\right)_{V,n}\bigg|_{\max}$$

$$\left(\frac{\partial S}{\partial E}\right)_{V,n}\bigg|_{\max} = \left(\frac{\partial S_r}{\partial E_r}\right)_{V,n}\bigg|_{\max} \qquad\qquad S = k_B \ln \Omega$$

$$\frac{1}{T} = \frac{1}{T_r} \qquad\qquad\qquad T = \left(\frac{\partial E}{\partial S}\right)_{V,n}$$

$$T = T_r \tag{9.14}$$

The condition that maximizes the overall number of microstates, that leads to the maximum total entropy, is when the sample and the reservoir have the *same temperature*. For the ideal gas that we used to obtain Eq. 9.6, this is the same as saying that the sample and reservoir have come to the *same temperature*. In any system where the temperature is well-defined, the canonical distribution requires that the average energy of the system increases with temperature. Therefore, when two interacting systems at different temperatures come into contact, the hot one must lose energy to lower its temperature, and the cold one must gain energy to raise its temperature. In other words,

Heat flows from a hot body to a cold body

because the maximum total entropy is found when the two arrive at the same temperature.

We can demonstrate the same principle more quickly by classical thermodynamics using the following scenario. If we bring any two subsystems A and B into thermal contact, allowing heat to transfer between them but forbidding any work to be done, then we expect the two following conditions to determine the outcome of their interaction: (*i*) the total energy $E_A + E_B$ must be conserved, and (*ii*) at equilibrium, the entropy of the combined system must be at its maximum value. We would express this mathematically as:

$$
\begin{aligned}
dE_T &= dE_A + dE_B \\
&= T_A dS_A - P_A dV_A + T_B dS_B - P_B dV_B && \text{Eq. 7.11} \\
&= T_A dS_A + T_B dS_B && \text{no work} \\
&= 0 && \text{conservation of energy}
\end{aligned}
$$

$$dS_B = -\frac{T_A}{T_B}dS_A \qquad\qquad\qquad \text{solve for } dS_B \quad (9.15)$$

$$dS_T = dS_A + dS_B = 0 \qquad\qquad \text{maximum total entropy}$$

$$dS_T = dS_A - \frac{T_A}{T_B}dS_A = dS_A\left(1 - \frac{T_A}{T_B}\right) = 0 \qquad \text{by Eq. 9.15}$$

$$T_A = T_B$$

As with Eq. 9.14, any two systems in thermal contact at equilibrium must be at the same temperature. Comparing our two methods, we see both how the classical thermodynamics and statistical mechanics are mutually consistent, and how the statistical mechanics carries more conceptual detail (in this case the link between the entropy and the number of ensemble states) at the expense of more demanding mathematics. We have increasingly taken the statistical mechanics approach for granted as we develop our tools for thermodynamics, but that approach remains available if we need to drop back down to the molecular scale to better understand a system's behavior.

The Second Law and Spontaneity

The **Clausius principle** is another way of phrasing the second law of thermodynamics:

$$(\Delta S_T)_{E,V,n} \geq 0. \tag{9.16}$$

The other extensive parameters E, V, and n are fixed because we know from Eq. 7.10 that we can write the energy as a function of S, V, and n,

$$E = E(S,V,n),$$

and therefore we must be able to write S in terms of E, V, and n. The equality in Eq. 9.16 holds only for the special case of a reversible process. For any irreversible process, the total entropy will increase.

This is a good time to stress the distinction between *reversible* and *adiabatic*. A reversible process is one in which the *total* entropy S_T, including all reservoirs, remains constant:

$$\text{reversible:} \quad \Delta S_T = 0 \tag{9.17}$$

An adiabatic process is one in which $q = 0$, and for a reversible process this requires that the entropy S of the *system* is constant:

$$\text{reversible and adiabatic:} \quad \Delta S = 0 \tag{9.18}$$

Reversible processes require that the system always be in a well-defined thermodynamic state, that is, in equilibrium. Adiabatic processes do not. Examples can be found of both non-adiabatic reversible processes (e.g., reversible isothermal expansion) and irreversible adiabatic processes (e.g., turbulent adiabatic expansion).

BIOSKETCH | **Denis Evans**

Denis Evans is a professor of chemistry at Australia National University in Acton, near Canberra. His work has centered around the thermodynamics of systems that are not in equilibrium, and in particular has looked at the second law of thermodynamics in the context of a finite system evolving over a short time period. He has been able to show from first principles that random fluctuations in the behavior of a non-equilibrium system can appear to violate the second law of thermodynamics over isolated periods of time, and that these local violations become more probable exponentially as we decrease either the size of the system or the time over which we measure the entropy. The mathematical form of this result is known as the *fluctuation theorem*. Remarkably, at the same time that the fluctuation theorem shows that the total entropy can decrease over very short periods of time in small systems, it also shows that in the limit of long times and macroscopic systems the entropy must *increase,* providing a mathematical proof of the second law (which we have justified only by a qualitative, intuitive argument). By mapping the trajectories of particles in an optical trap, Evans and coworkers were able to experimentally demonstrate entropy decreases in a system with a small ensemble size.

Applications of the Second Law

A process in which the total entropy increases is called **spontaneous.** If we know enough about the thermodynamic properties of a system, we can predict the direction in which change will occur when we adjust any of the system parameters, by seeing which processes make ΔS_T positive.

For example, we can derive an expression for the change in entropy of an ideal gas undergoing isothermal expansion from volume V_i to volume V_f. During the isothermal expansion of an ideal gas, the energy of the gas does not change because—in the absence of interactions between the particles—the energy depends only on the temperature. Therefore we can set Eq. 7.12 equal to zero,

$$dE = TdS - PdV = 0$$

to find the incremental change in the entropy dS:

$$dS = \frac{P}{T}dV = \frac{nR}{V}dV.$$

Integrating dS from the initial to final state gives us ΔS:

$$\Delta S = \int_i^f dS = nR \int_i^f \frac{1}{V}dV = nR\ln\frac{V_f}{V_i}. \tag{9.19}$$

For an expansion, $V_f > V_i$, so $\ln(V_f/V_i) > 0$, and therefore ΔS is positive for expansion, negative for compression. This is a manifestation of our expression for the ensemble size of the particle in a three-dimensional box: $\Omega \propto V^N$ (Eq. 9.7). The ensemble size increases as the volume increases, and therefore the entropy increases also. The expansion of a gas is a spontaneous process. Compression of the gas is not spontaneous, and somebody has to do work on the gas to reduce its volume. To make sure we're clear, let me point out that ΔS for the sample can be negative. Processes in which the sample loses entropy take place because the surroundings gain that much entropy, or more.

The energy, entropy, temperature, volume, pressure, and number of moles are all state functions. In the isothermal expansion, it does not matter whether we go between the initial and final states i and f by a reversible or irreversible process; those parameters will be the same for our sample at the end of the expansion. That means ΔE, ΔS, ΔT, ΔV, ΔP, and Δn are also the same no matter what path is chosen. However, the change in overall entropy ΔS_T does depend on the path. If the change in energy for a reversible process is ΔE_{rev} and for an irreversible process is ΔE_{irr}, then

$$q_{rev} + w_{rev} = \Delta E_{rev} = \Delta E_{irr} = q_{irr} + w_{irr}. \tag{9.20}$$

We showed for the isothermal expansion that the reversible process was the one that did the most work, and for *any* process

$$0 > w_{irr} > w_{rev}, \quad q_{irr} < q_{rev}, \tag{9.21}$$

where the sample is doing the work, so w is negative. For the reversible process, $đq_{rev} = TdS$, and this gives the **Clausius inequality**

$$đq \le TdS, \tag{9.22}$$

with the equality holding only for the reversible process. There is less heat flowing into the sample during an irreversible isothermal expansion because the sample does not have to do as much work. The overall entropy increases because the rapid expansion creates turbulence; the energy of the sample is expressed partly as violent motion of the reservoir gas, which contributes to the entropy of the reservoir.

The thermodynamic potentials can also give the direction of a spontaneous process under certain conditions. Measuring the total entropy change in Eq. 9.16 is a sure way of knowing, but we rarely measure S directly, and the other extensive parameters E and V are often not what we want to hold fixed. If we keep T and P constant (as indicated by subscripts T and P), the Gibbs energy is useful, because

$$G = E - TS + PV, \qquad \text{by Eq. 7.19}$$

$$dE = đq + đw$$

$$(dG)_{T,P} = (dE - TdS - SdT + PdV + VdP)_{T,P} \qquad d(xy) = xdy + ydx$$

$$= đq + đw - TdS + PdV.$$

From the Clausius inequality, we know that $đq \leq TdS$, and we also know that the maximum work is done in a reversible process (where $đw_{rev} = -PdV$), so $đw \geq -PdV$. Therefore, considering for now only PV work,

$$(dG)_{T,P} \leq 0, \tag{9.23}$$

where the equality holds for a reversible process. A spontaneous process under constant T and P will minimize G, and at equilibrium G is at its minimum value for those conditions.

Similarly, at constant T and V, the Helmholtz energy points out the direction of the spontaneous process:

$$(dF)_{T,V} = (dE - TdS - SdT)_{T,V} = đq + đw - TdS = đq - TdS \leq 0 \tag{9.24}$$

because $dV = 0$, so $đw = 0$. The spontaneous process is the one that minimizes F in this case, whereas $(dG)_{T,V} = đq - TdS + VdP$ may be positive. Note the sensitivity to the fixed conditions: under fixed volume and temperature, the equilibrium state of a system has constant F and G, but G need not be at its minimum value. Contrary to intuition, the spontaneous process is not necessarily the one that minimizes the energy E.

This begins to reveal the importance of the Gibbs energy in traditional bench-top chemistry, where reactions are carried out in a bath such as ice or boiling water (to fix T) and exposed to air or to an argon or nitrogen atmosphere (to fix P). The direction in which the reaction will move spontaneously is the one that leads to a negative $(\Delta G)_T$, where

$$(\Delta G)_T = \Delta(H - TS)_T$$

$$= (\Delta H)_T - (T\Delta S + S\Delta T)_T$$

$$= (\Delta H)_T - T(\Delta S)_T. \tag{9.25}$$

The temperature (and pressure as well) are often understood to be fixed when the Gibbs free energy is used, leading to the familiar form

$$\Delta G = \Delta H - T\Delta S. \tag{9.26}$$

The change in enthalpy in this equation is equal to the heat flowing into the surrounding temperature bath, which we can relate to the entropy change ΔS_r of the surrounding constant-temperature reservoir:

$$\Delta H = q = -q_r = -T\Delta S_r. \tag{9.27}$$

Combining this with Eq. 9.25, we have

$$(\Delta G)_{T,P} = -T\Delta S_r - T\Delta S = -T(\Delta S_r + \Delta S). \tag{9.28}$$

This number is negative for the spontaneous process, because the second law tells us that the total entropy change $\Delta S_r + \Delta S = \Delta S_T$ is positive. So when we say that negative ΔG corresponds to a spontaneous process, we're just restating the second law for the specific case of constant T and P.

9.3 Ideal Mixing

In our series of simple, non-chemical processes, let's consider one last application of entropy to a process: mixing. We have two containers, each holding an ideal gas at the same temperature T and the same pressure P, but the first container holds n_A moles of gas A and the other container holds n_B moles of gas B. The two containers are connected through a tube with a valve. What happens to the entropy and other parameters when we open the valve and let the two gases mix?

Entropy of Mixing

Intuitively, we know this to be a spontaneous process—the gases will not stay confined in their separate containers—and therefore the entropy must increase. How can that be, when the density of gas in any region of the container hasn't changed?

For ideal gases, there is no interaction between the molecules, whether they are the same composition or not. The process described earlier amounts to two independent isothermal expansions: the expansion of gas A from volume V_A to $V_A + V_B$ and the expansion of gas B from volume V_B to $V_A + V_B$. We know that the expansion of each gas alone increases its entropy (Eq. 9.19):

$$dS = \frac{P}{T}dV = \frac{nR}{V}dV,$$

and integrating both sides from initial state i to final state f gives

$$\Delta S = nR\ln\left(\frac{V_f}{V_i}\right).$$

The entropy is an extensive parameter, so the change in entropy of the entire sample is the sum of the changes of the two components, gases A and B, where

$$\Delta S_A = n_A R\ln\left(\frac{V_A + V_B}{V_A}\right)$$

$$\Delta S_B = n_B R\ln\left(\frac{V_A + V_B}{V_B}\right). \tag{9.29}$$

The volume ratios can be rewritten in terms of the mole fractions of the gases X_A and X_B, because at constant temperature and pressure,

$$\frac{V_A}{V_A + V_B} = \frac{n_A}{n_A + n_B} \equiv X_A,$$

$$\frac{V_B}{V_A + V_B} = \frac{n_B}{n_A + n_B} \equiv X_B. \tag{9.30}$$

Therefore, the change in entropy due to mixing for two samples at the same temperature and pressure is

$$\Delta_{\text{mix}}S = n_A R\ln\frac{1}{X_A} + n_B R\ln\frac{1}{X_B} = -R\left(n_A\ln X_A + n_B\ln X_B\right). \tag{9.31}$$

We didn't need any reservoirs for this process, and the total volume of the sample doesn't change, so in this example the sample is the entire system, and $\Delta S_T = \Delta S$. The mole fractions in Eq. 9.31 are less than 1 by definition, so $\ln X$ is negative and the entropy of mixing $\Delta_{mix}S$ is positive. Sure enough, the process is spontaneous.

Should we be satisfied with this? The equations predict mixing to be spontaneous, but the process rushes forward when there doesn't seem to be any compelling force. Unlike the solo ideal gas expansion used to find Eq. 9.19, gas A is not expanding into a vacuum. It's expanding into the second container, which already has its own molecules. *What drives this forward?* Furthermore, if we replace the gases with two liquids that have identical intermolecular potentials $u(R)$, we know intuitively that mixing will again occur spontaneously, even though no expansion takes place.

One way to show what's driving this forward is to approach exactly the same problem using a little statistical mechanics. We begin with N_A molecules of A in the first container and N_B molecules of B in the second, totaling $N_A + N_B = N$. Let's examine the entropies of A and B before mixing.

To isolate the main volume-dependence, we rephrase Eq. 9.7 this way:

$$\Omega_k^{trans} = V_k^{N_k} f(E_k, N_k)$$

$$f(E_k, N_k) = \frac{1}{N_k! \left[(3N_k/2) - 1 \right]!} \left(\frac{\pi}{4} \right) \left(\frac{2\pi m_k E_k}{h^2} \right)^{3N_k/2} \left(\frac{h^2}{2\pi m_k E_k V_k^{2/3}} \right). \tag{9.32}$$

We label this term "trans" because the translational term is what allows the particles to mix. We may also need to factor in degeneracies for the intermolecular vibrational states, and in general there will be degeneracy factors for rotation or other internal degrees of freedom. Including those contributions increases the number of available microstates by a factor Ω^{int}. Then our entropy before mixing is

$$\begin{aligned} S_i &= S_A + S_B = k_B(\ln\Omega_A + \ln\Omega_B) \\ &= k_B[\ln(\Omega_A^{int}\Omega_A^{trans}) + \ln(\Omega_B^{int}\Omega_B^{trans})] \\ &= k_B\{\ln[\Omega_A^{int}f(E_A, N_A)V_A^{N_A}] + \ln[\Omega_B^{int}f(E_B, N_B)V_B^{N_B}]\} \\ &= k_B\{\ln[\Omega_A^{int}f(E_A, N_A)] + \ln[\Omega_B^{int}f(E_B, N_B)]\} + N_A k_B\ln V_A + N_B k_B\ln V_B. \end{aligned} \tag{9.33}$$

When the substances are mixed, the total ensemble size Ω_{A+B} is the product of the degeneracies for A and B, where now they have both expanded into the full volume $V_A + V_B$, and we then obtain the final entropy S_f from the natural logarithm of that product:

$$\begin{aligned} S_f &= S_{A+B} = k_B\ln\Omega_{A+B} \\ &= k_B\ln\{[\Omega_A^{int}f(E_A, N_A)(V_A + V_B)^{N_A}][\Omega_B^{int}f(E_B, N_B)(V_A + V_B)^{N_B}]\} \\ &= k_B\ln\{[\Omega_A^{int}f(E_A, N_A)][\Omega_B^{int}f(E_B, N_B)]\} + N_A k_B\ln(V_A + V_B) + N_B k_B\ln(V_A + V_B) \\ &= k_B\{\ln[\Omega_A^{int}f(E_A, N_A)] + \ln[\Omega_B^{int}f(E_B, N_B)]\} + (N_A + N_B)k_B\ln(V_A + V_B). \end{aligned} \tag{9.34}$$

Finally, take the difference between the initial and final entropies:

$$\begin{aligned} \Delta S_{mix} &= S_f - S_i \\ &= [N_A k_B\ln(V_A + V_B) + N_B k_B\ln(V_A + V_B)] - [N_A k_B\ln V_A + N_B k_B\ln V_B] \\ &= R[(n_A + n_B)\ln(V_A + V_B) - n_A\ln V_A - n_B\ln V_B] \\ &= R[n_A[\ln(V_A + V_B) - \ln V_A] + n_B[\ln(V_A + V_B) - \ln V_B]] \\ &= R\left[n_A\ln\left(\frac{V_A + V_B}{V_A}\right) + n_B\ln\left(\frac{V_A + V_B}{V_B}\right)\right] \\ &= -R(n_A\ln X_A + n_B\ln X_B), \end{aligned} \tag{9.35}$$

which was the same result obtained classically for the ideal gas. This time, we can show that *as long as the A–B interaction is essentially the same as the A–A and B–B interactions,* it doesn't matter whether our substances are liquids or gases, monatomics or polyatomics.

This result, somewhat more general now than our classical derivation, shows that the driving force behind ideal mixing is the most basic one: the determination of our system to spread itself over as many ensemble states as possible. The number of distinct translational quantum states increases when mixing takes place by a factor of

$$\frac{(V_A + V_B)^{N_A + N_B}}{V_A^{N_A} V_B^{N_B}}.$$

Even with no net work done and no net heat flow, systems will still evolve so as to find more microstates of the ensemble.

In fact, in this idealized limit, the driving force behind the mixing has nothing to do with the mixing itself, but merely with each substance taking advantage of the opportunity to distribute its particles over a greater volume, increasing its own entropy.

Free Energy and Enthalpy of Mixing

Let's finish up, while we're on mixing, with a quick look at two thermodynamic potentials that we can evaluate for ideal mixing. Returning to the thermodynamic approach, we can evaluate $\Delta_{mix}G$ the same way we calculated $\Delta_{mix}S$. For the mixing of two ideal gases at constant T and P, our standard bench-top conditions, we find

$$\Delta_{mix}G = \Delta G_A + \Delta G_B = \int_i^f dG_A + \int_i^f dG_B$$

$$= \int_i^f (-S_A dT_A + V_A dP_A)_T + \int_i^f (-S_B dT_B + V_B dP_B)_T$$

$$= \int_i^f (V_A dP_A)_T + \int_i^f (V_B dP_B)_T$$

$$= n_A RT \int_i^f \frac{dP_A}{P_A} + n_B RT \int_i^f \frac{dP_B}{P_B}$$

$$= RT\left(n_A \ln \frac{P_A}{P} + n_B \ln \frac{P_B}{P} \right)$$

$$= RT(n_A \ln X_A + n_B \ln X_B) = -T\Delta_{mix}S. \tag{9.36}$$

Keep in mind that the initial pressures of both gases are equal to P, and P_A and P_B here are the partial pressures of A and B in the final mixture.

As expected, this is a negative number; X is less than 1, so $\ln X$ is negative. Mixing at constant temperature and pressure decreases the Gibbs energy and is a spontaneous process. How about the enthalpy of mixing, $\Delta_{mix}H$? We saw earlier (Eq. 9.26) that, *if the temperature is constant,* then

$$\Delta G = \Delta H - T\Delta S.$$

Solving for $\Delta_{mix}H$ using Eqs. 9.31 and 9.36 gives

$$\begin{aligned}
\Delta_{mix}H &= \Delta_{mix}G + T\Delta_{mix}S \\
&= RT(n_A \ln X_A + n_B \ln X_B) + T[-R(n_A \ln X_A + n_B \ln X_B)] \\
&= 0.
\end{aligned} \tag{9.37}$$

The enthalpy is just the energy with an additional term PV added in. We have not changed the temperature, so E is unchanged for these non-interacting substances, and the overall pressure and volume are constant, so the enthalpy must be constant as well. Or, more simply, because the process takes place at constant pressure, $\Delta H = q$, and no heat was flowing, so $\Delta H = 0$.

9.4   The Third Law of Thermodynamics

The entropy is therefore a useful concept if we want to find the direction that a process will move if left alone. This depends only on the *change* in overall entropy ΔS_T, not on the value of the entropy itself. However, as with E and H, it is often convenient to have a reference value for the entropy, and this is given by the **third law of thermodynamics**: the entropy of a substance approaches a minimum as T approaches zero, and we set $S = 0$ at $T = 0$.

Really, this is nothing new to us. Boltzmann's law (Eq. 2.11),

$$S = k_B \ln \Omega,$$

provides an absolute scale for the entropy. In the limiting case of $T = 0$, absolute zero, any system must be in its unique quantum ground state, so $\Omega(T = 0) = 1$. Therefore the entropy as determined by Boltzmann's law gives

$$S(T = 0) = k_B \ln(1) = 0, \tag{9.38}$$

in agreement with the third law.

An important experimental limitation exists, however, in establishing the third law. At absolute zero, the most stable form of a substance is a crystal that puts each particle in the most stable, reproducible environment possible. However, once any substance gets that cold, it loses the energy that it may need to cross the barrier from one crystalline configuration to the next, more stable configuration. If $T = 0$ K, then there is, by definition, no energy in any of the degrees of freedom relative to the ground state. We have been thinking of energy primarily as kinetic energy, because we have mainly been concerned with problems involving translational motion of gas molecules, in which case there is no potential energy except for small effects from intermolecular forces. In crystals, however, much of the energy is stored in the potential energy of the vibrational coordinates, and it is possible to cool a system down to a *frozen state*, a metastable state separated from the ground state by a substantial potential energy barrier.

A macroscopic example is the arrangement of marbles in a glass jar. If we start with a sample of 40 marbles and place them methodically into a cylindrical jar, using a perfect packing pattern, we may get them into the geometry that minimizes the average height, and therefore minimizes the gravitational potential energy. If we just pour the molecules into the jar, chances are good that we will instead get a metastable arrangement of the marbles. Maybe the gravitational potential energy would be lower if a few marbles shifted in the right directions, but the thermal energy that's left in the system is not nearly enough to push the marbles around and explore the neighboring microstates. The system will stay in that geometry until we reheat it, perhaps by shaking the container.[4] This commonly occurs in crystallization: local regions of order (**crystal domains**) may form simultaneously in different parts of the sample but with different orientations, leading to a metastable overall structure.

If cooling the substance traps us in any of several low-energy but not quite lowest-energy configurations of the substance, then there is a **residual entropy** that cannot, for practical purposes, be removed from the system. There are still several ensemble states possible, so $S = k_B \ln \Omega$ is not quite zero.

Adiabatic Demagnetization

Historically, experimental tests of the third law proved to be difficult because the law is specific to the unattainable zero of temperature. Measurements made at substantially higher temperature left too much room for uncertainty. There was great interest in getting to *very* low temperatures to test the convergence of the entropy toward zero.

Why is absolute zero classically unobtainable? Coolants work according to thermodynamics: to reach equilibrium, heat will flow spontaneously from one material (the sample) to a cooler material (the coolant), until the temperatures are the same. This means that the coolant is warming up, however, so even if the coolant had started at absolute zero, at equilibrium both coolant and sample must be warmer than absolute zero. Even using quantum mechanics, there is always a translational or (in solids) phonon degree of freedom that acts classically until very small numbers of particles and restricted spaces are used. Nevertheless, one can get arbitrarily close to absolute zero until quantum mechanics begins to set limits for practical solutions.

Eliminating the excitation in these low-energy degrees of freedom was a particular challenge in 20th-century attempts to measure properties of materials at very low temperatures, less than the 1 K obtainable using boiling liquid helium as the coolant.[5]

The motivation to achieve such low temperatures was largely our interest in studying and developing new magnetic materials. Iron is a **ferromagnetic** material at room temperature, meaning that the magnetic spins of unpaired electrons in the substance will spontaneously line up in parallel, generating a net, macroscopically detectable magnetic field. A block of iron can be a permanent magnet, maintaining a magnetic field without the benefit of external forces to line up the electron spins. Competing with the tendency of the spins to line up is the

---

[4]Quantum mechanical tunneling allows any sample to eventually move *through* potential energy barriers that separate microstates, but after crystallization the thermal energies are typically too low compared to the barrier heights to re-orient individual molecules, much less whole domains, tunneling or not.

[5]The normal boiling points of liquid helium are 3.19 K for ^{3}He and 4.22 K for ^{4}He, but that's at an ambient pressure of 1 bar. Temperatures near 1 K could be easily achieved by letting the helium boil into a vacuum of a few torr, and lower temperatures still (at much greater cost) by using ^{3}He instead.

External field on

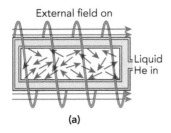

(a)

External field on

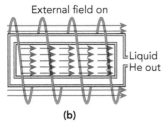

(b)

External field off

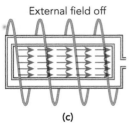

(c)

Induced field

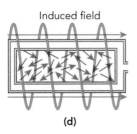

(d)

▲ FIGURE 9.4 **Schematic of W. F. Giauque's adiabatic demagnetization apparatus.** (a) Liquid helium is used to cool a crystal to near 1 K as it is magnetized by a high-field external solenoid magnet. (b) The helium is replaced by vacuum once the crystal is fully magnetized. (c) The external field is turned off. (d) The spins in the crystal equilibrate with the phonons. The net angular momentum of the spins is conserved by a current induced in the solenoid, which does work and further lowers the temperature of the crystal. (After a figure in *Nobel lectures, Chemistry,* Elsevier, Amsterdam, 1964.)

tendency of the spins to randomly reorient themselves as the thermal energy $k_B T$ increases. Consequently, iron ceases to function as a permanent magnet above a temperature of 1043 K. The thermal energy at that point is too great to keep the spins lined up. This temperature, where a substance ceases to be ferromagnetic, is called the **Curie temperature.** Many substances, including pure metals (such as dysprosium), alloys (such as Au_2MnAl), and ionic solids (such as EuS), become ferromagnetic at temperatures well below 298 K. By going to low temperature, researchers were able to greatly expand the range of materials that exhibit ferromagnetism and gather data on how the atomic and molecular structure affected the Curie temperature. Furthermore, at even lower temperatures, some materials were found to be superconducting (Section 4.4), encouraging efforts to attain even lower temperatures.

One way to reach these extremely low temperatures, on the order of 10^{-6} K, is a process called **adiabatic demagnetization.** This is an ingenious technique that turns magnetic crystals into their own coolants. The sample-coolant equilibration described earlier works not only when two different systems are brought into contact with each other, but also when different degrees of freedom in the same material are given different temperatures. This can be the case when bright ultraviolet light strikes a fluorescent liquid such as benzene in a flask: the translational and vibrational motions of the benzene molecules are in equilibrium with the walls of the flask and the temperature of the room, say 300 K, but the electronic states of the benzene are responding to the ultraviolet light, and their excitation may correspond to a temperature of over 2000 K. If the light is turned off, the different degrees of freedom will equilibrate, and the electronic temperature will quickly drop to the temperature of the room.

A similar process occurs during adiabatic demagnetization. We consider two degrees of freedom in magnetic crystals: the phonon or vibrational motions, and the magnetic spin. The magnetic state is determined by the orientation of the electron spins in the crystal. If they are all aligned, the crystal is in its lowest energy spin state,[6] and this can be accomplished using an intense magnetic field.

The sample, the magnetic crystal, is cooled to 1.8 K using liquid helium, and the magnet is activated to align all the spins in the crystal (in the language of classical thermodynamics, this corresponds to magnetic work done by the external field to prevent the spins from thermal fluctuations). The sample is then adiabatically isolated from the helium coolant, so there is no longer any heat flow, and the magnet is turned off. The spin temperature is then extremely low (effectively zero, which is possible because this particular degree of freedom is very non-classical), while the phonon temperature is still 1.8 K. Since there is no thermal contact to the surroundings, the two degrees of freedom equilibrate by heat flowing from the phonon states to the spin states. The spin temperature rises (the electron spins become more randomly aligned) as the phonon temperature drops (Fig. 9.5). Final temperatures below 10^{-6} K can be achieved.

---

[6]This is an idealized picture, because the fully magnetized state of a ferromagnetic material is metastable, not the global ground state of the crystal. Long-range interactions between the magnetic dipoles cause the crystal to be more stable when broken down into opposing magnetic domains.

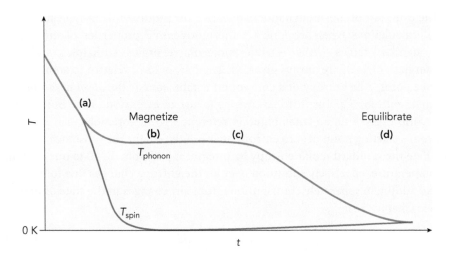

◄ FIGURE 9.5 **Adiabatic demagnetization.** The spin and phonon temperatures are initially equal because the system begins at equilibrium. The spins are then magnetized, reducing their effective temperature to zero. When the magnetic field is turned off, heat is transferred from vibrations to spin, increasing the spin temperature and decreasing the phonon temperature until the two are again equal.

The success of the technique depends very heavily on the material used for the sample, since the amount of heat that flows into the spin degree of freedom depends on the strength of the magnetic dipoles. The lowest temperatures attained as of 1994 used nuclear spin magnetization, because nuclei have much smaller magnetic energies than electrons. Bose-Einstein condensates have reached lower temperatures since then, but adiabatic demagnetization continues to be explored as an effective refrigeration technique.

Absolute Standard Entropies

In adiabatic demagnetization and similar experiments, the low temperature heat capacity $C_P(T)$ can be measured, and the absolute entropy can be obtained using

$$C_P(T) = \left(\frac{\partial q_{rev}}{\partial T}\right)_P = \left(\frac{T\partial S}{\partial T}\right)_P \tag{9.39}$$

so that at constant pressure,

$$S = \int_0^T \frac{C_P(T')}{T'}dT'. \tag{9.40}$$

The Debye theory for heat capacities of crystals predicts a low temperature heat capacity of (Eq. 7.73)

$$C_V \approx \frac{12\pi^4 N k_B^4 T^3}{5\omega_D^3}.$$

For crystals, the heat capacity at constant pressure is nearly equal to C_V because the coefficient of thermal expansion is relatively small (Eq. 7.52). In this limit, the equation for the entropy simplifies to

$$S \approx \frac{12\pi^4 N k_B^4}{5\omega_D^3}\int_0^T \frac{T'^3}{T'}dT' = \frac{12\pi^4 N k_B^4}{5\omega_D^3}\int_0^T T'^2 dT' = \frac{12\pi^4 N k_B^4}{5\omega_D^3}\left(\frac{T^3}{3}\right) \approx \frac{C_P}{3}. \tag{9.41}$$

Not all crystals obey the Debye theory, but this gives an illustration of how the absolute entropy is measurable at low temperatures.

The third law of thermodynamics appears most prominently in typical chemical calculations when we look up thermodynamic properties of molecules in standard tables. Unlike various molecular energies, enthalpies, and free energies— which are always given as ΔE, ΔH, or ΔG, relative to some reference point—the entropy of a compound is tabulated as the *absolute, standard molar entropy*, $S_m^{\ominus}$. The third law, which is just an extension of the Boltzmann equation, gives us an unambiguous reference point for molecular entropy. From a starting point of zero entropy at absolute zero for the substance, we calculate the standard molar entropy by integrating dS from $T = 0$ to our current temperature, effectively summing over all the entropy changes due to heating and adding in separately contributions from any changes in the state of matter (see Chapter 10):

$$S_m^{\ominus}(T) = \int_0^T dS. \tag{9.42}$$

Some example values of $S_m^{\ominus}$ are given in Table 9.2 so that we can examine the qualitative trends. The entropy tends to be fairly similar for compounds in the same phase, particularly with the same number of atoms. The gas-phase

CHECKPOINT Recall that the standard state (indicated by the superscript $\ominus$) refers to pure substances at a pressure of exactly 1 bar. It does not specify the temperature, so we have to state which temperature was used for any tabulated values.

TABLE 9.2 **Selected standard molar entropies.**

	$S_m^{\ominus}(\text{J K}^{-1}\,\text{mol}^{-1})$
diatomic gases	
Br_2(gas)	245.463
CO(gas)	197.674
F_2(gas)	202.78
HF(gas)	173.779
HI(gas)	206.594
KCl(gas)	239.10
O_2(gas)	205.138
SiN(gas)	216.76
tetratomic gases	
C_2H_2(gas)	200.94
$COCl_2$(gas)	283.53
ClF_3(gas)	281.61
SO_3(gas)	256.76
SO_2Cl_2(gas)	311.94
liquids	
H_2O(liq)	69.91
Br_2(liq)	152.231
CS_2(liq)	151.34
solids	
C(diamond)	2.377
C(graphite)	5.740
SiO_2(quartz)	41.84

diatomics Br_2, CO, F_2, HF, HI, KCl, O_2, and SiN all have standard entropies between $173.78 \, J \, K^{-1} \, mol^{-1}$ (HF) and $245.46 \, J \, K^{-1} \, mol^{-1}$ (Br_2), despite the variety of masses and bond strengths. The entropy increases with heavier compounds because the rotational and vibrational constants get smaller, so the density of states is bigger. If we move on to the gas-phase tetratomics C_2H_2, $COCl_2$, ClF_3, SO_3, and $SOCl_2$, the range is still $200.94 \, J \, K^{-1} \, mol^{-1}$ (C_2H_2) to $309.77 \, J \, K^{-1} \, mol^{-1}$ ($SOCl_2$), even though the number of vibrational coordinates per molecule has increased from one in the diatomics to six, and the number of rotational coordinates has increased from two to three. Compare the entropies of gas-phase SO_3 ($257 \, J \, K^{-1} \, mol^{-1}$) and $SO + O_2$ ($427 \, J \, K^{-1} \, mol^{-1}$) or $COCl_2$ ($284 \, J \, K^{-1} \, mol^{-1}$) and $CO + Cl_2$ ($421 \, J \, K^{-1} \, mol^{-1}$). Standard entropies of gases generally lie between $115 \, J \, K^{-1} \, mol^{-1}$ and $400 \, J \, K^{-1} \, mol^{-1}$. Very large molecules would have correspondingly large entropies, but at 298 K they are rarely found in the gas phase.

We saw how the Sackur-Tetrode equation successfully predicts the translational contribution to the entropy in monatomic gases. If we apply Eq. 7.35,

$$S = k_B \ln Q(T) + k_B T \left(\frac{\partial \ln Q(T)}{\partial T} \right)_{V,n},$$

we can also estimate the contributions from rotation and vibration. For example, for rotation in diatomics,

$$q_{rot} = \frac{k_B T}{B} \qquad \text{by Eq. 3.30}$$

$$S_{rot} = k_B \ln q_{rot} + k_B T \left(\frac{\partial \ln q_{rot}}{\partial T} \right)_{V,n} \qquad \text{by Eq. 7.35}$$

$$= k_B \ln q_{rot} + k_B T \frac{1}{q_{rot}} \left(\frac{\partial q_{rot}}{\partial T} \right)_{V,n} \qquad d \ln x = dx/x$$

$$= k_B \ln \frac{k_B T}{B} + k_B T \frac{B}{k_B T} \left(\frac{\partial k_B T / B}{\partial T} \right)_{V,n}$$

$$= k_B \ln \frac{k_B T}{B} + (B) \frac{k_B}{B}$$

$$= k_B \left(\ln \frac{k_B T}{B} + 1 \right)$$

$$S_{m,rot} = R \left(\ln \frac{k_B T}{B} + 1 \right) \qquad (9.43)$$

This predicts, for example, that the rotational motion of $^{12}C\,^{16}O$ ($B = 1.9313 \, cm^{-1}$) contributes $5.68R$ to the total entropy at 298.15 K. From the Sackur-Tetrode equation, the translational contribution is $(2.50 + 15.59)R$. The sum of the two terms comes to $197.6 \, J \, K^{-1} \, mol^{-1}$, which is correct to four digits (the actual value is $197.67 \, J \, K^{-1} \, mol^{-1}$). The vibrational constant of CO is too high to contribute substantially at 298 K. This is a remarkably successful example of using the microscopic, quantum mechanical properties of the material to predict the macroscopic properties. At higher temperatures or for heavier molecules, as vibrations become important, one can apply the same approach to calculating the vibrational contribution to the entropy.

In the condensed phases, the entropy is a much more sensitive function of the compound. Water has a very low liquid-phase standard entropy ($69.9 \, \text{J K}^{-1} \, \text{mol}^{-1}$) compared to less polar liquids (Br_2(liq) and CS_2(liq) $>$ $125 \, \text{J K}^{-1} \, \text{mol}^{-1}$). This results from the remarkable strength of water's hydrogen bonding interactions, which severely restrict motion of the molecules. The very sparse vibrational energy level structures of diamond and graphite cause these to have very low entropies (less than $6 \, \text{J K}^{-1} \, \text{mol}^{-1}$) compared to the entropies in excess of $100 \, \text{J K}^{-1} \, \text{mol}^{-1}$ for the heavier and more irregularly structured salts.

9.5 Computer Simulations

Entropy determines the direction that a process takes, but its control over our system lies behind the scenes. The laws of quantum and statistical and classical mechanics are all we need to simulate the behavior of thermodynamic systems, and this has been one of our major goals: to use the molecular-scale properties of our system to predict its macroscopic properties and dynamics. As we have seen, for example, in obtaining the van der Waals equation in Section 4.2, that can be a big job. Even simple gas properties require many approximations and assumptions, and the condensed phases are even more complicated. Add to that our goal of describing *how systems change over time,* and the algebraic equations that we can rely on become fewer and cruder approximations. We can bypass much of the intrinsic mathematical challenge (but much of the intuitive understanding as well) by using **computer simulations** of molecules moving over time to determine how any given chemical system will evolve over time.

In this section, we'll examine two important tools for modeling some basic dynamics of a system with a large number of particles. (We'll continue to assume no chemical reaction takes place, but these methods can be extended to include reactions.) As we'll see, these simulations are not programmed to obey the second law of thermodynamics, but they obey it nevertheless. The second law arises naturally from allowing particles to interact according to the laws that govern the microscopic scale.

In these simulations, we let the computer generate a large field of molecules, with properties programmed in, and then calculate the interactions on the molecular level. At any point we can average over the entire set of molecules to get the macroscopic properties. Because our job is only to establish the initial conditions and then to observe what happens, computer simulations of this sort are often considered to be computer-based experiments, rather than pure theory.

The challenges in computer simulation are similar to problems encountered in solving the quantum mechanics of a large molecule. The number of individual calculations necessary for a particular system increases as a power of the number of degrees of freedom, and it is an easy task to exceed the capabilities of any computer. In addition, only pairwise interactions are usually included explicitly; higher-order interactions are often ignored. Nevertheless, in many applications where analytical solutions aren't available, these simulations have enormous predictive power.

Molecular Dynamics Simulations

One straightforward and general form of computer simulation for a chemical system is the **molecular dynamics simulation,** applied to a fluid. The simplest such simulation for a gas or liquid could be programmed as follows:

1. Begin with a group of N molecules inside a container. The container is determined by minimum and maximum values of X, Y, and Z. The molecules will be described by some pair potential energy function $u(R_{ij})$, typically one of those described in Section 4.1. At a minimum, each molecule will have a hard sphere radius R_{hs}. Assign to each molecule i a random initial position $(X_i, Y_i, Z_i) = \vec{R}_i$ inside the container. If you are hoping for a very dense system, this can take some time while the simulation looks for an arrangement that prevents particles from overlapping with each other (in practice, check that all the distances between particles are greater than R_{hs}). Then assign values to the velocity components (v_{Xi}, v_{Yi}, v_{Zi}) such that each is also assigned randomly, but with the values weighted according to the Gaussian function

$$\mathcal{P}(v_{qi}) = e^{-mv_{qi}^2/(2k_BT)}, \tag{9.44}$$

where you have chosen the temperature T and particle mass m for the simulation, and where q is any of X, Y, or Z. This distribution has the shape of the canonical distribution of speeds in one dimension (see Eq. 3.61) and will lead to a Maxwell-Boltzmann distribution of translational energies.

2. Calculate for each molecule its new position after a small time interval δt:

$$\vec{R}_i(t + \delta t) = \vec{R}_i(t) + \vec{v}_i(t)\,\delta t. \tag{9.45}$$

The time interval δt should be small enough that the fastest molecule in the set moves only a fraction of its diameter. If δt is too large, molecules in the simulation will pass through one another or bury themselves in the walls of the container before you have a chance to correct for the interactions.

3. Check to see if any particles are about to hit the wall. For example, the wall at $Z = Z_{max}$ will be hit in the next time step by any molecule having $Z_{max} - Z_i < v_{Zi}\,\delta t$. If the wall is elastic, then each molecule that will hit the wall is given a new velocity vector, changing the sign of the velocity component perpendicular to the wall. After hitting the Z_{max} wall, the new velocity vector would be $(v_{Xi}, v_{Yi}, -v_{Zi})$.

4. Now comes the toughest part. Check the forces by finding the difference vector between every pair of molecules $\vec{R}_{ij} = \vec{R}_i - \vec{R}_j$ and then calculating a total force on molecule i of

$$\vec{F}_i = -\sum_{j \neq i}^{N} \frac{du(R_{ij})}{dR_{ij}} \frac{\vec{R}_{ij}}{|\vec{R}_{ij}|}. \tag{9.46}$$

This part of the calculation rapidly grows in computer time as the intermolecular potential becomes more complex, as the number of particles increases, and as internal degrees of freedom—electronic, vibrational, and rotational—are included. More clever algorithms for identifying nearest neighbor molecules and neglecting the other interactions become essential to keep the calculation manageable. Once the forces are computed for each particle, we use these to find the new velocity vectors:

$$\vec{v_i}(t + \delta t) = \vec{v_i}(t) + \frac{\vec{F_i}(t)\,\delta t}{m}. \tag{9.47}$$

5. Repeat from (2) for enough time intervals that the total time elapsed has allowed the average molecule to travel roughly the length of the sample.

6. Repeat the whole procedure from (1) several times to remove bias due to any single set of initial conditions.

EXAMPLE 9.1   Forces in a Lennard-Jones Fluid

CONTEXT Nanoparticles comprise a relatively new field of intense interest in materials research, partly because we have only recently developed the tools for isolating and characterizing them, but also because their sizes (which may extend from 1 nm in diameter to over 10 μm) cover a range comparable to wavelengths in the visible part of the spectrum, proteins and nucleic acid molecules in organisms, and dust grains in the air and in outer space. As such, they have untapped potential for applications in optics and medicine and may hold the key to central questions in atmospheric and interstellar chemistry. One way of obtaining high concentrations of nanoparticles is to suspend them in a *colloid*, a thick mixture of fine particles carried by a liquid. Within these colloids, however, we don't know whether the nanoparticles interact according to a Lennard-Jones potential (which is appropriate for small molecules) or according to a different potential energy function (*Bradley's formula*, which describes the van der Waals interaction of μm-scale particles). By carrying out molecular dynamics simulations, it is possible to predict properties of the colloid using the force laws of these two potential energy functions and see which better predicts the behavior of the substance in the laboratory.

PROBLEM Find an expression for the force acting between two particles in a fluid where the interactions are governed by the Lennard-Jones potential (Eq. 4.11),

$$u(R) = \varepsilon\left[\left(\frac{R_e}{R}\right)^{12} - 2\left(\frac{R_e}{R}\right)^6\right].$$

SOLUTION We take the derivative with respect to R, obtaining

$$\vec{F} = -\frac{du(R)}{dR}\frac{\vec{R}}{|\vec{R}|}$$

$$= -\varepsilon\left[-12\left(\frac{R_e^{12}}{R^{13}}\right) + 12\left(\frac{R_e^6}{R^7}\right)\right]\frac{\vec{R}}{|\vec{R}|}$$

$$= \frac{12\varepsilon}{R_e}\left[\left(\frac{R_e}{R}\right)^{13} - \left(\frac{R_e}{R}\right)^7\right]\frac{\vec{R}}{|\vec{R}|}.$$

The sign of the force is such that R tends to increase (the particles repel) if $R < R_e$, and R tends to decrease (the particles attract) if $R > R_e$.

The variations on the simulation are endless. The wall, for example, may be treated as a purely elastic surface, so molecules bounce off them with no loss of energy, or as an inelastic surface with some characteristic temperature that allows it to exchange energy with the particles. The wall may be eliminated entirely and the particle motions treated using **periodic boundary conditions.** In this case, we assign a domain along each coordinate, say from 0 to Z_{max} along Z. Then, if molecule i travels to some point (X_i, Y_i, Z_{max}), with velocity vector $\vec{v_i}$, the program immediately transfers the particle to the point $(X_i, Y_i, 0)$, still with velocity vector $\vec{v_i}$. This way, the simulation can mimic the properties of an unbounded system (reducing the impact of edge effects, discussed below).

By reducing the temperature, the system can be made to condense to a liquid, and even freeze into a solid. Properties of solutions may be calculated. Of particular interest to us in our studies of the condensed phases, the pair correlation function $\mathcal{G}(R)$ is easily computed from the distribution of particles at any point during the simulation.

From such a calculation, one can monitor *any* aspect of the system one wants to, as long as the computer will accommodate the complexity of the model. For example, the computer can count the number of molecular collisions per unit time, the average pressure against the walls over the total time of the simulation, and the extent to which the density fluctuates from one time or region to another during the simulation. In this way, the effect of the intermolecular potential on the pressure can be used to test the validity of approximations such as Eqs. 4.44 and 4.45 for the van der Waals coefficients,

$$b = \frac{2\pi \mathcal{N}_A R_{LJ}^3}{3} \qquad a = \frac{16\pi \mathcal{N}_A^2 \varepsilon R_{LJ}^3}{9},$$

that relate the observable non-ideality of a gas to its microscopic properties.

If one changes the external conditions (such as the volume or temperature), the evolution of the system in time can be monitored. Mass flow or other transport properties can be studied by initially grouping the molecules in one region of the container and then allowing them to flow or diffuse to other locations. This leads to countless engineering applications, including aerodynamics simulations in auto, aircraft, and weapons manufacture. Even more advanced simulations seek to predict the properties of dynamical systems involving chemical reactions, such as flames and combustion engines.

The molecular dynamics simulation, as described here, has some of the appeal of a genuine experiment: we merely set things up and measure the results, letting the physics do all the thinking. Of course it's often much easier to set up than a real experiment, and the physics is much simpler than the real physics; at best it approximates the behavior of the real system. Still, the simplified physics can function to our advantage, because there is hope of isolating chains of cause and effect in the simulation, which we might never accomplish in the experiment.

Figure 9.6 gives sample results from three 1000-particle simulations, starting from a random homogeneous distribution and then interacting through the Lennard-Jones potential. When the thermal energy $k_B T$ is ten times the well depth ε, the particles remain almost randomly distributed, and the pair correlation function has a single small peak. As the temperature decreases, the peak in $\mathcal{G}(R)$ becomes more pronounced, and you can see the molecules clustering toward the middle of the container in response to their mutual attractions. At $T = 0.1\ \varepsilon/k_B$, the attractions are strong enough that the simulation shows the sample starting to

▶ FIGURE 9.6 **Final particle distributions and pair correlation functions from molecular dynamics simulations.** Each simulation models 1000 spherical particles interacting for $2 \cdot 10^6$ steps in a container $20.0R_{LJ}$ on each side for selected values of the temperature. Distributions are the projections of all particle locations into the XZ plane. Distances are all in units of the molecular diameter R_{LJ}. The sloping tails of the pair correlation functions arise from the small size of the container.

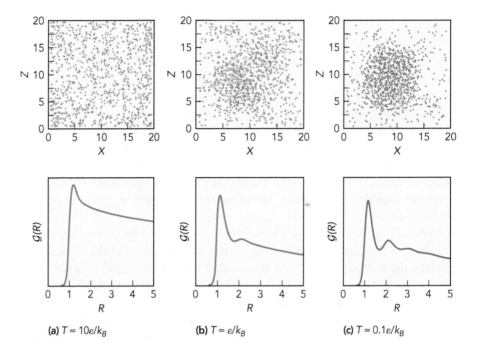

(a) $T = 10\varepsilon/k_B$ (b) $T = \varepsilon/k_B$ (c) $T = 0.1\varepsilon/k_B$

condense into a liquid in the middle of the container. All the major qualitative features of the fluid are reproduced by this simulation, and we control all the parameters of the system and can measure any variable we wish.

Unfortunately, a large molecular dynamics simulation demands extraordinary computer resources. For example, say we want to predict the properties of a liquid in a container shaped like a cube. How many molecules should we use? If we use 1000, then we'll have a distribution that's roughly ten particles on a side in three dimensions, and roughly 100 particles resting against each face of the container. With six sides to the cube, we'll have on average almost half of all the particles at the walls of our container, which is a poor approximation of the bulk liquid where only a tiny fraction of the molecules are at the boundaries. We say that the results will be contaminated by **edge effects,** because an unrealistically large fraction of the molecules lie at edges of the container. Setting $N = 10^6$ reduces the number of molecules at the walls to about 6%. Still not great but let's keep going anyway. At room temperature, the average speed of a small molecule is about 500 m s^{-1}, but many of the molecules are traveling faster than this, and we don't want to exclude them from a proper treatment. If we take into consideration all the molecules of speeds 10^3 m s^{-1} and less (which still leaves out roughly 30 molecules of our 10^6), and assume a molecular diameter of about 5 Å, then our minimum time step is on the order of

$$\delta t < \frac{(5\,\text{Å}/n)(10^{-10}\,\text{m Å}^{-1})}{10^3\,\text{m s}^{-1}} = \frac{1}{n}5 \cdot 10^{-13}\,\text{s,}$$

where $n = 5$, for example, means we are willing to let a fast molecule travel 1/5 of a molecular diameter before we check to see if it has hit anything. So our simulation will track the motions of 10^6 particles roughly every 100 fs. (In fact, a time step of less than 10 fs is more typical.) At each point we need to add up the forces and calculate new velocity vectors. To allow the simulation to evolve for one microsecond will require 10^7 time steps and $(10^6)(10^7) = 10^{13}$ individual velocity vector calculations. As fast as molecules move, it turns out that one microsecond gives our tiny sample of liquid barely enough time to respond to changes like the influx of a new set of molecules.

For very complicated, rapidly evolving systems, there may be no effective simulation technique that is more efficient than molecular dynamics. For relatively simple systems, however, a cleverer approach can save a lot of time.

Monte Carlo Methods

Let's phrase it this way. Our system has an enormous ensemble of possible states, indicated by the set of positions and velocities of the particles, and the simulation described earlier must move continuously from one state of the ensemble to one of its neighboring states. No matter how long we spend on the simulation, we can only hope to explore a set of connected states, all gathered near each other in some tiny part of the ensemble determined by the initial conditions.

Perhaps a more efficient method would be one that jumps around the ensemble, getting information from microstates that are widely separated from each other. This is the approach of **Monte Carlo simulations.** When the problem can be formulated for Monte Carlo, the simulation tends to be far less demanding than the brute-force molecular dynamics simulation.

Named for the gambling venue that made Monaco famous, a Monte Carlo simulation generates a series of configurations for the sample, each differing from the other by some random change in the microscopic parameters, but with the *probability* of each configuration determined by some non-random macroscopic law. What each configuration represents depends on the system being studied: if the structure of a liquid is being studied, each configuration is a different arrangement of the molecules; if it is a lattice of magnetic atoms, it is the set of quantum numbers m_s for the electron spins. Typically the configurations are forced to obey the canonical distribution, so the number of configurations with overall potential energy U is proportional to $e^{-U/(k_B T)}$. A typical goal of the calculation is to estimate the average of some parameter of the system over the entire ensemble of possible states, based on a limited number of carefully chosen states. The molecular dynamics simulation moves through microstates of the ensemble in tiny incremental steps, thoroughly sampling one small region of the ensemble. The Monte Carlo approach evolves more quickly than the molecular dynamics simulation, hopping around the ensemble, obtaining (we hope) a representative sampling of the *entire* ensemble in the time that it takes for the molecular dynamics simulation to wander one tiny corner.

What are the drawbacks? The model of Monte Carlo presented here does not incorporate the concept of time, so properties of the sample that gradually evolve process require a more advanced treatment. Related to this, properties that depend on the coupled motion of many particles can be difficult to model successfully by Monte Carlo, which typically changes from one configuration to

$\uparrow\downarrow$ $+D$

$\downarrow\downarrow$ $-D$

$\uparrow\uparrow$ $-D$

▲ FIGURE 9.7 **The Ising model.** The Ising model is intended to represent the behavior of a lattice of interacting spin 1/2 particles. We set an interaction energy D such that the interaction between like spins contributes $-D$ to the potential energy, and the interaction between unlike spins contributes $+D$.

another by manipulation of a single particle. Ultimately, the molecular dynamics simulation is more general, because we do not need to impose our own distribution function on the molecules. When they are in direct competition, however, Monte Carlo will normally be the more efficient technique.

To demonstrate the Monte Carlo method, we will explore a system known as the two-dimensional **Ising model,** intended to represent the properties of a layer of magnetic atoms on the surface of a non-magnetic crystal. Start with a square lattice of magnetic atoms. We wish to evaluate the average energy and other properties of this group of atoms as functions of temperature. Let's start with a small lattice of nine atoms in a 3 × 3 square and pretend that this is a small sample taken from the surface of a bulk crystal (Fig. 9.7). Each atom has an electron spin that may be "+" ($m_s = +\frac{1}{2}$) or "−" ($m_s = -\frac{1}{2}$). Let's start with the following array:

$$
\begin{array}{ccc}
+ & + & - \\
- & + & + \\
+ & - & +
\end{array}
\tag{9.48}
$$

There is an interaction energy between each pair of neighboring atoms that we set equal to $-D$ if the spins are both + or both −, and $+D$ if the signs oppose each other. Let's assume only the immediate neighbors interact with any given spin. The center atom, surrounded by eight neighboring atoms, has eight pair-interaction energies. For example, the center atom is +, and so is the atom directly above it. The contribution from these two atoms to the total potential energy is $-D$. There are four more atoms around the center atom with + spin, and five altogether, and three with − spin that will contribute $+D$ each to the total energy. Adding these contributions together, we find that the interaction energy due to all pairs that include the center atom is $(3 - 5)D = -2D$.

How about the atoms on the edges of our tiny lattice? They have fewer than eight neighbors, so their interaction energies are restricted to a narrower range than the center atom energies. If we want the simulation to give us some idea of what happens in a macroscopic planar array of magnetic atoms, we have a familiar problem: atoms on the perimeter of our simulated lattice will introduce edge effects into the calculations. While these effects exist in the macroscopic sample, the fraction of atoms on the edge is vanishingly small (if we have a *really* smooth surface). In our simulation, 8/9 of the atoms are on the edge, and they will skew the results of the simulation. Instead of magnifying the number of particles, as we did for the molecular dynamics simulation, this time let's use periodic boundary conditions.

To impose periodic boundary conditions in the Ising model, we pretend that the sample is surrounded on all sides by exactly identical samples so that the edge atoms on one side are neighbors to the edge atoms on the opposite side. For example, with our lattice, instead of just looking at the nine atoms drawn in Eq. 9.48, we would treat the lattice as a 5 × 5 lattice:

$$
\begin{array}{ccccc}
+ & + & - & + & + \\
- & + & + & - & + \\
+ & - & + & + & - \\
+ & + & - & + & + \\
- & + & + & - & +
\end{array}
\tag{9.49}
$$

The square in the middle is the original lattice, but now it is surrounded by eight identical lattices (only the sides and corners that neighbor the original lattice are drawn). Now we can continue to work with only nine atoms, but as if each atom is surrounded on all sides. If we change the spin of any of the atoms on the edge, we must also change the corresponding spin in the neighboring lattices. The spins surrounding our original lattice must continue to respond in a way appropriate to the temperature and potential energy. This does not eliminate the edge effects entirely. The results will still be skewed by the small number of active atoms, but the simulation with periodic boundary conditions approaches the behavior of the bulk more rapidly than if we had merely increased the size of the lattice.

We want to calculate the energy due to all the pairwise interactions among the N atoms in our sample. This may be written as

$$U = \tfrac{1}{2} \sum_{i=1}^{N} \sum_{r=1}^{8} (-1)^{m_s(i)+m_s(r)} D, \tag{9.50}$$

where the second sum is over the eight atoms neighboring each atom i in the first sum. The $\tfrac{1}{2}$ in front of the sum is to allow for the fact that each pair interaction should be counted only once, and the sum as written counts each twice. The argument of the sum is $-D$ when $m_s(i) = m_s(r)$ and $+D$ otherwise. Notice the impact of edge effects. If all the spins were aligned in our sample,

$$\begin{array}{ccc} + & + & + \\ + & + & + \\ + & + & + \end{array}$$

then the average energy per atom would be

$$\frac{1}{2}\left(\frac{[\,4(-3) + 4(-5) + 1(-8)\,]}{9} \right) = -2.22,$$

when edge effects are retained (four corners with -3 each, four edge atoms with -5 each, and one center atom with -8). In the macroscopic limit, however, this average would be $(-8)/2 = -4$, because almost every atom would be surrounded by eight aligned atoms. We obtain the correct result using periodic boundary conditions.

Once we have decided how we are going to model our system, the Monte Carlo calculation proceeds as follows:

1. Establish an initial state by randomly assigning spins ($+$ or $-$) to the atoms in the grid. Call this the "current" state.

2. Calculate the total potential energy for the current state, U_c.

3. Select an atom at random, and flip its spin. This makes a new state, which we'll call the "trial" state.

4. Calculate the total potential energy of the trial state, U_t.

5. Pick a random number between 0 and 1, and call it y.

   (a) If $\Delta U = U_t - U_c \leq 0$, then the trial state becomes the new current state.

   (b) If $\Delta U > 0$, the trial state becomes the new current state only if

   $$e^{-\Delta U/(k_B T)} > y. \tag{9.51}$$

CHECKPOINT The canonical distribution appears in red in Eq. 9.51 just to remind us that this is the same energy distribution that we derived in Chapter 2. This time we use the thermal energy $k_B T$ to gauge how likely is a particular *change* in energy (rather than a particular energy level alone).

Notice that $e^{-\Delta U/(k_B T)}$ approaches zero (keep *no* trial states) if $\Delta U \gg k_B T$, and approaches one (keep *all* trial states) if $\Delta U \ll k_B T$.

6. If the trial state is not kept, the spins return to the values they had in step 2. Whether we kept the trial state or not, the parameters of the state (such as U and the total spin) are then counted in a running average.

7. Return to step 3.

Let's walk through one cycle of a sample Monte Carlo calculation. For our lattice (Eq. 9.49), the energy for each atom is given in the following array.

$$
\begin{array}{ccc}
-2 & -2 & 4 \\
4 & -2 & -2 \\
-2 & 4 & -2
\end{array}
\tag{9.52}
$$

Adding this and dividing by 2 leaves a potential energy $U_c = 0$. This is the initial configuration, our *current* state.

We randomly select one of the spins, say the upper left-hand corner atom, and flip it so its spin is $-$ instead of $+$. This creates a *trial* lattice,

$$
\begin{array}{ccccc}
+ & + & - & + & + \\
- & - & + & - & - \\
+ & - & + & + & - \\
+ & + & - & + & + \\
- & - & + & - & -
\end{array}
\tag{9.53}
$$

with a different potential energy. Notice that we also flip the spin of the corresponding atoms in the boundaries. The trial lattice has the following energies at the atoms:

$$
\begin{array}{ccc}
2 & 0 & 2 \\
2 & 0 & 0 \\
0 & 2 & 0
\end{array}
\tag{9.54}
$$

which, added together and divided by 2, gives the total potential energy U_t equal to $4D$.

Now we come to step 5 and a decision—do we adopt this new configuration or not? The rule we impose here mimics the expected behavior of real systems subject to the canonical distribution. The potential energy has changed by $U_t - U_c = \Delta U$. A change in configuration that stabilizes the system, reducing the total potential energy, is always allowed; changes that increase the potential energy become less and less likely as the energy increase becomes larger and larger. In our example, let $D = 50\,\text{cm}^{-1}$ and $T = 298\,\text{K}$ ($k_B T = 207\,\text{cm}^{-1}$). Our potential energy increased from 0 to $4D$ or $200\,\text{cm}^{-1}$. Therefore ΔU is positive, and we must ask the computer for a random number y between 0 and 1.0. If the computer replies with $y = 0.592$, we find $e^{-\Delta U/(k_B T)} = 0.381$ is less than y, and we do *not* adopt the trial configuration.

Our criteria for selecting the next configuration introduce a clear temperature dependence. At $k_B T$ much larger than the interaction energy, $e^{-\Delta U/(k_B T)}$ is always close to 1, and therefore almost always greater than y. The system configuration at high temperature depends very little on the intermolecular potential, and this is what we expect for a real system.

This completes one cycle in the Monte Carlo simulation. This process is repeated over and over, occasionally reaching a new configuration and other times rejecting changes. The average properties calculated over all these configurations represent a canonical ensemble average. For the Ising model shown earlier, the average potential energy $\langle U \rangle$ and the average total spin $\langle M_S \rangle = \sum_i m_s(i)$ can be calculated, both as functions of the temperature T. These calculations on the two-dimensional Ising model can be used to illustrate the transition in a magnetic solid from an ordered phase, with the spins aligned at low temperature, to a disordered phase where randomized spins cause the average magnetization M (the ratio of $\langle M_S \rangle$ to the maximum possible $\langle M_S \rangle$) to approach zero. The results from a sample calculation are shown in Fig. 9.8.

The applications of Monte Carlo methods to chemical systems continue to expand, and they constitute a particularly important class of approaches to the difficult problem of chemical reaction dynamics.

CONTEXT *Where Do We Go From Here?*

The second law drives our system to *disperse* itself—to distribute its mass and energy more evenly and over a greater volume. This drive causes the molecules of a liquid to separate and form a gas when heated beyond the boiling point and causes the ions of sodium chloride to separate and dissolve in water. These fundamental changes that we observe and employ in chemistry on a routine basis cannot be justified on the basis of energetics alone. We need to understand how the entropy changes influence the process in order to justify what we see in the laboratory.

Now that we have invested some effort in understanding the entropy, we are ready to tackle these transformations: phase changes and solvation. These involve the most significant changes to the internal structure of our system that we've yet discussed, and they will bring us to the brink of chemical reactions in Part III.

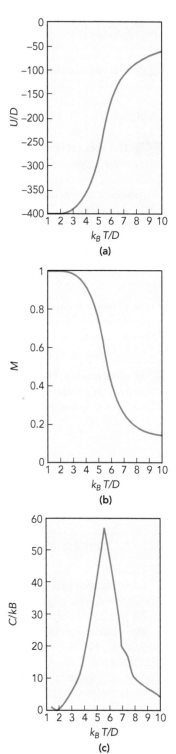

▶ FIGURE 9.8 **Results from Monte Carlo calculations on a 10 × 10 two-dimensional Ising model.** The phase transition from an ordered to a disordered spin system occurs at $k_B T = 5.5D$. We see the effects of the phase transition in (**a**) the steep increase in average potential energy, (**b**) the steep decrease in average magnetization M, and (**c**) the spike in the heat capacity at 5.5D, where energy absorbed by the system causes little temperature change.

KEY CONCEPTS AND EQUATIONS

9.1 **Entropy of a dilute gas.** The entropy of a monatomic gas is given accurately by the **Sackur-Tetrode equation:**

$$S_m = R\left\{\frac{5}{2} + \ln\left[\left(\frac{2\pi m k_B T}{h^2}\right)^{3/2} \frac{RT}{\mathcal{N}_A P}\right]\right\}. \quad (9.11)$$

9.2 **The second law of thermodynamics.** The second law of thermodynamics states that the *total* entropy (for the system and its surroundings) never decreases during a process and stays the same only for a reversible process. The second law may be written as the **Clausius principle**

$$(\Delta S_T)_{E,V,n} \geq 0. \quad (9.16)$$

The entropy change when an ideal gas changes volume at constant temperature is

$$\Delta S = RT \ln(V_f/V_i) \quad (9.19)$$

For a process at constant temperature and pressure, the total entropy increases if the Gibbs free energy change ΔG is negative, where

$$\text{constant } T : \Delta G = \Delta H - T\Delta S. \quad (9.26)$$

9.3 **Ideal mixing.** When we mix two ideal gases, at constant temperature and total pressure, the $\Delta_{mix}H$ of mixing is zero, but the change in entropy

$$\Delta_{mix}S = n_A R \ln\frac{1}{X_A} + n_B R \ln\frac{1}{X_B}$$

$$= -R(n_A \ln X_A + n_B \ln X_B) \quad (9.31)$$

is always positive. Therefore the free energy of mixing $\Delta_{mix}G = \Delta_{mix}H - T\Delta_{mix}S$ is negative, and so mixing

of ideal gases—when there are no intermolecular interactions—is always a spontaneous process.

9.4 **The third law of thermodynamics.** The third law of thermodynamics, that the entropy of a system is zero when $T = 0$, is a natural consequence of combining the Boltzmann entropy $S = k_B \ln \Omega$ with the notion that at absolute zero a system has one unique ground state, so $\Omega(T = 0\,\text{K}) = 1$ (however, see the residual entropy in Key Terms). The third law provides a reference point for the entropy, based on which we can define an absolute standard entropy for any given substance at a given temperature.

9.5 **Computer simulations.** A **molecular dynamics simulation** predicts the motion of molecules by breaking the process into tiny time steps. At each step, we calculate the forces acting on the set of model particles and use the corresponding acceleration vector to adjust all the velocity vectors. A **Monte Carlo simulation** mimics the canonical distribution of energies by randomly adjusting parameters (such as position or magnetic spin) in the simulation, calculating the change in energy ΔU from that adjustment, and then comparing $e^{-\Delta U/(k_B T)}$ to a random number to decide whether or not the new state contributes to the averaged properties of the system. The Monte Carlo method requires more careful planning than the molecular dynamics simulation, but when applicable it can converge to accurate results much more quickly by sampling a greater range of ensemble states.

KEY TERMS

- A **spontaneous process** is one that will occur without an external force required, because the overall entropy will increase during the process. A process that takes place *at constant temperature and pressure* is spontaneous if $\Delta G < 0$.
- In many real systems, there are multiple ensemble states too similar in energy for any practical experiment to isolate the ground state, even near 0 K. Because the

system has several ensemble states at this temperature, $\Omega > 1$ and the entropy does not fully reach zero near 0 K. The remaining entropy from these states is called the **residual entropy.**
- **Adiabatic demagnetization** is a technique for obtaining low temperatures of magnetic materials in order to measure properties such as heat capacity at extremely low temperatures.

OBJECTIVES REVIEW

1. *Calculate the entropy change for processes involving volume or temperature changes.*
   Find ΔS for the cooling of $100.\,\text{m}^3$ of air initially at 1.00 bar from $25.0°\text{C}$ to $23.0°\text{C}$.

2. *Estimate the absolute standard entropy of a gas as a function of temperature based on the mass and vibrational constants of the molecule.*
   Use the Sackur-Tetrode equation to estimate the standard molar entropy of radon gas at 298 K.

PROBLEMS

Discussion Problems

9.1 The number of moles in an ideal gas sample is increased adiabatically and at constant volume. During this process, which of the following occurs?

a. T increases and S decreases.

b. T decreases and S increases.

c. T decreases and S remains constant.

d. T and S remain constant.

9.2 Standard entropies for ions in solution are evaluated relative to the entropy of $H^+(aq)$; in other words, $S^{\ominus}[H^+(aq)] \equiv 0$. When H_2S is added to water, the solution contains $H_2S(aq)$, $HS^-(aq)$, and $S^{2-}(aq)$. The standard entropies (in $J\,K^{-1}\,mol^{-1}$) of these three species are written below; write the name of the correct compound next to its entropy.

a. -14.6;   b. 62.08;   c. 121

9.3 A rigid container initially at pressure P_1 is filled with gas to pressure P_2. During this time the temperature is kept constant. Explain whether the entropy of the sample in the container increases or decreases during this process.

9.4 Two containers are at exactly the same temperature, 298.15 K, and hold exactly 0.100 mol of gas, but container A holds helium gas whereas container B holds silane (SiH_4) gas. The internal energy of the silane must be higher, because silane has a higher heat capacity than helium. On average, does any energy transfer between the two containers?

9.5 The Ising model lattice is stable with all spins in the same direction for which of the following?

a. high temperature or high external magnetic field

b. low temperature or high external magnetic field

c. high temperature or low external magnetic field

d. low temperature or low external magnetic field

Heat Capacities and Expansions

9.6 Calculate the entropy change in $J\,K^{-1}$ for heating 1 mole of He gas ($C_{Pm} = 20.88\ J\,K^{-1}\,mol^{-1}$) from 300 K to 400 K at constant pressure.

9.7 Calculate the entropy change when a 56 g sample of pure iron is cooled from 373 K to 273 K at a pressure of 1.0 bar. For iron, $C_{Pm} = 25.1\ J\,K^{-1}\,mol^{-1}$.

9.8 Surprisingly, HCCH has a lower standard entropy at 298 K than HCN gas, even though acetylene has more vibrational modes and a lower rotational constant than hydrogen cyanide. For HCCH, $S_m^{\ominus} = 200.9\ J\,K^{-1}\,mol^{-1}$, $C_{Pm} = 43.93\ J\,K^{-1}\,mol^{-1}$ and for HCN $S_m^{\circ} = 201.8\ J\,K^{-1}\,mol^{-1}$, $C_{Pm} = 35.9\ J\,K^{-1}\,mol^{-1}$.

Find the temperature at which HCCH and HCN have the same standard entropy.

9.9 Two closed, rigid water containers with initial temperatures T_A and T_B are brought into thermal contact. At equilibrium, the two containers are at the same temperature. Prove that this process has positive ΔS and find an expression for the final temperature.

9.10 Find the entropy due to rotation in units of $J\,K^{-1}$ of a single molecule in the state $J = 5$.

9.11 Engineers sometimes model the propulsion of a bullet or an artillery shell through a gun barrel using the Nobel-Abel equation of state, which is the same as the van der Waals equation but without the attractive a term:

$$P(V_m - b) = RT. \qquad (9.55)$$

The justification for this is that the gases involved are very hot (so the attractive term in the intermolecular potential energy is negligible compared to the thermal energy) but extremely compressed (so the excluded volume is a significant fraction of the total volume). Find an expression for the entropy change $\Delta S = S_2 - S_1$ in a Nobel-Abel gas as it expands from volume and temperature V_1, T_1 to V_2, T_2, in terms of the number of moles n, the heat capacity C_V, and the excluded volume of the gas b. Assume that C_V is independent of T. THINKING AHEAD ▶ [What parameter can we use to link volume and temperature to C_V?]

9.12 A 25.0 L, 1.00 mole sample of air, composed in this case of 75% N_2 and 25% O_2, is separated into its components (by passing it over a liquid nitrogen trap, which condenses the oxygen). The separated N_2 and O_2 are then each allowed to return to the pressure and temperature of the initial air sample. Calculate ΔS.

9.13 Take our expression for the degeneracy of N particles in a box

$$\Omega = V^N f(E, N)$$

and find the change in entropy ΔS if the box expands from volume V_1 to V_2, keeping the total energy and number of particles constant.

9.14 One mole of an ideal monatomic gas is compressed isothermally at temperature T from pressure P_1 to P_2. What is the change in entropy?

9.15 If $\Omega = AV^N$, find $\Delta S = S(V_2) - S(V_1)$ in terms of R when $V_2 = 2V_1$ and $N = \mathcal{N}_A$.

9.16 During a reversible adiabatic expansion, one mole of an ideal monatomic gas cools by 100 K, and ΔG is -1000 J. What is the entropy of the final state in $J\,K^{-1}$?

9.17 In a temperature reservoir B at 298 K, a sample A undergoes an isothermal, isobaric process, for which $\Delta S_A = -327\ J\,K^{-1}$ and $\Delta H_A = -572$ kJ. Calculate the change in entropy of the sample plus its surroundings, ΔS_T under these conditions, and determine if the process is spontaneous.

9.18 When water is frozen at 3000 bar, it can be trapped in any of three crystalline forms, each corresponding to a different state of the water molecule: ice II, ice III, and ice V. Find the *apparent* molar entropy in $J K^{-1} mol^{-1}$ of water at $T = 0$ in such a case, assuming that each of these three forms is equally likely.

9.19 One prediction from the third law of thermodynamics is that the heat capacity C_P near absolute zero must not decrease with increasing temperature. Prove that this is the case by starting from the assumption that over a tiny increment in temperature, we can find some power x such that the heat capacity changes in proportion to T^x.

9.20 Calculate the entropy change when we heat calcium chloride from 298 K to 323 K, assuming that the heat capacity remains a constant $72.59 J K^{-1} mol^{-1}$ over this temperature range.

9.21 Find the entropy change ΔS for a 0.200 mol sample of N_2 heated and compressed from 298 K and 1.00 bar to 373 K and 3.00 bar.

9.22 We isolate n moles of an ideal gas with heat capacity C_{Pm} and then heat the gas and allow it to expand from a state P_1, V_1, T_1 to a final state P_2, V_2, T_2. Find a general formula for the ΔS of this process by treating the process as having two steps: an isothermal expansion step and a separate isobaric heating step. Show that the solution does *not* depend on which step (the expansion or the heating) is carried out first.

9.23 In adiabatic demagnetization, we use a magnetic field to couple away almost all of the residual energy of a crystal that has been cooled to 2.2 K by liquid helium. (We can get below the normal boiling point of liquid helium by boiling it at lower pressure.) Use Einstein's molar energy of the crystal,

$$E_m = 3\mathcal{N}_A \langle \varepsilon \rangle_{vib} = \frac{3\mathcal{N}_A \omega_E}{e^{\omega_E/(k_B T)} - 1},$$

in the low-temperature limit to estimate the residual molar energy of a crystal with an Einstein frequency of 105 cm^{-1} at 2.2 K. (The answer is smaller than you may expect.)

9.24 Figure 8.8 shows the P versus V graphs of three expansions: (a) the reversible isothermal, (b) the irreversible isothermal, and (c) the reversible adiabatic. Plot the corresponding curves for T versus S of these three processes, letting all three processes start from the same initial point (T_1, S_1) in the middle of your graph. Be quantitative if possible; otherwise, sketch an approximate curve. Label the curves a, b, and c. THINKING AHEAD ▶ [How should the areas under these curves compare?]

9.25 Evaluate the molar difference between the free energy and the energy, $G_m - E_m$ for argon at 298 K, assuming it is an ideal gas.

9.26 The container shown in the following figure separates an ideal gas into two compartments with a movable, thermally conducting wall between them. The conditions in each compartment are labeled.

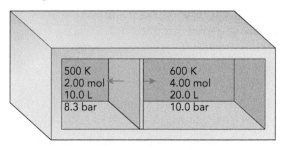

Circle **each** of the following statements that correctly describes how the system will change.

a. the wall will move to the right

b. the wall will move to the left

c. heat will flow to the right

d. heat will flow to the left

e. until the volume on both sides is the same

f. until the temperature on both sides is the same

g. until the pressure on both sides is the same

h. until the number of moles on both sides is the same

9.27 Show that the second law of thermodynamics applies to an isolated chamber, with two chambers A and B separated by a rigid, gas-tight piston. Both chambers are filled with equal amounts of the same gas, but gas A is initially at greater pressure than gas B. The piston slides from left to right until the two pressures are the same. Heat can flow freely between chambers A and B, so the process is isothermal. Find the total entropy change ΔS_T of the system, and show whether the value is always positive or always negative.

9.28 Use the Sackur-Tetrode equation to predict the standard molar entropy of *atomic* iodine I(g) at 573 K and 1.00 bar.

9.29 In many systems, the canonical distribution accurately describes populations in each degree of freedom (translations, rotations, and vibrations), but the effective temperature for each degree of freedom is different. For example, plasmas often heat vibrational motions more than they heat rotations or translations. Consider a sample of I_2 gas ($\omega_e = 214.5 cm^{-1}$, $B_e = 0.0559 cm^{-1}$) that initially has a vibrational temperature $T_{vib} = 653 K$, a rotational temperature $T_{rot} = 437 K$, and a translational temperature $T_{trans} = 298 K$. We then isolate the sample and wait for these different degrees of freedom to exchange energy until the temperatures are equal. What is the final temperature (now the same for all motions) of the sample?

Mixing

9.30 Find the entropy of mixing, $\Delta_{mix}S$, for two substances if they are mixed in the proportions that maximize $\Delta_{mix}S$.

9.31 Evaluate $\Delta_{mix}S$ for the mixing of two 1 L containers of water at 300 K. What is the change in $\ln \Omega$ for this process, given that $S_m(298 \text{ K})$ is 69.91 J K^{-1} mol^{-1}?

9.32 Calculate $\Delta_{mix}G$ in J for mixing three moles of helium and two moles of argon gas at 300 K, assuming they are ideal gases.

9.33 Calculate $\Delta_{mix}G$ in J at 298 K for mixing 0.01 mole each of H_2, Cl_2, and O_2, assuming they are ideal gases.

9.34 Calculate ΔG in the preparation of 1000 ml of 2.0 M $HClO_4$ solution, assuming ideal mixing and that each ion contributes individually to the entropy, at 298 K.

9.35 The desalinization of water attracts growing interest as potable water sources become more scarce and variable, but the process is not an easy one. Even assuming that the ions did not interact strongly with the solvent, desalinization must fight a substantial uphill battle against entropy. Assuming ideal mixing, calculate the minimum ΔS in J K^{-1} for the obtaining of pure water from 10 L of 1.0 M NaCl solution. Assume that the NaCl ionizes completely in solution and that each ion contributes independently to the entropy.

9.36 Calculate the entropy of mixing when we combine 0.100 mol of neon with 0.900 mol of argon at 298 K and 1.00 bar, assuming both are ideal gases.

9.37 Find an expression for $\Delta_{mix}F$ for the isothermal and isobaric mixing of two ideal gases.

9.38 For the isothermal mixing of two ideal substances at constant overall pressure P and volume $V_T = V_A + V_B$, we found

$$\Delta_{mix}S = -R(n_A \ln X_A + n_B \ln X_B).$$

If the initial isolated samples A and B are identical ($n_A = n_B$, $V_A = V_B$), and the total volume V_T is allowed to decrease, what is the final total volume $V_{T'}$ in terms of V_A if the overall process is adiabatic *and* isothermal?

9.39 Two ideal gases, n_A moles of A and n_B moles of B, initially occupy distinct volumes V_A and V_B at different temperatures, T_A and T_B. The molar heat capacities at constant pressure are C_A and C_B. A and B are combined to make a total volume $V_A + V_B$ at equilibrium. *First* write an equation for the total ΔS of this process in terms of the parameters above and the final temperature T_f. *Then* find the equation for T_f.

9.40 Consider the isothermal and isobaric mixing of the contents of two containers A and B holding identical samples of the same non-ideal gas, obeying the virial expansion (Eq. 4.39) to second order:

$$P = RT\left[\frac{n}{V} + B_2(T)\left(\frac{n}{V}\right)^2\right].$$

The combined volume of both containers is $V_A + V_B = V$, and the total number of moles of gas is $n_A + n_B = n$,

where $V_A = V_B$ and $n_A = n_B$. Find an expression for ΔS of this process in terms of n, V, and $B_2(T)$.

Computer Simulations

9.41 In a molecular dynamics calculation of the gas phase, after each time step the computer calculates new positions and orientations for each molecule. If it carries out one calculation for each translational coordinate and one for each rotational coordinate, how many calculations is this in a simulation of 1000 diatomic molecules over a period of 10^{-8} s with a resolution (time step-size) of 10^{-14} s?

9.42 Consider a molecular dynamics simulation using a square well potential where

$$u_{sq}(R) = \begin{cases} \infty & \text{if } R \leq R_1 \\ -\varepsilon & \text{if } R_1 < R \leq R_2 \\ 0 & \text{if } R > R_2 \end{cases}.$$

To correct the motion of the particles, we need to calculate the force of any interaction between two molecules A and B.

a. What is this force when A and B are separated by a distance $2R_2$?

b. What is this force when A and B are separated by a distance $(R_1 + R_2)/2$?

c. Use the change in potential energy to find an approximate expression for this force when the distance between A and B changes from $R_2 + (\Delta R/2)$ to $R_2 - (\Delta R/2)$, where $\Delta R \ll R_2 - R_1$.

9.43 A Monte Carlo simulation is carried out on the rotating dipole moment μ in an external electric field $\mathcal{E}$, which has potential energy

$$U(\theta) = \mathcal{E}\mu \cos\theta.$$

The following table shows a series of six values of U/k_B obtained during the simulation, whether they are included in the running average or not, and the corresponding values of the random number y. The previous energy in the simulation is given as "state 0" in the table, and the temperature is 300 K. Calculate $\langle U \rangle / k_B$ for this series, using only the correct states from the series.

trial state	U/k_B	y
0	388 K	
1	472 K	0.74
2	321 K	0.27
3	125 K	0.89
4	275 K	0.87
5	249 K	0.68
6	297 K	0.32

9.44 The following table gives averaged total energies for a series of Monte Carlo Ising model simulations at different temperatures. Calculate the approximate heat capacity of the system before and after the phase transition, and estimate the temperature of the phase transition.

T (K)	U (J)
4.4	−340.7
4.6	−324.2
4.8	−306.7
5.0	−285.8
5.2	−262.1
5.4	−237.3
5.6	−209.8
5.8	−186.8
6.0	−166.9
6.2	−152.6
6.4	−137.9

9.45 In a Monte Carlo simulation of a pure liquid at temperature T, we begin with some distribution of N particles, which interact according to a Lennard-Jones potential $u(R_{ij}) = 4\varepsilon\left[(R_{LJ}/R_{ij})^{12} - (R_{LJ}/R_{ij})^{6}\right]$. We select a molecule i and transfer it to a new location $\vec{r_i} = (x_i, y_i, z_i)$ in the sample to generate a trial state. A random number Y between 0 and 1 is generated.

a. List the conditions you should test to determine whether the trial state is to be kept in the simulation with another particle.

b. Write an equation for the total force at work on the relocated molecule.

9.46 Add the periodic boundary conditions and calculate the energy in terms of D for the following two-dimensional Ising model:

$$+ \quad - \quad + \quad -$$
$$- \quad + \quad - \quad +$$
$$+ \quad - \quad + \quad -$$
$$- \quad + \quad - \quad +$$

9.47 Calculate the total energy in terms of D of the 3 $\times$ 3 Ising model configuration drawn below, if the energy of interaction is $\pm D$ for any atom and each of its eight nearest neighbors. Include the periodic boundary conditions, which have not been drawn.

$$+ \quad - \quad -$$
$$- \quad - \quad -$$
$$+ \quad - \quad +$$

9.48 In a 1 cm^2 square layer of metal atoms arranged in a square lattice with lattice constant 2.5 Å, what fraction of atoms will be found at the very edge of the lattice?

(This problem justifies the need for elimination of edge effects in the Ising model.)

9.49 Find the maximum energy in terms of D of the 3 $\times$ 3 two-dimensional Ising model, and draw one of the configurations with that energy.

9.50 In terms of the interaction constant D, find the difference in the average energy per atom of a ground state 50 $\times$ 50 two-dimensional Ising model with and without the use of periodic boundary conditions.

9.51 For the following Ising model lattice, how much energy is necessary to flip the spin of the center atom, if $D = 2$ eV?

$$- \quad + \quad -$$
$$- \quad - \quad -$$
$$- \quad + \quad +$$

9.52 Draw any 4 $\times$ 4 grid for the Ising model that has zero energy, if periodic boundary conditions are included.

9.53 Derive a formula for the minimum energy in the two-dimensional Ising model, if the interaction energy between each spin and each of its eight nearest neighbors is $-D$. Assume that the model consists of a rectangular array of $i \times j$ spins.

9.54 A simpler Ising model than the one we have used counts only the interactions with the four nearest neighbor spins (above, below, left, and right) rather than the nearest eight. Apply the periodic boundary conditions and use this simpler system to find the potential energy in eV for the following array of spins, if the interaction constant $D = 1.5$ eV.

$$+ \quad - \quad -$$
$$- \quad + \quad -$$
$$+ \quad - \quad +$$

9.55 Shown below is a metastable configuration obtained from the Monte Carlo simulation of an Ising spin model under the condition $D = 4\,k_B T$. If the simulation proceeds by flipping the circled spin $\oplus$ from $+$ to $-$, find the probability that that new configuration will be kept.

$$+ \quad + \quad + \quad + \quad - \quad - \quad - \quad -$$
$$+ \quad + \quad + \quad \oplus \quad - \quad - \quad - \quad -$$
$$+ \quad + \quad + \quad + \quad - \quad - \quad - \quad -$$

9.56 If a magnetic spin system has an energy obeying the equation

$$E = E_0\left\{\left[\left(\frac{T}{T_0}\right) + \sqrt{2}\right]^{-4} - \left[\left(\frac{T}{T_0}\right) + \sqrt{2}\right]^{-2}\right\}$$

where E_0 and T_0 are constants, find the temperature (in terms of T_0) at which the heat capacity C_V reaches its maximum value, given that the maximum is the only critical point.

10 Phase Transitions and Phase Equilibrium

GOAL *Why Are We Here?*

The goal of this chapter is to describe phase transitions, both quantitatively and qualitatively. With the quantitative approach, we can calculate the dramatic effect that phase transitions have on our processes. However, we will first strive for a qualitative understanding of phase transitions, because the change from a liquid to a solid or from a liquid to a gas sorely tests our intuition about energy, entropy, and molecular structure. How can we add energy to a melting block of ice and yet see no change in temperature? Phase transitions are surprisingly incisive probes of the interactions between particles in our sample, providing an uncommonly accurate picture of the microscopic world through macroscopic properties.

CONTEXT *Where Are We Now?*

We've been inching toward genuine chemistry, looking for every opportunity to reduce or otherwise simplify molecular systems. Those opportunities are almost exhausted. This chapter and the next (on the thermodynamics of solutions) bring us to the threshold of our most ambitious goal: a basic understanding of chemical reactions. We still have a dodge or two left, however, before we arrive at full-blown chemistry. Here, we consider processes that significantly change the properties of a material without changing its chemical composition.

The thermodynamic properties of a substance can change discontinuously from one phase of matter to another, and crossing any such boundary defines the **phase transition.** When water goes from solid to liquid, molecular motions that were blocked in the solid become possible, causing a sudden increase in the heat capacity. On vaporization of the liquid, the density of water drops discontinuously by three orders of magnitude. The profound differences in these and other parameters make it necessary in classical thermodynamics to treat each phase of matter as a different substance. From the microscopic perspective, on the other hand, each phase corresponds to a different balance between the kinetic and potential energies of a *single* substance.

| SUPPORTING TEXT | *How Did We Get Here?*

The main qualitative preparation we need for the work ahead is an awareness of the balance between thermal energy and intermolecular attractions that separates solids from liquids from gases. For a more detailed background, we will draw on the following equations and sections of text to support the ideas developed in this chapter:

- The temperature T is defined by Eq. 2.24

$$T \equiv \left(\frac{\partial E}{\partial S}\right)_{V,N},$$

but a more intuitive (and less exact) understanding of the temperature can be expressed quantitatively using the equipartition principle (Eq. 3.20),

$$E = \frac{1}{2}N_{ep}Nk_{B}T = \frac{1}{2}N_{ep}nRT.$$

According to the equipartition principle, the temperature determines the average energy in each of the N_{ep} available degrees of freedom per molecule.

- In Section 4.1, we defined a solid as a system in which the forces between the particles are strong enough to stop translational motion, whereas in a liquid the thermal energy partly overcomes these forces, freeing the particles to travel throughout the system as long as the average density stays fixed. A gas is a system in which the thermal energy exceeds the capability of the attractive forces to hold the particles together, and the substance becomes free to expand and contract with the size of its container.

- The phase change from a liquid or solid to a gas is usually accompanied by an increase in volume, which we will find may be approximated using the expressions for an ideal gas expansion such as Eq. 9.19 for the entropy change:

$$\Delta S = nR\ln\left(\frac{V_f}{V_i}\right).$$

- In this chapter, we will begin to make use of the chemical potential μ, introduced as one of the thermodynamic intensive parameters in Eq. 7.6:

$$\left(\frac{\partial E}{\partial n_i}\right)_{S,V} = \mu_i,$$

which lets us write the incremental change in energy dE for a k-component system as (Eq. 7.11)

$$dE = TdS - PdV + \mu_1 dn_1 + \ldots + \mu_k dn_k.$$

The chemical potential serves in several equations as a molar Gibbs free energy for a particular chemical component, where the Gibbs free energy is given by Eq. 7.19,

$$G = E - TS + PV$$

and helps us determine the direction of a spontaneous process at constant T and P.

- In Section 7.3 we learn how to derive Maxwell relations by taking second derivatives of the thermodynamic potentials with respect to two different parameters. Also in Section 7.3, we find that we can express each of the thermodynamic parameters in terms of the system's N-particle partition function Q:

$$S = k_B \ln Q(T) + k_B T \left(\frac{\partial \ln Q(T)}{\partial T} \right)_{V,n} \tag{7.35}$$

$$P = k_B T \left(\frac{\partial \ln Q}{\partial V} \right)_{T,n}. \tag{7.37}$$

$$\mu = -k_B T \left(\frac{\partial \ln Q}{\partial n} \right)_{T,V}. \tag{7.38}$$

These expressions directly link the thermodynamic parameters to the properties of the individual molecules. In addition, the translational partition function Q_{trans} shows how the volume and number of particles explicitly affect these parameter values (Eq. 4.22):

$$Q_{trans}(T,V) = \frac{1}{N!} \left(\frac{2\pi m k_B T}{h^2} \right)^{3N/2} V^N.$$

10.1 Phase Transitions

The second law of thermodynamics may be the most important topic in physical chemistry, but not even quantum mechanics rivals phase transitions for a combination of sheer elegance and mystery. Phase transitions are the most fascinating processes that occur in thermodynamics, short of chemical reactions. So how do these work?

In a solid, each molecule vibrates about some average position, but the average positions and orientations are fixed. The liquid is almost completely nonrigid. The potential energy barriers that blocked rotation and translation of molecules in the solid have been exceeded by the available energy. Because we associate these microscopic degrees of freedom with energy flow in the form of heat, the substance must absorb heat in order to melt from a solid to a liquid. During the phase transition, however, the *temperature* does not change. But how can this be? How does the temperature *not* change if the liquid has been absorbing energy?

Following our technical definition of temperature in Eq. 2.24, we added a qualitative definition that temperature is roughly the *average kinetic energy in the easily excited degrees of freedom,* assuming that those energies have had time to equilibrate. For convenience, let us distinguish between translations, which are available to particles in the liquid but not in the solid, and the intermolecular vibrations, which are found in both solids and liquids. If two molecules are adjacent to each other, then their motion relative to each other is an intermolecular vibration. Once they stop being next to each other, we will simply call their motion translation, and that translation is what transforms the solid to a fluid.

CHECKPOINT From the standpoint of thermodynamics, the heat added to the substance during melting (and vaporization) doesn't increase the temperature because the added energy is going into transforming the substance from a solid to a liquid (or a liquid to a gas). In this respect we can see how a phase transition is equivalent to a chemical reaction as far as classical thermodynamics is concerned.

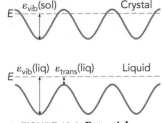

▲ FIGURE 10.1 **Potential energy curve for melting/freezing.** At the phase transition between solid and liquid, the *average* kinetic energy per degree of freedom can remain constant as the total energy increases because melting couples the added energy into translation—travel in new regions where the kinetic energy is very low.

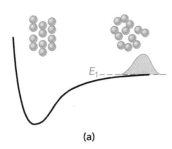

(a)

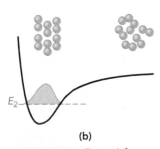

(b)

▲ FIGURE 10.2 **Potential energy curve for supercooling.** A supercooled liquid forms (a) when the sample has been cooled to an energy E_1 where very slow translational motions can occur among disordered molecular arrangements with high potential energy. The molecules will eventually locate a crystalline geometry (b) and stabilize into the solid, releasing their former potential energy as heat.

At the melting point, the solid has very energetic vibrations, but translations do not occur; if they did, we wouldn't have a solid. At the same temperature, the liquid has even more energetic vibrations, because we've had to add energy to the system, but it also has very low energy translations. These must be *low* energy translations, because at the melting point there is only exactly enough energy to reach those translational states (Fig. 10.1). At some point during melting, let the sample have two regions: (a) a solid region with an average kinetic energy per vibration of $\varepsilon_{vib}(sol)$, and (b) a liquid region with $N_{vib}(liq)$ vibrations of kinetic energy $\varepsilon_{vib}(liq)$ and $N_{trans}(liq)$ translations of kinetic energy $\varepsilon_{trans}(liq)$. The average kinetic energy *per degree of freedom* can remain equal in both regions, because the liquid has more degrees of freedom in which to distribute its energy:

$$\varepsilon_{vib}(sol) = \frac{N_{vib}(sol)\varepsilon_{vib}(sol)}{N_{vib}(sol)} = \frac{N_{vib}(liq)\varepsilon_{vib}(liq) + N_{trans}(liq)\varepsilon_{trans}(liq)}{N_{vib}(liq) + N_{trans}(liq)}, \quad (10.1)$$

where $\varepsilon_{vib}(liq) > \varepsilon_{vib}(sol) > \varepsilon_{trans}(liq)$. If the average kinetic energy is the same in the liquid and solid regions of the sample, the temperature will also be equal. Therefore, the liquid and solid can coexist at thermodynamic equilibrium, at the exact temperature of the phase transition. Adding or removing heat from the sample will only change the ratio of liquid to solid; it will not change the temperature until only one phase remains.

It is possible, in fact, to prepare a **supercooled liquid** by slowly lowering the temperature of the liquid below its standard freezing point (Fig. 10.2). The molecular motion slows down because of the loss in thermal energy, but the molecules continue to move with nearly random orientations, retaining their liquid character simply because they haven't arrived at the correct geometry for the more stable crystal. The *overall* energy is actually greater than what the solid has at its normal melting point, but much of that is potential energy due to unfavorable orientations of the liquid state molecules. The translations are so slow that the average kinetic energy *per degree of freedom,* and therefore the temperature, is lower than for the solid at the melting point. The substance is not at equilibrium, and it will eventually crystallize quite suddenly, releasing energy as the translational degrees of freedom shut down.

If we melt a solid by slowly applying heat at constant pressure, the temperature remains a continuous variable as a function of time, and so does the overall volume (Fig. 10.3). However, because the temperature remains constant during the phase transition from solid to liquid, the volume of a melting substance plotted as a function of the temperature changes discontinuously. A plot of V versus T shows a sudden jump or dip at the temperature of the phase transition. At constant pressure, the heat absorbed during the phase transition increases the enthalpy:

$$(\Delta H)_P = \int_{solid}^{liquid} (dH)_P$$

$$= \int_{solid}^{liquid} (TdS + VdP)_P$$

$$= \int_{solid}^{liquid} TdS = q_{rev}. \quad (10.2)$$

So graphs of H or S versus T also give discontinuous curves.

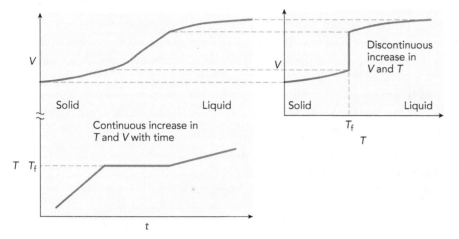

The same situation exists for the liquid–gas phase transition. During evaporation, the molecules go from one state, where their kinetic energy is measured relative to the attractive well of the intermolecular potential, to another state where the potential energy is nearly zero. To maintain the same average energy, and thus the same temperature, heat must be absorbed during evaporation (Fig. 10.4). Again the enthalpy, entropy, and volume are discontinuous functions of the temperature at constant pressure.

The difference that distinguishes solids, liquids, and gases as given in Section 4.1 is a difference in degrees of freedom: a solid has so little kinetic energy that it is trapped in a potential well corresponding to a single structure; a liquid has enough energy to allow the molecules to roll around each other but not enough energy to break them apart; the gas has enough energy for them to move independently. Phase transitions are processes in which energy flow changes the number or range of coordinates describing the system. These microscopic descriptions also illustrate why the entropy change is discontinuous. When a new kind of molecular motion becomes possible, the number of quantum states Ω in the ensemble grows exponentially, which appears as a sudden increase in the entropy, $S = k_B \ln \Omega$.

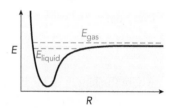

▲ FIGURE 10.4 **Potential energy curve for vaporization/ condensation.** During the transition from liquid to gas, the molecules must absorb heat to maintain the same average energy. New translations become possible during the phase transition, but only with very low kinetic energy.

EXAMPLE 10.1 Phase Transitions: A Microscopic Perspective

CONTEXT The latent heat of a phase transition is used in numerous applications. One area of current interest is the development of phase change materials that can be incorporated into the construction of buildings. For example, because melting absorbs much more heat than it takes to warm the substance up a few degrees to the melting point, phase transitions can be a very effective method for storing energy without changing the building temperature. Sunlight warms the structure during the day, but at the transition temperature the heat begins triggering endothermic phase transitions in the building material. At night, as the building cools off, the reverse, exothermic phase transition occurs, helping to maintain a constant temperature. In essence, the phase change materials greatly increase the heat capacity of the building, performing the same function that the ocean does in maintaining moderate temperatures in coastal areas. To identify what materials may serve this purpose, researchers are increasingly trying to predict phase transition properties by calculating potential energy surfaces and modeling the dynamics, as in this rather simplified problem.

PROBLEM The figure shows a crude model potential energy curve for one molecule above the surface of a liquid, all contained in a very small one-dimensional box. The energy u_0 corresponds to the strength of the intermolecular bonds binding the molecules together in the liquid. At low temperature, the molecule is bound by van der Waals forces to stay at a distance from the liquid surface of between 1.0 Å and 3.0 Å. At high temperature, the bond to the liquid is broken, and the molecule enters the gas phase, free to move between 1.0 Å and 18.0 Å (with the wall of the box keeping the molecule at $R < 18$ Å). Calculate the energy change necessary to vaporize the molecule at constant temperature. Assume the energy is a continuous variable.

SOLUTION As long as the molecule obeys the classical limits, the liquid can be warmed all the way to 200 K before the molecule leaves the liquid well. At that point, the potential energy of the molecule is still zero, so the kinetic energy $E - U$ is a constant 200 K (using temperature units for the energy for convenience). Once the molecule evaporates, it begins to bounce back and forth between the liquid surface at 1.0 Å and the wall at 18.0 Å. Between 1.0 and 3.0 Å the potential energy is still zero, but between 3.0 and 18.0 Å the potential energy is $u_0/k_B = 200$ K and the kinetic energy is $E_{gas} - u_0$. Let the minimum total energy of the molecule in the gas phase be E_{gas}. If we make a first approximation that the gas-phase molecule spends an equal time in every region between 1.0 and 18.0 Å, we estimate initially that the average kinetic energy of the gas is

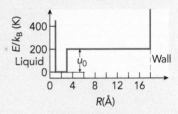

$$\langle K \rangle_{gas} = \frac{l_{liq} E_{gas} + l_{gas}(E_{gas} - u_0)}{l_{liq} + l_{gas}},$$

where the length of the liquid region l_{liq} is 2.0 Å and the length of the remaining region l_{gas} is 15 Å. Setting this equal to the average kinetic energy of the liquid at the melting point, $\langle K \rangle_{liq} = u_0$, we find

$$\langle K \rangle_{liq} = u_0 = \frac{l_{liq} E_{gas} + l_{gas}(E_{gas} - u_0)}{l_{liq} + l_{gas}}$$

$$= E_{gas} - \frac{u_0 l_{gas}}{l_{liq} + l_{gas}}$$

$$E_{gas} = u_0\left[1 + \frac{l_{gas}}{l_{liq} + l_{gas}}\right]$$

$$= k_B(200 \text{ K})\left(1 + \frac{15}{17}\right) = k_B(376 \text{ K}) = 5.19 \cdot 10^{-21} \text{ J}.$$

This means that the total energy required to keep the kinetic energy constant from liquid to gas is $k_B(376 \text{ K})$, so $\Delta E = k_B(176 \text{ K}) = 2.43 \cdot 10^{-21}$ J.

Because the molecule moves faster (having more kinetic energy) between 1.0 and 3.0 Å (call this the "liquid region"), the liquid region actually counts even less in our time-averaged kinetic energy, but this makes little difference in the qualitative result. Given that a typical gas container is much larger than the attractive region of the intermolecular potential, we would normally be at the limit $l_{gas} \gg l_{liq}$, for which

$$E_{gas} = 2u_0 = 2\langle K \rangle_{liq}. \qquad (10.3)$$

To go from the liquid to the gas at constant temperature, we should expect to have to add an energy equal to the kinetic energy of the liquid, or roughly the well depth u_0.

The most familiar phase transitions are among the three common phases of matter, and their conventional names are given in Table 10.1. In a fit of poor planning remarkable even in chemistry, the notation has evolved so that the phase transition we know as *melting* is officially named *fusion,* and its properties

TABLE 10.1   Conventional names for the common phase transitions.

from	to	phase transition
solid	liquid	fusion (melting)
liquid	solid	freezing
liquid	gas	vaporization
gas	liquid	condensation
solid	gas	sublimation
gas	solid	deposition

are evaluated at a temperature known as the standard **freezing point,** T_f. At the standard freezing point, the liquid and solid are in equilibrium while the total pressure (the *ambient* pressure) is exactly 1 bar, the standard state pressure. Faring only slightly better, the liquid-to-gas phase transition is officially *vaporization,* and is referenced to the **boiling point,** T_b.

However, unlike melting and fusion, there is a genuine distinction between *boiling* and *vaporization,* and to explain it we need to consider more carefully what happens when vaporization occurs. An unconfined liquid or solid sample loses at least a few molecules to the gas phase, except near absolute zero. The energy available from vibrations in the sample will occasionally be redistributed into kicking a particle out of the condensed phase and into the gas. So as long as any liquid or solid is present, vaporization takes place. As you'd expect, this happens more often as we heat up the sample and the vibrations have more energy available. As a result, we rarely have a sample present in just its solid phase or just its liquid phase: there's some of that substance in the gas phase as well. If the gas is allowed to escape, more of the sample will evaporate. If, on the other hand, the gas is allowed to remain and come into equilibrium with the liquid or solid, then the vaporization and condensation rates become equal when the gas reaches the **vapor pressure** of the sample at that temperature. The vapor pressure is the partial pressure of the pure sample in equilibrium with the condensed form of the material, and it increases with temperature.

Boiling is the particular case of vaporization taking place when the vapor pressure of the material *equals the ambient surrounding pressure.* Water, for example, doesn't need to be at its boiling point to evaporate. Hang a wet beach towel in the sun, and the water evaporates without the towel reaching scalding temperatures. What makes water *boil* at 373 K and 1.0 bar is that its vapor pressure is equal to 1.0 bar at that temperature. When that condition is satisfied, the gas forming within the liquid has enough pressure to push against the surrounding matter, and bubbles form in the liquid. The bubbles are what give boiling its name.

At constant pressure, the Gibbs energy G is a continuous variable for both melting and vaporization, because $(dG)_P = -S\,dT$. If dT is zero, then $(dG)_P$ must be zero also. These are examples of **first-order phase transitions** because, while G is continuous as a function of temperature, its first derivative $(\partial G/\partial T)_P = -S$ is not. The discontinuities occur in the energy, enthalpy, and in the variables V and S (assuming constant P) in these cases. For example, the plot of V versus T in Fig. 10.3 is discontinuous at T_f.

BIOSKETCH | Stephen Elliott

Stephen Elliott is a professor of chemical physics at the Department of Chemistry and Trinity College of Cambridge University who studies the glass phase transition in *phase change alloys,* combinations of metals and/or metalloids that undergo rapid phase transitions between crystalline and glassy forms. This phase transition can be induced over tiny areas using a focused laser, transforming an ordered crystalline structure into a disordered glass and vice versa. Because the glass and the crystal usually reflect and transmit light differently, phase change alloys provide a means of encoding data optically at high densities. This is the technology used, for example, to store data on optical disks. Similar effects can be applied to the electronics of flash memories as well. Professor Elliott studies the structure of the glassy phase of these materials and uses computer simulations to investigate the motions of the atoms during the phase transition. One goal of this work is to be able to reliably test candidate alloys by simulations so as to optimize the choice of materials prior to synthesizing and characterizing them in the laboratory.

There are also **second-order phase transitions,** in which $(\partial G/\partial T)_P$ is continuous but $(\partial^2 G/\partial T^2)_P$ is not. In these phase transitions, a graph of S versus T will be continuous, but plotting $C_P = (T\partial S/\partial T)_P$ versus T will yield a discontinuity at the phase transition temperature. These include phase transitions in which the number of degrees of freedom for molecular motion remains constant; for example, when enough energy is put into a magnetic crystal to suddenly allow the electron spins to become randomly oriented. The Monte Carlo methods and the Ising model of Section 9.5 simulate this phase transition effectively. Another common second-order phase transition is the **glass phase transition,** the transition between the brittle, amorphous glass and its corresponding fluid (or in the case of polymers, a transition from the crystalline to the rubbery form). Glass blowers take advantage of this phase transition to give glass the enormous elasticity required for shaping it into flasks and beakers and laser tubes, among a few other things. Because it is a second-order phase transition, there is no discontinuity in the sample volume, which allows the material to be smoothly and intricately shaped while in a viscous liquid state and then cooled to a solid with virtually no distortion or contraction.

10.2 Thermodynamics of Phase Transitions

As we move deeper into classical chemistry, which has older and closer ties to the laboratory than our initial microscopic perspective, we rely more heavily on differences in enthalpy ΔH rather than in energy ΔE to assess the relative stability of two chemical systems. We ascribe energy changes arising from the microscopic variables of the system to heat, and the heat is usually more easily evaluated as $q = (\Delta H)_P$ than as $q = (\Delta E)_V$. Again, it is a practical question: what parameters are easier to hold constant? For bench-top chemistry, we constrain the pressure more often than the volume.

We have previously shown how changes in thermodynamic parameters can be calculated in general by integrating the derivative; for example, to get ΔH we integrate dH. But for phase transitions, the functions we have to integrate have a discontinuity somewhere. A plot of H versus T for heating a substance through its melting or boiling point would look like Fig. 10.5, with a sudden rise in H at the temperature of the phase transition. To evaluate the overall ΔH for a process like the one shown, we are, in effect, integrating dH, but we break the process into distinct pieces:

- step ① for heating from the initial state i to the beginning of the phase transition α

- step ② for carrying out the phase transition from state α to state β at constant temperature

- step ③ for heating the substance to its final state f

The overall ΔH is just the sum of the ΔH's for each of the three steps:

$$\Delta H = \Delta H_1 + \Delta H_2 + \Delta H_3 = \int_i^f dH = \int_i^\alpha dH + \int_\alpha^\beta dH + \int_\beta^f dH. \quad (10.4)$$

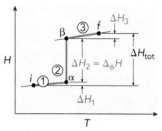

▲ FIGURE 10.5 **Enthalpy change for heating through a phase transition.** The enthalpy change for a temperature change that crosses a phase transition is the sum of the enthalpy change ΔH_1 of the initial phase, the latent enthalpy ΔH_2 of the phase transition, and the enthalpy change ΔH_3 of the final phase.

Analytical expressions may be obtainable for the parameters in one phase or the other, but the phase transition itself is defined by a discontinuity. Consequently, some of the thermodynamic parameters of the substance cannot be expressed across the phase transition by an equation for a well-behaved algebraic function. Instead, we measure the total change in that parameter's value from one phase to the other and then write the number in a table, to look up whenever we need it. For example, the **latent enthalpy of fusion** $\Delta_{\text{fus}} H^\ominus$ is the enthalpy of melting per mole for a particular substance, measured at its standard melting temperature T_f at 1 bar. For any phase transition ϕ—for example, fusion (fus), vaporization (vap), or sublimation (sub)—the latent enthalpy gives us the numerical value of the integral over the phase transition, from α to β, per mole of material:

$$\int_\alpha^\beta dH \equiv n\Delta_\phi H. \quad (10.5)$$

(The latent enthalpy is assumed to be a molar property, so we usually leave off the subscript "m" that goes with other molar properties.) The **latent entropy** of any phase transition, as we will see, can be calculated from the latent enthalpy and phase transition temperatures. Enthalpies and standard temperatures for the phase transitions of several substances are given in Table 10.2.

As with our other tables of molecular parameters, we can show how trends among these values reflect changes in the basic molecular properties. In this case, we observe principally that *more energy is required to melt or vaporize substances with stronger intermolecular bonds.* That's no surprise, if melting and boiling are the processes whereby we break those bonds. For example, helium, which has the weakest van der Waals forces of any chemical species, also has the lowest boiling point known and cannot even be frozen into a solid (except at pressures greater than about 10 bar). Melting points, boiling points, and the

TABLE 10.2 Standard molar enthalpies of vaporization and fusion at 1 bar, in order of increasing T_b.

Substance	T_f (K)	$\Delta_{fus}H^{\ominus}$ (kJ mol^{-1})	T_b (K)	$\Delta_{vap}H^{\ominus}$ (kJ mol^{-1})
He	—	—	4.22	0.0829
H$_2$	13.96	0.12	20.28	0.898
Ne	24.7	0.34	27.07	1.71
N$_2$	63.15	0.719	77.36	5.57
CO	68	0.83	81.6	6.04
F$_2$	53.5	0.51	85.03	6.62
Ar	83.8	1.12	87.30	6.43
O$_2$	54.5	0.44	90.20	6.820
CH$_4$	90.75	0.94	111.67	8.19
Kr	115.8	1.37	119.93	9.08
Xe	161.5	1.81	165.11	12.62
C$_2$H$_6$	90.3	2.86	184.55	14.69
C$_3$H$_8$	85.5	3.53	231.0	19.04
Cl$_2$	171.6	6.40	239.11	20.41
NH$_3$	195.41	5.66	239.82	23.33
CH$_3$Cl	176.1	6.431	249.06	21.40
CH$_2$Cl$_2$	178.01	6	313	28.06
Br$_2$	265.9	10.57	331.9	29.96
CHCl$_3$	209.6	8.8	334.32	29.24
CCl$_4$	250.0	3.28	349.9	29.82
H$_2$O	273.15	6.008	373.15	40.65
C$_{10}$H$_{22}$ (n-decane)	243.5	28.78	447.30	38.75
I$_2$ (rhombic)	386.8	15.52	457.6	41.57
Hg	234.32	2.29	629.88	59.11
TiI$_4$	423	19.8	650	58.4
S (monoclinic)	388.36	1.72	717.75	45
AgCl	728	13.2	1820	199
Pb	600.61	4.77	2022	179.5
Al	933.47	10.71	2792	294
B	2348	50.2	4300	480

corresponding latent enthalpies are lowest for the small, non-polar particles like Ne, N$_2$, CH$_4$, and CO$_2$. Larger molecules tend to have larger dispersion forces, and polar molecules form dipole–dipole bonds with each other. Both effects lead to higher phase transition temperatures and higher latent enthalpies. At the bottom of Table 10.2, having the highest phase transition temperatures and enthalpies, are the substances that form solids of chemical bond strength: the ionic (AgCl), metal (Al), and covalent crystals (B).

Let's try a crude, quantitative prediction. The change in entropy upon evaporation of the liquid at 1 bar is primarily due to the tremendous increase in volume. At T_b, most liquids have molar volumes between 14 and 250 cm³ mol⁻¹, whereas the corresponding gases have molar volumes ranging from 6400 to $3.6 \cdot 10^6$ cm³ mol⁻¹. The molar change in entropy ΔS_m we would predict from this isothermal expansion is (Eq. 9.19)

$$\Delta S_m = R \ln \left(\frac{V(\text{gas})}{V(\text{liq})} \right),$$

which is typically in the range of $6R$ to $10R$. Actual values of $\Delta_{\text{vap}} S^\ominus$ tend to be about 30% higher because the molecules in the liquid are constrained in their movement by their neighbors and therefore don't have as great an ensemble size as would an ideal gas at the same volume. This estimate that $\Delta_{\text{vap}} S^\ominus$ lies between about $8R$ and $13R$ is known as **Trouton's rule.**

The entropy change during any phase transition step is related to the corresponding latent enthalpy, because at constant pressure we can write $dH = TdS$, and at constant temperature we integrate both sides to find $\Delta H = T\Delta S$. Therefore, for a phase transition where both phases are in equilibrium at constant temperature and pressure,

$$\Delta_\phi S = \frac{\Delta_\phi H}{T_\phi}, \tag{10.6}$$

where T_ϕ is the phase transition temperature at which $\Delta_\phi H$ and $\Delta_\phi S$ are measured.

Returning to Trouton's rule, if $\Delta_{\text{vap}} S^\ominus$ is between $8R$ and $13R$, then we predict the latent enthalpy of vaporization

$$\Delta_{\text{vap}} H^\ominus = T_b \Delta_{\text{vap}} S^\ominus \approx T_b (8 \text{ to } 13)R.$$

For CCl_4, for example, this covers the range from $(350\,\text{K})(8R) = 23\,\text{kJ mol}^{-1}$ to $(350\,\text{K})(13R) = 38\,\text{kJ mol}^{-1}$, and the actual value is 30 kJ mol⁻¹. In Table 10.2, only He is inconsistent with Trouton's rule, having $\Delta_{\text{vap}} S^\ominus = \Delta_{\text{vap}} H^\ominus / T_b = 2.4R$, largely because of its very low boiling point.

Now that we know where the values in Table 10.2 come from, let's do an example showing how to use these values.

EXAMPLE 10.2 **Enthalpy and Entropy of a Phase Transition**

CONTEXT One of the questions in climate modeling is how warmer water and air will affect the annual formation and melting of the Arctic sea ice. Aside from factors involving the oceanography and meteorology, this question is fundamentally about the thermodynamics of the heat transfer between the sea ice and its surroundings. The ice acts as a buffer, limiting the temperature change possible in the surrounding sea by absorbing and releasing heat as required to maintain a relatively constant temperature. Here we will carry out a calculation on a small scale and with pure water to show how significant an effect the phase transition has on the heat transfer.

PROBLEM Calculate ΔH_m and ΔS_m when water ice at 263 K is warmed by 20 K at 1 bar. Use the constants $C_{Pm}(\text{sol}) = 37\,\text{J K}^{-1}\,\text{mol}^{-1}$, $C_{Pm}(\text{liq}) = 75\,\text{J K}^{-1}\,\text{mol}^{-1}$, and $\Delta_{\text{fus}} H^\ominus = 6.01\,\text{kJ mol}^{-1}$ for water, where (liq) and (sol) indicate the liquid and solid states, respectively.

SOLUTION There are three steps: warming the ice to 273 K, the phase transition to the liquid at 273 K, and warming the liquid water to 283 K (Fig. 10.6). For the first and last steps,

$$\Delta H_m = \int_{T_1}^{T_2} (dH_m)_P = \int_{T_1}^{T_2} (TdS_m)_P$$

$$= \int_{T_1}^{T_2} \left(\frac{\partial q_m}{\partial T}\right)_P dT = \int_{T_1}^{T_2} C_{Pm} dT$$

$$\approx C_{Pm}(T_2 - T_1), \qquad (10.7)$$

where we have approximated the heat capacity C_{Pm} to be a constant of the temperature. For the phase transition, the enthalpy change is, by definition, $\Delta_{fus}H^\ominus$, so for the molar enthalpy change we have

$$\Delta H_m = C_{Pm}(sol)(273 - 263) + \Delta_{fus}H^\ominus + C_{Pm}(liq)(283\,K - 273\,K)$$

$$= [37(10) + 6010 + 75(10)](J\,mol^{-1}) = 7130\,J\,mol^{-1}.$$

The entropy change can be calculated in a similar fashion. For the first and last steps, we can write

$$\Delta S_m = \int_{T_1}^{T_2} (dS_m)_P = \int_{T_1}^{T_2} \left(\frac{dH_m}{T}\right)_P$$

$$= \int_{T_1}^{T_2} \frac{C_{Pm}}{T} dT \approx C_{Pm} \ln \frac{T_2}{T_1}. \qquad (10.8)$$

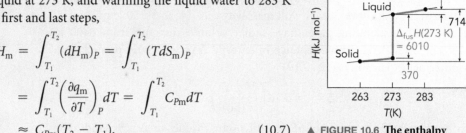

▲ **FIGURE 10.6 The enthalpy change for warming water from ice at 263 K to liquid at 283 K.** The overall ΔH is the same as for a series of three steps: warming the ice, melting the ice, and warming the liquid.

Equation 10.6 gives us the latent entropy of the phase transition itself in terms of $\Delta_{fus}H^\ominus$. Adding this to the two heating terms gives the overall change in molar entropy:

$$\Delta S_m^\ominus = C_{Pm}(sol)\ln\frac{273}{263} + \frac{\Delta_{fus}H^\ominus}{273} + C_{Pm}(liq)\ln\frac{283}{273}$$

$$= \left[37(0.037) + \frac{6010}{273} + 75(0.036)\right](J\,K^{-1}\,mol^{-1}) = (1.4 + 22.0 + 2.7)\,J\,K^{-1}\,mol^{-1}$$

$$= 26.1\,J\,K^{-1}\,mol^{-1}.$$

These results show that over this 20 K change in temperature, both ΔH and ΔS are dominated by the phase transition, not by the heating required for the temperature change.

EXAMPLE 10.3   Phase Transitions at Non-Standard Temperature

CONTEXT At an elevation of 5100 m in the Peruvian Andes, La Rinconada may be the town with the lowest atmospheric pressure in the world, averaging roughly 0.56 bar. Mount Evans in Colorado at 4300 m elevation has a paved road where the atmospheric pressure is somewhat higher, about 0.61 bar. At these elevations, the boiling point of water is significantly lower than at sea level. Coffee drinkers sometimes complain that coffee cannot be prepared properly at low pressure because the water simply doesn't get hot enough. To cook rice or beans in these locations, pressure cookers can be employed, building up the pressure above the food to raise the boiling point of water back to about 373 K. The temperature given below, 355 K, is a lower limit to the boiling point in La Rinconada.

PROBLEM Water boils at a lower temperature at high altitudes where the pressure is lower. What is ΔS_m for water boiling at 355 K? The heat capacities for H_2O are 33.6 J K^{-1} mol^{-1}(gas) and 75.3 J K^{-1} mol^{-1}(liq).

SOLUTION Although this is a one-step process, the phase transition at 355 K, ΔS_m must be the same as for the three-step process in which we first warm the water to 373 K, then boil it at its standard boiling temperature, and then cool the vapor back to 355 K (Fig. 10.7).

Evaluating ΔS_m in a fashion similar to that of the previous example, we have

$$\Delta S_m = C_{Pm}(\text{liq}) \ln \frac{373}{355} + \frac{\Delta_{vap}H}{373} + C_{Pm}(\text{gas}) \ln \frac{355}{373}$$

$$= \left[75.3(0.0495) + \frac{40{,}650}{373} + 33.6(-0.0495) \right] (\text{J K}^{-1}\text{mol}^{-1})$$

$$= (3.73 + 109 - 1.66)\,\text{J K}^{-1}\text{mol}^{-1} = 111\,\text{J K}^{-1}\text{mol}^{-1}.$$

The latent entropy at 373 K alone is $(40{,}650/373)\,\text{J K}^{-1}\text{mol}^{-1} = 109\,\text{J K}^{-1}\text{mol}^{-1}$. The change in temperature from 373 K to 355 K causes a relatively small shift (from 109 to 111 J K^{-1} mol^{-1}) in the entropy of vaporization. Pay careful attention to the signs; the direction of the process determines the sign of the change in enthalpy or entropy. The enthalpy and entropy will increase when the substance is heated, when it melts, and when it vaporizes.

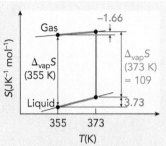

▲ FIGURE 10.7 **Latent entropy change at a non-standard temperature.** The ΔS for the boiling of water at 355 K can be determined using the entropy of vaporization $\Delta_{vap}S$ at 373 K.

Because H and S are state functions, it still does not matter how we get from the initial state to the final state; ΔH and ΔS must be the same for any path. If we induce a phase transition at some temperature other than the one at which the latent enthalpy and latent entropy are evaluated (say, by changing the pressure), we can still evaluate the change in enthalpy or entropy using the normal latent properties.

10.3 Chemical Potentials

In classical thermodynamics we treat different phases of the same compound as different substances, and it is the chemical potential μ that describes the change in energy or enthalpy as the number of molecules of the compound in one phase changes (Eq. 7.6):

$$\left(\frac{\partial E}{\partial n_i} \right)_{S,V} = \left(\frac{\partial G}{\partial n_i} \right)_{T,P} = \mu_i.$$

As an example, let's find an expression for the chemical potential of an ideal monatomic gas at constant entropy and volume. If S and V are constant, the chemical potential is the rate at which the energy increases as n increases. This is the same as the energy per mole of the substance—the molar energy:

$$\text{monatomic ideal gas: } E = \frac{3}{2}nRT$$

$$\mu = \left(\frac{\partial E}{\partial n} \right)_{S,V} = \frac{3}{2}RT.$$

Starting from the chemical potential, we can connect the chemical substances at the beginning of a process to the substances at the end of the process. In this chapter, we restrict the processes to phase transitions, but the results will be applicable to even more dramatic transformations later on.

CHECKPOINT Although the chemical potential can be described as a kind of molar energy, notice that we never simply set μ equal to E_m. That's because the chemical potential is defined not by the energy itself but by the *change* in energy (or free energy or enthalpy) under specific conditions, such as S and V held constant. We use the chemical potential because it automatically adapts to help us monitor the flow of chemical energy in our system in terms of whichever thermodynamic potential (E, H, G, or F) is most appropriate for the conditions that steer the process.

The Gibbs-Duhem Relation

The derivative of one extensive parameter (E) with respect to another (n_i) yields an intensive parameter (μ_i). The chemical potential depends on the concentration of each chemical component of the system, but not on the overall quantity of material. If we assume a system with k different chemical components, then (Eq. 7.20)

$$dE = TdS - PdV + \mu_1 dn_1 + \ldots + \mu_k dn_k.$$

Next, we use a sneaky trick to effectively integrate over the differentials.[1] Because μ does not depend on the amount of material, we can see easily how dE changes as we increase all the extensive parameters, including the total amount of material, by a small fraction $d\lambda$, without changing the *amount* of any of the components. For example, the change in V will be $Vd\lambda$, where we might set $d\lambda$ equal to 0.00001. Similarly the energy E, since it is an extensive parameter, must increase by an amount $Ed\lambda$. The pressure, temperature, and chemical potential are intensive parameters and do not depend on the amount of material. For this process, dE will obey the following equation:

$$dE = E\,d\lambda = TS\,d\lambda - PV\,d\lambda + \mu_1 n_1\,d\lambda + \ldots + \mu_k n_k\,d\lambda. \quad (10.9)$$

Dividing through by $d\lambda$ gives the following fascinating equation for the energy of our k-component sample:[2]

$$E = TS - PV + \mu_1 n_1 + \ldots + \mu_k n_k. \quad (10.10)$$

We derived Eq. 10.10 using only thermodynamics and some mathematical sleight-of-hand. We may justify the same result using statistical mechanics. Expressing our thermodynamic parameters in terms of the partition functions, we can write for a single-component sample (Eqs. 7.35, 7.37, 7.38):

$$TS - PV + \mu n =$$

$$k_{\mathrm{B}}T \ln Q + k_{\mathrm{B}}T^2\left(\frac{\partial \ln Q}{\partial T}\right)_{V,n} - k_{\mathrm{B}}TV\left(\frac{\partial \ln Q}{\partial V}\right)_{T,n} - n k_{\mathrm{B}}T\left(\frac{\partial \ln Q}{\partial n}\right)_{T,V}. \quad (10.11)$$

We need the partial derivatives of $\ln Q$ with respect to V and n, and we manage this by assuming a general form for the partition function, consistent with our results from Chapters 2 and 4:

$$Q = \frac{1}{N!}\zeta^N V^N, \quad (10.12)$$

where ζ includes all the dependence of the partition function on the temperature and the internal energies (such as vibrations and rotations). The logarithm then has the form

$$\ln Q = N - N\ln N + N\ln \zeta + N\ln V, \quad (10.13)$$

---

[1] The same trick forms the basis of perturbation theory, a general technique in physics but one with particular relevance to quantum mechanics. We use this approach in *QM* Section 4.2.

[2] We will not make direct use of this solution for the absolute energy E, but it is interesting to see how this derivation divides the energy of our sample into thermal (TS), mechanical ($-PV$), and chemical (μn) contributions.

where Stirling's approximation (Eq. 2.14) was used for the $N!$ term. The derivatives are

$$\left(\frac{\partial \ln Q}{\partial V}\right)_{T,n} = \frac{N}{V} \tag{10.14}$$

$$\left(\frac{\partial \ln Q}{\partial n}\right)_{T,V} = \mathcal{N}_A \left(\frac{\partial \ln Q}{\partial N}\right)_{T,V}$$

$$= \mathcal{N}_A \left(\frac{\partial[-N\ln N + N + N\ln \zeta + N\ln V]}{\partial N}\right)_{T,V} \quad \text{by Eq. 10.13}$$

$$= \mathcal{N}_A (-\ln N - 1 + 1 + \ln \zeta + \ln V)$$

$$= \left(\frac{\mathcal{N}_A}{N}\right)(-N\ln N + N\ln \zeta + N\ln V) \qquad N/N$$

$$= \frac{\ln Q - N}{n}. \tag{10.15}$$

Next, we substitute these results into Eq. 10.11:

$$TS - PV + \mu n = k_B T \ln Q + k_B T^2 \left(\frac{\partial \ln Q}{\partial T}\right)_{V,n} - k_B T V \left(\frac{N}{V}\right) - nk_B T \left(\frac{\ln Q - N}{n}\right)$$

$$= k_B T \ln Q + k_B T^2 \left(\frac{\partial \ln Q}{\partial T}\right)_{V,n} - Nk_B T - k_B T \ln Q + Nk_B T$$

$$= k_B T^2 \left(\frac{\partial \ln Q}{\partial T}\right)_{V,n}, \tag{10.16}$$

which is the same as the equation for the energy we obtained from Eq. 6.15. We thus obtain Eq. 10.10 for the case of one component:

$$E = TS - PV + \mu n,$$

and the result extends to the multi-component sample.

Whichever way we obtain Eq. 10.10, we learn something important about the chemical potential when we take the derivative to get dE again. This time, the derivative looks different:

$$dE = TdS + SdT - PdV - VdP + \mu_1 dn_1 + n_1 d\mu_1 + \ldots + \mu_k dn_k + n_k d\mu_k. \tag{10.17}$$

Substituting this expression into the left-hand side of Eq. 10.9, we have

$$TdS + SdT - PdV - VdP + \mu_1 dn_1 + n_1 d\mu_1 + \ldots + \mu_k dn_k + n_k d\mu_k$$
$$= TdS - PdV + \mu_1 dn_1 + \ldots + \mu_k dn_k,$$

and therefore

$$SdT - VdP + n_1 d\mu_1 + \ldots + n_k d\mu_k = 0. \tag{10.18}$$

This is called the **Gibbs-Duhem equation,** and it expresses neatly that the intensive degrees of freedom—T, P, and μ_i—are all interdependent parameters. For example, if we adjust the temperature of the system at constant pressure, then the chemical potentials must change so that the sum in Eq. 10.18 remains zero. Another interpretation of the Gibbs-Duhem equation is that it demonstrates how, at constant temperature and pressure ($dT = 0$ and $dP = 0$), the chemical potential gives the quantity of useful work available from each chemical component in the system. We will take advantage of this reasoning in Chapter 11.

The Clausius-Clapeyron Equation

We can go further still. Our definition of the Gibbs energy G from Eq. 7.19,

$$G = E - TS + PV,$$

can be combined with Eq. 10.10 for the energy to obtain a new expression for G:

$$G = TS - PV + \mu_1 n_1 + \ldots + \mu_k n_k - TS + PV$$

$$= \sum_{i=1}^{k} \mu_i n_i. \tag{10.19}$$

For a process at constant T and P at equilibrium $dG = 0$, so

$$\text{equilibrium:} \quad dG = \sum_{i=1}^{k} \mu_i dn_i = 0. \tag{10.20}$$

For a system consisting of only two phases, the number of moles lost from one phase B are gained by the other phase A, where A and B could be the gas and the liquid, for example. Therefore,

$$dn_A = -dn_B. \tag{10.21}$$

For two phases,

$$dG = \mu_B dn_B + \mu_A dn_A = dn_B(\mu_B - \mu_A). \tag{10.22}$$

At equilibrium, dG is zero, and

$$\mu_B = \mu_A. \tag{10.23}$$

At equilibrium, any two phases of the same material in contact with each other must have the same chemical potential. This is a powerful tool in the description of all sorts of bench-top chemistry when the reactants can be assumed to be in equilibrium with their vapor phases.

As our first application of Eq. 10.23, let's find how the phase change temperature shifts with pressure. For a single substance present in two phases A and B, we can write the Gibbs energy change of each phase individually (assuming T and P are the same for both substances):

$$dG_A = -S_A dT + V_A dP + \mu_A dn_A, \quad dG_B = -S_B dT + V_B dP + \mu_B dn_B. \tag{10.24}$$

The usual method of obtaining Maxwell relations (Section 7.3) gives us

$$\left(\frac{\partial S_A}{\partial n_A}\right)_{T,P} = -\left(\frac{\partial \mu_A}{\partial T}\right)_{P,n_A} \quad \left(\frac{\partial S_B}{\partial n_B}\right)_{T,P} = -\left(\frac{\partial \mu_B}{\partial T}\right)_{P,n_B}. \tag{10.25}$$

The entropy S is an extensive property, directly proportional to the number of moles, so S/n is the molar entropy S_m. Let the change in entropy ΔS be entirely due to the change in the number of moles Δn of molecules in that phase so that $\Delta S/\Delta n$ is also equal to S_m. Then we can write the molar entropies as derivatives of the chemical potential:

$$S_{mA} = -\left(\frac{\partial \mu_A}{\partial T}\right)_{P,n_A} \quad S_{mB} = -\left(\frac{\partial \mu_B}{\partial T}\right)_{P,n_B}. \tag{10.26}$$

Similarly,

$$V_{mA} = \left(\frac{\partial \mu_A}{\partial P}\right)_{T,n_A} \quad V_{mB} = \left(\frac{\partial \mu_B}{\partial P}\right)_{T,n_B}. \tag{10.27}$$

and, since $dG = \mu dn$ under constant temperature and pressure,

$$\text{constant } T \text{ and } P: \quad G_m = \mu. \tag{10.28}$$

Using the slope rule

$$dZ(X,Y) = \left(\frac{\partial Z}{\partial X}\right)_Y dX + \left(\frac{\partial Z}{\partial Y}\right)_X dY,$$

we can write $d\mu_A$ as follows:

$$d\mu_A = \left(\frac{\partial \mu_A}{\partial T}\right)_P dT + \left(\frac{\partial \mu_A}{\partial P}\right)_T dP = -S_{mA}dT + V_{mA}dP, \qquad (10.29)$$

and similarly

$$d\mu_B = -S_{mB}dT + V_{mB}dP.$$

If we change either the temperature or pressure, and the two phases are still in equilibrium, the chemical potentials must still be equal:

$$\mu_B = \mu_A.$$

As the conditions change, in order for the chemical potentials to remain equal, it is necessary that

$$d\mu_B = d\mu_A. \qquad (10.30)$$

Substituting in Eq. 10.29 for $d\mu$ gives

$$-S_{mB}dT + V_{mB}dP = -S_{mA}dT + V_{mA}dP,$$

$$\frac{dP}{dT} = \frac{S_{mB} - S_{mA}}{V_{mB} - V_{mA}} \equiv \frac{\Delta_\phi S}{\Delta_\phi V}, \qquad (10.31)$$

which is similar to one of our original Maxwell relations, Eq. 7.29:

$$\left(\frac{\partial S}{\partial V}\right)_T = \left(\frac{\partial P}{\partial T}\right)_V.$$

Using $\Delta_\phi S = \Delta_\phi H / T_\phi$, where T_ϕ is the phase transition temperature, this can be rewritten to give the **Clapeyron equation:**

$$\frac{dP}{dT} = \frac{\Delta_\phi H}{T_\phi \Delta_\phi V}. \qquad (10.32)$$

This equation is general to *any* phase transition.

We must pay attention to what pressure we're using here. These equations have been derived for a single substance, regardless of what other materials may be present. So the pressure here is the *vapor pressure of the pure sample,* not the overall ambient pressure.

For the special case of vaporization (setting A equal to liquid, B equal to gas, and ϕ equal to "vap"), the parameter $\Delta_{vap}V$ is essentially the molar volume of the gas: vaporization of 1 mole of water at 1 bar and 298 K, for instance, corresponds to a change in volume from 18 cm^3 = 0.018 L to 24.8 L, an increase of three orders of magnitude. Therefore,

$$\Delta_{vap}V \approx V_m(gas) \approx \frac{RT}{P}. \qquad (10.33)$$

If we plug this into the Clapeyron equation, we get

$$\frac{dP}{dT} = \frac{P\Delta_{vap}H}{RT^2} \qquad (10.34)$$

$$\frac{dP}{P} = \frac{\Delta_{vap}H}{R}\left(\frac{dT}{T^2}\right). \qquad (10.35)$$

This can be reorganized to get the **Clausius-Clapeyron equation:**

$$d\ln P = -\frac{\Delta_{vap}H}{R} d\left(\frac{1}{T}\right), \qquad (10.36)$$

which is generally used in its integrated form, with the standard boiling point T_b at 1 bar used as the lower limit of the temperature integral:

$$\int_{1\,bar}^{P} d\ln P' = -\frac{\Delta_{vap}H^{\ominus}}{R} \int_{T_b}^{T} d\left(\frac{1}{T'}\right)$$

$$\ln P(bar) = \frac{\Delta_{vap}H^{\ominus}}{R}\left(\frac{1}{T_b} - \frac{1}{T}\right). \qquad (10.37)$$

This form of the equation is convenient for finding the change in pressure necessary to shift the boiling temperature some fixed amount, or vice versa.

SAMPLE CALCULATION Clausius-Clapeyron equation. To find the pressure at which water's boiling point is shifted to 473 K, given $\Delta_{vap}H = 40.65$ kJ mol^{-1}, we substitute the new value of T into the integrated Clausius-Clapeyron equation:

$$\ln P\,(bar) = \frac{40{,}650\ J\ mol^{-1}}{8.3145\ J\ K^{-1}\ mol^{-1}}\left(\frac{1}{373\,K} - \frac{1}{473\,K}\right) = 2.77$$

$$P = e^{2.77} = 16\ bar.$$

EXAMPLE 10.4   Vapor Pressure

CONTEXT Chloroform was a common general anesthetic in the late 19th century, until it was discovered that, although it didn't irritate the lungs like ether, it killed the patient almost five times more often. One of the characteristics of a desirable inhaled anesthetic is that its vapor pressure helps to limit the patient's exposure so that it doesn't become lethal. But because vapor pressure is a sensitive function of the temperature, delivery devices for inhaled anesthetics require thermal monitors that compensate for changes in the ambient temperature.

PROBLEM Use the integrated Clausius-Clapeyron equation and the data in Table 10.2 to find the vapor pressure of chloroform ($CHCl_3$) at 298 K.

SOLUTION

$$\ln P\,(bar) = \frac{29{,}240\ J\ mol^{-1}}{8.3145\ J\ K^{-1}\ mol^{-1}}\left(\frac{1}{334.3\ K} - \frac{1}{298\ K}\right) = -1.28$$

$$P = e^{-1.28} = 0.28\ bar$$

When the gas is at equilibrium with the liquid, the pressure of the chloroform is over a quarter of an atmosphere, a finding consistent with its reputation as a volatile substance.

TOOLS OF THE TRADE   Differential Scanning Calorimetry

While calorimetry as a general technique dates back to the 18th century, precise measurements of the enthalpy changes in a substance were long hampered by ambiguous temperature measurements. Thermometers do not respond instantly to changes in temperature, and phase transitions can occur before a standard thermometer has a chance to respond. Certain phase transitions, particularly second order phase transitions, involve only slight changes in the heat capacity and are difficult to detect, particularly when the heat capacity of the calorimeter itself also has to be taken into account. One solution to the problem was differential thermal analysis (DTA), developed over several years from Henri Louis le Châtelier's work in 1887 to the working device invented by metallurgist William Chandler Roberts-Austen in 1899. In DTA, the same amount of heat is delivered to two heating vessels called *crucibles*. One crucible holds the sample of interest; the other holds a reference material. The temperature of each crucible is measured, and the difference between the temperatures assesses the thermal properties of the sample relative to the reference, while the effects of the calorimeter itself (in particular the crucibles) subtract out. To improve the sensitivity and precision of the technique, Michael J. O'Neill and Emmett S. Watson introduced the differential scanning calorimeter (DSC) in 1962, and this is now one of the most commonly used calorimetric techniques in industry.

What is differential scanning calorimetry (DSC)? A DSC measures the difference in heat delivered to a sample crucible and an empty reference crucible while it steadily increases the temperature in both crucibles. In contrast to DTA, which delivers the same *heat* to both crucibles, the DSC uses a feedback loop to maintain the same *temperature* in both crucibles. Since the sample has a higher heat capacity than the empty reference crucible, it has a higher heat capacity. As the DSC scans the temperature of the two crucibles, the DSC has to deliver more energy to heat the sample than the reference. Mathematically, the difference in the heat provided over a small temperature change δT may be expressed in terms of the difference in heat capacities:

$$q_{\text{sample}} - q_{\text{ref}} = (C_{P,\text{sample}} - C_{P,\text{ref}})\delta T.$$

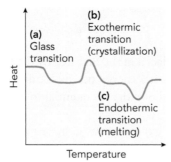

▲ **Sample DSC scan.** The scan shows **(a)** the change in offset during a glass phase transition, **(b)** a peak for an exothermic first-order phase transition (such as crystallization in a polymer), and **(c)** the dip corresponding to an endothermic first-order phase transition (melting).

This appears on the scan as an offset—a constant nonzero value. When the heat capacity changes, as during a glass phase transition, the amount of heat needed to change the temperature by an amount δT also changes, and that appears in the DSC scan as a change in the offset. For melting, on the other hand, because this is a first-order phase transition, the enthalpy changes discontinuously by an amount $\Delta_{\text{fus}}H^\ominus$ while the temperature (in principle) doesn't change at all. Therefore, the heat capacity $C_P = \left(\frac{\partial H}{\partial T}\right)_P$ should approach infinity, and the DSC scan would show a narrow spike at the phase transition temperature. In practice, the substance is never completely pure and the temperature is not exactly the same at all parts of the crucible, so the phase transition signal is spread over a range of temperatures and appears as a dip (if the transition is endothermic) or as a peak (if the transition is exothermic). The latent heat of the first-order phase transition can then be calculated by integrating the area of the peak or dip:

$$\Delta_\phi H = \int C_P dT.$$

Effectively by monitoring the heat capacity of the sample, the DSC becomes one of our most accurate and precise devices for characterizing phase transitions.

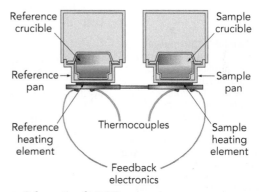

▲ **Schematic of a DSC.**

Why do we use differential scanning calorimetry? The DSC is a common instrument in analytical laboratories that support research into new compounds but DSC is also used in pharmaceutical, polymer, and biochemical research. Quantitative analysis of large molecules can be challenging, and the DSC provides a fast and convenient avenue for characterizing some of the most critical thermodynamic parameters of any reasonably pure sample. For example, pharmaceutical companies want to know at what temperature a new drug will decompose, because that value will affect the storage requirements. Materials scientists developing new polymers similarly need to know if the polymer will undergo a glass transition or crystallize at working temperatures. With a DSC, these determinations become routine.

How does it work? Briefly, the DSC uses a pair of matched thermocouples (see *Tools of the Trade*, Chapter 3)

to monitor the temperature simultaneously at the two crucibles and a feedback loop that directs the heat flow to the two crucibles to maintain the same signal from the two thermocouples. The DSC output is formed by plotting the heat required for the feedback versus T.

10.4 Statistical Mechanics of Vaporization

Here we demonstrate that the enthalpy of vaporization $\Delta_{vap}H^{\ominus}$ can be related to the intermolecular potential energy of the liquid. We already used Trouton's rule to predict $\Delta_{vap}H^{\ominus}$ from the boiling point, but with some statistical mechanics we can predict the value without knowing the boiling point. Instead, we need to know the nature of the intermolecular potential energy function for the liquid. We use the crude approximation that the liquid and gas differ only in two respects:

- There is no intermolecular potential energy in the gas, whereas there is a single average potential energy $-u_0$ experienced by every molecule in the liquid.

- The liquid before vaporization occupies a volume $V(\text{liq})$, whereas the gas occupies a much larger volume $V(\text{gas})$ after vaporization.

With these differences between the liquid and gas so clearly defined, we can write the partition function for either in the form

$$Q = Q_{vib}Q_{rot}Q_{K}Q_{U} \equiv Q_{1}Q_{U}',$$

where the configuration integral Q_U' contains all the dependence on V and on the intermolecular potential energy. For the gas, $Q_U'(\text{gas})$ is simply V^N. For the liquid, we get instead

$$Q_U'(\text{liq}) = \int_{-\infty}^{\infty} \cdots \int_{-\infty}^{\infty} e^{-U/(k_BT)} dX_1 \ldots dZ_N$$

$$= \int_{-\infty}^{\infty} \cdots \int_{-\infty}^{\infty} e^{Nu_0/(k_BT)} dX_1 \ldots dZ_N$$

$$= e^{Nu_0/(k_BT)} \int_{-\infty}^{\infty} \cdots \int_{-\infty}^{\infty} dX_1 \ldots dZ_N = e^{Nu_0/(k_BT)} V^N. \quad (10.38)$$

The overall potential energy U is replaced in the first step by the total number of molecules N times the average potential energy per molecule $-u_0$. This eliminates any distance dependence and allows the exponential to be factored out of the configuration integral.

Why does this help us? Because we can write μ in terms of the partition function (Eq. 7.38):

$$\mu = k_BT\left(\frac{\partial \ln Q}{\partial n}\right)_{T,V}.$$

In our case, $\ln Q$ is $\ln Q_1 + \ln Q_U'$. The configuration integrals give straightforward logarithms:

$$\ln Q_U'(\text{liq}) = \frac{Nu_0}{k_BT} + N\ln V(\text{liq}), \quad \ln Q_U'(\text{gas}) = N\ln V(\text{gas}).$$

CHECKPOINT The crucial step in this derivation is evaluating the configuration integral Q_U' for the liquid, obtaining Eq. 10.38, because this step connects the thermodynamic properties (in this case the enthalpy of vaporization) to the molecular properties (the well depth of the intermolecular potential). We see once again how the partition function serves to forge this link between the microscopic and macroscopic realms.

So for the liquid we have

$$\mu(\text{liq}) = k_BT\left[\left(\frac{\partial \ln Q_1}{\partial n}\right)_{T,V} + \left(\frac{\partial(Nu_0/(k_BT))}{\partial n}\right)_{T,V} + \left(\frac{\partial N\ln V(\text{liq})}{\partial n}\right)_{T,V}\right]$$

$$= k_BT\left[\left(\frac{\partial \ln Q_1}{\partial n}\right)_{T,V} + \frac{\mathcal{N}_Au_0}{k_BT} + \mathcal{N}_A\ln V(\text{liq})\right], \tag{10.39}$$

and for the gas we have

$$\mu(\text{gas}) = k_BT\left[\left(\frac{\partial \ln Q_1}{\partial n}\right)_{T,V} + \left(\frac{\partial N\ln V(\text{gas})}{\partial n}\right)_{T,V}\right]$$

$$= k_BT\left[\left(\frac{\partial \ln Q_1}{\partial n}\right)_{T,V} + \mathcal{N}_A\ln V(\text{gas})\right]. \tag{10.40}$$

Setting these chemical potentials equal at the phase transition, where $T = T_b$, we obtain a relationship among the volumes of the gas and liquid, the boiling point, and u_0:

$$\mu(\text{liq}) = \mu(\text{gas})$$

$$k_BT_b\left[\left(\frac{\partial \ln Q_1}{\partial n}\right)_{T,V} + \frac{\mathcal{N}_Au_0}{k_BT_b} + \mathcal{N}_A\ln V(\text{liq})\right] = k_BT_b\left[\left(\frac{\partial \ln Q_1}{\partial n}\right)_{T,V} + \mathcal{N}_A\ln V(\text{gas})\right]$$

$$\mathcal{N}_Au_0 + \mathcal{N}_Ak_BT_b\ln V(\text{liq}) = \mathcal{N}_Ak_BT_b\ln V(\text{gas}).$$

The latent entropy depends on the ratio of the gas and liquid volumes, so let's solve for that:

$$\ln V(\text{gas}) - \ln V(\text{liq}) = \ln\left(\frac{V(\text{gas})}{V(\text{liq})}\right) = \frac{u_0}{k_BT_b}$$

$$\Delta_{\text{vap}}S^{\ominus} = R\ln\left(\frac{V(\text{gas})}{V(\text{liq})}\right) = \frac{Ru_0}{k_BT_b} = \frac{\mathcal{N}_Au_0}{T_b}. \tag{10.41}$$

Finally, this gives us the desired equation for $\Delta_{\text{vap}}H^{\ominus}$:

$$\Delta_{\text{vap}}H^{\ominus} = T_b\Delta_{\text{vap}}S^{\ominus} = \mathcal{N}_Au_0. \tag{10.42}$$

In other words, our prediction is that the enthalpy of vaporization is exactly the energy necessary to overcome the average intermolecular attraction.

To be sure, our approximate treatment of the liquid keeps the math simple but oversimplifies the result. A more elaborate configuration integral could be used to yield increasingly accurate predictions of this fundamental thermodynamic quantity. Without this, the result we have is qualitatively correct. It accurately implies, for example, that because water is bound to the liquid by roughly one tenth the bond strength of a typical 400 kJ mol^{-1} chemical bond, it should have a vaporization enthalpy of about 40 kJ mol^{-1}.

10.5 Phase Diagrams

The Clapeyron equation was derived under the assumption that we had two phases of the same material in equilibrium. Since we can graph points of equilibrium as functions of the thermodynamic parameters, we can illustrate the equilibrium phase of a material on a **phase diagram** as a function of, for example, pressure P and temperature T. A phase transition is a discontinuity in one or more thermodynamic parameters along a certain path on this diagram, and therefore it occurs at a single point along that path. The temperature of the phase transition changes, in general, with the pressure or volume, as shown by the Clapeyron equation (Eq. 10.32).

Under the proper conditions, more than one phase can coexist at equilibrium, but the conditions may be very restrictive. Let's consider only the intensive properties of a sample that contains a mixture of k different compounds that can exist in p different phases. There are $p - 1$ equations similar to Eq. 10.23, relating the chemical potentials for each compound. For example, if we have a liquid mixture of H_2O, CH_3OH (methanol), and C_2H_5OH (ethanol) in a sealed container, and we allow the mixture to reach equilibrium with the gas phase, then

$$\mu_{H_2O(liq)} = \mu_{H_2O(gas)}, \quad \mu_{CH_3OH(liq)} = \mu_{CH_3OH(gas)}, \quad \mu_{C_2H_5OH(liq)} = \mu_{C_2H_5OH(gas)}.$$

We will always have a total of $k(p - 1)$ equations of this type. For our current example, there are $k = 3$ chemical components, $p = 2$ phases, and $k(p - 1) = 3$.

The intensive parameters we can adjust for the system are T, P, and the mole fractions X for the k components in each of the p phases, giving a total of $kp + 2$ intensive parameters. We have an additional p equations relating the mole fractions in each phase:

$$X_1 + X_2 + \ldots + X_k = 1. \tag{10.43}$$

For our example, we have $p = 2$ of these equations:

$$X_{H_2O(liq)} + X_{CH_3OH(liq)} + X_{C_2H_5OH(liq)} = 1,$$
$$X_{H_2O(gas)} + X_{CH_3OH(gas)} + X_{C_2H_5OH(gas)} = 1.$$

Given that we have $kp + 2$ parameters and $k(p - 1) + p$ equations, the number d of remaining degrees of freedom we will have for given values of p and k is the number of variables minus the number of constraining equations:

$$d = (kp + 2) - [k(p - 1) + p] = k - p + 2. \tag{10.44}$$

This is called the **Gibbs phase rule.** It predicts, for example, that for our example system of $k = 3$ components and $p = 2$ phases, there are $d = 3$ independent parameters. In other words, we could set T and P and the mole fraction $X_{H_2O(liq)}$ to arbitrary values, and still there would be some set of values for the other mole fractions such that *all three* compounds in both phases would be in equilibrium.

Let's take a simpler example, such as pure water. There is only one component ($k = 1$), and at standard pressure it can exist in up to three phases ($p = 3$): ice, liquid water, and water vapor. The Gibbs phase rule predicts that if we want to have all three of these phases in equilibrium at the same time, then the number of free parameters is

$$d = 1 + 2 - 3 = 0.$$

There are no free parameters, meaning that there is only one value of T and P for which the three phases can coexist *in equilibrium*. If we try something easier, requiring only two phases to coexist, then we have more freedom to choose our parameters. Water in a sealed jar coexists in equilibrium with water vapor only at its vapor pressure, which is a function of the temperature. The two phases (liquid and gas) come to equilibrium with one ($d = 1 + 2 - 2 = 1$) free parameter. If we change the pressure, there may still be some value of T for which the system is in equilibrium. This curve of values for P and T along which more than one phase can coexist is called a **phase boundary.** Values of P and T that do not lie on a phase boundary are values at which only one phase of water will exist at equilibrium.

In the phase diagram, we graph the phase boundaries that separate distinct phases at equilibrium. Assuming for now that we go from one phase to another only by a phase transition, there must be some point at which the liquid/vapor line meets the liquid/solid line. At that point, called the **triple point,** solid, liquid, and gas phases of the same substance coexist. This is a *unique* temperature and pressure, since it corresponds to a single point, not a line, on the P versus T graph. As shown earlier, there are no free parameters when all three phases coexist, so it can correspond to only a single point—we are not free to select the pressure or temperature.

The phase diagram for CO_2, shown in Fig. 10.8a, exemplifies several typical features of phase diagrams. The gas is the equilibrium phase at low pressure and high temperature, the solid at high pressure and low temperature. Isothermal compression of a gas (moving straight up on the phase diagram) normally results in a liquid if the temperature is high, and a solid if the temperature is low. Similarly, isobaric heating (moving straight to the right on the phase diagram) of a solid normally gives a liquid if the pressure is high and a gas if the pressure is low.

Do not forget that P in the phase diagram is the partial pressure of the substance, which is not necessarily the total pressure. For example, a block of dry ice (solid CO_2) left in a freezer at 200 K in air will evaporate. The solid is the equilibrium phase only if the partial pressure *of the carbon dioxide* is at least 1 bar.

A phase transition does not always take place when a material changes phase. That may seem bizarre, but our definition of phase transition requires a *discontinuity* in the parameters describing a substance. High temperature liquids have so much kinetic energy that the attractive forces binding the molecules are only barely sufficient to hold the liquid together. Similarly, high pressure gases are compressed nearly to the density of liquids. As long as both the temperature and pressure are high enough, the liquid-vapor transition can occur smoothly, with continuous changes in the volume, enthalpy, heat capacities, and all other parameters of the substance. The defining characteristics of liquids and gases are too blurred by the extreme conditions to separate one from the other.

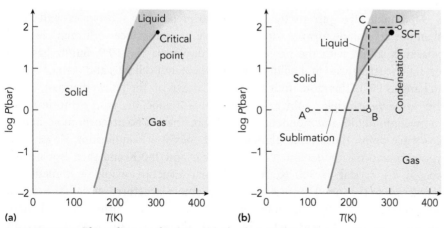

▲ **FIGURE 10.8 Phase diagram for CO_2.** (a) The three traditional states of matter and their phase boundaries. (b) A path described on the phase diagram as described in the text.

The pressure and temperature of the liquid-vapor line when the transition ceases to be a phase transition define the **critical point.** The line in the phase diagram is not drawn further than the critical point, because the equilibrium liquid-vapor transition is no longer truly a phase transition. When the temperature and pressure exceed the values at the critical point, the substance is called a **supercritical fluid,** rather than a gas or liquid. The supercritical fluid is like a liquid in that the molecules are in constant contact with each other, but like a gas in that it will expand if the external pressure is reduced.

We can use the phase diagram to predict what will happen to a substance as we change its environment. In Fig. 10.8b we trace a path across the phase diagram that corresponds to a process for obtaining the supercritical fluid (SCF) from dry ice. We begin by heating a sample of CO_2 from (A) 100 K at 1 bar to (B) 250 K at 1 bar. During this process, we cross the solid/gas phase boundary, and the dry ice sublimes to gas-phase CO_2. Next we increase the pressure of the gas up to (C) 100 bar (please notice that the vertical axis is $\log_{10} P$) at 250 K, and in doing so cross the liquid/gas phase boundary causing the sample to condense into the pressurized liquid CO_2. Finally, by heating the gas further, past room temperature to about 325 K at 100 bar (D), we obtain the supercritical fluid. During this step, however, we do not cross a phase boundary. We arrive at the SCF without any sudden changes in the properties of the substance.

Useful as they are, phase diagrams have their limitations:

- Under extreme conditions, the intramolecular structure of the sample changes. Compounds normally become unstable at temperatures of a few thousand K, where the thermal energy is sufficient to break molecular bonds. The temperature drives up the $T\Delta S$ contribution to ΔG, so the compound becomes more stable as a high-entropy mixture of atoms or of smaller compounds.

- Like all equilibrium thermodynamic treatments, the phase diagram is only as useful as the *equilibrium* system that it represents. We discussed the example of supercooled water; water can exist as a liquid at 263 K and 1 bar, but its phase diagram will not reveal this. Such a state, which can exist for long enough time to be observed but which is not at equilibrium, is called a **metastable state.**

Metastable states are particularly easy to prepare for substances with more than one crystalline form. A simple example of a substance with multiple solid phases is sulfur, with the phase diagram drawn in Fig. 10.9. Sulfur has two crystalline structures: monoclinic (two angles of unit cell 90°) and orthorhombic (all angles 90°). Therefore, four phases are drawn on the phase diagram. There are also *three* triple points: orthorhombic-monoclinic-gas, orthorhombic-monoclinic-liquid, and monoclinic-liquid-gas. There is no orthorhombic-liquid-gas triple point; those three phases cannot coexist at equilibrium. However, if you prepared the monoclinic crystal at 1 bar and 380 K and then isobarically cooled the crystal to 300 K, the molecular structure wouldn't immediately reorganize to the (now) more stable orthorhombic structure. Instead, the crystal will remain, for a long time, in the metastable monoclinic form. Standard phase diagrams *do not represent* metastable states. There are many ways that we can construct a non-equilibrium situation which does not correspond to a particular point on the phase diagram.

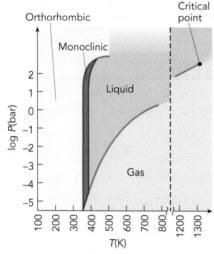

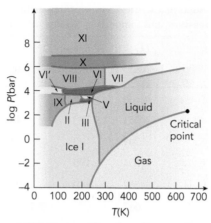

▲ FIGURE 10.10 **The phase diagram for H$_2$O.** Dotted lines indicate approximate phase boundaries.

▲ FIGURE 10.9 **Phase diagram for molecular sulfur.**

Perhaps the most thoroughly studied phase diagram is that for water, Fig. 10.10. This compound has at least 12 known crystalline structures.[3] Although the large number of known ice structures may be surprising, it is a function not only of water's structure, but also the enormous range of pressures over which the molecule has been studied. In ice X, identified at $6 \cdot 10^5$ bar, the molecules have been subjected to such great compression that they are no longer distinguishable. Unlike the ices at lower pressures, consisting of covalently bound H$_2$O molecules bound by hydrogen and dipole–dipole bonds, the H atoms in ice X occupy sites halfway between two O atoms. The structure is more characteristic of an ionic $(H^+)_2O^{2-}$ solid. These high-pressure forms of ice, although not naturally occurring on Earth, are believed to be present in the cores of certain icy moons of Saturn and Jupiter: Titan, Ganymede, and Callisto.

The most remarkable characteristic of the water phase diagram is the *negative* slope of the ice I/liquid phase boundary. An unusual consequence of this feature is that ice I *melts* when *compressed* at temperatures near 273 K, because the liquid form is more dense than the solid in that range of pressures. The ice I structure accommodates a very stable arrangement of the hydrogen and dipole–dipole bonds, including some long-range interactions, at the expense of a compact structure. The liquid, which is stable at higher temperatures, is insensitive to the long-range interactions and is slightly denser.

The critical point is a marvelous example of microscopic properties determining a macroscopic observable (Table 10.3). Carbon dioxide is a relatively small molecule with only the dispersion force to bind the liquid, water has strong hydrogen and dipole–dipole binding forces, and sulfur exists (at these temperatures) as the ring compound S$_8$, which forms strong enough intermolecular bonds to crystallize at 390 K. As these intermolecular forces increase, the binding energy

---

[3]The known forms of ice are numbered by Roman numerals, with some variations. Ice IV is not shown in the phase diagram, because it is a metastable structure. There is also a metastable cubic crystalline structure labeled ice Ic (the structure of normal ice I is hexagonal). Ice VI has sometimes been labeled ice X.

TABLE 10.3 Critical temperatures, critical pressures, ideal gas critical point number densities, and comparison to Lennard-Jones parameters of selected compounds.

	T_c (K)	ε/k_B (K)	P_c (bar)	ρ_c (cm^{-3}) $\approx \frac{N_A P_c}{RT_c}$	$\rho_c^{-1/3}/2$ (Å)	R_{LJ} (Å)
Ne	44	36	27.6	$4.50 \cdot 10^{21}$	3.0	2.95
Ar	151	120	4.09	$2.35 \cdot 10^{21}$	3.8	3.44
Kr	209	190	55.0	$1.90 \cdot 10^{21}$	4.0	3.61
CO_2	304	190	73.8	$1.75 \cdot 10^{21}$	4.1	4.00
C_6H_6	562	440	49.0	$6.31 \cdot 10^{20}$	5.9	5.27
H_2O	647		221	$2.47 \cdot 10^{21}$	3.7	
S_8	1314		207	$1.14 \cdot 10^{21}$	4.8	

of the liquid is more difficult to overcome, and higher temperatures are needed to reach the critical point: 304 K for CO_2, 647 K for H_2O, and 1314 K for S_8.

The critical number density ρ_c, in contrast, is determined largely by the size of the molecule. At the critical point, the molecules are shoulder-to-shoulder. You can't fit as many large molecules into a given volume as small molecules, so relatively large molecules (e.g., benzene and sulfur in Table 10.3) have smaller values of ρ_c. Taking the cube root of $1/\rho_c$ gives the average volume per molecule. If each molecule occupied a cubic volume of space, the length of one side of this cube would be $\rho^{-1/3}$. The critical density (calculated from P_c and T_c just using the ideal gas law) occurs when this distance is roughly twice the Lennard-Jones distance R_{LJ}.

The critical pressure increases with either T_c or ρ_c. As T_c climbs (to counteract a strong intermolecular binding force), the pressure necessary to compress the gas climbs as well. As the molecules get smaller, higher densities are required to achieve the density of the liquid, so again the critical pressure increases. The very high critical pressure of water (221 bar) is an example of this—the small size and strong intermolecular binding of water drive the critical pressure up.

Although we developed the Gibbs phase rule for the general case of k chemical components, these phase diagrams have been limited to only one chemical component. When we have two or more chemical components, the properties of each substance can be affected by the presence of the others. These effects merit special attention, so we consider the interactions between multiple chemical components in our next chapter and return then to phase diagrams of mixtures.

CONTEXT *Where Do We Go From Here?*

Phase transitions involve the changes in the bonding between particles of the same chemical composition. The formation of a solution involves similar changes to the intermolecular bonding, and solutions are the subject of the next chapter. It's then only one step from phase transitions and solvation to chemical reactions, where we change the bonding between the atoms within a molecule.

KEY CONCEPTS AND EQUATIONS

10.1 Phase Transitions. A phase transition occurs when some continuous change in one of the thermodynamic parameters of a substance, such as its temperature or pressure, results in a *discontinuous* change in one of its other thermodynamic parameters, such as its density or enthalpy. Phase transitions often correspond to some change in the balance between thermal energy, which we can control with the temperature, and intermolecular potential energy, which we can affect by means of the pressure, external magnetic field, or (as we find out in Chapter 11) mole fraction. Whereas thermal energy tends to randomize the distribution of particles, intermolecular forces tend to enforce structure on the distribution. Therefore, many phase transitions can be considered as transformations between ordered and disordered states.

10.2 Thermodynamics of Phase Transitions. Different phases of the same substance are treated as distinct entities in thermodynamics. Properties of the gas, liquid, and solid are listed separately, and the enthalpy change when one phase transforms into another is an empirically determined parameter called the **latent enthalpy** of the phase transition, $\Delta_\phi H^\ominus$. From $\Delta_\phi H^\ominus$ we can calculate the latent entropy change

$$\Delta_\phi S^\ominus = \frac{\Delta_\phi H^\ominus}{T_\phi}, \tag{10.6}$$

where T_ϕ is the temperature for the phase transition under standard conditions. **Trouton's rule** estimates that the latent entropy of vaporization lies between roughly $8R$ and $13R$, so $\Delta_{vap}H^\ominus$ is normally between roughly $8RT_b$ and $13RT_b$.

10.3 Chemical Potentials. The **Gibbs-Duhem equation**

$$SdT - VdP + n_1 d\mu_1 + \ldots + n_k d\mu_k = 0 \tag{10.18}$$

shows that all of the intensive parameters of the system are interrelated. A similar derivation shows that

the incremental change in the Gibbs free energy at equilibrium is given by

$$\text{equilibrium:} \quad dG = \sum_{i=1}^{k} \mu_i dn_i = 0.$$

In other words, at equilibrium, a change dn_i in the amount of any chemical component i must be balanced by changes in the amounts of the other chemical components. For the case of a single chemical species present in two different phases, this can be extended to give the **Clapeyron equation,**

$$\frac{dP}{dT} = \frac{\Delta_\phi H}{T_\phi \Delta_\phi V}, \tag{10.32}$$

which applies to any phase transition. For the specific case of vaporization, we can approximate $\Delta_{vap}V$ by the volume of the gas alone to get the **Clausius-Clapeyron equation**

$$\ln P(\text{bar}) = \frac{\Delta_{vap}H^\ominus}{R}\left(\frac{1}{T_b} - \frac{1}{T}\right). \tag{10.37}$$

10.4 Statistical Mechanics of Vaporization. By expressing the chemical potential in terms of the partition function, we can show that the enthalpy of vaporization of a substance should be roughly the energy needed to overcome the total intermolecular attraction of one particle for the bulk liquid.

10.5 Phase Diagrams. A **phase diagram** summarizes by way of a graph the equilibrium state or states of matter for a given substance as a function of different thermodynamic parameters, typically pressure and temperature when only one chemical species is present. The number of degrees of freedom d that may be independently adjusted is given by the **Gibbs phase rule,**

$$d = k - p + 2, \tag{10.44}$$

where k is the number of different chemical components present and p is the number of coexisting phases.

KEY TERMS

- A **phase transition** occurs when a continuous change in any thermodynamic parameter, such as temperature or pressure, causes a *discontinuous* change in other thermodynamic parameters, such as volume or heat capacity.
- In a **first-order phase transition,** the discontinuities appear in first derivatives of the thermodynamic

potentials such as S or V. In a **second-order phase transition,** the discontinuities do not appear until we reach second derivatives of the thermodynamic potentials, such as C_P.
- A **glass phase transition** is the second-order phase transition between a liquid and a glass (a solid that has the same microscopic structure as the liquid).

- The names of the various ordinary phase transitions are given in Table 10.1. **Boiling** is the special case of the vaporization of a liquid when its vapor pressure is equal to the total (or *ambient*) pressure. At a pressure of 1.00 bar, a liquid will freeze and a solid will melt at the standard **freezing point** T_f, and the liquid will boil and the gas will condense at the standard **boiling point** T_b.
- The **vapor pressure** of a liquid or solid is the equilibrium pressure of the substance in the gas phase at that temperature. It corresponds to the pressure of the liquid-gas or solid-gas phase boundary at the current temperature.
- A **triple point** is any location in the phase diagram of a single-component system at which three phases of matter may coexist at equilibrium. Normally this refers to the solid/liquid/gas triple point, although

other triple points may exist where two different crystalline forms of the substance are simultaneously at equilibrium.
- The **critical point** marks the end of the liquid-gas phase boundary on the phase diagram. At higher temperatures and pressures than the critical point, no phase transition (i.e., no discontinuity) is observed in going from the liquid to the gas. The **critical pressure** P_c and **critical temperature** T_c fix the location of the critical point on the phase diagram.
- A substance at higher pressure and higher temperature than the critical point is a **supercritical fluid.** The temperature is too high for the attractive forces between particles to bind them as in a liquid, but the pressure is so high that the density remains nearly that of the liquid anyway.

OBJECTIVES REVIEW

1. *Calculate enthalpy and entropy changes for processes involving phase transitions.*
   Calculate the overall $\Delta S^\ominus$ and $\Delta H^\ominus$ for condensing 105 g of carbon tetrachloride at 349.9 K and then cooling it to 323 K. The heat capacity of the liquid is 131.75 J K^{-1} mol^{-1}.

2. *Use the Clausius-Clapeyron equation to find the temperature or pressure of boiling under non-standard conditions.*

   Calculate the boiling point of CCl$_4$ if the pressure is 1.10 bar.

3. *Read a single-component phase diagram to identify the standard freezing and boiling points, critical and triple points, as well as which phase or phases exist at equilibrium at any given temperature and pressure.*
   Based on Fig. 10.9, estimate the standard freezing and boiling points for sulfur.

PROBLEMS

Discussion Problems

10.1 Consider the two liquids Cl$_2$ and NH$_3$, which according to Table 10.2 have nearly equal standard boiling points (239.11 K for Cl$_2$, 239.82 K for NH$_3$), but enthalpies of vaporization that differ by 14% (20.41 and 23.33 kJ mol^{-1}, respectively). Then, according to the Clausius-Clapeyron equation, at temperatures above 240 K, NH$_3$ has the higher vapor pressure. Qualitatively, how do you explain that NH$_3$, which *absorbs more heat* per mole to vaporize than Cl$_2$, then *exerts more pressure* after vaporizing than does Cl$_2$?

10.2 Values of $\Delta_{fus}H^\ominus$ for different compounds usually increase with T_f, but not always. Justify the fact that decane has a substantially lower freezing point (243 K) than water (273 K) but a much higher $\Delta_{fus}H^\ominus$ (28.78 kJ mol^{-1} versus 6.008 kJ mol^{-1}).

10.3 The boiling point of CCl$_4$ at 1 bar is 349.9 K. The boiling point at 2.00 bar is which of the following?

a. 200 K   b. 325 K   c. 375 K   d. 500 K

10.4 Why, in our look at Trouton's rule (Sec. 10.2), are we able to predict $\Delta_{vap}S^\ominus$ without knowing anything about u_0?

10.5 Between ^{1}H$_2$ and ^{2}H$_2$ (or D$_2$), which should have the higher melting point?

10.6 According to the phase diagram for water there is no phase boundary between ice and water vapor at pressures above about 0.006 bar. But ice has a measurable vapor pressure, and ice left in open air at a pressure of 1.00 bar and 250 K will eventually evaporate.

a. Which of the following is true?
   i. This is consistent with the phase diagram.
   ii. This is an example of the phase diagram's limitations.

b. Explain the source of the apparent inconsistency.

10.7 In a typical phase diagram for a single substance, what phase of matter do we end up with

a. if we start at the standard boiling point and increase P at constant T?

b. if we start at the triple point and decrease T at constant P?

c. if we start at the critical point and decrease P at constant T?

10.8 Draw a line connecting each statement on the left with the most appropriate explanation on the right.

$\Delta_{mix}S^{\ominus} > 0$ because	(A) C_{Pm} is bigger for the liquid than for the gas
$\Delta_{vap}H^{\ominus}$ decreases with T because	(B) dispersion forces are generally weaker than hydrogen bonds
vapor pressure increases with T because	(C) more energy per molecule is available to break intermolecular bonds
water has a higher T_f than propane because	(D) S increases with V for each substance
water melts at a constant T as heat is added because	(E) the average energy per degree of freedom remains constant

10.9 This is a cousin to Problem 9.4. Two containers are at exactly 273.15 K and hold exactly 0.100 mol of water, but container A holds water ice at its melting point, whereas container B holds liquid water at its freezing point. We bring the two containers into thermal contact. On average, does any energy transfer between the two containers?

Processes Involving Phase Transitions

10.10 Calculate the $\Delta_{fus}S^{\ominus}$ of Br_2 using Table 10.2.

10.11 The boiling point for isobutane (2-methylpropane) is 261.3 K. Use Trouton's rule to estimate the standard *enthalpy* of vaporization.

10.12 Iridium fluoride (IrF_6) has a standard boiling point of 326 K. Use Trouton's rule to estimate $\Delta_{vap}H^{\ominus}$ and then to estimate the vapor pressure of IrF_6 at 298 K.

10.13 Find $\Delta_{fus}H^{\ominus}$, $\Delta_{fus}S^{\ominus}$, and $\Delta_{fus}G^{\ominus}$ for water at its normal freezing point of 273.15 K.

10.14 Estimate $\Delta_{vap}E^{\ominus}$ for carbon tetrachloride at its standard boiling point. THINKING AHEAD ▶ [How is ΔE related to ΔH?]

10.15 It takes a 17.3 J K^{-1} change in entropy to heat 0.300 mol of glycerol, $C_3H_5(OH)_3$, from a solid at its melting point of 291 K to a liquid at 331 K, all at a pressure of 1 bar. The heat capacities are 150 J K^{-1} mol^{-1} for the solid and 221.9 J K^{-1} mol^{-1} for the liquid. Find $\Delta_{fus}H^{\ominus}$ for glycerol.

10.16 How many grams of water ice at 273.15 K are necessary to cool 1.000 L (55.56 mol) of water liquid from 298.15 K to 273.15 K?

10.17 We transfer 50.0 kJ of heat into a 36.0 g sample of liquid water initially at 350 K so that all the water heats up to the boiling point and some evaporates, all at 1 bar.

What is the mass of water remaining as liquid at the end of this process? $C_{Pm}(liq) = 75.3$ J K^{-1} mol^{-1}.

10.18 While brewing beer, a delinquent professor finds it necessary to rapidly cool the "wort," a fermented malt mixture, from near the boiling point of water to room temperature. A 5.7 L sample of the wort at 363 K is combined with 13.2 L of an equilibrium mixture of ice and liquid water at 1 bar pressure. What mass in g of the coolant must be ice if the final temperature of the combined system is to be 300 K? Assume the wort has the density and heat capacity of liquid water and that the heat capacity is constant with temperature.

10.19 If we add 20.0 kJ of heat to 90.0 g of water ice, initially at 250.0 K, find the number of grams of ice, the number of grams of liquid, and the final temperature of the sample. The heat capacities for the solid and liquid are 38.1 J K^{-1} mol^{-1} and 75.3 J K^{-1} mol^{-1}, respectively, and the process is carried out under standard conditions.

10.20 If at 1 bar we mix 10.0 mol of $H_2O(l)$ initially at 273.15 K with 10.0 mol of liquid nitrogen initially at 77.36 K and allow the mixture to come to equilibrium, what do we get? Give the phase or phases of each substance and how many moles in each phase, and the final temperature T_f.

10.21 A rigid vacuum chamber with volume 5.00 L initially contains 0.100 moles of liquid H_2O at 298 K. Air is also present at an initial pressure of 1.00 bar. How much energy is necessary to vaporize all of the water and obtain a final temperature for the sample of 405 K? Set the heat capacity of the air to 29.14 J K^{-1} mol^{-1}. THINKING AHEAD ▶ [What series of steps will accomplish the same net process but with phase changes under standard conditions?]

10.22 Calculate the total energy in kJ necessary to heat 18 g of ice at 273 K to water vapor at 373 K, all at 1 bar. Use densities of 0.917 g cm^{-3} for ice and 1.00 g cm^{-3} for the liquid.

10.23 Calculate $\Delta S^{\ominus}$ when 1 kg of dry ice at 194.65 K sublimes at that temperature (its normal sublimation temperature at 1 bar) and warms to CO_2 gas at 300 K. $\Delta H_{sub}^{\ominus} = 25$ kJ mol^{-1}.

10.24 Calculate the overall $\Delta S^{\ominus}$ for supercooling 2.00 moles of $H_2O(liq)$ from 273 K to 265 K and then freezing it at 265 K, all at 1.00 bar. For water, $T_f = 273.15$ K, $\Delta_{fus}H^{\ominus} = 6.008$ kJ mol^{-1}, $C_{Pm}(liq) = 75.3$ J K^{-1} mol^{-1}, and $C_{Pm}(sol) = 36$ J K^{-1} mol^{-1}

10.25 A 0.100 mol sample of liquid water at 298 K is injected into a sealed 5.00 L flask that was initially pumped to a vacuum. Some of the water evaporates, reaching its vapor pressure at 298 K. The flask is then heated to 373 K, and all the remaining liquid evaporates. The heat capacities of water are 35.6 J K^{-1} mol^{-1} for the gas and 75.3 J K^{-1} mol^{-1} for the liquid. Calculate the total entropy change ΔS for the water in this process.

10.26 Calculate ΔG for the heating of water ice at 263 K to liquid water at 283 K under a pressure of 1 bar.

10.27 A certain magnetic crystal loses its magnetization in a *second-order* phase transition when heated above 3000 K. Calculate ΔH_m in kJ mol^{-1} for the process of heating this crystal from 2000 K to 4000 K at constant pressure, if $C_{Pm}^{mag} = 100.0$ J K^{-1} mol^{-1} and $C_{Pm}^{non-mag} = 200.0$ J K^{-1} mol^{-1}.

10.28 For the graph shown of pressure versus molar volume of CO_2, estimate the values of T for the two isotherms drawn.

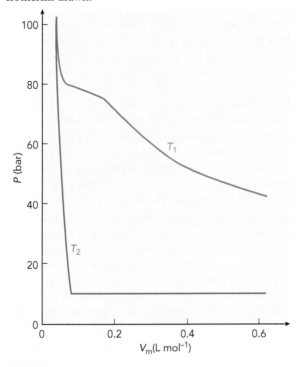

10.29 The heat capacities of N_2O_4 and H_2O for the liquid and gas phases are given in the appendix. The boiling point of N_2O_4 is 294.3 K and $\Delta_{vap}H^{\ominus} = 38.12$ kJ mol^{-1}. If 1.00 mole of H_2O(gas) at 373.15 K is combined with 1.00 mole of N_2O_4(liq) at 294.3 K, what is the temperature and composition (how much liquid and gas) of the resulting mixture, assuming for convenience that the vapor pressure of the liquids is zero?

10.30 An underground 180,000 L gasoline storage tank is exactly half-full with liquid octane at 298 K, with the rest of the volume occupied by air at a pressure of 1.00 bar. The tank is sealed, and the liquid and vapor phases of the octane come into equilibrium. What mole fraction of octane is still liquid, and what is the octane partial pressure? For pure octane V_m(liq) $= 0.164$ mol L, $\Delta_{vap}H^{\ominus} = 34.41$ kJ mol^{-1}, $T_b = 398.8$ K.

10.31 Sketch a graph for the function $\Delta H_{vap}^{\ominus}$ as a function of T for H_2O between 273 K and 700 K. Features you should be able to include are the value at the normal boiling point, the T-intercept, and the approximate curvature in different regions (linear, concave up, or concave down).

10.32 Estimate the melting point of ordinary ice under the skates of a person of mass 91 kg supported equally on two skates, each of which contacts the ice over an area of 2.2 cm^2. At 273.15 K, liquid water has a density of 1.000 g cm^{-3}, and the solid has a density of 0.917 g cm^{-3}. THINKING AHEAD ▶ [What happens if we push really hard against a block of ice?]

The Clapeyron and Clausius-Clapeyron Equations

10.33 Calculate the vapor pressure of decane at room temperature (298 K).

10.34 The vapor pressure of ethanol at 298 K is 0.320 bar, and the normal boiling point is 351 K. Find the standard enthalpy of vaporization for ethanol.

10.35 In Section 10.3 we estimate that the boiling point of water shifts to 473 K at a pressure of 16 bar, but this assumes that $\Delta_{vap}H^{\ominus}$ is constant.

a. Write a general equation for the vapor pressure of a liquid at temperature T that takes the temperature-dependence of $\Delta_{vap}H^{\ominus}$ into account.

b. Use this equation to calculate a more accurate boiling point for water at $P = 16$ bar, given the heat capacities C_{Pm} of 33.577 J K^{-1} mol^{-1} for the gas and 75.291 J K^{-1} mol^{-1} for the liquid.

10.36 Use the Clausius-Clapeyron equation and Table 10.2 to estimate the boiling temperature of CCl_4 at a pressure of 2 bar.

10.37 Find the freezing point (in K) of water at an altitude of 1500 m, where the atmospheric pressure is 0.85 bar. The densities of liquid water and water ice are 1.000 g cm^{-3} and 0.917 g cm^{-3}. $\Delta_{fus}H_m^{\ominus} = 6.01$ kJ mol^{-1} and $T_f = 273.15$ K.

10.38 Find the boiling temperature of NH_3 at 1 bar and at 2 bar.

10.39 Find the temperature at which CCl_4 and H_2O have the same vapor pressure.

10.40 Osmium tetroxide (OsO_4) is a highly toxic oxidizing agent used in organic synthesis. It is a solid at room temperature but sublimes ($\Delta_{sub}H^{\ominus} = 56.9$ kJ mol^{-1}) to a hazardous vapor that can stain corneas and cause blindness. The vapor pressure of OsO_4 at 298 K is 0.013 bar. Estimate the vapor pressure at 325 K.

10.41 Find an expression for each of the following in terms of $\Delta_{vap}H^{\ominus}$ and T_b or their derivatives:

a. $\Delta_{vap}E^{\ominus}$

b. $\Delta_{vap}F^{\ominus}$

c. $\Delta_{vap}\mu^{\ominus}$ at constant S

d. C_{Pm}(gas) $- C_{Pm}$(liq) at T_b and 1 bar

10.42 Estimate the triple point of H_2O using the Clapeyron equation and the boiling and freezing points at 1 bar. Use 1.000 g cm^{-3} for the density of liquid water and 0.917 g cm^{-3} for the solid.

10.43 Find the pressure such that $C_{10}H_{22}$ (n-decane) boils at the normal boiling point for water, 373.15 K.

10.44 Find $\Delta_{vap}H^{\ominus}$ of n-pentane if $T_b = 309.2\,K$ and the vapor pressure of the compound is 0.680 bar at 298.15 K.

10.45 Use the Clapeyron equation to predict the freezing point of neon at a pressure of 20.0 bar. The density of the solid is $1.444\,g\,cm^{-3}$ and of the liquid is $1.2073\,g\,cm^{-3}$.

Phase Diagrams

10.46 A sample contains a mixture of water and ammonia at a pressure and temperature such that the water is present in three phases (solid, liquid and gas) but the ammonia exists only as liquid and gas. Calculate how many degrees of freedom this system has at equilibrium, using the same method used to obtain the Gibbs phase rule.

10.47 Consider the phase diagram for ^{4}He.

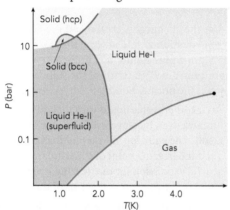

a. How many phases are shown?

b. How many triple points are shown?

c. What is the boiling point of helium at 0.01 bar?

d. What are the possible phase transitions?

10.48 A phase diagram for compound X is shown, with the phase boundaries shown in blue. Iodine has roughly the same triple point pressure and critical point pressure as compound X, but the corresponding temperatures are 300 K higher. Sketch in the approximate locations of the phase boundaries for iodine.

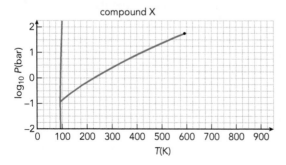

10.49 Using the phase diagram for isobutane, give T and P for (a) the critical point, (b) the triple point, and (c) the standard state melting point.

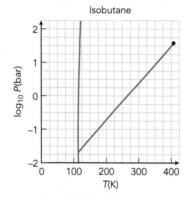

10.50 The phase diagram for mercury is shown, including four solid phases: α (rhombohedral), β (body-centered tetragonal), γ (orthorhombic), and δ (hexagonal close pack).

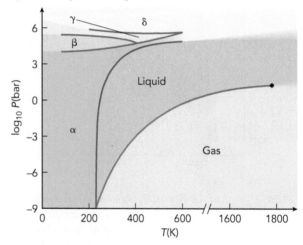

a. What is (approximately) the maximum pressure in bar at which mercury sublimes?

b. What is the maximum number of phases of pure mercury that can coexist simultaneously?

c. What is the maximum temperature at which vaporization can be observed as a phase transition?

d. Which of the phases shown has the greatest density?

e. Which of the phases shown has the greatest average interatomic bond strength?

10.51 Sketch a phase diagram for ethane consistent with the following data. Label the regions by phase, and identify any specific values of the coordinates that you can. (Suggestion: draw the right general graph first, and then identify any coordinate values you can for specific points on the graph.)

a. $T_f = 90.3\,K$ b. $T_b = 184.55\,K$

c. $\varepsilon = 230\,K$ d. $R_{LJ} = 4.42\,Å$

10.52 Draw any phase diagram that predicts the following behavior: (a) the liquid boils at 1 bar, 200 K and at 2 bar, 300 K; (b) below 200 K, the solid sublimes rather than melts; (c) no boiling point is determinable above 400 K.

10.53 Write the slope of the phase diagram gas/liquid phase boundary in terms of $\Delta_{vap}V$ and $\Delta_{vap}S$.

10.54 In a dewar filled with liquid nitrogen, we have an equilibrium mixture of N_2, O_2, and argon, with each substance present in both liquid and vapor phases. Give any *complete* set of intensive parameters that can be independently adjusted while maintaining this equilibrium (in other words, list as many as the number of free parameters).

10.55 Based on the phase diagram shown in the following figure for ethanol, estimate the ethanol–ethanol binding energy in cm^{-1} and the critical number density ρ_c in cm^{-3}. Pay close attention to all units, and note that $1\ MPa = 10^6\ Pa$.

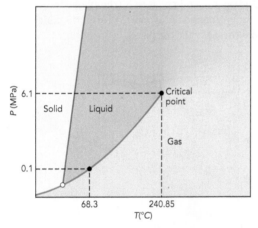

10.56 A graph of the ratio of the van der Waals coefficients a/b versus the thermal energy at the critical temperature RT_c for several substances yields a roughly linear relationship. Solve for an approximate numerical value for the slope of this line. If you have a linear regression package handy, check your prediction against the actual slope, based on the data in Table 4.2 and Table 10.3.

Derivations

10.57 Derive an expression for the change in Helmholtz free energy of vaporization, $\Delta_{vap}F^{\ominus}$, in terms of the boiling point T_b and any other relevant parameters. Simplify as much as possible, using any typical assumptions.

10.58 Use the density of solid I_2 ($4.933\ g\,cm^{-3}$) to estimate the number of molecules per unit area exposed on the surface of an iodine crystal. Furthermore, assume that the binding energy of one I_2 molecule to the surface is given approximately by its enthalpy of vaporization (Table 10.2). Using a Maxwell-Boltzmann distribution at 298 K, estimate the number of I_2 molecules per cm^2 of surface that have vibrational energies in the solid equal to or exceeding the binding energy.

10.59
a. Calculate the entropy of vaporization for ethane ($T_b = 184.55\ K$, $\Delta_{vap}H^{\ominus} = 14.69\ kJ\ mol^{-1}$) at its standard boiling point.

b. If the liquid has a molar volume of $0.054\ L\ mol^{-1}$, what percentage of this entropy is due to expansion from the liquid to the gas?

10.60 If the enthalpy of vaporization of some liquid at any temperature T is given by

$$\Delta_{vap}H^{\ominus}(J\ mol^{-1}) = 5.0 \cdot 10^5 - 20.0T + 1.0 \cdot 10^{-3}T^2,$$

and $C_P(gas) = 2.5nR$ for the gas and the normal boiling temperature is 700 K, find $C_{Pm}(liq)$ for the liquid as a function of T.

10.61 What is the numerical value of $(\partial\mu/\partial P)_{T,n}$ for liquid water?

10.62 An incompressible liquid at 5.0 bar boils at 50 K. Use the chemical potential to calculate the molar volume of the liquid.

10.63 Use the chemical potential and the Clausius-Clapeyron equation to find an equation for the molar entropy $S_m^{\bullet}$ of a pure liquid in equilibrium with the gas at a vapor pressure $P^{\bullet}$ in terms of the standard molar entropy $S_m^{\ominus}$, the normal boiling point T_b, and $\Delta_{vap}H^{\ominus}$.

10.64 An incompressible liquid with molar volume $V_m(liq)$ freezes to an incompressible solid with molar volume $V_m(sol)$. The enthalpy of fusion of the standard state is $\Delta_{fus}H^{\ominus}$, and the freezing point of the standard state is $T_f^{\ominus}$. Find a general expression for the freezing temperature T_f as a function of the pressure P in terms of these quantities.

10.65 Deviation from the ideal gas law is often written in terms of a compressibility factor $Z(P_R,T_R)$ such that

$$PV = Z(P_R,T_R)nRT,$$

where the reduced pressure $P_R = P/P_c$ and the reduced temperature $T_R = T/T_c$, where T_c and P_c define the critical point. A useful characteristic of $Z(P_R,T_R)$ is that its value at a given P_R and T_R is constant to within about 2% across a wide range of gases. This works because to a good approximation $\varepsilon = 0.80T_c$ and $R_{LJ} = 0.55\rho_c^{-1/3}$, where ρ_c is the number density at the critical point. Use these approximations and the virial expansion to find an equation for $Z(P_R,T_R)$ within our Chapter 4 approximations that is independent of the specific gas described (i.e., it should not depend on R_{LJ}, ε, P_c, or T_c).

10.66 Assume that liquid sulfur can be frozen at high pressure in such a way that there is an equal probability of forming the monoclinic or the orthorhombic crystal. In freezing a sample of 0.001 moles of sulfur under these conditions, what is the probability that less than 20% of the samples will freeze in the monoclinic form?

PART I
EXTRAPOLATING
FROM
MOLECULAR TO
MACROSCOPIC
SYSTEMS

PART II
NON-REACTIVE MACROSCOPIC SYSTEMS

PART III
REACTIVE
SYSTEMS

11   Solutions

LEARNING OBJECTIVES

After reading this chapter, you will be able to do the following:

❶ Express the chemical potential of a substance in solution in terms of a standard state potential and the activity.

❷ Use Raoult's law and Henry's law to estimate partial pressures of solute and solvent above a solution.

❸ Use a phase diagram to find vapor pressure, melting and boiling points, composition in each phase, and azeotropic and eutectic conditions for a mixture of substances.

❹ Estimate freezing point depression, osmotic pressure, and (for ionic solutes) membrane voltage based on the concentration.

GOAL *Why Are We Here?*

The goal of this chapter is to identify strategies for applying thermodynamics to systems of two or more interacting (but not chemically reacting) chemical components and to investigate a few ramifications of that interaction. The strategies we will use include finding a mathematical relation that lets us express thermodynamic quantities of a chemical in solution in terms of its equilibrium partial pressure above the solution and using phase diagrams to visualize the effects of solution concentration on the thermodynamic properties of the system.

CONTEXT *Where Are We Now?*

Phase transitions, under the right conditions, let us approach some key concepts of chemical change while avoiding the actual breaking of chemical bonds. Dissolving one compound in another can provide an additional example. We'll call any homogeneous mixture of two or more substances whose combined phase is liquid or solid (regardless of the equilibrium phases of the pure substances) a **solution.** We study solutions at this point because they allow us to extend our results from phase transitions in pure systems to multi-component systems, expand the concept of the chemical potential, and lead us to the doorstep of chemical reactions. Liquid solutions have historically dominated reaction chemistry in the laboratory because liquids provide unique stabilization mechanisms, gases require vacuum equipment, and solids do not mix as easily. In this chapter, we investigate the environment of the typical laboratory reaction before studying the reaction itself.

Why aren't gas mixtures included? The interactions between molecules are minimal in the gas phase, and it is the intermolecular forces that make solutions interesting to chemists. Those same interactions also demand a special effort to treat liquid solutions successfully. Unless specified otherwise, the solutions discussed in this chapter are liquids.

SUPPORTING TEXT *How Did We Get Here?*

This chapter relies on some understanding of liquid structure, primarily the notion that in a liquid the thermal energy is high enough that the particles can move anywhere in the system, but low enough that intermolecular attractions keep the average distance between the particles at a constant. At the macroscopic scale, these conditions mean that a liquid flows but does not expand to fill its container. A growing familiarity with the chemical potential, μ, will also prove helpful. As in the case of phase transitions, the relationship between chemical potentials of different components of the system at equilibrium will be our launching point into the thermodynamics. We will also draw on the following equations and sections of text in this chapter:

- In Eq 9.37 we find that when we combine two fluids A and B with no change in the intermolecular forces, the ΔH of the process is zero, but the entropy change (Eq. 9.31) and free energy change (Eq. 9.36) indicate that the process is spontaneous:

$$\Delta_{\text{mix}}S = -R(n_A \ln X_A + n_B \ln X_B) > 0 \quad \Delta_{\text{mix}}G = RT(n_A \ln X_A + n_B \ln X_B) < 0.$$

  In this chapter we will take a look at what happens when the intermolecular forces change as we move from the separated components to the mixture.

- From Section 10.3, we will borrow the expressions for the changes in entropy (Eq. 10.26), volume (Eq. 10.27) and free energy (Eq. 10.20) in terms of the chemical potential μ:

$$S_{\text{mA}} = -\left(\frac{\partial \mu_A}{\partial T}\right)_{P, n_A}$$

$$V_{\text{mA}} = \left(\frac{\partial \mu_A}{\partial P}\right)_{T, n_A}$$

$$\text{at equilibrium:} \quad (dG)_{T,p} = \sum_{i=1}^{k} \mu_i dn_i = 0.$$

- The Helmholtz free energy plays a cameo role in the presentation ahead, providing a convenient entrance for the statistical mechanics into the theory of solutions by means of Eq. 7.36:

$$F = -k_B T \ln Q(T).$$

- Chapter 4 presents derivations for the pressure of a non-ideal gas, arriving at a crude but useful approximation for the pair correlation function, $\mathcal{G}(R)$ (Eq. 4.59)

$$\mathcal{G}(R) \approx e^{-u(R)/(k_B T)}.$$

  In this chapter we will use this approximation to show how the microscopic intermolecular forces shape the behavior of bulk solutions.

- Section 10.5 introduces the phase diagram, which maps the equilibrium states of matter for a substance as a function of intensive parameters such as the temperature and pressure. In this chapter we will extend the use of phase diagrams to multi-component systems.

11.1 The Standard States

Choosing a reasonable reference point is an issue that stays with us from quantum mechanics all the way through thermodynamics, but nowhere is it more insidious than in the chemistry of solutions.

We previously defined the standard state as the case that all chemical components are pure and the total pressure is 1 bar. This is adequate for gases and some other ideal systems, but it doesn't suffice for solutions. Why should that be? We can derive many useful equations that describe solutions by starting from the chemical potential μ. The central importance of μ springs from its functioning effectively as a molar Gibbs free energy specific to each chemical species in the system, and the Gibbs free energy determines the direction of a process at constant T and P. As with the energy E, we measure *changes* in Gibbs energy rather than absolute energies, so it doesn't really matter what point we use as a reference for all our G measurements, right?

In fact, it doesn't, as long as we look only at *physically meaningful* values like ΔG, and not at numbers that depend on where we put the reference point. But many numbers that we use in calculations *do* depend on our choice of reference for G, and it is from these numbers that we eventually calculate the physically meaningful quantities. There is enough math in between for inconsistencies to ruin our calculations, so we tread carefully here.

We should consider how the standard state defines the reference chemical potential for several cases.

- *Gases.* The standard chemical potential of the ideal gas depends only on the ambient pressure being 1 bar (which we will also write as $P^{\ominus}$)

$$\mu_i^{\ominus} = \mu_i(P^{\ominus}). \quad \text{ideal gas}$$

Recall that we do not specify T for the standard state, so $\mu_i^{\ominus}$ changes with temperature.

- *Pure solids or liquids.* We set the standard state equal to that of the pure solid or liquid at an ambient pressure of 1 bar, so

$$\mu_i^{\ominus} = \mu_i^{\bullet}(P^{\ominus}). \quad \text{pure liquids and solids}$$

The superscript bullet in $\mu^{\bullet}$ is our symbol for a pure component.

- *Liquids in solution.* When the solution itself is a liquid, we can imagine starting from a pure liquid solvent and gradually adding solute until we achieve the concentration in our sample. The properties of the solution will then also change continuously, and the initial, pure solvent serves as an effective reference point for the chemical potential of that substance:

$$\mu_i^{\ominus} = \mu_i^{\bullet}(P^{\ominus}). \quad \text{liquid in solution} \qquad (11.1)$$

Furthermore, when one liquid dissolves in another, the relative concentrations can be varied so that we move either substance continuously from solute to solvent. For example, in a mixture of ethanol and methanol where the mole fraction of ethanol is 0.01, we would normally say that methanol is the solvent, ethanol the solute. The intermolecular interactions will primarily be methanol–methanol, with some ethanol–methanol, so the solution has properties very similar to those of pure methanol. If methanol were reduced to a 0.01 mole fraction, then we'd call ethanol the solvent, because now the solution's properties more closely resemble those of pure ethanol. If both liquids are at 0.5 mole fraction, then there is no clear distinction between solvent and solute. Therefore, we set the standard state of each liquid equal to the pure liquid, as in Eq. 11.1, not worrying about whether that liquid is the solvent or the solute.

- *Solid or gas solute in solution.* When solids or gases are dissolved in a liquid, let's assume that the liquid will be the solvent, surrounding (and often stabilizing) the other component while retaining the essential properties that make the mixture a liquid. The properties of a solid or a gas change so dramatically from the pure form to the dispersed form in solution that the pure substance has no direct bearing on the properties of the solution, and therefore the pure substance makes a poor reference point. Instead, the standard state of the solute is chosen to be the solute *in solution* at some specified concentration. For example, in biochemistry it is typical to set the standard state to an aqueous solution with a pH of 7 (which specifies the amount of H^+). For many solutes, the standard state is chosen to be 1 **molar** (1 M = 1 mole solute per liter of solution) or 1 **molal** (1 m = 1 mole solute per kilogram solvent).[1] We will set our standard state to be 1 M:

$$\mu_i^\ominus = \lim_{[X] \to 1M} \mu_i(\text{soln}). \quad \text{solid or gas in solution} \tag{11.2}$$

While these rules are applied to nearly all the examples in this text, any choice of reference state is possible, and many conventions exist for specific applications. Actual physical observables, like the maximum concentration of NaCl you can dissolve in water at 298 K, will not depend on the choice of reference state. As we will see, the reference state values often cancel out when we evaluate a measurable quantity. However, numbers that we use to calculate that concentration, such as the standard Gibbs energy of solvation of NaCl, *do* depend on the choice of standard state. The take-home message is "beware of comparing thermodynamic quantities for solutions from different sources unless you're sure that they are referring the values to the same standard state."

11.2 Statistical Mechanics of Solutions

Previously, we described several thermodynamic properties of liquids—heat capacity (Eq. 7.75), energy (Eq. 7.74), and pressure (see Problem 4.15)—in terms of integrals involving the pair distribution function $\mathcal{G}(R)$. If we wish to understand the origin of thermodynamic properties in liquid solutions, it is natural

---

[1]Why do both molarity and molality persist in solution measurements? Molarity has the advantage that volumes of liquids are more conveniently measured than masses: you can carry a volumetric flask anywhere in the lab, but you have to be at a scale to measure the mass. However, temperature changes alter the volume. Furthermore, when the solution becomes very concentrated, the solute molecules disrupt the structure of the solvent, and the volume of the solution is no longer directly proportional to the number of moles of solvent. Molality unambiguously specifies the amount of *both* solute and solvent, without our needing to know the change in volume upon heating, cooling, or mixing.

to return to the properties of the pure liquids and see how they are altered by the inclusion of a second chemical component.

Assumptions

To simplify our choice of the standard state, let's choose a solution of two liquids: a solvent A with a very dilute concentration of solute B. The standard state consists of the separated, pure solvent and solute at 1 bar. Although both of the components are liquids, we will take advantage of the simplicity of the gas phase whenever we can. The chemical potential forges this relationship between the solution and the gas, because at equilibrium, the chemical potential is the same for every phase of each chemical component. Before we can use that relationship, however, we have to fold into the chemical potential the effects of the *non-ideal* mixing of our two substances, the solvent A and the solute B.

To keep the mathematics simple, we shall assume the solution is very dilute, meaning that the mole fraction of solute X_B is much less than the mole fraction of solvent X_A, as in Fig. 11.1. This assumption allows the following simplifications:

1. We may neglect terms due to the B-B interactions between solute molecules.

2. We may neglect changes to the average thermodynamic properties of the solvent, such as its density, because most of the solvent molecules see only other solvent molecules and behave the same as if no solute were present.

Section 9.3 gives expressions for the change in enthalpy, entropy, and Gibbs energy of two samples when they are mixed together under ideal conditions—conditions that neglect differences among the A-B, A-A, and B-B interactions. But most of the interest in solution chemistry arises from those very differences, for example,

- ions may exist for relatively long periods in stabilizing or inert solvents, allowing chemical reactions to advance that depend on those intermediate molecules;

- isolation of compounds from a reaction mixture is often accomplished by combining two dissimilar solvents, which spontaneously separate into distinct phases, each carrying a different set of solutes with it; and

- aqueous solutions may promote the folding of proteins to reduce hydrophobic interactions with the solvent.

Enthalpy and Volume Changes with Solvation

To arrive at the thermodynamic properties of the solution that rely on the non-ideality of the mixing, we first find the shift in chemical potential of the two substances when they are mixed, $\Delta_{mix}\mu$:[2]

$$\Delta_{mix}\mu \equiv \mu(\text{soln}) - \mu^{\bullet}$$

---

[2]This property is similar to the **excess chemical potential,** but we will use $\Delta_{mix}\mu$ to keep our notation consistent with previous chapters.

A = ⚪   B = 🔵

▲ FIGURE 11.1 **The dilute solution of solute B in solvent A.**

$$= \mu_A(\text{soln}) + \mu_B(\text{soln}) - \mu_A^\bullet - \mu_B^\bullet$$

$$\approx \mu_B(\text{soln}) - \mu_B^\bullet. \qquad (11.3)$$

Here $\mu^\bullet$ denotes the chemical potential of the pure substance prior to mixing, and our dilute-mixing assumption lets us set $\mu_A \approx \mu_A^\bullet$.

How is $\Delta_{\text{mix}}\mu$ a useful quantity? As an example, at constant temperature and pressure, the chemical potential is the same as the molar Gibbs energy:

$$\mu = \left(\frac{\partial G}{\partial n} \right)_{P,T}. \qquad (11.4)$$

Therefore, for processes taking place at constant T and overall pressure P, the change in μ is the same as the change in molar Gibbs energy, ΔG_m. This goes for mixing a solution as well as any other process. Consequently, thermodynamic properties that we could determine from $\Delta_{\text{mix}}G$ we can also determine from $\Delta_{\text{mix}}\mu$. Some basic quantities of interest are the **excess volume of mixing**,

$$\Delta_{\text{mix}}V_m = \frac{V(\text{soln}) - (V_A^\bullet + V_B^\bullet)}{n_A + n_B}$$

$$= \left(\frac{\partial(\Delta_{\text{mix}}G_m)}{\partial P} \right)_{T,n_A,n_B}$$

$$= \left(\frac{\partial(\Delta_{\text{mix}}\mu)}{\partial P} \right)_{T,n_A,n_B}, \qquad (11.5)$$

and the molar **enthalpy of solvation**,

$$\Delta_{\text{mix}}H_m = \Delta_{\text{mix}}G_m + T\Delta_{\text{mix}}S_m \qquad\qquad \Delta G = \Delta H - T\Delta S$$

$$= \Delta_{\text{mix}}\mu - T\left(\frac{\partial(\Delta_{\text{mix}}\mu)}{\partial T} \right)_{P,n_A,n_B}. \qquad \text{by Eq. 10.26 } (11.6)$$

The excess volume is related to the **partial molar volumes** V_{mi} defined by

$$V_{mi} = \left(\frac{\partial V}{\partial n_i} \right)_{T,P,\, n_{j \neq i}}$$

As a quick example, if we have a vat of water, and add 58 ml (1.0 mol) of pure ethanol, the volume of the mixture will increase by only 54 ml. The partial molar volume of dilute ethanol in water is therefore 54 ml/mol. In general, the partial molar volume varies with the mole fraction of the solute: at a mole fraction of 0.10, the mole fraction of ethanol in water is even lower, about 53 ml/mol.

We can combine the terms in Eq. 11.6 into a single derivative by taking advantage of the relation

$$d\left(\frac{y}{x} \right) = y d\left(\frac{1}{x} \right) + \frac{1}{x} dy$$

$$\frac{d(y/x)}{d(1/x)} = y + \frac{1}{x}\frac{dy}{d(1/x)} \qquad\qquad \text{divide by } d(1/x)$$

$$= y - x\frac{dy}{dx}. \qquad\qquad d(1/x) = -x^{-2}dx$$

Setting $x = T$ and $y = \Delta_{mix}\mu$, we arrive at a more compact expression for the enthalpy of solvation (Eq. 11.6):

$$\Delta_{mix}H_m = \left(\frac{\partial(\Delta_{mix}\mu/T)}{\partial(1/T)} \right)_{P,n_A,n_B}. \qquad (11.7)$$

The chemical potential may be obtained by differentiating any one of our four principal thermodynamic potentials—E, H, G, or F—with respect to n. Which of these potentials we choose depends on which of the parameters T, S, P, and V we wish to hold constant. We choose to differentiate with the temperature and volume held constant, rather than the entropy or pressure, because the relevant results from Part I of this text are expressions in terms of V, N, and T. (This is a convenient choice for a liquid-phase problem anyway, because in that case we can often assume that the volume of the liquid remains constant during these processes.) The parameters we shall fix here are T, V, and n_A, and the thermodynamic potential we want in that case is the Helmholtz energy F. For a two component system,

$$\left(\frac{\partial F}{\partial n_B} \right)_{T,V,n_A} = \mu_B. \qquad (11.8)$$

Similarly, when we wish to find the chemical potential of mixing, we can simply differentiate the Helmholtz energy change with respect to n_B:

$$\Delta_{mix}\mu_B = \mu_B - \mu_B^\bullet = \left[\frac{\partial(F - F^\bullet)}{\partial n_B} \right]_{T,V,n_A}$$

$$= \left(\frac{\partial(\Delta_{mix}F)}{\partial n_B} \right)_{T,V,n_A}. \qquad (11.9)$$

Chemical Potential of Mixing

We've seen how $\Delta_{mix}\mu$ can be used to obtain some interesting properties of non-ideal solutions such as the enthalpy and volume changes of mixing. But these equations alone don't tell us, for example, what will be the numerical value of any of these parameters, such as $\Delta_{mix}H_m$. If we want to know that, we have to go back to the molecular scale. To show this, we will go ahead and derive an expression for $\Delta_{mix}\mu_B$ that gives an idea how that can be accomplished. If you'd rather skip the math, there's a **DERIVATION SUMMARY** after Eq. 11.15.

To find an expression for $\Delta_{mix}\mu_B$, we need an expression for F that connects to the molecular parameters. The partition function $Q(T)$ provides that connection, and we found F in terms of $Q(T)$ earlier (Eq. 7.36):

$$F = -k_B T \ln Q(T).$$

The Helmholtz energy change is a function of the potential energy U_{AB} for interactions between the solvent and solute, which we shall represent as the sum of the pairwise interactions for all N_A solvent molecules with all N_B solute molecules:

$$U_{AB} = \sum_{i=1}^{N_A} \sum_{j=1}^{N_B} u_{AB}(R_{ij}) = N_A N_B u_{AB}(R). \qquad (11.10)$$

Parameters Key: Chemical Potential of Mixing

symbol	parameter	SI units
F	Helmholtz free energy	J
Q_K, Q_U	partition functions (K for kinetic, U for potential energy)	unitless
V	volume	m^3
n_A	number of moles of substance A	mol
N_A	number of molecules of substance A	unitless
$dX'_{A1}dY'_{A1}dZ'_{A1}$	volume element for position of particle 1 of substance A, scaled by $1/V$	unitless
u_{AB}	pair potential energy function for one A particle interacting with one B particle	J
U_{AB}	total particle–particle potential energy	J
$\mathcal{G}(R)$	pair correlation function	unitless
ρ_A	number density of A $= N_A/V$	m^{-3}
ε	well depth of u_{AB}	J
R_{sq}, R'_{sq}	distance range for the attractive part of square well potential	m
$\Delta_{mix}\mu_B$	chemical potential of mixing	J mol^{-1}

In fact, we may write $\Delta_{mix}F$ in a compact form using the partition functions:

$$\Delta_{mix}F = F - F^{\bullet} = -k_B T(\ln Q - \ln Q^{\bullet}) = -k_B T \ln\frac{Q}{Q^{\bullet}}$$

$$\approx -k_B T \ln\left(\frac{Q_{K_A}Q_{K_B}Q_{U_{AA}}Q_{U_{BB}}Q_{U_{AB}}}{Q_{K_A}Q_{K_B}Q_{U_{AA}}Q_{U_{BB}}}\right)$$

$$= -k_B T \ln Q_{U_{AB}}, \tag{11.11}$$

where Q_{K_A} is the partition function for the kinetic energy of the solvent, $Q_{U_{AB}}$ is the partition function for the solvent–solute potential energy, and so on.[3] The partition function $Q^{\bullet}$ of our pure substances includes the kinetic and potential energies for the isolated samples and omits only the interaction term $Q_{U_{AB}}$:

$$Q_{U_{AB}}(T,V) = \int_0^1 \cdots \int_0^1 e^{-U_{AB}/(k_B T)}dX'_{A1}\ldots dZ'_{BN_B}. \tag{11.12}$$

The volume element is again composed of unitless scaled distances such that $dX'dY'dZ' = dXdYdZ/V$ (Eq. 4.29).

We shall also need a pair correlation function $\mathcal{G}_{BA}(R)$ that describes only the probability of locating a solvent molecule A at a distance R from a solute molecule B:

$$\mathcal{G}_{BA}(R) = \frac{\int_0^1 \cdots \int_0^1 e^{-U_{AB}/(k_B T)}dX'_{A2}\ldots dZ'_{AN_A}dX'_{B2}\ldots dZ'_{BN_B}}{Q_{U_{AB}}(T,V)}. \tag{11.13}$$

---

[3]The breakdown of the potential energy partition function into these three components is only approximate, because $Q_{U_{AB}}$ is an integral over the same coordinates as $Q_{U_{AA}}$ and $Q_{U_{BB}}$ and therefore cannot be exactly factored out. This is okay in the dilute solution limit because the effect of the approximation is to neglect the very small change in $Q_{U_{AA}}$ when a few of the A-A bonds are broken to accommodate the solute.

At last, we can solve for the chemical potential of mixing:

$$\Delta_{mix}\mu_B = \left(\frac{\partial(\Delta_{mix}F)}{\partial n_B}\right)_{T,V,n_A}$$

$$= -k_BT\left(\frac{\partial \ln Q_{U_{AB}}}{\partial n_B}\right)_{T,V,n_A} = -\frac{k_BT}{Q_{U_{AB}}}\left(\frac{\partial Q_{U_{AB}}}{\partial n_B}\right)_{T,V,n_A} \quad \text{by Eq. 11.11}$$

$$= -\frac{k_BT}{Q_{U_{AB}}}\frac{\partial}{\partial n_B}\int_0^1 \cdots \int_0^1 e^{-U_{AB}/(k_BT)}\, dX'_{A1}\cdots dZ'_{BN_B} \quad \text{by Eq. 11.12}$$

$$= -\frac{k_BT\mathcal{N}_A}{Q_{U_{AB}}}\frac{\partial}{\partial N_B}\int_0^1 \cdots \int_0^1 e^{-U_{AB}/(k_BT)}\, dX'_{A1}\cdots dZ'_{BN_B}$$

$$= \frac{\mathcal{N}_A}{Q_{U_{AB}}}\int_0^1 \cdots \int_0^1 e^{-U_{AB}/(k_BT)}\left(\frac{\partial U_{AB}}{\partial N_B}\right)_{T,V,n_A}\, dX'_{A1}\cdots dZ'_{BN_B} \quad de^{x/a} = e^{x/a}/a\, dx$$

$$= \frac{\mathcal{N}_A}{Q_{U_{AB}}}\int_0^1 \cdots \int_0^1 e^{-U_{AB}/(k_BT)}\left(\frac{\partial(N_AN_Bu_{AB})}{\partial N_B}\right)_{T,V,n_A}\, dX'_{A1}\cdots dZ'_{BN_B} \quad \text{by Eq. 11.10}$$

$$= \frac{\mathcal{N}_AN_A}{Q_{U_{AB}}}\int_0^1 \cdots \int_0^1 e^{-U_{AB}/(k_BT)}u_{AB}(R)\, dX'_{A1}\cdots dZ'_{BN_B}$$

$$= \frac{\mathcal{N}_AN_A}{Q_{U_{AB}}}\int_0^1 \cdots \int_0^1 u_{AB}(R)\left[\int_0^1 \cdots \int_0^1 e^{-U_{AB}/(k_BT)}dX'_{A2}\cdots dZ_{AN_A}dX_{B2}\cdots dZ'_{BN_B}\right]$$

$$\times\, dX'_{A1}dY'_{A1}dZ'_{A1}dX'_{B1}dY'_{B1}dZ'_{B1}$$

$$= \mathcal{N}_AN_A\int_0^1\int_0^1\int_0^1 u_{AB}(R)\mathcal{G}_{BA}(R)\, dX'_{A1}dY'_{A1}dZ'_{A1} \quad \text{by Eq. 11.13}$$

$$= \mathcal{N}_A\frac{N_A}{V}\int_0^a\int_0^b\int_0^c u_{AB}(R)\mathcal{G}_{BA}(R)\, dX_{A1}dY_{A1}dZ_{A1}$$

$$= 4\pi\mathcal{N}_A\rho_A\int_0^{V^{1/3}} u_{AB}(R)\mathcal{G}_{BA}(R)\, R^2dR. \quad \text{integrate over angles} \quad (11.14)$$

Admittedly, this equation is only useful if we know the pair correlation function for the solute–solvent interaction, and few of these are rigorously known. We can at least verify that this equation has the right behavior in the simplifying limit of a square well potential (Eq. 4.13) for $u_{AB}(R)$ and a pair correlation function (based on Eq. 4.59)

$$\mathcal{G}_{BA} \approx e^{-u_{AB}(R)/(k_BT)}.$$

Under these conditions,

$$\Delta_{mix}\mu = 4\pi\mathcal{N}_A\rho_A\left[\int_0^{R_{sq}}(\infty)e^{-\infty}R^2dR + \int_{R_{sq}}^{R'_{sq}}(-\varepsilon)e^{\varepsilon/(k_BT)}R^2dR + \int_{R'_{sq}}^\infty(0)e^0R^2dR\right]$$

$$= -\frac{4\pi}{3}\mathcal{N}_A\varepsilon\rho_A\, e^{\varepsilon/(k_BT)}\, (R'^3_{sq} - R^3_{sq}). \quad (11.15)$$

Unfortunately, Eq. 11.15 is not so successful at predicting quantitative values of $\Delta_{mix}H$ and other properties, because we have neglected the significant contribution of Q_{BB}. But qualitatively this is a reasonable result, in that $\Delta_{mix}\mu$ increases in magnitude as either the strength ε or the range $R'_{sq} - R_{sq}$ of the solute–solvent interaction increases. This predicts that $\Delta_{mix}\mu$ will normally be negative, which makes sense if the solute is stabilized by the solvent. We can see why this happens from the more accurate form in Eq. 11.14, where the integrand $u_{AB}(R)\mathcal{G}_{BA}(R)$ is likely to have greatest magnitude where the pair correlation function reaches its peak, which is in the attractive well where the potential energy is negative. On the other hand, where the repulsions are strong, at low values of R, $u_{AB}(R)$ becomes large and positive, but the pair correlation function approaches zero, and so the product $u_{AB}(R)\mathcal{G}_{BA}(R)$ also approaches zero. A negative value of $\Delta_{mix}\mu$ means a negative $\Delta_{mix}G$, implying (as we found in the ideal case) that the mixing is a spontaneous process.

DERIVATION SUMMARY Chemical Potential of Mixing. To obtain Eq. 11.14, we took Eq. 11.9, which writes $\Delta_{mix}\mu_B$ as a derivative of the Helmholtz free energy F with respect to the number of moles of solute, and we wrote F in terms of the partition function. The partition function depends explicitly on the number of particles present, so we could then evaluate the derivative, although this requires assuming the total potential energy is just the sum of the pair potentials u_{AB}. The resulting expression was rearranged to an integral over u_{AB} times the pair correlation function, indicating that the chemical potential of mixing can be positive or negative, depending on whether the intermolecular potential energy is dominated by the repulsive forces or the attractive forces. Equation 11.15 uses the square well potential to show that it will typically be negative.

BIOSKETCH | **Kim K. Baldridge**

Kim K. Baldridge is a professor of theoretical chemistry at the University of Zurich. Originally from North Dakota, Professor Baldridge made an early career choice for theoretical chemistry over medicine which led her to positions at the University of California and the San Diego Supercomputing Center before moving to Switzerland.

In the course of her work studying the quantum mechanics of fundamental organic reactions, Professor Baldridge develops methods in computational chemistry that apply to a broad spectrum of chemical problems, one highlight being the accurate modeling of solvation in chemical reactions. One method treats the solvent as a *polarizable continuum,* as a smooth shell that surrounds the reacting molecules with charges that mirror the charge distribution on the reactants. These calculations effectively average over the many possible orientations of the solvent molecules but lack the sensitivity to strong, local interactions such as hydrogen bonding between solvent and solute. The incorporation of a few solvent molecules creates a model with more accurate energies, but only a few of the thousands of possible geometries can be considered. Professor Baldridge and her research group study whether either approach or a combination of the two leads to more accurate predictions of the reaction properties. Accurate computational models of solvation are intensely sought after by the chemical industry in order to predict the characteristics of industrial-scale reactions and to guide research and development.

11.3 Thermodynamics of Solutions

Although the statistical definition of the activity allows us to directly relate the solution properties to the intermolecular potential, the difficulty in evaluating the pair correlation function and the intermolecular potential remains a substantial obstacle to applying Eq. 11.14 to real problems. More often, we rely on idealized models of solutions such as those described in this section.

Activities

The chemical potential μ shows up in our study of solutions because the equality of μ for different phases of the same component allows us to relate properties of the complex solution environment to simpler phases, particularly the vapor above the solution. However, μ rarely appears explicitly in our final equations. We use it temporarily to get other, more convenient formulas. When the chemical potential is evaluated, it is usually given relative to the chemical potential of the standard state, $\mu^{\ominus}$. Then we can write

$$\mu = \mu^{\ominus} + \int_{P^{\ominus}}^{P} \left(\frac{\partial \mu}{\partial P} \right)_{T,n} dP = \mu^{\ominus} + \int_{P^{\ominus}}^{P} V_m \, dP. \qquad (11.16)$$

There are two idealized cases in which the integral is easily evaluated: the ideal gas, in which case $V_m = RT/P$ and

$$\mu = \mu^{\ominus} + RT \ln \frac{P}{P^{\ominus}}, \qquad (11.17)$$

and the pure, incompressible liquid or solid, for which V_m is constant and

$$\mu = \mu^{\ominus} + V_m(P - P^{\ominus}). \qquad (11.18)$$

For non-ideal systems the chemical potential is usually given in a form similar to Eq. 11.17, but with the pressure ratio replaced by an empirical factor, most commonly the **activity** a:

$$\mu = \mu^{\ominus} + RT \ln a. \qquad (11.19)$$

The activity is unitless, equal to $P/P^{\ominus}$ for the ideal gas, for example. (We'll obtain some other expressions shortly.) We will have to keep careful track of the standard state, used to evaluate the standard chemical potential $\mu^{\ominus}$, because the standard state depends on the nature of the sample. For a gas, $\mu^{\ominus}$ is the chemical potential of the pure gas at a pressure of exactly 1 bar. For a liquid in solution, $\mu^{\ominus}$ is set to the value of the *pure* liquid's chemical potential at 1 bar.[4]

Raoult's Law

The most convenient thermodynamic potential to work with at constant temperature and pressure is G, and for a two-component solution (Eq. 10.20),

$$dG = \mu_A dn_A + \mu_B dn_B.$$

---

[4]This is why we define the standard state not only by the pressure, but also by insisting that all substances (except solutes) are in their pure forms.

In our derivation of the Gibbs-Duhem equation we showed that (Eq. 10.19)

$$G = \mu_A n_A + \mu_B n_B.$$

We can calculate the Gibbs energy of mixing under constant T and P, where G is the final Gibbs energy after mixing, and the initial Gibbs energy is

$$G_i = n_A \mu_A^{\bullet} + n_B \mu_B^{\bullet}. \tag{11.20}$$

Therefore,

$$\Delta_{mix}G = G_f - G_i$$
$$= n_A(\mu_A - \mu_A^{\bullet}) + n_B(\mu_B - \mu_B^{\bullet}). \tag{11.21}$$

We can write this in terms of the activity, but as long as the vapors of the two components behave like ideal gases, we use the ideal gas limit for the chemical potential of the vapor (Eq. 11.17):

$$\mu = \mu^{\ominus} + RT \ln \frac{P}{P^{\ominus}}.$$

If the solution is in equilibrium with the vapor phase of the components, then for each of A and B $\mu(\text{soln}) = \mu(\text{gas})$, and

$$\Delta_{mix}G = n_A \left(\mu_A^{\ominus} + RT \ln \frac{P_A}{P^{\ominus}} - \mu_A^{\ominus} - RT \ln \frac{P_A^{\bullet}}{P^{\ominus}} \right)$$
$$+ n_B \left(\mu_B^{\ominus} + RT \ln \frac{P_B}{P^{\ominus}} - \mu_B^{\ominus} - RT \ln \frac{P_B^{\bullet}}{P^{\ominus}} \right)$$
$$= RT \left(n_A \ln \frac{P_A}{P_A^{\bullet}} + n_B \ln \frac{P_B}{P_B^{\bullet}} \right), \tag{11.22}$$

where the pressures $P^{\bullet}$ are the vapor pressures of pure A or pure B at the same temperature as the solution.

So far we have had to assume only ideality of the gas phase, usually a good approximation. Now we make a more drastic assumption. **Raoult's law** states that the vapor pressure of the solvent is proportional to its mole fraction X_A:

$$P_A = X_A P_A^{\bullet}. \tag{11.23}$$

This equation has great intuitive appeal: it makes sense that the equilibrium amount of solvent escaping into the gas phase would be proportional to the amount in the solution. And in general, Raoult's law works very well, but we don't want to trust this relationship too much. It works best for dilute solutions, and only for the solvent. Solvation properties, such as the stabilization energy and cage structure, do not vary linearly with concentration, and the surface properties of the solution, where the interaction with the vapor takes place, may be much different from those of the rest of the solution.

Raoult's law provides a very convenient simplification. Substituting it into the Gibbs energy of mixing (Eq. 11.22) gives

$$\Delta_{mix}G = RT(n_A \ln X_A + n_B \ln X_B), \tag{11.24}$$

which is the same result we had for the mixing of two ideal gases. Similarly, the entropy of mixing is

$$\Delta_{mix}S = -R(n_A \ln X_A + n_B \ln X_B), \tag{11.25}$$

and the enthalpy of mixing $\Delta_{mix}H$ is 0. These hold strictly only for the **ideal solution,** where the potential energy function is the same for any pair of molecules, whether the molecules have the same identity or not.

We return now to the chemical potential. For a two-component system, we may use Eq. 10.20 to write the change in Gibbs energy in the following form:

$$(dG)_{T,P} = \mu_A dn_A + \mu_B dn_B. \tag{11.26}$$

Therefore, the chemical potential for component A can be written as

$$\mu_A = \left(\frac{\partial G}{\partial n_A}\right)_{T,P,n_B}. \tag{11.27}$$

To find the new chemical potential of component A after mixing it with component B, we compute $\Delta_{mix}\mu_A$:

$$\Delta_{mix}\mu_A = \left(\frac{\partial \Delta_{mix}G}{\partial n_A}\right)_{T,P,n_B}. \tag{11.28}$$

Combining this with our Eq. 9.36 for the Gibbs energy of mixing, we obtain

$$\Delta_{mix}\mu_A = \frac{\partial}{\partial n_A}\left[RT(n_A \ln X_A + n_B \ln X_B)\right] = RT \ln X_A. \tag{11.29}$$

For solutions, the reference chemical potential of the solvent is set to the potential of that compound when it is pure. Therefore,

$$\mu_A = \mu_A^{\bullet} + \Delta_{mix}\mu_A = \mu_A^{\bullet} + RT \ln X_A. \tag{11.30}$$

For any chemical component i in an ideal solution,

$$\text{ideal solution:} \quad a_i = X_i. \tag{11.31}$$

Another result of our selection of the pure substance as the reference is that the activity of pure substances becomes trivial to evaluate:

$$\mu = \mu^{\bullet} + RT \ln a = \mu^{\bullet}$$

$$RT \ln a = 0$$

$$\text{pure liquid or solid:} \quad a = X = 1. \tag{11.32}$$

The activity of a pure substance is equal to its fractional concentration, which is 1 by definition.

Henry's Law

Raoult's law is qualitatively accurate for solvents, because solvent molecules interact mostly with other solvent molecules, except in the most concentrated solutions. The solute, in contrast, interacts primarily with the solvent—not with other solute molecules. Therefore we should not expect the partial pressure of our solute B and its concentration in solution to be directly related to the vapor pressure of the pure solute. We can, however, reasonably hope to keep the same qualitative basis underlying Raoult's law: that the partial pressure of the solute above the solution is proportional to its concentration. When that is the case, we can assign a constant, k_X, and we have **Henry's law,**

$$P_B \approx k_X X_B, \tag{11.33}$$

where k_X is the Henry's law constant (or Henry's law coefficient), a function of *both* the solute and solvent, and varying with temperature as well. The approximation is given because Henry's law is defined to be correct in the limit of infinite dilution, and the Henry's law constant is therefore

$$k_X \equiv \lim_{X_B \to 0} \left(\frac{P_B}{X_B} \right). \tag{11.34}$$

By this definition, k_X has units of pressure because the mole fraction is unitless.[5]

Qualitatively, the stronger the attractive forces between the solvent and solute, the less solute can escape from the solution into the gas phase, and the lower the value of k_X. This trend appears in the selected Henry's law constants given in Table 11.1. Helium, with the weakest binding forces in all of chemistry, has the highest k_X values because a pressure of helium *above* the solution is necessary to keep enough helium *in* the solution. As we go to larger non-polar molecules, N_2 and CCl_4 and benzene, the value drops rapidly, reflecting the increasing stability of the solvent–solute interaction. Similarly, non-polar molecules are better stabilized by a polarizable solvent like benzene than by water, although water provides an excellent solvent for a polar molecule such as acetone. Decreases in temperature will tend to improve the stability in solution relative to the gas phase.

We can use Raoult's and Henry's laws to obtain the activities, which in turn give the chemical potentials (Eq. 11.19):

$$\mu = \mu^\ominus + RT \ln a,$$

where the standard state value $\mu^\ominus$ is a matter of convention and where the activity is determined by the intermolecular potential. The conventional reference states in solution are the pure liquid for the solvent A and the 1 M solution for the solute B:

$$\mu_A^\ominus = \mu_A^\bullet \tag{11.35}$$

$$\mu_B^\ominus = \lim_{[X] \to 1M} \mu_B. \tag{11.36}$$

TABLE 11.1 Values for the Henry's law constant k_X of selected solutions. The temperature is 298 K unless otherwise indicated.

solute	solvent	k_X (bar)	solute	solvent	k_X (bar)
He	H_2O	$1.45 \cdot 10^{5\,a}$	He	C_6H_6	$1.32 \cdot 10^{4\,a}$
H_2	H_2O	$7.17 \cdot 10^4$	H_2	C_6H_6	$3.93 \cdot 10^3$
N_2	H_2O	$8.64 \cdot 10^4$	N_2	C_6H_6	$2.27 \cdot 10^3$
N_2	H_2O	$5.44 \cdot 10^{4\,b}$			
C_2H_6	H_2O	$3.03 \cdot 10^4$	C_2H_6	C_6H_6	68
CCl_4	H_2O	$1.6 \cdot 10^3$	C_2Cl_4	H_2O	$9.6 \cdot 10^2$
C_6H_6	H_2O	$3.1 \cdot 10^2$	acetone	H_2O	2.2

[a] measured at 293 K.
[b] measured at 273 K.

---

[5] Pay close attention to units in other sources, however. Henry's law as we have written it expresses the *volatility* of the solute, how much escapes into the gas phase at a given concentration, but in many sources Henry's law is written in the inverse format as concentration/pressure, to reflect the *solubility* of the solute. Furthermore, the mole fraction in both definitions of the coefficient is often replaced in the experimental literature by the molal concentration, and of course you can never be sure what pressure units might be used.

Using these reference states, we define the activities for solution and solute as the mole fraction, using Raoult's law for the solvent,

$$\text{Raoult's law: } a_A = X_A = \frac{P_A}{P_A^\bullet}, \tag{11.37}$$

and Henry's law for the solute,

$$\text{Henry's law: } a_B = X_B = \frac{P_B}{k_X}. \tag{11.38}$$

We can illustrate how the Henry's law coefficient is related to the molecular properties by writing $\Delta_{\text{mix}}\mu$ in terms of the partial pressure of B over a pure sample $P_B^\bullet$ and over the solution P_B:

$$\Delta_{\text{mix}}\mu_B \approx \mu_B(\text{soln}) - \mu_B^\bullet$$

$$= \left[\mu_B^\ominus + RT\ln\left(\frac{P_B}{P^\ominus}\right) \right] - \left[\mu_B^\ominus + RT\ln\left(\frac{P_B^\bullet}{P^\ominus}\right) \right]$$

$$= RT\ln\left(\frac{P_B}{P_B^\bullet}\right). \tag{11.39}$$

Now we set Eqs. 11.39 and 11.14 equal to each other:

$$\Delta_{\text{mix}}\mu_B = 4\pi\mathcal{N}_A\rho_A \int_0^\infty u_{AB}(R)\mathcal{G}_{BA}(R)R^2 dR \qquad \text{by Eq. 11.14}$$

$$\approx RT\ln\left(\frac{P_B}{P_B^\bullet}\right) = RT\ln\left(\frac{k_X X_B}{P_B^\bullet}\right) \qquad \text{by Eq. 11.39}$$

$$k_X \approx \frac{P_B^\bullet}{X_B}\exp\left[\frac{4\pi\mathcal{N}_A\rho_A}{RT}\int_0^\infty u_{AB}(R)\mathcal{G}_{BA}(R)R^2 dR\right] \qquad \text{solve for } k_X$$

$$\approx \frac{P_B^\bullet}{X_B}\left[1 + \frac{4\pi\rho_A}{k_B T}\int_0^\infty u_{AB}(R)\mathcal{G}_{BA}(R)R^2 dR\right]. \; e^x \approx 1 + x \tag{11.40}$$

In the dilute solution limit, the first term can be expressed in terms of the number density of the solvent:

$$\lim_{X_B\to 0}\left(\frac{P_B^\bullet}{X_B}\right) = \lim_{n_B\to 0}\left[\left(\frac{n_B^\bullet RT}{V}\right)\left(\frac{n_A + n_B}{n_B}\right)\right]$$

$$\approx \frac{RTn_A}{V} = k_B T\rho_A. \tag{11.41}$$

So our expression for k_X becomes

$$k_X \approx k_B T\rho_A\left[1 + \frac{4\pi\rho_A}{k_B T}\int_0^\infty u_{AB}(R)\mathcal{G}_{BA}(R)R^2 dR\right]$$

$$\approx k_B T\rho_A + 4\pi\rho_A^2\int_0^\infty u_{AB}(R)\mathcal{G}_{BA}(R)R^2 dR. \tag{11.42}$$

The first term, $k_B T\rho_A = n_A RT/V$, is the ideal gas pressure of the solvent A. This is essentially a Raoult's law contribution to k_X in the limiting case that A and B have identical interactions. For that case, the partial pressures of A and B have the same proportionality constant, $P_A^\bullet$. The second term, which depends on the A-B pair correlation function and intermolecular potential, is the principal correction

CHECKPOINT Equation 11.41 cuts a corner here to save effort. The number of moles $n_B^\bullet$ when we rewrite the pressure $P_B^\bullet$ is for the number of moles in the gas phase, while n_B when we rewrite X_B corresponds to the solution. In the limit that we push n_B to zero, these two values do not necessarily approach zero at the same rate. More elegant derivations show how the leading term in Eq. 11.42 deviates from Raoult's law. In any case, it is the second term in Eq. 11.42 that shows how the A-B interactions shape Henry's law.

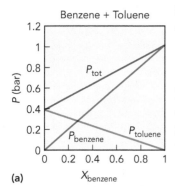

(a)

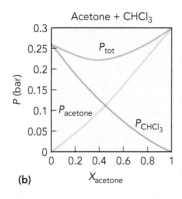

(b)

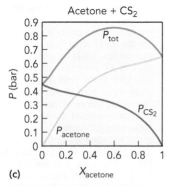

(c)

▲ **FIGURE 11.2** Vapor pressures as functions of mole fractions for three binary mixtures, illustrating **(a)** ideal behavior (benzene + toluene), **(b)** negative deviation from Raoult's law (acetone + $CHCl_3$), and **(c)** positive deviation from Raoult's law (acetone + CS_2).

arising from solvation effects. The second term is a negative contribution for favorable solvent–solute interactions, because it will be dominated by the attractive region where $u_{AB}(R)$ is negative and where $\mathcal{G}_{BA}(R)$ peaks. At lower temperature, both terms contribute to a reduction in the value of k_X: the Raoult's law term drops because the partial pressure of the gas drops in proportion to T, and the second term becomes more negative as the pair correlation function becomes more sharply defined. Although Eq. 11.42 is an approximation, you can see how a more rigorous application of our tools from statistical mechanics could allow us to estimate Henry's law constants and other parameters of real solutions.

Non-Ideal Solutions

Raoult's law and Henry's law together predict that we can set the activities of solvent and solute alike equal to their mole fractions. When that equality fails, we have a non-ideal solution. To express the chemical potential of a non-ideal substance in terms of mole fractions X, we must introduce the unitless **activity coefficient** γ:

$$\gamma_i \equiv \frac{a_i}{X_i}, \qquad (11.43)$$

which lets us replace the activity a_i with $\gamma_i X_i$. The non-ideality arises most frequently in concentrated solutions, where the density of solute disrupts the solvent structure, and where solute–solute interactions may invalidate the assumptions that went into Henry's law.

There is a straightforward, qualitative interpretation of γ. When $\gamma_i = 1$, the activity is equal to the mole fraction; in other words, the solution is ideal, obeying Henry's law for the solute or Raoult's law for the solvent. A value of γ_i greater than 1 indicates that the activity increases faster than the mole fraction, so the partial pressure of component i above the solution is greater than it would be for an ideal solution. If γ_i is less than 1, the partial pressure is lower. Therefore, a value of $\gamma_i > 1$ implies that component i is made less stable by its interactions with the other components of the solution, so it prefers to be in the gas phase (compared to the ideal case), whereas $\gamma_i < 1$ means the other components stabilize i, so it stays in the solution, reducing its partial pressure relative to the ideal case.

The three liquid/liquid solutions represented in Fig. 11.2 are worth examining to clarify these points.

- The benzene/toluene system in Fig. 11.2a is a well-known example of a relatively rare system: a solution of two substances that is nearly ideal at all mole fractions. From pure toluene ($X_{benzene} = 0$) on the left to pure benzene ($X_{benzene} = 1$) on the right, the equilibrium partial pressures of the two components vary in direct proportion to their concentrations, just as Raoult's law predicts. The total vapor pressure above the solution is the sum of the two partial pressures. Since the partial pressure curves are straight lines, their sum is also a straight line.

- Acetone, $(CH_3)_2C{=}O$, and chloroform, $CHCl_3$, form a solution with a **negative deviation** from Raoult's law, meaning that the actual partial pressure of the component is less than the Raoult's law partial pressure, as shown in Fig. 11.2b. This indicates that the dipole–dipole interaction between the acetone and the chloroform is stronger than the interactions between the molecules in pure acetone or pure chloroform. How can this be? The mixture provides an opportunity for a weak form of hydrogen bonding that is not available in the pure liquids to occur. The H — C bond in chloroform is not very strong so chloroform makes a good hydrogen donor. Acetone provides an oxygen atom, which readily accepts hydrogen to form hydrogen bonds. The pure liquids cannot form these bonds, because the C — H bonds in acetone are too strong to donate their H atoms, and the Cl atoms in chloroform are not as good hydrogen acceptors as the oxygen.

- In Fig. 11.2c, the pressures are shown for solutions of acetone and carbon disulfide, CS_2. In this case, the interaction between the highly polar acetone and the non-polar CS_2 is dominated by a dipole-induced dipole interaction that is weaker than either the dipole–dipole attraction in pure acetone or the dispersion forces in pure CS_2. The result is that the solution destabilizes the liquid relative to the gas, and the partial pressures are higher than the ideal case, giving a **positive deviation** from Raoult's law.

Multi-Component Phase Diagrams

Figure 11.2 shows graphs of the vapor pressure—which corresponds to the liquid–gas phase boundary in our phase diagrams from Section 10.5—as functions of intensive parameters of the system: the pressure and mole fraction. In other words, the graphs in Figure 11.2 are essentially phase diagrams themselves, but they illustrate the equilibrium states of matter for systems of more than one chemical component. Phase diagrams summarize a lot of data in one graphic, and they acquire particular importance in the analysis of mixtures, where every combination of different phases offers a new set of thermodynamic parameters.

We will consider only two examples of multi-component phase diagrams in this section to briefly illustrate how we can apply our understanding of solution thermodynamics to extend our phase diagrams beyond the case of a single substance.

Liquid–Liquid Equilibrium: Acetone and Chloroform

According to the Gibbs phase rule (Eq. 10.44)

$$d = k - p + 2,$$

when we have $k = 2$ chemical components present in $p = 2$ different phases, then we can independently adjust $d = 2 - 2 + 2 = 2$ intensive parameters. If only one phase is present, then there are three independent parameters.

The intensive parameters that we can control for a two-component system may be expressed as the temperature T, the pressure P, and the mole fraction of one component, say X_B. (Since the mole fraction of the other component X_A must equal $1 - X_B$ in a two-component system, it is never an independent parameter.) As an example, two phase diagrams for a mixture of acetone and chloroform ($CHCl_3$) are drawn in Fig. 11.3. Because there are up to three independent parameters, to get a two-dimensional graph we have fixed one of the intensive parameters—either P (Fig. 11.3a) or T (Fig. 11.3b)—while showing the equilibrium phase or phases of the mixture as we vary the other two parameters.

▶ **FIGURE 11.3 Phase diagrams for an acetone/chloroform mixtures.** These diagrams show the liquid–gas equilibrium as a function of mole fraction of chloroform with either **(a)** the pressure fixed to 1.013 bar or **(b)** the temperature fixed to 323.15 K.

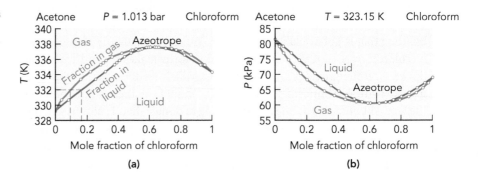

In a single-component phase diagram, we are free to pick the temperature (within a certain range), but at that temperature there can be only one pressure at which the gas and liquid are at equilibrium. The liquid–gas phase boundary in that case has only one degree of freedom. In our binary mixture, once we set the pressure in Fig. 11.3a to 1.013 bar, we only have one free parameter left if we want to find the liquid and gas both present. So, for example, if we set the temperature at 332 K, then we find that the mole fraction of chloroform X_{CHCl_3} has been determined.

Why then do we see *two* boundaries drawn between the pure liquid phase and the pure gas phase? The two curves appear because, for a non-ideal solution at a given temperature and pressure, the equilibrium value of X_B does not have to be the same in the liquid and gas phases. For example, in Fig. 11.3a, at 332 K we see that the gas phase has an equilibrium value for X_{CHCl_3} of about 0.10, whereas for the liquid at that temperature X_{CHCl_3} is roughly 0.17. In other words, at low concentrations of chloroform, the chloroform is less likely to go into the gas phase than the acetone.

The opposite happens at high chloroform concentrations. When the temperature reaches 336 K, one equilibrium exists where $X_{CHCl_3} = 0.90$ for the liquid and 0.94 for the gas. With excess chloroform, the proportion of chloroform in the gas phase is *larger* than its proportion in the liquid.

Let's examine Fig. 11.3a a little more closely. In Fig. 11.2b we saw that the binary mixture of chloroform and acetone has a negative deviation from Raoult's law, implying that the intermolecular attractions between acetone and chloroform are *stronger* than the attractions within the pure liquids. As a result, a higher temperature is required to vaporize the mixture than is required for vaporizing the acetone or the chloroform alone. This explains why the liquid–gas phase boundary in Fig. 11.3a reaches a maximum near the middle, where the mole fraction of chloroform is about 0.66. This corresponds to a maximally favorable interaction when there are roughly two chloroform molecules for each acetone, which is consistent with each O atom in acetone being able to accept the H atoms of two chloroforms (although we must keep in mind that these are not very strong hydrogen bonds, and in the liquid these bonds shift constantly as the molecules tumble).

For the general case of two components A and B coexisting in two phases α and β at equilibrium, the **lever rule** allows us to determine the amount of each substance n^α and n^β in each phase. We note first that the amount of A in the system can be written in terms of its overall mole fraction X_A:

$$n_A = nX_A = (n^\alpha + n^\beta)X_A,$$

and also in terms of its mole fraction in each phase:

$$n_A = n^\alpha X_A^\alpha + n^\beta X_A^\beta.$$

Setting the equations equal to one another, we obtain the lever rule:

$$n^\alpha(X_A - X_A^\alpha) = n^\beta(X_A^\beta - X_A). \tag{11.44}$$

From the total amounts of material in each phase, the lever rule lets us determine the proportion of A in each phase.

In between the low- and high-concentration limit there lies a mole fraction of chloroform such that the liquid and the gas have the *same* proportion of both compounds, a mixture called the **azeotrope.** Azeotropes are best known in chemistry for three reasons: (1) because the mole fraction that gives the azeotrope is unique for a given temperature, they provide a standard reference for concentrations in mixtures of liquids; (2) azeotropes that combine a flammable and nonflammable liquid often reduce the hazard of the flammable substance by preventing a buildup of the flammable substance either in the gas or liquid phase (the nonflammable component will always be present at a fixed proportion); and (3) azeotropic mixtures spoil an important separation technique, **fractional distillation,** precisely because they don't separate.

In fractional distillation, we try to separate one liquid from another by vaporizing the mixture in several stages. At each stage, the liquid is vaporized at some constant temperature T_i, the resulting vapor is condensed at a lower temperature T_{i+1}, and we arrive at a liquid that has a new composition: the composition of vapor at T_i. With each stage, the gas becomes higher in concentration of the more volatile component, while the liquid becomes more concentrated in the less volatile component. For the acetone/chloroform mixture, we can follow the chain of steps drawn in Fig. 11.4 and obtain a high-purity acetone at the end. The method doesn't work for azeotropic mixtures, however, because the concentration of both substances remains the same whether in the liquid or vapor phase. In this case, the mixture may be intentionally moved away from the azeotropic concentration ratio by adding one component or by changing the pressure to alter the phase diagram (which may be sufficient to significantly change the concentration ratio that gives the azeotrope).

If the deviation from Raoult's law is positive instead of negative, as for the acetone/CS_2 mixture in Fig. 11.2c, then the boiling point of the azeotrope appears at a lower temperature than the boiling points of the pure components.

Solid–Solid Equilibrium: Iron and Carbon

The iron-carbon phase diagram (Fig. 11.5) is one of the most important in all of industry, because it shows how temperature and composition determine the properties of the most basic steels. Iron is relatively abundant in nature, but only in combination with oxygen. To remove the oxygen, iron ore is smelted with carbon monoxide, shifting the oxygen from iron to CO to form CO_2. It was possibly during ancient smelting of iron that man discovered how the mixing of carbon with iron can form steel, an even stronger material than pure iron.

The term *steel* is applied to virtually any alloy of iron used to form a strong metal, but the most common component besides the iron is carbon. In the solid phase, the mixture may exist in many different forms. On the left side of the phase diagram, at $X_C = 0$, we have pure iron, which itself has four different stable structures or **allotropes** at different temperatures:

1. α-Fe or *ferrite* has a body-centered cubic crystal structure (see *QM* Chapter 13) and a density of $7.87\,\mathrm{g\,cm}^{-3}$. It is ferromagnetic, meaning it can be used to make permanent magnets.

2. β-Fe is an older designation for the paramagnetic form of α-Fe, above its Curie temperature of 1044 K where the thermal energy is too high for it to maintain a permanent magnetic field without the help of an external field.

3. γ-Fe has a face-centered cubic lattice and is 2% more dense than ferrite. In spite of the slightly higher density, γ-Fe can dissolve more carbon than ferrite, making it the preferred form of iron from which to make stainless steel. This allotrope of iron is not magnetic, and neither are the steels that it forms.

4. δ-Fe has a body-centered cubic lattice and is found over a narrow temperature range when molten iron is cooled.

▼ **FIGURE 11.4 Fractional distillation.** Starting from a 55:45 mixture of acetone and chloroform, we can increasingly isolate the acetone in the gas phase by vaporizing (which increases the concentration of the more volatile acetone) and then condensing (to lower the temperature and prepare the liquid for the next stage).

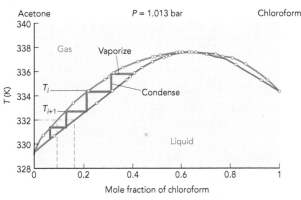

▶ **FIGURE 11.5 The iron-carbon phase diagram.**

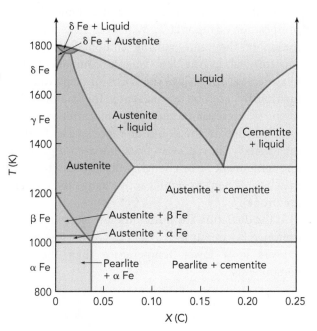

As we add carbon, we see many possible solid phases appear, differing in which iron allotrope is dominant and how much carbon is present:

1. *Austenite* is a common name for γ-iron and for its solid solutions of carbon.

2. *Cementite* is iron carbide, Fe_2C, a strong but inflexible material when in its pure form.

3. *Pearlite* is a combination of ferrite and cementite in clusters that link at the microscopic scale to form a strong steel.

The feature of this phase diagram common to solid–liquid binary phase diagrams is the curve that forms the solid–liquid phase boundary. The difference between the liquid and the solid is a matter of whether or not the particles are free to travel through the system. To form the solid from the liquid, we remove enough thermal energy from the system that direction-dependent forces (such as dipole–dipole forces or covalent bonding) fix the average positions of the particles. By locking the particles in place, we have lowered the entropy of the system. If the thermal energy is able to maintain the motion of the particles, a liquid is always favored by entropy over the solid. When we mix two different liquids together, the resulting solution has even a greater entropy than the pure components at the same temperature, as we found in showing with Eq. 9.31 that $\Delta_{mix}S$ is positive. In liquid form, the two substances can mix, something they can't do in the solid state. As a result, solutions tend to be stable as liquids at temperatures below the freezing points of the pure components.

In the iron-carbon phase diagram, this phenomenon appears as the decrease in the melting point of the mixture as we move away from the pure iron at the left. The solid–liquid phase boundary decreases as we approach a minimum called the **eutectic point.** A *eutectic alloy* is a mixture of components at the right composition to achieve a low melting point. This is desirable, for example, in soldering materials or obtaining alloys that can be easily melted for casting.

11.4 Ionic Solutions

Raoult's law and Henry's law allow us to estimate activities for the components of ideal solutions by referring to the chemical potential of the vapor phase. The solutes in Table 11.1 are all gases or liquids, because solids should be bound by fairly strong intermolecular forces, too strong to easily dissociate in solution. However, ionic solids provide a crucial exception to this. Often, the attractive forces between the ions and water are—surprisingly—comparable to the ion–ion binding forces of the solid. Add to that the increase in entropy obtained by allowing the ions to move about in solution, and ionic solids are very often found to be highly soluble in water. This is the one class of non-ideal solutions that we will consider carefully.

Aqueous solutions of ionic compounds tend to deviate far from ideal behavior, because (1) the charge–dipole interaction of the ions with the water is much stronger than the interaction between neutral solute and solvent, and

(2) the charge–charge interaction between the solvated ions is a relatively long-range interaction, influencing the motion of the solute even at relatively low concentrations. Ionic solutions play key roles in biochemistry, marine chemistry, and certain battery types. A detailed understanding of these systems from a microscopic level is fairly complicated, but ionic solutions are so important that many aspects of the thermodynamics are well-established.

TOOLS OF THE TRADE | Cyclic Voltammetry

The development of electrochemistry as an analytical technique took place gradually over roughly 150 years, from Alexander Volta's experiments with chemical batteries in 1800, through Michael Faraday's brilliant insights into the relationship between electrical current and numbers of particles in the 1830s, and well into the 20th century. Along the way, Walther Nernst published, in his 1887 dissertation, the relationship between ion activities and electrostatic potential, which accelerated the application of chemical thermodynamics to electrochemical reactions. In 1948, John E. B. Randles and A. Sevčik independently published their extensions of existing electrochemical methods to the technique of **cyclic voltammetry,** based on a new instrument—the three-electrode **potentiostat**—which had been invented in 1942. Cyclic voltammetry subsequently grew to become the principal electroanalytical technique in use today.

What is cyclic voltammetry? Cyclic voltammetry is the measurement of the current in an electrochemical cell as the voltage between the working and reference electrodes is varied up and down past both the reduction and oxidation potentials of the cell before returning to its original value. Ideally, this creates a closed curve that shows at what electrical potentials the various reduction and oxidation events take place.

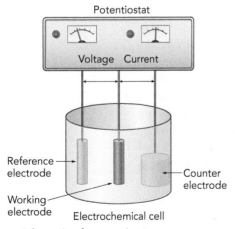

▲ **Schematic of a potentiostat.**

Why do we use cyclic voltammetry? From a cyclic voltammogram (CV) we can rapidly determine the redox potentials of a reaction under investigation and whether or not the reaction is reversible. These data alone often help to assess a proposed mechanism for the reaction, but in addition the CV is rich with information about the dynamics of the reaction. Quirks of all kinds in the reaction pathway appear in the CV and make its successful analysis a rewarding but challenging prospect.

How does it work? Cyclic voltammetry is carried out with a potentiostat, which controls the voltages and records the currents of a set of three electrodes:

1. The **working electrode,** where the reaction takes place.
2. The **reference electrode,** which is maintained at a constant potential. The reference potential is usually set by a well-calibrated redox reaction, such as the formation of AgCl from silver metal and chloride ions.
3. The **counter electrode** or **auxiliary electrode,** which provides the other half of the circuit needed for current to flow through the working electrode. The auxiliary electrode is given a much larger surface area than the reference electrode, to reduce the time required for ions to diffuse to the electrode surface and react. This strategy helps to prevent the dynamics measured at the working electrode from being significantly affected by the dynamics at the auxiliary electrode.

Cyclic voltammetry is used to describe many classes of redox reactions, but we'll assume for now a simple reaction that converts a group of positive ions M^+ to the neutral species M, and then converts them back to M^+ by the end of the scan. In a typical experiment, the potentiostat sets the voltage of the working electrode (relative to the reference electrode) at a high initial value, and then steadily decreases the voltage at a predetermined scan rate in the range of 10–100 mV s^{-1}. As the voltage at the working electrode decreases, positive ions are increasingly attracted to it, and when the voltage reaches the reduction potential, electrons will transfer from the electrode to the ions: the ions are *reduced* (the oxidation number of the ion goes down). This flow of electrons constitutes an electrical current, which is graphed by the potentiostat as a *voltammogram,* a plot of the current as a function of the voltage. The current registered in this section of the voltammogram is called the *cathodic current,* because the working electrode is acting as a cathode.

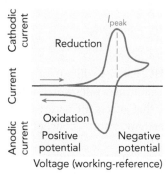

▲ **A sample cyclic voltammogram.**

Once most of the ions have been reduced, the current begins to drop. The potentiostat continues to decrease the voltage until it reaches the switching potential, and then it begins to increase the voltage. At this point, most of the ions have been reduced, so the current continues to drop. As the voltage rises, however, the working electrode becomes attractive to the electrons on the reduced species, and eventually current begins to flow in the *opposite direction,* as the reduced species become *oxidized* back to their original form. The signal from this *anodic current* appears in the voltammogram with the opposite sign during part of this sweep. When the voltage returns to its initial value, the current should be close to its original value.

Why does the current change gradually over this scan, instead of flowing all at once when we reach the appropriate reduction or oxidation potential? The current that flows is measured at the working electrode, and not all the ions can be present at the electrode at once. A negative voltage at the electrode will gradually attract the ions, but they travel largely by diffusion, which involves a high degree of random motion. Diffusion is slow compared to the timescale of the electron transfer itself. Stirring the solution will hasten the ions reaching the electrode, but even then the electrode becomes increasingly surrounded by a layer of reduced species, which increases the time it takes for the remaining ions to diffuse to the electrode surface.

As an example of how the CV can tell us about the dynamics of the system, the Randles-Sevčik equation gives the peak current I_{peak} in amperes as a function of the scan rate:

$$I_{\text{peak}}(\text{A}) = 0.4463AC\sqrt{\frac{\mathcal{F}^3 Dv}{RT}},$$

where A is the electrode surface area in cm^2, C the concentration of the species in mol cm^{-3}, D the diffusion constant in cm^2 s^{-1}, and v the scan rate in V s^{-1}. The amount of charge is quantified by the Faraday constant $\mathcal{F}$ (Eq. 11.70). We have assumed that only one electron is transferred per particle. (The concentration units are mol cm^{-3}, which ensures that the units work out to amperes.) The faster the scan (the greater v), the more rapidly charge is transferred to and from the electrode, and the greater the current (which is the *rate* of charge transfer). From the Randles-Sevčik equation, the diffusion constant of the species may be estimated, and in this way the CV serves as a probe not only of the reaction mechanism but also the mass transport properties of the solution.

Activities of Ionic Solutions

One of the defining characteristics of ionic solutions is that the solute dissociates to some extent when it is solvated, separating into ions of opposite charge. Unlike the solutes we have discussed so far, the concentration of individual solute species in the solution is not given by the molarity of the solution. For example, HCl dissociates to H^+ and Cl^- ions in water.[6] At equilibrium, the chemical potentials of solvated HCl and gas-phase HCl must be equal, but the solvated HCl corresponds (effectively) to *two different solutes:*

$$\mu_{\text{HCl(gas)}} = \mu_{\text{HCl(aq)}}$$

$$\mu_{\text{HCl(gas)}}^{\ominus} + RT\ln a_{\text{HCl(gas)}} = \mu_{\text{HCl(aq)}}^{\ominus} + RT\ln a_{\text{HCl(aq)}}$$

$$= \mu_{\text{HCl(aq)}}^{\ominus} + RT\big[\ln a_{\text{H}^+\text{(aq)}} + \ln a_{\text{Cl}^-\text{(aq)}}\big].$$

---

[6]Let us acknowledge that all ions in aqueous solution interact strongly with the water and not worry about whether or not aqueous H^+ is more properly written H_3O^+.

Using the activity coefficient γ, we find that the activity of aqueous HCl is related to the *square* of its mole fraction:

$$\ln a_{\text{HCl(aq)}} = \ln a_{\text{H}^+\text{(aq)}} + \ln a_{\text{Cl}^-\text{(aq)}}$$

$$= \ln\left[a_{\text{H}^+\text{(aq)}}a_{\text{Cl}^-\text{(aq)}}\right]$$

$$= \ln\left[\gamma_{\text{H}^+\text{(aq)}}X_{\text{H}^+\text{(aq)}}\gamma_{\text{Cl}^-\text{(aq)}}X_{\text{Cl}^-\text{(aq)}}\right] \qquad \text{by Eq. 11.43}$$

$$= \ln\left[\gamma_{\text{H}^+\text{(aq)}}\gamma_{\text{Cl}^-\text{(aq)}}X^2_{\text{HCl(aq)}}\right]$$

$$a_{\text{HCl(aq)}} = \gamma_{\text{H}^+\text{(aq)}}\gamma_{\text{Cl}^-\text{(aq)}}X^2_{\text{HCl(aq)}}. \qquad (11.45)$$

Even if we could set the γ values to 1, the HCl activity will not follow Henry's law. For those of us hoping for a simple and elegant theory, it's worse still, because we can expect the γ's to change with anything that affects the nature of the intermolecular forces in the solution, including the temperature, concentration, solvent, and the identity of the other ions present.

To make this general, an ionic compound with chemical formula $A_{v_+}B_{v_-}$ will have an activity that is the product of its single-ion activities,

$$a_{A_{v_+}B_{v_-}} = a_+^{v_+}a_-^{v_-} = (\gamma_+X_+)^{v_+}(\gamma_-X_-)^{v_-}, \qquad (11.46)$$

where v_+ and v_- are the coefficients of the cations and anions, respectively, released by dissociation of the ionic compound. The mole fractions of the ions are related to the mole fraction of compound by

$$X_+ = v_+X_{A_{v_+}B_{v_-}} \quad X_- = v_-X_{A_{v_+}B_{v_-}}, \qquad (11.47)$$

so we can write the activity in terms of the overall amount of compound:

$$a_{A_{v_+}B_{v_-}\text{(aq)}} = (\gamma_+v_+X_{A_{v_+}B_{v_-}})^{v_+}(\gamma_-v_-X_{A_{v_+}B_{v_-}})^{v_-} = (\gamma_+v_+)^{v_+}(\gamma_-v_-)^{v_-}X^{v_++v_-}_{A_{v_+}B_{v_-}}. \qquad (11.48)$$

The activity coefficients γ_+ for the cation and γ_- for the anion are rarely measured independently, because in typical solutions the cations and anions are both present at the same time.[7] Instead, we use a **mean ionic activity coefficient,** $\gamma_\pm$, such that

$$\gamma_\pm \equiv (\gamma_+^{v_+}\gamma_-^{v_-})^{1/(v_++v_-)}. \qquad (11.49)$$

Substituting Eq. 11.49 into Eq. 11.48 gives

$$a_{A_{v_+}B_{v_-}\text{(aq)}} = \left(\gamma_\pm X_{A_{v_+}B_{v_-}}\right)^{v_++v_-}. \qquad (11.50)$$

Although the mole fraction makes it easier to write this expression, the practical application will normally involve converting from the molarity (equal to $X/V_{\text{m,soln}}$) or the molality (equal to $Xn_{\text{soln}}/m_{\text{solvent}}$) instead.

For example, for HCl, $v_+ = v_- = 1$, because each HCl dissociates into one H^+ and one Cl^-. For this example,

$$\gamma_{\text{HCl}\pm} = (\gamma_{\text{H}^+\text{(aq)}}\gamma_{\text{Cl}^-\text{(aq)}})^{1/2}, \qquad (11.51)$$

and the activity of aqueous HCl can be written

$$a_{\text{HCl(aq)}} = \gamma_{\text{H}^+\text{(aq)}}\gamma_{\text{Cl}^-\text{(aq)}}X^2_{\text{HCl(aq)}} = \gamma^2_{\text{HCl}\pm}X^2_{\text{HCl(aq)}}. \qquad (11.52)$$

---

[7]Our ability to measure the **single-ion activity** of ions in solution is a discussion topic of continuing interest. The concept was historically shunned because single-ion activities are not generally determinable by experiment, but they play a key role in understanding the pH of concentrated acids and many other features of ionic solutions. In recent years, the availability of ion-selective electrodes and ion-exchange membranes has given the concept more traction.

The critical qualitative result for now is that we expect the activity of HCl to vary as the *square* of its concentration.

Unlike many other ionic compounds, HCl is quite volatile, and the pressure of HCl gas above the solution is straightforward to measure. From our analysis, we would predict that, at equilibrium, the partial pressure of the HCl varies as the square of its concentration in the solution:

$$\mu_{HCl(aq)} = \mu_{HCl(gas)}$$

$$\mu^{\ominus}_{HCl(aq)} + RT\ln a_{HCl(aq)} = \mu^{\ominus}_{HCl(gas)} + RT\ln a_{HCl(gas)}$$

$$\mu^{\ominus}_{HCl(aq)} + RT\ln\left[\gamma_{HCl\pm}X_{HCl(aq)}\right]^2 = \mu^{\ominus}_{HCl(gas)} + RT\ln\frac{P_{HCl(gas)}}{P^{\ominus}}$$

$$\frac{\mu^{\ominus}_{HCl(aq)} - \mu^{\ominus}_{HCl(gas)}}{RT} = \ln\frac{P_{HCl(gas)}}{P^{\ominus}} - \ln\left[\gamma_{HCl\pm}X_{HCl(aq)}\right]^2$$

$$= \ln\left\{\frac{P_{HCl(gas)}}{P^{\ominus}}\left[\gamma_{HCl\pm}X_{HCl(aq)}\right]^{-2}\right\}$$

$$\exp\left(\frac{\mu^{\ominus}_{HCl(aq)} - \mu^{\ominus}_{HCl(gas)}}{RT}\right) = \left\{\frac{P_{HCl(gas)}}{P^{\ominus}}\left[\gamma_{HCl\pm}X_{HCl(aq)}\right]^{-2}\right\}$$

$$\frac{P_{HCl(gas)}}{P^{\ominus}} = \left[\gamma_{HCl\pm}X_{HCl(aq)}\right]^2\exp\left(\frac{\mu^{\ominus}_{HCl(aq)} - \mu^{\ominus}_{HCl(gas)}}{RT}\right). \qquad (11.53)$$

Since the standard state chemical potentials and activity coefficients are all (at least roughly) constants, the partial pressure of HCl varies as $X^2_{HCl(aq)}$, which is proportional to the square of the concentration. Sure enough, this represents fairly well the variation in partial pressure of HCl with concentration, illustrated in Fig. 11.6.

SAMPLE CALCULATION **Activity of an Aqueous Ionic Compound.** To find an equation for the activity of aqueous Na_3PO_4 in terms of its mole fraction, we count the ions. Each sodium phosphate dissociates into three Na^+ ions and one PO_4^{3-} ion, so Eq. 11.50 gives

$$a_{Na_3PO_4(aq)} = \gamma^4_{Na_3PO_4\pm}X^4_{Na_3PO_4(aq)}. \qquad (11.54)$$

The activity in this case varies steeply, as the fourth power of the concentration.

We've been making things a little easier by looking first at ionic compounds that dissociate almost completely in water. What happens when we have a weak acid HA, for example, which only partially ionizes to H^+ and A^-? In that case, we can still equate the chemical potentials of solute in the gas and solution phases, but the activity in solution now consists of three terms: the activity of the cation, the activity of the anion, and the activity of the undissociated HA. We can still use the stoichiometry of the molecular formula to set the amount of H^+ equal to the amount of A^-, but the relative amount of HA will depend on the equilibrium between the neutral HA and the ions it forms. If HA is a neutral molecule, and the concentration is not too high, then we may be able to treat HA

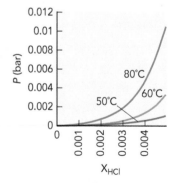

▲ FIGURE 11.6 **The vapor pressure of HCl above a solution.** The pressure is graphed as a function of mole fraction in aqueous solution at selected temperatures.

effectively using Henry's law and set γ_{HA} (the activity coefficient for undissociated HA) to 1. The resulting expression is

$$a_{HA(aq)} = \gamma_{HA\pm}^2 X_{H^+(aq)} X_{A^-(aq)} X_{HA}, \tag{11.55}$$

where X_{HA} is the mole fraction of undissociated HA. Typically, as we add a weak acid HA to solution, the product $X_{H^+(aq)} X_{A^-(aq)}$ is proportional to the mole fraction of HA. Therefore, the activity of the weak acid is still roughly proportional to X_{HA}^2.

11.5 Applications of the Activity

With these simplifying tools at our disposal, we can apply the activity to some basic properties of solutions. Collectively, these are known as the **colligative properties,** thermodynamic parameters of the system that vary with the relative *concentration* of the solute but *not* with the solute's chemical composition.

Freezing Point Depression

One fundamental impact of solvation is the shift in freezing and boiling points of the solution relative to the solvent. The solution is thermodynamically stable only if it is more stable than the separated solvent and solute, and this remains true as the solution approaches the phase transition temperatures of the solvent. For example, at the standard boiling point of the solvent, more energy is required to force the solvent into the gas phase from the solution than from the pure solvent; that is, the boiling point will increase. Similarly, the solution remains stable as a liquid at lower temperatures than the pure solvent; the freezing point *decreases*.[8]

At the freezing point of a solution,

$$\mu_A(\text{sol}) = \mu_A(\text{soln}) = \mu_A^\bullet(\text{liq}) + RT \ln a_A. \tag{11.56}$$

The chemical potential μ of the reference state for the solvent is $\mu_A^\bullet(\text{liq})$, the chemical potential of the pure liquid, in agreement with Raoult's law. Solving for the logarithm of the activity,

$$\ln a_A = \frac{1}{R}\left(\frac{\mu_A(\text{sol})}{T} - \frac{\mu_A^\bullet(\text{liq})}{T}\right), \tag{11.57}$$

and taking the derivative with respect to temperature (to find the shift in T as we change the activity), we get

$$\frac{\partial \ln a_A}{\partial T} = \frac{1}{R}\left[-\frac{\mu_A(\text{sol})}{T^2} + \frac{1}{T}\frac{\partial \mu_A(\text{sol})}{\partial T} - \left(-\frac{\mu_A^\bullet(\text{liq})}{T^2} + \frac{1}{T}\frac{\partial \mu_A^\bullet(\text{liq})}{\partial T}\right)\right]$$

$$= \frac{1}{R}\left[-\frac{G_A(\text{sol})}{T^2} - \frac{S_A(\text{sol})}{T} - \left(-\frac{G_A^\bullet(\text{liq})}{T^2} - \frac{S_A^\bullet(\text{liq})}{T}\right)\right]. \tag{11.58}$$

We took advantage of Eqs. 10.28 and 10.26 to eliminate the chemical potentials.

---

[8]Whereas solutes may be stabilized by the liquid solvent, they are less likely to be stabilized in the solid. Isolated solute molecules would tend to disrupt the solvent's crystal structure, and the components of the solution normally separate upon freezing. The solute, in other words, is not readily soluble in the solid.

Now we can find the change in Gibbs energy and entropy at the phase transition:

$$\Delta_{fus}G^{\bullet} = G^{\bullet}(liq) - G^{\bullet}(sol) \approx G^{\bullet}(liq) - G(sol) \quad (11.59)$$

$$\Delta_{fus}S^{\bullet} = S^{\bullet}(liq) - S^{\bullet}(sol) \approx S^{\bullet}(liq) - S(sol). \quad (11.60)$$

Although we are not dealing with the pure solid in this case, we will assume that the solvent dominates the phase transition properties. Therefore,

$$\frac{\partial \ln a_A}{\partial T} = \frac{1}{R}\left(\frac{\Delta_{fus}G_A^{\bullet}}{T^2} + \frac{\Delta_{fus}S_A^{\bullet}}{T}\right)$$

$$= \frac{1}{R}\left(\frac{\Delta_{fus}H_A^{\bullet} - T\Delta_{fus}S_A^{\bullet}}{T^2} + \frac{\Delta_{fus}S_A^{\bullet}}{T}\right) \quad \Delta G = \Delta H - T\Delta S$$

$$= \frac{\Delta_{fus}H_A^{\bullet}}{RT^2}. \quad (11.61)$$

What makes this useful is that we can put all the temperature dependence on one side and integrate to obtain the shift in freezing point. If we label the freezing points T_f for the solution and $T_f^{\bullet}$ for the pure solvent, we find

$$\ln a_A = \int_{T_f^{\bullet}}^{T_f} \frac{\Delta_{fus}H_A^{\bullet}}{RT^2} dT \quad (11.62)$$

$$= \frac{\Delta_{fus}H_A^{\bullet}}{R}\left(\frac{1}{T_f^{\bullet}} - \frac{1}{T_f}\right) = \frac{\Delta_{fus}H_A^{\bullet}}{R}\left(\frac{T_f - T_f^{\bullet}}{T_f^{\bullet}T_f}\right).$$

The shift in freezing point ΔT_f is

$$\Delta T_f \equiv T_f - T_f^{\bullet} = \frac{RT_f^{\bullet}T_f \ln a_A}{\Delta_{fus}H_A^{\bullet}}, \quad (11.63)$$

where we have assumed $\Delta_{fus}H_A^{\bullet}$ is constant. The assumption is a safe one, because the shift in freezing point ΔT_f is usually very small compared to T_f.

For a low concentration of solute, we can use the dilute solution limit where Raoult's law describes the solvent adequately, so

$$\ln a_A \approx \ln X_A = \ln(1 - X_B) \approx -X_B, \quad (11.64)$$

where $\ln x \approx (x - 1)$ (Eq. A.30) yields the final approximation. If we further assume $T_f \approx T_f^{\bullet}$, the result is

$$\Delta T_f \approx -\frac{RT_f^{\bullet 2}X_B}{\Delta_{fus}H_A^{\bullet}}, \quad (11.65)$$

where we have also made the substitution $T_f \equiv T_f^{\bullet}$, which is valid for small shifts in the freezing point. The value is negative, indicating that addition of solute lowers the freezing point. This result is often phrased in terms of the molality m of the solution and a freezing point depression constant K_f:

$$\Delta T = -K_f m = -\frac{RT_f^2 \mathcal{M}_A}{\Delta_{fus}H^{\ominus}(1000)} m$$

(where the factor of 1000 is normally included to remind us to convert the mass units from grams to kilograms, as required by the definition of the molality).

Freezing point depression was an important technique for obtaining approximate molecular weights of non-volatile materials (for gases one would use the ideal gas law to determine n), until sedimentation and electrophoretic techniques were developed. Once ΔT_f determined the mole fraction X_B, the molecular weight of B could be deduced from the mass of solute used to make the solution.

EXAMPLE 11.1   Freezing Point Depression

CONTEXT Seawater is full of dissolved ions, predominantly sodium and chloride ions at an effective NaCl molarity of roughly 0.60 M (about a tenth of the saturation limit). The high ion concentration lowers the freezing point of seawater relative to freshwater. This is one of the factors that prevents the Arctic Ocean from freezing completely in winter, although a much more important effect is the circulation of warm water from the Atlantic.
Nonetheless, the change in freezing point with salinity leads to some complex and elegant dynamics in the Arctic. As one example, consider what happens below the ice when seawater begins to freeze on the ocean surface. As the water freezes, salt that separates from the ice crystals increases the salinity of the water beneath the ice, forming *brine*. The brine is more salty than the water that froze initially, so it continues to cool below the freezing point of the seawater without freezing itself. And, because it's more dense than rest of the seawater that surrounds it, it tends to accumulate near the bottom of the ice floe. Once this accumulation reaches sufficient mass, the brine drops down toward the ocean floor in a narrow stream. As it does so, it *freezes* the seawater around it, forming icy tubes called *ice stalactites* or *brinicles* that stretch down from the ice floe above. The brine is warming as it extracts heat from the surrounding seawater, and it eventually rises above the freezing point and the brinicle ends. Sometimes, however, it reaches the seafloor. In much the way that cool air and warm air dictate the dynamics of the atmosphere, we see here an example of how differences in salinity contribute to the circulation of water in the ocean.

PROBLEM Estimate the freezing point of a 0.600 M aqueous solution of NaCl. Near the freezing point, the density of pure water is $0.997\,\mathrm{g\,cm^{-3}}$.

SOLUTION With water as the solvent, we use $T_f^{\bullet} = 273.15\,\mathrm{K}$ and $\Delta_{\mathrm{fus}}H_A^{\bullet} = 6008\,\mathrm{J\,mol^{-1}}$. The other number we need is the mole fraction of NaCl X_B. This is *not* the same as the molarity, which is moles of solute per liter of solution. We need moles of solute per mole of solution, and we use the molarity of pure water,

$$\frac{0.997\,\mathrm{g\,cm^{-3}}}{18.015\,\mathrm{g\,mol^{-1}}} = 0.0553\,\mathrm{mol\,cm^{-3}} = 55.3\,M,$$

to find $X_{\mathrm{NaCl}} = 0.600/55.3 = 0.0108$. However, because NaCl dissociates completely in water into $\mathrm{Na^+}$ and $\mathrm{Cl^-}$ ions, the mole fraction of particles in solution is twice the mole fraction of NaCl added, and the value we want is $X_B = 2(0.0108) = 0.0216$. We therefore estimate the change in the freezing point to be

$$\Delta T_f \approx -\frac{(8.3145\,\mathrm{J\,K^{-1}\,mol^{-1}})(273.15\,\mathrm{K})^2(0.0216)}{6008\,\mathrm{J\,mol^{-1}}} = -2.23\,\mathrm{K}.$$

The final freezing point we estimate is therefore

$$273.15\,\mathrm{K} - 2.23\,\mathrm{K} = 270.92\,\mathrm{K}.$$

So the freezing point of the seawater surrounding the brinicle is about 2°C cooler than for fresh water. The freezing point of the brine, which can have ion concentrations several times higher than this, will be several degrees lower.

Osmotic Pressure

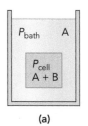

(a)

Another application of the chemical potential is to the analysis of **osmosis**, the flow of a solvent through a **semipermeable membrane** that blocks the flow of other substances. Consider one container (the cell) inside another (the bath), with the wall of the cell a semipermeable membrane (Fig. 11.7). Inside the cell we put a solute that cannot flow through the membrane, and put a freely flowing solvent outside. At equilibrium, the chemical potentials of the solvent inside and outside the cell must be equal:

$$\mu_A^{cell} = \mu_A^{bath}. \tag{11.66}$$

The solute cannot communicate with the rest of the sample across the membrane, and therefore its chemical potential needn't be balanced in the bath.

Returning to the solvent, the chemical potential in the bath is that of the pure solvent at the pressure P of the bath:

$$\mu_A^{bath} = \mu_A^\bullet(P), \tag{11.67}$$

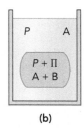

(b)

▲ FIGURE 11.7 **A cell with semipermeable membrane in a solvent bath.** Initially **(a)** the cell is at the same pressure as the solvent A. However, the presence of a solute B in the cell draws additional solvent into the cell **(b)** in order to balance the chemical potential of the cell and the bath. As a result, the pressure in the cell increases by an amount Π, the osmotic pressure.

but the chemical potential in the cell is a little more complicated. We cannot hold the pressure fixed, because as solvent flows into the cell, the pressure inside increases because the solute cannot flow out to compensate.

Having gotten used to pressures equilibrating in our systems, this could seem counter-intuitive: how can the system equilibrate by making the pressures *unequal*? The osmotic cell provides a perfect example of *chemical* work being done. The forces must balance, intuition tells us, and they do balance, but the forces are no longer limited to the elastic collisions of one group of molecules against another. In the osmotic cell, a strong attraction between solute and solvent molecules pulls molecules in from the surrounding bath, stopping when the excess pressure in the cell pushes back. This excess pressure in the cell we call the **osmotic pressure,** Π:

$$\mu_A^{cell} = \mu_A^\bullet(P + \Pi) + RT \ln a_A. \tag{11.68}$$

We still have to evaluate the impact of the pressure change on $\mu_A^\bullet$. If we assume the liquid to be incompressible, we can use the equation for the chemical potential of an incompressible substance derived earlier (Eq. 11.18) to get the chemical potential of the pure solvent at pressure P:

$$\mu_A^\bullet(P) = \mu_A^\bullet(P^\ominus) + V_m(P - P^\ominus).$$

The same equation leads to a similar equation for the potential of A at the elevated pressure in the cell:

$$\mu_A^\bullet(P + \Pi) = \mu_A^\bullet(P^\ominus) + V_m(P + \Pi - P^\ominus)$$

$$\mu_A^{cell} = \mu_A^\bullet(P^\ominus) + V_m(P + \Pi - P^\ominus) + RT \ln a_A.$$

Equating the chemical potentials inside and outside the cell wall gives

$$\mu_A^{bath} = \mu_A^\bullet(P)$$
$$= \mu_A^\bullet(P^\ominus) + V_m(P - P^\ominus)$$
$$= \mu_A^{cell} = \mu_A^\bullet(P^\ominus) + V_m(P + \Pi - P^\ominus) + RT \ln a_A.$$

The terms $\mu_A^\bullet(P^\ominus)$ and $V_m(P - P^\ominus)$ can be dropped from both sides. Raoult's law sets $\ln a_A$ equal to $\ln X_A \approx -X_B$ as before, and we solve for the osmotic pressure Π:

$$\Pi = \frac{RT X_B}{V_m} = RT[B], \tag{11.69}$$

where [B] is the molarity of the solute inside the cell. The excess or osmotic pressure inside the cell is directly proportional to the concentration of the solution.

The Nernst Equation

A more specific application of the osmosis problem applies to the case when the membrane allows some ions to pass but not others, allowing a charge difference to build up across the membrane. This mechanism is used by biological cells to receive and interpret signals from the rest of the organism.

To see how our recent exploration of the activities can give us useful tools for analyzing cell function, let's mimic the cell by putting an ionic solution inside the cell, separated again by a semipermeable membrane from a neutral solvent in the bath. This time, we allow the membrane to pass the solvent and only the *positive* ions from the solution. For this example, we'll allow the cell to expand to maintain a constant pressure on both sides of the cell (so there will be no osmotic pressure). What will happen? Entropy will drive the positive ions to disperse into the solvent on the other side of the membrane. But as this happens the imbalance between positive and negative charges across the membrane will generate a potential energy difference across the membrane that resists the loss of positive ions into the bath. At equilibrium, some balance must be struck between the drive for dispersion and the drive to keep the positive ions grouped with the negative ions in the cell. This equilibrium membrane potential helps to regulate the ion current into and out of real living cells, so we would like to be able to predict its magnitude. The chemical potential is again where we find our initial foothold on the problem.

For the ions that can flow across the membrane, we set the charge on each particle to ze, where z is the unitless ionic charge number: $z = +1$ for Na^+, $z = +2$ for Ca^{2+}, and so on. The charge *per mole* of these ions may therefore be written as $z\mathcal{F}$, where the **Faraday constant** $\mathcal{F}$,

$$\mathcal{F} = \mathcal{N}_A e = 96{,}485.34 \text{ C mol}^{-1}, \tag{11.70}$$

gives the charge in coulombs of one mole of protons (or -1 times the charge of a mole of electrons).

By allowing these positive ions to flow out of the cell, we build up a positive charge in the bath and a net negative charge in the cell (from the excess anions that are left behind). We've introduced another contribution to the energy of the system: the electrostatic potential energy. This term can be calculated by multiplying the voltage Φ at any given location by the charge at that location.[9] The molar energy of the system therefore acquires a new contribution of $\Phi\mathcal{F}z$. Since the chemical potential is the molar Gibbs free energy per particle under

---

[9]It would be more common to use V for voltage here, but we often need to use V for volume, so Φ for voltage may save some confusion. Voltage, just to remind ourselves, is not exactly equal to the electrostatic potential. In SI units, $1 \text{ V} = 1 \text{ J C}^{-1}$, so voltage is the electrostatic potential energy *per unit charge*. The work required to move an ion with charge ze through a voltage difference $\Delta\Phi$ is $ze\Delta\Phi$.

conditions of constant temperature and pressure, we add this new term to the activity-dependent chemical potential (Eq. 11.19), obtaining

$$\mu = \mu^{\ominus} + RT \ln a + \Phi \mathcal{F} z,$$

and we will make our usual approximation that the activity a of the solute can be replaced by the mole fraction X (Eq. 11.38):

$$\mu = \mu^{\ominus} + RT \ln X + \Phi \mathcal{F} z.$$

From this equation, we can solve for the voltage difference across the membrane by setting the chemical potentials for the ion on either side of the membrane equal to each other:

$$\mu^{\text{cell}} = \mu^{\text{bath}}$$

$$\mu^{\ominus} + RT \ln X^{\text{cell}} + \Phi^{\text{cell}} \mathcal{F} z = \mu^{\ominus} + RT \ln X^{\text{bath}} + \Phi^{\text{bath}} \mathcal{F} z$$

$$\Phi^{\text{cell}} \mathcal{F} z - \Phi^{\text{bath}} \mathcal{F} z = RT \ln X^{\text{bath}} - RT \ln X^{\text{cell}}$$

$$\Phi^{\text{cell}} - \Phi^{\text{bath}} = \frac{RT}{\mathcal{F} z} \left(\ln X^{\text{bath}} - \ln X^{\text{cell}} \right).$$

Rewriting $\Phi^{\text{cell}} - \Phi^{\text{bath}}$ as the equilibrium membrane voltage $\Delta \Phi_{\text{eq}}$ and combining the logarithms on the left side of the equation, we arrive at the **Nernst equation:**

$$\Delta \Phi_{\text{eq}} = \frac{RT}{\mathcal{F} z} \ln \frac{X^{\text{bath}}}{X^{\text{cell}}}. \tag{11.71}$$

In many references this value would be called the **membrane potential** for the ion, but keep in mind that it is expressed as a voltage (in V or J C^{-1}) rather than as a potential energy (in J).

SAMPLE CALCULATION **The Nernst Equation.** To find the equilibrium voltage difference for Ca^{2+} across a membrane at 303 K when the concentrations inside and outside the cell are 0.200 μM and 2.50 mM, respectively, we substitute these values into the Nernst equation, setting $z = +2$ for Ca^{2+}:

$$\Delta \Phi_{\text{eq}} = \frac{RT}{\mathcal{F} z} \ln \frac{X^{\text{bath}}}{X^{\text{cell}}}$$

$$= \frac{(8.3145 \, \text{J K}^{-1} \text{mol}^{-1})(303 \, \text{K})}{(96{,}485.34 \, \text{C mol}^{-1})(2)} \ln \frac{2.00 \cdot 10^{-7}}{2.50 \cdot 10^{-3}} = -0.123 \, V.$$

The Nernst equation predicts the membrane voltage for an individual ion, based on its concentrations on opposite sides of the cell membrane. If the channels permitting that ion to travel in and out of the cell are open, the ion will soon reach its equilibrium distribution across the membrane, and its ion flow will stop. If the concentration then changes on either side of the membrane, the ions or the solvent (as in osmosis) will flow until the voltage difference satisfies the Nernst equation. When other ions are present, the overall membrane potential must be obtained from the sum of their contributions.

11.6 The Kirkwood-Buff Theory of Solutions

We close this chapter by reviewing in brief the thinking behind an especially useful model for solutions as it applies to a two-component system of substances A and B. Arguably the principal aim of the Kirkwood-Buff theory is to incorporate the dynamic behavior of the solution into the classical thermodynamics that we would normally apply to the system. There are many related ways to assess these fluctuations, and K-B theory considers the fluctuations to be changes in the number of particles of A and/or B within any given volume V.

This approach indicates that we should work within the **grand canonical ensemble,** which is the set of all microstates of the system having fixed values of T, V, μ_A, and μ_B. In this ensemble, we fix the chemical potentials but *not* the numbers of particles N_A and N_B. As a result, our system is an *open* system: particles can enter and exit. The fluctuations in N_A and N_B cause variations in the number densities ρ_A and ρ_B, so we calculate instead the ensemble averages of the number densities, $\langle \rho_A \rangle$ and $\langle \rho_B \rangle$. These parameters in turn can be related through the volume (which is a constant) to the ensemble averaged values of N:

$$\langle N_A \rangle = \langle \rho_A \rangle V, \quad \langle N_B \rangle = \langle \rho_B \rangle V. \quad (11.72)$$

In K-B theory, we initially treat all the particles at once, whether each belongs to substance A or substance B. If we let i and j equal either A or B, then we can define the pair correlation function $\mathcal{G}_{ij}(R)$ to give the likelihood of finding molecules of substance i at a distance R from any randomly selected molecule of substance j. If $i = j$, then $\mathcal{G}_{ij}$ is a measure of the interactions between like molecules of A or B, and if $i \neq j$, then $\mathcal{G}_{ij}$ probes the interactions between different chemical species, such as solvent and solute. Recall that the pair correlation function in a typical liquid oscillates as it gradually approaches a value of 1, indicating that the relative positions of the molecules become more random as the distance between them increases.

It's worthwhile to keep track of the coordinates carefully here. The pair correlation functions depend only on the distances R between particles, not on the particular location of any particle in the system. On the other hand, the probability distribution functions $\mathcal{P}_V(X_i, Y_i, Z_i)$ are functions of the specific position of each particle i. For example, we can calculate the number of particles N_i of either substance by integrating over the corresponding probability distribution:

$$N_i = \int_{-\infty}^{\infty} \int_{-\infty}^{\infty} \int_{-\infty}^{\infty} \mathcal{P}_V(X_i, Y_i, Z_i) \, dX_i dY_i dZ_i. \quad (11.73)$$

In addition to the one-particle distribution function $\mathcal{P}_V(X_i, Y_i, Z_i)$, we can also define a two-particle distribution function $\mathcal{P}_{V^2}(X_i \ldots Z_j)$, which gives us the chance per unit volume squared that any particle of type i is at position X_i, Y_i, Z_i while any particle of type j is at X_j, Y_j, Z_j. We can integrate this two-particle distribution function over all six coordinates, but the results depend on whether or not $i = j$:

$$\int_0^{\infty} \ldots \int_0^{\infty} \mathcal{P}_{V^2}(X_i, \ldots, Z_j) \, dX_i \ldots dZ_j = \langle N_i N_j \rangle \, \text{if } i \neq j \quad (11.74)$$

$$\int_0^\infty \ldots \int_0^\infty \mathcal{P}_{V^2}(X_i, \ldots, Z_j)dX_i \ldots dZ_j = \langle N_i(N_i - 1)\rangle. \ if \ i = j \quad (11.75)$$

In other words, if we are counting molecules of both species in volume V, then we count N_i molecules of type i and N_j molecules of type j. When we count up the two-particle distribution for molecules of the *same* species, then for each of the N_i possible locations of the first molecule there are only $N_i - 1$ locations remaining for the second molecule. The triangular brackets $\langle \ \rangle$ remind us that in either case the value we obtain is averaged over the microstates of the grand canonical ensemble, in which the number of particles of each type can vary.

The **Kronecker delta function** δ_{ij} (where $\delta_{ij} = 1$ if $i = j$ and $\delta_{ij} = 0$ if $i \neq j$) lets us combine Eqs. 11.75 and 11.74 into a single equation:

$$\int_0^\infty \ldots \int_0^\infty \mathcal{P}_{V^2}(X_i, \ldots, Z_j)dX_i \ldots dZ_j = \langle N_i N_j\rangle - \langle N_i\rangle\delta_{ij}. \quad (11.76)$$

Because both the pair correlation function $\mathcal{G}_{ij}(R)$ and this two-particle distribution function $\mathcal{P}_{V^2}$ gauge the distribution of particles by pairs, it should be possible to relate the two. In fact, we've already done most of this work in Section 4.3 to get Eq. 4.56:

$$(N - 1)\,\mathcal{P}_{V^2}(X_1, \ldots, Z_2)dR = 4\pi\rho\mathcal{G}(R)R^2 dR.$$

In this expression, $\mathcal{P}_{V^2}(X_1, \ldots, Z_2)$ is the probability that specific particles 1 and 2 are located at the points X_1, Y_1, Z_1 and X_2, Y_2, Z_2. We can solve for the pair correlation function, obtaining

$$\mathcal{G}(R) = \frac{(N - 1)}{4\pi\rho R^2}\mathcal{P}_{V^2}(X_1, \ldots, Z_2).$$

Our next job is to generalize this to the case of *any* two molecules in the system, to change the coordinates of the probability function from $X_1 \ldots Z_2$ to $X_i \ldots Z_j$, where X_i, Y_i, Z_i are the coordinates of any molecule of type i and X_j, Y_j, Z_j are the coordinates of any molecule of type j. For molecule 1, we may swap in any of N possible molecules, and for molecule 2 we may swap in any of the $N - 1$ remaining molecules:

$$\mathcal{P}_{V^2}(X_i, \ldots, Z_j) = N(N - 1)\mathcal{P}_{V^2}(X_1, \ldots, Z_2).$$

Therefore, our pair correlation function becomes

$$\mathcal{G}(R) = \frac{(N - 1)}{4\pi\rho R^2}\left(\frac{1}{N(N - 1)}\right)\mathcal{P}_{V^2}(X_i, \ldots, Z_j)$$

$$= \frac{1}{4\pi N\rho R^2}\mathcal{P}_{V^2}(X_i, \ldots, Z_j).$$

But which ρ and which N are these? We derived Eq. 4.56 when we had only a single-component system. If the pair correlation function is now $\mathcal{G}_{ij}(R)$ for the average distribution of species j around any one molecule i, then the number density that normalizes $\mathcal{G}(R)$ in Eq. 4.56 is for the surrounding molecules j. On the other hand, the factor of N represents the number of molecules that we can choose as the reference molecule in the pair correlation function, which is N_i. We must also recall that we are studying the states of the system in the grand canonical ensemble, for which N_i and N_j are not constants, so we should use the ensemble averages $\langle N_i \rangle$ and $\langle \rho_j \rangle$. Making these adaptations gives the following equation:

$$\mathcal{G}_{ij}(R) = \frac{1}{4\pi \langle N_i \rangle \langle \rho_j \rangle R^2} \mathcal{P}_{V^2}(X_i, \ldots, Z_j).$$

To make the expression more symmetric, we may replace ρ_j by N_j/V:

$$\mathcal{G}_{ij}(R) = \frac{V}{4\pi \langle N_i \rangle \langle N_j \rangle R^2} \mathcal{P}_{V^2}(X_i, \ldots, Z_j). \tag{11.77}$$

This equation is worth looking at for a moment, just to see what the pair correlation function represents from a different perspective. Solving for the probability in Eq. 11.77 gives

$$\mathcal{P}_{V^2}(X_i, \ldots, Z_j) = \frac{4\pi R^2}{V} \langle N_i \rangle \langle N_j \rangle \mathcal{G}_{ij}(R). \tag{11.78}$$

If the positions of two molecules are utterly *uncorrelated,* the two-particle distribution function is just the product of the single-particle distribution functions:

$$\mathcal{P}_{V^2}(X_i, \ldots, Z_j) = \mathcal{P}_V(X_i, Y_i, Z_i) \mathcal{P}_V(X_j, Y_j, Z_j),$$

and the pair correlation function would be 1 everywhere except near the origin where the reference molecule resides. Under these conditions, Eq. 11.78 simplifies to

$$\mathcal{P}_{V^2}(X_i, \ldots, Z_j) = \mathcal{P}_V(X_i, Y_i, Z_i) \mathcal{P}_V(X_j, Y_j, Z_j) \approx \frac{4\pi R^2}{V} \langle N_i \rangle \langle N_j \rangle. \tag{11.79}$$

In other words, the chance that two particles of types i and j are separated by a distance R is given by the product of the number densities of i and j, scaled by the ratio of the surface area ($4\pi R^2$) to the volume V.

Now turn on the intermolecular forces. At separations R_{ij} where the interaction favors a molecule of type j appearing near a molecule of type i, $\mathcal{G}_{ij}(R) > 1$. According to Eq. 11.78, we will see a corresponding increase in the probability $\mathcal{P}_{V^2}$. Similarly, when the interactions make it less likely for two molecules to be separated by a distance R, then $\mathcal{G}_{ij}(R) < 1$ and $\mathcal{P}_{V^2}$ drops below the value for the random case. The pair correlation function is, in essence, the two-particle distribution function normalized by the random distribution.

Kirkwood-Buff theory relates the thermodynamic parameters of the solution to a set of integrals (the *K-B integrals,* G_{ij}) that are most commonly defined in terms of the pair correlation function:[10]

$$G_{ij} \equiv 4\pi \mathcal{N}_A \int_0^\infty \left[\mathcal{G}_{ij}(R) - 1 \right] R^2 dR. \tag{11.80}$$

By using the integrand $\mathcal{G}_{ij}(R) - 1$, Kirkwood and Buff isolate the favorable and unfavorable contributions of the intermolecular forces to the structure of the solution. Where molecules are *likely* to be found the integrand is positive, and where they are *unlikely* to be found the integrand is negative. The integral then averages all of these effects over the entire range of separations between two molecules, but with the integrand vanishing at large R where $\mathcal{G}_{ij}(R)$ approaches 1.

---

[10]The K-B integrals are more often defined without the factor of $\mathcal{N}_A$, but including that factor now simplifies the extension of the integrals to the molar quantities we use in thermodynamics.

With the relationships we've found, we can obtain the central equation to the Kirkwood-Buff theory of solutions:

$$G_{ij} = 4\pi \mathcal{N}_A \int_0^\infty \left[\mathcal{G}_{ij}(R) - 1 \right] R^2 dR \qquad \text{by Eq. 11.80}$$

$$= 4\pi \mathcal{N}_A \int_0^\infty \left[\frac{V}{4\pi N_i N_j R^2} \mathcal{P}_{V^2}(X_i, \ldots, Z_j) - 1 \right] R^2 dR \qquad \text{by Eq. 11.77}$$

$$= \mathcal{N}_A V \left[\frac{\langle N_i N_j \rangle - \langle N_i \rangle \delta_{ij}}{\langle N_i \rangle \langle N_j \rangle} - 1 \right] \qquad \text{by Eq. 11.76}$$

$$= \mathcal{N}_A V \left[\frac{\langle N_i N_j \rangle - \langle N_i \rangle \langle N \rangle_j - \langle N_i \rangle \delta_{ij}}{\langle N_i \rangle \langle N_j \rangle} \right]. \quad \text{combine terms} \quad (11.81)$$

Our derivation shows how the K-B integrals depend on the difference between $\langle N_i N_j \rangle$ and $\langle N_i \rangle \langle N_j \rangle$. In the same way that we can find the value of the chemical potential from the partition function in the canonical ensemble Q (Eq. 7.38),

$$\mu = -k_B T \left(\frac{\partial \ln Q}{\partial n} \right)_{T,V},$$

we can also find the value of the $\langle N_i \rangle$ from the partition function in the grand canonical ensemble. This partition function, which we will label Ξ for the short time it's with us, is different from Q because allowing the number of particles to change affects the number of available microstates of the system. In a two-component solution of A and B, the expressions for the average numbers of particles are

$$\langle N_A \rangle = -k_B T \left(\frac{\partial \ln \Xi}{\partial \mu_A} \right)_{T,V,\mu_B} \qquad \langle N_B \rangle = -k_B T \left(\frac{\partial \ln \Xi}{\partial \mu_B} \right)_{T,V,\mu_A}. \quad (11.82)$$

Substituting Eqs. 11.82 into Eq. 11.81 allows the K-B integrals to be related directly to thermodynamic parameters because V, T, S, and the thermodynamic potentials can all be expressed in terms of the partition function, as we saw in Eqs. 7.35–7.38. We've also seen how the chemical potential can be used to quantify properties dependent on interactions within the solution, such as the partial molar volume V_{Bm} (Eq. 11.5) and the enthalpy of solvation $\Delta_{mix}H_m$ (Eq. 11.7). When applied to a two-component solution, K-B theory requires the evaluation of three integrals: G_{AA}, $G_{AB} = G_{BA}$, and G_{BB}. These, in turn, can be expressed in terms of thermodynamic parameters: the isothermal compressibility κ_T of the solution, the partial molar volume V_{im} of each substance i, and the parameter D such that,

$$D \equiv \frac{X_i}{k_B T} \left(\frac{\partial \mu_i}{\partial X_i} \right)_{T,P} = X_i \left(\frac{\partial \ln (P_i/P^{\ominus})}{\partial X_i} \right)_{T,P},$$

which can be determined experimentally from measurements of the vapor pressure P_i at different mole fractions X_i of either substance. Although we will not prove these relationships here, it can be shown by writing each of these parameters as derivatives with respect to N_A or N_B that the following equations hold:

$$G_{AA} = RT\kappa_T - V_{mA}^{\bullet} \left[1 - \frac{V_{Bm}^2 \rho_{mix}}{V_{mB}^{\bullet} D} \right]$$

CHECKPOINT The grand canonical ensemble preserves terms that involve particle fluctuations, which would be lost in the canonical or microcanonical ensembles of Chapter 2. If N_i and N_j were constants, then $\langle N_i N_j \rangle$ would be equal to $\langle N_i \rangle \langle N_j \rangle$, so the K-B integral would vanish for $i \neq j$ and would always equal $-V/N_i = -1/\rho_i$ for $i = j$.

$$G_{AB} = RT\kappa_T - \frac{\rho_{mix}V_{Am}V_{Bm}}{D} \tag{11.83}$$

$$G_{BB} = RT\kappa_T - V_{mB}^{\bullet}\left[1 - \frac{V_{Am}^2\rho_{mix}}{V_{mA}^{\bullet}D}\right],$$

where the parameter ρ_{mix} is the molar density of the ideal solution,

$$\rho_{mix} = \frac{1}{V_{mA}^{\bullet}} + \frac{1}{V_{mB}^{\bullet}}.$$

Therefore, the values of the three K-B integrals can be determined solely on the basis of *experimentally measured* parameters.

Recall that each of the three K-B integrals are actually defined in terms of the pair correlation function (Eq. 11.80):

$$G_{ij} \equiv 4\pi \mathcal{N}_A \int_0^\infty \left[\mathcal{G}_{ij}(R) - 1\right] R^2 dR.$$

It's remarkable enough that the Kirkwood-Buff theory connects the pair correlation function directly to experimental data. More impressively, the theory accomplishes this feat without sacrificing accuracy for generality. Equations 11.83 are valid for solutions of electrolytes and for mixtures of three or more components.

As a check, let's examine one of these integrals in a limiting case. If we calculate the value of G_{AA} for a fluid of noninteracting particles (i.e., an ideal gas), then the partial molar volumes of A and B are equal to the molar volume of an ideal gas at that particular partial pressure P_i:

$$\text{ideal gas:}\quad V_{mi} = \frac{V}{n_i} = \frac{RT}{P_i}. \tag{11.84}$$

We also need to evaluate κ_T, ρ_{mix}, and D in the limit of the ideal gas and see if they can also be put in terms of T and P.

- The isothermal compressibility can be simplified using the ideal gas law:

$$\kappa_T = -\frac{1}{V}\left(\frac{\partial V}{\partial P}\right)_{T,n}$$

$$= -\frac{1}{V}(nRT)\left[\frac{\partial(1/P)}{\partial P}\right]_{T,n} \qquad\qquad PV = nRT$$

$$= \frac{1}{V}(nRT)\frac{1}{P^2} \qquad\qquad d(1/x) = -dx/x^2$$

$$= \frac{1}{V}\frac{V}{P}\frac{1}{P} = \frac{1}{P}. \qquad\qquad PV = nRT \quad (11.85)$$

- The molar density is the reciprocal of the molar volume:

$$\rho_{mix} = \frac{1}{V_{mA}^{\bullet}} + \frac{1}{V_{mB}^{\bullet}} = \frac{P_A}{RT} + \frac{P_B}{RT} = \frac{P}{RT}. \tag{11.86}$$

- The derivative term D can also be solved, because the partial pressure of each gas is proportional to its mole fraction:

$$P_A = X_A P_A^{\bullet}$$

$$D = X_A\left(\frac{\partial \ln(P_A/P^{\ominus})}{\partial X_A}\right)_{T,P}$$

$$= X_A \left(\frac{\partial \left[\ln(X_A) + \ln(P_A^\bullet / P^\ominus) \right]}{\partial X_A} \right)_{T,P}$$

$$= X_A \frac{1}{X_A} = 1.$$

Combining these gives us the ideal gas value of G_{AA}:

$$G_{AA} = RT\kappa_T - V_{mA}^\bullet \left[1 - \frac{V_{Bm}^2 \rho_{mix}}{V_{mB}^\bullet D} \right]$$

$$= \frac{RT}{P} - \frac{RT}{P_A} \left[1 - \frac{\left(\frac{RT}{P_B} \right)^2 \frac{P}{RT}}{\frac{RT}{P_B}(1)} \right]$$

$$= \frac{RT}{P} - \frac{RT}{P_A} \left[1 - \left(\frac{RT}{P_B} \right) \frac{P}{RT} \right]$$

$$= \frac{RT}{P} - \frac{RT}{P_A} \left[1 - \left(\frac{P}{P_B} \right) \right]$$

$$= \frac{RT}{P} - \frac{RT}{P_A} \left[\frac{P_B - P}{P_B} \right] = \frac{RT}{P} - \frac{RT}{P_A} \left[\frac{-P_A}{P_B} \right]$$

$$= \frac{RT}{P} + \frac{RT}{P_B} = \frac{RT}{P} \left[1 + \frac{1}{X_B} \right]. \qquad (11.87)$$

In this ideal case, this K-B integral is the molar volume with a correction based on the mole fraction of the second component.

EXAMPLE 11.2 Kirkwood-Buff theory and water–ethanol mixtures

CONTEXT The *Bartender's Conundrum* is the problem that mixing equal amounts of water and a high-proof liquor gives a final volume less than the sum of the two initial volumes. If we mix two noninteracting components with volumes V_A and V_B, then the final volume is $V_A + V_B$. The data in graph (a) show how the partial molar volumes of water and ethanol generally decrease as their mole fractions decrease, with a net result that (b) the volume of the final mixture V is less than the ideal volume $V_{H_2O} + V_{eth}$. These alcohol–water solutions serve other purposes, however, including the solvation of biochemicals. The same intermolecular forces that cause the density of molecules in the solution to increase (and the molar volume to therefore decrease) can also provide a favorable environment for hydrophobic proteins. The organic ends of the alcohols solvate the protein, while the alcohol OH groups form a network of hydrogen bonds to the water.

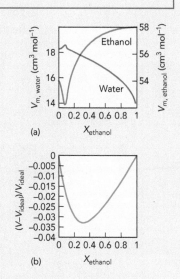

(a)

(b)

PROBLEM Use the data in the table to calculate the K-B integral G_{AA} for an aqueous solution of 40% ethanol by mole fraction at 298 K, where A is water and B is ethanol.

κ_T	$6.7 \cdot 10^{-10}\,\text{Pa}$
$V_{m,\text{eth}}^{\bullet}$	$5.8 \cdot 10^{-5}\,\text{m}^3\,\text{mol}^{-1}$
$V_{m,H_2O}^{\bullet}$	$1.8 \cdot 10^{-5}\,\text{m}^3\,\text{mol}^{-1}$
$V_{m,\text{eth}}$	$5.7 \cdot 10^{-5}\,\text{m}^3\,\text{mol}^{-1}$
V_{m,H_2O}	$1.7 \cdot 10^{-5}\,\text{m}^3\,\text{mol}^{-1}$
D	0.37

SOLUTION The data are all in SI units already—which would be rare in the real world, but let's not complain—so we don't need to be concerned about unit conversions. The values we obtain are

$$\rho_{\text{mix}} = \frac{1}{1.8 \cdot 10^{-5}\,\text{m}^3\,\text{mol}^{-1}} + \frac{1}{5.8 \cdot 10^{-5}\,\text{m}^3\,\text{mol}^{-1}}$$
$$= 7.3 \cdot 10^4 \,\text{mol}\,\text{m}^{-3}$$

$$G_{AA} = RT\kappa_T - V_{mA}^{\bullet}\left[1 - \frac{V_{Bm}^2\rho_{\text{mix}}}{V_{mB}^{\bullet}D}\right]$$
$$= (8.3145\,\text{J}\,\text{K}^{-1}\,\text{mol}^{-1})(298\,\text{K})(6.7 \cdot 10^{-10}\,\text{Pa}) - (1.8 \cdot 10^{-5}\,\text{m}^3\,\text{mol}^{-1})$$
$$\times \left[1 - \frac{(5.7 \cdot 10^{-5}\,\text{m}^3\,\text{mol}^{-1})^2(7.3 \cdot 10^4\,\text{mol}\,\text{m}^{-3})}{(5.8 \cdot 10^{-5}\,\text{m}^3\,\text{mol}^{-1})(0.37)}\right] = 0.000183\,\text{m}^3\,\text{mol}^{-1},$$

which is roughly the molar volume of liquid water.

From the values of the K-B integrals for this mixture based on experimental values such as these, it would be possible to directly check theoretical calculations of the ethanol–ethanol, water–ethanol, and water–water pair correlation functions. Kirkwood-Buff theory provides a valuable tool for assessing the quality of our fundamental understanding of solutions, among the most complex systems in chemistry.

Conclusion to Part II

Four functions that we concentrated on in the microscopic limit continue to govern the behavior of macroscopic matter: the total energy, the number of degrees of freedom, the potential energy, and the number of states.

1. The overall energy of the system E has been the focal point of our look at thermodynamics: it is the thermodynamic potential from which H and G and F are obtained; derivatives of E with respect to other extensive parameters give the intensive parameters T and P and μ.

2. The Gibbs phase rule tells us how many coordinates we need for a complete thermodynamic description of an equilibrium system, the macroscopic extension of our number of microscopic degrees of freedom N_{dof}. Once we have $k + 2 - p$ well-chosen parameters determined, our sample's location on the phase diagram and, in principle, all its other macroscopic parameters are known.

3. The potential energy function has played a subtle role, for it is the intermolecular potential that determines the nature of the activity. An especially strong solute–solvent interaction results in a large contribution of the activity to the chemical potential. We have seen only a few examples of how this is significant, but in Part III the potential energy term comes again to the forefront.

4. The ensemble size Ω travels through thermodynamics disguised in the form of the entropy S. In addition to its overwhelming importance in thermodynamics, this will be a pivotal parameter in the study of chemical reaction rates, as we find in Part III.

CONTEXT *Where Do We Go From Here?*

At last, we've covered all the fundamental principles that come into play when we study chemical reactions. These next chapters get to the heart of chemistry. How do molecules transform from one chemical structure to another? Where does the energy come from and where does it go? Our ability to integrate what we've learned about molecular structure, the canonical energy distribution, heat and work, entropy, and chemical potential all into a single model of chemical reactions is the greatest payoff available in studying physical chemistry.

KEY CONCEPTS AND EQUATIONS

11.1 **The Standard States.** For gases, pure liquids or solids, and solvents, we set the standard state chemical potential to the chemical potential of the pure substance at a pressure of 1 bar:

$$\mu_i^{\ominus} = \mu_i^{\bullet}(P^{\ominus}).\qquad(11.1)$$

For solutes, the reference state instead is the solute in a solution at a fixed concentration, typically 1 M.

11.2 **Statistical Mechanics of Solutions.** Non-ideal behavior in solutions, such as changes in the volume or enthalpy of the system on mixing, can be expressed in terms of the change in chemical potential of mixing:

$$\Delta_{mix}V_m = \left(\frac{\partial(\Delta_{mix}\mu)}{\partial P}\right)_{T,n_A,n_B}\qquad(11.5)$$

$$\Delta_{mix}H_m = \left(\frac{\partial(\Delta_{mix}\mu/T)}{\partial(1/T)}\right)_{P,n_A,n_B}.\qquad(11.7)$$

The chemical potential of mixing can be estimated from an integral over the product of the pair potential energy $u_{AB}(R)$ and the pair correlation function $\mathcal{G}_{BA}(R)$:

$$\Delta_{mix}\mu = 4\pi\mathcal{N}_A\rho_A\int_0^{V^{1/3}} u_{AB}(R)\mathcal{G}_{BA}(R)\,R^2 dR.\quad(11.14)$$

11.3 **Thermodynamics of Solutions.**
- The **activity** provides a catch-all correction to the chemical potential of a substance:

$$\mu = \mu^{\ominus} + RT\ln a.\qquad(11.19)$$

- For a solution that obeys **Raoult's law**, the partial pressure of the solvent above the solution at equilibrium is proportional to the mole fraction of solvent in the solution:

$$P_A = X_A P_A^{\bullet}.\qquad(11.23)$$

- According to **Henry's law,** the partial pressure of solute above the solution as we approach zero concentration is given by

$$P_B \approx k_X X_B,\qquad(11.33)$$

where k_X is the Henry's law constant.
- In either case, the activity is given by the mole fraction: $a_A = X_A$ and $a_B = X_B$.

11.4 **Ionic Solutions.** The activity of an ion in aqueous solution is poorly approximated by the mole fraction, so we often apply an empirical **activity coefficient** γ such that an ionic compound $A_{v_+}B_{v_-}$ has an activity

$$a_{A_{v_+}B_{v_-}(aq)} = \left(\gamma_{\pm}X_{A_{v_+}B_{v_-}}\right)^{v_++v_-}.\qquad(11.50)$$

11.5 **Applications of the Activity.**
- The freezing point of a solution decreases (relative to the pure solvent) as the solute concentration increases. We can explain this in terms of the activities and show that the shift in T_f is roughly proportional to the concentration of the solution:

$$\Delta T_f \approx -\frac{RT_f^{\bullet 2}X_B}{\Delta_{fus}H_A^{\bullet}}.\qquad(11.65)$$

- **The osmotic pressure** Π is also proportional to the concentration:

$$\Pi = \frac{RTX_B}{V_m} = RT[B], \qquad (11.69)$$

where $[B]$ is the molarity of the solute inside the cell.

- The Nernst equation,

$$\Delta\Phi_{eq} = \frac{RT}{\mathcal{F}z}\ln\frac{X^{bath}}{X^{cell}}, \qquad (11.71)$$

gives the equilibrium voltage difference between two parts of the system due to the diffusion of a particular ion from one region to the other.

KEY TERMS

- The **partial molar volume** of solute is the change in volume of the solution per mole of solute added.
- The **enthalpy of solvation** is the heat absorbed or emitted by the system as the solution is formed.
- An **azeotrope** is a mixture of two or more substances at concentrations such that the equilibrium composition of the vapor is identical to the equilibrium composition of the liquid.

- **Fractional distillation** is a technique for separating two fluids by repeatedly distilling the substance to isolate the more volatile component in the gas phase.
- The **eutectic point** in a multi-component phase diagram is the composition at which the liquid reaches a minimum melting point.
- **Osmosis** is the flow of materials across a membrane.
- The **Faraday constant** $\mathcal{F}$ is the charge per mole of electrons, equal to 96,485.34 C mol^{-1}.

OBJECTIVES REVIEW

1. *Express the chemical potential of a substance in solution in terms of a standard state potential and the activity.*
   Write an expression for the chemical potential of aqueous K_2SO_4 as a function of the mole fraction of solute and in terms of the activity coefficients γ for each ion. Assume that K_2SO_4 dissociates completely.

2. *Use Raoult's law and Henry's law to estimate partial pressures of solute and solvent above a solution.*
   Estimate the partial pressures of water and carbon tetrachloride above a 1.00 mM aqueous solution of carbon tetrachloride at 298 K. The vapor pressure of water at 25 °C is 0.032 bar.

3. *Use a phase diagram to find vapor pressure, melting and boiling points, composition in each phase, and azeotropic and eutectic conditions for a mixture of substances.*
   Identify the compositions of the gas and liquid phases (in terms of mole fraction) of the acetone/chloroform mixture characterized in Fig. 11.3 at equilibrium at 334 K, assuming acetone has the greater concentration.

4. *Estimate freezing point depression, osmotic pressure, and (for ionic solutes) membrane voltage based on the concentration.*
   Find the freezing point of a 0.50 M aqueous solution of K_2SO_4.

PROBLEMS

Discussion Problems

11.1 This is a quick review of a few concepts from thermodynamics. Indicate whether the following parameters will increase ("+"), decrease ("−") or remain unchanged ("0") during the process described.

parameter	process	+, −, or 0
S	reversible adiabatic expansion	
T	reversible adiabatic expansion	
T_b	increasing P	
$\Delta_{vap}H_m$	increasing P	
P_B	increasing X_B in a solution	

11.2 Explain whether the Henry's law constant for NH_3 in H_2O should be larger or smaller than the Henry's law constant for He in H_2O.

General Thermodynamic Properties of Solutions

11.3 Use the chemical potential and the activity to find an expression for the molar entropy of solute B in a solution of mole fraction X_B at temperature T.

11.4 Estimate $\Delta_{mix}H_m$ for a non-ideal solution having a square well potential for the interaction between solvent A and solute B at 298 K and constant pressure. Use $\varepsilon = 432\,K = 5.96 \cdot 10^{-21}$ J, $R_{sq} = 2.0$ Å, $R_{sq}{}' = 3.0$ Å, and $\rho_A = 3.3 \cdot 10^{22}$ cm^{-3}.

11.5 For ideal mixing, $\Delta_{mix}H = 0$. Write a general equation for $\Delta_{mix}H$ at 1 bar when two non-ideal substances A and B are mixed, in terms of their activities a_A and a_B.

11.6 Find an integral expression for $\Delta_{mix}F$ of a solution in terms of the pair correlation function $\mathcal{G}_{AB}(R)$ and potential energy $u_{AB}(R)$.

11.7 Try to approximate the freezing point of carbonated water by treating it as a 0.04 M solution of CO_2 in water.

11.8 Find an equation for the slope of the phase boundary of a two-component liquid as it approaches the eutectic point on a T versus X_B phase diagram, in terms of the temperature and the solvent enthalpy of fusion.

Raoult's and Henry's Laws

11.9 Plot the vapor pressure of water over a solution of mannitol as a function of the mole fraction of mannitol, assuming that the solution obeys Raoult's law and letting X_{water} vary from 1.00 to 0.80. The vapor pressure of pure water at 298 K is 23.8 torr.

11.10 In an **azeotropic** solution, the liquid and vapor phases have equal mole fractions of solute at equilibrium. Find the numerical value of the Henry's law coefficient of a solution that is azeotropic at its standard boiling point.

11.11 In Eq. 11.40,

$$k_X \approx \left(\frac{P_B^{\bullet}}{X_B}\right)\left[1 + \frac{4\pi\rho_A}{k_B T}\int_0^{V^{1/3}} u_{AB}(R)\mathcal{G}_{AB}(R)R^2 dR\right],$$

show whether k_X will generally be larger or smaller than $P_B^{\bullet}/X_B$.

11.12 Compound B is dissolved in water until the solution becomes saturated at a mole fraction X_B. From that point on, addition of B merely increases the amount of solid B lying at the bottom of the solution. Write an equation for X_B in the equilibrium, saturated solution at 1 bar in terms of the temperature T and ΔG_m for the process B(sol) $\rightarrow$ B(aq) (the Gibbs energy of solution).

11.13 An alternative to Henry's law is the power series expansion, $P_B = k_1 X_B + k_2 X_B^2$. Find an expression for the activity of B in a dilute solution in terms of the partial pressure and k constants in the given equation.

11.14 Use Eq. 11.40 to estimate the vapor pressure of B above a solution with a solvent–solute interaction u_{AB} described by the square well potential:

$$u_{AB}(R < R_1) = \infty \quad u_{AB}(R_1 \le R < R_2) = -u_0 \quad u_{AB}(R_2 \le R) = 0,$$

and where the pair correlation function obeys the equation

$$\mathcal{G}_{AB}(R) = X_B e^{-u_{AB}}.$$

Use the following parameters:

  a. $R_1 = 2.0\,\text{Å}$, $R_2 = 3.0\,\text{Å}$, $u_0 = 100\,\text{cm}^{-1}$
  b. solvent density $\rho_A = 3.35 \cdot 10^{22}\,\text{cm}^{-3}$

  c. temperature $T = 298$ K
  d. vapor pressure of pure solute $P_B^{\bullet} = 0.12$ bar
  e. mole fraction of B $X_B = 0.100$

11.15 We have used Raoult's law to obtain the expression for the freezing point shift of a solution:

$$\Delta T_f = -\frac{RT_f^{\bullet 2}X_B}{\Delta_{fus}H_{Am}^{\bullet}}.$$

If the activity of the solvent is given instead by

$$a_A = k_A X_A,$$

and the vapor pressures of the pure solvent and solute are $P_A^{\bullet}$ and $P_B^{\bullet}$, what is the new equation for ΔT_f in terms of the mole fraction of solute?

11.16 From the corresponding Henry's law coefficient $k_X = 8.64 \cdot 10^4\,\text{bar}$, estimate the **molarity** of N_2(gas) in water at 298 K when the water is in equilibrium with air at a total pressure of 1.10 bar. Assume that the air is 78% N_2 by volume (i.e., by mole number), and that the dissolved N_2 does not affect the volume of solution.

11.17 A supercritical fluid is a substance at higher temperature and higher pressure than its critical point on the phase diagram. Usually these pressures are above 100 bar. Solids and liquids can be dissolved in supercritical fluids. (a) Explain which of the following curves best represents the solubility curve for *any* liquid or solid in a supercritical fluid at constant pressure P as a function of temperature. (b) On that graph, show qualitatively what the solubility curve will look like for the pressure $2P$.

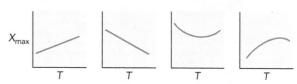

11.18 The Henry's law coefficient of 1-butanol at 298 K is 9.26 bar. Calculate the vapor pressure (in bar) of 1-butanol over a 0.0100M aqueous solution.

11.19 The Henry's law coefficient of CO_2 in water at 298 K is $1.6 \cdot 10^3$ bar. Calculate the *molarity* of CO_2 in water at 298 K when the solution is in equilibrium with the atmosphere, which has an average CO_2 abundance (by mole) of $3.94 \cdot 10^{-4}$ and a pressure of 1.07 bar.

11.20 For the phase diagram drawn in the following figure, identify (a) all the possible phase transitions indicated; (b) the freezing point of naphthalene at 0.5 fractional concentration in benzene; (c) the sequence of events as a 0.5 mole fraction mixture of benzene and naphthalene is cooled quasistatically from 350 K to 250 K.

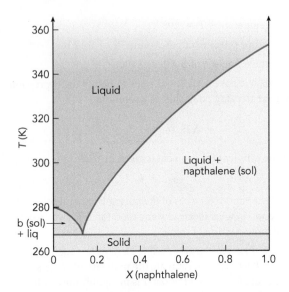

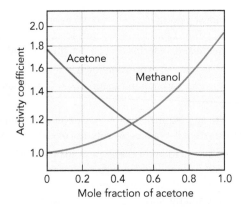

11.21 A partial phase diagram for the mixture of magnesium and zinc at 1 bar is shown in the following figure. Based only on this diagram, estimate the normal melting points of Zn, Mg, and MgZn$_2$.

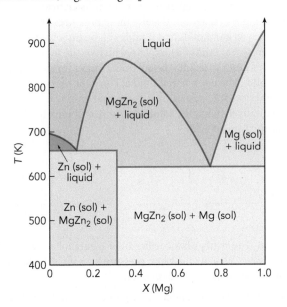

11.24 Find the osmotic pressure in bar of a cell where the solute has reached a concentration of 0.010 M at 298 K.

11.25 A rigid sample cell (with constant total volume) is divided by a flexible, semipermeable membrane into two initial volumes, V_1^i and V_2^i. Both volumes are filled with solvent A, which flows freely across the membrane. Then n_B moles of solute B are added to volume 1 and n_C moles of a different solute C to volume 2, at some constant temperature T. Neither B nor C may cross the membrane. Because the membrane is flexible, equilibrium is achieved when the pressures on both sides of the membrane are equal. What is the final value V_1^f of volume 1?

11.26 Consider the osmotic cell-within-a-cell design sketched in the figure. The membrane between 1 and 2 is permeable only to solvent A, and the membrane between 2 and 3 is permeable to both A and B but not C. Find an expression for the *total pressure* P_3 in terms of P_1 and the molarities [B] and [C] in cell 3. Assume that X_A is much greater than X_B and X_C and that the solution is incompressible.

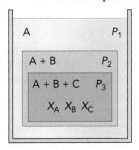

Activity

11.22 Find an equation for the activity a of a non-ideal gas whose behavior is given by the virial expansion:

$$\frac{PV_m}{RT} = 1 + B_2 V_m^{-1}.$$

11.23 A graph of the activity coefficients γ for methanol and acetone is shown below, where $a_i = \gamma_i X_i$. Circle the region or regions of the graph where either liquid obeys Raoult's law.

11.27 Start with an osmotic cell as in Fig. 11.7 where the volume of the cell is 10. ml and the volume of the bath (not including the cell) is 100. ml. We first add 0.100 mmol B to the cell but then find that the osmotic pressure is too high. We compensate by adding a solute C to the bath. Neither B nor C can penetrate the semipermeable membrane. Assume that Raoult's law is satisfied by the solvent. What is the minimum number of moles C that must be added to the bath to get a pressure difference across the membrane of 0.10 bar at 298 K?

PART I

EXTRAPOLATING
FROM
MOLECULAR TO
MACROSCOPIC
SYSTEMS

PART II

NON-REACTIVE
MACROSCOPIC
SYSTEMS

PART III
REACTIVE SYSTEMS

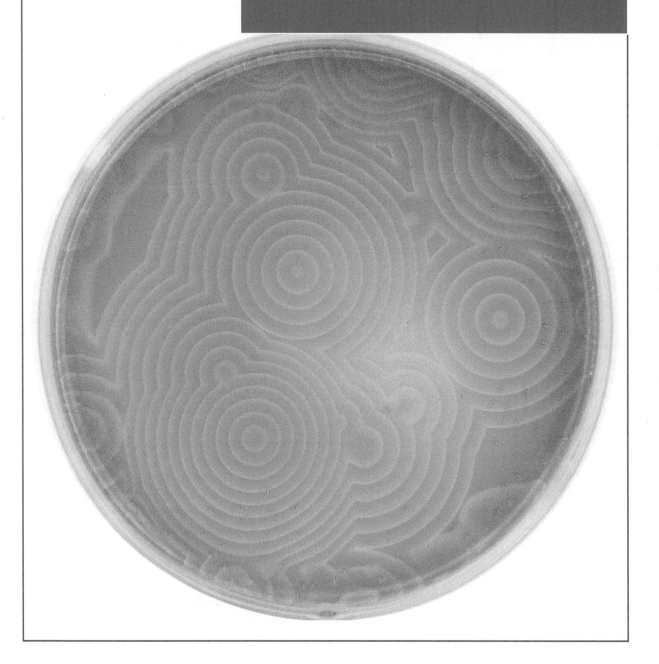

PART I
EXTRAPOLATING
FROM
MOLECULAR TO
MACROSCOPIC
SYSTEMS

PART II
NON-REACTIVE
MACROSCOPIC
SYSTEMS

PART III
REACTIVE SYSTEMS

12 The Thermodynamics of Chemical Reactions

13 Chemical Kinetics: Elementary Reactions

14 Chemical Kinetics: Multi-Step Reactions

12

The Thermodynamics of Chemical Reactions

LEARNING OBJECTIVES

After reading this chapter, you will be able to do the following:

❶ Identify the reactant, product, and transition state geometries on a reaction surface, and use the surface to estimate the energy of reaction and activation energy.

❷ Calculate the enthalpy of an isothermal reaction at arbitrary temperature and pressure from the enthalpies of formation and heat capacities.

❸ Calculate the adiabatic flame temperature of a compound.

❹ Calculate the equilibrium constant of a reaction from the Gibbs free energies of formation, and estimate the effects of temperature.

GOAL *Why Are We Here?*

Our goal in this chapter is to apply the principles of thermodynamics described in Part II to chemical reactions. What motivates us is the promise of understanding how we can predict the flow of energy in and out of reactive systems, both so we know what to expect and so we can look for opportunities to control the process. While our primary thesis in this textbook has been that molecular structure determines the macroscopic properties of our system, so too can we turn that concept around and use macroscopic parameters such as pressure and temperature to influence the direction and extent of chemical reactions—processes that are defined by activity at the microscopic scale.

CONTEXT *Where Are We Now?*

Finally we get to real chemistry, having developed all the tools of math and physics we need to describe chemical systems in the absence of reaction. Those tools will continue to serve us well in this final stretch. Thermodynamics, for example, continues to be useful, and we shall start with its application to chemical reactions.

Our brand of thermodynamics will tell us only about the equilibrium states of the system, but that's a good place to start. By understanding the endpoints of the reaction, we can begin to see whether or not the reaction will be spontaneous, and if so what the driving force is.

SUPPORTING TEXT *How Did We Get Here?*

This chapter focuses a good deal on enthalpy and entropy of reaction, so our most important preparation will be a familiarity with those two parameters and how we manipulate them in solving problems about thermodynamic processes. More specifically, we will draw on the following equations and sections of text:

- In Chapter 7, we define the enthalpy and Gibbs free energy in terms of the internal energy E (Eqs. 7.17 and 7.19):

$$H = E + PV$$
$$G = E - TS + PV = H - TS.$$

From the second expression, we are able to obtain an expression for ΔG for the process. At constant temperature, ΔG simplifies to the form (Eq. 9.26)

$$\Delta G = \Delta H - T\Delta S.$$

With T and P both held constant the incremental change in Gibbs free energy can also be expressed in terms of the chemical potentials μ_i of the k different chemical components (Eq. 7.26):

$$dG = \sum_{i=1}^{k} \mu_i dn_i.$$

In this chapter, we will connect these two expressions for the change in free energy as we explore the principle of chemical equilibrium, taking advantage of Eq. 11.19 to write the chemical potential in terms of the activity a:

$$\mu = \mu^{\ominus} + RT \ln a.$$

- To establish the link between the molecular and macroscopic scales, Eq. 6.15 allows us to calculate the energy of a system at any temperature T from its partition function Q:

$$E = k_B T^2 \left(\frac{\partial \ln Q(T,V)}{\partial T} \right)_{V,N}.$$

In similar fashion, we can also find the entropy from the partition function (Eq. 7.35):

$$S = k_B \ln Q(T) + k_B T \left(\frac{\partial \ln Q(T)}{\partial T} \right)_{V,n}.$$

12.1 Introduction to Chemical Reactions

We'll define a chemical reaction as a process in which the nature of the chemical bonds in our system is altered. That's only helpful after you've settled the argument about just what constitutes a chemical bond, however, and some ambiguities remain.

For example, the reaction of two carbon dioxide molecules to form the van der Waals molecule $(CO_2)_2$ is not generally considered to be a chemical reaction, because the chemical bonds within each monomer are essentially unchanged. During the formation of $(H_2O)_2$, bound by a 40 kJ mol^{-1} hydrogen bond, the perturbation is much more severe, but the water monomers may still be said to retain their original chemical identity. However, when we consider the condensation of Na gas to bulk sodium metal, we have to acknowledge that the two forms are dramatically different, not only in density but in electronic structure. Nevertheless, we usually count phase transitions as distinct from chemical reactions. Solvation offers a similar example.

The chemical natures of NaCl solid and NaCl in solution are undeniably different, but perhaps because the process is often easily reversible, solvation is not generally considered chemical reaction. Examples of stronger and weaker effects are again easy to come by.

How about processes in which there is only one chemical component and the arrangement of atoms *within* the molecule changes? As a class, these processes are known as **isomerization,** but of these we would usually consider only **structural** isomerization to be chemical reactions. Structural isomerizations involve changing the bonding sequence in the molecule, as in the transformation of cyclobutene to butadiene. Other isomerizations include **configurational** changes, such as the rotation around a double bond to change *cis* to *trans,* and the more rapid **conformational** isomerizations, which usually consist of rotations about a single bond.

Although we tend to refer to only a few of these processes as chemical reactions, they are all fundamentally the same kind of process: the motion of the atoms from one relative geometry to another. If the motion were all within one molecule, we would consider this motion along the vibrational coordinates, bond stretches and bends. In the same way that we analyzed molecular vibrations and intermolecular interactions using the potential energy curve, we begin our look at chemical reactions by seeing how the effective potential energy along the vibrational coordinates depends on the extent of the reaction.

12.2 Reaction Surfaces

When a chemical reaction takes place, compounds are converted from one structure to another. The bond lengths and bond angles change, and each of these corresponds to a vibrational coordinate—either an internal vibrational coordinate or a van der Waals vibrational coordinate—of the reactants. The chemical reaction can therefore be represented on a multidimensional potential energy surface called the **reaction surface,** a graph of the potential energy as a function of the vibrational coordinates of the entire set of atoms.

The simplest way to see this is for a reaction with very few coordinates, beginning with the unimolecular dissociation of a diatomic, such as

$$\text{HCl} + h\nu \longrightarrow \text{H} + \text{Cl}. \tag{12.1}$$

The $h\nu$ indicates that a photon, in this case in the ultraviolet, is necessary to break the chemical bond. We have already studied such processes—this is a photodissociation—and we know that the process can be represented on the potential energy curve of the H—Cl bond length, the vibrational coordinate of the reactant (Fig. 12.1).

For reactions involving more than two atoms, the surface becomes more complicated. The number of dimensions of the reaction surface is the same as the number of vibrational degrees of freedom of a single molecule made up of all the reactants: $3N_{atom} - 6$ where N_{atom} is the total number of atoms of all the reactants. To simplify the problem, chemists usually restrict the number of dimensions to those most relevant to the reaction. For example, the reaction

$$\text{F} + \text{H}_2 \longrightarrow \text{HF} + \text{H} \tag{12.2}$$

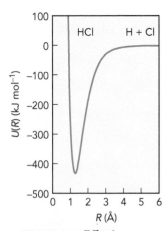

▲ FIGURE 12.1 **Effective vibrational potential energy curve for HCl.** The curve corresponds to the reaction surface for dissociation, with reactant (HCl) on the left proceeding towards product (H + Cl) on the right.

has been extensively studied by experiment and by theory. The reaction surface is a function of three coordinates, which may be chosen to be the HH bond length, the HF bond length, and the HHF angle. Of these, the angle may be held constant during the reaction, although the actual value does have a large impact on the results. Let's fix the HHF bond angle to 180° so the reaction takes place with all the atoms arrayed along a straight line.

At every value of the HH and the HF distances, there is a vibrational potential energy, which can be represented as a surface that varies up and down with the two coordinates. This reaction potential energy surface is presented in Fig. 12.2. We choose zero potential energy at the geometry of the reactants F and H_2. The lowest energy points on this curve are those where the H_2 or HF molecules are at their equilibrium bond lengths, about 1 Å, and the extra atom is at least 4 Å away.

In order for this reaction to take place, the three atoms can go through a **transition state,** a stationary point on the reaction surface, where the slope goes to zero, somewhere higher in energy than either the reactants and products. The transition state is the *highest energy molecular structure* along the *lowest energy path* along the reaction surface that connects the reactants to the products. In the case of F + H_2, the transition state is a structure about 30 kJ mol^{-1} above

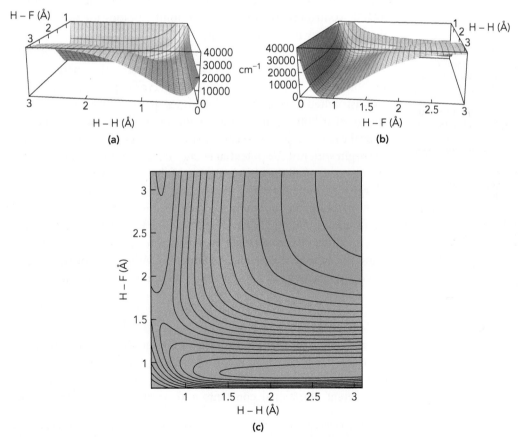

▲ FIGURE 12.2 **The reaction surface for the reaction F + H_2 at HHF = 180°.** Perspectives on the three-dimensional surface are chosen to illustrate the dependence on (a) R_{HH} and (b) R_{HF}. Also shown is (c) a contour plot of the surface. Each contour represents a curve of constant potential energy, and contours are given for intervals of 2500 cm^{-1} or 30 kJ mol^{-1}.

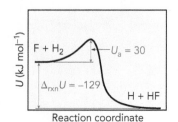

▲ FIGURE 12.3 **Relationships among various reaction energies.** Shown are the potential energy of activation U_a, the potential energy of reaction $\Delta_{rxn}U$, and the reaction potential energy surface for the $H_2 + F$ reaction. The reaction coordinate is the lowest energy path between reactants and products.

CHECKPOINT Chemical reactions are processes that change the chemical bonds in our system. The difference in potential energy between two separated atoms and two atoms in a chemical bond is typically about 400 kJ mol⁻¹, which is 160 times the thermal energy at room temperature. Therefore, chemical reactions often occur with energy changes much greater than the energy already available in random motion of the molecules. The reaction surface graphs the potential energy associated with the chemical bonds, showing how much energy can be released or absorbed for any given pathway.

the separated $F + H_2$ atom system. At that transition state, the H—F bond is quite weak, as indicated by the 1.5 Å bond length. The potential energy change from reactants to transition state is the potential energy of activation.

This gives a microscopic picture of what's happening as the reaction occurs, and it is directly related to the thermodynamics. As we move from one point on the surface near the reactants to another point near the products, the potential energy U changes. If we keep the temperature constant, so that the average kinetic energy in each degree of freedom stays the same during the reaction, then the change in energy of the reaction is roughly the same as the change in potential energy. We can define the reactants to be our initial thermodynamic state, the products to be our final thermodynamic state. Then we can use the tools we used for phase transitions and other thermodynamic processes to relate the change in energy during the reaction to other thermodynamic properties.

In principle, if we know the potential energy surface for the reaction, we can calculate the reaction properties under any conditions. In the same way, from the potential energy function for a quantum mechanical system, we can integrate the Schrödinger equation to completely describe the system. However, most chemical reactions do not receive this level of scrutiny, for the same reason that potential surfaces of stable molecules are rarely studied completely: there are simply too many coordinates. Furthermore, although the potential energy is convenient for a theoretical calculation, the experimental observations will include contributions from the zero-point vibrational energies and the average thermal energy in each coordinate.

This last point can be critical in some cases, so let's spell it out. When reaction properties are measured in bench-top chemistry, the energy is not the single variable that it is in a high-resolution spectroscopy experiment. The molecules are distributed over a range of energies at every step of the reaction, but the energy distribution can change as the molecular structure changes. When we draw the reaction potential energy surface, we ignore the nuclear motion entirely. We begin with just the potential energy $U(\overrightarrow{R})$, where $\overrightarrow{R}$ specifies any particular arrangement of all the nuclei. On top of that, the molecules will have zero-point energy E_{zp} from every vibrational mode. This gives us the ground state energy E_0 at every point on the surface, which is the bottom of our statistical energy distribution function.[1] Now, if we know the temperature and the system's partition function, we can calculate the average energy $\langle E \rangle$ in the thermal distribution using the canonical distribution function. Adding all of these up gives us the macroscopic energy of the system at that point on the reaction surface:

$$E(\overrightarrow{R}) = U(\overrightarrow{R}) + E_{zp}(\overrightarrow{R}) + k_B T^2 \left(\frac{\partial \ln Q(T, \overrightarrow{R})}{\partial T} \right)_{V,n}, \quad (12.3)$$

where the last term is the expression for the thermal correction $\langle E \rangle$ given in Eq. 6.15. The partition function Q is shown here to be a function of the position $\overrightarrow{R}$ on the reaction surface as well, because different geometries correspond to different vibrational constants and because the number of rotational and

---

[1]The zero-point energy, however, is not constant across the reaction surface, because as one bond forms or breaks, the force constants of all the other vibrational modes are affected. Therefore, our zero-point energy is now also a function of the geometry, $E_{zp}(\overrightarrow{R})$. For example, including the zero-point energy in our $F + H_2$ reaction greatly decreases the relative energy of the transition state.

translational coordinates may change as bonds are formed or broken. Once we know the energy $E(\vec{R})$, we can calculate the enthalpy and free energy and other experimental observables.

Currently, we can't hope to calculate, let alone read, reaction surfaces that describe all the vibrational modes of a reaction system with large molecules, and normally we wouldn't wish to. To a great extent, we can appreciate the challenges faced by a reactant molecule, as it struggles to become a product molecule, by examining the easiest path the molecule could take between reactant and product. Therefore, the reaction surface is usually drawn in a greatly simplified form, as a graph of the minimum energy path as a function of a variable called the **reaction coordinate** (Fig. 12.3). The reaction coordinate may be a single vibrational mode or a linear combination of vibrational modes, and it adjusts the geometry of the nuclei as necessary to take the system from reactants to products along a path of least resistance. We shall assume that the reaction diagram represents the experimentally observable energies, so the vertical axis is the energy $E(\vec{R})$ given in Eq. 12.3, not the potential energy. Henceforth, we'll call that the energy E of the reaction system, and from it we can always calculate the reaction enthalpy H by the definition of the enthalpy, $H = E + PV$ (Eq. 7.17).

The structure at the highest energy point of the reaction diagram is the transition state, and the energy of the transition state relative to the reactants is the **activation energy** E_a, the minimum energy of the reactants necessary for the reaction to occur by classically crossing over the barrier. Figure 12.4 shows how the zero-point and thermal corrections affect the reaction diagram for the $F + H_2$ system. In this case, the greater energy separation in the vibrational states of H_2 leads to a shift of 3 kJ mol^{-1} in the reaction energy $\Delta_{rxn}E$ measured at 298 K from the reaction potential energy $\Delta_{rxn}U$, which is temperature-independent.

A chemical reaction is analogous to the other thermodynamic processes that we have investigated in several respects:

1. The thermodynamic parameters of the initial (reactant) and final (product) states can be measured.

2. For identical sets of initial and final states, the thermodynamic parameters are independent of the reaction path.

3. The whole process can be reversible, in which case all the thermodynamic variables are well-determined at every point.

Just as we graphed H versus T for phase transitions, now we can graph H versus the reaction coordinate, where $H = E + PV$. The reaction coordinate corresponds to a macroscopic observable, in that it dictates the chemical composition of the system. Therefore, we can treat chemical reactions with the same thermodynamic formalism that we applied to expansions and mixing and phase transitions.

What will the thermodynamics tell us? No more than it told us in the other examples: properties of the reaction that relate to equilibrium states of the reactants and products. This includes the energy difference between reactants and products, which in turn gives the work and heat obtainable from chemical reactions and the course a reaction will take spontaneously, dictated as before by the entropy difference between reactants and products.

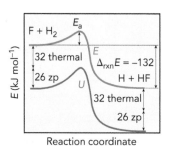

▲ FIGURE 12.4 **Reaction diagram for the $F + H_2$ reaction.** The reaction diagram (red line), is graphed in comparison to the potential energy curve (blue line) to show the relative impacts of the zero-point (zp) and the thermal energy corrections at 298 K.

12.3 Enthalpies of Reaction

Every chemical reaction involves a transformation between the potential energy of a chemical bond and the kinetic energy of the fragments when that bond is broken. To study this process in detail requires some of our most advanced experimental techniques, but a general picture of this exchange of energy is available from macroscopic measurements of the change in enthalpy as the reactants become products. For that reason, we devote this section to examples and methods of calculating reaction enthalpies.

Isothermal Reactions

We want to know how much heat is evolved during a particular reaction. The parameters of the initial and final states do not depend on the nature of the path between them, and so, as in other processes such as expansion or phase transitions, we can apply the first law:

$$\Delta E = q + w, \qquad dE = đq + đw,$$

where ΔE is the difference in internal energy between the initial and final states, and dE is the incremental energy change along the reaction coordinate, so that

$$\Delta E = \int_i^f dE.$$

The heat q that flows during a reversible reaction is related to the entropy:

$$đq_{rev} = TdS.$$

Returning to our thermodynamic potentials E, H, F, and G, we see that if we want the reversible heat, we can obtain it either at constant volume as

$$(q)_V = \int_i^f (TdS)_V = \int_i^f (TdS - PdV)_V = \int_i^f (dE)_V = \Delta E, \quad (12.4)$$

or at constant pressure as

$$(q)_P = \int_i^f (TdS)_P = \int_i^f (TdS + VdP)_P = \int_i^f (dH)_P = \Delta H. \quad (12.5)$$

Sticking for now to bench-top chemistry, which is at constant pressure, it is ΔH that we will generally look for.

CHECKPOINT We follow ΔH during a chemical reaction rather than ΔE because the energy change is divided into work (from expansion or contraction of the system, which we usually ignore) and heat. The enthalpy change gives the heat, which we can measure directly by calorimetry, and which is an important source of energy transfer that we use to power engines and other devices.

Equation 12.5 can be rewritten as

$$(q)_P = \Delta_{rxn}H = H(\text{products}) - H(\text{reactants}), \quad (12.6)$$

where $\Delta_{rxn}H$ is called the **enthalpy of reaction.** It is typical to evaluate the molar enthalpy of reaction $\Delta_{rxn}H$, but we will see that the moles in question depend on how the corresponding reaction has been balanced.

The enthalpy change is generally measured while the reaction remains at constant temperature and pressure. That measurement is quite feasible, because the heat evolved or absorbed during the reaction affects the temperature of the surroundings. Suppose the reaction takes place at constant P in a bath of temperature T, and the bath warms by an amount ΔT. Then the reaction has released an energy equal to $C_{P,\text{bath}}\Delta T$, where $C_{P,\text{bath}}$ is the heat capacity of the bath at constant pressure, so

$$\Delta_{rxn}H = -C_{P,\text{bath}}\Delta T. \quad (12.7)$$

This kind of measurement is called **calorimetry.** When $\Delta_{rxn}H$ is positive, heat is absorbed from the sample or its surroundings by the reaction, and the reaction is called **endothermic.** When $\Delta_{rxn}H$ is negative, heat is released by the reaction, and the reaction is called **exothermic.**

EXAMPLE 12.1 Calorimetry and Enthalpy of Reaction

CONTEXT Formic acid is the simplest carboxylic acid, having formula HCOOH, but its chemistry upon heating is surprisingly complex. In water and on metal surfaces, heated formic acid decomposes into H_2 and CO_2, but in the gas phase and in solution at low pH the following reaction occurs instead:

$$HCOOH(g) \rightarrow H_2O(g) + CO(g).$$

Calorimetry can help us understand how the interactions between HCOOH and its environment have such a substantial impact on the chemistry, beginning by telling us which reaction is more exothermic.

PROBLEM In the presence of a metal catalyst, we find that liquid formic acid decomposes at room temperature to form H_2 and CO_2. What is $\Delta_{rxn}H^{\ominus}$ for the decomposition of HCOOH into H_2 and CO_2 if the process warms 500. g of water by 1.48 K at 298.15 K when 0.100 moles of HCOOH react in a calorimeter at 1.00 bar? For liquid water at 298.15 K, the specific heat capacity is $4.186 \text{ J K}^{-1}\text{g}^{-1}$.

SOLUTION The heat gained by the water has been released by the reaction. Writing the mass of the water as M_{H_2O} and the specific heat capacity as c_{H_2O}, we set q for the reaction equal to $-q$ for the water:

$$q_{rxn} = -q_{H_2O}$$

$$\Delta H^{\ominus} = -C_{H_2O}\Delta T = -M_{H_2O}c_{H_2O}\Delta T$$

$$\Delta_{rxn}H^{\ominus} = \left(\frac{1}{n}\right)\Delta H^{\ominus} = -\frac{M_{H_2O}c_{H_2O}\Delta T}{n} = -\frac{(500.\text{ g})(4.186 \text{ J K}^{-1}\text{g}^{-1})(1.48 \text{ K})}{(0.100 \text{ mol})}$$

$$= -3.10 \cdot 10^4 \text{ J mol}^{-1} = -31.0 \text{ kJ mol}^{-1}.$$

This reaction is weakly exothermic. In contrast, the decomposition into H_2O and CO is endothermic with a $\Delta_{rxn}H^{\ominus}$ value of 102.6 kJ, but it is the faster reaction at high temperature in the gas phase.

Enthalpies of Formation

The enthalpy of reaction is predicted by using the enthalpies of the most stable forms of the component elements at 298.15 K and 10^5 Pa. The enthalpy of reaction for such a process is called the **enthalpy of formation** $\Delta_f H$ of the product. We'll need to remember that the $\Delta_f H$'s are molar quantities and that they are tabulated only for a specific pressure and temperature of the reaction, usually 298 K and 1.00 bar. Enthalpies of formation and other thermodynamic properties of various compounds are tabulated in the Appendix.

According to **Hess's law,** the enthalpy of reaction for our original reaction must be equal to the difference between the enthalpies of formation of the products and the enthalpies of formation of the reactants. In a general form, for the reaction

$$aA + bB \rightarrow cC + dD, \tag{12.8}$$

where a, b, c, and d are the stoichiometric coefficients for compounds A, B, C, and D, respectively, the enthalpy of reaction is given by

$$\Delta_{rxn}H = c\Delta_f H(C) + d\Delta_f H(D) - a\Delta_f H(A) - b\Delta_f H(B). \tag{12.9}$$

This is easily proved. The reaction must have the same number of moles of each element in the reactants as in the products. Assume that there are only two elements involved in the reaction, and their most stable forms are E and F, contributing e and f moles to each side of the reaction. The sum of the heats of formation for the reactants is $\Delta_{rxn}H$ for the reaction

$$e\mathrm{E} + f\mathrm{F} \rightarrow a\mathrm{A} + b\mathrm{B},$$

and the $\Delta_f H$'s for the products sum to the enthalpy of reaction for

$$e\mathrm{E} + f\mathrm{F} \rightarrow c\mathrm{C} + d\mathrm{D}.$$

The left-hand sides of these reactions correspond to the same thermodynamic state, so H of that state is a constant. The enthalpy of reaction we want can therefore be written as

$$\begin{aligned}
\Delta_{rxn}H &= H(c\mathrm{C} + d\mathrm{D}) - H(a\mathrm{A} + b\mathrm{B}) \\
&= \left[H(c\mathrm{C} + d\mathrm{D}) - H(e\mathrm{E} + f\mathrm{F})\right] - \left[H(a\mathrm{A} + b\mathrm{B}) - H(e\mathrm{E} + f\mathrm{F})\right] \\
&= \Delta_f H(\text{products}) - \Delta_f H(\text{reactants}).
\end{aligned} \tag{12.10}$$

This is sometimes written in slightly different form,

$$\Delta_{rxn}H = \sum_i v_i \Delta_f H_i, \tag{12.11}$$

where the v_i's are the stoichiometric coefficients and are positive for products and negative for reactants.

EXAMPLE 12.2 Heat of Reaction from Enthalpies of Formation

CONTEXT 1,2-Dibromoalkenes form a useful class of synthetic reagents in organic chemistry, allowing, for example, a reaction to close a chain of carbons up into a ring, driven partly by the elimination of Br_2. One way to make the dibromoalkenes is by the reaction of HBr with an *alkyne* (a molecule with a carbon–carbon triple bond). Because those reactions sacrifice one H—Br bond for two new C—Br and C—H bonds, the reaction is generally exothermic.

PROBLEM Find the enthalpy of reaction for

$$C_2H_2(g) + 2HBr(g) \rightarrow CH_2BrCH_2Br(g).$$

SOLUTION We need the enthalpies of the following reactions:

$$\begin{aligned}
H_2(g) + 2C(\text{graphite}) &\rightarrow C_2H_2(g) \quad 226.73 \text{ kJ mol}^{-1} \\
2\left[\tfrac{1}{2}H_2(g) + \tfrac{1}{2}Br_2(l) \rightarrow HBr(g) \right. &\left. -36.40 \text{ kJ mol}^{-1}\right] \\
2H_2(g) + 2C(\text{graphite}) + Br_2(l) &\rightarrow CH_2BrCH_2Br(g) \quad -38.91 \text{ kJ mol}^{-1}.
\end{aligned}$$

The most stable form of elemental hydrogen at 298 K and 1 bar is the gas, while for bromine it is the liquid and for carbon it is graphite.

For our particular case, we have

$$\begin{aligned}
\Delta_{rxn}H &= \Delta_f H(CH_2BrCH_2Br(g)) - \Delta_f H(C_2H_2(g)) - 2\Delta_f H(HBr(g)) \\
&= [-38.91 - 226.73 - 2(-36.40)] \text{ kJ mol}^{-1} \\
&= -192.84 \text{ kJ mol}^{-1}.
\end{aligned}$$

Because the enthalpy is an extensive parameter, its contribution from each reactant and product is proportional to the stoichiometric coefficients.

Similar results hold for the other extensive properties, in particular the Gibbs energy:

$$\Delta_{rxn}G = \sum_i v_i \Delta_f G_i. \qquad (12.12)$$

Gibbs energies of formation, $\Delta_f G$, are often tabulated, and $\Delta_f S$ is obtainable from these, since at constant temperature

$$\Delta S = \frac{\Delta H - \Delta G}{T}. \qquad (12.13)$$

Because the absolute molar entropies $S_m^\ominus$ can be evaluated using the third law of thermodynamics, tables of thermodynamic data generally list these instead of the entropies of formation, so $\Delta_{rxn}S$ will be evaluated directly from these:

$$\Delta_{rxn}S = \sum_i v_i S_{m,i}. \qquad (12.14)$$

There is an ambiguity in these definitions. If the enthalpies of formation are given per mole of product, are not the enthalpies of reaction molar quantities as well? The answer is yes, but molar in what? The enthalpy of reaction can be evaluated for any amount of material, but it is tabulated per mole of reactant or product. However, the reactants and products can have different stoichiometric coefficients. We could evaluate the enthalpy of our reaction in Example 12.2 per mole of HCCH or per mole of HBr, and we would get two different answers. In general, it is necessary to write the reaction for which we are evaluating the enthalpy change. The way we wrote the reaction, $\Delta_{rxn}H$ would be evaluated per mole of HCCH, but if we wrote the reaction instead as

$$\tfrac{1}{2}HCCH + HBr \rightarrow \tfrac{1}{2}CH_2BrCH_2Br, \qquad (12.15)$$

$\Delta_{rxn}H$ would be calculated per mole of HBr, giving half the value obtained per mole HCCH. This text adopts from here on the convention of giving values for the reaction enthalpy (and other reaction properties) per mole of "reaction," meaning that we take the stoichiometric coefficients to represent the number of moles of each substance. Our solution of -192.84 kJ mol^{-1} in Example 12.2, for instance, would mean that the reaction releases 192.84 kJ per mole of C_2H_2 and per 2 moles of HBr.

Temperature and Pressure Dependence

For calculating the temperature dependence of the enthalpy of reaction, we use the same method used for phase transitions at non-standard temperatures. Evaluate the change in enthalpy required to get the reactants to 298.15 K and the products from 298.15 K, and add these contributions to the $\Delta_{rxn}H^\ominus$ at 298 K. We combine these terms into a single term for the difference in heat capacities:

$$\Delta_{rxn}H^\ominus(T) = C_P(\text{reactants})(298.15 \text{ K} - T) + \Delta_{rxn}H^\ominus(298.15 \text{ K})$$
$$+ C_P(\text{products})(T - 298.15 \text{ K})$$
$$= \Delta_{rxn}H^\ominus(298.15 \text{ K}) + \Delta_{rxn}C_P(T - 298.15 \text{ K}), \qquad (12.16)$$

where

$$\Delta_{rxn}C_P = C_P(\text{products}) - C_P(\text{reactants}).$$

Similarly, for the reaction occurring at a non-standard pressure, the enthalpy of reaction can be obtained by adding the enthalpy changes as the reactants are taken to $P^\ominus$ and the products returned to the original pressure.

EXAMPLE 12.3   Pressure and Enthalpy of Reaction

CONTEXT Increasingly popular solvents include certain supercritical fluids—the hot, dense fluids at higher temperature and pressure than the critical point in the phase diagram (Section 10.5) providing the high density of a liquid without the strong intermolecular attractions that lead, for example, to high viscosities. Carbon dioxide is a popular substance for this application, because with a critical temperature of only 304 K, it does not need to be heated to temperatures that might damage the solute. However, the critical pressure of CO_2, at 74 bar, is often inconveniently high. In some applications requiring lower pressures, carbon monoxide is used instead, with a critical pressure of 35 bar (and a critical temperature of 133 K, so capable of acting as a supercritical fluid even at very low temperatures). However, researchers employ this strategy with caution for two reasons: (1) CO is toxic, and a leak into the laboratory could easily go undetected until the gas reaches hazardous concentrations; and (2) CO is flammable, and under pressure can react violently with oxygen.

PROBLEM Find an expression for the enthalpy of reaction in terms of the normal enthalpy of reaction for $CO + \frac{1}{2}O_2 \rightarrow CO_2$ at a pressure of 4.0 bar, assuming that T is maintained at 298.15 K and that these are all ideal gases. The $\Delta_{rxn}H^\ominus$ is -282.984 kJ mol^{-1}.

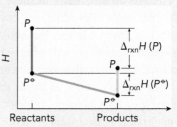

SOLUTION This enthalpy of reaction is the same as the enthalpy change for the three-step process (Fig. 12.5):

1. Reduce the CO and O_2 pressure to the standard state of 1.0 bar.

2. React the two compounds.

3. Compress the resulting CO_2 back to 4.0 bar.

▲ **FIGURE 12.5** Calculating the enthalpy of reaction at non-standard pressure.

The enthalpy change for the second step is the normal enthalpy of reaction, and ΔH for the other two steps is evaluated using the integral:

$$\Delta H = \int_{P_1}^{P_2} dH = \int_{P_1}^{P_2} (TdS + VdP)$$

$$= \int_{P_1}^{P_2} \left[\left(\frac{TdS}{dT} \right) dT + VdP \right] \qquad \text{times } dT/dT$$

$$= \int_{P_1}^{P_2} C_P dT + \int_{P_1}^{P_2} VdP. \tag{12.17}$$

The temperature is constant, so the first term is zero. (Although the pressure is changing, we used the heat capacity at constant pressure C_P under the assumption that the heat capacity of the gases does not vary significantly with pressure.) Assuming ideal gases, the second term is $nRT \ln(P_2/P_1)$, so the final value is

$$\Delta_{rxn}H(4.0\text{ bar}) = \int_{4.0\text{ bar}}^{1.0\text{ bar}} dH_{reactants} + \Delta_{rxn}H^\circ + \int_{1.0\text{ bar}}^{4.0\text{ bar}} dH_{products} \tag{12.18}$$

$$= \Delta_{rxn}H^\ominus + RT \ln\frac{4.0}{1.0}(n_{CO_2} - n_{CO} - n_{O_2})$$

$$= -282.984 \text{ kJ mol}^{-1} + (8.3145 \cdot 10^{-3} \text{ kJ mol}^{-1})(298.15 \text{ K})(\ln 4)[(1.0 - 1.0 - 0.5)\text{mol}]$$

$$= -284.70 \text{ kJ mol}^{-1}.$$

In this case the effect is a correction of about 2 kJ mol^{-1}. If the number of moles of products had been equal to the number of moles of reactants, the enthalpy of reaction would be constant at all pressures.

The same method as used in the example for gas phase chemistry applies to solution chemistry if the solvent can be assumed incompressible. In that case

$$\Delta_{rxn}H(P) = \int_{P}^{P^{\ominus}} dH + \Delta_{rxn}H(P^{\ominus}) + \int_{P^{\ominus}}^{P} dH$$

$$= \int_{P}^{P^{\ominus}} V dP + \Delta_{rxn}H(P^{\ominus}) + \int_{P^{\ominus}}^{P} V dP$$

$$= V(P - P^{\ominus}) + \Delta_{rxn}H(P^{\ominus}) + V(P^{\ominus} - P) = \Delta_{rxn}H(P^{\ominus}). \qquad (12.19)$$

Enthalpies of Combustion

Combustion is the rapid and exothermic oxidation of a compound.[2] The compound oxidized is the **fuel.** In addition to the enthalpies of formation, another standardized form of the enthalpy of reaction is made for combustible materials. These **enthalpies of combustion** are $\Delta_{rxn}H^{\ominus}$ tabulated for the reaction of one mole of the fuel with pure oxygen to yield complete conversion of all hydrogen into $H_2O(liq)$, all carbon into CO_2, all nitrogen into N_2, and other elements to their most stable oxidized form, *all at 298 K.* The idea is that we want to know how efficient the fuel is, starting from reactants still at room temperature, until the fuel is all consumed and the products have cooled back down to room temperature. Because enthalpy is a state function, the temperature of the reaction while it's occurring, even if it's thousands of degrees, will not affect the value of $\Delta_{rxn}H$ at 298 K.

> **SAMPLE CALCULATION** **Enthalpy of Combustion.** To calculate the enthalpy of combustion for gas-phase ethanol, $C_2H_5OH(g)$, we first write the balanced chemical equation,
>
> $$C_2H_5OH(g) + 3O_2(g) \rightarrow 2CO_2(g) + 3H_2O(l),$$
>
> and then use Hess's law:
>
> $$\Delta_{rxn}H = 2\Delta_f H(CO_2) + 3\Delta_f H(H_2O) - \Delta_f H(C_2H_5OH) - 3\Delta_f H(O_2)$$
> $$= [2(-393.509) + 3(-285.838) - (-235.10) - 3(0)] \text{ kJ mol}^{-1}$$
> $$= -1409.41 \text{ kJ mol}^{-1}.$$

These reaction heats are only a first, crude estimate of the net heat generated during combustion. The energy released by early combustion heats the remaining reactants, raising the temperature considerably over the 298 K reference point. Furthermore, combustion reactions normally involve many secondary reactions. In particular, with the large exothermicity of combustion, the temperature of the products may climb rapidly enough to induce thermal dissociation— so much kinetic energy is put into the vibrational coordinates that some of the bonds break. Water, for example, is one of the assumed products in hydrocarbon combustion, but it will begin to decay into hydrogen gas and OH radicals if temperatures climb over 2000 K. Many other chemical environments that are characterized by extremely energetic conditions—furnaces, the upper atmosphere, and interstellar space, for example—have this complexity.

---

[2]This is another definition open to argument, for the term combustion is often applied exclusively to reactions in which the oxidant is O_2, excluding other violent oxidations such as the $H_2(g) + Cl_2(g) \rightarrow 2HCl(g)$ reaction.

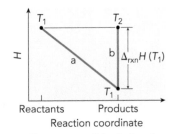

▲ **FIGURE 12.6 Calculating the adiabatic reaction temperature.** We assume a two-step process in which an exothermic reaction first occurs isothermally at temperature T_1 **(a)**, and the products are then heated from T_1 to T_2 to return the system to the original enthalpy **(b)**.

Adiabatic Reactions

As with gas expansions, the isothermal process is one convenient limit in which to study chemical reactions, the adiabatic is another. In the adiabatic process, no heat flows between the sample and its surroundings, so any heat generated during an exothermic reaction goes into heating the sample. At constant pressure under adiabatic conditions, the overall ΔH, which includes the enthalpy of reaction and the resulting heating or cooling of products, must be zero, since $(\Delta H)_P$ is equal to q, which must be zero for any adiabatic process. Figure 12.6 illustrates the enthalpy function for such a reaction. In this case, the products will be formed with the largest change in temperature, since the sample—which includes the products—cannot transfer the enthalpy of reaction to the surroundings.

This adiabatic reaction temperature (also called the **adiabatic flame temperature** for combustion reactions), can be obtained from the standard enthalpy of reaction and the heat capacities of the products. If the enthalpy of reaction at the initial temperature is $\Delta_{rxn}H(T_1)$, then

$$\Delta_{rxn}H(\text{adiabatic}) = \Delta_{rxn}H(T_1) + \int_{T_1}^{T_2} C_P(\text{products})dT = 0 \quad (12.20)$$

If the product heat capacities are constants of the temperature, or if an average value $\overline{C}_P$ for the heat capacities can be estimated, then

$$T_2 = T_1 - \frac{\Delta_{rxn}H(T_1)}{\overline{C}_P(\text{products})}. \quad (12.21)$$

One application of the adiabatic limit is to chemical reactions in ideal flames, for which the combustion reaction is too rapid for the heat to be carried away as the reaction occurs. However, the adiabatic flame temperature (like the enthalpy of combustion) is only a crude estimate, and values in excess of 3000 K are likely to significantly overestimate the temperature in the flame. Much of the heat evolved by the oxidation reactions is used up in endothermic dissociation reactions, so the final flame temperature is normally much lower.[3] For this application, it is usually a *terrible* approximation to use the heat capacity of the products at 298 K. The heat capacity of CO_2(gas) increases from 37.1 J K^{-1} mol^{-1} at 298 K to 58.4 J K^{-1} mol^{-1} at 1500 K, and for water the value changes from 33.6 J K^{-1} mol^{-1} to 47.4 J K^{-1} mol^{-1} over the same range. (Note that these are approaching the values of $15R/2$ and $7R$ expected for CO_2 and H_2O, respectively, from the equipartition heat capacity with vibrations *included*.) Failing to account for this can lead to T_2 estimates more than 50% too high.

EXAMPLE 12.4 Adiabatic Flame Temperatures

CONTEXT Carbon monoxide is one of the most abundant byproducts of hydrocarbon combustion, representing the incomplete oxidation of carbon atoms that didn't quite make it as far as carbon dioxide. The most common reason for large amounts of CO to appear among combustion products is a low oxygen concentration in the fuel mixture. With sufficient oxygen present, carbon monoxide itself will burn with a blue flame, which has been studied by thermodynamics and spectroscopy since the 19th century.

PROBLEM Calculate the adiabatic flame temperature of CO, first assuming that the CO_2 heat capacity is constant at its 298 K value of 37.11 J K^{-1} mol^{-1}, then assuming the value of 58.4 J K^{-1} mol^{-1} for 1500 K.

SOLUTION The balanced reaction is

$$CO(gas) \; + \; \frac{1}{2}O_2(gas) \; \longrightarrow \; CO_2(gas)$$

$\Delta_f H^\ominus$	-110.525	0	-393.509 kJ mol^{-1}
C_{P_m}(298 K)			37.11 J K^{-1} mol^{-1}
C_{P_m}(1500 K)			58.4 J K^{-1} mol^{-1}

From these we calculate a reaction enthalpy of

$$\Delta_{rxn}H^\ominus \; = \; -393.509 \; - \; (-110.525) \; = \; -282.984 \text{ kJ mol}^{-1}.$$

Together with the product heat capacity given for CO_2, we calculate an adiabatic flame temperature of

$$298 \text{ K} \; - \; \frac{(-282,984 \text{ J mol}^{-1})}{37.11 \text{ J K}^{-1} \text{mol}^{-1}} \; = \; 7924 \text{ K}$$

using the 298 K value of C_{Pm}, but a flame temperature of only

$$298 \text{ K} \; - \; \frac{(-282,984 \text{ J mol}^{-1})}{58.4 \text{ J K}^{-1} \text{mol}^{-1}} \; = \; 5144 \text{ K}$$

using the 1500 K value of C_{Pm}. In fact, typical CO flame temperatures are under 3000 K.

---

[3]The temperature in a flame can also be *higher* than the adiabatic flame temperature, although this is less common. One way this is accomplished is by designing a torch for the fuel that effectively circulates the heat back into the reaction mixture until a high steady-state temperature is achieved. That strategy can attain temperatures more than 400 K hotter than the adiabatic flame temperature.

12.4 Spontaneous Chemical Reactions

Most of the same questions we asked of other thermodynamic processes we can ask of chemical reactions. If we want to know which direction a chemical process will move and when it will stop (as far as our macroscopic parameters are concerned), we are dealing with the problem of spontaneity. The following rules still hold:

1. The reaction will spontaneously take the course that maximizes the *total* entropy: $\Delta S_T > 0$ (second law).

2. The heat transfer to the sample q at constant temperature is $q \leq T\Delta S$, with the equality holding only if the reaction is reversible—that is, if the total system entropy does not increase. Under these conditions, the entropy of the reservoir changes by an amount $\Delta S_{\text{res}} = -q/T = -\Delta H/T$ at constant pressure.

3. At constant temperature, $\Delta G = \Delta H - T\Delta S$.

Combining these rules, for the spontaneous reaction at constant P and T

$$\Delta G = \Delta H - T\Delta S = -T\Delta S_{\text{res}} - T\Delta S$$
$$= -T\Delta S_T \leq 0, \tag{12.22}$$

where the final result cannot be positive because T and ΔS_T must both be positive. Therefore, for a spontaneous reaction at constant temperature and under standard conditions, we have

$$\Delta_{\text{rxn}}G^{\ominus} = \Delta_{\text{rxn}}H^{\ominus} - T\Delta_{\text{rxn}}S^{\ominus} < 0, \tag{12.23}$$

and at equilibrium $\Delta_{\text{rxn}}G = 0$. This result brings together three important concepts: entropy of reaction, reversibility of reaction, and free energy of reaction. Let's look at these three concepts a little deeper in that order.

Entropy of Reaction

The first rule states that in spontaneous reactions the total entropy will increase. Recall Boltzmann's law (Eq. 2.11)

$$S \equiv k_B \ln \Omega;$$

the entropy increases with the number of states accessible to the system. There are two ways in which the entropy can increase that we have seen: the sample can remain unchanged in physical form, but its temperature or volume can increase, thereby allowing a greater density of internal states (for example, isothermal expansion of an ideal gas); or the sample itself can change to a form that has more states available (for example, vaporization of a liquid, which increases the effective number of degrees of freedom). In chemical reactions, both of these routes for changing the entropy are available, and generally both are important. The heat released or absorbed by the reaction affects the density of states of the products, and the products themselves will have different entropy than the reactants at the same temperature.

CHECKPOINT Let's briefly review the rules for counting degrees of freedom. In the gas phase, each particle—meaning an individual molecule or unbonded atom—has 3 translational degrees of freedom (or translational coordinates), each linear molecule has 2 rotational coordinates and $3N_{\text{atom}} - 5$ vibrational coordinates, and each non-linear molecule has 3 rotational coordinates and $3N_{\text{atom}} - 6$ vibrational coordinates. At room temperature, however, the vibrational coordinates rarely contribute a great deal to the available degrees of freedom because their excitation energies are so high.

An illustration is provided by the thermal decomposition of water into hydrogen and OH:

$$2H_2O(g) \rightarrow H_2(g) + 2OH(g).$$

Which side of the equation has greater entropy if the temperature of the compounds is the same on both sides? The right-hand side does. The enthalpies of formation in kJ mol^{-1} are H_2O(gas): -241.818, H_2: 0, and OH: $+38.95$, so the reaction is endothermic by $2[38.95 - (-241.82)] = +561.54$ kJ mol^{-1}, a considerable amount. However, the left-hand side has six atoms tied up in two molecules. Each water molecule has three vibrational, three rotational, and three translational degrees of freedom. The right-hand side has three diatomic molecules, each with one vibrational, two rotational, and three translational degrees of freedom. The total number of these degrees of freedom, 18, must be the same on both sides since it must be three times the number of atoms N_{atom}, and N_{atom} is a conserved quantity in chemical reactions. The difference is this: the energy levels of translational states are much closer together than rotational and vibrational states, by many orders of magnitude in any macroscopic gas-phase system. Therefore, the entropy is dominated by the number of *translational* degrees of freedom. More gas-phase molecules always means more translational degrees of freedom, and under any macroscopic conditions, this means greater entropy. For the water decomposition reaction, the entropy change is

$$[130.684 + 2(183.64) - 2(188.825)]\, JK^{-1}mol^{-1} = 120.32\, JK^{-1}mol^{-1}.$$

In chemical reactions, the entropy change is more subtle if the number of molecules is the same on both sides of the equation, because the effect of the reaction on the density of translational states is no longer obvious, and the effect of vibrational and rotational states becomes more important. For example, the reaction

$$H_2(g) + Cl(g) \rightarrow HCl(g) + H(g)$$

has the same number of each kind of degree of freedom on both sides. The entropy change is correspondingly small,

$$(-130 - 165 + 187 + 115)\, J\,K^{-1}\,mol^{-1} = +7\, J\,K^{-1}\,mol^{-1},$$

but slightly positive because the rotational and vibrational states of HCl are more dense than those of H_2.

CHECKPOINT The positive entropy for this reaction is the reason that the decomposition of water into H_2 and OH occurs in very hot flames, despite the reaction being very endothermic. As temperature increases, the entropy term in $\Delta G = \Delta H - T\Delta S$ becomes more important.

EXAMPLE 12.5 **Entropy of Reaction**

CONTEXT Sulfur dioxide is a product in the burning of fossil fuels, and if released to the atmosphere it will combine with H_2O to form sulfurous acid, one of the leading contributors to acid rain in heavily polluted areas. One method for removing SO_2 from exhaust gases is by reacting it with CO, formed during the same combustion process, to produce CO_2 and SO. A subsequent second oxygen shift to CO will convert the SO to elemental sulfur. Here we look at the entropy of that first step in the reaction.

PROBLEM Determine whether the reaction $CO(g) + SO_2(g) \rightarrow SO(g) + CO_2(g)$ should have a positive or negative $\Delta_{rxn}S$; then check the answer.

SOLUTION On the right we have two linear molecules (SO and CO_2); on the left, we have one nonlinear (SO_2) and one linear (CO). There are the same number of transitional coordinates but more rotational degrees of freedom on the left. Each of the linear molecules has 2 rotational degrees of freedom, but the SO_2 has 3. Therefore, the number of rotational coordinates—which are the most easily excited after translations—drops from 5 for the reactants to 4 for the products. Therefore, we can expect $\Delta_{rxn}S$ to be small (because the number of translation coordinates is unchanged) and negative (because the number of rotational coordinates decreases). The calculated value is

$$\underset{\text{CO}}{} \quad \underset{\text{SO}_2}{} \quad \underset{\text{SO}}{} \quad \underset{\text{CO}_2}{}$$
$$(-197.67 - 248.22 + 221.8 + 213.74) \text{ J K}^{-1} \text{ mol}^{-1} = -10.4 \text{ J K}^{-1} \text{ mol}^{-1}$$

Reversible Reactions

Chemical transformations occur by moving atoms along the coordinates of the reaction surface. There's no fundamental restriction on the direction that the atoms travel along that surface, though, so every reaction path where substance A turns into substance B is—at least theoretically—a two-way street, such that B can also be converted into A by moving backward along the same reaction coordinate. If A → B and A ← B are both observable, then we call the reaction **reversible** and write A ⇌ B. The term "reversible" when applied to reactions deviates somewhat from the precise meaning of a *reversible process* in thermodynamics, defined in Section 7.2. In particular, we can carry out a reversible reaction in such a way that the overall entropy of the system increases. For example, glacial acetic acid is pure CH_3COOH, and it dissociates spontaneously in water to form the acetate ion:

$$CH_3COOH(aq) \rightarrow CH_3COO^-(aq) + H^+(aq).$$

After a short time enough ions are present in the solution that the reverse reaction can also occur:

$$CH_3COO^-(aq) + H^+(aq) \rightarrow CH_3COOH(aq).$$

The reaction is reversible, but the overall process—mixing pure acetic acid and water to obtain acetic acid solution—is **irreversible** because we cannot separate the starting materials without driving down the entropy of our system.[4]

A number of cases arise in which, for practical purposes, the reaction is irreversible, including highly exothermic reactions, decompositions that result in several fragments, and photodissociations. The reverse of a highly exothermic reaction is a reaction that must *absorb* a huge amount of energy to proceed. If that energy isn't available—for example, because the reaction is carried out in an ice bath and heat from the forward reaction is removed—then the reverse reaction can't take place at a measurable rate. Decomposition reactions and photodissociations may be difficult to reverse because the fragments must

---

[4]We get away with using these two definitions of "reversibility" because we can distinguish between the bulk reaction (the total number of reactant molecules converted to product) and the microscopic reaction (a single collision, say, between two reactant molecules that results in products). If individual reactions at the microscopic level proceed in either direction, then the reaction is reversible. Thermodynamics doesn't care what happens at the microscopic level, however, and the reversibility of the bulk process depends on the overall change in entropy from all of these microscopic events.

come together again not only with enough energy but also with a specific orientation, and this becomes increasingly unlikely as the number and complexity of the fragments increase. In brief, an irreversible reaction is one where the overall entropy increase is so great (either because $\Delta_{rxn}S$ is big or because the reaction releases a lot of heat) that the reverse reaction becomes insignificant.

In a reversible reaction, one direction has a negative $\Delta_{rxn}S$. Let's not forget that it's still possible for the reaction to move in this direction, because the second law of thermodynamics, $\Delta S_T \geq 0$, only requires that the *entire* system entropy, including surroundings, does not decrease. We can reversibly compress a gas to a state of lower entropy if the drop in entropy of the sample is balanced by a rise in the entropy of the surroundings. Similarly, if a reaction occurs in a chamber and produces many moles of gas-phase product with positive $\Delta_{rxn}S$, we can compress the chamber and (at least partly) reverse the reaction. Again, the entropy of the surroundings must increase so that the second law is satisfied.

Free Energy of Reaction

At constant T and P, the simplest thermodynamic potential is the Gibbs energy. At constant temperature, we can write Eq. 9.25 for a chemical reaction:

$$(\Delta_{rxn}G)_T = G(\text{products}) - G(\text{reactants})$$

$$= \Delta_{rxn}H - T\Delta_{rxn}S. \tag{12.24}$$

The quantities $\Delta_{rxn}H$ and $\Delta_{rxn}S$ can be evaluated at any temperature and pressure from tabulated constants. Equation 12.24 lets us in turn evaluate $\Delta_{rxn}G$ over a large range of conditions. For standard reactions, such as formation and combustion, the Gibbs energy at the standard state and 298.15 K is often included among the tabulated constants.

But here's the real strength of the Gibbs energy, for our purposes: it depends only on the chemical potentials. We rewrite $\Delta_{rxn}G$ using Eq. 7.26 for a chemical reaction at constant temperature *and* pressure:

$$(\Delta_{rxn}G)_{T, P} = \int_{\text{reactants}}^{\text{products}} dG = \int_{\text{reactants}}^{\text{products}} \sum_i \mu_i dn_i. \tag{12.25}$$

For the reversible reaction, the chemical potentials are well-defined at every point, and the change in the number of moles dn of each chemical component is given by the stoichiometric coefficients, so

$$(\Delta_{rxn}G)_{T, P} = \sum_i v_i \mu_i, \tag{12.26}$$

where the v_i's are the stoichiometric coefficients, negative for reactants, positive for products. For the rest of this section, let's let T and P be fixed.

We can express this ΔG in terms of the Gibbs energy of reaction of the standard state, $\Delta_{rxn}G^{\ominus}$. This is the change in Gibbs energy when we begin with pure, separated reactants and end with pure, separated products. Although there is a Gibbs energy associated with any thermodynamic state during the reaction, the chemical potentials cause that Gibbs energy, which is for a mixture of products and reactants, to be different from the standard state

Gibbs energy. If we use our earlier empirical formula, Eq. 11.19, for the chemical potential,

$$\mu = \mu^{\ominus} + RT \ln a,$$

then we can rewrite the Gibbs energy as

$$\Delta_{rxn}G = \sum_i v_i \mu_i^{\ominus} + RT \sum_i v_i \ln a_i$$

$$\equiv \Delta_{rxn}G^{\ominus} + RT \ln \Xi, \tag{12.27}$$

where

$$\Delta_{rxn}G^{\ominus} = \sum_i v_i \mu_i^{\ominus} \tag{12.28}$$

and where the **activity quotient** Ξ is given by[5]

$$\ln \Xi = \sum_i v_i \ln a_i = \sum_i \ln a_i^{v_i} = \ln\left(\prod_i a_i^{v_i}\right), \tag{12.29}$$

or simply

$$\Xi = \prod_i a_i^{v_i}. \tag{12.30}$$

In fact, the activity quotient can help to describe our reaction when it travels an irreversible path, but in the next section we look at the value of Ξ only in the simplest case.

TOOLS OF THE TRADE | Reaction Calorimetry

What is reaction calorimetry? A **reaction calorimeter** measures the rate of heat flow $đq/dt$ in and out of a chemical reaction, in preparation for the scale-up to an industrial reactor.

Why do we use reaction calorimetry? In research laboratories, calorimetry is often carried out on tiny samples, a few grams or less, to conserve material and to manage the relatively large energy changes that may accompany chemical reactions. Small samples also allow the equipment to be designed to accommodate a large range of different types of materials and processes.

In contrast, industrial chemical reactors operate on scales up to hundreds of kilograms, and for that reason they are typically designed and built for a specific process. Optimizing the reactor design involves balancing competing concerns—for example, providing thermal contacts with the reaction mix so heat can be channeled away from an exothermic reaction, but without getting in the way of the agitator that mixes the reactants. Careful design of large reactors requires that the process has been thoroughly characterized at smaller scales. Furthermore, it is not so much the overall heat of the reaction that matters as much as the *rate* at which heat must be transferred in or out of the reactor.

How does it work? The sample volume may range from a few milliliters to about a liter. Once the reaction is initiated, heat flows across a threshold between the reactor and a temperature reservoir called the *jacket*. The jacket contains a fluid, usually several liters and often composed of water mixed with an agent to extend its useful temperature range. The fluid is circulated past the reactor, passing through a pump and temperature controller, which allows the jacket temperature to be regulated. The choice of fluid and the large jacket volume are designed to keep the heat capacity of the jacket much larger than the heat capacity of the mixture in the reactor.

---

[5]We see here one of the rare occurrences of the product symbol Π, which does for multiplication what Σ does for addition. It's interesting that these two fundamental symbols for sequence operations Σ and Π (sigma and pi, the Greek equivalents of s and p as in *sum* and *product*) also correspond to the two simplest angular momentum states of molecules, Σ and Π. Coincidence? I think so.

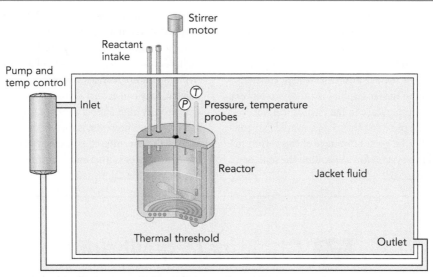

▲ **Schematic of a typical reaction calorimeter.**

Reaction calorimetry may be carried out under any of several sets of conditions, depending on the type of reaction to be studied, and sometimes just on the basis of what equipment happens to be available or affordable.

- **Isothermal heat flow calorimetry** maintains a constant temperature in the reactor mix by using an active feedback system to change the temperature of the jacket to compensate for heat flowing in or out of the reactor. The heat flow rate is then given by

$$\frac{đq_{rxn}}{dt} = -hA(T_{rxn} - T_{res}) + m_{rxn}c_{rxn}\frac{dT_{rxn}}{dt},$$

with the parameters defined in the following table.

parameter	definition	units
q_{rxn}	heat of the reaction	J
h	heat transfer coefficient of the threshold	$W\,m^{-2}\,K^{-1}$
A	area of the threshold	m^2
T_{rxn}	temperature of the reaction mixture	K
T_{res}	temperature of the jacket	K
m_{rxn}	mass of the reaction mixture	kg
c_{rxn}	specific heat capacity of the reaction mixture	$J\,K^{-1}\,kg^{-1}$

This technique is perhaps the most general, but it relies on careful calibration to determine the heat transfer coefficient h so that the temperature differential across the threshold is correctly converted to an energy flow rate.

- A variation is **isothermal heat balance calorimetry,** where the reactor is again kept at a constant temperature, but this time the temperature of the reactor is corrected by heating or cooling the jacket fluid. As the jacket fluid circulates, the temperature difference between the fluid before and after passing across the reactor is used to adjust the temperature of the fluid at the inlet. This method does not require calibrating the heat transfer rate across the threshold, but it instead requires a precise measurement of small temperature changes (perhaps hundredths of K) for the high-heat-capacity jacket fluid. In this case, the heat flow rate is

$$\frac{đq_{rxn}}{dt} = -\frac{dm_{res}}{dt}c_{res}\Delta T_{res} + m_{rxn}c_{rxn}\frac{dT_{rxn}}{dt},$$

where ΔT_{res} is the temperature change from inlet to outlet of the jacket fluid and $\frac{dm_{res}}{dt}$ is the mass flow rate through the jacket.

- **Isothermal power consumption calorimetry** fixes the jacket to a constant temperature some 10–20 K below the reactor so that heat is continually flowing from the reactor to the jacket. To compensate for the heat lost to the jacket, an external heater coil is submerged in the reaction mixture to keep the reactor temperature constant. The temperature of the heating coil can be adjusted at much faster timescales than the jacket fluid, so this method can characterize reaction systems with fast kinetics. Furthermore, this method does not require calibration of the heat flow rate across the threshold, because the recorded quantity is the *power delivered to the heater* while keeping a constant temperature, rather than the heat flow across the threshold. However, the heater can affect the chemistry of very reactive mixes, and uneven distribution of the heat can be a complicating factor.

- In **isoperibolic calorimetry,** it is the jacket temperature, not the reactor temperature, that is kept constant. The heat capacity of the jacket fluid is chosen to be very high so that the temperature in the reactor will not change drastically, provided that the threshold can conduct heat quickly. The great advantage of this technique is that no feedback system for controlling the reactor temperature is required.

- In **quasi-adiabatic calorimetry,** the jacket temperature is adjusted to remain equal to the reactor temperature so that the heat flow across the threshold is zero. This can be hazardous—reaction rates tend to increase rapidly as the temperature increases, so exothermic reactions run adiabatically can accelerate out of control, leading to dangerously high temperatures or explosion. One practical advantage, however, is that the jacket does not need to have the enormous heat capacity that's needed for the various types of isothermal calorimetry, so less fluid is needed and the calorimeter can be much smaller.

12.5 Chemical Equilibrium

Chemical equilibrium is important to us for two reasons: it is the state our system will attain if we wait forever, and it defines the *direction* that our system will be moving until then. Isolated systems approach equilibrium, and at equilibrium the values of the thermodynamic parameters—including the amount of each chemical component—stop changing. This principle allows us to take as dynamic a process as a chemical reaction and examine it without accounting for the time-dependence of the parameters. Here we consider what chemical equilibrium requires of the system, and this will serve as a reference point when we begin looking at the time-dependence of the reaction later.

Equilibrium Constants

At equilibrium, the activity quotient takes on the special value K_{eq}, because

$$\Delta_{rxn}G = \Delta_{rxn}G^{\ominus} + RT \ln K_{eq} = 0. \tag{12.31}$$

Solving for K_{eq}, we arrive at a unitless quantity called the **equilibrium constant** of the reaction,

$$K_{eq} = e^{-\Delta_{rxn}G^{\ominus}/(RT)}. \tag{12.32}$$

Therefore, to calculate K_{eq}, we may find the difference in the free energies of formation of products and reactants:[6]

$$\ln K_{eq} = -\frac{\Delta_{rxn}G^{\ominus}}{RT} = -\frac{\sum_i v_i \Delta_f G_i^{\ominus}}{RT}. \tag{12.33}$$

---

[6]Keep in mind that the *standard state* free energies of reaction $\Delta_{rxn}G^{\ominus}$ are *not* the free energies of the reactions as they take place in the laboratory. We calculate the standard state free energies by assuming that the reaction goes to completion. For any reaction at equilibrium (at constant T and P), the actual free energy change $\Delta_{rxn}G$ is zero, because the net reaction stops *before* it reaches completion.

Because K_{eq} is just a special case of Ξ, the equilibrium constant can also be expressed as

$$K_{eq} = \prod_i a_{i,eq}^{v_i}, \tag{12.34}$$

where the activities a_i take their equilibrium values. For the general reaction

$$aA + bB \rightarrow cC + dD,$$

the equilibrium constant takes on the familiar appearance

$$K_{eq} = a_A^{-a} a_B^{-b} a_C^c, a_D^d = \frac{a_C^c a_D^d}{a_A^a a_B^b}, \tag{12.35}$$

according to the definition of Ξ. By this definition, the equilibrium constant is *always unitless*, because the activities are unitless.

We will stick to this unitless definition of K_{eq} for convenience in linking the statistical and thermodynamic formulations, but the more common way of writing the equilibrium constant replaces the activities by the concentrations (for solutes) or partial pressures (for gases), to give the equilibrium constants commonly labeled K_c or K_P in general chemistry. Those constants, however, are also reported without units. At first sight, this seems worrisome, because the numerical value of the equilibrium constant is then affected by the units chosen for concentration or pressure, which, by Eq. 12.32, would in turn alter the value of the Gibbs energy $\Delta_{rxn}G^{\ominus}$. In practice, it doesn't matter, because we let the choice of units in our definition of K determine the standard state values used in $\Delta_{rxn}G^{\ominus}$. Say, for example, that we measured the partial pressures for a gas-phase reaction $A + B \rightleftharpoons 2C + D$ in bar initially:

$$K_P = \frac{P_C^2 P_D}{P_A P_B}.$$

If we then change all the partial pressures to units of atm, this will result in a new numerical value K'_P such that $K'_P = 0.987 K_P$. When we calculate the Gibbs energies from these two equilibrium constants, here's what we find:

$$\Delta_{rxn}G^{\ominus}_{bar} = -RT \ln K_P$$
$$\Delta_{rxn}G^{\ominus}_{atm} = -RT \ln K'_P$$
$$= -RT(\ln 0.987 + \ln K_P)$$
$$= -RT \ln 0.987 + \Delta_{rxn}G^{\ominus}_{bar}.$$

In other words, changing the units used to calculate K has the effect of shifting all the Gibbs energies by some added constant. This is equivalent to merely shifting the standard state used as a reference point for *all* the Gibbs energies, and it does not effect our conclusions, which are based only on *changes* in Gibbs energy. In practice, the standard state that is most often chosen puts all solutes at a concentration of 1 M and all gases at partial pressures of 1 bar, so concentrations in K_c are given as molarities and pressures in K_p as bar. For our general equilibrium constant, K_{eq}, we need the activities for the substances we're likely to encounter. We gather our results for the activities from Chapters 10 and 11 into Table 12.1.

TABLE 12.1 Activity laws for selected ideal systems.

substance	activity
ideal gas	$P_i/P^\ominus$
pure solid or liquid	1
Raoult's law (for solvent A)	$X_A = P_A/P_A^*$
Henry's law (for solute B)	$X_B = P_B/k_X$ or $[B]/1\,M$
non-ideal solution	$\gamma_i X_i$

If the ideal approximations are safe to apply to our reaction components, then we can replace the activities in Eq. 12.35 and predict the equilibrium constant. Once at equilibrium, the reaction has effectively ceased as far as macroscopic observables are concerned, so this gives the endpoint of the reaction. If we know the amounts of reactant and product present at some initial time, we can predict all the amounts once equilibrium is reached.

For example, for the reaction

$$I_2(aq) + Br_2(aq) \rightarrow 2IBr(aq)$$

in a Raoult's law solution, the equilibrium constant would be written as

$$K_{eq} = \frac{X_{IBr}^2}{X_{I_2} X_{Br_2}}. \tag{12.36}$$

We can calculate the equilibrium constant K_{eq} from the standard Gibbs energy of the reaction. If we start with equal concentrations c of the two reactants, then we can solve for the concentration $2x$ of product at equilibrium. The concentrations of product will each have diminished by an amount x, because there are two IBr molecules formed for each molecule of the two reactants consumed. Then we solve for x:

$$K_{eq} = \frac{X_{IBr}^2}{X_{I_2} X_{Br_2}}$$

$$= \frac{(2x)^2}{(c-x)^2} = \left(\frac{2x}{c-x}\right)^2$$

$$\sqrt{K_{eq}} = \left(\frac{2x}{c-x}\right)$$

$$(c-x)\sqrt{K_{eq}} = 2x$$

$$c\sqrt{K_{eq}} = (2 + \sqrt{K_{eq}})x$$

$$x = \frac{c\sqrt{K_{eq}}}{2 + \sqrt{K_{eq}}}. \tag{12.37}$$

Similarly, if we assume ideal gases for the reaction

$$2O_2(g) + Cl_2(g) \rightarrow 2ClO_2(g),$$

the equilibrium constant is

$$K_{eq} = \frac{P_{ClO_2}^2 P^\ominus}{P_{O_2}^2 P_{Cl_2}}. \tag{12.38}$$

If we started with no product, just O_2 and Cl_2, each in the amount c, then the partial pressure x of the product at any time is

$$P_{ClO_2} = x = c - P_{O_2} = 2(c - P_{Cl_2}).$$

At equilibrium, x can be determined in terms of these initial reactant pressures, because

$$x^2 = \frac{K_{eq}}{P^{\ominus}} P_{O_2}^2 P_{Cl_2}$$

$$= \frac{K_{eq}}{P^{\ominus}} (c - x)^2 \left(c - \frac{x}{2} \right).$$

This is a cubic equation in $x = P_{ClO_2}$, and it can be solved to get the pressure at equilibrium, but nobody said it would be easy. In this example and many others, solving for the equilibrium amounts of reactant and product often requires finding the roots of a high-order polynomial. These can always be solved numerically if necessary.

EXAMPLE 12.6 Activity Under Phase Equilibrium

CONTEXT In classical thermodynamics, a phase transition is no different from a chemical reaction. The enthalpy of vaporization for water is measured the same way we would measure the enthalpy of reaction for a chemical equation A → B at the same temperature. In this example, we show that the principles we are applying now to the concept of chemical equilibrium are the same as the principles of phase equilibrium that we derived in Chapter 10.

PROBLEM What is the activity of the water vapor in the water liquid–gas equilibrium at $T = 298$ K?

SOLUTION Since different phases of the same compound are treated as different compounds in classical thermodynamics, a phase transition is treated like a chemical reaction, in this case the reaction

$$H_2O(l) \rightarrow H_2O(g).$$

The Gibbs energies of formation are obtained from tables or from the enthalpies of formation and standard entropies. In this case,

$$\Delta_{rxn}G^{\ominus} = \Delta_f G[H_2O(g)] - \Delta_f G[H_2O(l)]$$

$$= (-228.572 \text{ kJ mol}^{-1}) - (-237.129 \text{ kJ mol}^{-1}) = 8.557 \text{ kJ mol}^{-1}.$$

The equilibrium constant is therefore

$$K_{eq} = e^{-\Delta_{rxn}G^{\ominus}/(RT)} = e^{-3.454} = 0.0317.$$

The equilibrium constant can also be written in terms of the activities:

$$K_{eq} = \frac{a[H_2O(g)]}{a[H_2O(l)]} = \frac{P}{P^{\ominus}},$$

where P is the water vapor pressure. Our equilibrium constant is the water vapor pressure in units of bar, so the normal vapor pressure of water is 0.0317 bar.

The equilibrium constant has another very convenient form that offers a different perspective on its physical significance. We know how to relate ΔG to the energy and other thermodynamic parameters, and we know how those parameters are related to the partition function Q. So we can write $\Delta_{rxn}G^{\ominus}$ in terms of the partition function, with help from Chapters 6 and 7:

$$G = E - TS + PV \qquad \qquad \text{by Eq. 7.19}$$

$$= k_B T^2 \left(\frac{\partial \ln Q(T)}{\partial T} \right)_{V,n} - T \left[k_B \ln Q(T) + k_B T \left(\frac{\partial \ln Q(T)}{\partial T} \right)_{V,n} \right]$$

$$+ k_B TV \left(\frac{\partial \ln Q}{\partial V} \right)_{T,n} \qquad \qquad \text{by Eqs. 6.15, 7.35, 7.37}$$

$$= k_B T \left[-\ln Q(T) + V \left(\frac{\partial \ln Q}{\partial V} \right)_{T,n} \right].$$

Let's assume for convenience that the system is an incompressible liquid, so we may neglect the volume derivative. Then we have

$$\Delta_{rxn}G^{\ominus} = \Delta_{rxn} \left[-k_B T \ln Q(T) \right] = -k_B T \left[\ln Q_{products} - \ln Q_{reactants} \right]$$

$$= -k_B T \ln \left(\frac{Q_{products}}{Q_{reactants}} \right) \qquad \qquad (12.39)$$

$$K_{eq} = e^{-\Delta_{rxn}G^{\ominus}/(RT)} = \frac{Q_{products}}{Q_{reactants}}. \qquad \qquad (12.40)$$

Notice two things here. First, this shows that the expression $K_{eq} = e^{-\Delta_{rxn}G^{\ominus}/(RT)}$ is nothing more than the canonical distribution $e^{-E(k_B/T)}$ showing up again.

Okay, maybe you're not impressed by that, but this next point really is amazing. The last equation for the equilibrium constant, $K_{eq} = Q_{products}/Q_{reactants}$, shows that equilibrium is effectively just the ratio of the number of product quantum states available to the number of reactant quantum states available. In the same way that two samples at different temperatures find the state of maximum entropy by exchanging heat to arrive at an intermediate temperature (Sec. 9.2), a chemical system finds a state of maximum entropy (and minimum free energy) by arriving at a mixture of reactants and products. The system is trying to distribute the molecular structures so as to obtain the greatest number of possible quantum states, the greatest ensemble size. This is the most fundamental, and arguably the simplest, of all pictures of chemical equilibrium. Look at the ensemble of states for the pure, separated reactants and at the ensemble for the pure, separated products. The ratio between the two is your equilibrium constant.

Temperature-Dependence of Equilibrium Constants

The equilibrium in a reaction will generally shift as a function of temperature. This makes sense, since the equilibrium state is one where $\Delta G = 0$, and ΔG depends on $T\Delta S$. We begin by writing the formula for $\ln K_{eq}$ from Eq. 12.32 for some new temperature T_2:

$$\ln K_{eq}(T_2) = -\frac{\Delta_{rxn}G^{\ominus}(T_2)}{RT_2}. \qquad \qquad (12.41)$$

But we must not stop here, because $\Delta_{rxn}G^{\ominus}$ is itself a function of the temperature:

$$\Delta_{rxn}G^{\ominus}(T_2) \;=\; \Delta_{rxn}H^{\ominus}(T_2) \;-\; T_2\Delta_{rxn}S^{\ominus}(T_2),$$

and even $\Delta_{rxn}H^{\ominus}$ and $\Delta_{rxn}S^{\ominus}$ are functions of temperature:

$$\Delta_{rxn}H^{\ominus}(T_2) = \Delta_{rxn}H^{\ominus}(T_1) \;+\; \int_{T_2}^{T_1} C_{Pm}(r)dT \;+\; \int_{T_1}^{T_2} C_{Pm}(p)dT$$

$$= \Delta_{rxn}H^{\ominus}(T_1) \;+\; (T_2 - T_1)\Delta_{rxn}C_{Pm}, \tag{12.42}$$

$$\Delta_{rxn}S^{\ominus}(T_2) = \Delta_{rxn}S^{\ominus}(T_1) \;+\; \int_{T_2}^{T_1}\left(\frac{C_{Pm}(r)}{T}\right)dT \;+\; \int_{T_1}^{T_2}\left(\frac{C_{Pm}(p)}{T}\right)dT$$

$$= \Delta_{rxn}S^{\ominus}(T_1) \;+\; \Delta_{rxn}C_{Pm}\ln\left(\frac{T_2}{T_1}\right), \tag{12.43}$$

where r = reactants and p = products. The importance of the temperature corrections for $\Delta_{rxn}S^{\ominus}$ and $\Delta_{rxn}C_{Pm}$ increases the further T_2 is from our reference temperature T_1, which will usually be 298.15 K. The enthalpy correction (which is proportional to the temperature difference) is usually much more important than the entropy correction (which varies more slowly as $\ln(T_2/T_1)$). Our result still neglects any temperature dependence of the heat capacities.

To get an explicit formula for $\ln K_{eq}(T_2)$, we need only to combine these equations:

$$\ln K_{eq}(T_2) = -\frac{\Delta_{rxn}G^{\circ}(T_2)}{RT_2}$$

$$= -\frac{1}{R}\left[\frac{\Delta_{rxn}H^{\circ}(T_2)}{T_2} \;-\; \Delta_{rxn}S^{\circ}(T_2)\right]$$

$$= -\frac{1}{R}\left[\frac{\Delta_{rxn}H^{\ominus}(T_1)}{T_2} \;+\; \frac{(T_2 - T_1)\Delta_{rxn}C_{Pm}}{T_2} \;-\; \Delta_{rxn}S^{\ominus}(T_1)\right.$$

$$\left. - \Delta_{rxn}C_{Pm}\ln\left(\frac{T_2}{T_1}\right)\right]. \tag{12.44}$$

Additional algebra shows that if the equilibrium constant is already known at temperature T_1, then $\ln K_{eq}(T_2)$ can be written in terms of that value as follows:

$$\ln K_{eq}(T_2) = \ln K_{eq}(T_1) \tag{12.45}$$

$$+ \frac{1}{R}\left[(\Delta_{rxn}H^{\ominus}(T_1) - T_1\Delta_{rxn}C_{Pm})\left(\frac{1}{T_1} - \frac{1}{T_2}\right) + \Delta_{rxn}C_{Pm}\ln\frac{T_2}{T_1}\right].$$

For a small deviation of the temperature from T_1, a simpler form suffices in which we eliminate the entropy correction:

$$\ln K_{eq}(T_2) \approx \ln K_{eq}(T_1) - \frac{\Delta_{rxn}H^{\ominus}}{R}\left(\frac{1}{T_2} - \frac{1}{T_1}\right). \tag{12.46}$$

Another useful expression of the temperature dependence appears if we rewrite Eq. 12.33 as $\ln K_{eq} = -(\Delta_{rxn}H^{\ominus} - T\,\Delta_{rxn}S^{\ominus})/(RT)$. If we take the derivative of both sides with respect to T, we obtain the **van't Hoff equation:**

$$\left(\frac{\partial \ln K_{eq}}{\partial T}\right)_P = \frac{\Delta_{rxn}H^{\ominus}}{RT^2}. \tag{12.47}$$

The van't Hoff equation is an application of **Le Châtelier's principle,** which states that when a stress is applied to a chemical reaction, equilibrium is shifted in the direction that will relieve the stress. In this case, K_{eq} is shifted in favor of the products when the temperature increases ($d\ln K_{eq}/dT$ is positive) if the reaction is endothermic ($\Delta_{rxn}H$ is positive), and vice versa if the reaction is exothermic. The temperature change is compensated by an opposite change in the heat generated by the reaction.

EXAMPLE 12.7 | Equilibrium Constant at Non-Standard Temperature

CONTEXT Silane, SiH_4, is a particularly exciting laboratory reagent to work with, being both a highly toxic gas and *pyrophoric*—igniting and burning on contact with room-temperature air. Silane is the silicon-based version of methane and has a similar tetrahedral molecular structure but with much weaker bonds. Low concentrations of impurities can catalyze the oxidation of silane, releasing enough heat to trigger the direct reaction of O_2 with SiH_4 to form SiO_2 and H_2O. Despite its hazards, silane is an important industrial compound, providing a quick route to get silicon atoms into the gas phase for numerous applications including the formation of integrated circuits by chemical vapor deposition.

PROBLEM Predict $\ln K_{eq}$ for the following reaction at 450 K:

$$SiH_4(g) + 2O_2(g) \rightarrow SiO_2(s) + 2H_2O(g).$$

SOLUTION We need the following constants:

	$SiH_4(g)$	$O_2(g)$	$SiO_2(s)$	$H_2O(g)$
$\Delta_f H^\ominus$ (kJ mol^{-1})	34.3	0.0	-903.49	-241.818
$\Delta_f G^\ominus$ (kJ mol^{-1})	56.9	0.0	-850.73	-228.572
C_{Pm} (J K^{-1} mol^{-1})	42.84	29.355	44.4	33.577

The equilibrium constant at 298.15 K is obtained in the usual way from the $\Delta_f G^\ominus$ values:

$$\Delta_{rxn}G^\ominus(298.15 \text{ K}) = \{[-850.73 + 2(-228.572)] - [56.9 + 2(0.0)]\} \text{ kJ mol}^{-1}$$

$$= -1364.7 \text{ kJ mol}^{-1}$$

$$\ln K_{eq}(298.152 \text{ K}) = -\frac{\Delta_{rxn}G^\ominus(298.152 \text{ K})}{RT}$$

$$= -\frac{(-1364.7 \text{ kJ mol}^{-1})}{(0.0083145 \text{ kJ K}^{-1} \text{ mol}^{-1})(298.15 \text{ K})}$$

$$= 550.5.$$

If we first assume that the enthalpy of reaction is constant at its 298.15 K value,

$$\Delta_{rxn}H^\ominus(298.15 \text{ K}) = \{[-903.49 + 2(-241.818)] - [34.3 + 2(0.0)]\} \text{ kJ mol}^{-1}$$

$$= -1421.4 \text{ kJ mol}^{-1},$$

then we can estimate the equilibrium constant using Eq. 12.46:

$$\ln K_{eq}(450\ \text{K}) = 550.5 - \frac{-1421.4\ \text{kJ mol}^{-1}}{0.0083145\ \text{kJ K}^{-1}\ \text{mol}^{-1}}\left(\frac{1}{450\ \text{K}} - \frac{1}{298.15\ \text{K}}\right)$$

$$= 550.5 - (193.5) = 357.0.$$

As required for an exothermic reaction, heat drives the equilibrium down, toward the reactants.

Now, to correct for the temperature dependence of the enthalpy of reaction, we combine the heat capacities

$$\Delta_{rxn}C_{Pm} = \{[44.4 + 2(33.577)] - [42.84 + 2(29.355)]\}\ \text{J K}^{-1}\ \text{mol}^{-1}$$

$$= 10.0\ \text{J K}^{-1}\ \text{mol}^{-1} = 0.0100\ \text{kJ K}^{-1}\ \text{mol}^{-1},$$

and use Eq. 12.45:

$$\ln K_{eq}(450\ \text{K}) = 550.5 + \frac{1}{0.0083145\ \text{kJ K}^{-1}\ \text{mol}^{-1}}\left\{\left[-1421.4\ \text{kJ mol}^{-1}\right.\right.$$

$$\left.-(298.15\ \text{K})(0.0100\ \text{kJ K}^{-1}\ \text{mol}^{-1})\right]\left(\frac{1}{298.15\ \text{K}} - \frac{1}{450\ \text{K}}\right)$$

$$\left. + (0.0100\ \text{kJ K}^{-1}\ \text{mol}^{-1})\ln\left(\frac{450}{298.15}\right)\right\}$$

$$= 550.5 + (-193.4) = 357.1.$$

In this case, the correction for the temperature dependence of $\Delta_{rxn}H^{\ominus}$ is small, and Eq. 12.46 works nearly as well.

CONTEXT *Where Do We Go From Here?*

In this chapter we've studied how a chemical reaction takes our system from one thermodynamic state to another, including how the entropy change directs this process and how the enthalpy change can inform us about this process. A major limitation of our approach has been that it allows us to describe only three distinct states of the system: the standard state reactants (pure and separated, before the reaction begins), the standard state products (which assume the reaction has gone to completion), and the equilibrium state (which lies in between). However, our reactions often don't reach equilibrium, either because the approach to equilibrium is too slow or because the system is not isolated, with material and energy entering or exiting the system as the reaction takes place. To describe what happens *as* the reaction occurs, we need a new approach—more flexible but also more demanding than the equilibrium thermodynamics we've applied here. For this job, we need kinetics, which will draw on our work in this chapter, but which will also rely more critically on our understanding of molecular structure and transport.

KEY CONCEPTS AND EQUATIONS

12.2 **Reaction Surfaces.** Chemical reactions involve changes in the relative positions of the atoms, which are accompanied by a redistribution of energy among the potential energies of the chemical bonds and the kinetic energies of the reactants and products. Looking at the process from the molecular level, we can graph the transformation of reactants into products as a path along a **reaction surface,** a plot of the potential energy of the particles involved in the reaction as a function of their positions. On a typical reaction surface, the reactants correspond to a range of geometries near one minimum, and the products to a range of geometries near another minimum. The lowest energy path from reactants to products requires crossing a saddle point at a geometry called the **transition state.** Graphing just the minimum energy path gives a **reaction diagram,** the potential energy of the molecular system as a function of the **reaction coordinate,** usually a combination of several bond length and bond angle changes that smoothly shift the reactant structure to the product structure.

12.3 **Enthalpies of Reaction.**
- **Hess's law** allows us to calculate the enthalpy of reaction at 1 bar and 298 K as a linear combination of the **enthalpies of formation** $\Delta_f H$ of the reactants and products:

$$\Delta_{rxn}H = \sum_i v_i \Delta_f H_i, \qquad (12.11)$$

where v_i is the stoichiometric coefficient for each component in the reaction, with the value for any reactant multiplied by -1.

- To calculate the enthalpy of an isothermal chemical reaction at a temperature other than 298 K, we can correct the enthalpy of reaction calculated at 298 K using Hess's law by a term for the difference in heat capacities in reactants and products needed to reach the new temperature:

$$\Delta_{rxn}H^\ominus(T) = \Delta_{rxn}H^\ominus(298.15\,\mathrm{K}) + \Delta_{rxn}C_P(T - 298.15\,\mathrm{K}).$$

- Using a similar strategy we find that the final temperature T_2 achieved by a reaction taking place under adiabatic conditions from an initial temperature T_1 is

$$T_2 = T_1 - \frac{\Delta_{rxn}H(T_1)}{\overline{C}_P(\text{products})}, \qquad (12.21)$$

assuming that the predicted temperature change is small enough that the reaction conditions do not change substantially.

12.4 **Spontaneous Chemical Reactions.** A chemical reaction is spontaneous if

$$\Delta_{rxn}G^\ominus = \Delta_{rxn}H^\ominus - T\Delta_{rxn}S^\ominus < 0. \,(12.23)$$

12.5 **Chemical Equilibrium.**
- The **activity quotient** Ξ may be expressed as the product of the activities of all the reactants and products, each raised to the power of the stoichiometric coefficient v_i:

$$\Xi = \prod_i a_i^{v_i}. \qquad (12.30)$$

- The equilibrium constant is the special case of the activity quotient when all the activities are at their equilibrium values,

$$K_{eq} = e^{-\Delta_{rxn}G^\ominus/(RT)}. \qquad (12.32)$$

- The **van't Hoff equation** uses the dependence of $\Delta_{rxn}G^\ominus$ on temperature to predict the temperature dependence of the equilibrium constant:

$$\left(\frac{\partial \ln K_{eq}}{\partial T}\right)_P = \frac{\Delta_{rxn}H^\ominus}{RT^2}. \qquad (12.47)$$

KEY TERMS

- The **activation energy** is effectively the energy difference between the transition state and the reactants.
- **Calorimetry** is the measurement of heat flow into or out of a system for the purposes of characterizing a process such as a phase change or chemical reaction.
- **Le Châtelier's principle** stipulates that the equilibrium state of a process shifts to reduce an applied stress. In chemical reactions, increasing the temperature, pressure, or amount of a particular compound triggers a response that partly offsets that increase.
- An **endothermic** reaction is one with a *positive* $\Delta_{rxn}H^\ominus$. An **exothermic** reaction has a *negative* $\Delta_{rxn}H^\ominus$.

- **Combustion** is the rapid oxidation of a compound.
- The **enthalpy of combustion** is the $\Delta_{rxn}H^\ominus$ for a combustion reaction that completely converts a compound to its individual elements in their most stable oxidized or pure elemental forms.
- The **adiabatic flame temperature** is the theoretical temperature that the products would reach if the combustion of a particular fuel were to be carried out under adiabatic conditions, with no other interconversion of energy taking place.

OBJECTIVES REVIEW

1. *Identify the reactant, product, and transition state geometries on a reaction surface, and use the surface to estimate the energy of reaction and activation energy.*
   Estimate to within 30 kJ mol^{-1} the activation energy for the reverse reaction $H + HF \rightarrow F + H_2$ from Fig. 12.2.

2. *Calculate the enthalpy of an isothermal reaction at arbitrary temperature and pressure from the enthalpies of formation and heat capacities.*
   Find $\Delta_{rxn}H^{\ominus}$ for the reaction $2NO_2(gas) \rightarrow N_2(gas) + 2O_2(gas)$ at 1.00 bar and 373 K.

3. *Calculate the adiabatic flame temperature of a compound.*
   Calculate the adiabatic flame temperature of hydrazine, $N_2H_4(liq)$.

4. *Calculate the equilibrium constant of a reaction from the Gibbs free energies of formation, and estimate the effects of temperature.*
   Find the equilibrium constant of the reaction $2NO_2(gas) \rightarrow N_2(gas) + 2O_2(gas)$ at 1.00 bar and 373 K.

PROBLEMS

Discussion Problems

12.1 Identify the following, using the reaction surface for $X + YZ \rightarrow XY + Z$ drawn in the following figure:

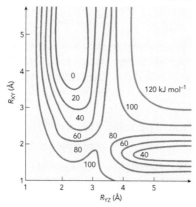

a. the equilibrium bond length of the R_{XY} molecule

b. the XY bond length of the transition state

c. $\Delta_{rxn}U$ ($\approx E_a$)

d. the YZ bond dissociation energy D_e

12.2 In Fig. 12.2, the reaction diagram using potential energy is corrected by the contributions from the zero-point energy ("zp") and from the thermal energy ("thermal"). If we change the H_2 to HD (replacing one H atom by a deuterium atom) on the reactant side of the graph, then (to a first approximation)

a. does the potential energy increase, decrease, or stay the same?

b. does the zero-point correction increase, decrease, or stay the same?

c. does the thermal energy correction increase, decrease, or stay the same?

12.3 A solid reactant decomposes to products in the gas phase. The $\Delta_{rxn}H^{\ominus}$ is -10.8 kJ at 273 K and $+8.7$ kJ at 325 K. Which of the following *must* be true?

a. The reaction is exothermic at all temperatures.

b. The reaction is endothermic at all temperatures.

c. The reaction is spontaneous at 273 K.

d. The reaction is spontaneous at 325 K.

e. The products have greater heat capacity than the reactant.

f. The products have greater entropy than the reactant.

g. The equilibrium pressure of the products decreases as the temperature increases.

12.4 For the following reaction, state whether $\Delta_{rxn}S^{\ominus}$ should be positive or negative and whether the magnitude should be large (>50 kJ K^{-1} mol^{-1}) or small (<50 kJ K^{-1} mol^{-1}).

$$C_6H_6(g) + Br(g) \rightarrow C_6H_5Br(g) + H(g).$$

12.5 For the reaction $Ag^+(aq) + Cl^-(aq) \rightarrow AgCl(s)$, which of the following is the correct $\Delta S^{\ominus}$ value?

a. -33 J K^{-1} mol^{-1} b. 33 J K^{-1} mol^{-1}

c. -33 kJ K^{-1} mol^{-1} d. 33 kJ K^{-1} mol^{-1}

12.6 Write "+" or "−" or "0" to indicate whether $\Delta_{rxn}S^{\ominus}$ of each of the following processes is likely to be positive, negative, or zero, respectively.

$2NO(g) + O_2(g) \rightarrow 2NO_2(g)$
formation reaction of NaCl(s)
formation reaction of $Cl_2(g)$
combustion of $CH_4(g)$
sublimation of $CO_2(s)$

12.7 The reaction $2SiN(g) + 2O_2(g) \rightarrow 2SiO_2(s) + N_2(g)$ has $\Delta_{rxn}H^\ominus$ of -2782.92 kJ mol^{-1}. Indicate whether the following values are greater than 1 (>1), between 0 and 1 ($0 - 1$), or less than zero (<0). (You can calculate the numbers but you shouldn't have to.)

parameter	$>1, 0-1, $ or <0
$\Delta_{rxn}S^\ominus$ (J K^{-1} mol^{-1}) at 298 K	
$\Delta_{rxn}G^\ominus$ (kJ mol^{-1}) at 298 K	
K_{eq} at 298 K	
$[K_{eq}$ at 350 K$]/[K_{eq}$ at 298 K$]$	

Reaction Surfaces and Reaction Diagrams

12.8 Sketch a reaction diagram (not a whole potential energy surface) for the minimum energy reaction path from reactants to products for the reaction $CH_3(g) + HCl(g) \rightarrow CH_4(g) + Cl(g)$, which has an activation energy of 6 kJ mol^{-1} and a $\Delta_{rxn}E$ of -8 kJ mol^{-1}. Label the reactants, products, and transition state (TS) on the diagram.

12.9 The heat of formation of $ClF_3 \cdot HF(g)$ is given in the Appendix. Estimate the bond strength (in kJ mol^{-1}) of the bond between ClF_3 and HF, and state whether this is consistent with a chemical bond, a van der Waals bond, or on the boundary between these limits.

12.10 According to Fig. 12.2, what is the approximate bond dissociation energy of H_2 in cm^{-1}?

12.11 Draw the geometry of the transition state for the $F + H_2$ reaction, according to Fig. 12.2, and indicate the bond lengths.

12.12 The contour surfaces given in the figure map the potential energy of the $NH + H$ reaction system. Figure (a) graphs the potential energy as a function of bond lengths, with the bond angle α fixed to $180°$, so the three atoms are in a line. Figure (b) is a function of α and the $H—H$ bond length with the NH bond length fixed to $1.963a_0$ (note that the distances are given in units of $a_0 = 0.529$ Å). In both graphs, contours are given every 0.3 eV [*J Chem. Phys.* **122** 114301 (2005)].

a. What is the equilibrium $H—H$ bond length in a_0?

b. At distances of $1.963a_0$ for the NH bond and $3.0a_0$ for the HH bond, what is the difference in potential energy between an NHH angle of $180°$ and $80°$?

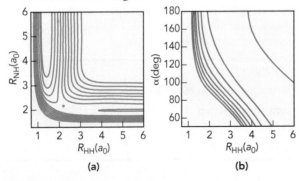

(a)   (b)

c. Draw the path on this diagram traveled by an NH molecule that is in an excited stretching state when no reaction takes place.

12.13 A reaction surface for the isomerization of the HC_4H_2 free radical is drawn in the figure, using energy contours at every 200 cm^{-1}. The lower energy regions are darker. Estimate ΔE for this isomerization.

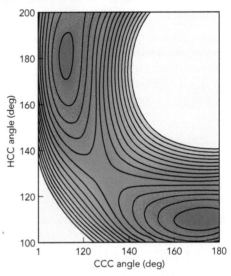

12.14 The figure shows a crude reaction surface for the isomerizations of the pentadienyl radical (A, B, C) and cyclopentenyl radical (D). The contour lines are drawn every 25 kJ mol^{-1}, and darker regions indicate lower energy.

a. Which form of pentadienyl (A, B, or C) is the least stable?

b. Estimate the ΔE for the isomerization A $\rightarrow$ D.

c. Estimate the activation energy E_a—the minimum energy necessary to add to the reactant—for the isomerization A $\rightarrow$ D to proceed.

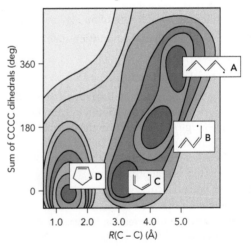

12.15 The vibrational constant for HF is 4138 cm^{-1}, and the vibrational energy is $\omega_e(v + 1/2)$. (For this you *do*

need to include the zero-point energy.) Using the $F + H_2$ reaction surface, Fig. 12.2, determine the following:

a. the equilibrium bond length of HF
b. the minimum bond length of HF during vibration in the $v = 2$ state
c. the maximum bond length of HF during vibration in the $v = 2$ state
d. the minimum H—H separation when HF is in the $v = 2$ state (assuming no other energies contribute)

Enthalpies and Entropies of Reaction

12.16 Calculate $\Delta_{vap}H^{\ominus}$ and the normal boiling temperature for CS_2. Assume the heat capacities and ΔS are constants over this temperature range.

12.17 Calculate $\Delta_{rxn}H^{\ominus}$ for the formation of three moles of $H_2(gas)$ from atomic hydrogen at 298 K.

12.18 Calculate the enthalpy of reaction for $SO_3(g) + 2NO(g) \rightarrow SO_2(l) + N_2O_3(g)$ at 263 K. The enthalpy of formation for $SO_2(l)$ is not in the Appendix, but the $\Delta_{vap}H^{\ominus}$ value is 24.9 kJ mol^{-1}.

12.19 Several products are possible in the gas-phase reaction: $NO + aO_2 \rightarrow bN_xO_y$, where a and b are stoichiometric coefficients. Find the single gas-phase product $N_xO_y(gas)$ that results from the most exothermic reaction (per mole of NO) of this form.

12.20 Calculate the heat of formation Δ_fH of NO gas at 2.00 bar and 373.15 K.

12.21 Calculate the standard enthalpy of combustion for benzene.

12.22 For the isothermal combustion of C_2H_4O at 298 K, find the enthalpy of combustion. (Keep in mind that this gives us the heat released if we start with everything at 298 K and then let the products return to 298 K, regardless of the actual temperature *during* the combustion.)

12.23
a. Write the chemical reaction for the combustion of ketene gas, H_2CCO, at 298 K.
b. If the enthalpy of combustion for ketene is -1025.3 kJ mol^{-1}, what is the enthalpy of formation for ketene?
c. Estimate the adiabatic flame temperature of ketene, using the heat capacities at 298 K. For this calculation, assume water is a gas.

12.24 Calculate $\Delta_{rxn}H^{\ominus}$ and $\Delta_{rxn}E^{\ominus}$ for the adiabatic isobaric combustion of C_2H_6.

12.25 For the reaction $Br_2(g) + Cl_2(g) \rightarrow 2BrCl(g)$ at 298 K, $\Delta_{rxn}H^{\ominus} = 29.28$ kJ, $C_P(\text{reactants}) = 68.07$ J K^{-1} mol^{-1}, and $C_P(\text{products}) = 69.96$ J K^{-1} mol^{-1}. Assuming the heat capacities remain constant, find the temperature at which $\Delta_{rxn}H^{\ominus}$ would be 30.00 kJ.

12.26 Calculate $\Delta_{rxn}H^{\ominus}$ for the reaction $SO(g) \rightarrow S_8(\text{rhombic}) + 1/2\ O_2(g)$ at 1.00 bar and 343 K.

12.27 Calculate the overall change in volume of the gases for the reaction $N_2O_4(g) \rightarrow N_2(g) + 2O_2(g)$ carried out adiabatically at 1.00 bar, starting from 0.100 mol N_2O_4 at 298 K.

12.28 Calculate the change in temperature of 5.00 L of water in a calorimeter from the complete reaction of 1.00 g of iodine in the reaction $1/2\ I_2(s) + 5/2\ F_2(g) \rightarrow IF_5(g)$ at 298 K and 1.00 bar.

12.29 Estimate the mean heat capacity C_{Pm} of CO_2 and N_2 between 300 K and 4800 K, given that the adiabatic flame temperature of C_2N_2 is 4800 K and $\Delta H_f^{\ominus}$ of C_2N_2 is 307.0 kJ mol^{-1}. What would the equipartition value of $C_{Pm}(CO_2)$ be?

12.30 Identify the number of degrees of freedom for each reactant and product in the two following reactions, and indicate whether the overall $\Delta_{rxn}S$ is likely to be positive

	$2O_2$ (gas)	$\rightarrow$	O_3 (gas)	$+$	O (gas)
translational					
rotational					
vibrational					
$\Delta_{rxn}S$ + or −					

	GeH_4 (gas)	$+$	$2O_2$ (gas)	$\rightarrow$	GeO_2 (sol)	$+$	$2H_2O$ (gas)
translational							
rotational							
vibrational							
$\Delta_{rxn}S$ + or −							

or negative:

12.31 Calculate the Helmholtz free energy of reaction $\Delta_{rxn}F^{\ominus}$ for the formation of 1.00 mol $IF_5(gas)$ from solid iodine and fluorine gas at 320 K and 1.00 bar. For IF_5 at 298 K, $\Delta_fH^{\ominus} = -822.49$ kJ mol^{-1} $\Delta_fS^{\ominus} = -237.32$ J K^{-1} mol^{-1}, and $\Delta_fC_{Pm} = -6.27$ J K^{-1} mol^{-1}.

12.32 If the activation energy of the reaction $2NO(g) + Cl_2(g) \rightarrow 2NOCl(g)$ is 15 kJ mol^{-1}, what is the *enthalpy* of reaction at 298 K?

12.33 Determine whether the entropy of reaction is positive or negative for the two following reactions. NOCl is non-linear.

(a) $2NO(g) + Cl_2(g) \rightarrow 2NOCl(g)$

(b) $NOCl(g) + Cl(g) \rightarrow NO + Cl_2(g)$.

12.34 Balance the following chemical equation and then calculate the $\Delta_{rxn}S$ at 298.15 K of the combustion of $HCN(g)$ ($\Delta_fH^{\ominus} = 135.1$ kJ mol^{-1}, $S_m^{\ominus} = 201.8$ J K^{-1} mol^{-1} at 298.15 K) at a pressure of 2.00 bar: $HCN(g) + O_2(g) \rightarrow CO_2(g) + N_2(g) + H_2O(l)$. Assume the liquid is incompressible.

12.35 Identify the number of degrees of freedom contributed by each term in the following reaction of linear diacetylene (HC_4H), and indicate whether the overall $\Delta_{rxn}S$ is likely to be positive or negative:

	$9O_2$ (g)	+	$2HC_4H$ (g)	→	$8CO_2$ (g)	+	$2H_2O$ (g)
translational							
rotational							
vibrational							
$\Delta_{rxn}S$ + or −							

12.36 Evaluate the entropy change for the combustion of C_2H_6 (ethane) at 1000 K.

12.37 At what temperature does the reaction of ethylene oxide (C_2H_4O) to ethanol, $C_2H_4O(g) + H_2(g) \rightarrow C_2H_5OH(g)$, have a $\Delta_{rxn}S$ of -100 J K^{-1} mol^{-1}? The following possibly useful parameters are evaluated at 298.15 K:

$$\Delta_{rxn}S^{\ominus} = -90.5 \text{ J K}^{-1}\text{ mol}^{-1}$$
$$\Delta_{rxn}H^{\ominus} = -182.47 \text{ kJ mol}^{-1}$$
$$\Delta_{rxn}G^{\ominus} = -155.48 \text{ kJ mol}^{-1}$$
$$\Delta_{rxn}C_P = -11.29 \text{ J K}^{-1}\text{ mol}^{-1}$$

12.38 The gas-phase reaction of N_2O and O_2 to form N_2O_3 is endothermic by 1.67 kJ mol^{-1}. If the reaction is carried out adiabatically and at a constant pressure of 1 bar, and it reaches a final temperature of 298.15 K, what was the initial temperature? The heat capacities in J K^{-1} mol^{-1} are 38.45 for N_2O, 29.355 for O_2, and 65.61 for N_2O_3.

12.39 If 0.0100 mol of ClO_3F gas, initially at 298 K and 1.00 bar and mixed with 0.400 mol of Ar, decomposes completely in an *adiabatic* balloon at constant pressure to F_2, Cl_2, and O_2, what is the final volume of the balloon? The total heat capacity C_P of the initial mixture is 8.96 J K^{-1} and of the final mixture is 9.08 J K^{-1}.

12.40 Calculate the adiabatic flame temperature of gas-phase heptanoic acid, $C_7H_{14}O_2$ ($\Delta_f H^{\ominus} = -536.2$ kJ mol^{-1}), and find the initial average speed $\langle v \rangle$ of the CO molecules at 298 K before the reaction and $\langle v \rangle$ of the CO_2 molecules after the reaction. Use the heat capacities of CO_2 and H_2O at 1500 K: 58.4 and 47.4 J K^{-1} mol^{-1}, respectively.

12.41 A sample of 0.500 mol H_2S gas reacts with O_2 gas at 1.00 bar to yield H_2O and SO_3, both in the gas phase. If the reactants start at 298 K and the final temperature is 458 K, what is the total change in entropy for this sample?

12.42 Estimate to within 10 J K^{-1} mol^{-1}, showing your reasoning, the standard molar entropies of *n*-propyl alcohol (1-propanal, C_3H_7OH) in liquid and in gas phase.

Free Energies of Reaction and Chemical Equilibrium

12.43 What is the minimum temperature at which graphite and chlorine gas would spontaneously react to form CCl_4(gas) at standard pressure?

12.44 Find the temperature at which the direction of the spontaneous reaction changes for the system

$$CH_2Cl_2 + Cl_2 \rightleftharpoons CCl_4 + H_2.$$

12.45 A particular reaction at constant volume and temperature has positive ΔG and increases the pressure. Find the remaining requirement for which the reaction is spontaneous under these conditions.

12.46 A reaction at constant T and V is spontaneous if $\Delta_{rxn}F$ is negative. Calculate $\Delta_{rxn}F$ for the complete combustion of 0.0020 moles of C_6H_6 (benzene) to $H_2O(g)$ and $CO_2(g)$ at 800 K and in 0.5 L.

12.47 Calculate the work done during the explosive combustion of 0.01 L of acetylene at 1 bar, 298 K. Use any approximations you think are appropriate.

12.48 At 1.00 bar and 298 K, N_2O_4 gas exists simultaneously in equilibrium with $NO_2(g)$ and with $N_2O_4(l)$. If we have excess $N_2O_4(l)$ present in a sealed container, calculate the equilibrium partial pressures of $N_2O_4(g)$ and $NO_2(g)$.

12.49 For the oxidation of $ClO(g)$ to $ClO_2(g)$, $\Delta_{rxn}H = 0.66$ kJ and $\Delta_{rxn}S = -72.36$ J K^{-1} mol^{-1}. Estimate the value of K_{eq} for this reaction at 373 K. (Don't worry about the heat capacity corrections.)

12.50 Find the $\Delta_{rxn}G^{\ominus}$ and K_{eq} at 298 K for the reaction $N_2(g) + 3H_2(g) \rightarrow 2NH_3(g)$.

12.51 For the metabolism of methylamine to formaldehyde and ammonia, the enthalpy of reaction is -201.8 kJ mol^{-1} and the entropy of reaction is 148.8 J K^{-1} mol^{-1}. Find the value of the equilibrium constant at 325 K.

12.52 For the reaction $A + NO_2(g) \rightarrow B + NO(g)$, identify compounds A and B if the equilibrium constant at 298 K is 12.87, and if A and B contain only nitrogen and/or oxygen.

12.53 Evaluate the equilibrium constant at 298 K for the reaction

$$C_2H_2(g) + 2HCl(g) \rightarrow C_2H_4Cl_2(g).$$

12.54 For the reaction of ClO_3F and H_2O to form ClO, HF, and O_2, calculate $\Delta_{rxn}H$ and $\Delta_{rxn}G$ per mole of ClO_3F at 298.15 K and at 500 K. Calculate the equilibrium constant at both temperatures.

12.55 For the reaction in which diamond oxidizes to form CO_2, find K_{eq} (298 K) and K_{eq} (398 K).

12.56 Calculate the equilibrium constant for the ethane combustion reaction at 1000 K and at 2000 K, assuming the heat capacities are constant.

12.57 For a gas-phase chemical reaction carried out at constant volume and temperature, write an equation for $\ln K_{eq}$ in terms of ΔG, ΔP, V, and T.

PART I
EXTRAPOLATING
FROM
MOLECULAR TO
MACROSCOPIC
SYSTEMS

PART II
NON-REACTIVE
MACROSCOPIC
SYSTEMS

PART III
REACTIVE SYSTEMS

12 The Thermodynamics of Chemical Reactions
13 Chemical Kinetics: Elementary Reactions
14 Chemical Kinetics: Multi-Step Reactions

13 Chemical Kinetics: Elementary Reactions

LEARNING OBJECTIVES

After reading this chapter, you will be able to do the following:

❶ Write the rate law for an elementary reaction based on the chemical equation.

❷ Estimate the rate constant of a reaction using simple collision theory or transition state theory, given the appropriate parameters.

❸ Calculate the half-life of the reactant in an elementary first-order reaction from the Arrhenius constants.

GOAL *Why Are We Here?*

Our goal in this chapter is to identify the factors that determine the *time-dependence* of the simplest, single-step chemical reactions. In addition, we will consider two derivations for the rate constant of a reaction between any two gas-phase molecules.

CONTEXT *Where Are We Now?*

After looking at several aspects of chemical reaction thermodynamics, there's a haunting feeling that something's been missing. If we never move beyond thermodynamics alone, we find these omissions:

1. While we've described equilibrium initial and final states adequately, we can approach the system *during* the reaction only in the limiting case of the reversible chemical reaction, when the system is always in equilibrium.
2. The thermodynamic approach fails to describe the time-scale for the reaction.

These two statements are related. Thermodynamics dictates that at room temperature diamond will spontaneously convert to graphite, but we know that this process is too slow to be of any practical concern. Diamond at room temperature and standard pressure is a *non-equilibrium* situation, and equilibrium thermodynamics cannot describe it adequately. Consider also a chemical explosion: the chemical process occurs too rapidly for the mechanical and thermal processes of the surroundings to remain in equilibrium with the chemical products. In either case, the slow phase transition of solid carbon or the rapid combustion of an explosive, we want tools to describe the rate at which chemical change occurs.

SUPPORTING TEXT | *How Did We Get Here?*

The work ahead will rely primarily on our grasp of two concepts: the canonical distribution of energies introduced in Section 2.4 and the dynamics of colliding molecules as covered in Section 5.1. In particular, we will draw on the following equations and sections of text to support the ideas developed in this chapter:

- Among the collision parameters we've defined are the *collision frequency* (Eq. 5.12):

$$\gamma = \rho\sigma\langle v_{AA}\rangle,$$

the *collision cross-section*, the effective cross-sectional area of the colliding molecules, (Eq. 5.10):

$$\sigma_{AB} = \frac{1}{4}(\sigma_A + 2\sqrt{\sigma_A\sigma_B} + \sigma_B),$$

and the *average relative speed* (Eq. 5.4):

$$\langle v_{AB}\rangle = \sqrt{\frac{8k_BT}{\pi\mu}},$$

where μ is the reduced mass $(m_Am_B)/(m_A + m_B)$.

- Some reaction rates are limited by diffusion, and in this case we will turn to Fick's first law (Eq. 5.42) to help quantify the dynamics:

$$J(Z_0) = -D\left(\frac{d\rho}{dZ}\right)\bigg|_{Z_0}.$$

- In Section 12.2 we define the *activation energy* to be the maximum energy above the reactants that the reaction pathway encounters as we travel along the minimum energy path. The geometry at that point is called the *transition state*.

- The equilibrium constant for a reaction can be expressed in terms of the partition function (Eq. 12.40):

$$K_{eq} = \frac{Q_{products}}{Q_{reactants}}.$$

We can also write the equilibrium constant in thermodynamic terms, using the free energy change (Eq. 12.32):

$$K_{eq} = e^{-\Delta_{rxn}G^\circ/(RT)},$$

and the free energy change in turn can be broken into contributions from the enthalpy and entropy (Eq. 9.26):

$$\Delta G = \Delta H - T\Delta S.$$

The enthalpy for any process is defined as (Eq. 7.17)

$$H = E + PV.$$

13.1 Reaction Rates

Chemical kinetics is the study of chemical reaction rates. The thermodynamic properties are still important, but there are additional parameters that must be determined to predict the kinetics of a reaction. As in thermodynamics, we can determine those parameters empirically (from experiment), or we can try to predict them from the molecular structure using quantum mechanics. In kinetics as in thermodynamics, both the microscopic and macroscopic parameters of the system are crucial to understanding its behavior. Let's take a look now at the parameters that most clearly affect the time-dependence of the reaction.

Microscopic Parameters

In Section 12.4, we saw that we could refer back to the molecular structures of the reactants and products in order to predict the sign of $\Delta_{rxn}S^{\ominus}$. The time-dependence of the reaction is even more sensitive to the molecular structure, depending on the geometry at the reactants and products and the transition state in between.

Orientation of Reactants

Consider the reaction

$$Cl(g) + HF(g) \rightarrow HCl(g) + F(g), \qquad (R13.1)$$

with

$$\Delta_{rxn}H^{\ominus} = 165.198 + 173.779 - 186.908 - 158.754 = -6.685 \text{ kJ mol}^{-1}.$$

This reaction is less probable if the Cl atom attacks from the F end of the HF molecule than from the H end; the Cl is more likely to bounce away from the HF molecule (or react with F instead of H) when it approaches from the end *opposite* to the hydrogen (Fig. 13.1a). The probability becomes even lower if we replace the fluorine atom with a much larger functional group, as in the reaction

$$Cl(g) + (CF_3)_3CH(g) \rightarrow HCl(g) + (CF_3)_3C(g). \qquad (R13.2)$$

The $(CF_3)_3$ group is large enough compared to the single F atom that the Cl is much more likely to be bumped away during its approach to the H atom. Consequently, we would expect the reaction rate to be slower for Reaction R13.2 than for Reaction R13.1. (We should note that a second possible reaction results instead in H + ClF, but these are less thermodynamically favored products and contribute little under typical conditions.)

Energy Barrier to Reaction

Recall the reaction potential energy surface. During a chemical reaction, *potential energy* tends to increase as the reactants change their chemical composition, because the bond structure often becomes unfavorable. However, the *total energy* is still constant. Do not imagine the energy of the molecules increasing; the reaction system is merely exchanging kinetic energy (translational and vibrational, principally) for potential energy as it climbs the hill toward the transition state.

The bottom line: a chemical reaction must generally overcome the activation energy, a peak in the potential energy between the reactants and products. The height of this barrier depends strongly on the molecular structure and quantum mechanics of the system.

Macroscopic Parameters

While events at the molecular level define the chemical reaction, the rate at which we observe changes in our system depends also on the system's thermodynamic parameters.

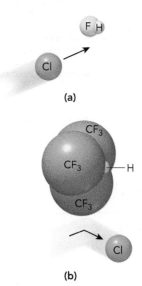

(a)

(b)

▲ FIGURE 13.1 **Steric effects in the reaction probability.** More angles of collision allow Cl to react with H from HF **(a)** than from $(CF_3)_3CH$ **(b)**.

Concentrations

The language of chemical kinetics grew mainly out of studies in solution chemistry, with the amount of reactant and product expressed in units of molarity. Therefore, we tend to represent the amount of each molecule in formulas using the square brackets for molarity, [X]. The kinetics of gas phase reactions tend also to be described in concentration units (rather than partial pressures) but using the number density in particles per cm^3 rather than molarity.

To report the kinetics of a single reaction unambiguously when different molecules are formed or consumed at different rates, a useful definition is the **reaction velocity:**

$$v = \frac{1}{v_i}\frac{d[i]}{dt} = \frac{1}{Vv_i}\frac{dn_i}{dt}, \tag{13.1}$$

where $[i]$ is the molar concentration of product or reactant i, v_i again the stoichiometric coefficient in the reaction, and n_i the number of moles of component i contained in volume V. The reason that v_i appears in the definition is so that the reaction velocity is equal for species in the reaction that have different stoichiometries. For example, the reaction

$$2H_2 + O_2 \longrightarrow 2H_2O \tag{R13.3}$$

has a reaction velocity that is the same whether we look at H_2 ($v = -2$), O_2 ($v = -1$), or H_2O ($v = +2$):

$$-\frac{1}{2}\frac{d[H_2]}{dt} = -\frac{d[O_2]}{dt} = \frac{1}{2}\frac{d[H_2O]}{dt}. \tag{13.2}$$

CHECKPOINT We'll say more about these exponents *x* and *y* shortly, but in general their values are found by experiment. They determine in what way the rate of the reaction depends on the amount of reactant. The values are not necessarily positive, and don't even have to be integers. However, the assumptions we use in this chapter will result in values of 1 or 2.

Often, the reaction velocity can be written in terms of a **rate law,** a power law in the reactant concentrations, with a concentration-independent coefficient called the **rate constant** k:

$$v = k[A]^x[B]^y.\ldots \tag{13.3}$$

The **reaction order** is the sum of the exponents $x + y + \ldots$

A **net reaction** is a reaction written to show only the initial and final chemical components. For example, the net reaction R13.3 of H_2 with O_2 to form water occurs through a series of reactions, such as the following:

$$H_2 + O_2 \rightarrow H + HO_2$$
$$HO_2 + H_2 \rightarrow H_2O + OH \tag{R13.4}$$
$$H + OH + M \rightarrow H_2O + M.$$

The "M" in the last step is a shorthand for any particle that participates in the reaction without being changed, normally to provide or (as in this case) carry away energy. Each of these individual steps is called an **elementary reaction,** in which the reactants transform into products with a single molecular collision or decay. Reaction R13.3, for example, is not an elementary reaction, because the H_2 and O_2 bonds do not all break at the same time. Intermediate steps are involved. Even when two H_2 molecules and one O_2 molecule do collide all at once, the probability of forming two water molecules is negligible.

The series of elementary reactions leading to a net equation is called the **reaction mechanism.** If we add up Reactions R13.4, the HO_2, H, and OH cancel because they each appear as both a reactant and a product, and what remains is the net change from $2H_2 + O_2$ to $2H_2O$. Reactions R13.4 in this case form the reaction mechanism.

In this chapter, we will focus on elementary reactions. These may be unimolecular, if the initial reactant is unstable:

$$XeCl \rightarrow Xe + Cl, \tag{R13.4}$$

or if the molecule has been previously energized, say by a collision or by a laser:

$$C_2H_4O_2\,(\text{dioxetane}) \rightarrow 2H_2CO.$$

An elementary reaction may also be bimolecular, as we have already considered:

$$Cl + HCl \rightarrow Cl_2 + H$$

or even termolecular,

$$He + H + Cl \rightarrow He + HCl.$$

This last reaction cannot occur at any significant rate unless the helium is present to take away the energy released by the H — Cl bond formation.[1] In the gas phase, the probability of three reactants colliding simultaneously is very small compared to the probability of a bimolecular collision, but many reactions cannot proceed by any other mechanism. Elementary reactions involving more than three reactants are unknown.

Temperature

In the rate law, Eq. 13.3, we can anticipate that the rate constant k is a function of temperature T. Each reaction between two gas-phase molecules occurs only when the two molecules collide. The temperature, which is the thermodynamic parameter associated with the average molecular speed, will affect both the average collision rate and the average collision energy. In the next section, we will apply the laws of mass transport to find explicitly what this temperature-dependence could be.

13.2 Simple Collision Theory

We've listed the parameters that we expect to have the greatest influence on the speed of the reaction. Now we need a theory that folds these together in an effort to predict the value, or at least the trends, for the rate constant k.

For our simplest model of chemical reactions, we fix the temperature and assume the following:

1. The reaction is bimolecular, meaning that two reactant molecules collide for the reaction to proceed:

$$A + B \xrightarrow{k} \text{products.}$$

---

[1] The two-body reaction $H(g) + Cl(g) \rightarrow HCl(g)$ can only occur by radiative association, an unlikely mechanism except in one case. When a neutral fragment A collides with an ionic fragment B^+, the dipole derivative of the neutral-ion complex can be large enough to make the infrared vibrational transitions along that coordinate quite strong. In that case, there may be a good chance of a photon being emitted during the picosecond or less that A and B^+ are the right distance from each other to form the stable AB^+.

2. The reactants must be close enough for intermolecular forces to be important before the reaction can take place (this is what we will call a collision).

3. A single distribution function describes the probability of the reaction occurring for any such collision group.

4. There will be some activation barrier to overcome.

The activation barrier is *not* always present, but reactions without activation barriers will be a special application of our more general results. With only these assumptions, and some results from statistical mechanics, we can obtain a semi-empirical rate law based on **simple collision theory**. If you'd rather not wait for the result, there's a DERIVATION SUMMARY after Eq. 13.16.

We can count the total number of collisions N_{coll} between A and B molecules in a volume V of our sample over a period of time δt like this:

1. The rate at which *each* molecule A collides with B molecules is the heterogeneous collision frequency, based on Eq. 5.12,

$$\gamma_{A:B} = \sigma_{AB} \langle v_{AB} \rangle \rho_B, \qquad (13.4)$$

Parameters Key: Simple Collision Theory

symbol	parameter	SI units
$\gamma_{A:B}$	collision frequency of A with B per molecule A	s^{-1}
σ_A, σ_{AB}	collision cross-section for A-A and A-B collisions, respectively	m^2
$\langle v_{AB} \rangle$	average relative speed of A and B	$m\,s^{-1}$
N_A, N_B	number of molecules A and B	unitless
V	volume	m^3
ρ_A, ρ_B	number densities of A and B	m^{-3}
$[A], [B]$	concentrations of A and B (normally in units of $mol\,L^{-1}$)	$mol\,m^{-3}$
$\mathcal{N}_A$	Avogadro's number	mol^{-1}
k_B	Boltzmann constant	$J\,K^{-1}$
R	gas constant $= \mathcal{N}_A k_B$	$J\,K^{-1}\,mol^{-1}$
T	temperature	K
m_A, m_B	masses of molecule A, molecule B	kg
μ	reduced mass $= m_A m_B / (m_A + m_B)$	kg
δt	small time increment	s
N_{coll}	number of A-B collisions in time δt	unitless
$\mathcal{P}_{rxn}$	probability that an A-B collision leads to products	unitless
p	steric factor, the probability that A and B have the right orientation at the collision to form products	unitless
E_a	activation energy	$J\,mol^{-1}$
E_{AB}	kinetic energy of the A-B collision	J
k_{SCT}	rate constant for the A+B reaction, based on simple collision theory	$m^3\,mol^{-1}\,s^{-1}$

where (Eqs. 5.10 and 5.4)

$$\sigma_{AB} = \frac{1}{4}(\sigma_A + 2\sqrt{\sigma_A \sigma_B} + \sigma_B), \qquad \text{by Eq. 5.10}$$

$$\langle v_{AB} \rangle = \sqrt{\frac{8k_B T}{\pi \mu}}, \qquad \text{by Eq. 5.4}$$

and where μ is the reduced mass of the collision pair:

$$\mu = \frac{m_A m_B}{m_A + m_B}.$$

2. Equation 13.4 gives the rate per A molecule, so to get the *total* collision rate, we multiply by the number of A molecules in the volume, $N_A = \rho_A V$.

3. Finally, to get the total number of collisions, we multiply this rate by the time interval δt, arriving at

$$N_{coll} = N_A \gamma_{A:B} \delta t = \rho_A V \sigma_{AB} \langle v_{AB} \rangle \rho_B \delta t. \qquad (13.5)$$

The number of A+B collisions in which A is converted to product is $N_{coll} \mathcal{P}_{rxn}$, and this gives us the amount of A consumed in the reaction over a time δt:

$$dN_A = -N_{coll} \mathcal{P}_{rxn}. \qquad (13.6)$$

We can now write the reaction velocity as

$$v = -\frac{d[A]}{dt} = -\frac{d[N_A/(\mathcal{N}_A V)]}{dt} \qquad \text{def. of molarity}$$

$$= -\frac{1}{\mathcal{N}_A V} \frac{dN_A}{dt} = \frac{1}{\mathcal{N}_A V} \frac{N_{coll} \mathcal{P}_{rxn}}{dt} \qquad \text{by Eq. 13.6}$$

$$\approx \frac{1}{\mathcal{N}_A V} \frac{\rho_A V \sigma_{AB} \langle v_{AB} \rangle \rho_B \delta t \, \mathcal{P}_{rxn}}{\delta t} \qquad \text{by Eq. 13.5}$$

$$= \sigma_{AB} \langle v_{AB} \rangle \mathcal{N}_A^{-1} \mathcal{P}_{rxn} \rho_A \rho_B. \qquad \text{rearrange} \quad (13.7)$$

Like all our probability functions, $\mathcal{P}_{rxn}$ is limited to values between 0 and 1, where a value of 1 would give an upper limit to the reaction rate—every collision leading to products—and $\mathcal{P}_{rxn} = 0$ would mean that the reaction never takes place. We can break the reaction probability into two contributions to improve the quantitative and intuitive utility of our model:

$$\mathcal{P}(rxn) \approx p \mathcal{P}(E_{AB} > E_a). \qquad (13.8)$$

These are the two microscopic contributions to the reaction rate discussed previously: the orientation term and the activation energy term. The **steric factor** p is determined by the fraction of reactant orientations that lead to products. It is difficult to estimate, and is usually determined empirically (from experiment). The activation energy term $\mathcal{P}(E_{AB} > E_a)$ is the fraction of reactant collisions with collision energy greater than the activation energy E_a, and this we can obtain from statistical mechanics.

We use the canonical distribution to assume that the probability of two molecules having a collision energy E_{AB} is proportional to $e^{-E_{AB}/(k_B T)}$. This is a crucial assumption, because it means that we effectively ignore the contribution to the reaction from internal rotational and vibrational energies of the reactants, which can total much more than the translational energy. However, it is also a necessary assumption at this stage, because the ability of a molecule to channel energy into the reaction coordinate from rotations and vibrations is a sensitive function of the nature of the reaction. We can apply the canonical distribution to rotations and vibrations also, but without knowing how much of that energy can be channeled into the reaction we cannot write a general formula for that contribution. Our approximation will work well for reactions of diatomic molecules with single atoms (such as the H+HF reaction) but breaks down increasingly as the number of atoms in each molecule increases.

You may notice that we appear to be ignoring *electronic* excitation in the reactants as well, and there are many examples of reactions that depend on excited state atoms and molecules to go forward. However, in practice we effectively treat these as being distinct from the same species in their ground states. For example, we would draw a different potential energy surface for the reaction of an excited state molecule A^* than for the ground state molecule A. And in the chemical equation, excited electronic states are normally indicated as part of the name of the reactant.

Treating the collision energy E_{AB} as we are, as a relative translational energy, it is not a discrete variable like the quantized vibrational and rotational energies. Therefore, we need a continuous distribution function $\mathcal{P}_E(E_{AB})$, which can be normalized (using one of our non-unitless partition functions q') as follows:

$$1 = \int_0^\infty \mathcal{P}_E(E_{AB}) dE_{AB} = \int_0^\infty \frac{e^{-E_{AB}/(k_B T)}}{q'} dE_{AB} = -\frac{k_B T}{q'} e^{-E_{AB}/(k_B T)} \Big|_0^\infty = \frac{k_B T}{q'}$$

$$q' = k_B T$$

$$\mathcal{P}_E(E_{AB}) = \frac{e^{-E_{AB}/(k_B T)}}{k_B T}. \tag{13.9}$$

If we want the fraction of collisions with E_{AB} greater than E_a, we integrate this distribution function from E_a to infinity:

$$\mathcal{P}(E_{AB} > E_a) = \int_{E_a}^\infty \mathcal{P}_E(E_{AB}) dE_{AB}$$

$$= \frac{1}{k_B T} \int_{E_a}^\infty e^{-E_{AB}/(k_B T)} dE_{AB}$$

$$= \left[-e^{-E_{AB}/(k_B T)} \right]_{E_a}^\infty$$

$$= e^{-E_a/(k_B T)}. \tag{13.10}$$

For macroscopic systems, the activation energy is usually given in molar units, so $1/(k_B T)$ is replaced by $1/(RT)$:

$$\mathcal{P}(E > E_a) = e^{-E_a/(RT)}. \tag{13.11}$$

Folding this into Eq. 13.8, we may write

$$\mathcal{P}_{rxn} \approx p e^{-E_a/(RT)}. \tag{13.12}$$

Plugging this result into our reaction velocity in Eq. 13.7 gives

$$v = \sigma_{AB} \langle v_{AB} \rangle \mathcal{N}_A^{-1} p e^{-E_a/(RT)} \rho_A \rho_B. \tag{13.13}$$

To get macroscopic units, we relate the number densities to the molarities by Avogadro's number:

$$\rho_A = \mathcal{N}_A[A] \quad \rho_B = \mathcal{N}_A[B], \tag{13.14}$$

and arrive at a reaction velocity

$$v = \sigma_{AB}\langle v_{AB}\rangle \mathcal{N}_A p e^{-E_a/(RT)}[A][B] = k[A][B]. \tag{13.15}$$

We add the second expression, $k[A][B]$, to put the reaction velocity in the common form of Eq. 13.3. This at last gives the rate constant from simple collision theory:

$$k_{SCT} = \sigma_{AB}\langle v_{AB}\rangle \mathcal{N}_A p e^{-E_a/(RT)}$$

$$k_{SCT} = \sigma_{AB}\sqrt{\frac{8k_B T}{\pi\mu}}\mathcal{N}_A p e^{-E_a/(RT)}. \tag{13.16}$$

DERIVATION SUMMARY Simple Collision Theory. To obtain Eq. 13.16, we counted the number of collisions per unit time between reactant molecules A and B over the entire system. Then we multiplied this number by the fraction of those collisions in which the kinetic energy exceeds the activation energy, which brings in the canonical factor of $e^{-E_a/(RT)}$. Because the collision frequency depends on the number of molecules A and B within the volume V, the reaction velocity ends up being proportional to the concentrations $[A]$ and $[B]$.

The first major result from simple collision theory is its accurate prediction of the experimentally measured temperature dependence that was discovered by Svante Arrhenius, who devised the empirical **Arrhenius equation,**

$$k_{Arr} = Ae^{-E_a/(RT)}. \tag{13.17}$$

The pre-exponential factor A and the activation energy E_a vary from one reaction to another, but nearly all reaction rate constants follow the general behavior predicted by Eq. 13.17: k_{Arr} climbs rapidly from zero as the temperature increases, slowing down as the thermal energy RT approaches the activation energy E_a and asymptotically approaching $k_{Arr} \approx A$ as RT becomes much larger than E_a (Fig. 13.2).

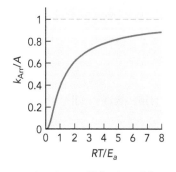

▲ **FIGURE 13.2 Behavior of the Arrhenius rate constant k_{Arr} as a function of temperature.** The value of k_{Arr} converges on A at high T. At $RT = E_a$, k_{Arr} is about one third of its maximum value.

EXAMPLE 13.1 Estimate Arrhenius Pre-Exponential Factor A

CONTEXT If we listed all the chemical species that appear in the net reaction of (for example) the combustion of gasoline or the formation of nitric acid in the atmosphere, we would find that the list is dominated by *reactive intermediates,* molecules with short lifetimes that form while the reactants at the beginning of the reaction are transforming into the products we will be left with at the end. The road from reactants to products passes through these intermediates along the way, and the direction we take as we move through the middle of this mechanism is determined not by the reactants (which are only a memory) and not by the products (which aren't yet on the scene) but by the next intermediate.

Paradoxically, although these intermediates determine the direction of the reaction at each step, they often pass much too quickly to be characterized easily in the laboratory. In order to predict how we can modify a chemical reaction to make it more to our liking, we need to understand the reaction mechanism and the reaction rate of each step. In this example we see how this is often feasible, as we successfully estimate the pre-exponential factor of a rate constant relevant to atmospheric HNO_3 production, using only fundamental molecular parameters.

PROBLEM Estimate the value of A in $cm^3 mol^{-1} s^{-1}$ for the reaction between oxygen atom and molecular nitrogen, $O(gas) + N_2(gas)$, at 298 K. Let the collision cross-sections be 30Å^2 for O and 37Å^2 for N_2.

SOLUTION Removing the exponential factor from Eq. 13.16 leaves A:

$$A = \sigma_{AB}\sqrt{\frac{8k_BT}{\pi\mu}}\mathcal{N}_A p.$$

For this simple reaction, we can estimate that the steric factor p is close to 1, because almost any relative orientation of the atom and the N_2 molecule should make reaction possible. We calculate σ_{AB} and μ and then plug away:

$$\sigma_{AB} = \frac{1}{4}(30 + 2\sqrt{30\cdot37} + 37)(\text{Å}^2) = 33.4\,\text{Å}^2$$

$$\mu = \frac{16.00\cdot28.00}{(16.00 + 28.00)}(\text{amu}) = 10.18\,\text{amu}$$

$$\langle v_{AB}\rangle = \sqrt{\frac{8k_BT}{\pi\mu}} = 787.3\text{ m s}^{-1}$$

$$A \approx (33.4\cdot10^{-20}\text{ m}^2)(787.3\text{ m s}^{-1})(6.022\cdot10^{23}\text{ mol}^{-1})(1)$$
$$= 1.6\cdot10^8\text{ m}^3\text{ mol}^{-1}\text{ s}^{-1} = 1.6\cdot10^{14}\text{ cm}^3\text{ mol}^{-1}\text{ s}^{-1}.$$

This agrees well with the value of 10^{14} $cm^3 mol^{-1} s^{-1}$ in Table 13.1.

Simple collision theory doesn't assist us in predicting E_a, because the activation energy depends too specifically on the chemical structures involved, but it does let us estimate A.

Our rate constant in Eq. 13.16 predicts the general form of the Arrhenius equation but with an added temperature dependence to the pre-exponential constant A. In general the pre-exponential term does have some temperature dependence, and this is usually written in the form

$$A(T) = A'T^w, \tag{13.18}$$

where in almost all cases $-2 \le w \le 2$. The simple collision theory picture shows one way in which this temperature dependence can come about: the relative speed of the molecules increases at $\sqrt{T}$, so the collision rate (and hence the reaction rate) should increase by the same factor:[2] $A_{SCT} = A'T^{1/2}$.

---

[2]Then why didn't Arrhenius include this in his expression? The Arrhenius equation is empirical, based on experimental measurements of reaction rates. And measuring reaction rates to high precision has never been easy. The $e^{-E_a/(RT)}$ term is one big problem. For a reaction at 298 K with an activation barrier of 100 kJ mol^{-1}, a change of only 0.1 K causes a 10% change in the reaction rate. Rate constants for fast reactions can be difficult to measure because the reaction starts before you're done combining the reactants, and you need precise measurements of the reaction time as well as the concentrations. Slow reactions may involve concentrations that are too low or that change too little for precise measurements. Furthermore, many reactions maintain their identity only over a temperature range much less than 100 K, because many reactions are studied in liquid solutions where the temperature range must normally lie between the solvent's freezing and boiling points and must support the required solubilities of the reactants.

The T-dependence of $A(T)$ is much less dramatic than the exponential term $e^{-E_a/(k_B T)}$ in determining the rate constant. For most reactions over the range of easily accessible temperatures, the Arrhenius equation is a suitable approximation. Even when a pre-exponential temperature dependence is observed, it's common to report the reaction's activation energy E_a as *defined* by the temperature dependence of the Arrhenius rate constant:

$$\frac{k}{A} = \exp\left(-\frac{E_a}{RT}\right). \qquad \text{by Eq. 13.17}$$

$$\ln \frac{k}{A} = -\frac{E_a}{RT}. \qquad \text{ln of both sides}$$

$$\frac{d}{dT}\left(\ln \frac{k}{A}\right) = -\frac{d}{dT}\frac{E_a}{RT} \qquad d/dT \text{ of both sides}$$

$$\frac{d}{dT}(\ln k - \ln A) = \frac{d}{dT}\ln k = -\left(-\frac{E_a}{RT^2}\right) = \frac{E_a}{RT^2} \qquad \text{assume } A \text{ constant}$$

$$E_a \equiv RT^2 \frac{d\ln k}{dT}. \qquad \text{solve for } E_a \ (13.19)$$

To determine the activation energy experimentally, a typical approach is to measure the rate constant at various temperatures, and then to graph $\ln k$ vs. $1/T$, an **Arrhenius plot.**

We can rearrange Eq. 13.19 to show that the slope of this graph gives the activation energy (Fig. 13.3):

$$d\ln k = \frac{E_a}{R}\frac{dT}{T^2} = -\frac{E_a}{R}d\left(\frac{1}{T}\right).$$

The Arrhenius constants for several selected reactions are given in Table 13.1.

TABLE 13.1 **Arrhenius factors for various gas-phase elementary reactions at 298.15 K.**

First-Order	A (s^{-1})	E_a (kJ mol^{-1})
$C_2H_5SH \rightarrow C_2H_4 + H_2S$	1.0×10^{13}	215
$C_2H_5I \rightarrow C_2H_4 + HI$	2.5×10^{13}	209
$CH_3NC \rightarrow CH_3CN$	4×10^{13}	160
$N_2O_5 \rightarrow NO_2 + NO_3$	6.3×10^{14}	88
$CH_3CO \rightarrow CH_3 + CO$	1×10^{15}	43
$C_2H_6 \rightarrow 2CH_3$	2.5×10^{17}	384
Second-Order	A (cm^3 mol^{-1} s^{-1})	E_a (kJ mol^{-1})
$O_3 + C_3H_8 \rightarrow C_3H_7O + HO_2$	10^9	51
$NO + O_3 \rightarrow NO_2 + O_2$	7.9×10^{11}	10.5
$NO + Cl_2 \rightarrow NOCl + Cl$	4.0×10^{12}	84.9
$S + N_2 \rightarrow SN + N$	4.0×10^{12}	459
$S_2 + H \rightarrow SH + S$	7.9×10^{12}	69.5
$2NOCl \rightarrow 2NO + Cl_2$	1×10^{13}	103.6
$CH_3 + CH_3 \rightarrow C_2H_6$	2×10^{13}	≈ 0

(continued)

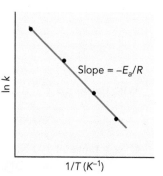

▲ FIGURE 13.3 **Arrhenius plot.** If the Arrhenius equation holds, then a graph of $\ln k$ vs. $1/T$ yields a straight line with negative slope equal to $-E_a/R$.

TABLE 13.1 continued **Arrhenius factors for various gas-phase elementary reactions at 298.15 K.**

Second-Order	A (cm^3 mol^{-1} s^{-1})	E_a (kJ mol^{-1})
$2NOBr \rightarrow 2NO + Br_2$	4.15×10^{13}	58.1
$OH + H_2 \rightarrow H_2O + H$	8×10^{13}	42
$Cl + H_2 \rightarrow HCl + H$	8×10^{13}	23
$O + N_2 \rightarrow NO + N$	10^{14}	315

Third-Order	A (cm^6 mol^{-2} s^{-1})	E_a (kJ mol^{-1})
$2NO + O_2 \rightarrow 2NO_2$	1.05×10^9	-4.6
$2NO + Br_2 \rightarrow 2NOBr$	3.2×10^9	≈ 0
$2NO + Cl_2 \rightarrow 2NOCl$	1.7×10^{10}	15
$O + O_2 + Ar \rightarrow O_3 + Ar$	3.2×10^{12}	-9.6
$O + O + O_2 \rightarrow O_2 + O_2$	1×10^{15}	≈ 0
$I + I + Ar \rightarrow I_2 + Ar$	6.3×10^{15}	5.4
$H + H + H_2 \rightarrow H_2 + H_2$	1×10^{16}	≈ 0
$SO_3 + O + He \rightarrow SO_2 + O_2 + He$	5.0×10^{16}	-6.5

EXAMPLE 13.2 **Temperature Dependence in A**

CONTEXT When a given reaction can take place over a temperature range of hundreds of degrees, it can be difficult to tabulate the data so that the rate constant can be reliably calculated to high precision. A common solution is shown in Eq. 13.18: set $A(T) = A'T^w$ to include a temperature dependence in the pre-exponential factor. But, as we'll see, theory doesn't always predict the same temperature dependence, the same value of the exponent w. How big a difference does this make?

PROBLEM For the reaction $2NOBr \rightarrow 2NO + Br_2$ in Table 13.1, calculate k at 298 K assuming only the Arrhenius equation. Then predict the values of k at 400 K if (a) the Arrhenius equation continues to hold *and* (b) if the Arrhenius value at 298 K is correct but A is actually of the form $A'T^{1/2}$.

SOLUTION At 298 K:

$$k = Ae^{-E_a/(RT)} = (4.15 \cdot 10^{13})e^{-58.1/(.008314 \cdot 298)} = 2.7 \cdot 10^3 \text{ cm}^3 \text{ mol}^{-1} \text{ s}^{-1}$$

At 400 K:

$$k = (4.15 \cdot 10^{13})e^{-58.1/(.008314 \cdot 400)} = 1.07 \cdot 10^6 \text{ cm}^3 \text{ mol}^{-1} \text{ s}^{-1}.$$

With the added $T^{1/2}$ dependence, at 400 K:

$$k = (4.15 \cdot 10^{13})\sqrt{\frac{400}{298}}e^{-58.1/(.008314 \cdot 400)} = 1.24 \cdot 10^6 \text{ cm}^3 \text{ mol}^{-1} \text{ s}^{-1}.$$

Over a range of 100 K, the pre-exponential temperature dependence (as predicted by simple collision theory) contributes only a 20% change, which is dwarfed by the factor of 400 increase from the exponential term. This extreme temperature dependence, by the way, means that if we wanted to measure the rate constant to within a 1% error, we would need the temperature to be stable to within about 0.1 K, which is difficult to maintain in the middle of a chemical reaction and becomes only more challenging at higher temperatures. Partly for this reason, rate constants are relatively poorly determined constants, often given to only one or two significant figures.

Simple collision theory is straightforward and instructive, but not very accurate. The steric factor may have to be empirically determined, and often it is much smaller than true steric arguments can justify. The fault is in the omission of more subtle barriers to reactivity than the energy of the transition state and the orientation of the reactants, and in the neglect of rotational and vibrational energy contributions. In essence, the appeal of simple collision theory is that it allows us to describe reaction kinetics using primarily macroscopic parameters. A more accurate treatment will require a more detailed consideration of what happens at the *molecular* level.

A chemical reaction involves a change between different thermodynamic states, but these states must be described by quantum mechanics, the same as for non-reacting compounds. At every point during the breaking and forming of chemical bonds, the chemical system must be in a quantum state. The likelihood of a quantum state of the reactants becoming a quantum state of the products is determined by the wavefunction for each state. The overall bulk reaction rate is determined by the sum over the reaction probabilities for all those wavefunctions for all possible distributions of the translational, vibrational, and other energies in the system. The following treatment for reactive collisions simplifies this picture somewhat, but takes an approach to the reaction probability that can be adapted to incorporate, to varying degrees, the influence of the quantum mechanics.

BIOSKETCH | Robert E. Continetti

Robert E. Continetti is Professor of Chemistry at the University of California at San Diego. Professor Continetti's research is directed at fundamental studies of chemical reactions and reactive compounds. One method developed in the laboratory allows his team to study single-step reactions initiated by the transfer of an electron to a positive ion. One of these studies focused on the dissociation of the *sym*-triazine molecule ($H_3C_3N_3$: benzene with every other CH replaced by N). The molecule was formed as the cation, $H_3C_3N_3^+$, and mass spectrometric methods were used to isolate it in a molecular beam. The beam of $H_3C_3N_3^+$ is channeled through a cell containing neutral atomic cesium vapor. Cesium has the

lowest ionization energy in the periodic table (at 3.9 eV, slightly lower even than francium's) and easily surrenders an electron to the $H_3C_3N_3^+$, which dissociates according to the reaction $H_3C_3N_3^+ + e^- \rightarrow 3\ HCN$. The dissociation into *three* fragments instead of two is highly unusual, and the Continetti group was able to determine the distribution of energies of the HCN product molecules by measuring their velocities at a time- and position-sensitive detector. These techniques allow sophisticated reaction surfaces such as the one in Fig. 12.2 to be mapped experimentally, including the region near the transition state.

Ion trap chamber　　Path of molecular beam　　Source chamber

▲ **Experimental apparatus used for dissociative charge exchange experiments such as the *sym*-triazine study.**

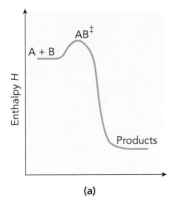

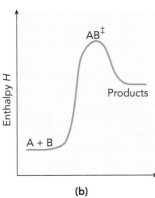

▲ **FIGURE 13.4 Reaction diagram for a bimolecular transition state theory reaction.** The reactants A and B form the activated complex AB‡ on their way to forming the reaction products. The reaction may be exothermic (**a**) or endothermic (**b**).

13.3 Transition State Theory

Many kinetic models take advantage of **transition state theory** or **activated complex theory.** Transition state theory actually encompasses a range of chemical reaction models, some specific to particular classes of reaction. What they all share is the description of the chemical reaction as a process in which reactants combine to form the activated complex, AB‡, which can then decay into the products (Fig. 13.4):

$$A + B \rightarrow AB^{\ddagger} \rightarrow \text{products.}$$

We will go over the simplest form of transition state theory as applied to bimolecular reactions, but we will be able to see qualitatively how the model can be made more sophisticated and extended to other systems.

Isn't the activated complex the same as the transition state? Definitions of the activated complex vary, and for our purposes, the two may as well be the same. In more rigorous implementations of transition state theory, the activated complex corresponds to a *region* of the reaction surface, including the unique transition state geometry, that the reactants must cross as they transform into products. The transition state, after all, corresponds to a single geometry on the reaction surface: the highest point on the minimum energy path connecting reactants to products. But when reactants collide with more than enough energy to overcome the activation barrier, the reaction path can take a higher energy path that goes around the transition state. In transition state theory, it is common to define not so much a single point on the reaction surface, but a **dividing surface** that separates the reactants from products. As shown in Fig. 13.5, the dividing surface contains the transition state but extends upward in energy from that saddle point and perpendicular to the reaction pathway. On one side of the dividing surface, the potential energy slopes down toward the reactants. On the other side, it slopes down toward the products. The activated complex is whatever geometry the reacting particles have as they cross the dividing surface. This broader definition of the activated complex recognizes that each combination of reactants will find a

▶ **FIGURE 13.5 Dividing surface for the H + HF reaction.** The activated complex may correspond to any geometry along the dividing surface (shown here as the red curve).

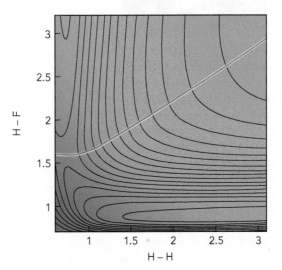

slightly different path across the barrier. The thermal and zero-point energies make the geometry of each reaction system even more dynamic than the reaction diagram suggests, and the molecules do not actually form a single, well-defined transition state structure during the reaction.

In any case, the activated complex is the only route to the products for this reaction. We would consider routes to the same products through different activated complexes to be distinct reactions. It's always possible for the products to recombine to form the reactants via the activated complex, but for irreversible reactions this process is negligible. At present, we will ignore reverse reactions because we are studying only a single step in the reaction.

The central assumption in transition state theory is that the reaction dynamics are dominated by three sets of states: the reactant states, the activated complex states, and the product states. By determining the energy levels and degeneracies of each set, the reaction rate for a given set of conditions can be determined. We will go through one possible derivation of the rate constant now, and a DERIVATION SUMMARY follows Eq. 13.33.

Parameters Key: Transition State Theory

symbol	parameter	SI units
$K^{\ddagger}$	equilibrium constant for formation of $AB^{\ddagger}$ from $A+B$	unitless
X_A	mole fraction of A	unitless
$[A], [B]$	concentrations of A and B (normally in units of mol L^{-1})	mol m^{-3}
$[AB^{\ddagger}]$	concentration of $AB^{\ddagger}$	mol m^{-3}
n_T	sum of moles for all components in volume V, including any solvent	mol
V	volume	m^3
C	overall concentration $= n_T/V$	mol m^{-3}
f	transmission factor, probability per unit time that activated complex becomes product	s^{-1}
$\langle v_{rxn} \rangle$	average speed with which activated complex crosses the dividing surface	m s^{-1}
R_{rxn}	distance to travel across the dividing surface	m
k_B	Boltzmann constant	J K^{-1}
R	gas constant $= \mathcal{N}_A k_B$	J K^{-1} mol^{-1}
T	temperature	K
m_A, m_B	masses of molecule A, molecule B	kg
μ	reduced mass $= m_A m_B/(m_A + m_B)$	kg
$Q^{\ddagger}_{tot}$	overall partition function for activated complex	unitless
$Q^{\ddagger}_{rxn}$	translational partition function for activated complex along reaction path	unitless
$Q^{\ddagger}$	partition function for activated complex neglecting motion along reaction path, $Q^{\ddagger}_{tot}/Q^{\ddagger}_{rxn}$	unitless
$\Delta G^{\ddagger}, \Delta H^{\ddagger}, \Delta S^{\ddagger}$	free energy, enthalpy, and entropy for formation of activated complex from reactants	J mol^{-1} J K^{-1} mol^{-1}

At this point many methods diverge to treat the quantum mechanics of the transition state explicitly. We shall only treat it thermodynamically. The first step in our reaction, the formation of the activated complex, has equilibrium constant

$$K^{\ddagger} = \frac{a^{\ddagger}}{a_A a_B}. \tag{13.20}$$

If we use the ideal gas law for gases, or Henry's or Raoult's law for solutions, the activity is directly proportional to the density of the substance, written either as a mole fraction or partial pressure:

$$K^{\ddagger} = \begin{cases} = \dfrac{X^{\ddagger}}{X_A X_B} & \text{solution-phase} \\[2mm] = \dfrac{P^{\ddagger} P^{\ominus}}{P_A P_B} & \text{gas-phase} \end{cases} \tag{13.21}$$

Although it is usually easier to investigate details of gas-phase reactions, kinetics equations are conventionally written in terms of molarities, as appropriate for solution-phase chemistry. To rewrite Eq. 13.21 in terms of molarities, we need to convert the mole fraction into molarity using the total number of moles n_T:

$$K^{\ddagger} = \frac{X^{\ddagger}}{X_A X_B} = \frac{\left(\dfrac{n^{\ddagger}}{n_T}\right)}{\left(\dfrac{n_A}{n_T}\right)\left(\dfrac{n_B}{n_T}\right)} = \frac{\left(\dfrac{n^{\ddagger}}{V}\right)\left(\dfrac{n_T}{V}\right)}{\left(\dfrac{n_A}{V}\right)\left(\dfrac{n_B}{V}\right)} = \frac{[AB^{\ddagger}]C}{[A][B]}, \tag{13.22}$$

where C is the fraction n_T/V, the overall molarity of the solution. For dilute solutions, where our approximations for the activities are valid, C is accurately estimated by the molarity of the solvent. Therefore, we can write the concentration of the transition state in terms of the concentrations of the reactants:

$$[AB^{\ddagger}] = \frac{K^{\ddagger}[A][B]}{C}. \tag{13.23}$$

The reaction velocity v is the rate of product formation, and we now assume this to be proportional to the concentration of the activated complex:

$$v = f[AB^{\ddagger}] = \frac{fK^{\ddagger}[A][B]}{C} = k_{TST}[A][B]. \tag{13.24}$$

The **transmission factor** f is the fraction per second of activated complexes that will convert to products. The rate constant is therefore

$$k_{TST} = \frac{fK^{\ddagger}}{C}. \tag{13.25}$$

We have two terms to solve for: the transmission factor f and the equilibrium constant $K^{\ddagger}$.

The transmission factor f may be written the following way:

$$f = \frac{\langle v_{rxn}\rangle}{2R_{rxn}}, \tag{13.26}$$

where $\langle v_{rxn}\rangle$ is the average rate at which the reactants move across the threshold of the transition state to become products, and R_{rxn} is the short distance they must travel to cross this threshold. The factor of 2 appears because it is assumed that the reactants are traveling with equal probability in either direction along the reaction coordinate, and therefore only half of them are moving in the right direction for the reaction to go forward. It would be tricky to quantify R_{rxn}, but we will find that our final expression does not depend on this value. We can write an expression for $\langle v_{rxn} \rangle$ based on our analysis of characteristic speeds in Section 5.1:

$$\langle v_{rxn} \rangle = \sqrt{\frac{2k_B T}{\pi \mu}}. \tag{13.27}$$

This is Eq. 5.4 written for motion in one dimension instead of three dimensions.

The equilibrium constant $K^{\ddagger}$ can be rewritten in terms of the partition functions of the reactants and the transition state. Equation 12.40 lets us write

$$K^{\ddagger} \approx \frac{Q^{\ddagger}_{tot}}{Q_A Q_B}. \tag{13.28}$$

There is a problem with this form of $K^{\ddagger}$, however: it contains a contribution to the partition function $Q^{\ddagger}_{tot}$ from the reaction coordinate that we have to treat more carefully. The transition state is not bound along the reaction coordinate—it falls apart either to create reactants or products—and therefore this coordinate is not a vibrational coordinate, but an *extra* translational coordinate, counted in addition to the center of mass translational motion of the transition state. Therefore, we break $Q^{\ddagger}_{tot}$ up into two pieces:

$$Q^{\ddagger}_{tot} = Q^{\ddagger} Q^{\ddagger}_{rxn}, \tag{13.29}$$

where $Q^{\ddagger}_{rxn}$ is the translational partition function for motion along the reaction coordinate. Equation 3.44 gives us the translational partition function for motion in three dimensions, which is the product of the three independent partition functions for motion along X, Y, and Z. We take the cube root of that to find the partition function for motion along only one coordinate. We also divide the result by a factor of 2, because we are changing a vibrational coordinate into a translational one. While we would integrate the bond length R for vibration from 0 to ∞, we integrate translational coordinates from $-\infty$ to ∞, which introduces the factor of 2. We are left with

$$Q^{\ddagger}_{rxn} = \frac{1}{2}\left(\frac{8\pi \mu k_B T}{h^2}\right)^{1/2} V^{1/3} = \left(\frac{\sqrt{2\pi \mu k_B T}}{h}\right)R_{rxn}, \tag{13.30}$$

where we have limited the partition function to the same distance R_{rxn} over which we are evaluating the reactant speed $\langle v_{rxn} \rangle$.

Combining these results, we obtain

$$
\begin{aligned}
k_{TST} &= \frac{f K^{\ddagger}_{tot}}{C} && \text{by Eq. 13.25} \\[2mm]
&= \left(\frac{\langle v_{rxn} \rangle}{2CR_{rxn}}\right)\frac{Q^{\ddagger} Q^{\ddagger}_{rxn}}{Q_A Q_B} && \text{by Eqs. 13.26, 13.28, 13.29} \\[2mm]
&= \left(\frac{1}{2CR_{rxn}}\right)\sqrt{\frac{2k_B T}{\pi \mu}}\left(\frac{\sqrt{2\pi \mu k_B T}}{h}\right)(R_{rxn})\frac{Q^{\ddagger}}{Q_A Q_B} && \text{by Eq. 13.30} \\[2mm]
&= \left(\frac{k_B T}{Ch}\right)\frac{Q^{\ddagger}}{Q_A Q_B} = \left(\frac{k_B T}{Ch}\right)K^{\ddagger}, && (13.31)
\end{aligned}
$$

where $K^{\ddagger}$ no longer contains the partition function from the reaction coordinate.

This equilibrium constant $K^{\ddagger}$ can also be rewritten in terms of the entropy and enthalpy change in formation of the transition state (Eq. 12.41):

$$\ln K^{\ddagger} = -\frac{\Delta G^{\ddagger}}{RT} = -\frac{\Delta H^{\ddagger} - T\Delta S^{\ddagger}}{RT}. \tag{13.32}$$

With that, we can finally write the rate constant according to the **Eyring equation:**

$$k_{\text{TST}} = \frac{k_{\text{B}}T}{Ch}e^{\Delta S^{\ddagger}/R}e^{-\Delta H^{\ddagger}/(RT)}. \tag{13.33}$$

DERIVATION SUMMARY Transition State Theory. We imagined a dividing surface that separates reactants from products, including the transition state but also higher energy geometries that the reaction could pass through, and we called the geometries on or near this dividing surface the activated complex, $AB^{\ddagger}$. Next we assumed that an equilibrium exists between the reactants and $AB^{\ddagger}$, and we wrote the equilibrium constant $K^{\ddagger}$ in terms of their activities. We then reasoned that the activity of $AB^{\ddagger}$ would be proportional to the rate at which product was formed, because the reaction would have to pass through $AB^{\ddagger}$ to get to product. Therefore, we solved for the activity of $AB^{\ddagger}$ and converted the reactant activities to concentrations, getting the expected dependence on $[A][B]$. Finally, we used the thermodynamics of chemical equilibrium to replace the equilibrium constant $K^{\ddagger}$ by $e^{-\Delta G^{\ddagger}/(RT)}$, which breaks up into an enthalpy contribution $e^{-\Delta H^{\ddagger}/(RT)}$ and an entropy contribution $e^{-\Delta S^{\ddagger}/R}$.

There is still an Arrhenius-like term $e^{-E_a/(RT)}$. Although $\Delta H^{\ddagger}$ is not precisely the same as the energy difference $\Delta E^{\ddagger}$ between the reactants and the transition state, it is usually fairly close. Starting with our definition of the enthalpy (Eq. 7.17), we find that $\Delta H = \Delta E + \Delta(PV)$ for any thermodynamic system. For ideal solutions, the volume is independent of the chemistry, and the $\Delta(PV)$ term is zero; in that case, the distinction between E_a and $\Delta H^{\ddagger}$ can be ignored. For ideal gases at constant temperature, $\Delta(PV) = RT\Delta n$. Therefore, we can quickly convert between energy and enthalpy for any nearly ideal system by counting the change Δn_{gas} in moles of gas from reactants to product:

$$\Delta H = \Delta E + RT\Delta n_{\text{gas}}. \tag{13.34}$$

At 298 K, RT is 2.5 kJ mol^{-1}. Because chemical bond changes typically involve activation energies in the range of $40 - 400$ kJ mol^{-1}, the $\Delta(PV)$ term is often a correction of less than 10%.

The temperature dependence of the pre-exponential factor differs from the factor of $T^{1/2}$ that we saw in simple collision theory, but it still pales in comparison to the T-dependence in the exponential term.

A more significant difference between the Eyring and simple collision theory rate constants is the replacement of the steric factor by a term $\Delta S^{\ddagger}$, the entropy change during formation of the transition state. Steric effects will be included in this because fewer orientations of the reactants in the transition state corresponds to a lower entropy transition state. However, it also includes any other factors that reduce the number of states accessible to the transition state, such as highly concerted motions, where many atoms must be moving in the right direction at the same time. An example of a highly concerted motion is the S_N2 displacement reaction

$$CH_3I + H \rightarrow CH_4 + I,$$

which involves an inversion of the methyl group.

It is $\Delta S^{\ddagger}$ that poses the greatest challenge in calculating the Eyring rate constant. We can relate this term to the ensemble size in statistical mechanics by application of Boltzmann's law (Eq. 2.11) for the entropy,

$$S = k_B \ln \Omega.$$

Keeping in mind that S in Boltzmann's law differs from the molar entropy S_m by a factor of Avogadro's number $\mathcal{N}_A$, we obtain an expression for $e^{\Delta S^{\ddagger}}$ in terms of the ensemble size as follows:

$$
\begin{aligned}
\exp\left[\Delta S^{\ddagger}/R\right] &= \exp\left[(S^{\ddagger} - S_{mA} - S_{mB})/R\right] \\
&= \exp\left\{\mathcal{N}_A k_B\left[\ln \Omega^{\ddagger}(E^{\ddagger}) - \ln \Omega_A(E_A) - \ln \Omega_B(E_B)\right]/R\right\} \\
&= \exp\left\{\ln \Omega^{\ddagger}(E^{\ddagger}) - \ln \Omega_A(E_A) - \ln \Omega_B(E_B)\right\} \\
&= \exp\left\{\ln\left[\frac{\Omega^{\ddagger}(E^{\ddagger})}{\Omega_A(E_A)\Omega_B(E_B)}\right]\right\} \\
&= \frac{\Omega^{\ddagger}(E^{\ddagger})}{\Omega_A(E_A)\Omega_B(E_B)}.
\end{aligned}
\tag{13.35}
$$

In order to evaluate this term, chemical theoreticians must calculate the number of quantum states accessible by the reactants and the corresponding number of states of the transition state at every energy that is likely to contribute substantially to the reaction probability.

TOOLS OF THE TRADE | **Ultrafast Lasers**

In a laser, energy is transferred into a substance (the laser medium), usually using an electric current or a light source (a lamp or another laser), promoting the particles of the laser medium into excited quantum states. Specific combinations of laser medium and excitation method allow for a population inversion, in which more particles are in the excited energy level than in some lower energy level (accounting for differences in degeneracy). The energized laser medium emits radiation as it relaxes to the lower energy state, and some of that radiation is reflected back through the laser medium to stimulate more emission at the same wavelength, phase, and photon trajectory. In this way, the original beam of light is effectively amplified: the laser medium invests much of its energy into magnifying the intensity of the original beam, rather than emitting the light in all directions and over a wide wavelength range.

If the upper quantum state of the laser transition can be continuously supplied *and* if the lower state can be continuously depleted, then the laser can operate continuously. However, the intense emission of many lasers delivers molecules out of the excited state to the lower state faster than the system can respond to maintain the population inversion required for light amplification. The laser emits light only in short pulses, usually on the order of 10 ns, because after each burst of light the population inversion has to be built up again.

In many applications, pulsed operation of a laser has advantages over continuous wave (cw) operation. By concentrating the emission into a short time, pulsed lasers often achieve much greater energy densities in the beam than cw lasers, allowing some pulsed lasers to vaporize small quantities of materials with very high boiling points. Furthermore, by pushing these pulses to extremely short durations, pulsed lasers can serve as strobe lights on chemical processes, permitting glimpses of atomic rearrangements *on the time scale of the reaction.*

What is an ultrafast laser? The term *ultrafast laser* is now generally applied to lasers with pulse durations of less than roughly 1 picosecond (1 ps = 10^{-12} s). Efforts to shorten the pulse duration have extended from the 1980s to the present, when lasers with pulse widths of about 100 femtoseconds (1 fs = 10^{-15} s) are found in hundreds of laboratories, and the shortest pulses attained are 100 attoseconds (1 as = 10^{-18} s).

Why do we use ultrafast lasers? Femtosecond lasers have been employed in chemistry primarily to examine motions of the atomic nuclei during a chemical process such as reaction or phase transition. A typical chemical bond stretches back and forth once over a period of between 10 and 40 femtoseconds. Chemical reactions involve motions along the same vibrational coordinates (one can imagine, for example, a stretch so extreme that it breaks the bond), and single steps in a reaction occur on the same time scale as the vibrational motions—a few tens of femtoseconds. Therefore, to see the motions of the atoms during that step requires pulse widths of 10 fs or less.

How do they work? If the medium is capable of emitting laser light over a broad range of wavelengths, then the spacing l between the principal laser optics determine a set of possible standing wave laser wavelengths $\lambda = l/n$ where n must be an integer. These allowed wavelengths are called the *longitudinal modes* of the laser. In principle, all wavelengths λ within the range of emissions of the laser medium may lase simultaneouly. However,

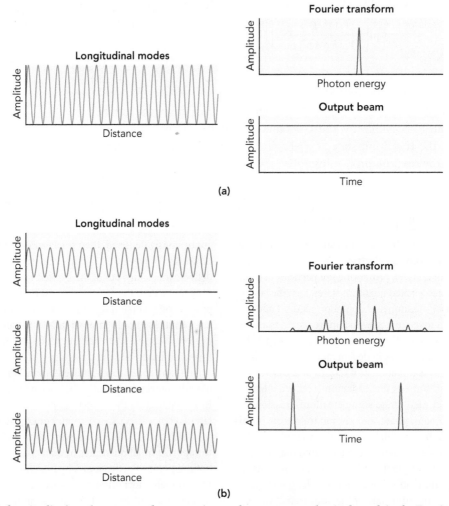

▲ **(a) A single longitudinal mode corresponds to a continuous laser output and a single peak in the Fourier transform. (b) By combining many longitudinal modes with a well-defined relative phase, the laser output can be concentrated into a series of short pulses.**

light amplification at each of these modes is stimulated by a different set of photons, and in general the oscillation of one mode is not coordinated with the oscillations of other modes. Small instabilities in the laser cause the relative phases of the modes to shift randomly.

The key to getting short pulses in femtosecond lasers has been **mode-locking,** defining a relative phase among the different longitudinal modes. The combination of many modes oscillating with well-defined phases leads to a series of pulses. The wider the range of modes (and therefore of photon energies) that can be combined, the shorter the pulse that can be generated. This effect is the manifestation of a Fourier transform (Section A.1) between the frequency and the duration of the radiation. The Fourier transform establishes a trade-off: the shorter the pulse, the less precise the energy of the radiation.

Several different techniques have been developed to mode-lock the laser. One of the oldest methods is the use of a *saturable absorber,* a substance (often a semiconductor) that absorbs radiation less as the radiation's intensity increases. This approach causes the laser action to cease when the power is too low, leaving the amplification to occur only when the signal is intense. Narrowing the window of time in which the different modes can successfully lase forces the relative phases to come into alignment. A second method uses an optical modulator within the laser to vary either the intensity or frequency of the laser as a function of time, enforcing a specific relative phase on the different modes. More recent experiments have used *high harmonic generation* to achieve sub-femtosecond pulses. In this technique, high intensity laser pulses ionize the laser medium, and when the electrons recombine with the ions the energy is released in a burst of coherent radiation. To form a pulse of only 10^{-16} s, we must combine photon frequencies over a range of roughly 10^{16} s^{-1}, which requires photon energies extending into the x-ray region of the spectrum.

As an example of the application of this technology, Ahmed Zewail (1999 Nobel Prize in Chemistry) was able to measure the time for formation of the transition state in the dissociation of ICN to form the I and CN radicals, using one femtosecond laser pulse to initiate the dissociation and a second pulse to probe the reaction complex spectroscopically.

13.4 Diffusion-Limited Rate Constants

All of the theory involved in predicting rate constants by simple collision or transition state theory comes to nothing if the only substantial barrier to the reaction is in simply getting the reactants to find each other. In many reactions, the activation barriers and other factors are so favorable that the exponential terms in Eqs. 13.16 or 13.33 are essentially 1, predicting that the reaction occurs as soon as the reactants collide. Our derivations have taken into account the collision rate in the gas phase, but in solution we need to allow also for the diffusion of the reactants through the solvent to form the activated complex.

In the diffusion-limited reaction between reactants A and B, molecule A is consumed at the rate that molecule B diffuses into contact with A. The diffusion rate we can estimate from Fick's first law (Eq. 5.42),

$$ J = -D\left(\frac{d\rho}{dZ}\right), $$

where J is the diffusion-limited flux of molecules per unit area per unit time. If we make the area the surface area of a sphere of radius R_{AB} (Fig. 13.6), then any molecule B diffusing into that sphere comes in contact with A, and the reaction occurs. The rate, in molecules per unit time, of B diffusing into any sphere of radius r we'll call γ_D. A negative sign accounts for this flow being in the negative r direction:

$$ \gamma_D = -J_B(4\pi r^2) $$

$$ = D_{AB}(4\pi r^2)\left(\frac{d\rho_B}{dr}\right). $$

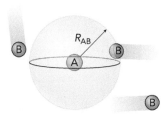

▲ FIGURE 13.6 **Diffusion-limited reaction.** To estimate the diffusion-limited rate constant, we assume a sphere centered on each molecule A such that any molecule B that gets inside the sphere will react.

Assuming that the concentration of B is not changing too rapidly with time, we can set γ_D to be constant and integrate over the volume outside the sphere around A to get the total rate of diffusion through the sphere:

$$\gamma_D = D_{AB}(4\pi r^2)\left(\frac{d\rho_B}{dr}\right)$$

$$\frac{\gamma_D}{r^2} dr = D_{AB}(4\pi)d\rho_B \qquad\qquad \text{divide by } r^2$$

$$\int_{R_{AB}}^{\infty} \frac{\gamma_D}{r^2} dr = D_{AB}(4\pi) \int_0^{\rho_B} d\rho_B' \qquad \text{integrate outside sphere}$$

$$\frac{\gamma_D}{R_{AB}} = D_{AB}(4\pi)\rho_B \qquad\qquad \int x^{-2} dx = -x^{-1}$$

$$\gamma_D = 4\pi R_{AB}D_{AB}\rho_B$$

This last expression gives us the rate at which molecules of B diffuse into contact with any molecule A, and we can set that equal to the rate at which that particular A molecule is consumed. To find the total rate of consumption of A, we multiply γ_D by the concentration of A, adding a factor of Avogadro's number to convert the number density ρ_B to a molarity:

$$-\frac{d[A]}{dt} = \frac{1}{\mathcal{N}_A}\gamma_D[A] = 4\pi R_{AB}D_{AB}[A][B].$$

We can calculate the diffusion-limited rate constant k_D by solving for k in the bimolecular rate law, $-d[A]/dt = k[A][B]$:

$$k_D = \frac{-d[A]/dt}{[A][B]}$$

$$= \frac{4\pi R_{AB}D_{AB}[A][B]}{[A][B]}$$

$$= 4\pi R_{AB}D_{AB}. \qquad\qquad (13.36)$$

For a typical molecular collision radius of 5 Å and diffusion constant of $1 \cdot 10^{-5}\,\text{cm}^2\,\text{s}^{-1}$, this predicts a diffusion limited rate of about $6 \cdot 10^{-12}\,\text{cm}^3\,\text{s}^{-1}$ per molecule, which converts to $4 \cdot 10^9\,\text{L mol}^{-1}\,\text{s}^{-1}$. Bimolecular reactions in solution that are predicted to have rate constants of this order or higher are likely to be limited by diffusion.

13.5 Rate Laws for Elementary Reactions

Both the simple collision and transition state theories find that the reaction rate depends only on the rate constant and the product of the concentrations of the reactants:

$$\text{elementary reaction:} \quad v = k \prod_{\text{reactants}} [i]^{v_i}. \qquad (13.37)$$

The application to various cases is straightforward, bearing in mind that these rate laws hold only for *elementary reactions,* such as those in Table 13.1:

$$A \rightarrow \text{products} \quad v = k[A]$$
$$A + B \rightarrow \text{products} \quad v = k[A][B]$$
$$2A \rightarrow \text{products} \quad v = k[A]^2$$
$$A + B + C \rightarrow \text{products} \quad v = k[A][B][C].$$

With these rate laws, to the extent that the reaction we are interested in can be considered an elementary reaction, the product concentrations can be obtained as a function of time and the initial reactant concentrations. The simplest example is the unimolecular decomposition, where one molecule falls apart irreversibly into fragments. To find the concentration [B] in the reaction

$$A \rightarrow B + C,$$

one needs to solve the rate equation:

$$v = \frac{d[B]}{dt} = -\frac{d[A]}{dt} = k[A]. \tag{13.38}$$

This is the fundamental problem in classical reaction kinetics: we know the rate law, which is a first-order differential equation, and we wish to find the concentrations of the chemical components as a function of time and the initial concentrations.

In this case, if we know the initial concentration of reactant $[A]_0$, then we can multiply both sides by dt and divide by [A] and integrate, obtaining

$$-\int_{[A]_0}^{[A]} \frac{d[A]'}{[A]'} = k \int_0^t dt'$$

$$-\ln \frac{[A]}{[A]_0} = kt$$

$$[A] = [A]_0 \, e^{-kt} \tag{13.39}$$

if $[B]_0 = [C]_0 = 0$: $\quad [B] = [C] = [A]_0 (1 - e^{kt}). \tag{13.40}$

Equations 13.39 and 13.40 are examples of **integrated rate laws,** solutions to the rate law differential equation that explicitly describe how the concentrations of reactants and products vary with time.

If we have a fixed amount of A at the start of the reaction, then [A] drops steadily as the reaction progresses, slowing as the amount of A diminishes (Fig. 13.7). The time it takes for half of the initial compound to decompose is a useful characteristic time for the reaction, known as the **half-life.** From Eq. 13.39, we find $t_{1/2}$ by setting $[A]/[A]_0 = 0.5$ and solving for t:

$$t_{1/2} = -\frac{1}{k}\ln\tfrac{1}{2} = \frac{\ln 2}{k}. \tag{13.41}$$

This is the direct measurement of a reaction rate: measuring the concentration of reactant and/or product as a function of time. Although it sounds straightforward, concentrations in reaction mixtures are rarely completely straightforward to measure, particularly for an elementary reaction

CHECKPOINT In addition to the first-, second-, and third-order elementary reactions used as examples here, there are zero-order reactions, for which the reaction rate is independent of the reactant concentration. These are often reactions that are limited by the availability of a required catalyst or other factor, such that only a relatively small number of reactant molecules are able to convert to products at any given time.

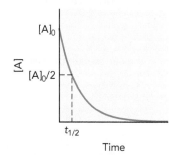

▲ FIGURE 13.7 **Concentration vs. time for a first-order elementary reaction.** The half-life is the time at which the concentration of reactant has dropped to half its original value.

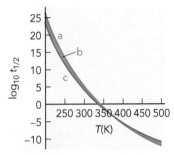

▲ **FIGURE 13.8 Logarithms of predicted half-lives as a function of temperature.** The values are given for a reaction with activation energy $E_a = 100.0$ kJ mol^{-1}, assuming
(a) $A = (1.25 \cdot 10^{10} \, T^2) \, \text{s}^{-1}$,
(b) $A = 1.0 \cdot 10^{14} \, \text{s}^{-1}$, and
(c) $A = (3.2 \cdot 10^{20} \, T^{-2}) \, \text{s}^{-1}$.

that usually involves at least one eager-to-react molecule. Short lifetimes for reactive molecules, adequate mixing of the reactants and products, control of the temperature throughout the reaction mixture (given that the reaction will absorb or release at least a little heat), and effects from competing reactions: all of these complicate the seemingly simple problem of measuring how fast a single-step reaction takes place. Given that, one understands why the temperature dependence of the pre-exponential factor appears to be such a mild effect. Figure 13.8 graphs $\log_{10}$ of the predicted half-life for a reaction as a function of temperature, assuming that the pre-exponential factor is of the form (a) $A' T^2$, (b) A', and (c) $A' T^{-2}$. Despite this wide range of temperature dependence, the curves appear very similar, because the overwhelming T-dependence is in the exponential term $e^{-E_a/(k_B T)}$. An experimentalist attempting to determine the temperature dependence in both the pre-exponential and exponential terms must be able to measure the reaction rates very precisely over a wide range of temperatures.

EXAMPLE 13.3   Integrated Rate Laws

CONTEXT Polystyrene is one of the most popular plastics in the world, used for packing materials, CD cases, laboratory plasticware, and so on. In contrast, the compound 1,3-butadiene is produced during the cracking of petroleum in greater quantities than industry has a use for. However, at high temperature butadiene will dimerize to form vinylcyclohexene, which can be dehydrogenated to produce styrene, the compound from which polystyrene is made. Using catalysts to lower the temperature for the dimerization reaction, this synthesis may prove to be a clever and economical use of the excess butadiene.

PROBLEM Find the integrated rate law for the elementary reaction 2A → B, where A is butadiene and B is vinylcyclohexene. Use this to obtain an expression for the half-life of butadiene in terms of the rate constant k and the initial concentration $[A]_0$.

SOLUTION The rate law is

$$-\frac{d[A]}{dt} = k[A]^2.$$

We can isolate the two variables $[A]$ and t and integrate both sides, setting $[A]$ at $t = 0$ equal to $[A]_0$:

$$\frac{d[A]}{[A]^2} = -k \, dt$$

$$\int_{[A]_0}^{[A]} \frac{d[A]'}{[A]'^2} = -k \int_0^t dt'$$

$$-\frac{1}{[A]} + \frac{1}{[A]_0} = -kt$$

$$[A] = \left(\frac{1}{[A]_0} + kt \right)^{-1} = \frac{[A]_0}{1 + [A]_0 kt}.$$

To get the half-life, we set $[A]$ to $[A]_0/2$ and solve for t:

$$[A]_{1/2} = \frac{[A]_0}{2} = \frac{[A]_0}{1 + [A]_0 k t_{1/2}}$$

$$1 + [A]_0 k t_{1/2} = 2$$

$$t_{1/2} = \frac{1}{[A_0] k}.$$

Unlike our result for the unimolecular elementary reaction (Eq. 13.41), this half-life is a function of the initial concentration and is not a constant.

CONTEXT *Where Do We Go From Here?*

In this chapter we've examined single-step chemical reactions. While many reactions can be controlled in the laboratory to effectively take place one step at a time, few natural processes occur with that degree of isolation. In a flame, for example, dozens if not hundreds of elementary reactions are occurring simultaneously. The final touch left for us to add to our model of chemistry is a more complete treatment of a system as it undergoes *multi*-step reactions. We will assume that the theory we've just built up in this chapter is sufficient to understand the origin of the reaction rate constants for each step, and will focus on how to combine the individual steps to accurately describe how the composition of the system will evolve over time.

KEY CONCEPTS AND EQUATIONS

13.1 **Reaction Rates.** The reaction velocity v can be expressed in terms of the rate of consumption of any reactant or generation of any product in the reaction:

$$v = \frac{1}{v_i} \frac{d[i]}{dt}, \qquad (13.1)$$

where v_i is the stoichiometric coefficient, negative for reactants and positive for products. For an **elementary reaction** (one that results from a single collision or other event at the molecular level), we can normally set this reaction velocity equal to a **rate constant** k multiplied by the product of the reactant concentrations raised to the power of their (positive) stoichiometric coefficients. The resulting equation is called the **rate law** and is a first-order differential equation. For example, for the second-order elementary reaction $A + B \rightarrow$ products, one form of the rate law is

$$-\frac{d[A]}{dt} = k[A][B].$$

13.2 **Simple Collision Theory.** One approach to predicting the rate constant of a bimolecular gas-phase reaction breaks the value of k into (1) a rate at which the reactants collide with the right orientation and (2) a probability that the collision occurs at an energy greater than the activation barrier, arriving at the rate constant from **simple collision theory:**

$$k_{SCT} = \sigma_{AB} \sqrt{\frac{8 k_B T}{\pi \mu}} \mathcal{N}_A p e^{-E_a/(RT)}. \qquad (13.16)$$

In this expression, the **steric factor** p is an empirically determined parameter for the fractional probability that the reactant orientation is correct for the reaction to occur. Simple collision theory successfully reproduces the exponential dependence on the activation energy that led to the empirical **Arrhenius equation** for the rate constant:

$$k_{Arr} = A e^{-E_a/(RT)}. \qquad (13.17)$$

13.3 **Transition State Theory.** In **transition state theory,** the reaction rate is calculated based on the rate at which the reactants can form an **activated complex,** usually in the neighborhood of the transition state, which crosses over to products. Our version of transition state theory arrives at the **Eyring equation,**

$$k_{TST} = \frac{k_B T}{Ch} e^{\Delta S^{\ddagger}/R} e^{-\Delta H^{\ddagger}/(RT)}, \qquad (13.33)$$

which divides the rate constant into contributions from the enthalpy change $\Delta H^{\ddagger}$ and entropy change $\Delta S^{\ddagger}$ on forming the activated complex.

13.4 **Diffusion-Limited Rate Constants.** For a reaction in which the rate is limited by diffusion rather than by the activation energy or reactant orientation once the collision occurs, the reaction rate constant may be approximated as

$$k_D = 4\pi R_{AB} D_{AB}. \qquad (13.36)$$

13.5 **Rate Laws for Elementary Reactions.** For the simplest elementary chemical reactions, we can integrate the rate law to arrive at an algebraic form of the **integrated rate law,** which gives the concentration of a given reactant or product as a function of time. One important example is the case of the elementary first-order reaction $A \xrightarrow{k}$ products, which gives the integrated rate law for the reactant

$$[A] = [A]_0 e^{-kt}. \qquad (13.39)$$

The **half-life** $t_{1/2}$ is the time required for half of the reactant A to be consumed, and in this first-order example it is given by

$$t_{1/2} = \frac{\ln 2}{k}. \qquad (13.41)$$

KEY TERMS

- The **reaction order** is the sum of the exponents on the concentrations in the rate law. For example, if $v = k[A][B]^2$, then the reaction order is 3, and we call this a third-order reaction.
- The **reaction mechanism** is the series of individual chemical transformations that take place at the molecular scale as the reactants are converted to products.

- A **net reaction** is a chemical equation that shows only the initial and final species of the reaction mechanism, neglecting any molecules that may be formed and then destroyed over a series of steps.
- In transition state theory, the **transmission coefficient** f is the probability per unit time that an activated complex will become product.

OBJECTIVES REVIEW

1. *Write the rate law for an elementary reaction based on the chemical equation.*
   Write the rate law for the elementary reaction
   $2NO(g) + Cl_2(g) \rightarrow 2NOCl(g)$.

2. *Estimate the rate constant of a reaction using simple collision theory or transition state theory, given the appropriate parameters.*
   Use simple collision theory to estimate the activation energy of the elementary reaction
   $H(g) + O_2(g) \rightarrow OH(g) + O(g)$ at 2000 K if the

   rate constant is $2.92 \cdot 10^{11}$ cm^3 mol^{-1} s^{-1}. Assume the steric factor is 1, and set the collision cross-section to the value for O_2 of 36 Å^2.

3. *Calculate the half-life of the reactant in an elementary first-order reaction from the Arrhenius constants.*
   Using the values in Table 13.1, estimate the half-life of ethane (C_2H_6) at 1100 K.

PROBLEMS

Reactions in these problems are elementary reactions.

Discussion Problems

13.1 This problem includes some simple thermodynamics review. State whether each of the following parameters is intensive (I) or extensive (E).

  a. the excluded volume b in the van der Waals equation

  b. the collision cross section σ

  c. the Joule-Thomson coefficient $\left(\dfrac{\partial T}{\partial P}\right)_H$

  d. the chemical potential μ

  e. $d[A]/dt$ for a chemical reaction

13.2 Explain why the $O_3(g) + C_3H_8(g)$ reaction has such a small A value in Table 13.1.

13.3 For some three-body reactions between free radicals, such as $O(g) + O_2(g) + Ar(g) \rightarrow O_3(g) + Ar(g)$ in Table 13.1, the Arrhenius activation energy is *negative*. What does this mean for experimental measurements of the reaction rate? What could cause this effect?

Theory of Chemical Rate Constants

13.4 The activation energy for the forward reaction $2OH(g) \rightarrow H_2O(g) + O(g)$ is 0.42 kJ mol^{-1}. Use the table of thermochemical data in the appendix to calculate the activation energy for the *reverse* reaction.

13.5 Find the steric factor p for the reaction of Cl with H_2, based on simple collision theory, using the data in Table 13.1. The collision cross-sections for H_2 and Cl_2 are 27 Å^2 and 93 Å^2, respectively; use the value for Cl_2 to estimate the cross section of Cl.

13.6 In simple collision theory, we might assume that the steric factor p for the $O + N_2$ reaction was nearly 1. Use this assumption to estimate the pre-exponential factor A (in cm^3 mol^{-1} s^{-1}) for that reaction at 298 K. Use a collision cross-section of 28 Å^2 for atomic oxygen.

13.7 Using assumptions similar to those used in simple collision theory, find an equation for the rate constant of a photodissociation reaction $AB + h\nu \xrightarrow{k} A + B$. What do the steric factor and activation energy represent in this case?

13.8 The Arrhenius activation energy E_a is defined to be the best fit value to the equation

$$E_a = RT^2 \frac{d(\ln k)}{dT}.$$

Use this definition to show that for a gas-phase trimolecular elementary reaction within the assumptions of transition state theory,

$$E_a = \Delta H^{\ddagger} + 3RT \text{ and } A = \frac{e^3 k_B T}{h(C^{\ominus})^2} e^{\Delta S^{\ddagger}/R}, \text{ where } k = \frac{fK^{\ddagger}}{C^{\ominus}}.$$

13.9 Find $\Delta G^{\ddagger}$ at 298 K for the reaction $2I + Ar \rightarrow I_2 + Ar$.

13.10 Rewrite the Eyring equation,

$$k = \frac{k_B T}{Ch} e^{\Delta S^{\ddagger}/R} e^{-\Delta H^{\ddagger}/(RT)},$$

using the vibrational partition functions Q_A and Q_B for the reactants and $Q^{\ddagger}$ for the transition state.

13.11 Estimate $S^{\ddagger}$ (not $\Delta S^{\ddagger}$) for the reaction of NO with O_3 from the Arrhenius constants in Table 13.1.

13.12 Calculate $\Delta H^{\ddagger}$ and $\Delta S^{\ddagger}$ for the gas-phase reaction at 298 K

$$Cl(g) + H_2(g) \rightarrow HCl(g) + H(g)$$

from the Arrhenius constants. Based on your results and the enthalpies of formation, plot the reaction diagram (energies of reactants, transition state, products) as a function of reaction coordinate.

13.13 Calculate $\Delta_f H^{\ominus}$ for the CH$_3$ radical.

13.14 Using the table in the Appendix and Table 13.1 for the reaction $NO(g) + O_3(g) \rightarrow NO_2(g) + O_2(g)$, do the following:

  a. find $\Delta_{rxn} H^{\ominus}$ at 298 K

  b. find $\Delta_{rxn} S^{\ominus}$ at 298 K

  c. find $\Delta_{rxn} G^{\ominus}$ at 298 K

  d. find K_{eq} at 298 K

  e. find $\Delta H^{\ddagger}$ at 298 K

  f. find $\Delta S^{\ddagger}$ at 298 K

  g. find k_f and k_r at 298 K

h. draw the reaction diagram

i. find $\Delta_{rxn}H^{\ominus}$ at 180 K

j. find K_{eq} at 180 K

k. if the Arrhenius A constant for this reaction actually varies as $T^{1/2}$, and the given in Table 10.2 is for 298 K only, find k_f at 500 K, and find k_f at 500 K when A is independent of temperature.

Experimental Kinetics

13.15 Calculate $[CH_3NC]/[CH_3CN]$ after 10 hours at 298 K and at 600 K if the initial sample is pure CH_3NC.

13.16 Given an initial solution of 0.01 M NOBr at 298 K, calculate the concentration $[Br_2]$ after 1 hour and after 100 hours. Neglect the reverse reaction.

13.17 The half-life of $SO_2Cl_2(g)$, as it decomposes by first-order reaction into SO_2 and Cl_2, is $3.15 \cdot 10^4$ s at 320 °C. What is the rate constant of the reaction at this temperature?

13.18 Calculate the half-life of N_2O_5 at 298 K.

13.19 The half-life of a gas-phase unimolecular decay is 362 hours at 500 K and 86 seconds at 600 K. Find the Arrhenius constants E_a and A.

13.20 The half-life of a unimolecular decomposition is 3 hours at 298 K and 2 minutes at 500 K. Find the half-life at 200 K.

13.21 Several studies of fundamental combustion reactions have focused on the reaction

$$CH_3(g) + H(g) \xrightarrow{k_1} CH_4(g).$$

Such a reaction normally requires a third body to absorb the energy released by the bond formation:

$$CH_3(g) + H(g) + M \xrightarrow{k_2} CH_4(g) + M.$$

Otherwise the CH_4 molecule is formed with too much energy, and one of the CH bonds breaks, giving back the original reactants, CH_3 + H. The experiments monitor the density of CH_3 (by mass spectrometry or infrared spectroscopy) to measure the reaction rate, and no net change in $[CH_3]$ is measured under the first reaction if the CH_4 is stable for too short a time. The second reaction is much more difficult to study theoretically, however, so the first reaction has been studied in experiments using the following reaction system:

$$CH_3(g) + D(g) \xrightarrow{k_1'} CH_3D(g).$$

Why does this make it possible to measure the rate constant k_1' with fairly high accuracy, using the same experimental technique just described?

14 Chemical Kinetics: Multi-Step Reactions

LEARNING OBJECTIVES

After reading this chapter, you will be able to do the following:

❶ Write the differential rate law for any reactant, product, or intermediate of a multi-step reaction in terms of the concentrations and rate constants, given the full reaction mechanism.

❷ Verify that analytical solutions to integrated rate laws give the correct solution in various limits.

❸ Apply steady-state, fast equilibrium, and other approximations to the kinetics analysis of a multi-step reaction.

GOAL *Why Are We Here?*

The goal of this chapter is to analyze the time-dependence of systems involving more than one elementary reaction with shared chemical components. We will find exact solutions for some of the basic cases and useful approximations for more complicated systems. We will wrap up the book by applying these methods to a few examples drawn from natural systems.

CONTEXT *Where Are We Now?*

Although there are laboratory conditions under which we may study a single-step reaction, most chemistry involves numerous component elementary reactions. Hundreds of individual chemical species have been studied in an effort to understand the network of interactions that forms the chemistry of the atmosphere, for example. Nature loves to explore possibilities in chemistry as much as anything. In the same way that a river changes from its source to its mouth and changes with the seasons, the chemistry of the river water—and of the soil beneath it and the air above it—constantly shift with time and place. As rich and complex as the chemistry of natural systems can be, we can break it down into a set of elementary steps, and each step we can understand with the tools from Chapter 13. What we're going to do now is look at how small groups of connected elementary steps build up to form complex reaction networks and examine some common approximations that seek to make the problem more tractable mathematically and intuitively.

SUPPORTING TEXT *How Did We Get Here?*

The main qualitative preparation that will be helpful for the work ahead is covered by the material in Chapter 13 on the chemical kinetics of single-step reactions. However, there are few specific equations that we will draw from that chapter, because the focus now shifts to the time-dependence of the macroscopic concentrations, away from the molecular-scale parameters that determine the rate constants. We will draw on the following equations and sections of text to support the ideas developed in this chapter:

- To obtain the integrated rate laws, we will need to be able to integrate the differential form of the rate laws. In those places, we will take advantage of the general solutions offered by Table A.7.

- Solving the integrated rate law for the elementary first-order reaction gave us the following result for the concentration of reactant as a function of time (Eq. 13.39):

$$[A] = [A]_0\, e^{-kt}.$$

- In our brief look at molecular astrophysics, we will refer back to the principles of blackbody radiation, covered in Section 6.2, specifically the profile of the radiation density ρ) as a function of radiation frequency ν (Eq. 6.20):

$$\rho(\nu)d\nu = \frac{8\pi h\nu^3 d\nu}{c^3(e^{h\nu/(k_B T)} - 1)}.$$

- The effect of molecular properties on the kinetics will appear in our study of the atmosphere and interstellar space, drawing on the collision properties defined in Section 5.1 and then implemented in Section 13.2.

14.1 Elements of Multi-Step Reactions

In principle, if we understand the mechanism of a multi-step reaction, no matter how complicated, it can be reduced to its elementary reactions. These elementary reactions are joined because they share chemical components, either as reactants or products. The combustion of methane at 2000 K, for example, yields the chemical density profiles shown in Fig. 14.1 for a only few selected compounds. The methyl radical, CH_3, is one of several highly reactive species formed during preliminary steps of the overall reaction and consumed by later reactions. The CO profile shows that, although CO is a stable molecule formed initially as a product, it too is largely consumed by later reactions. The curves shown are predicted by a system of 23 reactions involving 14 different chemical species. A more recent model combines 52 species in over 320 distinct steps. The ways that these species connect the individual elementary steps break down into 3 types of reaction coupling: reversible reactions, parallel reactions, and sequential reactions. We'll now look at the simplest examples of each of these types of coupled reactions to see how the relationship between the elementary steps affects the outcome.

Reversible Steps

When considering an elementary reaction

$$A \xrightarrow{k_1} B,$$

we ignore the reverse reaction

$$B \xrightarrow{k_{-1}} A.$$

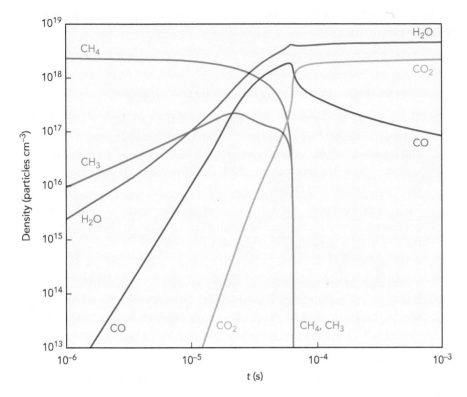

◀ **FIGURE 14.1 Log-log plot of density versus time for several compounds involved in the combustion of methane at 2000 K.** After a figure in J.I. Steinfield, J.S. Francisco, and W.L. Hase, *Chemical Kinetics and Dynamics*, Prentice-Hall (1989).

Although the two reactions share the same transition state, the change in direction makes these distinct elementary reactions. If the reverse reaction is extremely slow, it can be ignored, depending on the level of precision required. Commonly, however, reverse reactions must be included in the analysis of a complete reaction system.

It is possible to obtain the integrated rate law (which gives expressions for the reactant and product concentrations) by methods similar to those used in Section 13.5 for the single elementary equation. We choose the simplest case: the reversible unimolecular reaction

$$A \underset{k_{-1}}{\overset{k_1}{\rightleftarrows}} B,$$

with rate constants k_1 and k_{-1} for the forward and reverse reactions, respectively. Whenever we encounter a multistep reaction such as this, with more than one elementary reaction involved, the change in concentration of *any* species in the mechanism can be expressed as the sum over the rate law expressions for that species in each elementary step where it plays a part. In our current reaction, for example, species A participates in two reactions: as the reactant in the forward reaction with rate constant k_1, and as a product in the reverse reaction with rate constant k_{-1}. The change in concentration of A as the two reactions proceed is therefore given by the rate of loss (from the forward reaction) plus the rate of gain (from the reverse reaction):

$$\frac{d[A]}{dt} = -k_1[A] + k_{-1}[B].$$

In the same way, we can write a rate law for [B] that combines the contributions from the two reactions:

$$\frac{d[B]}{dt} = k_1[A] - k_{-1}[B].$$

There are a few points worth noting about these two equations:

- The number of terms on the right side of the rate law is equal to the number of elementary reactions involving that species, where the forward and reverse reactions count as *separate* elementary steps. In this case, each species (A and B) participates in both the forward and reverse reactions, so each rate law has two terms.

- Each term multiplies the rate constant for the reaction by the *reactant* concentrations of that step. The product concentration does not influence the rate of the elementary reaction that forms it.

- If the species is a reactant in the reaction, then that term appears with a minus sign in the rate law, because the species is consumed in that step. If the species is a product, then the term appears with a plus sign. That is why the first term in $d[A]/dt$ appears with a negative sign ($-k_1[A]$) whereas the second term, for the reverse reaction that forms A, appears with a plus sign ($k_{-1}[B]$).

Given initial concentrations $[A]_0$ and $[B]_0$, we define the extent of reaction x to be the amount of B generated or the amount of A lost since the reaction started:

$$x = [B] - [B]_0 = [A]_0 - [A]. \tag{14.1}$$

Keep in mind that any initial concentrations such as $[A]_0$ and $[B]_0$ are treated as constants, so their derivatives are zero. The concentrations at any later time can then be solved by separating t and x in the rate law and integrating both sides of the equation as follows:

$$v = \frac{d[B]}{dt} = -\frac{d[A]}{dt} = \frac{dx}{dt}$$

$$= k_1[A] - k_{-1}[B] = k_1([A]_0 - x) - k_{-1}([B]_0 + x)$$

$$dt = \frac{dx}{k_1[A]_0 - k_{-1}[B]_0 - (k_1 + k_{-1})x} \qquad \text{isolate } x \text{ and } t$$

$$-(k_1 + k_{-1})dt = \frac{dx}{\left(\dfrac{1}{k_1 + k_{-1}}\right)(k_{-1}[B]_0 - k_1[A]_0) + x}$$

$$-(k_1 + k_{-1})t = \ln\left[\frac{\left(\dfrac{1}{k_1 + k_{-1}}\right)(k_{-1}[B]_0 - k_1[A]_0) + x}{\left(\dfrac{1}{k_1 + k_{-1}}\right)(k_{-1}[B]_0 - k_1[A]_0)}\right] \qquad \text{integrate}$$

$$x = \left(\frac{k_1[A]_0 - k_{-1}[B]_0}{k_1 + k_{-1}}\right)\left(1 - e^{-(k_1 + k_{-1})t}\right). \tag{14.2}$$

This result can then be solved for [A] or [B] as a function of time using our definition of x:

$$[A] = [A]_0 - \left(\frac{k_1[A]_0 - k_{-1}[B]_0}{k_1 + k_{-1}}\right)(1 - e^{-(k_1+k_{-1})t}) \qquad (14.3)$$

$$[B] = [B]_0 + \left(\frac{k_1[A]_0 - k_{-1}[B]_0}{k_1 + k_{-1}}\right)(1 - e^{-(k_1+k_{-1})t}). \qquad (14.4)$$

We can verify that these equations make sense in various extreme limits. For example, if $k_{-1} = 0$, then we obtain the same result as for the forward reaction only (Eq. 13.39):

$$\lim_{k_{-1} \to 0}[A] = [A]_0 - \left(\frac{k_1[A]_0}{k_1}\right)(1 - e^{-k_1t}) = [A]_0 e^{-k_1t}.$$

We can also check that at time $t = 0$ the concentration of A is given by its initial concentration:

$$\lim_{t \to 0}[A] = [A]_0 - \left(\frac{k_1[A]_0 - k_{-1}[B]_0}{k_1 + k_{-1}}\right)(1 - 1) = [A]_0.$$

The same general approach we just used to get the integrated rate laws rarely works out so pleasantly. Often, there is no algebraic solution to the integrated rate laws, but we can always obtain solutions by numerical methods, starting from a particular set of initial concentrations. However, we would like to get analytical solutions whenever possible. Unlike the numerical integrals, we can use analytical solutions to solve the whole kinetics problem at once for any set of initial conditions and at the same time see some of the trends in the reaction rate. It's worth it to derive the algebraic forms of these integrated laws when we can.

If the forward and reverse reactions share the same transition state, how independent can the forward and reverse reaction rates be? The reaction velocity depends *only on the nature of the reactants and the transition state, not on the products*. In contrast, the thermodynamic parameters of the system (also called the **thermochemistry**) depend only on the reactants and products, not on the transition state, so there is a symmetry to the forward and reverse reactions that our analysis of the kinetics does not make apparent. In particular,

$$\Delta_{rxn}H_1(A \to B) = -\Delta_{rxn}H_{-1}(B \to A),$$

but $k_1 \neq k_{-1}$.

Nevertheless, there *is* a thermodynamic link between the forward and reverse rate constants k_1 and k_{-1}. The origin of this link is that the forward and reverse reactions share the same transition state, so E_a for the forward reaction differs from E_a for the reverse reaction by ΔE for the reactants and products. For the reaction

$$A \underset{k_{-1}}{\overset{k_1}{\rightleftharpoons}} B,$$

we can set the reaction rate to zero at equilibrium:

$$\text{at equilibrium:} \quad -\frac{d[A]}{dt} = \frac{d[B]}{dt} = k_1[A]_{eq} - k_{-1}[B]_{eq} = 0$$

$$K_{eq} = \frac{[B]_{eq}}{[A]_{eq}} = \frac{k_1}{k_{-1}}. \qquad (14.5)$$

This relation can be seen by checking the integrated rate laws in the limit of infinite time, at which point the system must have reached equilibrium:

$$\lim_{t \to \infty} [A] = [A]_0 - \left(\frac{k_1[A]_0 - k_{-1}[B]_0}{k_1 + k_{-1}} \right)(1 - 0) = \frac{k_{-1}[A]_0 + k_{-1}[B]_0}{k_1 + k_{-1}}$$

$$\lim_{t \to \infty} [B] = [B]_0 + \left(\frac{k_1[A]_0 - k_{-1}[B]_0}{k_1 + k_{-1}} \right)(1 - 0) = \frac{k_1[A]_0 + k_1[B]_0}{k_1 + k_{-1}}$$

$$\lim_{t \to \infty} \frac{[B]}{[A]} = \frac{k_1}{k_{-1}} = K_{eq}. \tag{14.6}$$

The concentration ratio, equal to the equilibrium constant, is given by the ratio of the forward and reverse reaction rates.

Parallel Reactions

Competing reactions draw from the same reactant or set of reactants to form different products:

$$A \xrightarrow{k_1} B, \quad A \xrightarrow{k_2} C. \tag{R14.1}$$

This occurs, for example, in an organic synthesis that yields two distinct isomers of the same product. The overall reaction velocity for this example is

$$v = -\frac{d[A]}{dt} = k_1[A] + k_2[A] = (k_1 + k_2)[A]. \tag{14.7}$$

Solving for the concentration of reactant A as a function of time gives

$$[A] = [A]_0 e^{-(k_1 + k_2)t}, \tag{14.8}$$

which we can use to find the concentrations of the products:

$$\frac{d[B]}{dt} = k_1[A] = k_1[A]_0 e^{-(k_1 + k_2)t} \tag{14.9}$$

$$[B] = \int_0^t k_1[A]_0 e^{-(k_1 + k_2)t} dt$$

$$= \frac{k_1[A]_0}{k_1 + k_2}[1 - e^{-(k_1 + k_2)t}] \tag{14.10}$$

$$[C] = \frac{k_2[A]_0}{k_1 + k_2}[1 - e^{-(k_1 + k_2)t}]. \tag{14.11}$$

Therefore, the product ratio is the ratio of their respective reaction rate constants:

$$\frac{[B]}{[C]} = \frac{k_1}{k_2}. \tag{14.12}$$

This product ratio is valid for the reaction mechanism R14.1, which does not permit the products B and C to come into equilibrium with each other. In this case, the reaction is said to be under **kinetic control,** meaning that the product yields are determined only by the kinetics, not by the thermodynamics of the competing reactions.

EXAMPLE 14.1   Parallel Reactions

CONTEXT Hops are a flavoring agent in beer, and one of the major chemical components responsible for that flavor is β-pinene, a bicyclic compound with molecular formula $C_{10}H_{16}$. Although a stable compound at room temperature, at high temperature β-pinene can decompose into myrcene, which is used to make perfumes. However, a second product competes with the formation of myrcene.

PROBLEM The compound β-pinene isomerizes at 650 K to either 4-isopropenyl-1-methylcyclohexene (call it IMC), with a rate constant of $k_1 = 0.022$ s^{-1}, or to myrcene, with a rate constant of $k_2 = 0.13$ s^{-1}. If the reaction is carried out under kinetic control, what is the ratio of IMC to myrcene, and what is the half-life of the β-pinene?

SOLUTION The product ratio is given by the ratio of the rate constants:

$$\frac{[\text{IMC}]}{[\text{myrcene}]} = \frac{k_1}{k_2} = 18\%.$$

The decay rate of the β-pinene is the same as for a first-order reaction with overall rate constant $k_1 + k_2$, (Eq. 14.8), so its half-life is $t_{1/2} = \ln 2 / (k_1 + k_2) = 4$ s.

Sequential Reactions

The sequential reaction is the basic element of the multi-step reaction. Products of one reaction are available to be reactants for the next step. Molecules generated in one step of the reaction only to be consumed in a later step, and therefore not appearing in the net reaction, are called **chemical intermediates.**

The simplest set of sequential reactions is

$$A \xrightarrow{k_1} B \xrightarrow{k_2} C.$$

In this example, species B would be the chemical intermediate. The first reaction forms B, but the second reaction converts B to the product C. Therefore, the net reaction, $A \rightarrow C$ does not include B. This sequence is appropriate for the radioactive decay of uranium-239,

$$^{239}U \xrightarrow{-e^-} {}^{239}Np \xrightarrow{-e^-} {}^{239}Pu,$$

and for unimolecular reactions that involve an intermediate structure, such as this structural isomerization sequence in C_5H_7 (see Problem 12.14):

2-vinylcyclopropyl $\rightarrow$ 1,4-pentadienyl $\rightarrow$ cyclopenten-3-yl.

How bad is the math for finding the final product concentration [C] as a function of time? Not bad for [A], if we start by assuming initial concentrations $[A]_0$ for A and 0 for B and C:

$$-\frac{d[A]}{dt} = k_1[A] \quad [A] = [A]_0\, e^{-k_1 t}. \tag{14.13}$$

Finding [B] is a little tougher because the variables [B] and t don't separate cleanly:

$$\frac{d[B]}{dt} = k_1[A] - k_2[B]$$

$$= k_1[A]_0\, e^{-k_1 t} - k_2[B]. \tag{14.14}$$

However, this can be solved by applying one of our differential equation solutions from Table A.7. Equation 14.14 is in the form

$$\frac{dx}{dy} + g_1(y)x = g_2(y)$$

with these substitutions:

$$x = [B], \quad y = t, \quad g_1(y) = k_2, \quad g_2(y) = k_1[A]_0\, e^{-k_1 t}.$$

The integrated rate law is the solution to that differential equation:

$$x \exp\left(\int g_1(y)dy\right) = \int g_2(y)\exp\left(\int g_1(y)dy\right)dy.$$

Making our substitutions and integrating from 0 to t, we find

$$[B]e^{\int_0^{t'} k_2 t'} = \int_0^t [A]_0 k_1\, e^{-k_1 t'}\, e^{\int_0^{t'} k_2 t''}\, dt'$$

$$[B]e^{k_2 t} = \int_0^t [A]_0 k_1\, e^{-k_1 t'}\, e^{k_2 t'} dt'$$

$$= \frac{[A]_0 k_1}{k_2 - k_1}\left(e^{(k_2 - k_1)t} - 1\right)$$

$$[B] = \frac{[A]_0 k_1}{k_2 - k_1}\left(e^{-k_1 t} - e^{-k_2 t}\right). \tag{14.15}$$

This equation for [B] can be verified by substituting it into both sides of Eq. 14.14, the rate equation for $d[B]/dt$:

$$\frac{d[B]}{dt} = \frac{[A]_0 k_1}{k_2 - k_1}\left(-k_1 e^{-k_1 t} + k_2 e^{-k_2 t}\right)$$

$$k_1[A] - k_2[B] = k_1[A]_0 e^{-k_1 t} - [A]_0 \frac{k_1 k_2}{k_2 - k_1}\left(e^{-k_1 t} - e^{-k_2 t}\right)$$

$$= \frac{[A]_0 k_1}{k_2 - k_1}\left[(k_2 - k_1)e^{-k_1 t} - k_2 e^{-k_1 t} + k_2 e^{-k_2 t}\right]$$

$$= \frac{[A]_0 k_1}{k_2 - k_1}\left(-k_1 e^{-k_1 t} + k_2 e^{-k_2 t}\right).$$

To find [C] as a function of time, we start from an equation for the conservation of mass:

$$[C] = [A]_0 - [A] - [B]$$

$$= [A]_0 - [A]_0 e^{-k_1 t} - \frac{[A]_0 k_1}{k_2 - k_1}\left(e^{-k_1 t} - e^{-k_2 t}\right)$$

$$= [A]_0\left(1 - \frac{k_2 e^{-k_1 t} - k_1 e^{-k_2 t}}{k_2 - k_1}\right). \tag{14.16}$$

In determining integrated rate laws, we typically assume that the reactants and products that appear in the net reaction are sufficiently stable that the initial concentrations (in this case $[A]_0$ and $[C]_0$) may appear in the rate law, although we often set the initial concentration of product to zero. However, any intermediates that appear in the mechanism are taken to be too reactive to have any significant initial concentration, and therefore often do not appear in the integrated rate law.

Equation 14.16 can be tested in a couple of limiting cases. For example, what if $k_1 \gg k_2$? In that case, the concentration [C] simplifies to

$$[\text{C}] = [\text{A}]_0\left(1 - \frac{-k_1 e^{-k_2 t}}{-k_1}\right) = [\text{A}]_0(1 - e^{-k_2 t}), \qquad (14.17)$$

which is the same as for the reaction $\text{A} \rightarrow \text{C}$ with rate constant k_2. Similarly, if $k_1 \ll k_2$, then

$$[\text{C}] = [\text{A}]_0\left(1 - \frac{k_2 e^{-k_1 t}}{k_2}\right) = [\text{A}]_0(1 - e^{-k_1 t}). \qquad (14.18)$$

This illustrates mathematically the **rate-limiting step.** In a sequential reaction, the kinetics are dominated by the slowest step in the chain.

14.2 Approximations in Kinetics

In a few cases, we know the elementary steps of a multi-step reaction and can integrate to write useful integrated rate laws. More often, even if we could get analytical solutions, they would be so lengthy that the qualitative appreciation we're looking for would be lost.

Take, for example, the reversible bimolecular reaction of A and B to form an intermediate C, which can decay irreversibly to product D. If we attempt to solve this analytically, we can only get as far as the differential equations:

$$\text{A} + \text{B} \underset{k_{-1}}{\overset{k_1}{\rightleftharpoons}} \text{C} \overset{k_2}{\rightarrow} \text{D} \qquad (\text{R14.2})$$

$$-\frac{d[\text{A}]}{dt} = k_1[\text{A}][\text{B}] - k_{-1}[\text{C}]$$

$$\frac{d[\text{C}]}{dt} = k_1[\text{A}][\text{B}] - k_{-1}[\text{C}] - k_2[\text{C}]$$

$$\frac{d[\text{D}]}{dt} = k_2[\text{C}].$$

We cannot solve for the appearance rate of D without simultaneously solving for [C] as a function of time. Matrix methods can yield analytical solutions for complex reaction schemes, but the more complex the reaction, the less enlightening the equation for the concentration. Frequently, what we would prefer is a fairly accurate but *approximate* solution, one that correctly describes the major terms contributing to the rate of product formation, without overburdening our brains with the math.

The Steady-State Approximation

The intermediates in a series of sequential elementary reactions are often extremely reactive, meaning that the rate constant for their conversion to product is fast compared to the rate constant for their formation. This corresponds to the case illustrated in Fig. 14.2a, and the concentration [C] of the intermediate stabilizes much more rapidly than the principal reactants and products. Under these conditions, the **steady-state approximation** may be useful.

▶ FIGURE 14.2 **The steady-state and fast equilibrium approximations.** Concentration (in arbitrary units) is graphed as a function of time for the two-step reaction $A + B \overset{k_1}{\underset{k_{-1}}{\rightleftharpoons}} C \overset{k_2}{\rightarrow} D$ under various combinations of rate constants. In all cases, $[B] = [A]$. **(a)** $k_1 = 0.1$, $k_{-1} = 0.0001$, $k_2 = 10.0$. Concentrations $[A]'$ and $[D]'$ based on the steady-state approximation are indistinguishable from the correct concentrations in this graph. **(b)** $k_1 = 10.0$, $k_{-1} = 0.0001$, $k_2 = 0.1$. Concentrations $[A]'$ and $[D]'$ based on the steady-state approximation illustrate deviations from steady state when $k_2 \ll k_1$. **(c)** $k_1 = 10.0$, $k_{-1} = 10.0$, $k_2 = 0.1$. Concentrations $[A]'$ and $[D]'$ are based on the fast-equilibrium assumption.

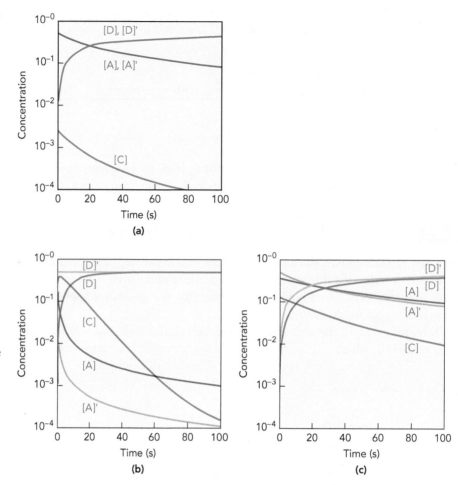

CHECKPOINT In Fig. 14.2a the concentration of the intermediate C is always much lower than the concentrations of reactants or products (note the logarithmic scale). For that reason, the derivative $d[C]/dt$ is also a small number compared to the changes in [A] and [D] early in the reaction. That is the condition that allows us to apply the steady-state approximation.

Steady state assumes that the concentration of the intermediate changes so slowly, compared to the reactant and product concentrations, that it can be safely treated as a constant over a short time. This is *not* assuming that the reaction has reached equilibrium, because the reactant and product concentrations are permitted to vary.

For example, the second order Reaction R14.2

$$A + B \overset{k_1}{\underset{k_{-1}}{\rightleftharpoons}} C \overset{k_2}{\rightarrow} D$$

has a reaction velocity for the concentration [C] of the intermediate

$$\frac{d[C]}{dt} = k_1[A][B] - (k_{-1} + k_2)[C]. \tag{14.19}$$

Under steady state we assume that [C] becomes constant shortly after we initiate the reaction. The point of the assumption is to set $d[C]/dt = 0$ so that we can solve Eq. 14.19 for the steady-state concentration of intermediate:

$$[C]_{ss} = \frac{k_1[A][B]}{k_{-1} + k_2}. \tag{14.20}$$

This is the foothold that lets us move through the equations for the other concentrations, simplifying as we go. For example, the rate law for A becomes solvable.

Let's use $[A]'$ and $[B]'$ to represent the approximate value of $[A]$ and $[B]$ under the steady-state approximation.[1] We substitute the steady-state concentration of C into the usual rate law:

$$-\frac{d[A]'}{dt} = k_1[A]'[B]' - k_{-1}[C]_{ss}$$

$$= \left(1 - \frac{k_{-1}}{k_{-1} + k_2}\right)k_1[A]'[B]'. \qquad \text{by Eq. 14.20}$$

As long as A and B react at the same rate, the two concentrations are related:

$$[B]_0 - [B]' = [A]_0 - [A]', \qquad [B]' = [B]_0 - [A]_0 + [A]'. \quad (14.21)$$

Therefore, we can write our rate law for A in terms of the concentration $[A]'$ alone:

$$-\frac{d[A]'}{dt} = \left(1 - \frac{k_{-1}}{k_{-1} + k_2}\right)k_1[A]'([B]_0 - [A]_0 + [A]')$$

$$-\frac{d[A]'}{[A]'([B]_0 - [A]_0 + [A]')} = \left(1 - \frac{k_{-1}}{k_{-1} + k_2}\right)k_1 dt. \quad \text{isolate } [A]' \text{ and } t \quad (14.22)$$

This can be integrated to obtain the time-dependent concentration of A:

$$\frac{1}{[B]_0 - [A]_0}\left[\ln\left(\frac{[B]_0 - [A]_0 + [A]'}{[A]'}\right) - \ln\left(\frac{[B]_0}{[A]_0}\right)\right] = k_1\left(1 - \frac{k_{-1}}{k_{-1} + k_2}\right)t$$

$$\frac{1}{[B]_0 - [A]_0}\left[\ln\left(\frac{[A]_0([B]_0 - [A]_0 + [A]')}{[B]_0[A]'}\right)\right] = k_1\left(1 - \frac{k_{-1}}{k_{-1} + k_2}\right)t$$

$$[A]' = [A]_0\left(1 - \frac{[A]_0}{[B]_0}\right)\left\{\exp\left[([B]_0 - [A]_0)\left(1 - \frac{k_{-1}}{k_{-1} + k_2}\right)k_1 t\right] - \frac{[A]_0}{[B]_0}\right\}^{-1}. \quad (14.23)$$

It takes a lot of algebra to obtain Eq. 14.23 after solving the integral, but once done, the equation is manageable. One can show, for example, that when $t = 0$, the concentration $[A]'$ is accurately predicted to be $[A]_0$. Similarly, when t becomes very large, $[A]'$ approaches either 0 (if $[B]_0 \geq [A]_0$) or $[A]_0 - [B]_0$ (if $[B]_0 < [A]_0$).

As the first step increases in reaction rate over the second step (Fig. 14.2b), the concentration of the intermediate becomes more strongly time-dependent, and the steady-state approximation becomes invalid.

EXAMPLE 14.2 Approximations and Limits in Kinetics

PROBLEM Show that Eq. 14.23 correctly predicts the rate law of an irreversible bimolecular reaction $A + B \xrightarrow{k_1} D$ in the limit that $k_{-1} \ll k_1$.

SOLUTION To solve this problem, we need to (*i*) find the correct rate law for the reaction $A + B \xrightarrow{k_1} D$ and (*ii*) show that Eq. 14.23 yields the same result. (*i*) The concentrations of A and B are connected by the extent of reaction x, where

$$x = [A]_0 - [A] = [B]_0 - [B],$$

---

[1] We use $[A]'$ and not $[A]_{ss}$ because this approximate concentration is obtained by applying the steady-state approximation to C, not to A. The time-dependencies of $[A]$, $[B]$, and $[D]$ are not explicitly affected and (we hope) not seriously shifted from their correct forms.

which allows us to write $[B]$ in terms of $[A]$: $[B] = [A] + [B]_0 - [A]_0 \equiv [A] + a$ where we have set $a = [B]_0 - [A]_0$. The rate law we need to integrate is

$$\frac{d[A]}{dt} = -k_1[A][B] = -k_1[A]([A] + a).$$

We can separate terms in the two variables, $[A]$ and t, and then integrate to solve the differential equation:

$$\frac{d[A]}{[A]([A] + a)} = -k_1 dt$$

$$\int_{[A]_0}^{[A]} \frac{d[A]'}{[A]'([A]' + a)} = -\int_0^t k_1 dt'$$

From a table of integrals, we learn that there is an analytical solution to the integral on the left side. The general formula is

$$\int \frac{dx}{(x + a)(x + b)} = \frac{1}{b - a} \ln \frac{x + a}{x + b} + C.$$

Applying this formula with $x = [A]$ and $b = 0$ gives the following:

$$-\frac{1}{a} \ln \frac{[A]' + a}{[A]'} \bigg|_{[A]_0}^{[A]} = -k_1 t$$

$$-\frac{1}{a} \left\{ \ln \frac{[A] + a}{[A]} - \ln \frac{[A]_0 + a}{[A]_0} \right\} = -k_1 t.$$

The remaining work is to solve for $[A]$, which we will summarize here in a few steps:

$$-\frac{1}{a} \left\{ \ln \left[\left(\frac{[A] + a}{[A]} \right) \left(\frac{[A]_0}{[A]_0 + a} \right) \right] \right\} = -k_1 t \qquad \ln x - \ln y = \ln(x/y)$$

$$\ln \left[\left(\frac{[A] + a}{[A]} \right) \left(\frac{[A]_0}{[A]_0 + a} \right) \right] = ak_1 t \qquad \text{multiply by } -a$$

$$\left(\frac{[A] + a}{[A]} \right) \left(\frac{[A]_0}{[A]_0 + a} \right) = e^{ak_1 t} \qquad e^x \text{ of both sides}$$

$$[A] = \frac{[A]_0 a}{[A]_0 + a} \left(e^{ak_1 t} - \frac{[A]_0}{[A]_0 + a} \right)^{-1} \qquad \text{rearrange}$$

$$= [A]_0 \left(1 - \frac{[A]_0}{[B]_0} \right) \left\{ \exp[([B]_0 - [A]_0)k_1 t] - \frac{[A]_0}{[B]_0} \right\}^{-1}. \qquad a = [B]_0 - [A]_0$$

(*ii*) Setting $k_{-1} = k_2 = 0$, equation 14.23 simplifies to

$$\lim_{k_{-1} \ll k_1} [A]' = [A]_0 \left(1 - \frac{[A]_0}{[B]_0} \right) \left\{ \exp[([B]_0 - [A]_0)(1 - 0)k_1 t] - \frac{[A]_0}{[B]_0} \right\}^{-1}$$

$$= [A]_0 \left(1 - \frac{[A]_0}{[B]_0} \right) \left\{ \exp[([B]_0 - [A]_0)k_1 t] - \frac{[A]_0}{[B]_0} \right\}^{-1}$$

So the steady-state approximation, in the limit that the reverse reaction in the first step is negligible, predicts the correct time-dependent concentrations for the reactants.

The Fast-Equilibrium Approximation

If $k_{-1} \gg k_2$ in Reaction R14.2, then **fast-equilibrium approximation** may be applicable. This is a specific case within the steady-state approximation, where the forward and reverse reactions of step 1 (A + B $\rightleftharpoons$ C) are both fast compared to step 2 (C $\rightarrow$ D). Then the concentrations of A, B, and C approach their equilibrium ratio for step 1 right away:

$$\frac{[C]}{[A][B]} \approx K_1 = \frac{k_1}{k_{-1}}. \tag{14.24}$$

Meanwhile, the overall concentration of this near-equilibrium mixture decreases as some C slowly leaks through the irreversible step 2 to form product. Because K_1 is a constant, we can solve again for [C] in terms of the reactant concentrations:

$$[C]_{fe} = \frac{k_1}{k_{-1}}[A][B]. \tag{14.25}$$

We can see that this is a special case of the steady-state approximation by comparing this result to Eq. 14.20,

$$[C]_{ss} = \frac{k_1}{k_{-1} + k_2}[A][B].$$

The fast-equilibrium approximation is the steady-state case where k_2 is negligible compared to k_{-1}. Concentration curves are shown in Fig. 14.2c for an example case in which the forward and reverse rate constants for the first step are equal. The fast-equilibrium approximation is fairly successful under the selected conditions.

The Initial Rate Approximation

One of the simplest approximations in kinetics is based on the basic definitions underlying all this calculus: the derivative of the concentration $d[A]/dt$ is the change in the concentration [A] per unit time, in the limit that the unit of time gets really short. If we measure [A] at two times, separated by a tiny interval δt, then we can estimate the derivative at that point by subtraction, a method of numerical differentiation called **finite differences:**

$$\frac{d[A]}{dt} \approx \frac{[A]_{t+\delta t} - [A]_t}{\delta t}. \tag{14.26}$$

Therefore, if we measure the reactant concentrations *very soon* after starting the reaction, we can solve for the rate law without having to work through any difficult calculus. Better yet, if δt is small enough that none of the concentrations have changed by more than a few percent, *we can pretend that those concentrations are fixed.*

For example, a reaction takes place between two reactants A and B. All we know at the outset about this rate law is that it should have the form

$$-\frac{d[A]}{dt} = k[A]^n[B]^m,$$

but we don't know what the exponents n and m are. For a complex reaction, we cannot even assume that these exponents should be integers.

Let's say that we can monitor an electronic transition of A using a UV–vis absorption spectrometer so that we know when its concentration drops by 0.10 mol L^{-1}. To find the initial rate law, we run three experiments:

1. The initial concentrations are each 1.0 mol L^{-1}. The concentration of A drops to 0.90 mol L^{-1} at 22.4 seconds after starting the reaction.

2. We double the initial concentration of B to 2.0 mol L^{-1} and now find that A reaches 0.90 mol L^{-1} at 11.1 seconds.

3. We double the initial concentration of A to 2.0 mol L^{-1}, (with B back at 1.0 mol L^{-1}) and find that A reaches 1.90 mol L^{-1} at 5.3 seconds.

We replace $d[A]$ in the rate law by the change in concentration of $[A]$, -0.10 mol L^{-1}, we replace dt by the time required for that change, and we approximate the concentrations by their initial values. This is enough to give us the approximate rate law for the net reaction:

$$-\frac{d[A]}{dt} = k[A]^n[B]^m$$

$$-\frac{\Delta[A]}{\Delta t} \approx k[A]_0^n[B]_0^m$$

$$\frac{0.10}{22.4} \approx k(1.0)^n(1.0)^m$$

$$\frac{0.10}{11.1} \approx k(1.0)^n(2.0)^m$$

$$\frac{0.10}{5.3} \approx k(2.0)^n(1.0)^m$$

Dividing the first equation into the second and third, we obtain

$$2 \approx (2.0)^m \quad 4 \approx (2.0)^n,$$

which means m is 1 and n is 2, at least at early times in the reaction. So the rate law may be written as

$$-\frac{d[A]}{dt} = k[A]^2[B],$$

and we can even estimate the rate constant:

$$k = \frac{0.10 \text{ mol } L^{-1}}{22.4 \text{ s } (1.0 \text{ mol } L^{-1})^3} = 0.0045 \text{ L}^2 \text{ mol}^{-2} \text{ s}^{-1}. \tag{14.27}$$

These results were evaluated for the complex reaction

$$2A + B \underset{k_{-1}}{\overset{k_1}{\rightleftharpoons}} C + D \quad A + B \underset{k_{-2}}{\overset{k_2}{\rightleftharpoons}} D + E,$$

with $k_1 = 2.54 \cdot 10^{-3} \text{ L}^2 \text{ mol}^{-2} \text{ s}^{-1}$, $k_2 = 6.57 \cdot 10^{-5} \text{ L}^2 \text{ mol}^{-2} \text{ s}^{-1}$, and small reverse rates. The initial rate approximation allows us to pull out the correct molecularity of the fast first reaction, even though the overall kinetics will be dominated by the slow rate of the second step.

The Pseudo-Lower Order Approximation

Finally, consider the case when Reaction R14.2

$$A + B \underset{k_{-1}}{\overset{k_1}{\rightleftharpoons}} C \xrightarrow{k_2} D$$

is carried out in a concentrated solution of B or in an atmosphere of B, so the concentration [B] is very large and remains essentially constant at value $[B]_0$. Then the reaction velocity is

$$\frac{d[C]}{dt} \approx k_1[A][B]_0 - (k_{-1} + k_2)[C], \qquad (14.28)$$

and the steady-state concentration of C can be solved for directly:

$$[C]_{(1)} = \frac{k_1[A][B]_0}{k_{-1} + k_2}. \qquad (14.29)$$

The subscript (1) on the left-hand side indicates that this concentration for C assumes the reaction is a **pseudo-first order reaction.** When [B] is kept constant, the conversion of A and B into C becomes a first-order reaction, proportional to [A] and the pseudo-first order rate constant $k_1[B]_0/(k_{-1} + k_2)$. This limiting case is often used to distinguish the rate law dependence on different reactants.

BIOSKETCH | Enrique Iglesia

Enrique Iglesia holds the Theodore Vermeulen Chair in Chemical Engineering at the University of California at Berkeley, where he studies the reaction mechanisms for catalytic reactions of interest in a wide variety of applications. Recent work in the Iglesia group has included a detailed study of the mechanism for reactions between O_2 and small hydrocarbons found in natural gas, using platinum nanoclusters as the catalyst. These reactions convert highly volatile organic compounds into a more manageable oxidized form, but the reaction mechanism is complex and largely unknown. Using precise conditions, Professor Iglesia and his coworkers are able to show, for example, that the mechanism for reaction of methane and ethane with O_2 are identical, but the reaction rates and the optimal organic/O_2 ratio differ in part because the two hydrocarbons have distinct $C-H$ bond strengths. The researchers have quantified the effect of the platinum cluster size, temperature, and O_2 partial pressure—among other factors—in order to determine what the most effective conditions are for these complex reactions.

14.3 Chain Reactions

We have only been dealing with some of the most general aspects of kinetics. There are many more specific applications of kinetics that require considerable additional attention if they are to be treated with any quantitative accuracy. We narrow our focus now with a brief look at one class of multi-step reactions that appears in many contexts.

The **chain reaction** is a self-propagating reaction series in which intermediate reaction steps generate the species necessary to initiate the multi-step reaction elsewhere. The general chain reaction consists of four categories of reactions, illustrated next for the combustion of H_2:

initiation: $H_2 \rightarrow 2H$ non-radical forms radicals

              $O_2 \rightarrow 2O$

branching: $H + O_2 \rightarrow OH + O$ radical forms more radicals

$O + H_2 \rightarrow OH + H$

propagation: $OH + H_2 \rightarrow H_2O + H$ radical is regenerated

termination: $2H + M \rightarrow H_2 + M$ radicals are destroyed

$H + O_2 + M \rightarrow HO_2 + M$

$HO_2 + OH \rightarrow H_2O + O_2$

$HO_2 + H \rightarrow H_2 + O_2.$

CHECKPOINT The chain carriers in the H_2 combustion example are free radicals, atoms or molecules with at least one unpaired electron. The odd electron makes the radical very reactive, but reactions with other even-electron species still leave an odd electron somewhere, and the chain reaction continues. Chain carriers in other processes include ions and (in nuclear reactions) neutrons.

The chain reaction cannot begin until after one of the initiation reactions results in the atomic free radicals H or O. The initiation reactions are highly endothermic; ΔH is the bond dissociation energy for the reactant. Many processes can drive these reactions, however, such as a match flame or an electrical spark. The chain branching steps get their name because they produce more free radical chain carriers (such as OH, O, or H) than are consumed in the reaction. This distinguishes the branching reactions from the propagation steps in which one chain carrier is consumed for each one produced. As long as there is sufficient fuel, the chain reaction can continue, ending only with various termination reactions. The symbol M in the termination reactions is commonly used to indicate any third molecule (for example, H_2, O_2, H_2O or even a solid surface). The third molecule is necessary to remove the excess energy as the other reactants form a bond. The termination steps in this series of reactions are much slower than the branching and propagation steps,[2] so this particular chain reaction usually ends only when all the O_2 or H_2 is consumed.

The chain reaction is only effective at self-propagation when the **kinetic chain length** is large. The kinetic chain length is the ratio of the reaction velocities, $v_{propagation}$ to $v_{initiation}$. We find that the propagation velocity depends on the termination velocity, so all three reaction steps figure in the kinetic chain length.

The kinetic analysis of chain reactions therefore involves a sort of catalysis by intermediates, which can complicate the integration for the concentrations, and the results are often extremely sensitive to the relative reaction rates.

Thermal explosions are another example of how specific cases pose specific kinetic problems. In this process, the heat generated by an extremely exothermic reaction is channeled initially into a very small volume. This causes the temperature of other reactants to rise, and most rate constants increase with temperature, so the reaction proceeds even faster. Shortly the production of products is rapid enough and the local heating intense enough that the gas expands violently. This is a difficult problem in kinetics, since not only the concentrations but the rate constants vary with time.

---

[2]Unless a container wall is nearby, that is. Chain reactions can sometimes be defeated by the geometry of the vessel in which the reaction takes place. The chain carriers, free radicals in this example, need to be able to travel far enough through the reaction mixture to find their targets. If they are more likely to meet with a quenching body, such as a wall, the chain reaction may never take off.

14.4   Reaction Networks

In this final section, we test our chemical model in a few complex environments. Consider it a measure of considerable accomplishment that after all the effort it takes to get here, we could now examine contemporary research topics in the chemistry of the atmosphere, biological systems, interstellar space—indeed, *anywhere*. Physical chemistry provides the basis for understanding any chemical process.

We shall apply our chemical model to a few reaction networks in nature that have attracted the curiosity of chemists. Much effort in laboratory chemistry is devoted to simplifying the system of interest by reducing the number of parallel reactions and enforcing controlled conditions. In nature, chemical reactions rarely take place in isolation or at constant temperature and pressure. Reactions occur in networks of parallel and reverse and sequential elementary steps, in environments where the pressures and temperatures and densities all vary with time and location. Although we shall take into account only the dominant reactions for part of the network, even a cursory description of such a system is a rigorous test of the chemical model.

Whole textbooks are devoted to the following topics. These short sections merely pull a few illustrations from each area. Rather than limit ourselves to the kinetics of these chemical systems, we will show how these areas integrate quantum mechanics, statistical mechanics, and thermodynamics as well as kinetics. A single question in each of these areas can require a synthesis of many aspects of physical chemistry for its solution.

Atmospheric Chemistry

The Atmosphere

We must first describe the environment in which the chemistry is taking place. The earth's atmosphere is a perfect example of how reaction conditions in nature are functions of time and location. For instance, the chemical composition of the air changes drastically with the altitude. Water and other polar molecules are found almost entirely in the warm lower atmosphere because they condense at lower temperatures than the non-polar O_2, N_2, and CO_2. The relative abundances of N_2 and O_2 in air remain fairly constant throughout the lower atmosphere, at 0.78 and 0.21, respectively (Table 14.1), but molecules with low and extremely variable concentrations, such as O_3 (ozone), may still dominate the chemistry.

TABLE 14.1 **The principle chemical components of dry air in the lower atmosphere, and their abundances by mole fraction.**

N_2	0.7808	He	$5.24 \cdot 10^{-6}$
O_2	0.2095	CH_4	$2 \cdot 10^{-6}$
Ar	$9.34 \cdot 10^{-3}$	Kr	$1.14 \cdot 10^{-6}$
CO_2	$3.14 \cdot 10^{-4}$	H_2	$5 \cdot 10^{-7}$
Ne	$1.82 \cdot 10^{-5}$	Xe	$8.7 \cdot 10^{-8}$

▶ **FIGURE 14.3 The atmosphere.**
Approximate curves are drawn
representing temperature (red
curve) and number density
(blue curve) of the atmosphere as
functions of altitude z.

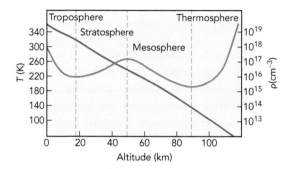

In Fig. 14.3, approximate graphs are given of the temperature and gas number density as functions of the altitude above the earth's surface. These functions are only approximate; the temperature and density depend on several other parameters, including the latitude of the measurement and time of day.

The number density, which is proportional to pressure when the temperature is constant, varies qualitatively as one might expect: the greater the height above the earth, the less dense the air and the lower the pressure. We have the tools to obtain a much more quantitative prediction than this, however. If we consider the atmosphere to be a statistical sample of gas molecules, we can use the probability distribution for the *location* of these molecules to get the number density as a function of altitude. All we have, after all, is a large number of molecules with a distribution determined by a potential energy field—in this case, the earth's gravitational field. Equation 2.31 restates the canonical distribution:

$$\mathcal{P}(E) = \frac{\Omega(E)e^{-E/(k_B T)}}{Q(T)}.$$

For the case at hand, E depends both on the speed v of the molecule and on the altitude z:

$$E = \tfrac{1}{2}mv^2 + mgz, \tag{14.30}$$

where m is the mass of *one molecule* and g is the gravitational acceleration near the earth's surface ($g = 980 \text{ cm s}^{-2}$). We shall simplify the system with the approximation that these two variables v and z are independent (they are not, because the temperature changes with altitude). Then our canonical distribution becomes

$$\mathcal{P}(E) = \mathcal{P}(v,z) = \frac{Cv^2 e^{-(mv^2/2 + mgz)/(k_B T)}}{Q_{\text{trans}}(T)Q_{\text{grav}}}. \tag{14.31}$$

We only want the z-dependence, so we integrate over all speeds:

$$\mathcal{P}(z) = \int_0^\infty \mathcal{P}(v,z)dv$$

$$= \frac{e^{-mgz/(k_B T)}\int_0^\infty Cv^2 e^{-mv^2/(2k_B T)}}{Q_{\text{trans}}(T)Q_{\text{grav}}} = \frac{e^{-mgz/(k_B T)}Q_{\text{trans}}(T)}{Q_{\text{trans}}(T)Q_{\text{grav}}}$$

$$= \frac{e^{-mgz/(k_B T)}}{Q_{\text{grav}}}. \tag{14.32}$$

Because we assumed z and v to be independent, we could have neglected the translational term completely. Our probability distribution drops exponentially with altitude. The number density, the number of molecules per unit volume, must be proportional to the probability function at any altitude z. We can't predict the proportionality factor—it depends on how many gas molecules are available, an independent parameter of the atmosphere. However, we can set the number density at some reference altitude, sea level, to $\rho(0)$ and evaluate the number density at any other point in terms of this reference value:

$$\rho(z) = \rho(0)e^{-mgz/(k_B T)} \equiv \rho(0)e^{-z/Z_0}. \tag{14.33}$$

The coefficient Z_0 in the exponent is the **scale height** and is defined by this equation. Equation 14.33 predicts $Z_0 = k_B T/mg = 8.3$ km if an average temperature of 280 K is used. The number density of the atmosphere near sea level $\rho(0)$ fluctuates but has an average of about $2.57 \cdot 10^{19} \, \text{cm}^{-3}$.

The temperature curve, on the other hand, is not so easily explained. Different gas densities and distances from the earth's surface favor different mechanisms for heating and cooling of the atmosphere. The following designations are given to those regions of the atmosphere closest to the earth's surface:

1. **Troposphere** (0 to ~18 km above the earth's surface): T decreases with altitude. The atmosphere is largely transparent to the visible and infrared radiation from the sun, but the earth's surface is not. It absorbs much of this radiation and warms to roughly 300 K. This in turn heats the lowest level of the atmosphere, the troposphere, partly by convection and partly by thermal radiation in the infrared. Molecules that absorb the infrared, particularly water, are most abundant in this region of the atmosphere. Throughout the troposphere, the trend is for the air to be cooler at greater distances from the earth, its source of heat.

2. **Stratosphere** (~18 to ~50 km): T increases with altitude. The next layer is the stratosphere and is heated by chemical reactions. In particular, the formation and destruction cycle of O_3 contributes to the heating of this region. These reactions are initiated by the sun's ultraviolet radiation and absorb much of the radiation along the way. Therefore, the heating from this mechanism increases as the altitude increases; there is more ultraviolet light available on the upper end of the stratosphere than the lower end.

3. **Mesosphere** (~50 to ~90 km): T decreases with altitude. Chemical reactions require a sufficient density of reactant to take place, and in the atmosphere the densities tend to drop at higher altitudes. Above 50 km, the ozone approaches such low concentrations that the atmosphere begins to cool off again as altitude increases. Additional cooling is provided by the infrared radiation of thermally excited CO_2 into the infrared-transparent upper atmosphere. This is the mesosphere, where the atmosphere reaches its coldest temperatures, as low as 100 K under extreme conditions.

4. **Thermosphere** (~90 to ~500–1000 km): T increases with altitude. Above 90 km, the atmosphere is heated by the direct photodissociation and photoionization of O_2 and N_2. These very endothermic processes convert high-frequency ultraviolet light into kinetic energy of separated atoms and atomic ions, resulting in temperatures on the order of 1000 K. Hence the thermosphere, particularly the lower sections in which the ions interact most strongly with ground-based radio transmission, is also known as the **ionosphere.**

The Chapman Mechanism

Having discussed how variable the reaction conditions are in the atmosphere, let's content ourselves with a simplified analysis of the ozone cycle at one set of conditions. At the temperature and radiation intensity characteristic of an altitude of 30 km, the following reactions contribute to the formation and destruction of O_3 with the reaction rates shown:

$$O_2 + h\nu \xrightarrow{j_1} 2O \qquad\qquad j_1 = 5.2\ 10^{-11}\,\text{s}^{-1} \qquad\qquad \text{(R14.3)}$$

$$O + O_2 + M \xrightarrow{k_2} O_3 + M \quad k_2 = 5.6\ 10^{-34}\,\text{cm}^6\,\text{mol}^{-2}\,\text{s}^{-1} \qquad \text{(R14.4)}$$

$$O_3 + h\nu \xrightarrow{j_3} O + O_2 \qquad j_3 = 9.5\ 10^{-4}\,\text{s}^{-1} \qquad\qquad \text{(R14.5)}$$

$$O + O_3 \xrightarrow{k_4} 2O_2 \qquad\qquad k_4 = 1.0\ 10^{-15}\,\text{cm}^3\,\text{mol}^{-1}\,\text{s}^{-1}. \quad \text{(R14.6)}$$

(We have written the rate constants for the photodissociation reactions R14.3 and R14.5 using j instead of k to emphasize that these depend on the radiation intensity.) In Reaction R14.4, M is any third molecule, in this case usually N_2 or O_2. The third molecule is not chemically changed by the reaction, but it is necessary for the reaction to occur because two-body association reactions are normally forbidden.

This set of reactions constitutes a cycle in which O_3 is formed from O_2 in a two-step process and then is returned to O_2 in another two-step process. The cycle is driven by the input of ultraviolet light in Reactions R14.3 and R14.5. It is Reaction R14.3, requiring radiation at wavelengths shorter than 320 nm, which gives O_3 its reputation for blocking the sun's most harmful ultraviolet light. Ionization of N_2 and O_2 requires radiation of wavelengths shorter than 79 nm and 102 nm, respectively. Radiation of 127 nm can dissociate N_2 into atomic nitrogen, and 240 nm radiation can dissociate O_2. Therefore ultraviolet solar radiation with wavelengths shorter than 240 nm is reduced as it passes through the atmosphere by the ionization and dissociation of N_2 and O_2, but the 240–320 nm radiation would pass unattenuated if it weren't for Reaction R14.3.

This model for the atmospheric O_3 cycle, called the **Chapman mechanism,** neglects numerous additional reactions for the sake of simplicity. Nonetheless, it yields correct order-of-magnitude results for the O_3 and O densities and is useful for understanding several other related cycles involving nitrogen, hydrogen, and chlorine.

We can estimate the steady-state abundances of O_3 and O at 30 km using the Chapman mechanism and the rate constants listed earlier. First, we need the rate laws for the relevant chemical components, where the steady-state approximation has been invoked to set all three rate laws to zero:

$$\frac{d[O]}{dt} = 2j_1[O_2] - k_2[O][O_2][M] + j_3[O_3] - k_4[O][O_3] = 0 \quad (14.34)$$

$$\frac{d[O_2]}{dt} = -j_1[O_2] - k_2[O][O_2][M] + j_3[O_3] + 2k_4[O][O_3] = 0 \quad (14.35)$$

$$\frac{d[O_3]}{dt} = k_2[O][O_2][M] - j_3[O_3] - k_4[O][O_3] = 0. \quad (14.36)$$

We have four concentrations, but do we really need to solve all four of them? The abundance of N_2, the dominant compound in the atmosphere, is not affected by this series of reactions; we may therefore reasonably assume that [M] is a constant. Using Eq. 14.33, we estimate the number density of M at $6.7 \cdot 10^{17}\,\text{cm}^{-3}$ or

$1.1 \cdot 10^{-6} \, \text{mol cm}^{-3}$ when the altitude is 30 km. Similarly, the O_2 density is too great at these altitudes to be substantially affected by this series of reactions. Multiplying the air density of $6.7 \cdot 10^{17} \, \text{cm}^{-3}$ by 0.21, we predict the O_2 density to be about $1.4 \cdot 10^{17} \, \text{cm}^{-3}$. If we calculate the abundances of O_3 and O, for now assuming [M] and $[O_2]$ are constant, we can easily check those assumptions afterward.

Adding Eqs. 14.35 and 14.36, we obtain

$$-j_1 [O_2] + k_4 [O][O_3] = 0,$$

which allows us to express the concentration [O] in terms of $[O_2]$ and $[O_3]$:

$$[O] = \frac{j_1[O_2]}{k_4[O_3]}. \tag{14.37}$$

Then, from Eq. 14.34, we can solve for the concentration $[O_3]$ in terms of [M], $[O_2]$, and the rate constants:

$$[O_3] = \frac{1}{j_3 - k_4[O]}(k_2[O][O_2][M] - 2j_1[O_2])$$

$$= \frac{1}{j_3 - (j_1[O_2]/[O_3])}\left(\frac{k_2 j_1[O_2]_2[M]}{k_4[O_3]} - 2j_1[O_2]\right).$$

Multiplying through to remove $[O_3]$ from the denominators leaves the equation in quadratic form, which can be solved to obtain $[O_3] = 1.8 \cdot 10^{13} \, \text{cm}^{-3}$. Using this value to solve for [O] in Eq. 14.37 gives the value $3.6 \cdot 10^8 \, \text{cm}^{-3}$. Both of these are at least three orders of magnitude lower than the O_2 number density, so our initial assumptions are acceptable.

The actual O_3 density is lower by a factor of 2 to 4 because the nitrogen, hydrogen, and chlorine cycles also participate. The Chapman mechanism is confined to molecules that do not condense under atmospheric conditions. Consider, in contrast, the following chemical cycles:

1. The HO_x cycle is dominated by two systems, one below an altitude of 30 km:

$$OH + O_3 \rightarrow HO_2 + O_2$$
$$HO_2 + O_3 \rightarrow OH + O_2 + O_2$$

Net reaction: $\qquad 2O_3 \rightarrow 3O_2$

and another above 40 km:

$$OH + O \rightarrow H + O_2$$
$$H + O_2 + M \rightarrow HO_2 + M$$
$$HO_2 + O \rightarrow OH + O_2$$

Net reaction: $\qquad 2O \rightarrow O_2$

2. The NO_x cycle:

$$NO + O_3 \rightarrow NO_2 + O_2$$
$$NO_2 + O \rightarrow NO + O_2$$
$$NO_2 + OH + M \rightarrow HNO_3 + M$$

Net reaction: $\quad O_3 + O + NO_2 + OH \rightarrow 2O_2 + HNO_3$

3. The ClO_x cycle:

$$CCl_2F_2 + h\nu \rightarrow CF_2Cl + Cl$$

$$Cl + O_3 \rightarrow ClO + O_2$$

$$ClO + O \rightarrow Cl + O_2$$

$$Cl + CH_4 \rightarrow HCl + CH_3$$

Net reaction: $CCl_2F_2 + O_3 + O + CH_4 + h\nu \rightarrow 2O_2 + CF_2Cl + HCl + CH_3$

In the ClO_x cycle, CCl_2F_2 is chosen as a representative chlorofluorocarbon. The final step listed in each of these cycles results in a condensate, such as H_2O, HNO_3, or HCl. These precipitate back to Earth, terminating the chain reactions.

EXAMPLE 14.3   Steady-State in the NO_x Cycle

PROBLEM Use the concentrations of atomic oxygen and ozone as determined by the Chapman mechanism to find an expression for the steady-state concentration of NO, which appears as an intermediate in the NO_x cycle, in terms of $[NO_2]$, $[O_2]$, $[O_3]$, and relevant rate constants. Assume for simplicity that no other reactions contribute significantly to the NO concentration.

SOLUTION The NO concentration appears in the rate laws for steps 1 and 2 in the NO_x reactions listed above. If we write those rate constants as k_{1N} and k_{2N}, then the steady-state concentration of NO can be found as follows:

$$\frac{d[NO]}{dt} = -k_{1N}[NO][O_3] + k_{2N}[NO_2][O] = 0$$

Solving for the steady-state NO concentration yields

$$[NO]_{ss} = \frac{k_{2N}[NO_2][O]}{k_{1N}[O_3]}.$$

From the Chapman mechanism, we have

$$[O] = \frac{j_1[O_2]}{k_4[O_3]},$$

and substituting this expression in for the O atom concentration above, we find the NO concentration in terms of the NO_2, O_2, and O_3 concentrations:

$$[NO]_{ss} = \frac{k_{2N}j_1[NO_2][O_2]}{k_{1N}k_4[O_3]^2}.$$

An important point illustrated by the Chapman mechanism is that O_3 is not easily formed except by the three-body reaction of O atom with O_2. The photolysis of O_3 leaves behind an odd O atom from which O_3 can be formed again. As long as these "odd" oxygen species (O atom or O_3) are abundant, there will be abundant ozone. It is reactions that remove the net number of "odd" oxygens from the upper atmosphere that account for the depleted O_3 layer. Reaction R14.4 in the Chapman mechanism is very slow. We can compare the product $k_2[M] = 6.1 \cdot 10^{-40} \, cm^3 \, mol^{-1} \, s^{-1}$, the pseudo-second order rate constant for the step that forms ozone, to $k_4 = 1.0 \cdot 10^{-15} \, cm^3 \, mol^{-1} \, s^{-1}$ for the destruction of ozone by O atom in Reaction R14.6. Even accounting for the difference in reactant concentrations,

---

TOOLS OF THE TRADE | **Flow Tube Kinetics**

In the 1900s, an apparatus for studying electrical discharges through gases as they flowed through a tube was developed by Robert W. Wood (who would later become better known for developing the first techniques in infrared and ultraviolet photography). By the 1920s, researchers were using similar devices to measure the time-dependence of chemical reactions, capitalizing on the contemporary development of spectroscopic techniques.

What is a flow tube and why do we use it? A flow tube in reaction kinetics is a channel that combines reactants while the mixture flows continuously towards a pump, so that the extent of reaction increases with distance along the tube. By knowing the average flow rate through the tube, the concentration of any given chemical species can be monitored as a function of time by using an appropriate detection method at different locations downstream from the point where the reactants first mix.

The technique provides a versatile means of characterizing gas-phase reactions over timescales longer than about 10 μs. Because the reaction is being carried out continuously, with each location along the tube corresponding to some steady state of the reaction, the detection scheme does not have to respond at the time scale of the reaction. For example, the measurement of weak fluorescence signals may take seconds of acquisition time, while the half-life of the reactants may be only milliseconds. In a flow tube, this discrepancy in time scales is not a problem, because the reaction mixture at any given location remains fixed.

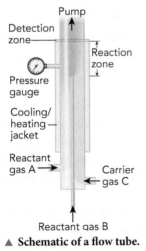

▲ **Schematic of a flow tube.**

How does it work? The schematic of a typical flow tube for kinetics studies is shown in the figure. The main tube is normally about 2.5 cm in diameter and 1 m long. The reaction is initiated by the introduction of a reactive species A, typically a radical or ion, to a second reactant, with both reactants being carried at low concentrations in a non-reactive gas such as helium or argon. The reactive species is generated just before entry into the flow tube, usually by electric discharge or thermal decomposition of a less reactive precursor molecule. For example, F_2 gas in an AC discharge provides an abundant source of fluorine atoms, whereas I_2 may be dissociated at high temperature to provide iodine atoms. Upon entering the flow tube, the reactive gas A mixes with the carrier gas C, which has been added just upstream, and is carried down the flow tube by the action of a vacuum pump attached to the opposite end. The temperature of the reaction mixture is controlled by heating or cooling a fluid that circulates through a jacket surrounding the flow tube.

The second reactant, gas B, enters the flow tube through a section of small-diameter tubing that can slide back and forth along the axis of the flow tube to control the point at which gases A and B begin to mix. The reaction zone extends from this point to the detector.

The reaction kinetics are then measured by varying the length of the reaction zone. In a typical experiment, the concentration [B] is much greater than [A], so the reaction may be treated as pseudo-first order, such that

$$\frac{d[A]}{dt} = -k_1[A],$$

where $k_1 = k[B]_0$, with k being the second-order rate constant. The detector is often a mass spectrometer, although fluorescence and other spectroscopic techniques are also used.

The low cost and simplicity of the flow tube design led to its use in a large number of the experiments used to determine rate constants for atmospheric and combustion processes.

Reaction R14.4 is much slower. All three of the cycles given result in a net reduction of O_3 production, as shown in the net reactions. The HO_x cycle is not very fast, but the NO_x cycle, which has the distinction of requiring no photo-initiation, is very efficient. The ClO_x cycle is extremely fast, but there is relatively little Cl in the upper atmosphere. It is estimated that roughly two thirds of the O_3 depletion mechanism is accounted for by the NO_x cycle and one sixth by the HO_x cycle. The cause of the remaining sixth is disputed at this writing. Considerable concern arose in the 1980s that the extremely efficient ClO_x reactions, and a set of even more efficient Br-mediated reactions, would irreversibly reduce the O_3 content of the stratosphere if man-made emissions of these halogens into the atmosphere continued unchecked. A massive reduction in the use of chlorofluorocarbons for household and industrial applications resulted.

Enzyme Catalysis

Enzymes are macromolecules that catalyze biochemical reactions, often increasing reaction rate constants by over eight orders of magnitude compared to the uncatalyzed reactions. These reactions include the decomposition of polysaccharides into simple sugars, the hydrolysis of peptide bonds in proteins, and numerous exchange reactions, isomerizations, oxidations, hydrogenations, and dehydrogenations. Most enzymes are composed of over 1000 atoms, and they may be pure proteins or part protein (the **apoenzyme**) and part non-protein (the **coenzyme**). In many cases the enzymes are much larger than the principal reactants, called **substrates,** and their role in the reaction is similar to that of a solid surface catalyst: the reactants form a complex (through van der Waals forces) with the catalyst and in this environment react much more swiftly than the free reactants. The reaction takes place *on* the enzyme.

This pivotal role and the rich variety and sophistication of enzymes have drawn scientists to enzyme research from a wealth of diverse fields, including fundamental physical chemistry. Some central questions directly addressed by the methods we have examined in this text include the following:

- How can we describe the structure of molecules with so many stable conformers?
- How does this structure influence the enzyme's biochemical function?
- What are the chemical thermodynamics and kinetics governing the enzyme's biochemical function?

Biomolecular Structure

Spectroscopy and diffraction have been the chemist's workhorses for determining molecular structure. We can show how the vibrationally averaged geometry of a molecule can be precisely determined from its rotational spectrum, how its bond strengths influence the vibrational spectrum, and how spectroscopy of the electronic states can be used to determine the nature of the molecular orbitals and therefore distribution of the electrons.

However, very large biomolecules are rarely confined to a single geometry, and we would expect to learn little from the rotational spectrum of a large protein, assuming that we could even get it into the gas phase to free up the rotations. Furthermore, the vibrational spectrum of a 3000-atom molecule is nearly a continuous series of absorptions from 3000 cm^{-1} on down to the low-energy torsional motions. Some absorptions are stronger than others, permitting some limited analysis when these are easily enough separated from the surrounding transitions. Worse yet, biomolecules are generally found in water, which is a very strong and broad absorber of infrared radiation, blocking out large frequency windows from study when the molecule is in its normal environment.

However, there are several other means for analyzing biomolecular structure: electronic and NMR spectroscopy, optical rotation studies, and x-ray diffraction of crystallized or fibrous biochemicals. A brief outline follows of only one of these methods, as an illustration of the application of quantum mechanics to macromolecules.

Electronic spectroscopy of biomolecules. The spectroscopy of electronic transitions is most easily carried out on those molecules with relatively low-lying excited states. Equipment costs (and potential health hazards) rise rapidly as the photon energy of a spectrometer climbs above 6 eV (50,000 cm^{-1} or 0.22 E_h). Recall the energy levels of a particle in a one-dimensional box—the energy gap decreases rapidly as the length a of the box increases:

$$E_n = \frac{n^2 \pi^2 \hbar^2}{2ma^2}.$$

That suggests that these low-lying excited electronic states are easiest to find when molecular orbitals extend over a large distance. The MOs of conjugated π electrons found in polyacetylene or benzene are good examples of this, as are the relatively unfettered valence electron orbitals of the more massive elements. Consequently, ultraviolet and visible wavelength spectroscopy of biomolecules is usually sensitive to those regions of the molecule occupied by a metal or sulfur atom, or by some aromatic functional group (Table 14.2).

TABLE 14.2 Chromophores of selected amino acids and typical absorption wavelengths.

λ (nm)	group	amino acid
210, 250	$-S-S-$	cysteine dimer
235	$-S^-$	cysteine anion
210, 250	$-C-S-CH_3$	methionine
206, 261	$-C_6H_5$	phenylalanine
222, 270	phenol	tyrosine
235, 287	phenol anion	tyrosine anion
195, 220, 280, 286	indole	tryptophan
211	imidazole	histidine

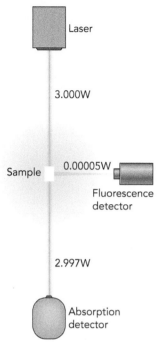

▲ FIGURE 14.4 Spectroscopy of a protein sample. In one possible setup, spectroscopy of the sample takes place with one detector set to monitor the absorbed power of the beam passing through the sample and a second detector that receives only the fluorescent radiation.

These sites on the molecule that absorb UV or visible radiation easily are called **chromophores.** They offer the opportunity to probe a large macromolecule at a specific location. As shown in Table 14.2, the environment of the site can have a significant effect on the absorption wavelength. For example, the absorption spectrum can be used to study the effect of pH on the ionization and dimerization of proteins at a cysteine amino acid.

Among the strongest amino acid chromophores are the aromatic rings of phenylalanine, tyrosine, and tryptophan. These same functional groups have the distinction among amino acids of offering these highly polarizable π-electron systems for binding to substrates through particularly strong dispersion and charge-induced dipole interactions. For this reason, spectroscopic studies of these sites often permit direct measurements of an enzyme's biochemical activity.

It is actually more common to measure the emitted fluorescence from one of these chromophores than to measure the absorbed power, however, because the fluorescence is often easier to detect than absorbance and carries some additional information besides. It may seem odd that fluorescence is easier to measure than absorbance, given that we won't see more light emitted as fluorescence than was absorbed in the first place. Say that a 3-watt laser beam is focused into a dilute solution of tryptophan, and 3 mW of the power is absorbed by the indole rings of the sample (Fig. 14.4). Fluorescence is never completely effective, so when the indole fluoresces, it emits only 0.3 mW, and this is emitted in all directions so only 0.05 mW actually reaches our detector. It is still much easier to see 0.05 mW of light against a dark background than to see the difference between 3.000 watts of light and 2.997 watts of light.

Common applications of fluorescence spectroscopy to protein structure include analysis of fluorescence wavelengths, intensities, or decay times to identify the folding or unfolding rate of the protein or the site at which a particular molecule (the **ligand**) will bind. For example, the phenyl group on tryptophan may form dispersion bonds with another molecule, and as a result the electronic wavefunction of the phenyl π system is subtly changed. We don't expect the change to be very dramatic if the bond formed is relatively weak, but there will be some effect on the tryptophan fluorescence:

- If the ligand shifts the phenyl π antibonding orbitals higher in energy, the fluorescence signal shifts to a shorter wavelength.

- If the ligand is itself a chromophore, it may absorb some of the fluorescence from the phenyl ring, decreasing the intensity of the observed fluorescence.

- If the ligand blocks rotation of the phenyl ring, it can absorb some energy from the excited electronic state by collisions, accelerating the loss of energy from the excited state and decreasing the duration of the fluorescence signal.

Similar effects can be seen when the protein is folded and the phenyl ring of the tryptophan is influenced by proximity to some other functional group on the same protein.

Enzyme Kinetics: The Michaelis-Menten Equation

The enzyme-catalyzed reaction has a general form

$$\text{E} + \text{S} \underset{k_{-1}}{\overset{k_1}{\rightleftharpoons}} \text{ES} \underset{k_{-2}}{\overset{k_2}{\rightleftharpoons}} \text{EP} \underset{k_{-3}}{\overset{k_3}{\rightleftharpoons}} \text{E} + \text{P}, \tag{R14.7}$$

where E is the enzyme, S the substrate or substrates, and P the product or products. We shall obtain an approximate equation for the initial reaction rate of this process in terms of the substrate concentration [S] and the overall enzyme concentration $[E]_0$.

The mechanism by which the reaction rate increases will vary, but there are two general categories of catalysis: the enthalpy of activation $\Delta H^{\ddagger}$ may be decreased by stabilizing the transition state through intermolecular binding, or the entropy of activation $\Delta S^{\ddagger}$ may be increased by forcing the reactants into a suitable orientation for the reaction to proceed (effectively increasing the steric factor p).

In organisms, the enzyme concentration is usually less than 1% the substrate concentration, so the concentration [ES] is essentially in a steady state. Furthermore, the third step in Reaction R14.7 is extremely fast compared to the first and second steps, allowing us to treat the sequence as a two-step process:

$$E + S \underset{k_{-1}}{\overset{k_1}{\rightleftharpoons}} ES \underset{k_{-2}}{\overset{k_2}{\rightleftharpoons}} E + P.$$

If we concentrate on early stages of the reaction, the product concentration [P] is small enough that the second step is essentially irreversible. This reduces the problem to a reaction scheme of the form

$$E + S \underset{k_{-1}}{\overset{k_1}{\rightleftharpoons}} ES \overset{k_2}{\rightarrow} E + P.$$

The rate law for the product P would then be written

$$\frac{d[P]}{dt} = k_2[ES]. \tag{14.38}$$

A solution for the product rate law can be obtained with help from the examples in Section 14.2. The enzyme exists as either free enzyme in concentration [E] or complexed enzyme with concentration [ES]:

$$[E]_0 = [E] + [ES]. \tag{14.39}$$

The rate law for the ES complex is

$$\frac{d[ES]}{dt} = k_1[E][S] - k_{-1}[ES] - k_2[ES],$$

and under the steady-state approximation we can set this derivative to zero and solve for the ES concentration:

$$\frac{d[ES]}{dt} = k_1\{[E][S]\} - k_{-1}[ES]_{ss} - k_2[ES]_{ss} = 0$$

$$[ES]_{ss} = \frac{k_1[E][S]}{k_{-1} + k_2}.$$

To simplify the notation, we introduce the **Michaelis-Menten constant,**

$$K_M \equiv \frac{k_{-1} + k_2}{k_1}, \tag{14.40}$$

so the ES steady-state concentration becomes

$$[ES]_{ss} = \frac{k_1[E][S]}{k_{-1} + k_2} = \frac{[E][S]}{K_M}.$$

The Michaelis-Menten constant is itself informative in that, in the limit that k_2 is very small, K_M approaches the equilibrium constant for dissociation of the ES complex to $E + S$.

Our goal is an expression for the rate law, which we can write using the steady-state concentration of the ES complex, but the most convenient form would put the rate in terms of the concentrations that we can monitor most easily by experiment: the initial concentration of enzyme $[E]_0$ and the time-dependent concentration of substrate $[S]$. To get to this form, we replace $[E]$ by $[E]_0 - [ES]_{ss}$ and solve for $[ES]_{ss}$:

$$
\begin{aligned}
[ES]_{ss} &= \frac{[E][S]}{K_M} \\
&= \frac{([E]_0 - [ES]_{ss})[S]}{K_M} \\
&= \frac{[E]_0[S]}{K_M} - \frac{[ES]_{ss}[S]}{K_M} \\
[ES]_{ss}\left(1 + \frac{[S]}{K_M}\right) &= \frac{[E]_0[S]}{K_M} \\
[ES]_{ss} &= \frac{[E]_0[S]/K_M}{(1 + [S]/K_M)} \\
&= \frac{[E]_0[S]}{K_M + [S]}.
\end{aligned}
\tag{14.41}
$$

Combining Eqs. 14.38 and 14.41 gives the equation we want:

$$
\frac{d[P]}{dt} = k_2[ES]_{ss} = k_2\frac{[E]_0[S]}{K_M + [S]}.
\tag{14.42}
$$

Equation 14.42 is one form of the **Michaelis-Menten equation.** All the values on the right-hand side are constants except the substrate concentration $[S]$. Consider two limits:

$$
\lim_{[S]\to 0}\frac{d[P]}{dt} = \frac{k_2[E]_0[S]}{K_M}
\tag{14.43}
$$

$$
\lim_{[S]\to\infty}\frac{d[P]}{dt} = \frac{k_2[E]_0[S]}{[S]} = k_2[E]_0.
\tag{14.44}
$$

When $[S]$ is small, the reaction velocity is roughly first order in $[S]$. If more substrate is added, more of the enzyme becomes tied up in complexes, and the reaction velocity approaches a maximum value of $v_{max} = k_2[E]_0$, independent of $[S]$. The Michaelis-Menten equation is often rewritten to use v_{max} as a reference point:

$$
\frac{d[P]}{dt} = v_{max}\frac{[S]}{K_M + [S]} = k_2[E]_0\frac{[S]}{K_M + [S]}.
\tag{14.45}
$$

This relationship is plotted in Fig. 14.5. The value of our rate constant k_2 is also known as the enzymatic **turnover number**—the number of substrate molecules converted to product per unit time, and is often written as k_{cat}.

The Michaelis constant is often used as an indicator of the kinetic efficiency of an enzyme. Representative values are given in Table 14.3.

The Michaelis constant represents the substrate concentration at which half the enzyme reactive sites are occupied. For a given substrate concentration, the fraction of occupied reactive sites f_{ES} can then be calculated from K_M:

$$
f_{ES} = \frac{[ES]}{[E]_0} = \frac{[S]}{[S] + K_M}.
\tag{14.46}
$$

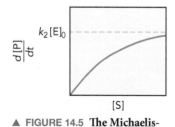

▲ **FIGURE 14.5 The Michaelis-Menten reaction velocity.** The predicted reaction velocity $d[P]/dt$ for early times in an enzyme-catalyzed reaction is graphed as a function of substrate concentration $[S]$.

TABLE 14.3 Selected Michaelis constants.

enzyme	substrate	$K_M (\mu M)$
carbonic anhydrase	CO_2	8000
threonine deaminase	threonine	5000
pyruvate carboxylase	pyruvate	400
pyruvate carboxylase	HCO_3^-	1000
pyruvate carboxylase	ATP	60
arginine-tRNA synthetase	arginine	3
arginine-tRNA synthetase	tRNA	0.4
arginine-tRNA synthetase	ATP	300

One common way to find the value of K_M is from a **Lineweaver-Burk plot** of the reciprocal rate versus $1/[S]$, which gives a slope proportional to K_M:

$$\frac{1}{d[P]/dt} = \frac{1}{v_{max}} + \frac{K_M}{v_{max}[S]}. \tag{14.47}$$

This is a fairly simple picture of the kinetics, but it serves as an excellent starting point for a large fraction of the enzyme reaction systems that have been studied. One of the complications that may arise is the presence of an inhibitor, a substance that blocks the catalytic activity of the enzyme. In the simplest case, the inhibitor binds to the enzyme at the same site as the substrate would, tying up that location so the substrate can't get access and can't get catalyzed. If the formation of the EI complex has the reversible reaction

$$E + I \underset{k_{-3}}{\overset{k_3}{\rightleftharpoons}} EI$$

with equilibrium constant

$$K_3 = \frac{k_3}{k_{-3}} = \frac{[EI]}{[E][I]},$$

then we need to include EI as one of the possible forms in which the enzyme may be found in Eq. 14.39:

$$[E]_0 = [E] + [ES] + [EI] = [E] + [ES] + K_3[E][I]$$

$$[E] = \frac{[E]_0 - [ES]}{1 + K_3[I]}.$$

This changes our solution for the steady-state concentration of ES (Eq. 14.41) to

$$[ES]_{ss} = \frac{[E][S]}{K_M} = \frac{([E]_0 - [ES]_{ss})[S]}{K_M(1 + K_3[I])}$$

$$[ES]_{ss}\left[1 + \frac{[S]}{K_M(1 + K_3[I])}\right] = [ES]_{ss}\left(\frac{K_M + K_M K_3[I] + [S]}{K_M(1 + K_3[I])}\right) = \frac{[E]_0[S]}{K_M(1 + K_3[I])}$$

$$[ES]_{ss} = \left(\frac{[E]_0[S]}{K_M + K_M K_3[I]}\right)\left(\frac{K_M + K_M K_3[I] + [S]}{K_M + K_M K_3[I]}\right)^{-1}$$

$$= \left(\frac{[E]_0[S]}{K_M + K_M K_3[I] + [S]}\right). \tag{14.48}$$

And this gives us the Michaelis-Menten equation corrected for the presence of the inhibitor:

$$\frac{d[P]}{dt} = \frac{k_2[E]_0[S]}{K_M + K_M K_3[I] + [S]}. \tag{14.49}$$

As the inhibitor concentration exceeds that of the substrate, the reaction rate generally plummets at a rate inversely proportional to [I].

The function of the enzyme is sufficiently similar to a solid surface that the same general reaction sequence is appropriate to surface catalysis. In typical surface reactions, a reactant molecule forms a bond with one or more molecules of the surface and is held in place during the chemical transformation, then released. The location on the surface where the bond forms is called an **active site.** Each enzyme molecule in the previous analysis can be used to represent one active site on a catalytic surface. The substrate forms a complex with the active site or enzyme, undergoes a chemical transformation, and emerges from the catalyst as product.

The efficiency of chemical reactions catalyzed by surfaces vary with the **coverage** θ, the ratio of the number of active sites occupied to the total number of active available. In our analogy with enzyme chemistry, this number is the ratio of the concentrations $[ES]/[E]_0$, and is quickly evaluated from the results just given:

$$\theta = \frac{[ES]}{[E]_0} = \frac{k_1[S]}{k_{-1} + k_2 + k_1[S]} \tag{14.50}$$

$$= \frac{\left(\dfrac{k_1}{k_{-1} + k_2}\right)[S]}{1 + \left(\dfrac{k_1}{k_{-1} + k_2}\right)[S]}$$

$$= \frac{K_{ads}[S]}{1 + K_{ads}[S]},$$

where

$$K_{ads} \equiv \frac{k_1}{k_{-1} + k_2}$$

is approximately the equilibrium constant for adsorption of the substrate to the surface.

Combustion Chemistry

We defined combustion as rapid and exothermic oxidation in Section 12.3. There are numerous variations on this process. One valuable distinction is made by comparing the state of the sample immediately after combustion to the original state. If the final state is characterized by lower pressure and density, the process is a **deflagration.** Flames are usually weak deflagrations, with pressures and densities kept close to those of the initial state by the continuous flow of product gases away from the reaction. If the combustion is extremely rapid, the pressure and density skyrocket, and the process is called a **detonation.** This is characteristic of the reactions that lead to explosion: the product gases do not have time to flow away from the reaction before the fuel is all consumed, enormous pressures are generated in a small volume, and this pressure is released as shock waves through the surrounding medium.

Laminar Flames

Laminar flow is fluid motion in a single direction without **turbulence,** and turbulence is any discontinuity in the density or flow velocity. An ideal Bunsen burner would provide a laminar flame in which the CH_4 entering the combustion region is completely oxidized into H_2O and CO_2. The principal elementary reactions relevant to this process are given in Table 14.4, and there are more besides. Detailed analysis of such a system would require accounting not only for the kinetics of over thirty simultaneous and reversible reactions, but also for the variance of those reaction rates over the large range of temperatures and densities characteristic of flames, to say nothing of the common instabilities and sources of turbulence. What's more, experimental measurements for the rate constants are very difficult to obtain. Of all the reactants and products in Table 14.4, only seven are considered chemically stable: CH_4, O_2, H_2O, CO_2, CO, H_2, and CH_2O.

TABLE 14.4 **Important reactions in the combustion of CH_4 in O_2.** The reaction rates are determined at 2000 K in units of $(cm^3 \, mol^{-1})^{n-1} \, s^{-1}$, where n is the molecularity of the reaction. Where k_r is not given, the reverse reaction is negligible.

	reaction	k_f	k_r
1.	$H + O_2 \rightleftharpoons OH + O$	$2.92 \cdot 10^{11}$	$1.15 \cdot 10^{13}$
2.	$H + O_2 + M \rightleftharpoons HO_2 + M$	$5.26 \cdot 10^{15}$	$2.17 \cdot 10^{10}$
3.	$H_2 + O \rightleftharpoons H + OH$	$8.97 \cdot 10^{12}$	$7.02 \cdot 10^{12}$
4.	$H_2 + OH \rightleftharpoons H + H_2O$	$8.34 \cdot 10^{12}$	$8.36 \cdot 10^{11}$
5.	$2OH \rightleftharpoons H_2O + O$	$8.48 \cdot 10^{12}$	$1.20 \cdot 10^{12}$
6.	$HO_2 + H \rightleftharpoons 2OH$	$1.17 \cdot 10^{14}$	$6.00 \cdot 10^{8}$
7.	$HO_2 + H \rightleftharpoons H_2 + O_2$	$2.10 \cdot 10^{13}$	
8.	$HO_2 + H \rightleftharpoons H_2O + O$	$1.95 \cdot 10^{13}$	
9.	$HO_2 + OH \rightleftharpoons H_2O + O_2$	$1.30 \cdot 10^{13}$	
10.	$CO + OH \rightleftharpoons CO_2 + H$	$4.74 \cdot 10^{11}$	$2.01 \cdot 10^{11}$
11.	$CH_4 + H \rightleftharpoons H_2 + CH_3$	$1.95 \cdot 10^{13}$	$9.40 \cdot 10^{11}$
12.	$CH_4 + OH \rightleftharpoons H_2O + CH_3$	$7.37 \cdot 10^{12}$	$1.05 \cdot 10^{11}$
13.	$CH_3 + O \rightleftharpoons CH_2O + H$	$7.00 \cdot 10^{13}$	$5.30 \cdot 10^{7}$
14.	$CH_3 + OH \rightleftharpoons CH_2O + 2H$	$1.83 \cdot 10^{13}$	
15.	$CH_3 + OH \rightleftharpoons CH_2O + H_2$	$8.00 \cdot 10^{12}$	
16.	$CH_3 + OH \rightleftharpoons CH_2 + H_2O$	$4.26 \cdot 10^{12}$	
17.	$CH_3 + H \rightleftharpoons CH_2 + H_2$	$4.07 \cdot 10^{12}$	
18.	$CH_3 + H + M \rightleftharpoons CH_4 + M$	$1.11 \cdot 10^{17}$	$4.30 \cdot 10^{7}$
19.	$CH_2 + H \rightleftharpoons CH + H_2$	$4.00 \cdot 10^{13}$	$1.31 \cdot 10^{13}$
20.	$CH_2 + OH \rightleftharpoons CH_2O + H$	$2.50 \cdot 10^{13}$	
21.	$CH_2 + OH \rightleftharpoons CH + H_2O$	$2.11 \cdot 10^{13}$	
22.	$CH_2 + O_2 \rightleftharpoons CO_2 + 2H$	$4.45 \cdot 10^{12}$	
23.	$CH_2 + O_2 \rightleftharpoons CO + OH + H$	$4.45 \cdot 10^{12}$	
24.	$CH + O_2 \rightleftharpoons CHO + O$	$3.00 \cdot 10^{13}$	
25.	$CH + OH \rightleftharpoons CHO + H$	$3.00 \cdot 10^{13}$	
26.	$CH_2O + H \rightleftharpoons CHO + H_2$	$9.16 \cdot 10^{12}$	$4.20 \cdot 10^{9}$
27.	$CH_2O + OH \rightleftharpoons CHO + H_2O$	$2.22 \cdot 10^{13}$	$8.60 \cdot 10^{9}$
28.	$CHO + H \rightleftharpoons CO + H_2$	$2.00 \cdot 10^{14}$	$2.80 \cdot 10^{5}$
29.	$CHO + OH \rightleftharpoons CO + H_2O$	$1.00 \cdot 10^{14}$	$1.30 \cdot 10^{4}$
30.	$CHO + M \rightleftharpoons CO + H + M$	$1.04 \cdot 10^{13}$	$3.63 \cdot 10^{12}$
31.	$CHO + O_2 \rightleftharpoons CO + HO_2$	$3.00 \cdot 10^{12}$	

The remaining eight species are extremely reactive: H, O, OH, HO_2, CH, CH_2, CH_3, and CHO. Consequently, they cannot be easily isolated and they often participate in several simultaneous reactions.

The goal of this kinetic analysis justifies the profound effort, however. Imagine being able to test different designs of internal combustion engines, rocket engines, and furnaces without having to actually build and test the prototypes. With even a qualitatively accurate computer simulation, critical properties such as fuel efficiency, explosion potential, and exhaust pollution could be evaluated at a tiny fraction of the time, expense, and hazard associated with the physical models.

So take a deep breath and look again at Table 14.4. We will simplify the problem by making some drastic but rational approximations. Assume a fixed temperature of 2000 K and fixed overall density for these reactions, as though all the combustion took place within a very narrow region of the flame. This primarily neglects many soot-forming reactions in the cooler part of the flame downstream from the initial combustion. Assume also that the flow rate is constant throughout the flame, so that the position of a molecule in the flame corresponds directly to the time elapsed since that molecule was involved in the combustion reaction. Already, this makes the problem tractable for a computer using numerical integration. Within minutes, a moderately fast computer could determine the densities of all the reactant and product species as functions of location and initial gas mixture. However, it is difficult to understand a system from such a computer output. Given those results, how would you know what to change in order to improve the fuel efficiency or reduce CO production?

We would like to simplify the set of equations in Table 14.4, but we must do so carefully. It is not sufficient to look at the rate constants and dismiss those that are lowest. Reaction rates depend also on reactant concentrations. A reaction involving CH_4 will be much more important initially than a much faster reaction whose reactants are all radical intermediates formed later in the reaction. In fact, a careful analysis of the numerical results can be used to identify the most important series of reactions and break them up into different categories. The dominant chemistry of the methane flame can be described by the reduced set of multi-step reactions given in Table 14.5.

The net reaction from Table 14.5 is the expected

$$CH_4 + 2O_2 \rightleftharpoons CO_2 + 2H_2O. \tag{14.51}$$

In this sketch of the methane flame, the initiation step in the reaction took place at some earlier time, producing free-radical chain carriers such as OH and H radicals. During the flame, H_2O and H atoms are formed from collisions between H- and O-containing molecules. These are the species that transform the fuel CH_4 into CO and H_2. The CO becomes CO_2 only in the subsequent shift reaction. The recombination reaction keeps the free radical concentration in check; the unpaired electrons of a free radical usually remain unpaired after reaction with a closed-shell molecule, so the radical concentration is reduced only by radical–radical reactions. All four steps in Table 14.5 are exothermic, and all are spontaneous at temperatures greater than 1000 K.

One qualitative result predicted by this reaction sequence, but not very obvious from Table 14.5, is the structure of the flame, illustrated in Fig. 14.6. As the fuel/air mixture flows into the combustion region, the CH_4 and O_2

TABLE 14.5 A reduced set of multi-step reactions used to describe the methane flame.
Reaction numbers correspond to those in the previous table. The stoichiometry of the O_2 consumption reactions is doubled to provide the correct net reaction.

a. O_2 consumption		b. CH_4 consumption	
1.	$2(H + O_2 \rightleftharpoons OH + O)$	11.	$CH_4 + H \rightleftharpoons H_2 + CH_3$
3.	$2(H_2 + O \rightleftharpoons H + OH)$	13.	$CH_3 + O \rightleftharpoons CH_2O + H$
4.	$4(H_2 + OH \rightleftharpoons H + H_2O)$	26.	$CH_2O + H \rightleftharpoons CHO + H_2$
net	$2(3H_2 + O_2 \rightleftharpoons 2H_2O + 2H)$	30.	$CHO + M \rightleftharpoons CO + H + M$
		3.	$H + OH \rightleftharpoons O + H_2$
		4.	$H + H_2O \rightleftharpoons OH + H_2$
		net	$CH_4 + 2H + H_2O \rightleftharpoons CO + 4H_2$
	$\Delta_{rxn}H = -42.7 \text{ kJ mol}^{-1}$		$\Delta_{rxn}H = -229.8 \text{ kJ mol}^{-1}$
	$\Delta_{rxn}S = 9.9 \text{ kJ K}^{-1} \text{ mol}^{-1}$		$\Delta_{rxn}S = 115.9 \text{ kJ K}^{-1} \text{ mol}^{-1}$

c. CO shift		d. H_2 recombination	
10.	$CO + OH \rightleftharpoons CO_2 + H$	2.	$O_2 + H + M \rightleftharpoons HO_2 + M$
4.	$H + H_2O \rightleftharpoons H_2 + OH$	9.	$OH + HO_2 \rightleftharpoons H_2O + O_2$
		4.	$H + H_2O \rightleftharpoons OH + H_2$
net	$CO + H_2O \rightleftharpoons CO_2 + H_2$	net	$2H + M \rightleftharpoons H_2 + M$
	$\Delta_{rxn}H = -41.2 \text{ kJ mol}^{-1}$		$\Delta_{rxn}H = -453.9 \text{ kJ mol}^{-1}$
	$\Delta_{rxn}S = -42.1 \text{ kJ K}^{-1} \text{ mol}^{-1}$		$\Delta_{rxn}S = -98.7 \text{ kJ K}^{-1} \text{ mol}^{-1}$

encounter an increasing density of H and OH free radicals left over from previous reactions. The CH_4 and O_2 consumption steps take place in this early region of the flame, called the **preheat zone.** The net products of these reactions are primarily CO and H_2O. Excess H atoms produced during these reactions are recombined into H_2, with O_2 acting as a catalyst. The CH_4 fuel is used up in a narrow boundary called the **fuel consumption layer,** and thereafter the focus of the chemistry changes. Now it is the CO that is consumed, converted to the more stable CO_2 in the shift reactions catalyzed by OH in the **oxidation layer.** It is possible to decrease the amount of CO in the exhaust gases by increasing the initial O_2 concentration or by seeding the fuel mix with other sources of OH.

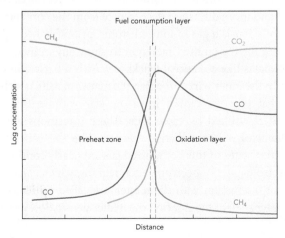

◀ FIGURE 14.6 **A schematic of the regions in the methane-air flame.** The fuel/air mixture flows into the combustion region from the left, passing through a preheat zone that warms the mixture and initiates reactions, through the fuel consumption layer in which the methane is effectively used up, and into an oxidation layer. The CO shift reaction and oxidation of H_2 to H_2O occur primarily in the oxidation layer.

Molecular Astrophysics

Astrophysics and Spectroscopy

Physical chemistry and astrophysics share many common interests. Part of this overlap results from the origins of each field in classical physics. Predicting the motion of groups of planets or stars that interact through their gravitational fields requires integration of a many-body classical Hamiltonian, just as scary a prospect as integrating the quantum mechanical Hamiltonian of a polyatomic molecule. The thermodynamics of gas expansions finds application to stellar radiation problems and black holes.

However, the greatest similarities in modern research in astrophysics and physical chemistry rest with the remarkable coincidence of observational methods. Both sciences are faced with the dilemma that the object of study is difficult to probe directly: in physical chemistry it is often a chemical sample too small or too reactive to isolate; in astrophysics it may be a star one thousand light years away. Yet a precise, general, and sensitive technique exists that has worked well for both sciences: spectroscopy.

Astronomers are hampered by rarely being in the same place as what they study. They rely on *any* data that reaches them about distant planets, stars, galaxies, comets, and other extraterrestrial phenomena. The richest such data arrive in the form of radiation emitted by these objects in the radio, microwave, infrared, visible, ultraviolet, x-ray, and gamma-ray regions of the spectrum, carrying information about its sources across the largest distances known. This radiation carries with it spectra of atoms and molecules in any of the countless extraterrestrial environments observable from Earth. The chemical composition of the sun was determined by spectroscopy, and from this information the power source of the sun and most other stars was deduced to be hydrogen fusion. The distances to the farthest galaxies and the expansion rate of the universe are generally based on measurements of the hydrogen atom spectrum. How does the universe come to be a gigantic spectrometer? There are many mechanisms that generate extraterrestrial radiation with valuable spectroscopic information; we shall consider only two examples, stellar emission and radiative cooling of molecular clouds.

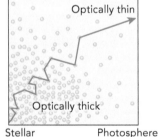

Stellar core   Photosphere

▲ FIGURE 14.7 **Optical thickness of a stellar interior.** In the denser regions of a star, matter is so compressed that photons are absorbed, scattered, and emitted many times before they work their way toward the surface, where they can at last travel unimpeded. The arrow in the figure represents a photon emitted near the core of the star, interacting with many atoms in the central, optically thick region of the star, and eventually finding its way to the optically thin photosphere.

Stellar emission. A star is a cloud of gas so massive and so dense that its own gravity compresses the matter at its core with the necessary force to enable nuclear fusion to take place. The core is the hottest and densest region of the star. The star becomes cooler and more diffuse as the distance from the core increases.

The interior of a star is a gas of ionized atoms, primarily hydrogen and helium, too hot and dense for the emitted radiation to show any individual spectroscopic transitions; it radiates like a classical blackbody, with a fairly continuous range of emitted energy. In the inner part of the star, the matter density is so high over such a large distance that photons cannot leave the star without first being scattered or absorbed and re-emitted by these ions. These transitions are not specific to particular frequencies of the radiation because, like the conduction band of a metal, the quantum states of this hot ionized gas are nearly continuous throughout much of the electromagnetic spectrum. Such regions, which can't be penetrated by radiation without interaction with the matter, are called **optically thick.** Photons emitted in the outer shell of the star find the matter density low enough that there is a good chance of their leaving the star without further scattering or absorption. This outer shell, called the **photosphere** of the star, is **optically thin,** essentially transparent to the radiation (Fig. 14.7). At this distance from the stellar core, the temperature has dropped to the point that neutral atoms and even some molecules may be stable. The radiation released by stars is generally consistent with the emission profile of a blackbody at the temperature of the star where it first becomes optically thin. Absorptions of the blackbody radiation from the optically thick interior of the sun by atoms and molecules in the stellar photosphere account for much of what we know of the chemical composition of stars, including our own sun.

EXAMPLE 14.4 Solar Blackbody Radiation

PROBLEM The sun's emission spectrum is characteristic of a blackbody at a temperature of 5800 K. Find the peak emission wavelength.

SOLUTION The blackbody emission spectrum is given by Eq. 6.20:

$$\rho(\nu)d\nu = \frac{8\pi h\nu^3 d\nu}{c^3(e^{h\nu/(k_B T)} - 1)},$$

where the radiation density $\rho(\nu)$ is proportional to the intensity of the light emitted at frequency ν. To express the radiation density in terms of the wavelength, we replace ν by c/λ and $d\nu$ by

$$|d\nu| = \left|d\left(\frac{c}{\lambda}\right)\right| = |-c\lambda^{-2}d\lambda| = c\lambda^{-2}d\lambda$$

so our equation becomes

$$\rho(\lambda)d\lambda = \frac{8\pi hcd\lambda}{\lambda^5(e^{hc/(k_B T\lambda)} - 1)},$$

In Problem 6.5, we find that the derivative of ρ is given by

$$\frac{d\rho(\lambda)}{d\lambda} = \frac{8\pi hc}{\lambda^6(e^{hc/(k_B T\lambda)} - 1)}\left[\frac{hc/(k_B T\lambda)}{1 - e^{-hc/(k_B T\lambda)}} - 5\right].$$

Setting the derivative to zero and assuming that $\frac{hc}{\lambda} \gg k_B T$ at λ_{max} leads (to a very good approximation) to the following expression for the wavelength of highest intensity:

$$\lambda_{max}\,(\text{nm}) = \frac{2.88 \cdot 10^6}{T(\text{K})}.$$

Substituting $T = 5800$ K, we obtain $\lambda_{max} = 496$ nm, at the blue end of the visible region of the spectrum. The hottest stars have effective temperatures of roughly 40,000 K, corresponding to a peak emission wavelength of about 72.4 nm, deep in the ultraviolet.

EXAMPLE 14.5 Solar Photon Diffusion

PROBLEM Heat is carried to the optically thin layer of a star primarily by convection. Radiation is just too inefficient a process in the dense core. Use the Einstein diffusion equation (Eq. 5.36) to estimate the time required for a photon generated at the sun's center to diffuse out of the sun's core, a distance of $1.7 \cdot 10^8$ m. Although the actual number density in the core approaches 10^{26} cm^{-3}, the photons interact weakly with the matter, and have a mean free path of about 0.1 cm. Use the mean free path to estimate the diffusion constant.

SOLUTION We may approximate the time required by solving for t in the Einstein equation:

$$t \approx \frac{R^2}{6D},$$

where we have replaced the root mean square distance with R. Using this equation assumes that the diffusion constant D is a constant throughout the sun's core, which is a poor assumption but will suffice for now. To estimate the diffusion constant, we use Eq. 5.33:

$$D = \frac{\lambda^2 \gamma}{2} = \frac{\lambda^2 (c/\lambda)}{2} = \frac{\lambda c}{2} = 1.5 \cdot 10^5 \, \text{m}^2 \, \text{s}^{-1}.$$

We used the speed c of the photon to solve for the collision frequency γ. With this data, we can solve (very approximately) for the time:

$$t \approx \frac{(1.7 \cdot 10^8 \, \text{m})^2}{6(1.5 \cdot 10^5 \, \text{m}^2 \, \text{s}^{-1})}$$

$$= 3 \cdot 10^{10} \, \text{s},$$

which is about 1000 years. In more sophisticated treatments [*Astrophys. J.* **401** 759 (1999)], the photon diffusion time is actually longer by nearly two orders of magnitude, which is why convection is a much more important mechanism for heat transfer.

EXAMPLE 14.6 Solar H_2/H Equilibrium

PROBLEM Calculate K_{eq} for the dissociation of H_2 at 5800 K.

SOLUTION For this reaction, $\Delta_{rxn}H(298 \text{ K})$ is twice the enthalpy of formation for H(g), or 435.93 kJ mol^{-1}, and similarly $\Delta_{rxn}G(298 \text{ K})$ is 406.494 kJ mol^{-1}. The difference in heat capacities is

$$\Delta_{rxn}C_P = 2(20.786) - 28.824 = 12.748 \, \text{J K mol}^{-1}.$$

The room temperature equilibrium constant is

$$K_{eq}(298 \text{ K}) = e^{-\Delta_{rxn}G/(RT)}$$

$$= e^{-164} = 6 \cdot 10^{-72}.$$

Indeed, the equilibrium lies on the side of H_2. Now we calculate the constant at 5800 K, using Eq. 12.45:

$$\ln K_{eq}(T_2) = \ln K_{eq}(T_1) + \frac{1}{R}\left[\left(\Delta_{rxn}H°(T_1) - T_1\Delta_{rxn}C_{Pm} \right)\left(\frac{1}{T_1} - \frac{1}{T_2} \right) + \Delta_{rxn}C_{Pm} \ln \frac{T_2}{T_1} \right]$$

$$= -164 + \frac{1}{8.3145}\left[(435930 - 298 \cdot 12.748)\left(\frac{1}{298} - \frac{1}{5800} \right) \right.$$

$$\left. + 12.478 \ln\left(\frac{5800}{298} \right) \right]$$

$$= 5.90.$$

$$K_{eq}(5800 \text{ K}) = e^{5.90} = 365.$$

The equilibrium now lies on the product side of the reaction, so most of the hydrogen *will* be dissociated.

There are many tools of classical physics used to describe the structure of stars, for example, how the density and temperature vary with distance from the center. Our own tools let us assess some of the chemical properties of stars, however. For example, at our sun's "effective surface temperature" of 5800 K, what is the equilibrium constant for dissociation of H_2 into hydrogen atoms? Certainly at room temperature, the equilibrium state of the reaction

$$H_2(g) \rightarrow 2H(g)$$

lies far to the left. Because the reaction is endothermic, raising the temperature will shift the equilibrium to the right. Furthermore, the dissociation of the molecule has a positive $\Delta_{rxn}S$, which will further shift the equilibrium toward products as the temperature increases.

The interstellar medium. The universe is composed of much more than stars, however. Most of the observable matter in the universe is believed to be in a less obvious form: interstellar gas and dust. Within our galaxy, the number density of matter between stars generally varies between 10^3 and 10^6 cm^{-3}, almost all of it in one of three forms of hydrogen: H^+ ions, neutral H atoms, and H_2 molecules. This matter is known collectively as the **interstellar medium**. The ionized gas corresponds to the largest, hottest, and sparsest regions of interstellar space. By contrast, the molecular hydrogen exists only in compact, cold, relatively dense clumps known as **molecular clouds**. However, even in the molecular clouds, number densities rarely exceed 10^5 cm^{-3}, which is 10^{13} times less dense than our atmosphere and comparable to the densities obtained by the best laboratory vacuum pumps.[3] Temperatures are roughly in the range 10–100 K, generally lower for denser clouds.

What provides the heat for the molecular clouds? These border a warmer region of atomic hydrogen gas, which in turn is heated by cosmic rays (charged particles accelerated to near the speed of light), shock waves from supernovae, and most importantly stellar radiation.[4] Then a logical question to ask is "what keeps the molecular clouds from heating to the same temperature as the surrounding gas?" The answer lies in one of the clearest differences between atoms and molecules in the opening parts of this text: molecules have more

---

EXAMPLE 14.7 **Atomic Fine Structure Transitions**

PROBLEM What is the lowest atomic number for a neutral atom capable of fine structure transitions within its ground state term?

SOLUTION The ground state terms of H, He, Li, Be are $^2S_{1/2}$, $^1S_0$, $^2S_{1/2}$, and $^1S_0$, respectively. Each of these terms has but a single value of J because $L = 0$ for each. For each of these terms, $J = S$. We need electrons with orbital angular momentum (e.g., p orbitals) before there can be fine structure splitting. Boron, with electron configuration $1s^2 2s^2 2p$, has ground state terms $^2P_{1/2}$ and $^2P_{3/2}$ and is the first atom in the periodic table with ground state fine structure transitions. However, boron has a very low cosmic abundance (less than 10^{-10} that of hydrogen) and makes no significant contribution to cooling of the atomic gas. The next three elements—carbon, nitrogen, and oxygen—are the most abundant of the elements after He, and they contribute the most to the cooling. (Ground state $^4S_{3/2}$ nitrogen also has no fine structure splitting, but its ions N^+ and N^{2+} both contribute substantially to the cooling.)

---

[3]Many experiments and industrial applications rely on "ultra-high vacuum" or UHV, working at pressures of about 10^{-12} torr, which corresponds to a density of about $3 \cdot 10^4$ cm^{-3}.

[4]The photochemistry of our upper atmosphere, described earlier (*Atmospheric Chemistry*), absorbs enormous amounts of stellar ultraviolet radiation. The majority of the galaxy is very unfriendly to molecules; the high ultraviolet intensity dissociates chemical bonds too efficiently.

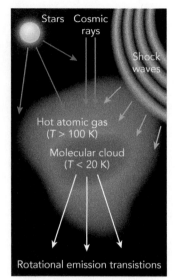

Stars   Cosmic rays

Shock waves

Hot atomic gas (T > 100 K)

Molecular cloud (T < 20 K)

Rotational emission transitions

▲ FIGURE 14.8 **Energy transfer in molecular clouds.** Molecular clouds are cooler than the surrounding atomic gas, as low as 5 K, because they constantly emit radiation from rotational transitions of polar molecules into the optically thin surroundings. No significant cooling transitions are available for the atomic gas until its temperature exceeds about 100 K.

degrees of freedom. To be cooler than the surrounding atomic gas, molecular clouds need to be able to release heat in a way that the surrounding gas cannot— by rotational and vibrational *emission* transitions.

The atomic gas is in the temperature range 100–4000 K, which corresponds to less than 3000 cm^{-1}. This is too cold for atomic hydrogen to be bumped by collision into any of its excited electronic states, the lowest of which is $n = 2$ at approximately 80,000 cm^{-1}. Radiative cooling is not an option for the atomic gas except through the forbidden fine structure transitions of the heavier elements, for example, the $^3P_1 \rightarrow {}^3P_0$ transition of atomic carbon at 16.4 cm^{-1}. In fact, these transitions are the dominant cooling mechanism for the atomic gas, but the molecules have a much more efficient mechanism.

Hydrogen molecules on the fringe of a molecular cloud occasionally collide with atoms from the neighboring warm regions. Those molecules in turn collide with other molecules and heat up the molecular cloud. However, when one of the fast-moving H_2 molecules collides with CO or HCN, these will often be left in excited rotational states. Unlike H_2, which has no permanent dipole moment, CO and HCN have allowed rotational transitions. The $J = 1 \rightarrow 0$ transition of CO, for example, is at 3.3 cm^{-1}. What this implies is that emission lines are available for the molecular clouds even at energies corresponding to only a few K in temperature. Consequently, the molecular clouds can get much colder than the atomic gas (Fig. 14.8).

These emission lines provide the bulk of our knowledge of the molecular clouds, although additional information is available in some cases from the way in which light from a star on its way to Earth is blocked or scattered by an intervening cloud. Carried in the rotational emission spectrum of a molecular cloud are clues to the cloud's chemical composition, density, depth, temperature, energy sources, direction and speed of motion, magnetic and electrical properties, and virtually every other physical aspect of the cloud an astrophysicist would want to know.

The drawback is that these are often only clues; for some properties the astrophysicist must solve a puzzle lacking several critical pieces. The chemical composition is a perfect example. The dominant elements in our galaxy, and presumably all galaxies, are hydrogen, helium, oxygen, carbon, and nitrogen, in that order. The abundances for the atomic interstellar gas are given in Table 14.6.

TABLE 14.6 Cosmic abundances of the dominant elements, relative to hydrogen, and some of their likely common forms in the molecular clouds.

element	relative abundance	species in molecular clouds
H	1	H, H_2, H_2O, HCN
He	$6.3 \cdot 10^{-2}$	He
O	$6.9 \cdot 10^{-4}$	CO, SiO, H_2O, OH, HCO, MgO, CO_2, OCS, O_2
C	$4.2 \cdot 10^{-4}$	CO, HCN, CS, HCO, CO_2, CH_4, OCS
N	$8.7 \cdot 10^{-5}$	N_2, HCN, NH_3, NH_2
Si	$4.5 \cdot 10^{-5}$	Si, SiS, SiO, SiO_2
Mg	$4.0 \cdot 10^{-5}$	Mg, MgH, MgS, MgO
Ne	$3.7 \cdot 10^{-5}$	Ne
Fe	$3.2 \cdot 10^{-5}$	Fe
S	$1.6 \cdot 10^{-5}$	SiS, CS, S, HS, H_2S, MgS

These values are determined almost entirely from spectroscopic data. However, in the molecular clouds, spectra are not available for every compound. Some of the most common molecules in these clouds, such as H_2, N_2, CO_2, and CH_4, have no permanent dipole moment and therefore no allowed rotational transitions. Nearly as difficult to detect is H_2O, because it is so abundant in our own atmosphere that the very weak signals due to extraterrestrial H_2O are usually completely obscured. Nonetheless, the spectra that can be measured and interpreted have shown a fascinating and unexpected chemistry in this environment.

Interstellar Chemical Synthesis

The elements in Table 14.6 may be organized into three categories: the noble gases He and Ne, which do not form chemical bonds; the **refractory elements** Mg, Fe, and Si, which easily condense to form large particles called **dust grains;** and the non-metals H, O, C, N, and S, which tend to form small gas-phase molecules. The division between the last two categories is not by any means rigid, since small molecules of the refractory elements such as SiS have been observed in the molecular clouds, and much of the H_2O is expected to be condensed as water ice on the dust grains. Nonetheless, it appears clear that the actual chemistry, apart from the formation and dissociation of H_2, overwhelmingly lies in the domain of the non-metals H, O, C, and N. Of this chemistry, the greatest number of reactions are those that involve the highly flexible bonding capabilities of carbon; in short, organic chemistry.

Numerous organic molecules have been detected in the molecular clouds. Table 14.7 comprises only a partial list, but it demonstrates that these include several of the familiar varieties of organic compounds: nitriles (CH_3CN), alkynes (CH_3CCH), alcohols (CH_3OH), aldehydes (H_2CO), and organic acids (HCOOH). The list is highly biased toward molecules that are easier to observe by virtue of having large permanent dipole moments and, therefore, more intense rotational spectra. This explains the conspicuous absence of the alkanes, which tend to have small or zero dipole moments.

In addition to the familiar organic compounds, however, several unusual molecular classes are detected: molecular ions (HCO^+), isocyanides (HNC), long-chain unsaturated nitriles (HC_9N), carbenes ($c-C_3H_2$), and free radicals (C_3H) have all been detected in the interstellar medium, including several examples of each.

TABLE 14.7 A partial list of carbon-containing compounds observed in the interstellar molecular clouds. The "c-" prefix indicates a cyclic molecule.

CH	C_2	CN	CO	CS
CH^+	HOC^+	HCS^+	$HCNH^+$	HCO^+
HCN	HNC	HNCO	HNCS	
HC_3N	HC_5N	HC_7N	HC_9N	$HC_{11}N$
CH_3CN	CH_3NC	C_2H_3CN	C_3H_5CN	
CH_3CHO	HC_2CHO	HC_3HOH	CH_3CH_2OH	CH_3OH
C_3O	HC_2O	H_2CS	HCOOH	CH_3COOH
CH_3SH	NH_2CHO	NH_2CN	$c-SiC_2$	$c-C_3H_2$
C_3N	CH_3C_3N	CH_3C_4H	CH_3CCH	
C_2H	C_3H	C_4H	C_5H	

What accounts for the fact that these extremely reactive compounds—much too reactive to isolate under typical laboratory conditions—are so abundant in the molecular clouds that they account for roughly half of the known interstellar molecules?

Observational bias is certainly part of the answer, because that eliminates many symmetric and stable compounds, such as CH_4 and HCCH, in favor of more asymmetric molecules with very large dipole moments, such as the unsaturated carbon chains with a nitrile or carbonyl group at the end (HC_5N has a dipole moment of 4.3 D). Nevertheless, the environment does not discriminate so strongly against reactive molecules as our denser, hotter environment does. There are several reasons for this: molecular collisions are much less frequent in the molecular clouds, there is less kinetic energy available to overcome activation barriers, and there are few solid surfaces to act as catalysts for radical recombination.

When a reactive compound such as CCH or HCO^+ is formed, days may elapse before it even strikes another molecule. When it does, the other molecule is 99% likely to be H_2 or He. With thermal energies of less than 100 K, activation barriers of 10 kJ mol^{-1} are virtually insurmountable, and even if a reaction does take place, chances are that the products themselves are still very reactive. A reaction between a molecular ion and a neutral molecule must still produce a charged product, for example, and a molecule with an unpaired electron reacts with a closed shell molecule to produce a new free radical.

CHECKPOINT We can explain the relatively high abundance of reactive species in the interstellar gas in terms of the chain reaction steps outlined in Section 14.4. Collisions between molecules are likely to be chain propagation steps, in which a net charge or an unpaired electron is shifted from one species to another. But chain termination steps are rare in these low densities. The chance of two ions or two radicals running into one another charged is almost negligible at these low densities.

Interstellar c-C_3H_2. A series of reactions is often initiated when one of these reactive compounds is formed. Reactive molecules are first formed by reactions at the boundaries between the molecular clouds and the hotter atomic gas, by shock waves from exploding stars, by cosmic rays, by ultraviolet and x-ray photons, and by free electrons accelerated along magnetic fields. These events are relatively rare inside the cloud, and the chemistry of the cloud's interior must generally proceed through a series of steps with very low activation barriers.

EXAMPLE 14.8   **Interstellar Collision Frequencies**

PROBLEM Calculate the collision frequency for H_2 in a cold interstellar molecular cloud with a temperature of roughly 10 K and a number density of about 10^5 cm^{-3}.

SOLUTION Section 5.1 on collision parameters includes Eq. 5.12 for the collision frequency γ:

$$\gamma = \rho\sigma\langle v_{AA}\rangle$$

$$= \rho\sigma\sqrt{16\frac{k_B T}{\pi m}} \qquad \text{by Eq. 15.5}$$

$$= (10^5\,\text{cm}^{-3})(100\ \text{cm/m})^3\left(27\cdot 10^{-20}\,\text{m}^2\right)\sqrt{\frac{16(1.381\cdot 10^{-23}\,\text{J K}^{-1})(10\text{K})}{\pi(2)(1.661\cdot 10^{-27}\,\text{kg amu}^{-1})}}\ \text{m s}^{-1}$$

$$= 1.24\cdot 10^{-5}\,\text{s}^{-1}$$

$$= 1\ \text{collision every 22 hours.}$$

Consider the following proposed mechanism for the synthesis of interstellar c-C_3H_2 (Fig. 14.9):

$$C^+ + C_2H_2 \rightarrow c\text{-}C_3H^+ + H$$

$$c\text{-}C_3H^+ + H_2 \rightarrow c\text{-}C_3H_3^+ + h\nu$$

$$c\text{-}C_3H_3^+ + e^- \rightarrow c\text{-}C_3H_2 + H$$

▲ FIGURE 14.9
Cyclopropenylidene.

The chain of events is spurred by the entrance of a carbon atomic ion into a region of stable molecules such as C_2H_2 and H_2.

The first step, formation of the cyclic c-C_3H^+ ion, has an advantage over reactions between neutral molecules: the monopole-induced dipole interaction is powerful enough at these low temperatures to increase the collision frequency of ions and neutrals. The C^+ ion induces a slight charge separation in the C_2H_2, creating an appreciable attractive force between the two species. Molecules that would otherwise miss each other may be pulled together by this mechanism. The ion-molecule bond, which may be nearly as strong as a chemical bond, also helps reduce the activation barriers for these reactions. The formation mechanisms proposed for most interstellar polyatomic molecules are driven by ion-neutral reactions at every step.

The second step looks at first as though it may proceed rapidly by comparison simply because the very next collision partner is 90% likely to be H_2; however, there is a significant obstacle. This reaction forms the c-$C_3H_3^+$ ion by **radiative association.** A single product is formed from two reactants, and the excess energy is carried away by a photon, rather than by the third body we would use to make this reaction occur under terrestrial conditions. The very low densities in the interstellar medium make three-body reactions unheard of. The only way for the excess energy of the activated complex $(c\text{-}C_3\,H_3^+)^\ddagger$ to be removed before it falls apart is for the complex to emit a photon. The low temperature of the molecular cloud plays an important role here by lengthening the lifetime of the activated complex, increasing the chances that it will radiate before decaying back into products. Even so, this is a slow step, made feasible largely by the high abundance of H_2 and by the enhanced collision rates for ion-neutral reactions. The reaction rate constant at 10 K is estimated to be of the

---

EXAMPLE 14.9 Ion-Molecule Reaction Rates

PROBLEM Based on the parameters given in Table 14.8, estimate the rates of formation of HCl^+ and HCl at 80 K in the Orion Molecular Cloud from the reactions given.

$$Cl^+ + H_2 \xrightarrow{k_1} HCl^+ + H$$
$$Cl + H_2 \xrightarrow{k_2} HCl + H$$

TABLE 14.8 **Selected parameters for Orion Molecular Cloud and the $Cl^+ + H_2$ and $Cl + H_2$ reactions.**

$[Cl^+]$	$1 \cdot 10^{-12}$ cm^{-3}	$[Cl]$	$4 \cdot 10^{-3}$ cm^{-3}
$[H_3^+]$	$1 \cdot 10^{-4}$ cm^{-3}	$[H_2]$	$1 \cdot 10^5$ cm^{-3}
$[e^-]$	$1 \cdot 10^{-12}$ cm^{-3}	T	80 K
$A_1(80\text{ K})$	$1 \cdot 10^{-9}$ cm^3 s^{-1}	$A_2(80\text{ K})$	$1 \cdot 10^{-10}$ cm^3 s^{-1}
E_{a1}	0 kJ mol^{-1}	E_{a2}	23 kJ mol^{-1}

SOLUTION

$$k_1 = A_1 e^{-E_{a1}/RT} = 1 \cdot 10^{-9} \text{ cm}^3 \text{ s}^{-1}$$

$$k_2 = A_2 e^{-E_{a2}/RT} = 1.0 \cdot 10^{-25} \text{ cm}^3 \text{ s}^{-1}$$

Note that the ion-molecule reaction has no activation energy because the transfer of an electropositive H atom to the Cl^+ ion is so exothermic that any progress along the reaction coordinate stabilizes the system. The Cl^+ ion also polarizes the H_2 molecule, using monopole-induced dipole forces to enhance the collision rate. The added attractive forces account for the factor of ten increase in A_1 over A_2, because other contributions to the pre-exponential factor are expected to be similar for both reactions.

We use the standard rate law equations to find the formation rates:

$$\frac{d[HCl^+]}{dt} = k_1 [Cl^+] [H_2]$$

$$= (1 \cdot 10^{-9})(1 \cdot 10^{-12})(1 \cdot 10^5) = 1 \cdot 10^{-16} \text{ cm}^{-3} \text{s}^{-1}$$

$$\frac{d[HCl]}{dt} = k_2 [Cl] [H_2]$$

$$= (1.0 \cdot 10^{-25})(4 \cdot 10^{-3})(1 \cdot 10^5) = 4 \cdot 10^{-23} \text{ cm}^{-3} \text{s}^{-1}.$$

So based on reactions 1 and 2 alone, the formation rate of HCl is seven orders of magnitude slower than the formation rate of HCl^+. This is not the whole story, however, as the next example shows.

EXAMPLE 14.10   Steady State in the Interstellar Medium

PROBLEM Contrary to the apparent results of the previous example, the $[HCl]/[HCl^+]$ ratio in the Orion Molecular Cloud is estimated at $2 \cdot 10^9$. There is a more efficient way to produce the HCl than by reaction 2 in the mechanism given above, and the HCl^+ is itself consumed in one of those reactions:

$$Cl + H_3^+ \xrightarrow{k_3} HCl^+ + H_2 \qquad k_3 = 1.0 \cdot 10^{-9} \text{ cm}^3 \text{s}^{-1}$$

$$HCl^+ + H_2 \xrightarrow{k_4} H_2Cl^+ + H \qquad k_4 = 1.3 \cdot 10^{-9} \text{ cm}^3 \text{s}^{-1}$$

$$H_2Cl^+ + e^- \xrightarrow{k_5} HCl + H \qquad k_5 = 1.5 \cdot 10^{-7} \text{ cm}^3 \text{s}^{-1}$$

Reaction 3 is actually about four times more efficient at forming HCl^+ than reaction 1 (but doesn't offer such a straightforward comparison with neutral-neutral kinetics). The net reaction is

$$Cl + H_3^+ + e^- \rightarrow HCl + 2H.$$

Find the steady-state number density of HCl^+ in this mechanism, using the parameters in Table 14.7.

SOLUTION Treating HCl^+ as an intermediate in the mechanism, we apply the steady-state approximation:

$$\frac{d[HCl^+]}{dt} \approx k_3 [Cl] [H_3^+] - k_4 [HCl^+]_{ss} [H_2] = 0$$

$$[HCl^+]_{ss} = \frac{k_3 [Cl] [H_3^+]}{k_4 [H_2]}$$

$$= \frac{(1.0 \cdot 10^{-9})(4 \cdot 10^{-3})(1 \cdot 10^{-4})}{(1.3 \cdot 10^{-9})(1 \cdot 10^5)} = 3 \cdot 10^{-12} \text{ cm}^{-3}.$$

The steady-state concentration is roughly the measured value of $1 \cdot 10^{-12} \text{ cm}^{-3}$.

order $10^{12} \text{ cm}^3 \text{ mol}^{-1} \text{ s}^{-1}$, which compares favorably to some of those tabulated in Table 14.4 for the methane flame. This is a rate constant that will *decrease* with temperature, because the lifetime of the activated complex becomes shorter.

In the third step, the molecular ion is neutralized by a free electron. Again, conservation of energy must be maintained, in this case by release of one of the hydrogen atoms, leaving behind the cyclic carbene c-C_3H_2.

Interstellar carbon chains. The unsaturated carbon chains observed in the interstellar medium appear with unexpectedly high densities in a variety of environments. The C_4H radical, for example, has been detected in very cold molecular clouds and stellar photospheres and also in the laboratory in acetylene flames and electrical discharges. A proposed mechanism for the formation of C_4H in the interstellar medium is

$$C^+ + CH_4 \rightarrow C_2H_2^+ + H_2$$

$$C^+ + CH_4 \rightarrow C_2H_3^+ + H$$

$$C_2H_3^+ + e^- \rightarrow C_2H_2 + H$$

$$C_2H_2^+ + C_2H_2 \rightarrow C_4H_2^+ + H_2$$

$$C_4H_2^+ + e^- \rightarrow C_4H + H.$$

This reaction mechanism proceeds through a series of either ion-neutral or ion-electron reactions, without requiring the relatively slow radiative association step called for in the c-C_3H_2 synthesis. Note that longer carbon chains, as found in $HC_{11}N$, can be formed by additional ion-neutral steps of the form

$$C_{2n}H_2^+ + C_2H_2 \rightarrow C_{2n+2}H_2^+ + H_2,$$

before the final recombination with an electron that ends the reaction sequence.

Conclusion to the Text

This chapter is intended to illustrate both the power of the modern chemical model and the daunting challenges to which it is applied. Although we have been able to describe the reaction dynamics of a few selected environments in this chapter, we can only master the chemistry of these problems with a global understanding of the physics of the chemical system: its quantum mechanics, transport properties, and thermodynamics. The chemical model unites all these.

Several conceptual landmarks, many of them familiar from introductory chemistry, lie within the chemical model, providing snapshots of the logic underlying specific chemical properties. The Bohr model of the atom, the Lewis dot model of molecular bonding, the non-ideal gas laws, the Arrhenius equation of chemical kinetics, and numerous other equations and techniques simplify the formidable problems presented in the study of chemistry. These snapshots reduce the far-reaching and intricate assembly of our chemical theory to a more manageable size for application to individual problems. We are in the fortunate position, however, of having a theoretical framework that unites *all* chemistry and that can be drawn upon when new measurements test our understanding of molecular structure and dynamics. That framework forms the basis of an intuition that will guide you as you expand the frontiers of chemistry.

KEY CONCEPTS AND EQUATIONS

14.1 **Elements of Multi-Step Reactions.** The simplest multi-step reaction has two steps, and there are three ways that the two steps can be related to each other:

- A **reversible step** consists of a forward reaction and the corresponding reverse reaction, as in the simplest case, $A \underset{k_{-1}}{\overset{k_1}{\rightleftharpoons}} B$. By introducing the reverse reaction, we allow equilibrium to be reached at a point where the forward and reverse reactions occur at the same rate. This relationship lets us express the equilibrium constant as the ratio of the forward and reverse rate constants:

$$K_{eq} = \frac{[B]_{eq}}{[A]_{eq}} = \frac{k_1}{k_{-1}}. \qquad (14.5)$$

- **Parallel reactions** occur when two distinct reaction steps share one or more of either the same reactants or the same products, as in the reaction series $A \overset{k_1}{\rightarrow} B$, $A \overset{k_2}{\rightarrow} C$. In this mechanism, if we begin with zero concentration of the products B and C, then the ratio $[B]/[C]$ is determined by the ratio of the rate constants:

$$\frac{[B]}{[C]} = \frac{k_1}{k_2}. \qquad (14.12)$$

  This illustrates the principle of **kinetic control** of a reaction: if we don't wait for equilibrium to be reached, the distribution of products is initially determined only by the ratio of the forward rate constants.

- A series of **sequential reactions** are steps chained to each other such that a product of one reaction step becomes a reactant in the next: $A \overset{k_1}{\rightarrow} B \overset{k_2}{\rightarrow} C$. The shared species (B in the example) is called a chemical **intermediate**. Unlike a transition state, which is a saddle point on the reaction surface, a chemical intermediate corresponds to a local minimum on the surface. By solving for the time-dependent concentrations in this mechanism, we are able to show that when one of the two reaction rates is much smaller than the other, then the overall reaction rate is effectively determined by the rate constant of the *slower* or **rate-limiting step.**

14.2 **Approximations in Kinetics.**

- The **steady-state approximation** assumes that a chemical intermediate has a much shorter lifetime than the reactants from which it is formed. This condition is satisfied, for example, in the reaction mechanism $A + B \underset{k_{-1}}{\overset{k_1}{\rightleftharpoons}} C \overset{k_2}{\rightarrow} D$ when $k_2 \gg k_1$. Under these conditions, the concentration of the intermediate varies at a slower rate than the reactant and product concentrations, and we can solve for a steady-state concentration of the intermediate.

- When the forward and reverse reactions of one step are fast in comparison to other steps in the mechanism, it may be possible to assume that the reactant and product concentrations of that step satisfy the **fast equilibrium approximation**. In our sample mechanism, if $A + B$ is in fast equilibrium with the intermediate C, then the concentration of the intermediate can be related to the reactant concentrations through the equilibrium ratio:

$$[C]_{fe} = \frac{k_1}{k_{-1}}[A][B]. \qquad (14.25)$$

- The **initial rate approximation** allows the rate law to be determined empirically by measuring the changes $\Delta[X]$ of the reactant concentrations in the sample and dividing by the time interval Δt. Changing the starting concentration of the each reactant and repeating the measurements then allows us to estimate the exponent of that reactant in the rate law.

- In the **pseudo-lower order approximation** one or more of the reactants are present in excess, so that their concentrations remain essentially constant during the reaction. This strategy reduces the number of concentrations that need to be treated as variables, sometimes allowing an analytical solution to the integrated rate law when it would not be possible otherwise.

KEY TERMS

- A **chain reaction** occurs when a single reaction at the molecular scale leads to a product while *also* leaving behind one or more reactive intermediates that initiate the same reaction with other reactants. As a result, the reaction propagates through the system at a rapidly increasing rate.

- The **Chapman mechanism** is a series of reactions that model the ozone cycle, including the destruction of O_3 by UV radiation and its regeneration by O_2 and atomic oxygen.

- An **enzyme** is a protein or protein-based molecule that catalyzes biochemical reactions.
- The **Michaelis-Menten** constant is effectively the equilibrium constant for dissociation of an enzyme-substrate complex under steady-state conditions, and it is used as a measure of the enzyme activity.

- Combustion reactions occur as **deflagrations,** which are relatively slow processes in which the heat is efficiently carried away from the reaction mixture, and **detonations,** in which the energy is released at an accelerating rate, producing a shock wave.

OBJECTIVES REVIEW

1. *Write the differential rate law for any reactant, product, or intermediate of a multi-step reaction in terms of the concentrations and rate constants, given the full reaction mechanism.*
   Write the differential rate law for [Cl] in the ClO_x cycle described in Section 14.4.

2. *Verify that analytical solutions to integrated rate laws give the correct solution in various limits.*

Show that Eq. 14.23 gives the correct result in the limit $t = 0$.

3. *Apply the steady-state, fast equilibrium, and other approximations to the kinetics analysis of a multi-step reaction.*
   For the reaction sequence $A + B \xrightarrow{k_1} C \xrightarrow{k_2} D$, find the integrated rate law for [D] if $[B]_0 \gg [A]_0$.

PROBLEMS

Discussion Problems

14.1 The following drawing represents the reaction diagram for a sequential reaction of three steps. Explain which of the four steps is likely to be the rate-limiting step at very low temperature and which at very high temperature, and explain why.

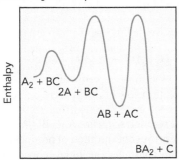

14.2 For the reaction sequence $A + B \underset{k_{-1}}{\overset{k_1}{\rightleftharpoons}} C + D$ $C + E \underset{k_{-2}}{\overset{k_2}{\rightleftharpoons}} A + E$, what would be a suitable approximation to use in obtaining the concentration of C if the rate constants are given by the values (all in the same units) $k_1 = k_{-2} = 10^{14}$, $k_{-1} = 10^{18}$, and $k_2 = 10^7$?

Integrated Rate Laws

14.3 Equation 14.16 gives the concentration of C as a function of time in the two-step, unimolecular sequential reaction:

$$[C] = [A]_0 \left(1 - \frac{k_2 e^{-k_1 t} - k_1 e^{-k_2 t}}{k_2 - k_1} \right).$$

What happens to the concentration [C] if $k_1 = k_2$, making the denominator in the equation vanish?

14.4 For the consecutive reactions

$$A \xrightarrow{k_1} B \xrightarrow{k_2} C,$$

find [B] as a function of t when $k_1 = k_2$.

14.5 For the sequential reaction, $A \xrightarrow{k_1} B \xrightarrow{k_2} C$, when $[B]_0 = [C]_0 = 0$ the concentration of product C is given by the amount of A converted to C:

$$[C]_A = [A]_0 \left(1 - \frac{k_2 e^{-k_1 t} - k_1 e^{-k_2 t}}{k_2 - k_1} \right).$$

Modify this equation for the case when $[B]_0 \neq 0$ and $[C]_0 \neq 0$.

14.6 Find the algebraic solution for the concentration of the intermediate [C] in the reaction $A \xrightarrow{k_1} B \xrightarrow{k_2} C \xrightarrow{k_3} D$, assuming all initial concentrations are zero except for A, with initial concentration $[A]_0$.

14.7 For the reaction $A + B \underset{k_{-1}}{\overset{k_1}{\rightleftharpoons}} C$, the concentration of C obeys the following equation in the limit $t \rightarrow \infty$:

$$[C] = [C]_0 + \frac{1}{2a} \left[\sqrt{b^2 - 4ac} - b \right]$$

where

$$a = k_1 \quad b = -k_1([A]_0 + [B]_0) - k_{-1}$$
$$c = k_1([A]_0 [B]_0) - k_{-1}[C]_0.$$

Show that this is the correct answer for infinite time. You do not need to use or solve the integrated rate law.

14.8 For the reaction $A \underset{k_{-1}}{\overset{k_1}{\rightleftharpoons}} B + C$, obtain an expression for $[B]$ as a function of t, assuming that $[B]_0 = [C]_0 = 0$. Verify that this solution is valid in the limits $t = 0$ and $t \to \infty$.

14.9 Find an equation for the concentrations $[A]$ and $[C]$ in the reaction $A + B \underset{k_{-1}}{\overset{k_1}{\rightleftharpoons}} C$ in terms of the initial concentrations $[A]_0$, $[B]_0$, and $[C]_0$. Verify that these results are valid at $t = 0$ and $t = \infty$.

14.10 Find a single expression for the concentration $[B]$ in terms of the rate constant and initial concentrations for the parallel reactions:

$$A + B \overset{k_1}{\rightarrow} D,$$
$$A + C \overset{k_2}{\rightarrow} E,$$

for the case $k_1 = k_2$. For all initial concentrations $[A]_0$, $[B]_0$, and $[C]_0$, prove that your answer is valid at (a) $t = 0$, and (b) $t = \infty$.

14.11 For the reaction system $A \overset{k_1}{\rightarrow} C$, $B \overset{k_2}{\rightarrow} C$, find an expression for $[C]$ in terms of the rate constants and initial concentrations $[A]_0$, $[B]_0$, and $[C]_0$.

14.12 For the reactions $A + B \overset{k}{\rightarrow} D$, $A + C \overset{k}{\rightarrow} E$, a possible solution is

$$[B] = [B]_0 \left(1 - 2\frac{[B]_0}{[A]_0}\right)\left[e^{([A]_0 - 2[B]_0)kt} - 2\frac{[B]_0}{[A]_0}\right]^{-1}.$$

Verify that this solution is valid in the limits $t \to 0$ and $t \to \infty$. Under what initial conditions is the expression valid?

14.13 Given the equation in Problem 14.12, write equations for the concentrations of C, D, and E as functions of time, assuming $[D]_0 = [E]_0 = 0$.

14.14 Consider the simplest catalysis reaction, involving a molecule A that isomerizes to a form B, with greatly enhanced reaction velocity in the presence of catalyst C. There is the fast catalyzed reaction

$$A + C \underset{k_{-1}}{\overset{k_1}{\rightleftharpoons}} B + C \quad v_1 = k_1[A][C] \quad v_{-1} = k_{-1}[B][C]$$

and the slow, non-catalyzed reaction

$$A \underset{k_{-2}}{\overset{k_2}{\rightleftharpoons}} B \quad v_2 = k_2[A] \quad v_{-2} = k_{-2}[B].$$

These two reactions occur simultaneously in a mixture of A and C. Find an expression for the ratio of $[B]$ to $[A]$ at equilibrium, in terms of the rate constants. At equilibrium, the ratio should be independent of $[C]$, because the catalyst changes only the chemical kinetics, not the thermodynamics.

14.15 Ethenyl cyclopropene isomerizes at 640 K to yield several products. If the product ratio is determined entirely by kinetics (so the products are not allowed to come to equilibrium), then the relative yields are as follows: (1) 3-methylcyclobutene, 96.2%; (2) 1,4-pentadiene, 1.5%; (3) *cis*-1,3-pentadiene, 1.2%; (4) *trans*-1,3-pentadiene, 1.1%. If the half-life of ethenyl cyclopropene overall is 1570 s, estimate the rate constants k_1, k_2, k_3, and k_4 for formation of the four products.

Complex Reactions and Approximations

14.16 Write the equations for the production rates $d[A]/dt$ of each chemical component in the chain reaction network

$$Cl_2 \underset{k_{-1}}{\overset{k_1}{\rightleftharpoons}} 2Cl$$
$$Cl + H_2 \overset{k_2}{\rightarrow} HCl + H$$
$$Cl_2 + H \overset{k_3}{\rightarrow} HCl + Cl.$$
$$2Cl + H_2 \overset{k_4}{\rightarrow} Cl_2 + H_2$$

14.17 The principal elementary reactions in the oxidation of methane that involve CH_4 itself are these:

a. $CH_4 + M \overset{k_1}{\rightarrow} CH_3 + H + M$
b. $CH_4 + OH \overset{k_2}{\rightarrow} CH_3 + H_2O$
c. $CH_4 + H \overset{k_3}{\rightarrow} CH_3 + H_2$
d. $CH_4 + O \overset{k_4}{\rightarrow} CH_3 + OH$

First, write the rate law for the disappearance of CH_4 in this oxidation. Then use this to obtain a differential equation for the rate of appearance of OH. Don't worry about solving the equation or isolating all the dependence on $[OH]$.

14.18 Write the net reaction that predicts the following rates, if A, B, C, D, and E represent all the species in the mechanism.

$$\frac{d[A]}{dt} = -k_1[A][B] + k_{-1}[C][B] + k_2[C][D]$$
$$\quad - k_{-2}[A][E]$$
$$\frac{d[B]}{dt} = 0$$
$$\frac{d[C]}{dt} = k_1[A][B] - k_{-1}[C][B] - k_2[C][D]$$
$$\quad + k_{-2}[A][E]$$
$$\frac{d[D]}{dt} = -k_2[C][D] + k_{-2}[A][E]$$
$$\frac{d[E]}{dt} = k_2[C][D] - k_{-2}[A][E]$$

14.19 For the reaction sequence $A + B \underset{k_{-1}}{\overset{k_1}{\rightleftharpoons}} C + D$, $C \overset{k_2}{\rightarrow} E$, express the concentration of product E at time t in terms of the rate constants, the initial concentrations $[A]_0$ and $[B]_0$ (of A and B), and the concentration $[D]$ at time t. Use the steady-state approximation for $[C]$. Initial concentrations of C, D, and E are zero.

14.20 One function of organometallic reagents in chemical synthesis is the catalysis of hydrogen transfer reactions that would be very difficult in isolated organic compounds. For example, a chemical complex of rhodium and an alkyne is believed to form an intermediate of the form HMCCH (where the M represents the Rh and some additional ligands). This intermediate then transfers the metal-bound H atom to the other end of the alkyne, forming the alkene complex $MCCH_2$. This elementary step was proposed to be bimolecular, with two complexes meeting to form a square-bridged dimer so that H atoms were traded from one complex to the next. To test this, the reaction was carried out with an equimolar mixture of normal reagent and deuterium *and* ^{13}C doubly-isotopically-labeled reagent, leading to the following possible reactions:

$$2HM^{12}C{\equiv}^{12}CH \xrightarrow{k_1} 2M{=}^{12}C{=}^{12}CH_2$$

$$HM - {}^{12}C{\equiv}^{12}CH + DM - {}^{13}C{\equiv}^{13}CD \xrightarrow{k_2}$$
$$M{=}^{12}C{=}^{12}CHD + M{=}^{13}C{=}^{13}CHD$$

$$2DM - {}^{13}C{\equiv}^{13}CD \xrightarrow{k_3} 2M{=}^{13}C{=}^{13}CD_2.$$

Assume that the initial concentration of both reactants is 0.020 M, and that the isotopic masses do not influence the reaction rate.

a. Predict the final concentrations of the four products, without solving the integrated rate laws.

b. Now solve for the integrated rate laws that predict the concentrations of the four products as functions of time, and predict the final concentrations of each product in the long-time limit. (Note that without reverse reactions the final concentrations need not be the equilibrium concentrations.)

If your two answers for the final product yields do not agree, find the incorrect assumption.

14.21 The reaction system is

$$A \xrightarrow{k_1} B,$$
$$B \underset{k_{-2}}{\overset{k_2}{\rightleftharpoons}} C$$
$$B \xrightarrow{k_3} D.$$

Assume that fast equilibrium applies to the second step, and let all initial concentrations be zero except set $[A] = [A]_0$ initially. Find an equation for the concentration of D as a function of time and the fast equilibrium constant $K_2 \equiv k_2/k_{-2}$.

14.22 Hypochlorite ion, used as the chlorinating agent in swimming pools, can decompose exothermically, in rare occasions even causing fires. The principal decomposition mechanism is believed to be

$$OCl^- + OCl^- \xrightarrow{k_1} ClO_2^- + Cl^-,$$
$$OCl^- + ClO_2^- \xrightarrow{k_2} ClO_3^- + Cl^-.$$

a. Find the steady-state concentration of ClO_2^- ion.

b. Use this to obtain a differential (*not integrated*) rate law for the concentration of ClO_3^-, such that only $[ClO^-]$ and t appear as variables.

14.23 An organometallic catalyst converts the (E,E) isomer of a dialkenyl ester to the (E,Z) isomer in a bimolecular reaction of the form $A + C \xrightarrow{k_1} B + C$. However, the reaction slows with time because the catalyst slowly degrades according to the first order reaction $C \xrightarrow{k_2} D$. The reaction rate without the catalyst is negligible, and the initial concentration of B, the (E,Z) isomer, is zero.

a. Use a symbolic math program to find the integrated rate law for $[B]$.

b. Verify that the solution yields reasonable results in the following limits: (*i*) $t = 0$, (*ii*) $t \to \infty$ and $k_1[C]_0 \gg k_2$, (*iii*) $t \to \infty$ and $k_1[C]_0 \ll k_2$.

c. If the experimentalists graph $\ln([A]_0/[A])$ versus t, what will be the shape of the resulting curve? How is that shape affected by the value of k_2?

Reaction Networks

14.24 Hot air rises, but what happens to it as it rises through our atmosphere? Given the number density function Eq. 14.33 and the temperature curve plotted in Fig. 14.3 to the surrounding atmosphere, calculate the final temperature of a bubble of air that rises adiabatically from a state $z_1 = 0$, T 300 K to an altitude $z_2 = 20$ km.

14.25 The $^1D_2$ state of atomic oxygen lies 1.968 eV above the ground state. Calculate the enthalpies of reaction for the HO_x cycle reaction $^1DO + H_2O \longrightarrow 2OH$ and the corresponding ground state reaction $^3PO + H_2O \longrightarrow 2OH$.

14.26 In this problem we show that simple gas diffusion from the ground into the stratosphere is a very slow process. For this reason, convection is the dominant mechanism for delivering pollutants into the upper atmosphere. Find an equation for the diffusion constant for CCl_2F_2 in air as a function of altitude z above the earth, assuming a collision cross section of 60 $Å^2$, a constant temperature of 240 K, and a zero-altitude number density ρ_0 of $2.43 \cdot 10^{19}$ cm^{-3}. Then use the Einstein equation for diffusion to estimate how long it would take for CCl_2F_2 emitted at the earth's surface to diffuse into the stratosphere. Estimate the altitude z traveled in time t as

$$z \approx \sqrt{6Dt},$$

and find the time necessary for z to reach a value of 18 km. Because the number density of the atmosphere is a function of altitude, you will need to integrate over z.

14.27 How long does it take a flame obeying the rate laws in Table 14.4 to consume 10% of the methane if the concentrations [H] and [OH] are steady state and correspond to partial pressures of 1 Pa for each radical and if the temperature is 2000 K?

14.28 For the combustion of acetylene, give an example of a likely reaction in each of the following categories: (a) initiation; (b) propagation; (c) chain branching; (d) termination.

14.29 For the combustion reaction $H + O_2 \rightarrow OH + O$, the rate constant at 2000 K is $2.92 \cdot 10^{11}$ cm^3 mol^{-1} s^{-1}. Using simple collision theory, estimate the activation energy in $kJ \, mol^{-1}$. Use $\sigma_H = 21 \, Å^2$ and $\sigma_{O_2} = 36 \, Å^2$. Let p be equal to 1.

14.30 Estimate $\Delta_{rxn}H$ and $\Delta_{rxn}G$ at 2000 K (per mole CH_4 in the overall reaction) for each of the four *net* reactions in Table 14.5. Assume that all the heat capacities obey the ideal gas relation

$$C(T) = \text{constant} \times T.$$

Which reactions, if any, are entropy-favored?

14.31 In a molecular cloud with an overall number density of 10^5 cm^{-3}, and fractional abundances $X[CCl^+] = 2.4 \cdot 10^{-13}$ and $X[NH_3] = 1.0 \cdot 10^{-7}$, calculate the number of forward reactions $CCl^+ + NH_3 \longrightarrow HCl + HCNH^+$ occurring in a volume of one cubic light year over a period of 100 years, if the rate constant is $1.30 \cdot 10^{-9}$ cm^3 s^{-1}. One light year (ly) = $9.46 \cdot 10^{17}$ cm.

14.32 The power detected at a radio telescope is often given as the **antenna temperature** T_A, such that

$$\text{Power} = k_B T_A \delta \nu,$$

where $\delta \nu$ is the frequency range of the radiation admitted to the detector. Consider a 100 m diameter dish antenna radio telescope set to detect the $J = 2 \rightarrow 1$ rotational transition of CO ($B_0 = 1.9313$ cm^{-1}) with a frequency range $\delta \nu$ of 1 MHz, pointed at a molecular cloud 10 light years (10 ly) distant at temperature 10 K.

(a) Find the flux of photons per second striking the antenna from the cloud if the measured antenna temperature is 0.5 K. (b) The lifetime τ of CO in the $J = 2$ state is roughly $1.5 \cdot 10^6$ s before emission. If the cross-sectional area observed by the telescope is a circle 0.001 light years in diameter, and the number density of CO can safely be assumed to be 10 cm^{-3}, what is the total volume sampled by the telescope and the average depth in light years of the cloud along the telescope's line of sight? One light year = $9.46 \cdot 10^{17}$ cm.

14.33 Find an equation for $d[P]/dt$ equivalent to the Michaelis-Menten equation when the third step of the reaction (Reaction R14.7) is *not* required to be fast compared to the second step, and when k_{-2} is significant. Assume only that both intermediates ES and EP are in steady-state concentrations and that the reaction is at early enough a stage that $[P] \approx 0$. The equation should depend on the concentrations $[E]_0$ and $[S]$ and on the rate constants. Verify that your solution is consistent with the Michaelis-Menten equation in the appropriate limit.

Thermochemical Properties

All values evaluated at 298.15 K, $P = 1$ bar

	S_m° (J K^{-1} mol^{-1})	$\Delta_f H^\circ$ (kJ mol^{-1})	$\Delta_f G^\circ$ (kJ mol^{-1})	C_{Pm} (J K^{-1} mol^{-1})
He(gas)	126.153	0.0	0.0	20.786
Ne(gas)	146.219	0.0	0.0	20.786
Ar(gas)	154.846	0.0	0.0	20.786
Kr(gas)	164.08	0.0	0.0	20.786
Xe(gas)	169.573	0.0	0.0	20.786
Br(gas)	175.022	111.884	82.396	20.786
Br$_2$(gas)	245.463	30.907	3.110	36.02
Br$_2$(liq)	152.231	0.0	0.0	75.689
BrCl(gas)	240.10	14.64	−0.98	34.98
C(diamond)	2.377	1.895	2.900	6.113
C(graphite)	5.740	0.0	0.0	8.527
CO(gas)	197.674	−110.525	−137.168	29.142
CO$_2$(gas)	213.785	−393.509	−394.359	37.11
COCl$_2$(gas)	283.53	−218.8	−204.6	57.66
CS$_2$(gas)	237.84	117.36	67.12	45.40
CS$_2$(liq)	151.34	89.70	65.27	75.7
CCl$_4$(gas)	309.85	−102.9	−60.59	83.30
CCl$_4$(liq)	216.40	−135.44	−65.21	131.75
CF$_4$(gas)	261.61	−925.	−879.	61.09
CH$_2$Cl$_2$(gas)	270.23	−92.47	−65.87	50.96
CH$_3$Cl(gas)	234.58	−80.83	−57.37	40.75
CH$_3$OH(gas)	239.81	−200.66	−161.96	43.89
CH$_3$OH(liq)	126.8	−238.66	−166.27	81.6
CH$_4$(gas)	186.264	−74.81	−50.72	35.309
CHCl$_3$(liq)	201.7	−134.47	−73.66	113.8
HCOOH(liq)	128.95	−424.72	−361.35	99.04
C$_2$H$_2$(gas)	200.94	226.73	209.20	43.93
C$_2$H$_4$(gas)	219.56	52.26	68.15	43.56
CH$_2$ClCH$_2$Cl(gas)	308.39	−129.79	−73.87	78.7
CH$_2$BrCH$_2$Br(gas)	329.7	−38.9	−10.6	86.2
c-CH$_2$CH$_2$O(gas)	242.53	−52.63	−13.01	47.91
C$_2$H$_5$OH(gas)	282.70	−235.10	−168.49	65.44
C$_2$H$_5$OH(liq)	160.7	−277.69	−174.78	111.46
C$_2$H$_6$(gas)	229.60	−84.68	−32.82	52.63
C$_6$H$_6$(gas)	269.31	82.927	129.72	81.67
CaCl$_2$(sol)	104.6	−795.8	−748.1	72.59

(continued)

(continued) All values evaluated at 298.15 K, $P = 1$ bar

	S_m° (J K^{-1} mol^{-1})	$\Delta_f H^\circ$ (kJ mol^{-1})	$\Delta_f G^\circ$ (kJ mol^{-1})	C_{Pm} (J K^{-1} mol^{-1})
Cl(gas)	165.198	121.679	105.680	21.840
Cl$_2$(gas)	223.066	0.0	0.0	33.907
ClF(gas)	217.89	−54.48	−55.94	32.05
ClF$_3$(gas)	281.61	−163.2	−123.0	63.85
ClF$_3 \cdot$ HF(gas)	360.	−450.6	−384.0	(unavailable)
ClO(gas)	226.63	101.84	98.11	31.46
ClO$_2$(gas)	256.84	102.5	120.5	41.97
ClO$_3$F(gas)	278.97	−23.8	48.2	64.94
F(gas)	158.754	78.99	61.91	22.744
F$_2$(gas)	202.78	0.0	0.0	31.30
H(gas)	114.713	217.965	203.247	20.786
H$_2$(gas)	130.684	0.0	0.0	28.824
H$_2$O(gas)	188.825	−241.818	−228.572	33.577
H$_2$O(liq)	69.91	−285.830	−237.129	75.291
H$_2$S(gas)	205.79	−20.63	−33.56	34.23
HF(gas)	173.779	−271.1	−273.2	29.133
HCl(gas)	186.908	−92.307	−95.299	29.12
HBr(gas)	198.695	−36.40	−53.45	29.142
HI(gas)	206.594	26.48	1.70	29.158
I(gas)	180.791	106.838	70.250	20.786
I$_2$(gas)	260.69	62.438	19.327	36.90
I$_2$(sol)	116.135	0.0	0.0	54.438
IF$_5$(gas)	327.7	−822.49	−751.73	99.2
K(gas)	160.336	89.24	60.59	20.786
K(sol)	64.18	0.0	0.0	29.58
KCl(gas)	239.10	−214.14	−233.0	36.48
KCl(sol)	82.59	−436.747	−409.14	51.30
NH$_3$(gas)	192.45	−46.11	−16.45	35.06
NO(gas)	210.761	90.25	86.55	29.844
NO$_2$(gas)	240.06	33.18	51.31	37.20
N$_2$(gas)	191.61	0.0	0.0	29.125
N$_2$H$_4$(gas)	238.47	95.40	159.35	49.58
N$_2$H$_4$(liq)	121.21	50.63	149.34	98.87
N$_2$O(gas)	219.85	82.05	104.20	38.45
N$_2$O$_3$(gas)	312.28	83.72	139.46	65.61
N$_2$O$_4$(gas)	304.29	9.16	97.89	77.28
N$_2$O$_4$(liq)	209.2	−19.50	97.54	142.7
N$_2$O$_5$(gas)	355.7	11.3	115.1	84.5
Na(gas)	153.712	107.32	76.761	20.786
Na(sol)	51.21	0.0	0.0	28.24
NaCl(gas)	229.81	−176.65	−196.66	35.77
NaCl(sol)	72.13	−411.153	−384.138	50.50
O(gas)	161.055	249.170	231.731	21.912
O$_2$(gas)	205.138	0.0	0.0	29.355
O$_3$(gas)	238.93	142.7	163.2	39.20
OH(gas)	183.64	38.95	34.23	29.89

(continued)

(continued) All values evaluated at 298.15 K, $P = 1$ bar

	S_m° (J K^{-1} mol^{-1})	$\Delta_f H^\circ$ (kJ mol^{-1})	$\Delta_f G^\circ$ (kJ mol^{-1})	C_{Pm} (J K^{-1} mol^{-1})
S_8(rhombic)	31.8	0.0	0.0	22.6
SO(gas)	221.8	6.26	–19.84	30.17
SO_2(gas)	248.22	–296.830	–300.194	39.87
SO_2Cl_2(gas)	311.94	–364.0	–320.0	77.0
SO_3(gas)	256.76	–395.72	–371.06	50.67
$SOCl_2$(gas)	309.77	–212.5	–198.3	66.5
SF_4(gas)	292.03	–774.9	–731.3	73.01
SF_6(gas)	291.82	–1209.	–1105.3	97.28
Si(sol)	18.83	0.0	0.0	20.00
$SiCl_4$(gas)	330.73	–657.01	–616.98	90.25
$SiCl_4$(liq)	239.7	–687.	–619.84	145.31
SiF_4(gas)	282.76	–1614.94	–1572.65	73.64
SiH_4(gas)	204.62	34.3	56.9	42.84
SiN(gas)	216.76	486.52	456.08	30.17
SiO_2 (quartz)	41.84	–910.94	–856.64	44.43
SiO_2 (cristobalite)	42.68	–909.48	–855.46	44.18
SiO_2 (amorphous)	41.9	–903.49	–850.73	44.4
SiS(gas)	223.66	112.47	60.89	32.26

Solutions to Objectives Review Questions

Numerical answers to problems are included here. Complete solutions to selected problems can be found in the *Student's Solutions Manual*.

Chapter 1

P1.1 -743 kJ

P1.2 $\left[\int_0^\infty \mathcal{P}_v(v)(mv)^2 dv\right]^{1/2}$, with $\mathcal{P}_v(v)$ given by Eq. 1.27

Chapter 2

P2.1 $1.33 \cdot 10^{-22}$ J K^{-1} (Boltzmann) and $1.24 \cdot 10^{-22}$ J K^{-1} (Gibbs)

P2.2 6.98

P2.3 0.00080

Chapter 3

P3.1 trans: 15.5 kJ; rot: 10.3 kJ; vib: 10.3 kJ

P3.2 0.0475

P3.3 0.155

P3.4 Any $\int_{-\infty}^\infty v^2 e^{-mv^2/(2k_BT)} v\, dv$ is zero because the integrand has canceling positive and negative values

Chapter 4

P4.1 Use canonical distribution to get $\mathcal{P}_Z(Z)$; then integrate $\int_0^\infty \mathcal{P}_Z(Z)Z\, dZ$

P4.2 333 K, versus 289 K for ideal gas

P4.3 $\exp\left\{-\varepsilon\left[\left(\dfrac{R_e}{R}\right)^{12} - 2\left(\dfrac{R_e}{R}\right)^6\right]/(k_BT)\right\}$, $\varepsilon/k_B = 230$ K, $R_e = 4.42$ Å

P4.4 a and c

Chapter 5

P5.1 $1.1 \cdot 10^{-7}$ m

P5.2 $D \approx 0.20$ cm^2 s^{-1}, $t \approx 8.4 \cdot 10^3$ s $= 2.3$ hr

P5.3 $1.0 \cdot 10^{-8}$ mol s^{-1} m^{-2}

Chapter 6

P6.1 convection and conduction

P6.2 13.7 μm

P6.3 $2.8 \cdot 10^5 M^{-1}$ cm^{-1}

P6.4 Doppler: 0.010 cm^{-1}, collision: 0.87 cm^{-1}

Chapter 7

P7.1 $\left(\dfrac{\partial V}{\partial S}\right)_{T,n} = \left(\dfrac{\partial V}{\partial P}\right)_{T,n}\left(\dfrac{\partial P}{\partial S}\right)_{T,n}$

$= -\left(\dfrac{\partial V}{\partial P}\right)_{T,n}\left(\dfrac{\partial T}{\partial V}\right)_{P,n} = \dfrac{\kappa_T}{\alpha}$

P7.2 $N_{ep} \approx 3 + 3 + 2 \times 27 = 60$; $C_{Vm} \approx N_{ep} R/2$; $C_{Pm} = C_{Vm} + R = 31\,R = 258$ J K^{-1} mol^{-1}

P7.3 7.1 J

Chapter 8

P8.1 171 J

P8.2 0.00160 L K J^{-1} = 0.160 K bar^{-1}

P8.3 430 K

Chapter 9

P9.1 -792 J K^{-1}

P9.2 176 J K^{-1} mol^{-1}

Chapter 10

P10.1 $\Delta H = -22.8 \text{ kJ}, \Delta S = -65.4 \text{ J K}^{-1}$

P10.2   353 K

P10.3 $T_f \approx 390 \text{ K}, T_b \approx 700 \text{ K}$

Chapter 11

P11.1 $\mu = \mu^\circ + 2RT \ln \gamma_{K^+} X_{K^+} + RT \ln \gamma_{SO_4^{2-}} X_{SO_4^{2-}}$

P11.2 $P_{H_2O} \approx 0.032 \text{ bar}, P_{CCl_4} \approx 0.029 \text{ bar}$

P11.3   gas: 0.20 chloroform, 0.80 acetone; liquid: 0.30 chloroform, 0.70 acetone

P11.4   270.4 K

Chapter 12

P12.1 $\approx 180 \text{ kJ mol}^{-1}$

P12.2 -65.36 kJ

P12.3   3460 K

P12.4 $\Delta_{rxn} G^\circ = -111.73 \text{ kJ}, K_{eq} = 4.43 \cdot 10^{15}$

Chapter 13

P13.1 $$-\frac{d[NO]}{2dt} = -\frac{d[Cl_2]}{dt} = \frac{d[NOCl]}{2dt}$$
$$= k[NO]^2[Cl_2]$$

P13.2   141 kJ mol^{-1}

P13.3   4.8 s

Chapter 14

P14.1 $$\frac{d[Cl]}{dt} = j_1[CCl_2F_2] - k_2[Cl][O_3]$$
$$+ k_3[ClO][O] - k_4[Cl][CH_4]$$

P14.2 $\lim_{t \to 0} [A]' = [A]_0$

P14.3 $[D] \approx [A]_0 \left(1 - \dfrac{k_2 e^{-[B_0]k_1 t} - [B]_0 k_1 e^{-k_2 t}}{k_2 - [B]_0 k_1} \right)$

CREDITS

Cover Richard Megna/Fundamental Photographs

Front Matter **Page v:** John Donne (1572–1631), Satire III.

Part I **Page 37:** AIST-NT, Inc.

Chapter 1 **Page 45:** J.W. Gibbs, "A Method of Geometrical Representation of the Thermodynamic Properties of Substances by Means of Surfaces," Transactions of the Connecticut Academy of Arts and Sciences 2, Dec. 1873. **Pages 45–47:** Clausius, Rudolf. (1850). "On the Motive Power of Heat, and on the Laws which may be deduced from it for the Theory of Heat", Communicated in the Academy of Berlin, Feb.; Published in Poggendorff's Annalen der Physick, March-April. LXXIX, 368, 500. (b) Translated in: the Philosophical Magazine, July 1851, Vol. ii, pgs, 1, 102. **Pages 48–54:** James Clerk Maxwell, "Molecules". Nature, 8(204) (25 September 1873). **Page 54:** Joseph Francisco.

Chapter 2 **Page 71:** Courtesy of Juan de Pablo

Chapter 3 **Page 108:** Source: The Theory of Rotating Diatomic Molecules by Masataka Mizushima, John Wiley & Sons 1975. CRC Handbook of Chemistry and Physics (77th edition) CRC Press 1996. **Page 116:** Photo by David A. Harvey, courtesy of David A. Harvey Photography.

Chapter 5 **Page 191:** Desorption Electrospray Ionization (DESI), Courtesy of Prosolia, Inc., Indianapolis, IN, http://www.prosolia.com/desi.php.

Chapter 6 **Page 208:** Stephen Klippenstein.

Part II **Page 231:** ilumus photography/Fotolia.

Chapter 7 **Page 245:** L. Brian Stauffer, University of Illinois News Bureau. **Page 260:** "Die Plancksche Theorie der Strahlung und die Theorie der spezifischen Wärme", A. Einstein, Annalen der Physik, volume 22, pp. 180–190, 1907.

Chapter 8 **Page 283:** Lavoisier, Antoine; Elements of Chemistry, In a New Systematic Order, Containing all the Modern Discoveries, Translated by Robert Kerr, Edinburgh; Marcellin Berthelot, Sur la force de la poudre et des matières explosives (1872). **Page 295:** Nicolas Léonard Sadi Carnot, Reflections on the Motive Power of Fire, 1824.

Chapter 9 **Page 314:** Denis Evans. **Page 315:** Clausius, Rudolph. The Mechanical Theory of Heat. London: Taylor & Francis, 1867. eBook. **Page 322:** After a figure in "William F. Giauque - Nobel Lecture: Some Consequences of Low Temperature Research in Chemical Thermodynamics". Nobelprize.org. 31 Oct 2012 http://www.nobelprize.org/nobel_prizes/chemistry/laureates/1949/giauque-lecture.html.

Chapter 11 **Page 382:** Kim Baldridge.

Part III **Page 415:** Viktor Horvath and Irving Epstein, Brandeis University.

Chapter 12 **Page 428:** William Lester.

Chapter 13 **Page 459:** Source: Physical Chemistry by Joseph H. Noggle, Little, Brown & Company 1984, Table 10.2. **Page 461:** Robert Continetti

Chapter 14 **Page 479:** Steinfeld, Jeffrey I.; Francisco, Joseph S.; Hase, William L., Chemical Kinetics and Dynamics, 2nd Ed., ©1999. Reprinted and Electronically reproduced by permission of Pearson Education, Inc., Upper Saddle River, New Jersey. **Page 496:** Sydney Chapman; J. Bartels (1940). Geomagnetism, Vol. II, Analysis and Physical Interpretation of the Phenomena. Oxford Univ. Press.

See end of Solutions Manual for additional citations and annotations.

INDEX

Note: Page numbers followed by n indicate footnotes.